Organic Reactions

Organic Reactions

VOLUME 40

JOHN WILEY & SONS, INC.

New York • Chichester • Brisbane • Toronto • Singapore

In recognition of the importance of preserving what has been written, it is a policy of John Wiley & Sons, Inc., to have books of enduring value published in the United States printed on acid-free paper, and we exert our best efforts to that end.

Published by John Wiley & Sons, Inc.

Library of Congress Catalog Card Number 42-20265

ISBN 0-471-53841-8

Printed in the United States of America

10 9 8 7 6 5 4 3 2 1

PREFACE TO THE SERIES

In the course of nearly every program of research in organic chemistry the investigator finds it necessary to use several of the better-known synthetic reactions. To discover the optimum conditions for the application of even the most familiar one to a compound not previously subjected to the reaction often requires an extensive search of the literature; even then a series of experiments may be necessary. When the results of the investigation are published, the synthesis, which may have required months of work, is usually described without comment. The background of knowledge and experience gained in the literature search and experimentation is thus lost to those who subsequently have occasion to apply the general method. The student of preparative organic chemistry faces similar difficulties. The textbooks and laboratory manuals furnish numerous examples of the application of various syntheses, but only rarely do they convey an accurate conception of the scope and usefulness of the processes.

For many years American organic chemists have discussed these problems. The plan of compiling critical discussions of the more important reactions thus was evolved. The volumes of *Organic Reactions* are collections of chapters each devoted to a single reaction, or a definite phase of a reaction, of wide applicability. The authors have had experience with the processes surveyed. The subjects are presented from the preparative viewpoint, and particular attention is given to limitations, interfering influences, effects of structure, and the selection of experimental techniques. Each chapter includes several detailed procedures illustrating the significant modifications of the method. Most of these procedures have been found satisfactory by the author or one of the editors, but unlike those in *Organic Syntheses* they have not been subjected to careful testing in two or more laboratories.

Each chapter contains tables that include all the examples of the reaction under consideration that the author has been able to find. It is inevitable, however, that in the search of the literature some examples will be missed, especially when the reaction is used as one step in an extended synthesis. Nevertheless, the investigator will be able to use the tables and their accompanying bibliographies in place of most or all of the literature search so often required.

Because of the systematic arrangement of the material in the chapters and the entries in the tables, users of the books will be able to find information desired by reference to the table of contents of the appropriate chapter. In the interest of economy the entries in the indices have been kept to a minimum, and, in particular, the compounds listed in the tables are not repeated in the indices.

The success of this publication, which will appear periodically, depends upon the cooperation of organic chemists and their willingness to devote time and effort to the preparation of the chapters. They have manifested their interest already by the almost unanimous acceptance of invitations to contribute to the work. The editors will welcome their continued interest and their suggestions for improvements in *Organic Reactions*.

Chemists who are considering the preparation of a manuscript for submission to *Organic Reactions* are urged to write either secretary before they begin work.

CUMULATIVE CHAPTER TITLES BY VOLUME

Volume 1 (*1942*)

1. **The Reformatsky Reaction**: Ralph L. Shriner
2. **The Arndt-Eistert Reaction**: W. E. Bachmann and W. S. Struve
3. **Chloromethylation of Aromatic Compounds**: Reynold C. Fuson and C. H. McKeever
4. **The Amination of Heterocyclic Bases by Alkali Amides**: Marlin T. Leffler
5. **The Bucherer Reaction**: Nathan L. Drake
6. **The Elbs Reaction**: Louis F. Fieser
7. **The Clemmensen Reduction**: Elmore L. Martin
8. **The Perkin Reaction and Related Reactions**: John R. Johnson
9. **The Acetoacetic Ester Condensation and Certain Related Reactions**: Charles R. Hauser and Boyd E. Hudson, Jr.
10. **The Mannich Reaction**: F. F. Blicke
11. **The Fries Reaction**: A. H. Blatt
12. **The Jacobsen Reaction**: Lee Irvin Smith

Volume 2 (*1944*)

1. **The Claisen Rearrangement**: D. Stanley Tarbell
2. **The Preparation of Aliphatic Fluorine Compounds**: Albert L. Henne
3. **The Cannizzaro Reaction**: T. A. Geissman
4. **The Formation of Cyclic Ketones by Intramolecular Acylation**: William S. Johnson
5. **Reduction with Aluminum Alkoxides (The Meerwein-Ponndorf-Verley Reduction)**: A. L. Wilds

6. **The Preparation of Unsymmetrical Biaryls by the Diazo Reaction and the Nitrosoacetylamine Reaction**: Werner E. Bachmann and Roger A. Hoffman

7. **Replacement of the Aromatic Primary Amino Group by Hydrogen**: Nathan Kornblum

8. **Periodic Acid Oxidation**: Ernest L. Jackson

9. **The Resolution of Alcohols**: A. W. Ingersoll

10. **The Preparation of Aromatic Arsonic and Arsinic Acids by the Bart, Béchamp, and Rosenmund Reactions**: Cliff S. Hamilton and Jack F. Morgan

Volume 3 (1946)

1. **The Alkylation of Aromatic Compounds by the Friedel-Crafts Method**: Charles C. Price

2. **The Willgerodt Reaction**: Marvin Carmack and M. A. Spielman

3. **Preparation of Ketenes and Ketene Dimers**: W. E. Hanford and John C. Sauer

4. **Direct Sulfonation of Aromatic Hydrocarbons and Their Halogen Derivatives**: C. M. Suter and Arthur W. Weston

5. **Azlactones**: H. E. Carter

6. **Substitution and Addition Reactions of Thiocyanogen**: John L. Wood

7. **The Hofmann Reaction**: Everett S. Wallis and John F. Lane

8. **The Schmidt Reaction**: Hans Wolff

9. **The Curtius Reaction**: Peter A. S. Smith

Volume 4 (1948)

1. **The Diels-Alder Reaction with Maleic Anhydride**: Milton C. Kloetzel

2. **The Diels-Alder Reaction: Ethylenic and Acetylenic Dienophiles**: H. L. Holmes

3. **The Preparation of Amines by Reductive Alkylation**: William S. Emerson

4. **The Acyloins**: S. M. McElvain

5. **The Synthesis of Benzoins**: Walter S. Ide and Johannes S. Buck

6. **Synthesis of Benzoquinones by Oxidation**: James Cason

7. **The Rosenmund Reduction of Acid Chlorides to Aldehydes**: Erich Mosettig and Ralph Mozingo

8. **The Wolff-Kishner Reduction**: David Todd

Volume 5 (1949)

1. **The Synthesis of Acetylenes**: Thomas L. Jacobs

2. **Cyanoethylation**: Herman A. Bruson

3. **The Diels-Alder Reaction: Quinones and Other Cyclenones**: Lewis W. Butz and Anton W. Rytina

4. **Preparation of Aromatic Fluorine Compounds from Diazonium Fluoborates: The Schiemann Reaction**: Arthur Roe

5. **The Friedel and Crafts Reaction with Aliphatic Dibasic Acid Anhydrides**: Ernst Berliner

6. **The Gattermann-Koch Reaction**: Nathan N. Crounse

7. **The Leuckart Reaction**: Maurice L. Moore

8. **Selenium Dioxide Oxidation**: Norman Rabjohn

9. **The Hoesch Synthesis**: Paul E. Spoerri and Adrien S. DuBois

10. **The Darzens Glycidic Ester Condensation**: Melvin S. Newman and Barney J. Magerlein

Volume 6 (1951)

1. **The Stobbe Condensation**: William S. Johnson and Guido H. Daub

2. **The Preparation of 3,4-Dihydroisoquinolines and Related Compounds by the Bischler-Napieralski Reaction**: Wilson M. Whaley and Tuticorin R. Govindachari

3. **The Pictet-Spengler Synthesis of Tetrahydroisoquinolines and Related Compounds**: Wilson M. Whaley and Tuticorin R. Govindachari

4. **The Synthesis of Isoquinolines by the Pomeranz-Fritsch Reaction**: Walter J. Gensler

5. **The Oppenauer Oxidation**: Carl Djerassi

6. **The Synthesis of Phosphonic and Phosphinic Acids**: Gennady M. Kosolapoff

7. **The Halogen-Metal Interconversion Reaction with Organolithium Compounds**: Reuben G. Jones and Henry Gilman

8. **The Preparation of Thiazoles**: Richard H. Wiley, D. C. England, and Lyell C. Behr

9. **The Preparation of Thiophenes and Tetrahydrothiophenes**: Donald E. Wolf and Karl Folkers

10. **Reductions by Lithium Aluminum Hydride**: Weldon G. Brown

Volume 7 (*1953*)

1. **The Pechmann Reaction**: Suresh Sethna and Ragini Phadke

2. **The Skraup Synthesis of Quinolines**: R. H. F. Manske and Marshall Kulka

3. **Carbon-Carbon Alkylations with Amines and Ammonium Salts**: James H. Brewster and Ernest L. Eliel

4. **The von Braun Cyanogen Bromide Reaction**: Howard A. Hageman

5. **Hydrogenolysis of Benzyl Groups Attached to Oxygen, Nitrogen, or Sulfur**: Walter H. Hartung and Robert Simonoff

6. **The Nitrosation of Aliphatic Carbon Atoms**: Oscar Touster

7. **Epoxidation and Hydroxylation of Ethylenic Compounds with Organic Peracids**: Daniel Swern

Volume 8 (*1954*)

1. **Catalytic Hydrogenation of Esters to Alcohols**: Homer Adkins

2. **The Synthesis of Ketones from Acid Halides and Organometallic Compounds of Magnesium, Zinc, and Cadmium**: David A. Shirley

3. **The Acylation of Ketones to Form β-Diketones or β-Keto Aldehydes**: Charles R. Hauser, Frederic W. Swamer, and Joe T. Adams

4. **The Sommelet Reaction**: S. J. Angyal

5. **The Synthesis of Aldehydes from Carboxylic Acids**: Erich Mosettig

6. **The Metalation Reaction with Organolithium Compounds**: Henry Gilman and John W. Morton, Jr.

7. **β-Lactones**: Harold E. Zaugg

8. **The Reaction of Diazomethane and Its Derivatives with Aldehydes and Ketones**: C. David Gutsche

Volume 9 (*1957*)

1. **The Cleavage of Non-enolizable Ketones with Sodium Amide**: K. E. Hamlin and Arthur W. Weston

2. **The Gattermann Synthesis of Aldehydes**: William E. Truce

3. **The Baeyer-Villiger Oxidation of Aldehydes and Ketones**: C. H. Hassall

4. **The Alkylation of Esters and Nitriles**: Arthur C. Cope, H. L. Holmes, and Herbert O. House

5. **The Reaction of Halogens with Silver Salts of Carboxylic Acids**: C. V. Wilson

6. **The Synthesis of β-Lactams**: John C. Sheehan and Elias J. Corey

7. **The Pschorr Synthesis and Related Diazonium Ring Closure Reactions**: DeLos F. DeTar

Volume 10 (*1959*)

1. **The Coupling of Diazonium Salts with Aliphatic Carbon Atoms**: Stanley M. Parmerter

2. **The Japp-Klingemann Reaction**: Robert R. Phillips

3. **The Michael Reaction**: Ernst D. Bergmann, David Ginsburg, and Raphael Pappo

Volume 11 (*1960*)

1. **The Beckmann Rearrangement**: L. Guy Donaruma and Walter Z. Heldt

2. **The Demjanov and Tiffeneau-Demjanov Ring Expansions**: Peter A. S. Smith and Donald R. Baer

3. **Arylation of Unsaturated Compounds by Diazonium Salts**: Christian S. Rondestvedt, Jr.

4. **The Favorskii Rearrangement of Haloketones**: Andrew S. Kende

5. **Olefins from Amines: The Hofmann Elimination Reaction and Amine Oxide Pyrolysis**: Arthur C. Cope and Elmer R. Trumbull

Volume 12 (*1962*)

1. **Cyclobutane Derivatives from Thermal Cycloaddition Reactions**: John D. Roberts and Clay M. Sharts

2. **The Preparation of Olefins by the Pyrolysis of Xanthates. The Chugaev Reaction**: Harold R. Nace

3. **The Synthesis of Aliphatic and Alicyclic Nitro Compounds**: Nathan Kornblum

4. **Synthesis of Peptides with Mixed Anhydrides**: Noel F. Albertson

5. **Desulfurization with Raney Nickel**: George R. Pettit and Eugene E. van Tamelen

Volume 13 (*1963*)

1. **Hydration of Olefins, Dienes, and Acetylenes via Hydroboration**: George Zweifel and Herbert C. Brown

2. **Halocyclopropanes from Halocarbenes**: William E. Parham and Edward E. Schweizer

3. **Free Radical Additions to Olefins to Form Carbon-Carbon Bonds**: Cheves Walling and Earl S. Huyser

4. **Formation of Carbon-Heteroatom Bonds by Free Radical Chain Additions to Carbon-Carbon Multiple Bonds**: F. W. Stacey and J. F. Harris, Jr.

Volume 14 (1965)

1. **The Chapman Rearrangement**: J. W. Schulenberg and S. Archer

2. **α-Amidoalkylations at Carbon**: Harold E. Zaugg and William B. Martin

3. **The Wittig Reaction**: Adalbert Maercker

Volume 15 (1967)

1. **The Dieckmann Condensation**: John P. Schaefer and Jordan J. Bloomfield

2. **The Knoevenagel Condensation**: G. Jones

Volume 16 (1968)

1. **The Aldol Condensation**: Arnold T. Nielsen and William J. Houlihan

Volume 17 (1969)

1. **The Synthesis of Substituted Ferrocenes and Other π-Cyclopentadienyl-Transition Metal Compounds**: Donald E. Bublitz and Kenneth L. Rinehart, Jr.

2. **The γ-Alkylation and γ-Arylation of Dianions of β-Dicarbonyl Compounds**: Thomas M. Harris and Constance M. Harris

3. **The Ritter Reaction**: L. I. Krimen and Donald J. Cota

Volume 18 (1970)

1. **Preparation of Ketones from the Reaction of Organolithium Reagents with Carboxylic Acids**: Margaret J. Jorgenson

2. **The Smiles and Related Rearrangements of Aromatic Systems**: W. E. Truce, Eunice M. Kreider, and William W. Brand

3. **The Reactions of Diazoacetic Esters with Alkenes, Alkynes, Heterocyclic, and Aromatic Compounds**: Vinod David and E. W. Warnhoff

4. **The Base-Promoted Rearrangements of Quaternary Ammonium Salts**: Stanley H. Pine

Volume 19 (1972)

1. **Conjugate Addition Reactions of Organocopper Reagents**: Gary H. Posner
2. **Formation of Carbon-Carbon Bonds via π-Allylnickel Compounds**: Martin F. Semmelhack
3. **The Thiele-Winter Acetoxylation of Quinones**: J. F. W. McOmie and J. M. Blatchly
4. **Oxidative Decarboxylation of Acids by Lead Tetraacetate**: Roger A. Sheldon and Jay K. Kochi

Volume 20 (1973)

1. **Cyclopropanes from Unsaturated Compounds, Methylene Iodide, and Zinc-Copper Couple**: H. E. Simmons, T. L. Cairns, Susan A. Vladuchick, and Connie M. Hoiness
2. **Sensitized Photooxygenation of Olefins**: R. W. Denny and A. Nickon
3. **The Synthesis of 5-Hydroxyindoles by the Nenitzescu Reaction**: George R. Allen, Jr.
4. **The Zinin Reduction of Nitroarenes**: H. K. Porter

Volume 21 (1974)

1. **Fluorination with Sulfur Tetrafluoride**: G. A. Boswell, Jr., W. C. Ripka, R. M. Scribner, and C. W. Tullock
2. **Modern Methods to Prepare Monofluoroaliphatic Compounds**: William A. Sheppard

Volume 22 (1975)

1. **The Claisen and Cope Rearrangements**: Sara Jane Rhoads and N. Rebecca Raulins
2. **Substitution Reactions Using Organocopper Reagents**: Gary H. Posner
3. **Clemmensen Reduction of Ketones in Anhydrous Organic Solvents**: E. Vedejs
4. **The Reformatsky Reaction**: Michael W. Rathke

Volume 23 (1976)

1. **Reduction and Related Reactions of α,β-Unsaturated Compounds with Metals in Liquid Ammonia**: Drury Caine
2. **The Acyloin Condensation**: Jordan J. Bloomfield, Dennis C. Owsley, and Janice M. Nelke
3. **Alkenes from Tosylhydrazones**: Robert H. Shapiro

Volume 24 (*1976*)

1. **Homogeneous Hydrogenation Catalysts in Organic Synthesis**: Arthur J. Birch and David H. Williamson
2. **Ester Cleavages via S_N2-Type Dealkylation**: John E. McMurry
3. **Arylation of Unsaturated Compounds by Diazonium Salts (The Meerwein Arylation Reaction)**: Christian S. Rondestvedt, Jr.
4. **Selenium Dioxide Oxidation**: Norman Rabjohn

Volume 25 (*1977*)

1. **The Ramberg-Bäcklund Rearrangement**: Leo A. Paquette
2. **Synthetic Applications of Phosphoryl-Stabilized Anions**: William S. Wadsworth, Jr.
3. **Hydrocyanation of Conjugated Carbonyl Compounds**: Wataru Nagata and Mitsuru Yoshioka

Volume 26 (*1979*)

1. **Heteroatom-Facilitated Lithiations**: Heinz W. Gschwend and Herman R. Rodriguez
2. **Intramolecular Reactions of Diazocarbonyl Compounds**: Steven D. Burke and Paul A. Grieco

Volume 27 (*1982*)

1. **Allylic and Benzylic Carbanions Substituted by Heteroatoms**: Jean-François Biellmann and Jean-Bernard Ducep
2. **Palladium-Catalyzed Vinylation of Organic Halides**: Richard F. Heck

Volume 28 (*1982*)

1. **The Reimer-Tiemann Reaction**: Hans Wynberg and Egbert W. Meijer
2. **The Friedländer Synthesis of Quinolines**: Chia-Chung Cheng and Shou-Jen Yan
3. **The Directed Aldol Reaction**: Teruaki Mukaiyama

Volume 29 (*1983*)

1. **Replacement of Alcoholic Hydroxy Groups by Halogens and Other Nucleophiles via Oxyphosphonium Intermediates**: Bertrand R. Castro

2. **Reductive Dehalogenation of Polyhalo Ketones with Low-Valent Metals and Related Reducing Agents**: Ryoji Noyori and Yoshihiro Hayakawa

3. **Base-Promoted Isomerizations of Epoxides**: Jack K. Crandall and Marcel Apparu

Volume 30 (*1984*)

1. **Photocyclization of Stilbenes and Related Molecules**: Frank B. Mallory and Clelia W. Mallory

2. **Olefin Synthesis via Deoxygenation of Vicinal Diols**: Eric Block

Volume 31 (*1984*)

1. **Addition and Substitution Reactions of Nitrile-Stabilized Carbanions**: Simeon Arseniyadis, Keith S. Kyler, and David S. Watt

Volume 32 (*1984*)

1. **The Intramolecular Diels-Alder Reaction**: Engelbert Ciganek

2. **Synthesis Using Alkyne-Derived Alkenyl- and Alkynylaluminum Compounds**: George Zweifel and Joseph A. Miller

Volume 33 (*1985*)

1. **Formation of Carbon-Carbon and Carbon-Heteroatom Bonds via Organoboranes and Organoborates**: Ei-Ichi Negishi and Michael J. Idacavage

2. **The Vinylcyclopropane-Cyclopentene Rearrangement**: Tomáš Hudlický, Toni M. Kutchan, and Saiyid M. Naqvi

Volume 34 (*1985*)

1. **Reductions by Metal Alkoxyaluminum Hydrides**: Jaroslav Málek

2. **Fluorination by Sulfur Tetrafluoride**: Chia-Lin J. Wang

Volume 35 (*1988*)

1. **The Beckmann Reactions: Rearrangements, Elimination-Additions, Fragmentations, and Rearrangement-Cyclizations**: Robert E. Gawley

2. **The Persulfate Oxidation of Phenols and Arylamines (The Elbs and the Boyland-Sims Oxidations)**: E. J. Behrman

3. **Fluorination with Diethylaminosulfur Trifluoride and Related Aminofluorosulfuranes**: Miloš Hudlický

Volume 36 (1988)

1. **The [3 + 2] Nitrone-Olefin Cycloaddition Reaction**: Pat N. Confalone and Edward M. Huie

2. **Phosphorus Addition at sp^2 Carbon**: Robert Engel

3. **Reduction by Metal Alkoxyaluminum Hydrides. Part II. Carboxylic Acids and Derivatives, Nitrogen Compounds, and Sulfur Compounds**: Jaroslav Málek

Volume 37 (1989)

1. **Chiral Synthons by Ester Hydrolysis Catalyzed by Pig Liver Esterase**: Masaji Ohno and Masami Otsuka

2. **The Electrophilic Substitution of Allylsilanes and Vinylsilanes**: Ian Fleming, Jacques Dunoguès, and Roger Smithers

Volume 38 (1990)

1. **The Peterson Olefination Reaction**: David J. Ager

2. **Tandem Vicinal Difunctionalization: β-Addition to α,β-Unsaturated Carbonyl Substrates Followed by α-Functionalization**: Marc J. Chapdelaine and Martin Hulce

3. **The Nef Reaction**: Harold W. Pinnick

Volume 39 (1990)

1. **Lithioalkenes from Arenesulfonylhydrazones**: A. Richard Chamberlin and Steven H. Bloom

2. **The Polonovski Reaction**: David Grierson

3. **Oxidation of Alcohols to Carbonyl Compounds via Alkoxysulfonium Ylides: The Moffatt, Swern, and Related Oxidations**: Thomas T. Tidwell

CONTENTS

CHAPTER 1

THE PAUSON–KHAND CYCLOADDITION REACTION FOR SYNTHESIS OF CYCLOPENTENONES

NEIL E. SCHORE

Department of Chemistry, University of California, Davis, California

CONTENTS

Organic Reactions, *Vol. 40*, Edited by Leo A. Paquette et al.
ISBN 0-471-53841-8

ACKNOWLEDGMENTS

I wish to thank the following individuals for providing reprints and preprints describing their work in this area and for making helpful comments and suggestions in the preparation of this chapter: Profs. R. Caple, A. de Meijere, D. H. Hua, B. Kerr, M. E. Krafft, P. D. Magnus, P. L. Pauson, F. Serratosa, and W. A. Smit. Financial support from the National Institutes of Health, the University of California, and the Chevron Research Corporation is gratefully acknowledged. Finally, I wish to thank Prof. R. G. Bergman, who introduced me to organometallic chemistry in general, and the Pauson–Khand reaction in particular, for his continuing support and his encouragement.

INTRODUCTION

The Pauson–Khand reaction is a cocycloaddition of alkynes, alkenes, and carbon monoxide to generate cyclopentenones in a formal [2+2+1] cycloaddition process. The reaction was discovered and first reported in detail by Ihsan U. Khand and Peter L. Pauson in 1973 in the course of a study aimed principally at the preparation and characterization of various alkene and alkyne complexes derived from $Co_2(CO)_8$.[1] The generality of the reaction, typically carried out by heating a mixture of the alkene and the readily formed $Co_2(CO)_6$ complex of the alkyne in hydrocarbon or ethereal solvent, was established primarily by an extensive series of studies carried out by the Pauson group throughout the 1970s.

$$R^1C{\equiv}CR^2 + R^3R^4C{=}CR^5R^6 \xrightarrow{Co_2(CO)_8} \text{cyclopentenone } (R^1, R^2, R^3, R^4, R^5, R^6)$$

The earliest studies established that synthetically reasonable yields, usually in the 40–60% range, and significant regio- and stereoselectivity could be expected even for relatively simple examples. Development of this reaction has reached the point where predictable success and control of selectivity are possible. The variety of successful reactions and systems accessible by means of this cycloaddition is now quite substantial.

This chapter addresses the scope and generality of this reaction as well as the current state of the art with regard to control of regio- and stereochemistry. Both inter- and intramolecular versions of the cycloaddition are presented, and current models that have been put forth concerning the reaction mechanism are noted. Synthetic applications of the Pauson–Khand reaction have taken many forms, and representative examples of all major types of systems accessed are presented. The process exemplifies a nearly ideal merging of organometallic with synthetic chemistry, in which advances in each area have promoted development in the other in an almost symbiotic manner. Excellent shorter reviews on the Pauson–Khand reaction have been published.[2–4] We hope that this chapter will succeed in providing the reader with both an overview of the reaction as well as a sufficiently detailed understanding of its complexities to permit meaningful evaluation as a possible solution to current or future research needs.

MECHANISM

The only direct evidence that bears on the mechanism of the Pauson–Khand reaction is the unambiguous observation that the alkyne complex $Co_2(CO)_6{\cdot}R^1C{\equiv}CR^2$ is involved in the first stage of the process. No intermediates have been detected beyond this alkyne complex. The current level of mechanistic understanding is instead inferred from observations of regio- and stereochemistry in a large number of examples, and is illustrated schematically in Eq. 1 (for clarity the alkene substituents have been omitted).[5–7] It is usually assumed that complexation of the alkene to one cobalt atom takes place via a dissociative mechanism involving initial loss of CO. This process is almost certainly reversible. Subsequently, irreversible insertion of the complexed face of the alkene π bond into one of the formal cobalt–carbon bonds of the alkyne complex occurs, in the step that is probably both rate-

and product-determining and is followed by addition of CO to the coordinatively unsaturated cobalt atom. The metallocycle that forms may proceed to product by a standard sequence of steps beginning with migratory insertion of a cobalt-bound CO, addition of a ligand (e.g., another CO molecule), and reductive elimination of the $Co(CO)_3$ moiety. The structure obtained is simply the $Co_2(CO)_6$ complex of the final enone; loss of the $Co_2(CO)_6$ fragment, either before or after attachment of an additional ligand, completes the process.

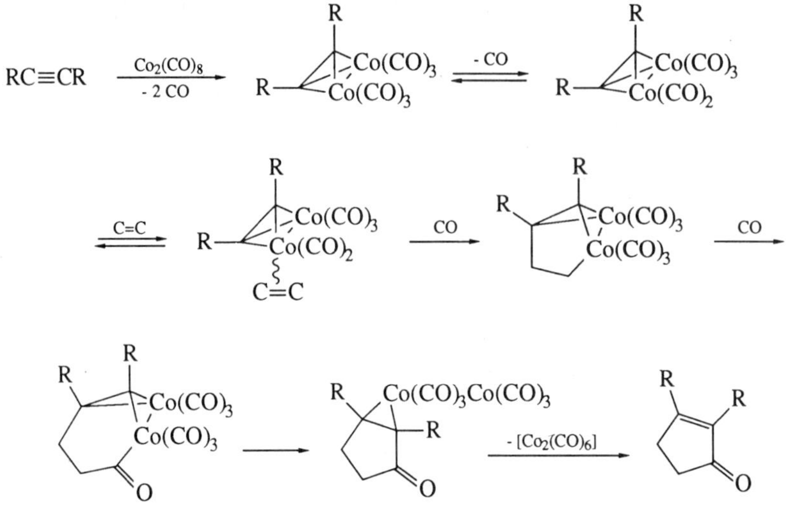

(Eq. 1)

The structure of the product is influenced both by steric interactions associated with the alkene insertion as well as by the structure of the favored configurational and conformational isomers of the precursor alkene complex. When both the alkene and alkyne are unsymmetrically substituted, eight such structures capable of subsequent insertion reaction may be drawn. These are shown as **1–8** for the case in which both the alkene and alkyne are terminal. In structures **1–6** the alkene is complexed *cis* to the bond between cobalt and the substituted alkyne carbon, and these complexes presumably suffer steric destabilization. Of these, isomers **1** and **2**, in which the alkene eclipses the bond between cobalt and the unsubstituted alkyne carbon, are probably preferred over **3–6**, where interaction with R^1 is most severe. However, the isomers most likely to lead to insertion are **7** and **8**. The alkene is complexed *trans* to the bond between cobalt and the substituted alkyne carbon, and eclipses the bond between cobalt and the unsubstituted alkyne carbon. Insertion occurs into this cobalt–carbon bond, forming the first new carbon–carbon bond with the least sterically hindered alkyne carbon. This fixes the regiochemistry of the alkyne component, placing the larger of its substituents in the 2 position of the final cyclopentenone. This result is general for alkynes containing substituents of different sizes.

Depending on the size of the substituent(s) on the alkene and on the less hindered alkyne carbon, there may be a significant conformational preference for isomer **7**, in which the more heavily substituted end of the alkene is oriented toward a carbonyl group, and away from the bond to the alkyne carbon. When this occurs, it leads to regioselectivity in incorporation of the alkene component as well, resulting in a preference for the 5-substituted cyclopentenone **9**. However, the degree of steric interaction necessary for high selectivity is present only in more heavily substituted situations, such as reactions with internal alkynes. Most Pauson–Khand cycloadditions involving terminal alkynes and terminal alkenes are unselective in the incorporation of the alkene, although regioselectivity involving the alkyne remains high.

Electronic effects are also observed in the Pauson–Khand reaction. Alkynes conjugated to electron-withdrawing groups do not undergo the cycloaddition. Alkenes bearing electron-withdrawing groups react anomalously, giving 1,3-dienes. This reaction is completely regioselective, with the new carbon–carbon bond forming between the less-hindered alkyne carbon and the less-hindered alkene carbon.[8,9] It is reasonable to assume that complexation of the alkene and subsequent insertion occur as in the normal cycloaddition sequence. The π-conjugating, electron-withdrawing group EWG on the alkene apparently renders a β-hydrogen elimination–reductive elimination sequence competitive with CO insertion, leading to the diene product. Intermediate behavior is observed with styrene derivatives, most of which give both diene and cyclopentenone products, both with complete regioselectivity. Discussion of electronic effects on regioselectivity is presented in a subsequent section.

R Co(CO)$_3$ Co(CO)$_3$ H EWG → R Co(CO)$_3$ Co(CO)$_3$ H EWG → R H Co(CO)$_3$Co(CO)$_3$ EWG

- [Co$_2$(CO)$_6$] → R H EWG

SCOPE AND LIMITATIONS

The Pauson–Khand reaction is quite tolerant of substrate structure. The most satisfactory alkynes are acetylene and simple terminal alkynes, including arylalkynes. Internal alkynes typically give lower yields of cyclopentenones. The scope of the reaction with respect to the alkene is somewhat more limited. Strained cyclic alkenes are generally good substrates, frequently giving yields in excess of 50%. However, steric hindrance around the double bond reduces cycloaddition reactivity considerably. This result is apparently due to a reduction in the ability of the alkene to compete with additional molecules of alkyne for reaction with the initially formed $Co_2(CO)_6 \cdot RC{\equiv}CR^1$ complex. As a result, side reactions such as alkyne trimerization and multicomponent cycloadditions involving only alkyne and carbon monoxide become dominant.[10] Simple acyclic alkenes and unstrained cyclic alkenes are less satisfactory, although ethylene itself is an exception, reacting smoothly with a variety of alkynes.

The Pauson–Khand reaction tolerates a wide range of remote functionality including ethers, alcohols, ketones, ketals, esters, tertiary amines, tertiary amides, thioethers, and aromatic and heteroaromatic rings. Complications do arise with substrates bearing allylic or propargylic functionality, and the effects of conjugation of either substrate π system with another carbon–carbon π bond vary with the specific nature of the substrate.

THE INTERMOLECULAR PAUSON–KHAND CYCLOADDITION REACTION

The intermolecular Pauson–Khand reaction of simple acylic alkenes is generally limited by both low reactivity and lack of regiocontrol in incorporation of the alkene, although incorporation of the alkyne remains highly or totally regioselective regardless of the structure of the alkene.[11] Arylalkenes and certain heteroatom-substituted systems offer better opportunities for alkene regiocontrol. Cyclic alkenes present a quite different situation. Cycloaddition yields are good to excellent in many cases, and considerable stereo- and regioselectivity is obtained with respect to both the alkyne and the alkene components. Polycyclic molecules in which the alkene is strained provide the most favorable results, both in yield and selectivity. Reaction conditions vary widely, mainly as a function of the structure of the alkene; details are presented in the appropriate sections that follow.

Acyclic Alkenes

Ethylene itself reacts readily with the $Co_2(CO)_6$ complexes of terminal alkynes. The reaction proceeds slowly even at room temperature, although forcing conditions (toluene, 80–160°, 50–120 atm initial ethylene pressure, autoclave) are required for best results. Yields of cyclopentenones typically fall in the 30–60% range, and may be improved somewhat by the addition of tri-*n*-butylphosphine oxide to the reaction mixture (Eq. 2).[12–15]

$$n\text{-}C_5H_{11}C{\equiv}CH\cdot Co_2(CO)_6 + CH_2{=}CH_2 \xrightarrow[\text{toluene, 36 h}]{85°,\ 120\ \text{atm}} \text{2-}(C_5H_{11}\text{-}n)\text{cyclopent-2-enone}\ (55\%) \qquad \text{(Eq. 2)}$$

Internal alkynes have also been used with some success. The example shown in Eq. 3 is one of the very few cases of incomplete regioselectivity in alkyne incorporation.[16] In this regard it is noted that no systematic examination of regioselectivity has been carried out for alkynes in which the substituents are more similar in size than methyl vs. primary alkyl.

$$C_2H_5C{\equiv}CCH_3\cdot Co_2(CO)_6 + CH_2{=}CH_2 \xrightarrow[\text{toluene, 36 h}]{110°,\ 35\ \text{atm}} \text{2-}C_2H_5\text{-3-}CH_3\text{-cyclopent-2-enone}\ (24\%) + \text{2-}CH_3\text{-3-}C_2H_5\text{-cyclopent-2-enone}\ (3\%) \qquad \text{(Eq. 3)}$$

Pauson–Khand cycloaddition reactions of gaseous substituted alkenes require conditions similar to those used for ethylene itself. Reactions involving liquid alkenes are typically carried out in solution under nitrogen. A variation that occasionally gives superior results uses a catalytic amount of the $Co_2(CO)_6$·alkyne complex, with the reaction carried out under an atmosphere of free alkyne and carbon monoxide. As mentioned earlier, terminal aliphatic alkenes usually give modest yields, but alkene incorporation occurs without regioselectivity in reactions with $Co_2(CO)_6$ complexes of terminal alkynes (Eq. 4).[7]

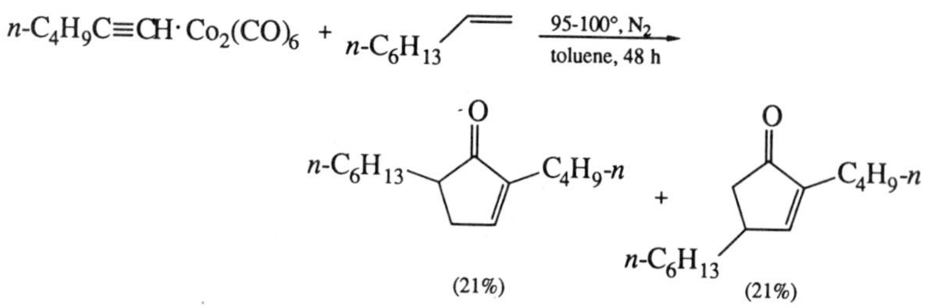

$$n\text{-}C_4H_9C{\equiv}CH\cdot Co_2(CO)_6 + n\text{-}C_6H_{13}CH{=}CH_2 \xrightarrow[\text{toluene, 48 h}]{95\text{-}100°,\ N_2}$$

2-n-butyl-5-n-hexylcyclopent-2-enone (21%) + 2-n-butyl-4-n-hexylcyclopent-2-enone (21%) (Eq. 4)

Increased alkene regioselectivity but reduced chemical yields are observed in reactions with internal alkynes (Eq. 5).[7] As described above, the site of coordination of the alkene determines alkyne regioselectivity, while the conformation of the coordinated alkene prior to insertion determines alkene regioselectivity.

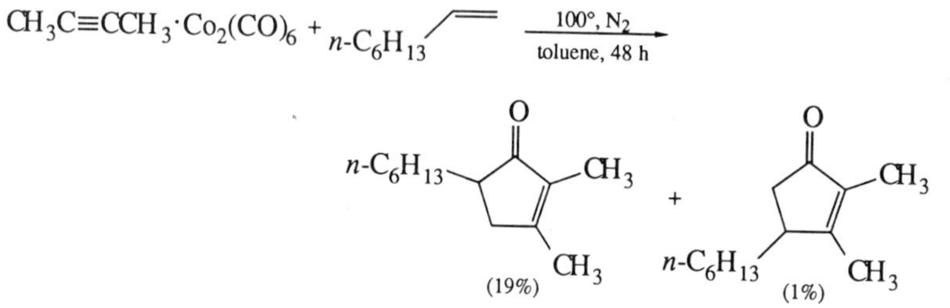

$$CH_3C{\equiv}CCH_3\cdot Co_2(CO)_6 + n\text{-}C_6H_{13}CH{=}CH_2 \xrightarrow[\text{toluene, 48 h}]{100°,\ N_2}$$

2,3-dimethyl-5-n-hexylcyclopent-2-enone (19%) + 2,3-dimethyl-4-n-hexylcyclopent-2-enone (1%) (Eq. 5)

Examples of unexpectedly high, but solvent-dependent regiocontrol have been reported for allyl ethers.[4,17] In a very interesting and possibly related observation by Krafft, alkenes containing groups at a homoallylic position capable of acting as soft ligands give higher yields and often very high regioselectivities. This result is thought to result from coordination of the heteroatom to cobalt prior to insertion, thereby fixing the conformation of the alkene to favor the 5-substituted product.[6] An example is given in Eq. 6.

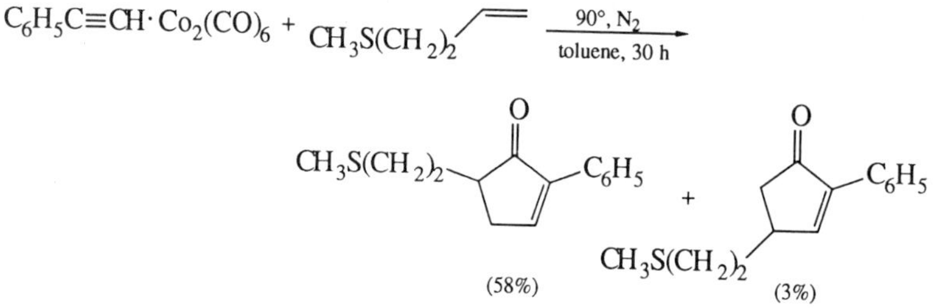

$$C_6H_5C{\equiv}CH\cdot Co_2(CO)_6 + CH_3S(CH_2)_2CH{=}CH_2 \xrightarrow[\text{toluene, 30 h}]{90°,\ N_2}$$

2-phenyl-5-[$CH_3S(CH_2)_2$]cyclopent-2-enone (58%) + 2-phenyl-4-[$CH_3S(CH_2)_2$]cyclopent-2-enone (3%) (Eq. 6)

Alkenes bearing electron-withdrawing groups give rise entirely to conjugated dienes via the mechanism already described (Eq. 7).[8] Conjugated acyclic dienes, in contrast to *cyclic* dienes (vide infra), also give only linear oligomerization, resulting in stereoisomeric mixtures of acyclic polyene products.[2]

$$C_6H_5C{\equiv}CH\cdot Co_2(CO)_6 + CH_3CH{=}CHCHO \longrightarrow C_6H_5CH{=}CHCH{=}C(CH_3)CHO \quad \text{(Eq. 7)}$$

major stereoisomer

Styrene derivatives have been extensively studied and are intermediate, giving comparable yields of dienes and 5-arylcyclopentenones, both with complete regioselectivity.[11,18,19] The example in Eq. 8 is typical, both in terms of overall yield as well as chemoselectivity.[18]

$$C_6H_5C{\equiv}CH\cdot Co_2(CO)_6 + 4\text{-}CH_3OC_6H_4CH{=}CH_2 \xrightarrow[\text{toluene, 7 h}]{110°,\ N_2} 4\text{-}CH_3OC_6H_4CH{=}CHCH{=}CHC_6H_5\ (42\%) + \text{5-(4-}CH_3OC_6H_4\text{)-2-}C_6H_5\text{-cyclopent-2-enone}\ (27\%) \quad \text{(Eq. 8)}$$

More heavily substituted acyclic alkenes are not useful substrates unless homoallylic heteroatom substitution is present to facilitate complexation to cobalt. Alkene stereochemistry is generally lost in the cycloaddition process.[6]

Vinyl and allyl halides cyclize in low yield with apparent hydrogenolysis of the carbon–halogen bond.[11] Although there are examples of partial success in Pauson–Khand cycloadditions of vinyl ethers and esters, neither the substrates nor the products (especially 4-alkoxycyclopentenones) tolerate the reaction conditions very well.[20]

Methylenecyclopropane and methylenecyclobutane give poor results under typical stoichiometric conditions. However, Smit finds that adsorption of a mixture of alkene and $Co_2(CO)_6$-complexed alkyne on any of several solid supports (e.g., silica, alumina, Zeolite) and heating of the *dry solid* leads to good-to-excellent yields of cycloaddition products.[21] It is thought that adsorption may promote ligand exchange and, therefore, facilitate alkene complexation. Reactions of methylenecyclopropane with internal alkynes give only 5-spiroannulated cyclopentenones, indicating steric control of alkene complex conformation (Eq. 9; cf. Eq. 5). Regioselectivity is lower and favors the 4-spiroannulated product from terminal alkynes and acetylene, where large steric interactions upon complexation are absent (Eq. 10). Instead, alkene orientation is probably controlled upon insertion, with the more substituted end of the alkene preferring to bond to an unsubstituted alkyne carbon rather than to a $Co(CO)_3$ moiety. A small electronic effect, which is discussed in a later section on bridged bicyclic alkenes, may also contribute to this result.

$$C_2H_5C{\equiv}CC_2H_5{\cdot}Co_2(CO)_6 + \text{methylenecyclopropane} \xrightarrow[50^\circ,\ 4\ h]{\text{Zeolite NaX}}$$

(<3%) + (54%) (Eq. 9)

$$C_6H_5C{\equiv}CH{\cdot}Co_2(CO)_6 + \text{methylenecyclopropane} \xrightarrow[50^\circ,\ 2\ h]{\text{Zeolite NaX}}$$

(65%) + (14%) (Eq. 10)

Monocyclic Alkenes

Simple cyclopropenes and cyclobutenes have not been studied as Pauson–Khand substrates although based on results with bicyclic systems (vide infra) at least the latter should give good results. Cyclopentene reacts with terminal alkynes to give 30–70% yields of bicyclo[3.3.0]octenones.[11,13,14,22] Better yields with gaseous alkynes are often obtained under "catalytic" conditions, in which a benzene solution of the alkene is heated in the presence of ca. 0.2 equivalent of the $Co_2(CO)_6{\cdot}RC{\equiv}CH$ complex under an atmosphere consisting of a ca. 1:1 mixture of the alkyne and carbon monoxide. Several turnovers may be obtained, although the improvements in yield are only occasionally dramatic. Reported yields in these reactions are based either on starting alkene or on starting $Co_2(CO)_6{\cdot}RC{\equiv}CH$ complex. With less volatile alkynes reaction with a stoichiometric amount of complex is carried out in alkane or arene solvent at 70–110°, or in heptane at 110–120° (sealed tube). Pauson has found that addition of 1 equivalent of tri-*n*-butylphosphine oxide often improves results in these cases, perhaps by facilitating loss of CO (Eq. 11).[15] Catalytic conditions give much better results than stoichiometric conditions in reactions of trisubstituted cycloalkenes such as 1-methylcyclopentene with acetylene (Eq. 12).[22] The alkene regioselectivity is similar to that of methylenecyclopropane. Ring-containing vinyl esters and ethers give poor results.[4]

$n\text{-}C_5H_{11}C\equiv CH\cdot Co_2(CO)_6$ + [cyclopentene] $\xrightarrow[\text{hexane, 24 h}]{70°,\ N_2}$ [bicyclic enone]$-C_5H_{11}\text{-}n$ (Eq. 11)

(41%)
[with $(n\text{-}C_4H_9)_3PO$, 70%]

$HC\equiv CH$ + CH_3-[cyclopentene] $\xrightarrow[\text{benzene, 24 h}]{\text{catalytic } HC\equiv CH\cdot Co_2(CO)_6,\ 80°,\ CO}$ [O, CH_3 product] (61%) + [CH_3, O product] (8%) (Eq. 12)

Unlike acyclic dienes, cyclopentadienes and fulvenes react with alkynes to give bicyclo[3.3.0]octenones in good-to-excellent yield. As seen with reactions of styrenes, the favored position of conjugated unsaturation is the 5 position of the product (Eq. 13).[23]

$C_6H_5C\equiv CH\cdot Co_2(CO)_6$ + [cyclopentadiene] $\xrightarrow[\text{toluene, 5 h}]{110°,\ N_2}$ [O]$-C_6H_5$ (≈50%) + [O]$-C_6H_5$ (≈10%) (Eq. 13)

Excellent yields are obtained in cycloadditions of 2,5-dihydrofuran with gaseous alkynes under catalytic reaction conditions. As in the case of cyclopentene, reactions with less volatile alkynes under stoichiometric conditions give good results which are usually improved by addition of tri-*n*-butylphosphine oxide.[15] Substituted dihydrofurans give poorer yields and low regioselectivity.[2,24]

Cyclohexene itself gives very poor results, although the presence of a homoallylic amine in the substrate leads to some improvement.[6,11] Cycloheptene and cyclooctene give moderate yields of cyclopentenones, but only with phenylacetylene, not with alkyl acetylenes.[11] Cyclohexadienes undergo Diels–Alder reaction with alkynes under Pauson–Khand conditions giving bicyclo[2.2.2]octa-1,4-dienes, which then react with additional alkyne to give tricyclic cyclopentenones (Eq. 14) (vide infra).[25]

$C_6H_5C\equiv CH\cdot Co_2(CO)_6$ + [cyclohexadiene] $\xrightarrow[\text{toluene, 5-6 h}]{60\text{-}80°,\ N_2}$ [C_6H_5, O tricyclic product]$-C_6H_5$ (Eq. 14)

(major isomer, 65%)

Ring-Fused Bicyclic and Polycyclic Alkenes

Bicyclo[3.2.0]hept-6-enes, containing a fused cyclobutene ring, react with both terminal and internal alkynes to form *cis*,*anti*,*cis*-tricyclo[5.3.0.0^{2,6}]dec-4-en-3-ones. With bicyclo[3.2.0]hepta-3,6-dienes reaction occurs entirely at the cyclobutene double bond. The cycloaddition is completely stereoselective, taking place exclusively on the less hindered *exo* face of the bicyclic alkene.[1] Alkene regiochemistry is directed by steric interactions involving allylic substituents. Insertion to give metallocycle **11** avoids a 1,3-pseudodiaxial interaction between the allylic substituent and a $Co(CO)_3$ moiety, which would occur in the regioisomeric intermediate **12**. This leads via the mechanism shown earlier exclusively to enone **13**, with the larger allylic substituent farther from the newly formed cyclopentenone carbonyl. Cycloadditions of bicy-

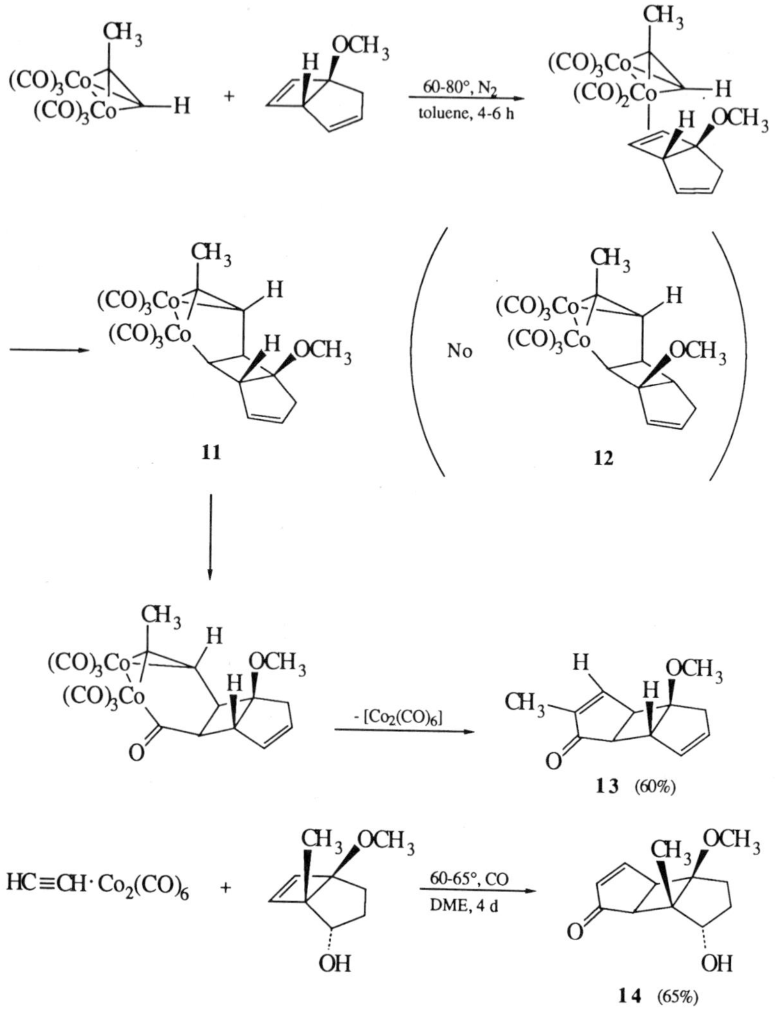

CH3
(CO)3Co
(CO)3Co
H
+
OCH3
60-80°, N2
toluene, 4-6 h
(CO)2Co
No
11
12
- [Co2(CO)6]
13 (60%)
HC≡CH·Co2(CO)6
60-65°, CO
DME, 4 d
OH
O
14 (65%)

clo[3.2.0]hept-6-enes are completely regioselective. Upon replacement of a ring fusion hydrogen with methyl, which is effectively larger than methoxy, enone **14** is the only product formed.[26,27]

Bicyclo[3.3.0]oct-2-enes undergo isomerization to bicyclo[3.3.0]oct-1-enes prior to cycloaddition; the product is therefore an angularly fused triquinane, rather than the linear isomer (Eq. 15).[28] (Exceptions are found in reactions of silyl- and cyclopropylacetylenes.[29]) Metal-mediated hydride transfer via a π-allyl complex is probably involved. Evidence for the presence of metal hydrides is found in the observation of products of both hydrogenation and hydrogenolysis under more forcing conditions. The regioselectivity is similar to that observed for 1-methylcyclopentene. Similar yields, with a 1:1 mixture of isomeric products, are found with 2-methylbicyclo[3.3.0]oct-2-ene, the only example of a tetrasubstituted alkene successfully undergoing Pauson–Khand cycloaddition.[22]

$t\text{-}C_4H_9(CH_3)_2SiOCH_2C{\equiv}CH \cdot Co_2(CO)_6$ + [bicyclo[3.3.0]oct-2-ene] $\xrightarrow[\text{benzene, 24 h}]{80°,\ CO}$ [angular triquinane enone bearing $t\text{-}C_4H_9(CH_3)_2SiOCH_2$] (18%) (Eq. 15)

Indenes and acenaphthalene are similar to cyclopentadiene, reacting with alkynes to give largely or exclusively cyclopentenones; the indenes react with complete regioselectivity.[21,23] Dihydronaphthalene gives similar results, also with the expected regioselectivity for a styrene analog.[18]

Bridged Bicyclic and Polycyclic Alkenes

Norbornene and a number of its polycyclic derivatives give cyclopentenones in yields of 20–40% from internal alkynes, and 30 to >90% from terminal alkynes, including cyclopropyl- and trimethylsilylacetylene (Eq. 16).[29,30] Stoichiometric, catalytic, and dry state adsorption conditions have all been used successfully. Stereoselectivity in formation of the *exo* ring fusion is always 100%, as is regioselectivity in incorporation of the alkyne but not necessarily the alkene.[3,14,21]

$(CH_3)_3SiC{\equiv}CH \cdot Co_2(CO)_6$ + [norbornadiene] $\xrightarrow[\text{toluene, 18 h}]{80\text{-}90°}$ [tricyclic enone]–$Si(CH_3)_3$ (93%)

(Eq. 16)

Phenylacetylene-$Co_2(CO)_6$ reacts with (*R*)-(+)-2,3,*O*-isopropylideneglycerine-1-diphenylphosphine [(*R*)-(+)-glyphos] to give two separable diastereomers of phenylacetylene-$Co_2(CO)_5$-(*R*)-glyphos. Cycloaddition of the $(-)_{589}$ diastereomer with norbornene at a temperature at which diastereomer equilibration is slow (≤60°) gives the enone in only 31% yield, but 100% enantiomerically pure (Eq. 17). This result does *not* require direct steric interaction between the chiral ligand and the complexed alkene. Complexation and insertion of the alkene at exclusively one of the two diastereotopic cobalt atoms [presumably at the $Co(CO)_3$ rather than at the $Co(CO)_2$-(phosphine)] is sufficient, as the reaction is already inherently face-selective. It is fitting that this first demonstration of optical induction in Pauson–Khand cycloaddition comes in part from the Pauson group, where the intermolecular version of this reaction was explored to the greatest extent.[31]

$C_6H_5C{\equiv}CH \cdot Co_2(CO)_5$[(*R*)-(+)-glyphos] + [norbornene] —(45°, ultrasound; toluene, 6 h)→ [enone]–C_6H_5 (Eq. 17)

(31%, 100% ee)
absolute configuration unknown

If other types of double bonds are present only the norbornene undergoes cycloaddition.[32] Cycloaddition succeeds with 1-arylnorbornenes as well. An electronic regioselectivity effect is evident by comparing cycloadditions of norbornen-2-ols, which are not regioselective, with those of norbornen-2-ones. The double bond in the latter, polarized by homoconjugation, reacts preferentially to give **15** via metallocycle **16**, in which the partially positive C-5 is bonded to a carbon of the complexed alkyne, rather than to a partially positive cobalt center.[33]

$CH_3C{\equiv}CH \cdot Co_2(CO)_6$ + [norbornenone] —(60-65°, CO; isooctane, 20 h)→ [enone]–CH_3 (15%) + **15** (43%) via **16** [$(CO)_3Co$, $(CO)_3Co$, CH_3, H, O]

Reactions with certain functionalized alkynes such as 4-pentyn-1-ol give low yields (<25%), in part a result of competing alkyne trimerization. Suppression of this side reaction is achieved by covalent attachment to a functionalized polystyrene. Pauson–Khand reaction of the polymer-bound alkyne followed by cleavage of the polymer linkage affords excellent yields of the cycloaddition product (Eq. 18).[34]

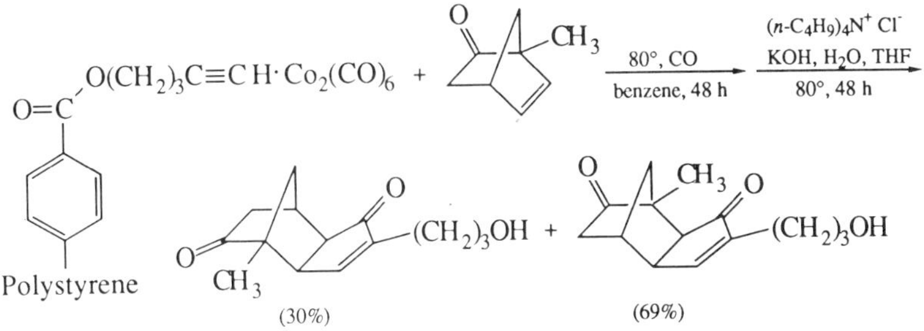

(Eq. 18)

Norbornadiene may react at one or both double bonds. Yields are moderate in the former case, lower in the latter; both are regioselective.[1,14,32] Appreciable amounts of *endo*-fused products result in the reaction with acetylene, but other alkynes give nearly exclusively *exo* products (Eq. 19).[15,35] Polymer-linkage of alkynols gives improved yields of both single and double cycloaddition.[34] Adsorption on silica also improves yields.[21] A heterocyclic analog, 2,3-diaza-5-norbornene, undergoes cycloaddition, but 7-oxanorbornadiene deoxygenates.[2,3]

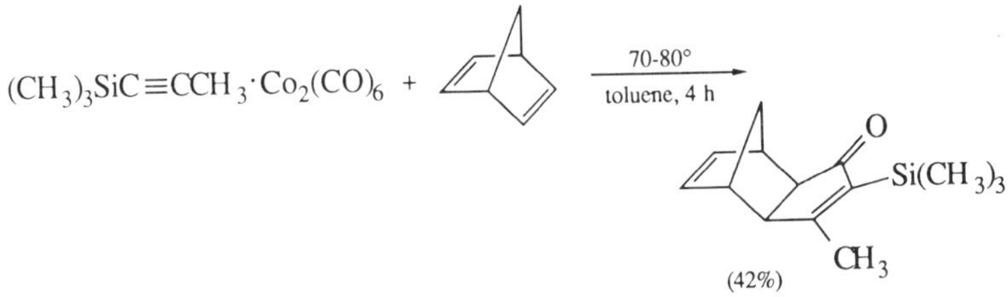

(Eq. 19)

Derivatives of bicyclo[2.2.2]octene, 8-oxabicyclo[3.2.1]oct-6-ene, and 8-azabicyclo[3.2.1]oct-6-ene all cycloadd readily to acetylene and terminal alkynes.[5,25,36–38] However, addition of a methyl group to the double bond of the oxabicyclooctene or even bulky bridgehead substitution eliminates cycloaddition reactivity, and smaller bridgehead substitutents result in only low regioselectivity (Eq. 20).[5]

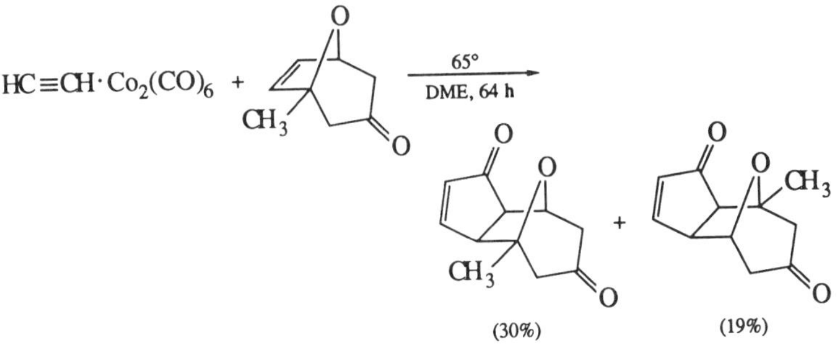

(Eq. 20)

THE INTRAMOLECULAR PAUSON–KHAND CYCLOADDITION REACTION

Intramolecular cycloadditions occur upon complexation of derivatives of hept-1-en-6-ynes and oct-1-en-7-ynes to $Co_2(CO)_8$ and subsequent heating, giving bicyclic enones. Intramolecularity permits satisfactory results with terminal, internal, and even trisubstituted alkenes, although reactions of trisubstituted alkenes are limited to terminal alkynes by steric hindrance. Hex-1-en-5-yne undergoes alkyne trimerization instead, avoiding four-membered ring formation.[39]

All-Carbon Enynes

Stoichiometric solution-phase conditions are most frequently used for these systems. Best yields are generally obtained by heating the reactants in a sealed tube. Substitution effects in cycloadditions of hept-1-en-6-ynes to give bicyclo[3.3.0]oct-1-en-3-ones have been well studied, and have been accommodated by Magnus into the mechanistic scheme previously described. In particular, a stereochemical preference for substituents at the allylic (C-3) and propargylic (C-5) positions of the substrate to be on the *exo* face of the bicyclic product is observed, and this is enhanced by bulky substitution on the alkyne terminus (Eqs. 21 and 22).[40,41] Steric interactions between the *endo* allylic and propargylic positions and the alkyne substituent are responsible. Structures **17** and **18** show the relevant conformations for the substrate in Eq. 21 with the ring fusion bond that would form upon alkene insertion shown by a dashed line. The development of a severe pseudo-1,3-diaxial interaction upon insertion from conformation **17** (the "Felkin–Ahn" conformation, with the double bond terminus closest to the medium-size group) diverts the reaction through **18**, leading to the product with the allylic substituent *cis* to the ring fusion hydrogen and on the *exo* face of the molecule.

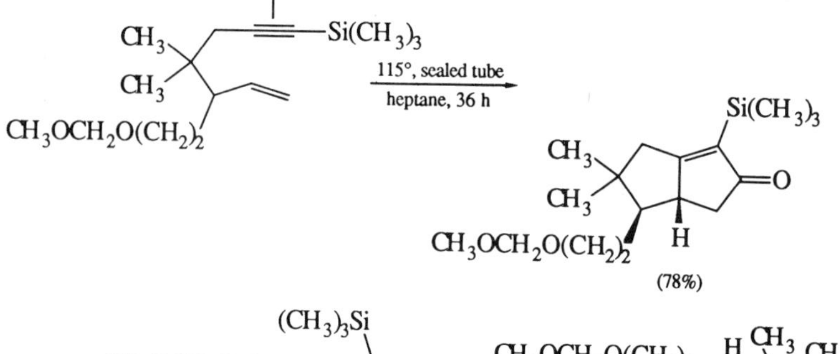

$Co_2(CO)_6$ $Si(CH_3)_3$ CH_3 CH_3 $CH_3OCH_2O(CH_2)_2$

115°, sealed tube / heptane, 36 h

$Si(CH_3)_3$ CH_3 CH_3 O $CH_3OCH_2O(CH_2)_2$ H (78%) (Eq. 21)

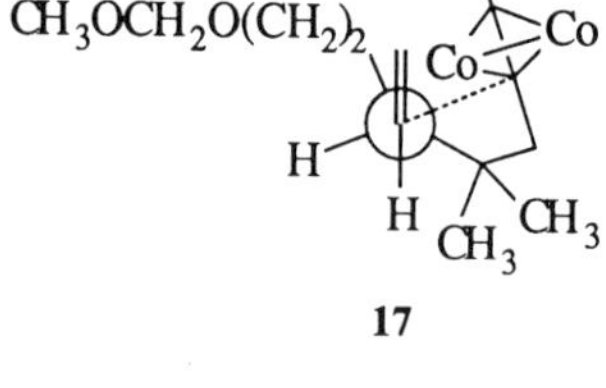

$(CH_3)_3Si$ $CH_3OCH_2O(CH_2)_2$ Co Co H H CH_3 CH_3

17

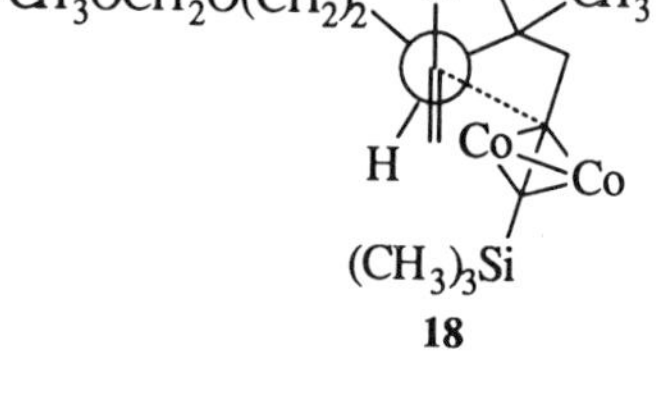

$CH_3OCH_2O(CH_2)_2$ H CH_3 CH_3 H Co Co $(CH_3)_3Si$

18

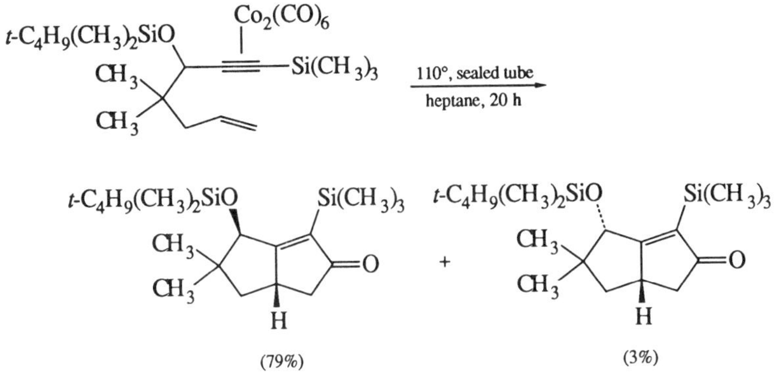

110°, sealed tube, heptane, 20 h

(79%) + (3%) (Eq. 22)

Substitution at C-4 has no stereochemical consequences, but improves yields and may shorten reaction times: compare Eqs. 23 and 24.[42,43] Heavy substitution at C-3 and C-5 is detrimental in the absence of C-4 substitution.[44] More forcing conditions are required, and double bond reduction becomes a problem. For systems with free or protected hydroxy groups, dry conditions on silica give the best results.[45–47] Dry state conditions have not been examined with enynes lacking polar functionality.

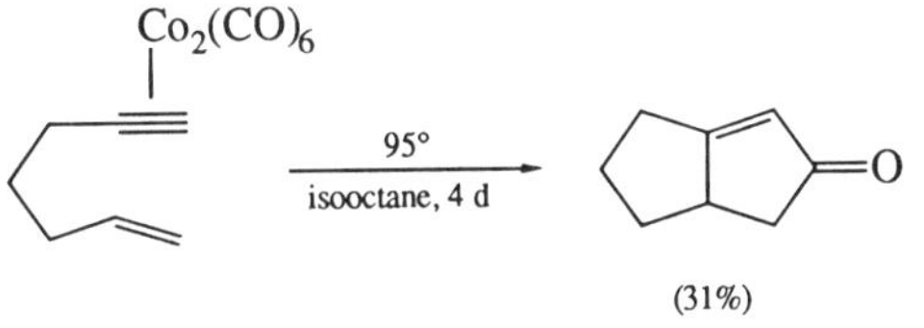

95°, isooctane, 4 d (31%) (Eq. 23)

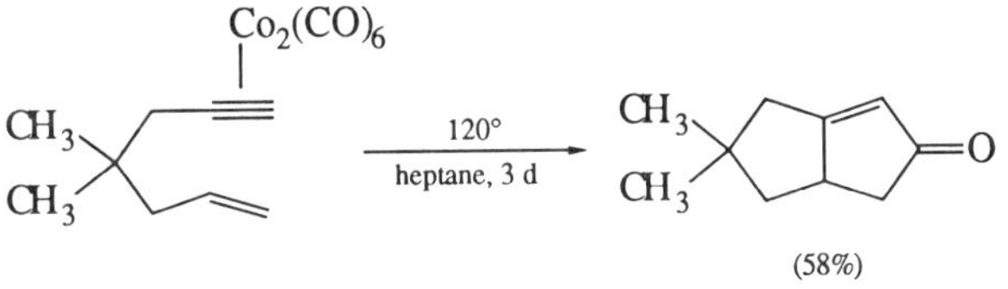

120°, heptane, 3 d (58%) (Eq. 24)

Cycloadditions of 1-(4-pentynyl)cyclopentenes and 3-(3-butynyl)cyclopentenes lead to the angularly fused triquinane (tricyclo[6.3.0.0.^{1,5}]undecane) and the triquinacene (tricyclo[5.2.1.0^{4,10}]undecane) ring systems, respectively.[48–51] Useful stereoselectivity is observed in triquinane formation when an allylic methyl group is present on the cyclopentene ring (Eq. 25).[52]

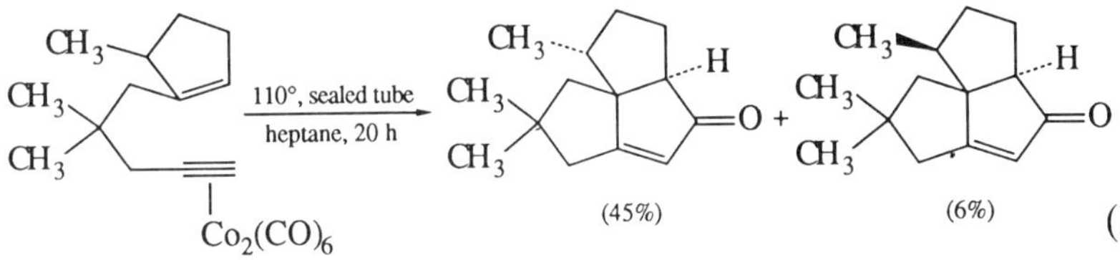

110°, sealed tube, heptane, 20 h (45%) + (6%) (Eq. 25)

Insertion intermediates **19** and **20**, which lead to the major and minor products respectively, suggest that a pseudo-1,3-diaxial interaction is again responsible.

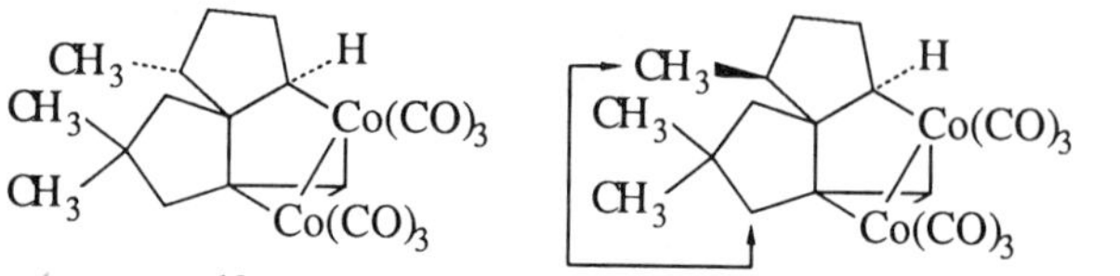

19 **20**

In the 3-(3-butynyl)cyclopentene series the alkene is only *cis*-1,2-disubstituted, and therefore cycloaddition is compatible with and actually benefits from substitution on the alkyne terminus. Not only is stereocontrol with respect to propargyl substituents obtained, but equilibration is observed with labile substituents. Beginning with a mixture of diastereomeric substrates, reversible ionization of the propargylic leaving group occurs, facilitated by stabilization of the cationic intermediate by complexation to cobalt. As the precursor to the less sterically hindered 5-*exo* substituted product is formed in this equilibration it cyclizes readily, forming a single triquinacene stereoisomer in high yield (Eq. 26).[53] The reaction tolerates unsubstituted hydroxy functionality when carried out under dry conditions.

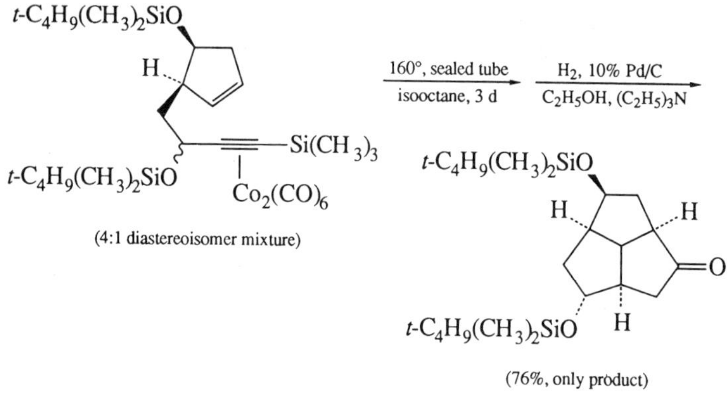

(4:1 diastereoisomer mixture)

160°, sealed tube, isooctane, 3 d; H_2, 10% Pd/C, C_2H_5OH, $(C_2H_5)_3N$

(76%, only product)

(Eq. 26)

Heteroatom-Linked Enynes

Intramolecular Pauson–Khand cycloaddition of allyl propargyl ethers, readily prepared from $Co_2(CO)_6$-complexed propargyl cations, gives only moderate yields under the usual solution conditions.[54–56] In their studies of the effects of silica adsorption on these reactions, Smit and Caple found that reaction times drop from days to hours, reaction temperatures are reduced, and yields are often doubled.[45,57,58] In addition to facilitation of ligand exchange, adsorption of the ether oxygen atom to silica may act like bulky substitution to restrict conformational motion, favoring intramolecular reaction. Substitution is well tolerated in these reactions, although little stereochemical work has been done.[59] A side reaction, hydrogenolysis of the allylic carbon–oxygen bond, is suppressed by heating the adsorbed substrate under oxygen, presumably to scavenge reducing species such as cobalt hydrides. This hydrogenolysis process may be promoted by heating the substrate complex on alumina under argon. Good yields of 3-alkyl-4-(hydroxyalkyl)cyclopentenones are obtained (Eq. 27).[60]

$Co_2(CO)_6$; 60°, CO, isooctane, 24 h (29%); 45°, O_2, SiO_2, 30 min (76%); 90°, Ar, Al_2O_3, 70 min (69%) (Eq. 27)

Pauson–Khand cycloaddition reactions of *N*-allyl-*N*-propargyl amides are also best carried out under dry conditions on silica gel. However, unless the alkyne terminus is substituted the expected 7-azabicyclo[3.3.0]octenone is reduced to a saturated ketone under the reaction conditions. Use of a chlorocarbon solvent suppresses reduction, but yields are lower. In contrast, cycloaddition of an *N*-allyl-*N*-(3-butynyl) amide gives the 7-azabicyclo[4.3.0]nonenone without reduction.[61]

APPLICATIONS TO SYNTHESIS

The Pauson–Khand cycloaddition has been used as a key step in numerous syntheses. In this section a selection is presented. Only the cycloaddition reactions themselves are illustrated. The reader is referred to the original references for details concerning the role the cycloaddition plays in each synthesis. A number of cyclopentenones formed in intermolecular cycloadditions of ethylene, cyclopentene, and dihydrofuran with terminal alkynes have been used in the synthesis of prostanoid analogs (Eq. 28).[62] Cycloaddition products of norbornadiene with alkynes give 4,5-disubstituted-2-cyclopentenones after conjugate addition followed by retro-Diels–Alder elimination of cyclopentadiene.[35] Natural products or natural product precursors prepared from products of dihydrofuran cycloadditions include methylenomycin B,[17] cyclomethylenomycin A, cyclosarkomycin (Eq. 29),[63] and Japanese hop ether.[24]

$$CH_3O_2C(CH_2)_3CH{=}CHCH_2C{\equiv}CH\cdot Co_2(CO)_6 \xrightarrow[\text{toluene, 3 h}]{CH_2{=}CH_2,\ 160°,\ 50\text{ atm}}$$

(57%) (Eq. 28)

$$CH_3C{\equiv}CCH_3\cdot Co_2(CO)_6 + \text{(dihydrofuran)} \xrightarrow[\text{isooctane, 8 d}]{85°,\ CO,\ CH_3C{\equiv}CCH_3}$$

(70%) (Eq. 29)

A potential precursor to the sesquiterpene illudin M has obtained via a remarkably regioselective cycloaddition of acetylene to a densely functionalized 7-oxanorbornene (Eq. 30).[64] Cycloaddition regioselectivity with this alkene varies with alkyne substitution, similar to that seen with simple acyclic alkenes (cf. Eqs. 4 and 5).

$$HC{\equiv}CH\cdot Co_2(CO)_6 + \xrightarrow[\text{toluene, 4 h}]{60°}$$

(53%) + (11%) (Eq. 30)

Potential precursors to hydrazulenoid natural products in the guaianolide and pseudoguaianolide families of natural products have been prepared from cycloaddition products of bicyclo[3.2.0]heptenes with acetylene (e.g., **14**).[26] The alkene serves as a synthetic equivalent to the less reactive cycloheptene.[11] The lactarane furanether B has been prepared in two complementary ways using cycloaddition reactions of 8-oxabicyclo[3.2.1]oct-6-enes, one of which is shown in Eq. 31.[36,37]

$CH_3C{\equiv}CH \cdot Co_2(CO)_6$ + [bicyclic furan substrate] $\xrightarrow[\text{benzene, 44 h}]{64°, CO}$ (43%) + (21%) (Eq. 31)

Cycloadditions of cyclopropylacetylenes with cyclopentene and norbornene give 2-cyclopropylcyclopentenones whose vinylcyclopropane functionality may be rearranged thermally to give a new cyclopentene ring (Eqs. 32 and 33). De Meijere has applied this sequence to synthesize linearly fused triquinanes characteristic of the hirsutane class of natural products.[29,63,65] Cycloadditions of both bicyclo[3.3.0]oct-1-enes and bicyclo[3.3.0]oct-2-enes with alkynes to give angularly fused triquinanes have already been described (Eq. 15).[28]

$C{\equiv}CH \cdot Co_2(CO)_6$ / $Si(CH_3)_3$ + cyclopentene $\xrightarrow[\text{heptane}]{85°, 19 d;\ 100°, 11 d;\ \text{sealed tube}}$ product with $Si(CH_3)_3$ (45%) (Eq. 32)

$C{\equiv}CH \cdot Co_2(CO)_6$ / OC_4H_9-*t* + norbornene $\xrightarrow[\text{toluene, 16 h}]{80\text{-}90°}$ product with O, H, OC_4H_9-*t* (79%) (Eq. 33)

The intramolecular Pauson–Khand preparation of the bicyclo[3.3.0]oct-1-en-3-ones was first applied to the synthesis of complex natural products by Magnus. Solutions to problems arising in the syntheses of coriolin,[40] hirsutic acid,[42] and quadrone[66] included ring fusion equilibration to control remote stereochemistry (hirsutic acid) and addition of base to scavenge cobalt hydrides responsible for a hydrogenolysis problem (quadrone) (Eq. 34). Stereoselective syntheses of optically pure carbocycline analogs have employed intramolecular cycloadditions of enynes derived from D-(+)-ribonolactone (Eq. 35)[67] as well as nonracemic glyceraldehyde derivatives.[68] Bicyclo-[3.3.0]oct-1-en-3-ones have also been used for syntheses of pentalenene and pentalenolactone E methyl ester.[43,69]

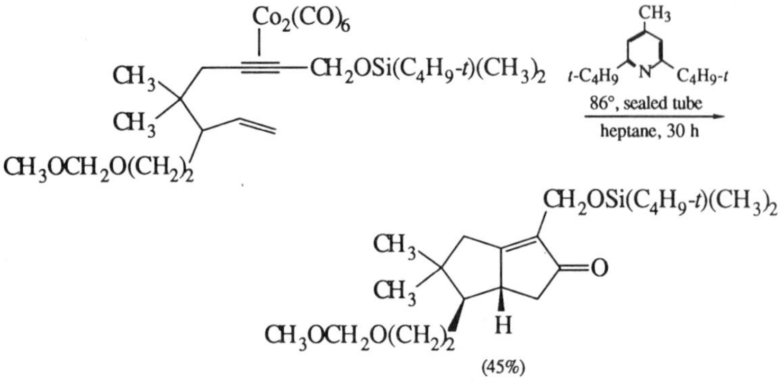

(Eq. 34)

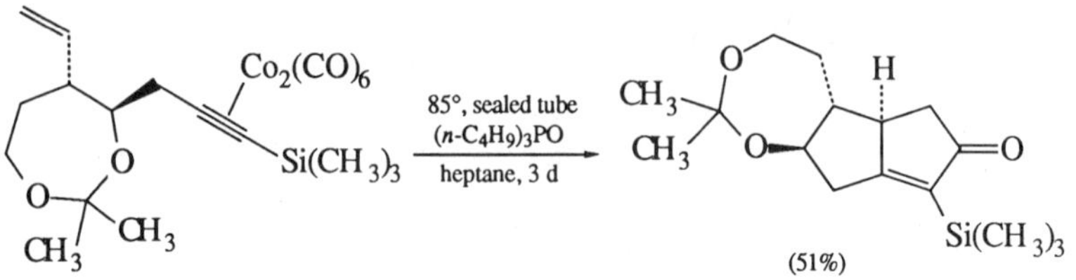

(Eq. 35)

Novel carbopolycyclic systems have been made from inter- as well as intramolecular Pauson–Khand cycloadditions of medium ring alkynes formed initially by alkylation of the stabilized propargyl cation complexes.[70] Through Pauson–Khand cycloaddition of an allyl propargyl ether Billington completed a formal synthesis of the natural product aucubigenone.[55] Smit and Caple have extended this chemistry to include ring-substituted ethers whose cycloadditions result in novel heteropolycyclics (Eq. 36)[45].

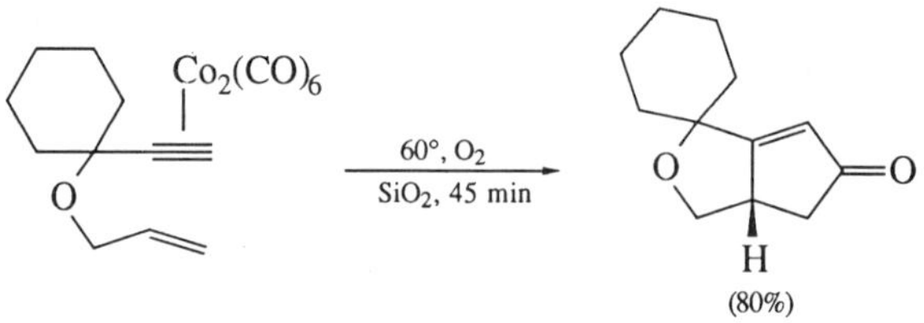

(Eq. 36)

Cycloaddition of a trimethyl substituted 1-(4-pentynyl)cyclopentene was the key step in a stereocontrolled synthesis of pentalenene (Eq. 25).[52] However, isocomene, possessing methyl groups at the ring fusions, could not be prepared by this general route because of the failure of cycloaddition to occur with a tetrasubstituted alkene.[48,49]

A variety of possible entries to the triquinacene tricyclo[5.2.1.$0^{4,10}$]decane-2,5,8-trione were evaluated as being of value in syntheses of dodecahedrane and its derivatives (Eq. 26).[51] Optically active substrates were prepared and cyclized to the corresponding chiral trione with ≥97% stereoselectivity.[71]

Smit prepared an 8-butenyloxybicyclo[3.3.0]oct-1-en-3-one and a 9-oxa-8-

butenylbicyclo[3.3.0]oct-1-en-3-one by dry state Pauson–Khand cycloaddition, and photocyclized each to a tetracyclic fenestrane derivative (Eq. 37).[72]

$$\text{CH}_3\text{-C(O-allyl(2-methyl))(CH}_2\text{CH}_2\text{CH=CH}_2\text{)-C}\equiv\text{CH}\cdot\text{Co}_2\text{(CO)}_6 \xrightarrow[\text{SiO}_2]{\text{heat}} \text{product } (52\%) \quad \text{(Eq. 37)}$$

EXPERIMENTAL CONDITIONS

The Pauson–Khand cycloaddition reaction is most frequently carried out under stoichiometric conditions. The alkyne is allowed to react with commercially available $Co_2(CO)_8$ at room temperature for 2–4 hours in hydrocarbon or ether solvent, forming $Co_2(CO)_6 \cdot RC{\equiv}CR^1$.[73] Moderate heating of this solution with alkene, usually under nitrogen, but occasionally under carbon monoxide (and alkyne, if the latter is gaseous), generates the cyclopentenone product.[23,25] Improvements in yields have occasionally been found when the reaction is carried out in a sealed tube, or in the presence of added phosphine oxide, or under ultrasonic irradiation.[15,61] In situ generation of $Co_2(CO)_8$ for Pauson–Khand chemistry is also known.[74]

With gaseous alkynes a catalytic variation often results in substantially improved yields with less reactive alkene substrates. A mixture of the alkene and ca. 10 mol % $Co_2(CO)_8$ in an inert solvent is heated under a 1:1 alkyne–carbon monoxide atmosphere. Several turnovers are observed in favorable cases, all involving intermolecular cycloadditions.

Dry state adsorption conditions often offer the most dramatic improvements, and are effective for both intermolecular and intramolecular cycloadditions. The cobalt-complexed enyne, or a mixture of the alkene and the cobalt-complexed alkyne, is applied to the adsorbent, the solvent is removed by evaporation, and the solid is then warmed until the color of the complex fades. Cycloadditions of allyl propargyl ethers are best done on silica gel under oxygen to suppress hydrogenolysis of the propargylic carbon–oxygen bond.[59] For other substrates a variety of adsorbents may be used including alumina and Zeolites, and an inert atmosphere is preferred.[21,60]

Note added in proof. Of the several papers to have appeared between the submission of this chapter and the preparation of the proofs, we note the following three that describe significant qualitative advances in the state of the art of Pauson–Khand (PK) cycloaddition: A. L. Veretenov, A. S. Gybin, and V. A. Smit, *Izv. Akad. Nauk SSSR, Ser Khim.*, 495 (1989) describes successful intramolecular PK cycloaddition involving an electron-poor double bond (that of a 1,4-heptadien-6-yn-3-one). S. Shambayati, W. E. Crowe, and S. L. Schreiber, *Tetrahedron Lett.*, **31**, 5289 (1990) describes promotion of intramolecular PK cycloaddition at room temperature, with concomitant improvement in stereoselectivity, by tertiary amine oxides. Finally, J. Castro, H. Sörenson, A. Riera, C. Morin, A. Moyano, M. A. Pericàs, and A. E. Greene, *J. Am. Chem. Soc.*, **112**, 9388 (1990) describes an enantioselective intramolecular PK cycloaddition mediated by a chiral auxiliary.

EXPERIMENTAL PROCEDURES

2-Pentylcyclopent-2-en-1-one [Cycloaddition of an Alkynehexacarbonyldicobalt Complex with a Gaseous Alkene; Reaction in the Presence and Absence of a Phosphine Oxide] (cf. Eq. 2).[15] The alkyne complex was prepared in 80–85% yield by stirring octacarbonyldicobalt (34.2 g, 0.10 mol) under nitrogen with an equimolar quantity or slight excess of 1-heptyne in dry, olefin-free light petroleum (b.p. 40–60°) (150–200 mL) for 3–4 hours and purified by distillation. The 1-heptynehexacarbonyldicobalt complex was isolated as a dark red oil, b.p. 120°, 0.1 torr; IR ν(CO) (film) 2005, 2025, 2042 cm^{-1}; ^{1}H NMR ($CDCl_3$) δ 0.9 (t, J = 6 Hz, 3*H*), 1.55 (m, 6*H*), 2.9 (t, J = 7 Hz, 2*H*), 6.0 (s, 1*H*). Anal. Calcd for $C_{13}H_{12}Co_2O_6$: C, 40.9; H, 3.2 Found: C, 41.8; H, 3.5%. A solution of 1-heptynehexacarbonyldicobalt (7.5 g, 19.6 mmol) in toluene (200 mL) was placed in a 200-mL steel autoclave which was pressurized with ethylene (60 atm). The autoclave was shaken and heated to 110° for 36 hours, then cooled and the contents filtered. After removal of toluene the residue was chromatographed on alumina. Light petroleum eluted unchanged complex (0.2 g), and ether/light petroleum (1:1) eluted the product which was further purified by flash chromatography[75] and distillation at 100° (bath temp.)/0.4 torr yielding a very pale yellow oil (1.06 g, 36%); IR ν_{max} (film) 1633, 1700 cm^{-1}; ^{1}H NMR ($CDCl_3$) δ 0.90 (t, J = 6 Hz, 3*H*), 1.31 (m, 4*H*), 1.48 (m, 2*H*), 2.16 (br t, J = 7 Hz, 2*H*), 2.40 (m, 2*H*), 2.57 (m, 2*H*), 7.30 (m, 1*H*). The reaction in the presence of tri-*n*-butylphosphine oxide (4.1 g, 18.8 mmol) was carried out in the same manner yielding 1.4 g (49%) of the ketone. Anal. Calcd for $C_{10}H_{16}O$: C, 78.8; H, 10.6. Found: C, 78.9; H, 10.6%.

***exo*- and *endo*-3a,4,7,7a-Tetrahydro-4,7-methanoinden-1-ones [Cycloaddition of an Alkynehexacarbonyldicobalt Complex with a Liquid Alkene under Stoichiometric Conditions]** (cf. Eq. 19).[15] Ethynehexacarbonyldicobalt was prepared by stirring octacarbonyldicobalt in light petroleum for 3–4 hours under an acetylene atmosphere. The complex was purified by distillation as a dark red oil, b.p. 64–66°, 3.5–4 torr.[76] Reaction of ethynehexacarbonyldicobalt (4 g, 12.8 mmol) and norbornadiene (2.4 g, 26 mmol) in toluene ($\approx$150 mL) at 70° for 4 hours under nitrogen gave a product mixture which was freed from metal-containing products and from the diketones ($\approx$0.6 g) by chromatography on neutral alumina and elution with ether. Separation of the isomers was then effected by flash chromatography on MN-Kieselgel using 1:4 ethyl acetate/light petroleum as eluent. The *exo* isomer (0.822 g, 44%)

was eluted before its *endo* isomer (0.192 g, 10%). For the *exo* isomer, IR ν_{max} (film) 1700 cm^{-1}; ^{1}H NMR ($CDCl_3$) δ 1.22 (m, 2*H*), 2.12 (d, 1*H*), 2.60 (br s, 1*H*), 2.78 (m, 2*H*), 6.12 (m, 3*H*), 7.44 (dd, 1*H*). For the *endo* isomer, IR ν_{max} (film) 1700 cm^{-1}; ^{1}H NMR ($CDCl_3$) δ 1.67 (q, 2*H*), 2.77 (t,1*H*), 2.95 (m, 1*H*), 3.19 (br s, 1*H*), 3.40 (m, 1*H*), 5.77 (dd, 1*H*), 5.94 (br d, 2*H*), 7.38 (dd, 1*H*).

***cis, anti, cis*-1-Methoxy-7-methyl-*endo*-8-hydroxytricyclo[5.3.0.0^{2,6}]dec-4-en-3-one (14) [Preparation in situ of an Alkynehexacarbonyldicobalt Complex and its Cycloaddition with a Liquid Alkene under Stoichiometric Conditions].**[26] To 20 mL of dry 1,2-dimethoxyethane under nitrogen was added 0.250 g (0.719 mmol) of octacarbonyldicobalt and the solution was then stirred under an acetylene atmosphere for 1 hour at 25°. The formation of ethynehexacarbonyldicobalt was accompanied by vigorous CO evolution and a color change from yellowish brown to reddish violet. A solution of 0.154 g (1.00 mmol) of 1-methyl-5-methoxybicyclo[3.2.0]hept-6-en-*endo*-2-ol in 5 mL of 1,2-dimethoxyethane was added and the mixture heated to 60–65° for 4 days under CO. After cooling and removal of solvent the residue was precoated on silica gel and chromatographed. Hexane elution removed residual organometallics and 1:1 hexane/ether eluted a small amount of the cyclopentadienone Diels–Alder dimer. Upon further elution 0.135 g (65%) of **14** was isolated as a white crystalline solid: m.p. 136.0-136.5°; IR ν_{max} ($CHCl_3$) 1690-1705, 3350-3550 cm^{-1}; ^{1}H NMR ($CDCl_3$) δ 1.00 (s, 3*H*), 1.66 (m, 2*H*), 2.06 (m, 2*H*), 2.31 (d, J = 4.2 Hz, 1*H*), 2.78 (d, J = 5.0 Hz, 1*H*), 3.22 (m, 1*H*), 3.30 (s, 3*H*), 4.00 (m, 1*H*), 6.42 (dd, J = 1.0, 5.6 Hz, 1H), 7.68 (dd, J = 3.2, 5.6 Hz, 1*H*). Anal. Calcd for $C_{12}H_{16}O_3$: C, 69.21; H, 7.74. Found: C, 69.19; H, 7.73%.

***exo*-4-(2′-*tert*-Butoxycyclopropyl)tricyclo[5.2.1.0^{2,6}]dec-4-en-3-one [Preparation in situ of an Alkynehexacarbonyldicobalt Complex and its Cycloaddition with Excess Liquid Alkene under Stoichiometric Conditions]** (Eq. 33).[64] To a solution of 4.00 g (11.7 mmol) of octacarbonyldicobalt in 30 mL of dry toluene kept in a screw cap vessel at room temperature (22°) with the exclusion of light was added 1.55 g (11.2 mmol) of 1-*tert*-butoxy-2-ethynylcyclopropane ($E/Z \approx 11{:}1$), and the mixture was stirred for 2 hours at room temperature. Then 5.47 g (58.1 mmol) of norbornene was added and the mixture stirred for 16 hours at 90°. The cobalt complexes in the mixture were removed by chromatography on silica gel (200 g), eluting with 60/80° petroleum ether until the eluent was colorless. Elution with ether removed the organic products which were rechromatographed using 200 g of silica gel. Elution with 8:1 petroleum ether/ether gave 2.29 g (79%) of the product (R_F = 0.2) as a mixture of diastereoisomers ($E/Z \geq 11{:}1$); ^{1}H NMR ($CDCl_3$) δ 0.95 (m, 2*H*), 1.08 (ddd, 2*H*), 1.23 (s, 9*H*), 1.26 (m, 2*H*), 1.44-1.67 (m, 2*H*), 1.72 (ddd, 1*H*), 2.11 (m, 1*H*), 2.17 (d, 1H), 2.37 (m, 1H), 2.51 (m, 1H), 3.28 (ddd, J = 2.6, 3.8, 6.8 Hz, 1H), 6.75 (d, J = 2.8 Hz, 1*H*); ^{13}C

NMR (62.90 MHz, $CDCl_3$) δ 14.46, 16.34, 28.20, 28.45, 29.07, 31.03, 38.27, 39.08, 47.90, 54.41, 75.21, 148.4, 154.55, 209.6. High resolution MS: Calcd for $C_{17}H_{24}O_2$: 260.1776. Found: 260.1781.

***exo*-3a,4,5,6,7,7a-Hexahydro-4,7-methanoinden-1-one [Preparation in situ of an Alkynehexacarbonyldicobalt Complex and its Cycloaddition with a Liquid Alkene under Catalytic Conditions]** (Cf. Eq. 16).[1] A solution of octacarbonyldicobalt (1 g, 3 mmol) and norbornene (3 g, 32 mmol) in isooctane was stirred first under acetylene, and then under 1:1 acetylene/carbon monoxide at 60–70° until gas absorption ceased. The mixture was then concentrated and the residue chromatographed on neutral alumina. Light petroleum/benzene (1:1) eluted ethynehexacarbonyldicobalt (≈70 mg). Benzene/chloroform (1:1) then eluted a yellow oil which was distilled at 101–102° and 15 torr to give the ketone (3.54 g, 74%), which solidified on prolonged storage at 0°: m.p. 32° (from pentane); IR ν_{max} (film) 1695 cm^{-1}; ^1H NMR ($CDCl_3$) δ 1.00 (m, 2*H*), 1.23–1.72 (m, 4*H*), 2.16 (m, 2*H*), 2.38 (br s, 1*H*), 2.69 (m, 1*H*), 6.26 (dd, 1*H*), 7.52 (dd, 1*H*).

***cis*-4,5,6,6a-Tetrahydro-1(3a*H*)-pentalenone [Cycloaddition of an Alkyne with a Liquid Alkene under Stoichiometric vs. Catalytic Conditions]** (cf. Eq. 11).[22] A mixture of cyclopentene (17 g, 250 mmol), ethynehexacarbonyldicobalt (17.1 g, 55 mmol), and benzene (350 mL) was heated to reflux under nitrogen for 2 days, cooled to 20°, and filtered through Kieselguhr, the residue being washed with chloroform. The filtrate was evaporated and the residue chromatographed on an alumina column. Light petroleum eluted residual organocobalt compounds and chloroform eluted the product, which was further purified by flash chromatography or distillation (90–100°, 15 torr) to give the ketone (3.25 g, 49%) having IR and ^1H NMR data in agreement with the literature[77] and giving a single peak on GLC (5% FFAP at 80°).

The reaction was repeated using acetylene/carbon monoxide (1:1) instead of a nitrogen atmosphere, cyclopentene (13.76 g, 200 mmol), ethynehexacarbonyldicobalt (3.12 g, 10 mmol), and benzene (100 mL) at 65° for 2 days and the product isolated as above to give the ketone (855 mg, 70% based on cobalt complex).

6-Methylspiro[2.4]hept-6-en-5-one and 5-Methylspiro[2.4]hept-5-en-7-one [Cycloaddition of an Alkyne with a Gaseous Alkene under Dry State Adsorption Conditions] (cf. Eq. 9).[21] A solution of propynehexacarbonyldicobalt[1,8] (0.63 g, 1.9 mmol) in hexane (15 mL) was mixed with chromatography grade silica gel (10.0 g). Solvent was removed on a rotary evaporator and the resulting dry powder was added to a precooled (−78°) ampule charged with 0.3 mL (4 mmol) of methylenecyclopropane. The ampule was sealed, vigorously shaken, and heated at 50° for 2 hours (color changed from pink to gray). The contents were thoroughly extracted with ether and, after removal and TLC separation, pure 6-methylspiro[2.4]hept-6-en-5-one (0.125

g, 53%) and 5-methylspiro[2.4]hept-5-en-7-one (0.025 g, 11%) were isolated as colorless liquids.

Tricyclo[6.3.0.0^{1,5}]undec-7-en-6-one [Intramolecular Enyne Cycloaddition in Refluxing Solvent](cf. Eq. 25).[49] To 1.20 g (8.96 mmol) of 1(4-pentynyl)cyclopentene in 150 mL of benzene under a nitrogen atmosphere was added 3.77 g (11.0 mmol) octacarbonyldicobalt. The solution was blanketed with CO and stirred at room temperature for 5 hours, and then heated to reflux and allowed to stir for 4 days. The crude reaction mixture was concentrated onto neutral alumina and placed on top of a neutral alumina column. Elution with hexane removed nonpolar organometallic species. Elution with ether gave 0.59 g of crude product, which was purified by elution with ether using a Chromatotron (radially accelerated preparative TLC apparatus, silica gel) to give 0.51 g (35% yield) of enone; IR ν_{max} (CCl_4) 1630, 1690 cm^{-1}; ^{1}H NMR ($CDCl_3$) δ 1.30 (septet, J = 6.1 Hz, 1H), 1.40 (dt, J_d = 5.6 Hz, J_t = 12.3 Hz, 1H), 1.59-17.2 (m, 3H), 1.76 (dd, J = 4.8, 6.5 Hz, 1H), 1.86 (dd, J = 6.5, 11.3 Hz, 1H), 1.93–2.16 (m, 3H), 2.42 (br d, J = 8.3 Hz, 1H), 2.54 (br dt, J_d = 18.4 Hz, J_t = 7.0 Hz, 1H), 2.65 (dddd, J = 2.0, 3.8, 11.0, 18.4 Hz, 1H), 5.82 (t, J = 2.0 Hz, 1H). Anal. Calcd for $C_{11}H_{14}O$: C, 81.44; H, 8.70. Found: C, 81.29; H, 8.52%.

5β*H*-2-(Trimethylsilyl)-6β-[2-(methoxymethoxy)ethyl]-7,7-dimethylbicyclo[3.3.0]oct-1-en-3-one [Intramolecular Enyne Cycloaddition in Solvent in a Sealed Tube] (Eq. 21).[66] To a solution of the methoxymethyl ether of 4,4-dimethyl-7-trimethylsilyl-3-vinyl-6-heptyn-1-ol (166 mg, 0.59 mmol) in 4 mL of dry heptane, purged with CO for 2 hours, was added $Co_2(CO)_8$ (230 mg, 0.67 mmol), and the CO purging continued for 3 hours in a screw-top resealable Pyrex tube. The tube was then sealed and heated at 115° for 36 hours. The contents were evaporated in vacuo and chromatographed over Florisil, eluting with 2% ether/petroleum ether, followed by petroleum ether/ethyl acetate (4:1) to afford 143 mg (78% yield) of the ketone as a mobile, slightly yellow liquid; IR ν_{max} (film) 1600, 1680, 2900 cm^{-1}; ^{1}H NMR ($CDCl_3$) δ 0.16 (s, 9H), 1.02 (s, 3H), 1.11 (s, 3H), 1.32 (dt, J_d = 3 Hz, J_t = 12 Hz, 1H), 1.55 (m, 1H), 1.85 (m, 1H), 2.50 (AB q, J = 18 Hz, 2H), 2.53 (dd, J = 4, 16.5 Hz, 1H), 2.80 (m, 1H), 3.35 (m, 3H), 3.51 (m, 1H), 3.61 (m, 1H), 4.60 (s, 2H). High resolution MS (chemical ionization): Calcd for $C_{17}H_{30}O_3Si$ (M^+ − 1): 309.1841. Found: 309.1812.

4,4-Dimethyl-3-oxabicyclo[3.3.0]oct-5-en-7-one [Intramolecular Cycloaddition of an Allyl Propargyl Ether under Dry State Adsorption Conditions] (Eq. 27).[59] To a solution of allyl 1,1-dimethylpropynyl ether hexacarbonyldicobalt complex (351 mg, 1 mmol)[55] in 30 mL of pentane was added silica gel (10 g, ≈10% H_2O content w/w). After 30 minutes the solvent was removed in vacuo (rotary evaporator) and the residue heated to 45° for 30 minutes in the rotating flask under a slow stream of oxygen. Subsequent extraction of the silica gel with ether (5 × 50 mL) and removal of solvent gave the enone, which was purified by TLC on silica gel eluting with 1:2 hexane/ether, and isolated as a colorless oil (115 mg, 76% yield); ^{1}H NMR ($CDCl_3$) δ 1.42 (s, 3*H*), 1.55 (s, 3*H*), 2.18 (dd, 1*H*), 2.62 (dd, 1*H*), 3.44 (m, 1*H*), 3.55 (dd, *J* = 8, 10 Hz, 1*H*), 4.35 (t, *J* = 8 Hz, 1*H*), 5.96 (d, *J* = 2.5 Hz, 1*H*).

TABULAR SURVEY

We have attempted to cover the literature thoroughly up to the end of 1988, and we have also covered all references that were available to us through the end of 1989.

The tables are arranged, as is the text, by cycloaddition type. Tables I–III present tables of intermolecular cycloadditions of alkynes with acyclic alkenes. Cycloadditions of ethylene are cited in order of complexity of alkyne structure in the following way: terminal alkynes with alkyl substituents, by carbon number and then by increasing functionalization, then terminal alkyl substituents with aryl substituents, and finally internal alkynes. Cycloadditions of mono- and disubstituted alkenes are ordered similarly, first by increasing alkene complexity, and then by increasing alkyne complexity. This order in particular optimizes proximity of examples with similar structural characteristics to facilitate easy comparison. Tables IV–VII present cycloadditions of alkynes with carbo- and heteromonocyclic alkenes, and followed by cycloadditions with carbo- and heteropolycyclic alkenes. In order to keep cyclic alkenes of similar structural types together the entries in these tables are grouped by alkene ring size, as this is a primary factor in the success of the cycloaddition reaction. For a given alkene, alkyne cycloaddition partners are cited in the order described above. Alkenes are ordered by number of substituents, then carbon number, then degree of functionalization. Polycyclic alkenes are presented with ring-fused systems first, followed by bridged systems. Entries are ordered first by ring size of the cycloaddition-reactive double bond, second by number of atoms in the bicyclic system containing that double bond, third by number of additional nonfunctionalized rings, fourth by number of additional functionalized rings, fifth by functionalization of the bicyclic system containing the reactive double bond, sixth by number of substituents, seventh by carbon number, eighth by degree of functionalization, and ninth by alkyne structure. Some exceptions are made with entries that cannot readily be incorporated in tabulated subgroups of reactions within the tables. Tables VIII–XI present intramolecular cycloadditions. Table VIII lists enynes in which the shortest linkage between the double and triple bonds consists of carbon atoms only, while Table IX lists enynes in which a heteroatom is present in the linkage. Order of citation is, first by number of atoms in this linkage, second by increasing atomic number of the heteroatom (Table IX only), third by position of substituents in relationship to the double bond, fourth by number of substituents, fifth by carbon number, and sixth by degree of functionalization. Tables X and XI list two exceptional groups of cycloadditions.

Reactions are presented with principal details of reaction conditions (omitting workup) including solvent, temperature, time, and atmosphere, as they are available from the original references. Yields are in parentheses. Where the same reaction under apparently identical conditions is reported by the same group more than once, only the most recent citation is tabulated. Where multiple examples of the same reaction under only slightly varying conditions are reported by the same group, only enough examples are tabulated to give the reader guidance to the presence or absence of a pattern between reaction conditions and product yields.

The following abbreviations are used in the tables when necessary to save space:

Ac	acetyl
Bn	benzyl
cat. Co	catalytic amount of $Co_2(CO)_8$ or alkyne·$Co_2(CO)_6$
c-C_3H_5	cyclopropyl
DME	1,2-dimethoxyethane
DMPS	dimethylphenylsilyl
ee	enantiomeric excess
glyphos	2,3-*O*-isopropylideneglycerine-1-diphenylphosphine
HMPA	hexamethylphosphoric triamide
hν	ultraviolet/visible irradiation
isooctane	2,2,4-trimethylpentane
MOM	methoxymethyl
PCC	pyridinium chlorochromate
TBDMS	*tert*-butyldimethylsilyl
THF	tetrahydrofuran
THP	tetrahydropyranyl
TMS	trimethylsilyl
Ts	*p*-toluenesulfonyl

TABLE I. ALKYNES WITH ETHYLENE

Alkyne[a] $R^1C{\equiv}CR^2$		Reaction Conditions	Product(s) and Yield(s) (%)		Refs.
R^1	R^2		I	II	
TMSCH$_2$	H	Benzene, 90°, 60 atm[b], 36 h	(29)	(0)	78
n-C_5H_{11}	"	Toluene, 15-20°, 50 atm[b], 14 d	(47)	(0)	15
"	"	Toluene, (n-C_4H_9)$_3$PO, 15-20°, 50 atm[b], 16 d	(49)	(0)	15
"	"	Toluene, 80°, 50 atm[b], 36 h	(31)	(0)	15
"	"	Toluene, (n-C_4H_9)$_3$PO, 80°, 50 atm[b], 36 h	(47)	(0)	15
"	"	Toluene, 85°, 120 atm[b], 36 h	(55)	(0)	15
"	"	Toluene, 110°, 50 atm[b], 36 h	(36)	(0)	15
"	"	Toluene, (n-C_4H_9)$_3$P, 110°, 50 atm[b], 36 h	(26)	(0)	15
"	"	Toluene, (n-C_4H_9)$_3$PO, 110°, 50 atm[b], 36 h	(70)	(0)	15
cis-$C_2H_5CH{=}CHCH_2$	"	Toluene, 100°, 30 atm[b], 36 h	(32)	(0)	16
$CH_3O_2C(CH_2)_6$	"	Toluene, 160°, 15 atm[b], 7 h	(46)	(0)	13
cis-$CH_3O_2C(CH_2)_3CH{=}CHCH_2$	"	Toluene, 160°, 50 atm[b], 3 h	(57)	(0)	13

$CH_3O_2CCH_2$-(2,5-thienyl)-CH_2	H	Toluene, 130°, 60 atm[b], 10 h	(33)	(0)	12
$CH_3O_2C(CH_2)_2$-(2,5-furyl)-CH_2	"	Toluene, 130°, 60 atm[b], 10 h	(32)	(0)	12
C_6H_5	"	Toluene, 110°, 50 atm[b], 36 h	(31)	(0)	15
"	"	Toluene, $(n\text{-}C_4H_9)_3PO$, 110°, 50 atm[b], 36 h	(45)	(0)	15
"	"	Toluene, 160°, 80 atm[b], 7 h	(30)	(0)	11
C_6H_5S	"	Toluene, 110°, 5 h	(11)	(0)	14
C_2H_5	CH_3	Toluene, 110°, 35 atm[b], 36 h	(24)	(3)	16
n-C_5H_{11}	"	Toluene, 110°, 60 atm[b], 36 h	(22)	(2)	16
"	"	Toluene, 160°, 80 atm[b], 7 h	(17)	(0)	11
trans-$CH_3CH{=}CHCH_2$	"	Toluene, 160°, 80 atm[b], 7 h	(25)	(0)	11
cis-$C_2H_5CH{=}CHCH_2$	"	Toluene, 100°, 30 atm[b], 36 h	(22)	(0)	16

[a]This refers to the alkyne portion of the preformed alkyne·$Co_2(CO)_6$ complex.

[b]The reaction was carried out in an autoclave under the indicated pressure of ethylene prior to heating.

TABLE II. ALKYNES AND MONOSUBSTITUTED ALKENES

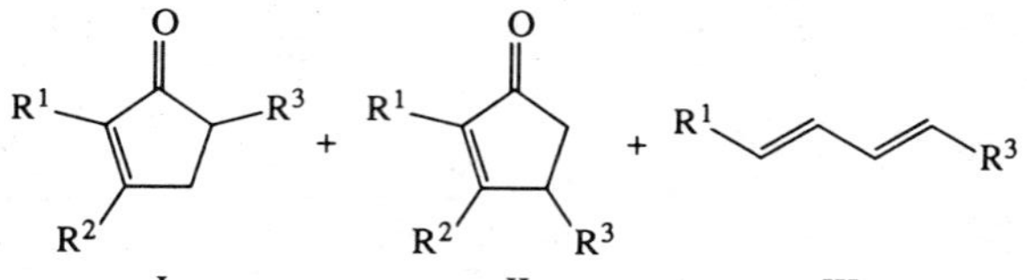

Alkyne[a] $R^1C{\equiv}CR^2$ R^1	R^2	Alkene $R^3CH{=}CH_2$ R^3	Reaction Conditions	Product(s) and Yield(s) (%) I	II	III	Refs.
H	H	t-C_4H_9	Benzene, 80°, HC≡CH/CO, cat. Co	(28)	(<2)	(0)	4
"	"	"	Benzene, 80°, N_2	(4)	(<1)	(0)	4
"	"	CH_3CO_2	–	(17)	(0)	(0)	4
"	"	"	Isooctane, 65°, 60 h, HC≡CH/CO, cat. Co	(4)	(2)	(0)	20
"	"	t-C_4H_9O	Isooctane, 55°, 72 h, HC≡CH/CO, cat. Co	(15)	(15)	(0)	20
"	"	$THPOCH_2$	Benzene, heat	(48)	(0)	(0)	4
"	"	"	Benzene, $h\nu$	(42)	(0)	(0)	4
"	"	"	$(n$-$C_4H_9)_2O$, heat	(4)	(2)	(0)	4
"	"	"	$(n$-$C_4H_9)_2O$, $h\nu$	(10)	(6)	(0)	4
"	"	"	Petroleum ether, heat	(20)	(25)	(0)	4
"	"	"	Petroleum ether, $h\nu$	(14)	(9)	(0)	4
CH_3	"	C_6H_5	Toluene, 110°, 6 h, N_2	(24)	(7)	(0)	11
n-C_4H_9	"	n-C_6H_{13}	Toluene, 95-100°, 48 h, N_2	(21)	(21)	(0)	7
"	"	$CH_3S(CH_2)_2$	Toluene, 90°, 24-36 h, N_2	(53)	(7)	(0)	6
"	"	$CH_3OCH_2O(CH_2)_2$	Toluene, 95-100°, 48 h, N_2	(25)	(16)	(0)	7
n-C_5H_{11}	"	n-C_6H_{13}	THF, 65°, 24 h, N_2[b]	(30) (I and II)		(0)	74

C_6H_5	H	Br	Toluene, 160°, 80 atm[c], 7 h	(19)(R^3=H)	(0)	(0)	11
"	"	CN	–	(0)	(0)	(-)	8
"	"	CH_3	Toluene, 160°, 80 atm[c], 7 h	(12)	(11)	(0)	11
"	"	CH_2Br	Toluene, 160°, 80 atm[c], 7 h	(5)(R^3=CH_3)	(0)	(0)	11
"	"	n-C_6H_{13}	Toluene, 110°, 6 h, N_2	(5)	(9)	(0)	11
"	"	"	Toluene, 95°, 24 h, N_2	(25)	(25)	(0)	7
"	"	n-C_8H_{17}	THF, 65°, 24 h, N_2[b]	(26) (I and II)		(0)	74
"	"	Cyclohexyl	Toluene, 95-100°, 48 h, N_2	(32)	(13)	(0)	7
"	"	$(CH_3)_2N(CH_2)_2$	Toluene, 90°, 24-36 h, N_2	(60)	(12)	(0)	6
"	"	$(CH_3)_2N(CH_2)_3$	Toluene, 90°, 24-36 h, N_2	(52)	(17)	(0)	6
"	"	$CH_3S(CH_2)_2$	Toluene, 90°, 30 h, N_2	(58)	(3)	(0)	6
"	"	$CH_3S(CH_2)_3$	Toluene, 90°, 24-36 h, N_2	(39)	(13)	(0)	6
"	"	C_6H_5	Toluene, 110°, 6 h, N_2	(12)	(0)	(39)	11
"	"	4-FC_6H_4	Toluene, 110°, 7 h, N_2	(35)	(0)	(0)	18
"	"	2-ClC_6H_4	Toluene, 110°, 7 h, N_2	(4)	(0)	(16)	18
"	"	4-ClC_6H_4	Toluene, 110°, 7 h, N_2	(16)	(0)	(13)	18
"	"	4-$CH_3C_6H_4$	Toluene, 110°, 7 h, N_2	(13)	(0)	(26)	18
"	"	4-$CH_3OC_6H_4$	Toluene, 110°, 7 h, N_2	(27)	(0)	(42)	18
"	"	2-Furyl	Toluene, 110°, 7 h, N_2	(0)	(0)	(15)	18
"	"	Ferrocenyl	Toluene, 110°, 7 h, N_2	(trace)	(0)	(49)	18
"	"	$(CO)_3Cr{\cdot}C_6H_5$	Toluene, 110°, 5 h, N_2	(37)	(0)	(32)	19
"	"	$(CO)_3Cr{\cdot}4\text{-}FC_6H_4$	Toluene, 110°, 5 h, N_2	(29)	(0)	(44)	19
"	"	$(CO)_3Cr{\cdot}4\text{-}ClC_6H_4$	Toluene, 110°, 5 h, N_2	(0)	(0)	(29)	19

TABLE II. ALKYNES AND MONOSUBSTITUTED ALKENES (*Continued*)

Alkyne[a]		Alkene	Reaction Conditions	Product(s) and Yield(s) (%)			Refs.
R^1	R^2	R^3		I	II	III	
C_6H_5	H	$(CO)_3Cr{\cdot}4\text{-}CH_3C_6H_4$	Toluene, 110°, 5 h, N_2	(29)	(0)	(44)	19
"	"	$(CO)_3Cr{\cdot}4\text{-}CH_3OC_6H_4$	Toluene, 110°, 5 h, N_2	(47)	(0)	(35)[d]	19
"	"	$C_6H_5CH_2$	Toluene, 110°, 7 h, N_2	(25)	(0)	(0)	18
"	"	$4\text{-}CH_3OC_6H_4CH_2$	Toluene, 110°, 7 h, N_2	(29)[e]	(0)	(0)	18
"	"	$C_6H_5CH{=}CH$	Toluene, 110°, 7 h, N_2	(0)	(0)	(17)	18
"	"	$(CO)_3Cr{\cdot}C_6H_5CH_2$	Toluene, 100°, 5 h, N_2	(32)	(0)	(0)	19
CH_3	CH_3	$n\text{-}C_6H_{13}$	Toluene, 100°, 48 h, N_2	(19)	(1)	(0)	7
"	"	Cyclohexyl	Toluene, 100°, 48 h, N_2	(23)	(0)	(0)	7
"	"	$THPOCH_2$	Toluene, 110°, 8 h, N_2	(32)	(0)	(0)	17
"	"	$CH_3OCH_2O(CH_2)_2$	Toluene, 100°, 48 h, N_2	(25)	(<1)	(0)	7
C_6H_5	"	$n\text{-}C_6H_{13}$	Toluene, 100°, 48 h, N_2	(17)	(1)	(0)	7
"	"	Cyclohexyl	Toluene, 100°, 48 h, N_2	(22)	(0)	(0)	7
"	"	$CH_3OCH_2O(CH_2)_2$	Toluene, 100°, 48 h, N_2	(40)	(1)	(0)	7

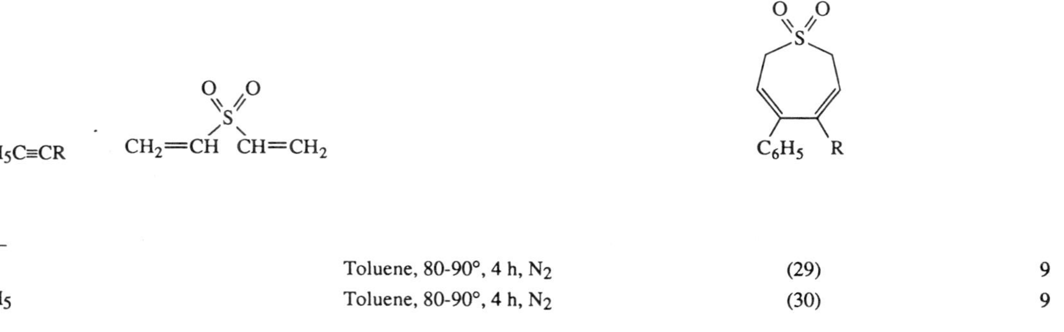

R	Conditions	Product	Ref.
H	Toluene, 80-90°, 4 h, N_2	(29)	9
C_6H_5	Toluene, 80-90°, 4 h, N_2	(30)	9

[a]This refers to the alkyne portion of the preformed alkyne·$Co_2(CO)_6$ complex.
[b]The alkyne·$Co_2(CO)_6$ complex was prepared in situ from $CoBr_2$, Zn, CO, and alkyne.
[c]The reaction was carried out in an autoclave under the indicated pressure of alkene prior to heating.
[d]Some loss of the $Cr(CO)_3$ group is observed.
[e]The rearranged diene $C_6H_5CH{=}CHC(CH_3){=}CHC_6H_4OCH_3$-4 is formed as a mixture of isomers in 5% yield.

TABLE III. ALKYNES WITH DISUBSTITUTED ALKENES

Alkyne[a]	Alkene				Reaction Conditions	Product(s) and Yield(s) (%)						Refs.
$R^1C{\equiv}CH$	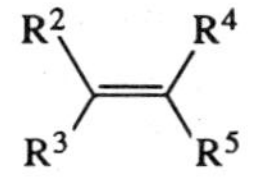 $R^2R^3C{=}CR^4R^5$					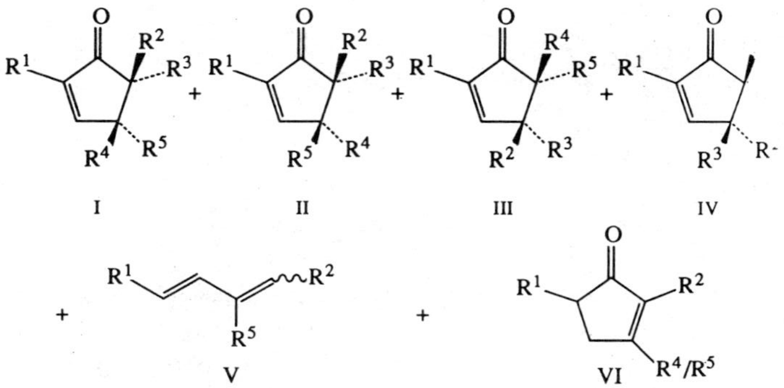 I + II + III + IV + V + VI						
R^1	R^2	R^3	R^4	R^5		I	II	III	IV	V	VI	
H	$-(CH_2)_2-$		H	H	SiO_2, 50°, 2 h[b]	(7)		(39)		(0)	(0)	21
CH_3	"		H	H	SiO_2, 50°, 2 h[b]	(11)		(53)		(0)	(0)	21
"	"		H	H	Al_2O_3, 50°, 2 h[b]	(13)		(64)		(0)	(0)	21
"	"		H	H	Zeolite NaX, 50°, 2 h[b]	(25)		(66)		(0)	(0)	21
"	$-(CH_2)_3-$		H	H	SiO_2, 70°, 2 h[b]	(24)		(24)		(0)	(0)	21
$CH_2{=}CH$	$-(CH_2)_2-$		H	H	Zeolite NaX, 50°, 1 h[b]	(≤1)		(25)		(0)	(0)	21
CH_3OCH_2	"		H	H	Zeolite NaX, 50°, 1 h[b]	(13)		(66)		(0)	(0)	21
"	$-(CH_2)_3-$		H	H	$MgO{\cdot}SiO_2$, 70°, 2 h[b]	(16)		(16)		(0)	(0)	21
c-C_3H_5	$-(CH_2)_2-$		H	H	Zeolite NaX, 50°, 1.5 h[b]	(17)		(50)		(0)	(0)	21

$CH_2{=}C(CH_3)$	$-(CH_2)_2-$		H	H	Zeolite NaX, 50°, 2 h[b]	(15)		(44)		(0)	(0)	21
"	"		H	H	Zeolite H-ZSM, 50°, 30 min[b]	(11)		(58)		(0)	(0)	21
$(CH_3)_2C(OH)$	"		H	H	SiO_2, 50°, 6 h[b]	(12)		(35)		(0)	(0)	21
TMS	$CH_3S(CH_2)_2$	H	CH_3	H	Toluene, 90°, 24-36 h, N_2	(3)	(59)	(3)(III+IV)		(0)	(0)	6
n-C_4H_9	$(CH_3)_2N(CH_2)_2$	CH_3	H	H	Toluene, 90°, 24-36 h, N_2	(57)		(3)		(0)	(0)	6
C_6H_5	CHO	H	H	CH_3	–	(0)	(0)	(0)	(0)	(-)	(0)	8
"	$CO_2C_2H_5$	H	H	CH_3	–	(0)	(0)	(0)	(0)	(45)[c]	(0)	8
"	C_6H_5	H	H	CH_3	Toluene, 110°, 7 h, N_2	(15)	(0)	(0)	(0)	(0)	(0)	18
"	$(CH_3)_2N(CH_2)_2$	CH_3	H	H	Toluene, 90°, 30 h, N_2	(73)		(4)		(0)	(0)	6
"	"	H	CH_3	H	Toluene, 90°, 24-36 h, N_2	(68)(I+II+VI)		(0)	(0)	(0)		6
"	"	H	H	CH_3	Toluene, 90°, 24-36 h, N_2	(68)(I+II+VI)		(0)	(0)	(0)		6
"	$CH_3S(CH_2)_2$	H	CH_3	H	Toluene, 90°, 24-36 h, N_2	(4)	(67)	(2)(III+IV)		(0)	(0)	6
"	"	H	H	CH_3	Toluene, 90°, 24-36 h, N_2	(32)	(3)	(5)(III+IV)		(0)	(0)	6
"	$-(CH_2)_2-$		H	H	SiO_2, 50°, 2 h[b]	(15)		(64)		(0)	(0)	21
"	$-(CH_2)_3-$		H	H	SiO_2, 70°, 2 h[b]	(45)		(45)		(0)	(0)	21
$C_2H_5C{\equiv}CC_2H_5$	$-(CH_2)_2-$		H	H	Zeolite NaX, 50°, 4 h[b]	(54)		(<2)		(0)	(0)	21
$TMSC{\equiv}CCH_3$	"		H	H	Zeolite NaX, 50°, 2 h[b]	(81)		(<4)		(0)	(0)	21

[a]This refers to the alkyne portion of the preformed alkyne·$Co_2(CO)_6$ complex.

[b]This reaction was carried out in a sealed tube.

[c]The product consisted of a 3.5:1 mixture of *E,E* and *2Z,4E* stereoisomers.

TABLE IV. ALKYNES WITH MONOCYCLIC ALKENES

Alkene		Alkyne[a]	Reaction Conditions	Product(s) and Yield(s) (%)		Refs.
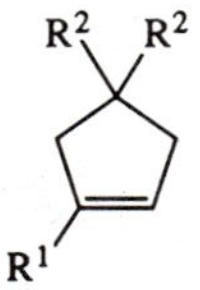		$R^3C{\equiv}CH$		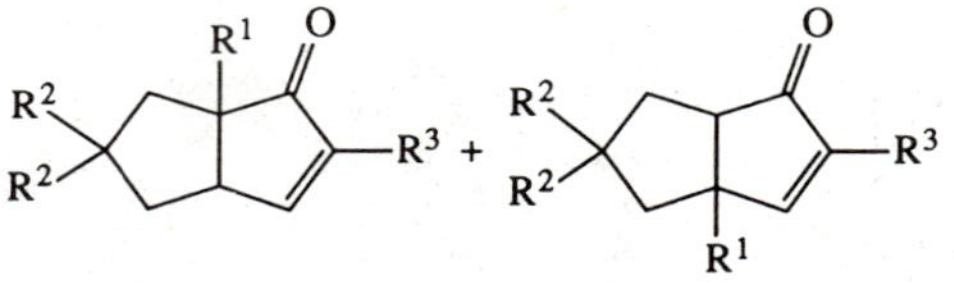 I	II	
R^1	R^2	R^3		I	II	
H	H	H	Toluene, 110°, 36 h	(49)		15
"	"	"	Toluene, $(n\text{-}C_4H_9)_3PO$, 110°, 36 h	(53)		15
"	"	"	Benzene, 80°, 2 d	(49)		22
"	"	"	Benzene, 65°, 2 d, $HC{\equiv}CH/CO$, cat. Co	(70)		22
"	"	$c\text{-}C_3H_5$	120°, 2 d[b,c]	(42)		30
"	"	$1\text{-}Cl\text{-}c\text{-}C_3H_4$	45°, 12 h[b]	(18)		30
"	"	$1\text{-}TMS\text{-}c\text{-}C_3H_4$	45°, 50 h[b]	(20)		30
"	"	"	85°, 19 d, then 100°, 11 d, CO[b,c]	(45)		30
"	"	$2\text{-}(C_2H_5O)\text{-}c\text{-}C_3H_4$[d]	Hexane, 120°, 2 d, CO[c]	(34)		65
"	"	$2\text{-}TMS\text{-}c\text{-}C_3H_4$[d]	128°, 36 h[b,c]	(30)		29
"	"	$2\text{-}DMPS\text{-}c\text{-}C_3H_4$[d]	130°, 36 h[b,c]	(11)		29
"	"	$TMSCH_2$	Benzene, 80°, 24 h, N_2	(28)		78
"	"	$n\text{-}C_5H_{11}$	Hexane, 70°, 24 h	(41)		15
"	"	"	Hexane, $(n\text{-}C_4H_9)_3PO$, 70°, 24 h	(70)		15

H	H	C_6H_5	Hexane, 70°, 36 h	(40)		15
"	"	"	Hexane, $(n\text{-}C_4H_9)_3PO$, 70°, 36 h	(70)		15
"	"	"	Toluene, 70°, 2 d	(25)		15
"	"	"	Toluene, 110°, 36 h	(49)		15
"	"	"	Toluene, ultrasound, 70°, 3 h	(59)		15
"	"	"	Toluene, $(n\text{-}C_4H_9)_3PO$, 110°, 36 h	(68)		15
"	"	"	Toluene, 150-160°, 7 h[c]	(47)		11
"	"	"	THF, 65°, 12 h, N_2[e]	(47)		74
"	"	C_6H_5S	Toluene, 110°, 2 h	(53)		15
"	"	$n\text{-}C_8H_{17}$	THF, 65°, 12 h, N_2	(58)		74
"	"	$CH_3O_2C(CH_2)_6$	Benzene, 80°, 7 h	(33)		13
"	"	$Cl(CH_2)_3CH{=}CHCH_2$[f]	–	(33)		13
AcO	"	H	–	(26)	(0)	4
CH_3	"	"	Benzene, 80°, 3 h	(2)	(17)	22
"	"	"	Benzene, 65°, 3 d, HC≡CH/CO, cat. Co	(9)	(60)	22
H	CH_3	$c\text{-}C_3H_5$	Heptane, 125°, 34 h[c]	(12)		30
"	"	1-TMS-$c\text{-}C_3H_4$	Heptane, 85°, 19 d[c]	(16)(28)[g]		30

TABLE IV. Alkynes with Monocyclic Alkenes (*Continued*)

Alkene	Alkyne[a]	Reaction Conditions	Product(s) and Yield(s) (%)	Refs.
 TMSO, TMSO	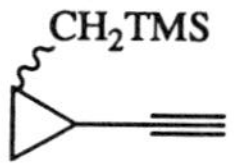 CH_2TMS	100°, 5 d[b]	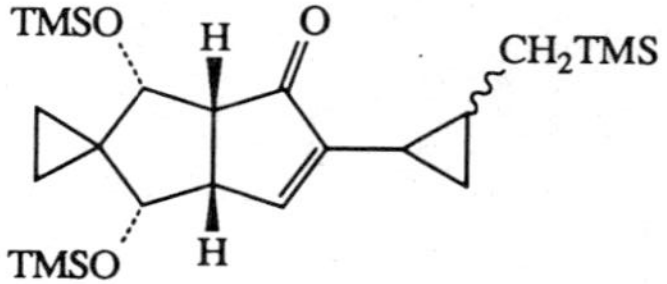 TMSO, H, O, CH_2TMS, TMSO, H (19)	29
	RC≡CH		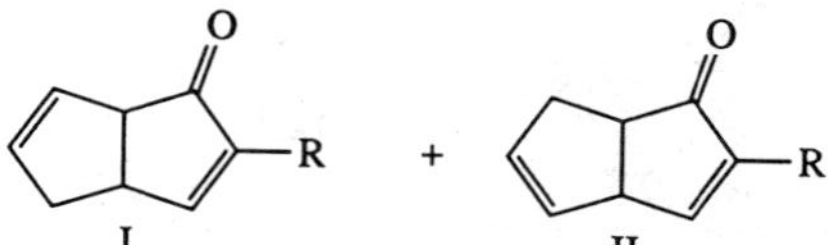 I + II	
R			I, II	
CH_3		Toluene, 110°, 5 h, N_2	(≈45), (≈10)	23
C_6H_5		Toluene, 110°, 5 h, N_2	(≈49), (≈11)	23
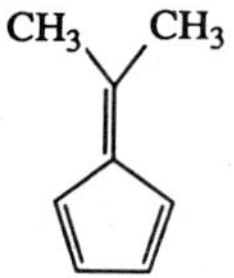 CH_3, CH_3	$C_6H_5C{\equiv}CH$	Toluene, 110°, 5 h, N_2	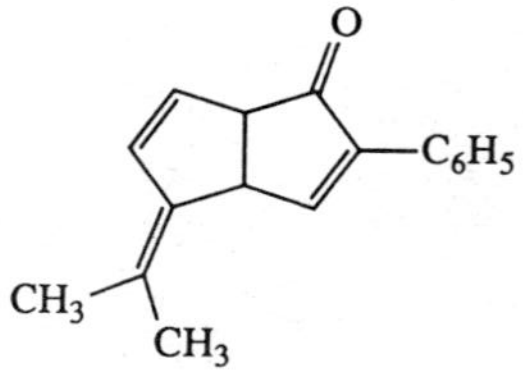 O, C_6H_5, CH_3, CH_3 (23)	23

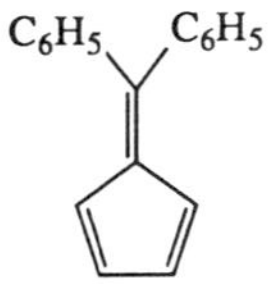

C_6H_5 C_6H_5 (fulvene structure) + $RC{\equiv}CH$ → C_6H_5, C_6H_5, O, R (product structure)

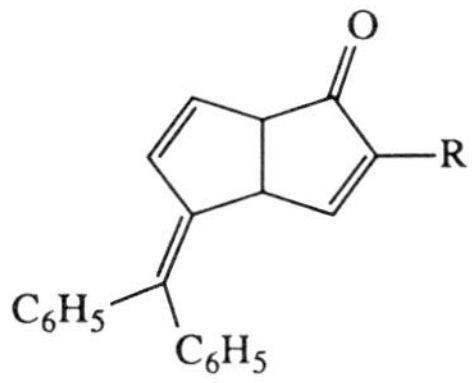

R	Conditions	Yield	Refs.
H	Toluene, 110°, 5 h, N_2	(15)	23
CH_3	Toluene, 110°, 5 h, N_2	(15)	23

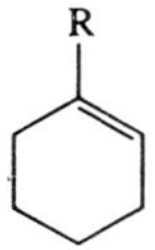

R (cyclohexene structure) + $C_6H_5C{\equiv}CH$ 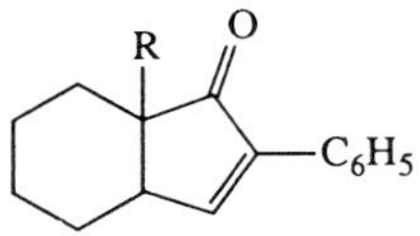 → R, O, C_6H_5 (product structure)

R	Conditions	Yield	Refs.
H	Toluene, 110°, 6 h, N_2	(3)	11
$(CH_3)_2N(CH_2)_2$	Toluene, 90°, 24-36 h, N_2	(28)	6

TABLE IV. ALKYNES WITH MONOCYCLIC ALKENES (*Continued*)

Alkene	Alkyne[a]	Reaction Conditions	Product(s) and Yield(s) (%)			Refs.
(1,3-cyclohexadiene)	$RC\equiv CH$		I + II + III			
R			I	II	III	
H		Toluene, 60-80°, 5-6 h, N_2	(20)[h]		(0)	25
CH_3		Toluene, 60-80°, 5-6 h, N_2	(38)[h] (I or II)		(0)	25
C_6H_5		Toluene, 60-80°, 5-6 h, N_2	([i]) (I or II)		(65)[h]	25
 (1,3-cyclohexadiene)	$C_6H_5C\equiv CH$	Toluene, 70-80°, 5 h, N_2	(4) (I or II)		(19)[h]	25
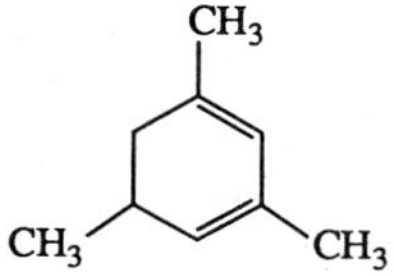 (1,3,5-trimethyl-1,3-cyclohexadiene)	$HC\equiv CH$	Toluene, 110°, 5 h, N_2	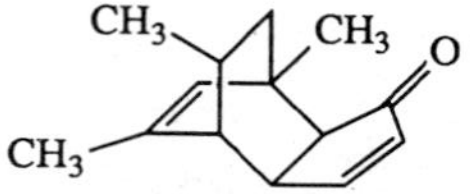 (trimethyl bicyclic enone) (15)[h]			25
 (cycloheptene)	$R^1C\equiv CR^2$		(bicyclic cyclopentenone with R^1, R^2)			

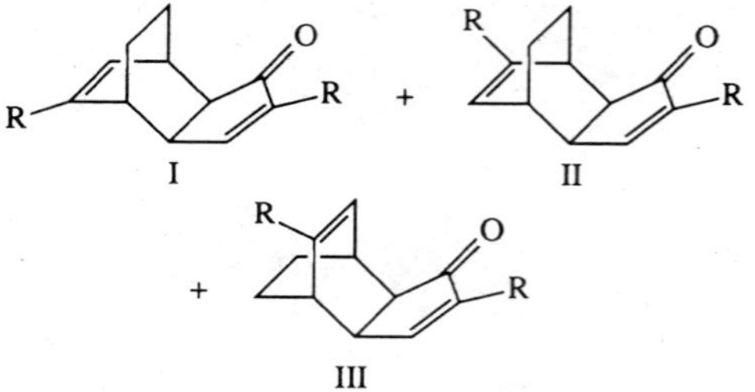

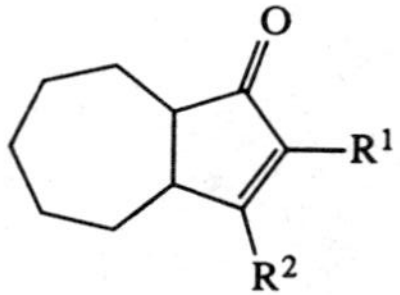

Alkene	Alkyne	Conditions	Product(s) and Yield(s) (%)	Refs.
	$R^1C \equiv CR^2$			
	R^1 / R^2			
	CH_3 / H	Toluene, 110°, 6 h, N_2	(17)	11
	$CH_2=C(CH_3)$ / H	Toluene, 110°, 6 h, N_2	(5)	11
	C_6H_5 / H	Toluene, 110°, 6 h, N_2	(41)[j]	11
	C_6H_5 / C_6H_5	Toluene, 110°, 6 h, N_2	(7)	11
 (structure)	$RC \equiv CH$, R = H, CH_3, or C_6H_5	Toluene, 110°, N_2	No cyclic ketone products	11
 (structure)	$C_6H_5C \equiv CH$	Toluene, 110°, 6 h, N_2	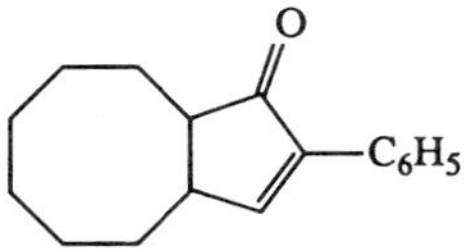 (structure with O, C_6H_5) (35)[j]	11

[a]This refers to the alkyne portion of the preformed alkyne·$Co_2(CO)_6$ complex.
[b]The alkene was the solvent for this reaction.
[c]This reaction was carried out either in a sealed tube or in an autoclave.
[d]Both the starting alkyne and the product were mixtures of stereoisomers.
[e]The alkyne·$Co_2(CO)_6$ complex was prepared in situ from $CoBr_2$, Zn, CO, and alkyne.
[f]The *cis* isomer was used.
[g]The second yield is based on unrecovered starting complex.
[h]This is the most likely structure of the major product.
[i]The yield of this isomer was variable and low.
[j]The 3-phenyl regioisomer was thought to be formed in trace amounts.

TABLE V. ALKYNES WITH HETEROCYCLIC ALKENES

Alkene	Alkyne[a]		Reaction Conditions	Product(s) and Yield(s) (%)		Refs.
2,5-dihydrofuran bearing R^1, R^1 at C-2	$R^2C{\equiv}CR^3$			I + II (bicyclic cyclopentenones bearing R^1, R^1, R^2, R^3)		
R^1	R^2	R^3		I	II	
H	H	H	Isooctane, 65°, 1 d, $HC{\equiv}CH/CO$, cat. Co	(150)[b]		63
"	n-C_5H_{11}	"	Hexane, 70°, 36 h, N_2	(37)		15
"	"	"	Hexane, $(n\text{-}C_4H_9)_3PO$, 70°, 2 d, N_2	(69)		3
"	"	"	Isooctane, $(n\text{-}C_4H_9)_3P$, 60°, 36 h, N_2	(48)		15
"	"	"	Toluene, 100°, 1 d, N_2	(47)[c]		15
"	"	"	Toluene, HMPA, 100°, 1 d, N_2	(56)		15
"	"	"	Toluene, $(n\text{-}C_4H_9)_3PO$, 100°, 1 d, N_2	(64)		15
"	$CH_3O_2C(CH_2)_3CH{=}CHCH_2$[d]	"	Benzene, 80°, 3 h, N_2	(65)		62
"	C_6H_5	"	Hexane, 70°, 36 h, N_2	(37)		15
"	"	"	Hexane, $(n\text{-}C_4H_9)_3PO$, 70°, 36 h, N_2	(69)		15
"	CH_3	CH_3	Toluene, 70-100°, 8 h, N_2	(20)		63
"	"	"	Isooctane, 85°, 8 d, $CH_3C{\equiv}CCH_3/CO$, cat. Co	(70)		63
CH_3	H	H	Benzene, 60°, 4 d, N_2	(13)	(9)	24
"	"	"	Benzene, 65°, 6 d, $HC{\equiv}CH/CO$, cat. Co	(32)	(21)	24

Alkene	Alkyne	Conditions	Product(s) and Yield(s) (%)	Refs.
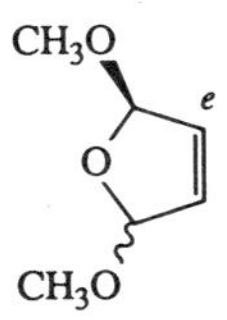 CH_3O, CH_3O [e]	$C_6H_5C{\equiv}CH$	Toluene, 110°, 6 h 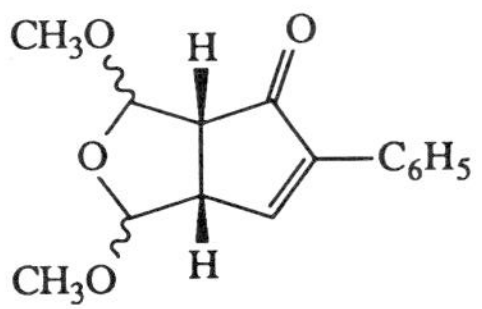	CH_3O, CH_3O, H, H, O, C_6H_5 (10)	2
	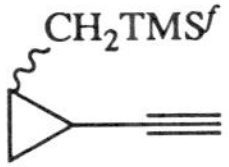 CH_2TMS [f]	Toluene, 80°, 3 d	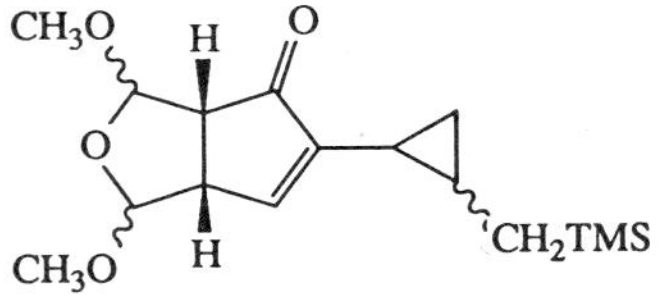 CH_3O, CH_3O, H, H, O, CH_2TMS (35) [g]	29
	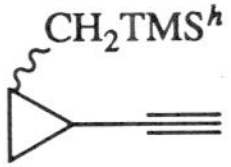 CH_2TMS [h]	Toluene, 80°, 5 d 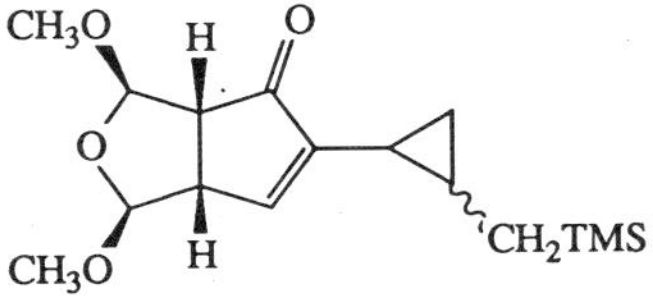	CH_3O, CH_3O, H, H, O, CH_2TMS (46) [i] + CH_3O, CH_3O, H, H, O, CH_2TMS (6) [j]	29
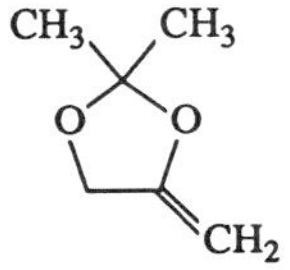 CH_3, CH_3, CH_2	$HC{\equiv}CH$	Benzene, 65°, 16 h 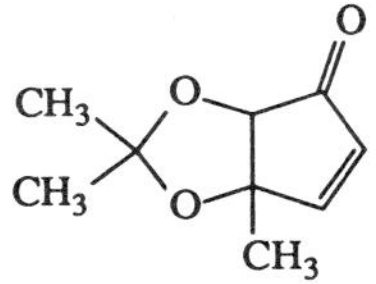	CH_3, CH_3, CH_3, O (23)	22

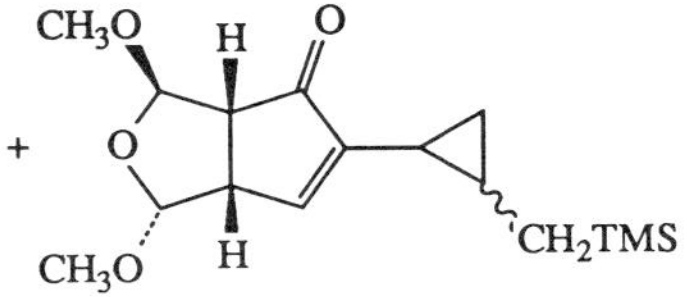

TABLE V. ALKYNES WITH HETEROCYCLIC ALKENES (*Continued*)

Alkene	Alkyne[a]	Reaction Conditions	Product(s) and Yield(s) (%)	Refs.
 2,5-dihydrothiophene 1,1-dioxide (structure)	$C_6H_5C\equiv CH$	Toluene, 110°, 6 h	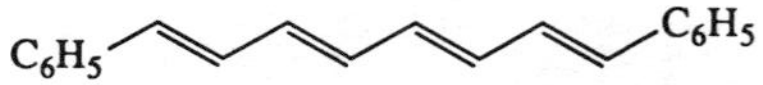 C_6H_5–(CH=CH)$_4$–C_6H_5 (19)	2
	$C_6H_5C\equiv CC_6H_5$	Toluene, 110°, 6 h	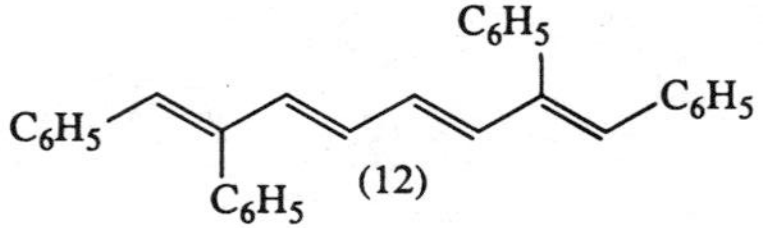 C_6H_5CH=C(C_6H_5)–CH=CH–CH=CH–C(C_6H_5)=CHC_6H_5 (12)	2
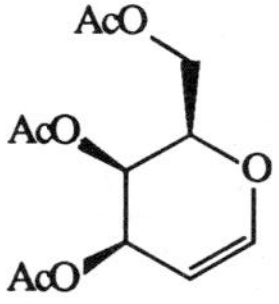 tri-O-acetyl glucal (AcO, AcO, AcO; structure)	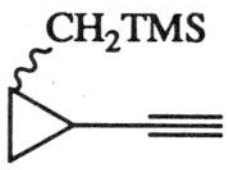 1-ethynyl-2-(CH$_2$TMS)cyclopropane (structure)	Toluene, 80°, 2 d	No cyclic ketone products	29

[a]This refers to the alkyne portion of the preformed alkyne·$Co_2(CO)_6$ complex.

[b]The yield is based on available cobalt complex.

[c]This is an average of many experiments under slightly differing conditions.

[d]The *cis* isomer was used.

[e]A 1:1 mixture of *cis* and *trans* isomers of the alkene was used in large excess.

[f]A 97:3 *cis*:*trans* mixture was used.

[g]A mixture of 5 diastereoisomers was obtained.

[h]A 92:8 *trans*:*cis* mixture was used.

[i]This was formed as a 1:1 mixture of diastereoisomers.

[j]This was formed as a 4:1 mixture of diastereoisomers.

TABLE VI. ALKYNES WITH BICYCLIC OR POLYCYLIC ALKENES

Alkene	Alkyne[a]	Reaction Conditions	Product(s) and Yield(s) (%)	Refs.
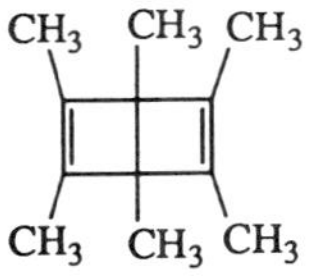 CH_3 CH_3 CH_3 / CH_3 CH_3 CH_3	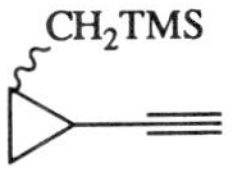 CH_2TMS	Toluene, 70°, 8 h	No cyclic ketone products	29
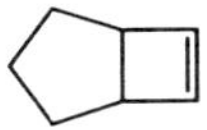	$R^1C{\equiv}CR^2$		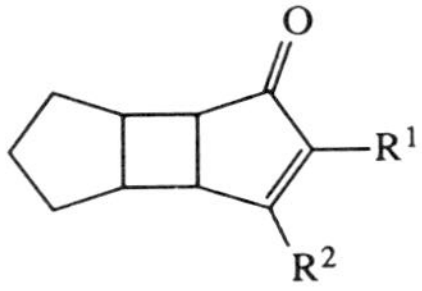 O, R^1, R^2	
R^1 R^2				
H H		Toluene, 60-80°, 4-6 h, N_2	(32)	27
CH_3 "		Toluene, 60-80°, 4-6 h, N_2	(43)	27
C_6H_5 "		Toluene, 60-80°, 4-6 h, N_2	(43)	27
" C_6H_5		Toluene, 60-80°, 4-6 h, N_2	(30)	27

TABLE VI. ALKYNES WITH BICYCLIC OR POLYCYLIC ALKENES (*Continued*)

Alkene		Alkyne[a]	Reaction Conditions	Product(s) and Yield(s) (%)	Refs.
CH_3O		$R^1C{\equiv}CR^2$		CH_3O, R^1, R^2, O	
R^1	R^2				
H	H		Toluene, 60-80°, 4-6 h, N_2	(46)	27
CH_3	"		Toluene, 60-80°, 4-6 h, N_2	(60)	27
C_6H_5	"		Toluene, 60-80°, 4-6 h, N_2	(30)	27
C_2H_5	C_2H_5		Toluene, 60-80°, 4-6 h, N_2	(44)	27
C_6H_5	CH_3		Toluene, 60-80°, 4-6 h, N_2	([b])	27
"	C_6H_5		Toluene, 60-80°, 4-6 h, N_2	(35)	27
HO, CH_3, CH_3O		HC≡CH	DME, 65°, 4 d, HC≡CH/CO	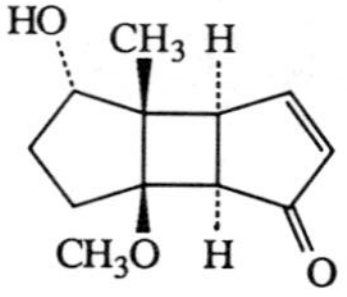 HO, CH_3, H, CH_3O, H, O (65)	26

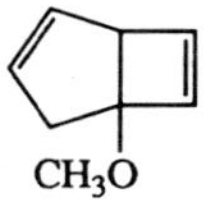

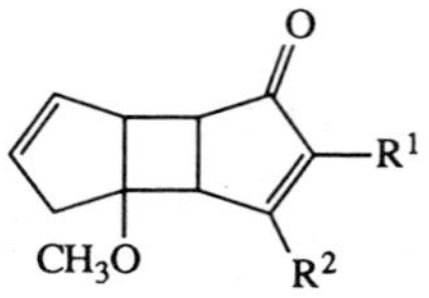

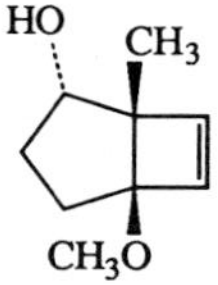

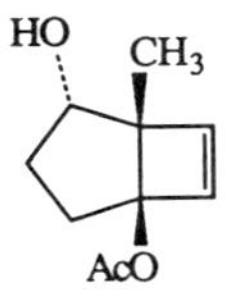 HO, CH_3, AcO	HC≡CH	DME, 65°, 4 d, HC≡CH/CO	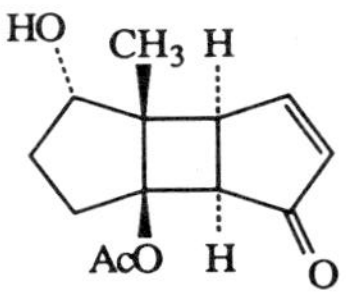 HO, CH_3, H, AcO, H, O (≈20)	26
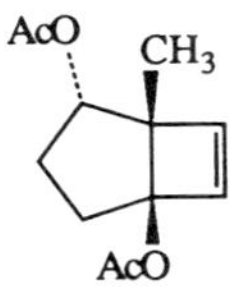 AcO, CH_3, AcO	HC≡CH	DME, 65°, 4 d, HC≡CH/CO	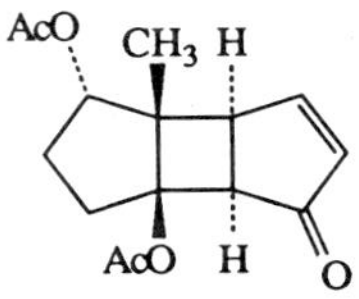 AcO, CH_3, H, AcO, H, O (54)	26

TABLE VI. ALKYNES WITH BICYCLIC OR POLYCYLIC ALKENES (*Continued*)

Alkene	Alkyne[a]	Reaction Conditions	Product(s) and Yield(s) (%)			Refs.
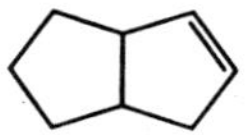	RC≡CH		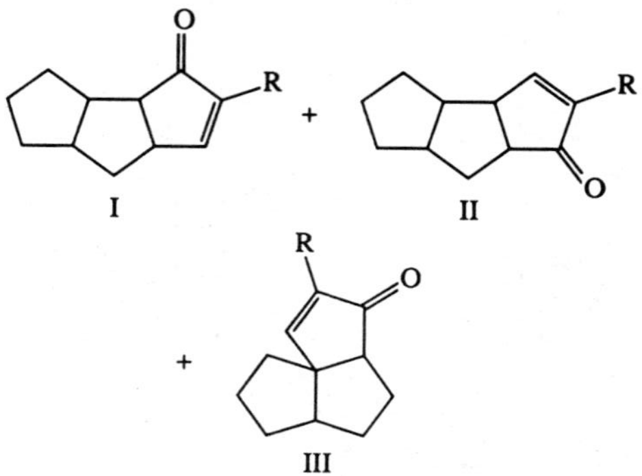 I + II + III			
R			I	II	III	
TMS		Toluene, 100°, 14 d	I:II:III = 1:1:1.5			29
1-TMS-*c*-C_3H_4		Heptane, 85°, 19 d, 100°, 11 d, CO[c]	(24)	(14)	(0)	29

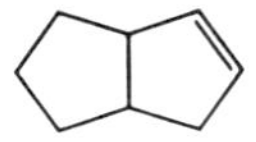

$R^1C{\equiv}CR^2$

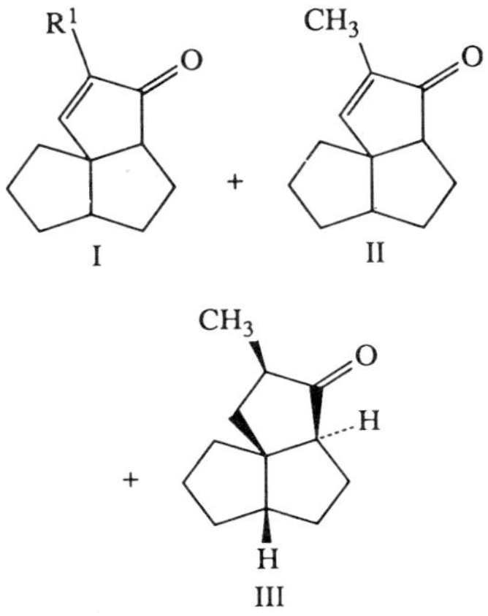

I + II + III

R^1	R^2	Conditions	I	II	III	Refs.
$THPOCH_2$	H	220°, 5 d, CO[c,d]	(0)	(10)	(33)	28
"	"	Tetralin, 220°, 18 h, CO[c]	(0)	(10)	(40)	28
$TBDMSOCH_2$	"	Benzene, 80°, 1 d, CO	(18)	(0)	(0)	28
"	"	134°, 4 h, CO[d]	(15)	(0)	(0)	28
"	"	220°, 7 d, CO[c,d]	(0)	(11)	(15)	28
"	"	t-$C_4H_9C_6H_5$, 140°, 8 h, CO[e]	(30)	(10)	(0)	28
"	TMS	134°, 21 h, , CO[d]	No cyclic ketone products			28

TABLE VI. ALKYNES WITH BICYCLIC OR POLYCYLIC ALKENES (*Continued*)

Alkene	Alkyne[a]	Reaction Conditions	Product(s) and Yield(s) (%)		Refs.
[structure]	HC≡CH		I	II	
		148°, N_2[d]	(3)	(3)	22
		Benzene, 65°, 10 d, HC≡CH/CO, cat. Co	(12)	(12)	22
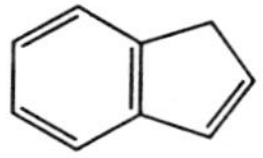 [structure]	RC≡CH		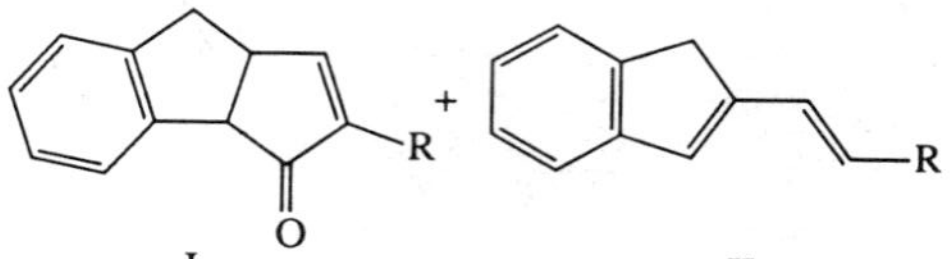 I	II	
R					
H		Toluene, 110°, 5 h, N_2	(31)	(0)	23
CH_3		Toluene, 110°, 5 h, N_2	(41)	(0)	23
"		SiO_2 (dry), 55°, 2 h	(43)	(0)	21
CH_3OCH_2		SiO_2 (dry), 70°, 1 h	(21)	(0)	21
C_6H_5		Toluene, 110°, 5 h, N_2	(52)	(4)	23

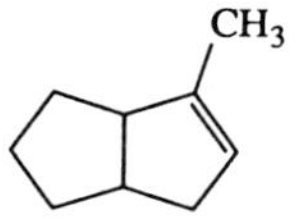

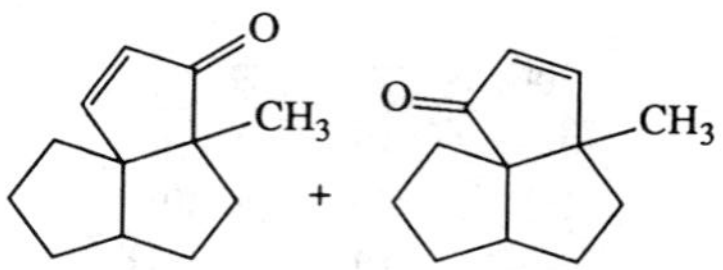

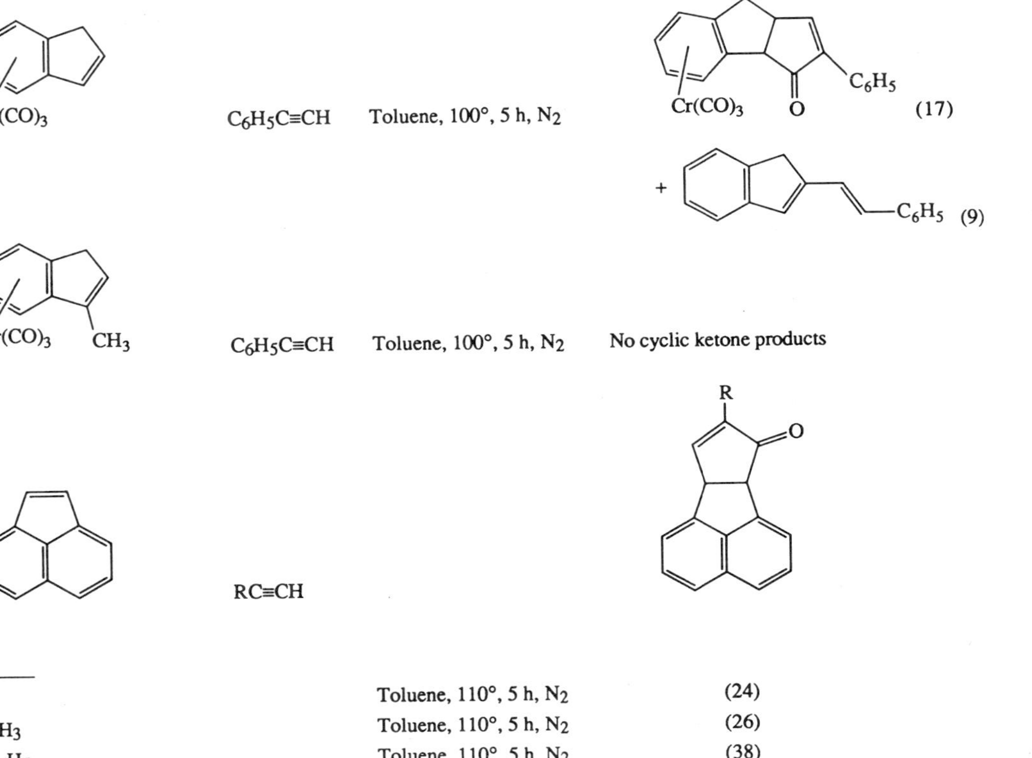

Reactant	Alkyne	Conditions	Product(s) and Yield(s) (%)	Refs.
Cr(CO)$_3$	$C_6H_5C{\equiv}CH$	Toluene, 100°, 5 h, N_2	Cr(CO)$_3$, O, C_6H_5 (17) + C_6H_5 (9)	19
Cr(CO)$_3$, CH_3	$C_6H_5C{\equiv}CH$	Toluene, 100°, 5 h, N_2	No cyclic ketone products	19
	RC≡CH		R, O	
R				
H		Toluene, 110°, 5 h, N_2	(24)	23
CH_3		Toluene, 110°, 5 h, N_2	(26)	23
C_6H_5		Toluene, 110°, 5 h, N_2	(38)	23

TABLE VI. ALKYNES WITH BICYCLIC OR POLYCYLIC ALKENES (*Continued*)

Alkene	Alkyne[a]	Reaction Conditions	Product(s) and Yield(s) (%)		Refs.
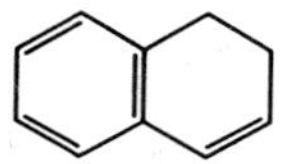 1,2-Dihydronaphthalene	$RC\equiv CH$		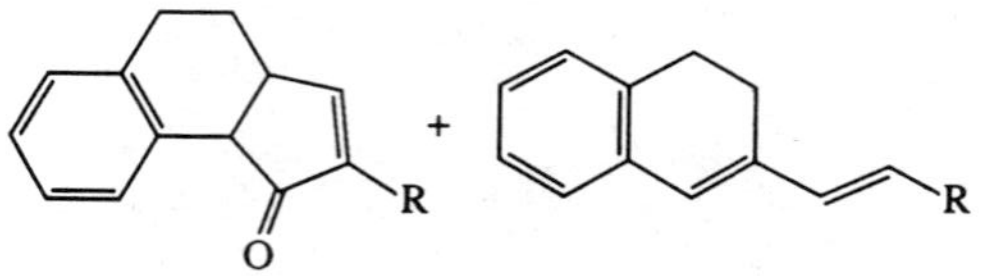 I + II		
R			I	II	
CH_3		Toluene, 110°, 7 h, N_2	(35)	(0)	18
C_6H_5		Toluene, 110°, 7 h, N_2	(38)	(4)	18
 Norbornene	$R^1C\equiv CR^2$		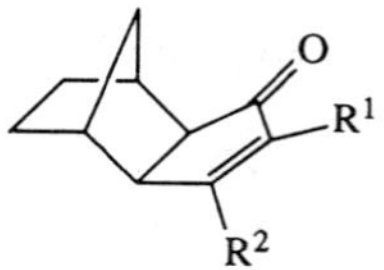		
R^1	R^2				
H	H	Benzene, 60-70°, 4 h, N_2	(55)		1
"	"	Isooctane, 60-70°, $HC\equiv CH/CO$, cat. Co	(74)		1
C_2H_5O	"	–	(*f*)		3
C_6H_5S	"	Toluene, 60-70°, 5 h, N_2	(59)		14
TMS	"	Isooctane, 80-90°, 18 h	(93)		30

R^1	R^2	Conditions	Yield (%)	Refs.
CH_3	H	Toluene, 60-70°, 4 h, N_2	(33)	1
"	"	SiO_2 (dry), 60°, 2 h	(74)	21
CH_2OH	"	–	(0)	3
CH_3OCH_2	"	Hexane, 60°	(37)	21
"	"	SiO_2 (dry), 55°, 2 h	(74)	21
CH_3O_2C	"	–	(0)	3
c-C_3H_5	"	Toluene, 80-90°, 18 h	(83)	64
1-Cl-*c*-C_3H_4	"	70-75°, 3 h, Ar[d]	(30)	29
1-TMS-*c*-C_3H_4	"	74°, 4 h[d]	(50)	64
1-[$C_2H_5OCH(CH_3)O$]-*c*-C_3H_4	"	Toluene, 80°, 18 h	(28)	29
2-(CH_3O_2C)-*c*-C_3H_4	"	Toluene, 80°, 45 h, then 20°, 16 h	(58)	29
2-($CH_3O_2CCH_2$)-*c*-C_3H_4	"	Toluene, 80°, 1 d, then 20°, 3 d	(72)	29
2-(*t*-C_4H_9O)-*c*-C_3H_4	"	Toluene, 80-90°, 18 h	(79)	64
1-Cl-*c*-$C_3(CH_3)_4$	"	Toluene, 70°, 5 h	(18)	29
i-$C_4H_9OCH{=}CH$	"	Toluene, 80°, 2 d	(45)	29
C_6H_5	"	Mesitylene, 60-70°, 4 h, N_2	(59)	1
"	"	THF, 65°, 4 h, N_2[g]	(80)	74
"	"	SiO_2 (dry), 55°, 2 h	(86)	21
"	"	Toluene, (*n*-C_4H_9)$_3$P, 111°, 3 h	(51)	31
"	"	Toluene, (+)-Glyphos, 111°, 3 h	(49)	31
"	"[h]	Toluene, 111°, 1 d	(28)[i]	31
"	"[j]	Toluene, 45°, ultrasound, 6 h	(31)[k]	31
n-C_6H_{13}	"	THF, 65°, 4 h, N_2[g]	(88)	74
n-C_8H_{17}	"	THF, 65°, 4 h, N_2[g]	(92)	74

TABLE VI. ALKYNES WITH BICYCLIC OR POLYCYCLIC ALKENES (*Continued*)

Alkene	Alkyne[a]	Reaction Conditions	Product(s) and Yield(s) (%)	Refs.
R^1	R^2			
TMS	CH_3	Toluene, 70-80°, 4 h, N_2	(38)	32
$CH_3CH{=}CHCH_2$	"	–	([l])	32
$CH_3C{\equiv}CCH_2$	"	–	([l])	32
C_2H_5	C_2H_5	Toluene, 60-70°, 4 h, N_2	(23)	1
n-C_3H_7	n-C_3H_7	THF, 65°, 12 h, N_2[g]	(78)	74
C_6H_5	CH_3	Toluene, 60-70°, 4 h, N_2	(35)	1
C_6H_5	C_6H_5	Benzene, 60-70°, 4 h, N_2	(65)	32
"	"	THF, 65°, 12 h, N_2[g]	(38)	74
 [structure]	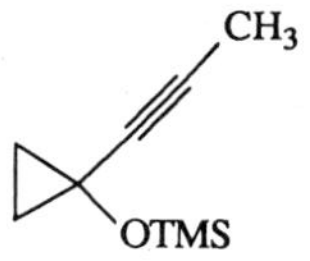 [structure] CH_3, OTMS	Toluene, 80°, 3 d	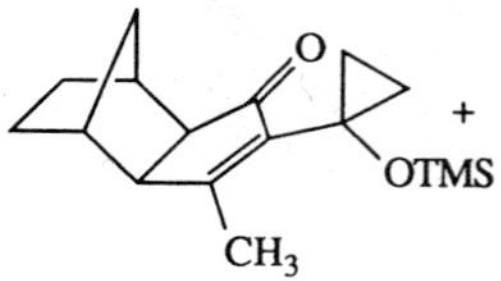 [structure] O, OTMS, CH_3 (1) 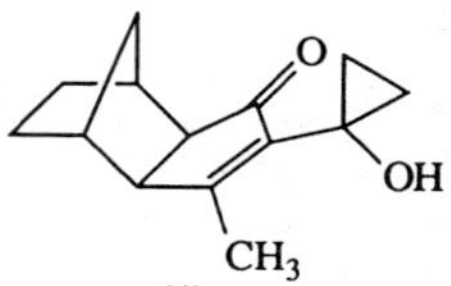 + [structure] O, OH, CH_3 (4) 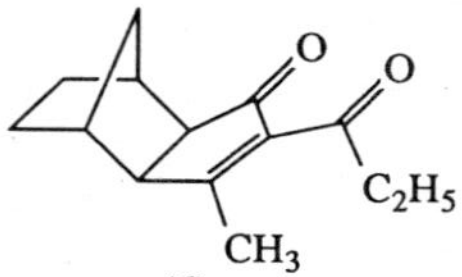 + [structure] O, O, C_2H_5, CH_3 (6)	29

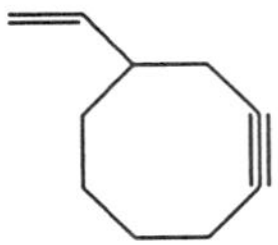

Benzene, 80°, CO — (34) + (34) — 70

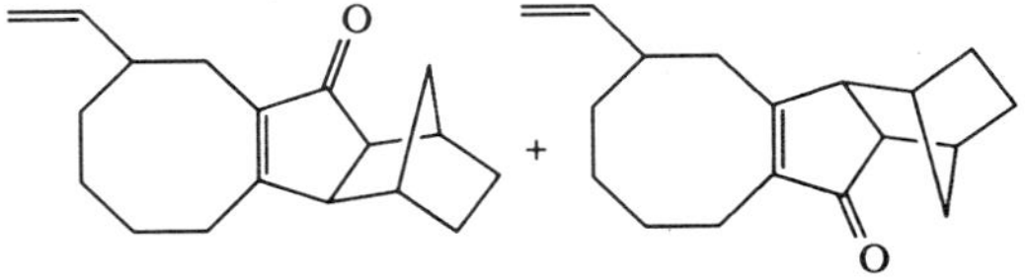

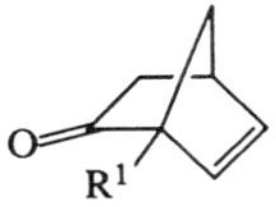

$R^2C\equiv CH$ — I + II

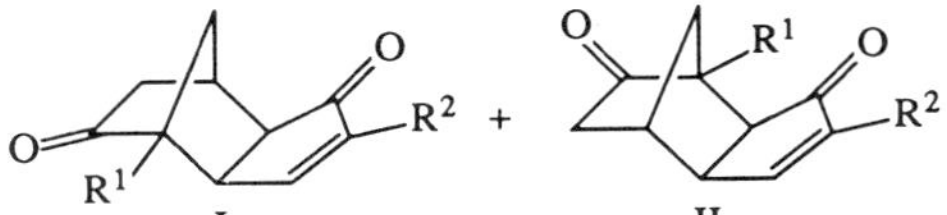

R^1	R^2	Conditions	I	II	Refs.
H	CH_3	Isooctane, 70°, 2 d, CO, cat. Co	(10)	(31)	33
"	C_6H_5	Isooctane, 65°, 20 h, CO, cat. Co	(16)	(42)	33
CH_3	CH_3	Isooctane, 70°, 2 d, CO, cat. Co	(16)	(39)	33
"	$HO(CH_2)_3$	Benzene, 80°, 2 d, CO	(<4)	(<6)	34
"	Polymer-$O(CH_2)_3$[m]	Benzene, 80°, 2 d, CO	(29)[n]	(70)[n]	34

TABLE VI. ALKYNES WITH BICYCLIC OR POLYCYLIC ALKENES (*Continued*)

Alkene		Alkyne[a]	Reaction Conditions	Product(s) and Yield(s) (%)		Refs.
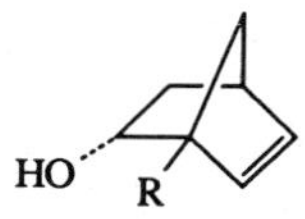		$CH_3C \equiv CH$		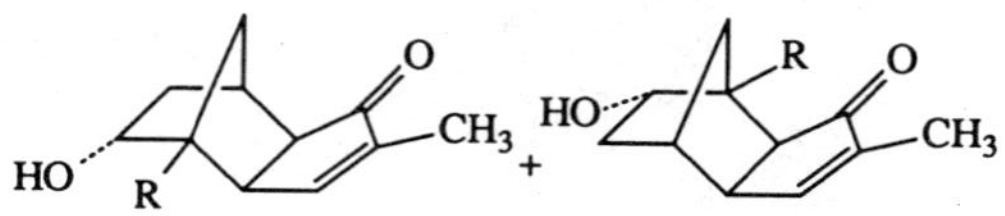 I + II		
R				I	II	
H			Isooctane, 70°, 2 d, CO, cat. Co	(21)	(17)	33
CH_3			Isooctane, 70°, 2 d, CO, cat. Co	(35)	(24)	33
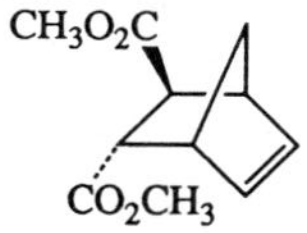		$R^1C \equiv CR^2$		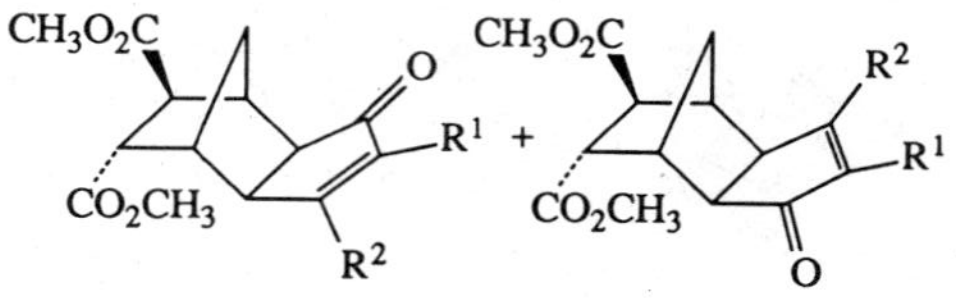		
R^1	R^2			Major	Minor	
H	H		Toluene, 60-70°, 4 h, N_2	(25)[o]	(8)[o]	32
CH_3	"		Toluene, 60-70°, 4 h, N_2	(53)[o]	(9)[o]	32
C_6H_5	"		Toluene, 60-70°, 4 h, N_2	(44)[o]	(14)[o]	32
C_6H_5	CH_3		Toluene, 60-70°, 4 h, N_2	(28)[o]	(14)[o]	32

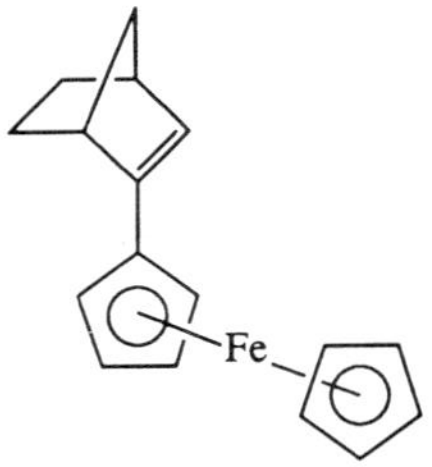	HC≡CH	Toluene, 70-80°, 4 h, N_2	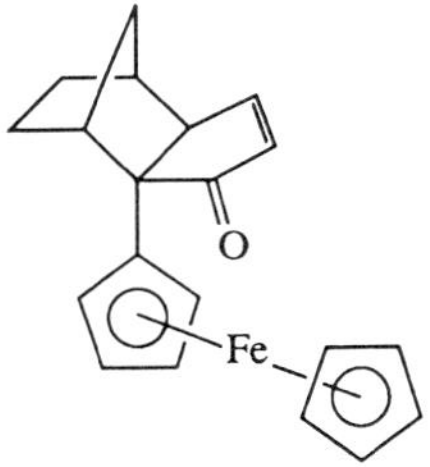 (18)	32
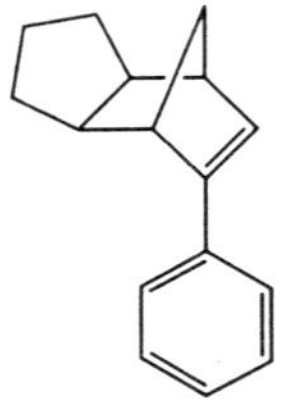	HC≡CH	Toluene, 70-80°, 4 h, N_2	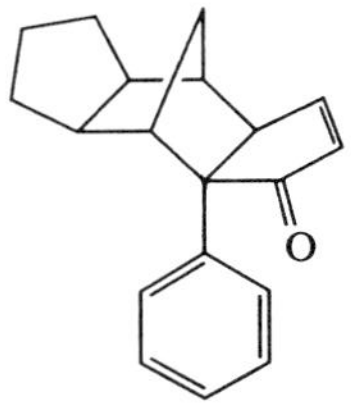 (60)	32

TABLE VI. ALKYNES WITH BICYCLIC OR POLYCYLIC ALKENES (*Continued*)

Alkene	Alkyne[a]	Reaction Conditions	Product(s) and Yield(s) (%)	Refs.
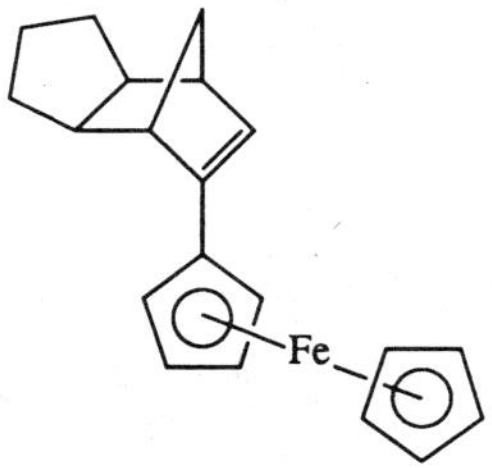 Fe	HC≡CH	Toluene, 70-80°, 4 h, N_2	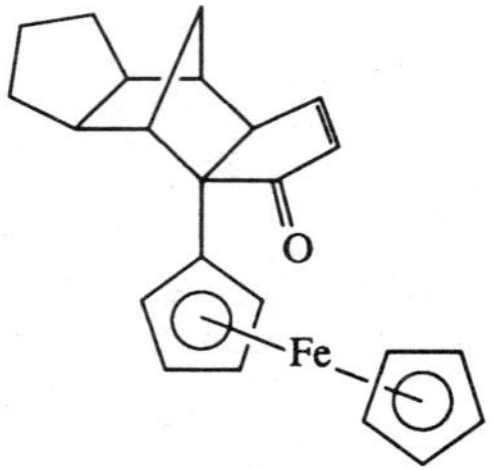 O, Fe (65)	32
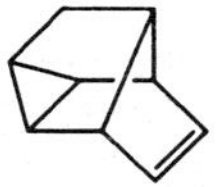	RC≡CH		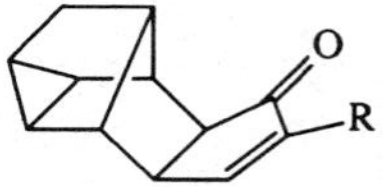 O, R	
R				
TMS		Toluene, 80-90°, 18 h	(85)	30
c-C_3H_5		Toluene, 80-90°, 18 h	(85)	64
1-TMS-c-C_3H_4		Toluene, 80-90°, 18 h	(83)	30
1-[$C_2H_5OCH(CH_3)O$]-c-C_3H_4		Toluene, 80-90°, 18 h	(8)	64
2-(t-C_4H_9O)-c-C_3H_4		Toluene, 80-90°, 18 h	(65)	64

RC≡CH

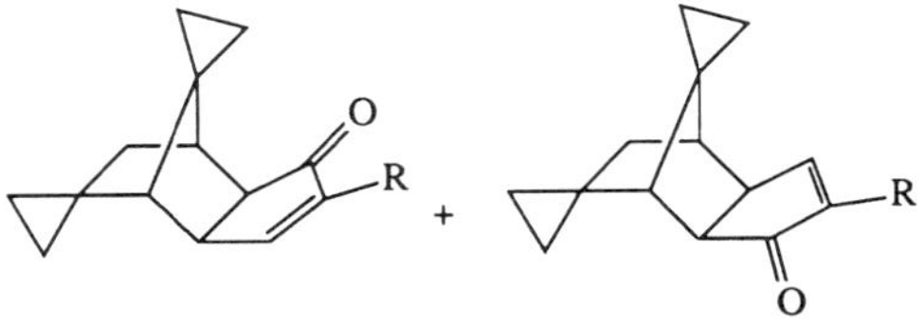

R	Conditions	Yield (%)	Refs.
$(CH_3)_3Si$	Toluene, 80-90°, 18 h	(84)^p	30
1-TMS-*c*-C_3H_4	Toluene, 80-90°, 18 h	(58)^p	30

$R^1C≡CR^2$

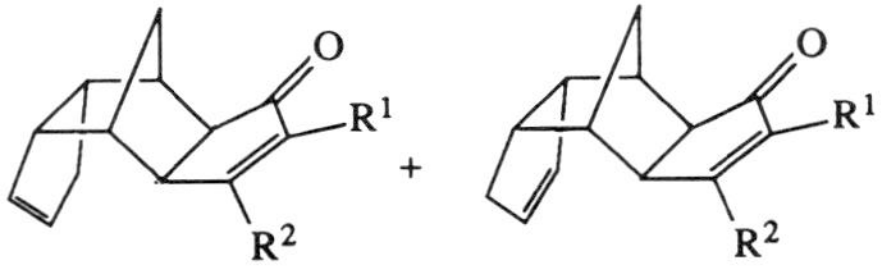

R^1	R^2	Conditions	Yield (%)	Refs.
H	H	Toluene, 70-80°, 4 h, N_2	(74)^p	32
CH_3	"	Toluene, 70-80°, 4 h, N_2	(51)^p	32
C_6H_5	"	Toluene, 70-80°, 4 h, N_2	(57)^p	32
C_6H_5	CH_3	Toluene, 70-80°, 4 h, N_2	(32)^p	32
C_2H_5	C_2H_5	Toluene, 70-80°, 4 h, N_2	(18)^p	32
C_6H_5	C_6H_5	Toluene, 70-80°, 4 h, N_2	(23)^p	32

TABLE VI. Alkynes with Bicyclic or Polycyclic Alkenes (*Continued*)

Alkene	Alkyne[a]	Reaction Conditions	Product(s) and Yield(s) (%)		Refs.
 [structure]	RC≡CH		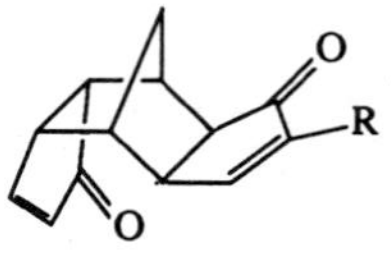 [structure] I + [structure] II	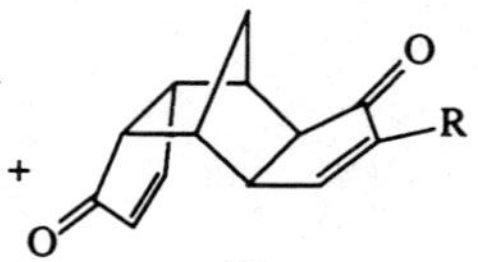	
R			I	II	
H		Toluene, 70-80°, 4 h	(18)	(33)	32
CH_3		Toluene, 70-80°, 4 h	(10)	(24)	32
C_6H_5		Toluene, 70-80°, 4 h	(9)	(32)	32
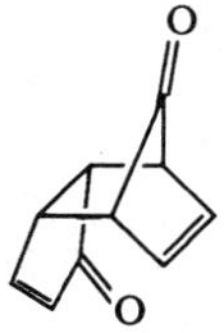 [structure]	HC≡CH	Benzene, 65°, 5 d, HC≡CH/CO	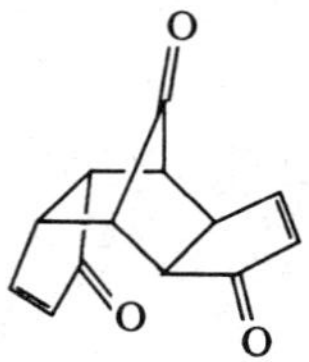 [structure] (14)		10

$RC{\equiv}CH$

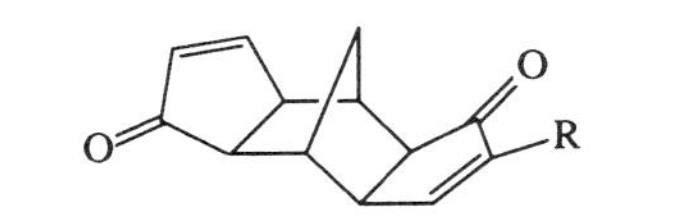

R			
CH_3	Toluene, 90°, 8 h	(21)	1
C_6H_5	Toluene, 90°, 8 h	(23)	1

$R^1C{\equiv}CR^2$

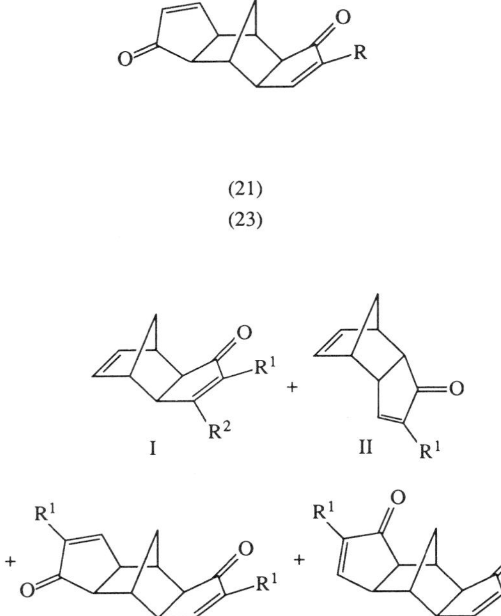

TABLE VI. ALKYNES WITH BICYCLIC OR POLYCYLIC ALKENES (*Continued*)

Alkene	Alkyne[a]	Reaction Conditions	Product(s) and Yield(s) (%)				Refs.
R^1	R^2		I	II	III	IV	
H	H	Toluene, 70°, 4 h, N_2	(44)	(10)	(≈45) (III and IV)		1,15
"	"	Toluene, $(n\text{-}C_4H_9)_3PO$, 70°, 4 h, N_2	(34)	(7)	(-)	(-)	15
"	"	Toluene, 60-70°, 10 h, N_2	(-)	(-)	(29)	(4)	1
"	"	Toluene, 100°, 24 h, N_2	(33) (I and II)		(-)	(-)	15
"	"	Toluene, $(n\text{-}C_4H_9)_3PO$, 100°, 24 h, N_2	(54) (I and II)		(-)	(-)	15
"	"	Isooctane, 70°, HC≡CH/CO, cat. Co	(47)	(-)	(q)	(f)	1
CH_3	"	Toluene, 60-70°, 4 h, N_2	(33)	(-)	(17)	(4)	1
$n\text{-}C_4H_9$	"	Isooctane, 65°, 36 h, N_2	(53)	(2)	(0)	(0)	35
$n\text{-}C_5H_{11}$	"	Toluene, 55-60°, 3 h, CO	(37)	(1)	(0)[r]	(0)[r]	15
"	"	Toluene, 100°, 48 h, N_2	(52)	(-)	(0)	(0)	15
"	"	Toluene, $(n\text{-}C_4H_9)_3PO$, 100°, 48 h, N_2	(62)	(-)	(0)	(0)	15
"	"	Toluene, ultrasound, 70°, 3 h, N_2	(65)	(-)	(0)	(0)	15
$HO(CH_2)_3$	"	Benzene, 80°, 10 h, CO	(26)	(0)	(0)	(0)	34
C_6H_5	"	Benzene, 60-70°, 4 h, N_2	(45)	(0)	(0)	(0)	1
"	"	Toluene, 80-90°, 7-8 h, N_2	(-)	(0)	(13)	(-)	1
"	"	Isooctane, 70°, 5 h, CO, cat. Co[s]	(26)	(0)	(10)	(<1)	1
C_6H_5S	"	Toluene, 60-70°, 5 h, N_2	(57)	(0)	(0)	(0)	14
Polymer-$O(CH_2)_3$[m]	"	Benzene, 80°, 6 h, CO	(59)[n]	(0)	(0)[t]	(0)	34
TMS	CH_3	Toluene, 70-80°, 4 h, N_2	(42)	(0)	(0)	(0)	32
C_2H_5	C_2H_5	Toluene, 60-70°, 4 h, N_2	(23)	(0)	(0)	(0)	1
C_6H_5	C_6H_5	Toluene, 60-70°, 4 h, N_2	(28)	(0)	(0)	(0)	1

Reactant	Alkyne	Conditions	Product(s) and Yield(s) (%)	Refs.
(norbornadiene)	$THPO(CH_2)_4C{\equiv}CCH_2C{\equiv}CH$	Toluene, 60-70°, 4 h, N_2	$(CH_2)_4OTHP$ (bis-cyclopentenone adduct) + $(CH_2)_4OTHP$ (regioisomeric bis-cyclopentenone adduct) (11 total)	62
CH_3O_2C, CH_3O_2C (dimethyl norbornadiene-2,3-dicarboxylate)	$R^1C{\equiv}CR^2$	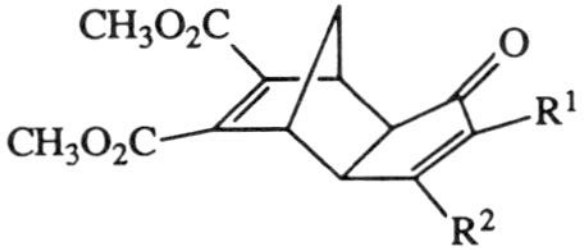	CH_3O_2C, CH_3O_2C, O, R^1, R^2 (cyclopentenone adduct)	

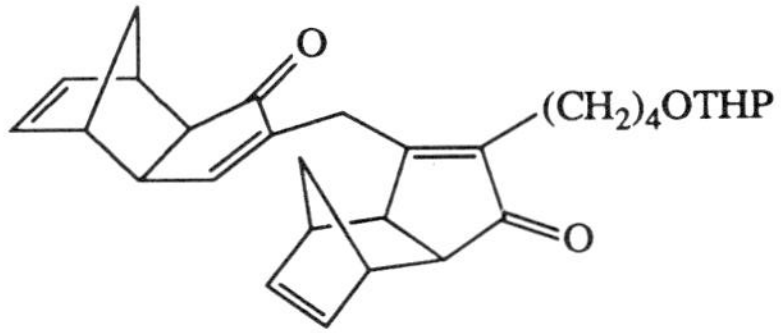

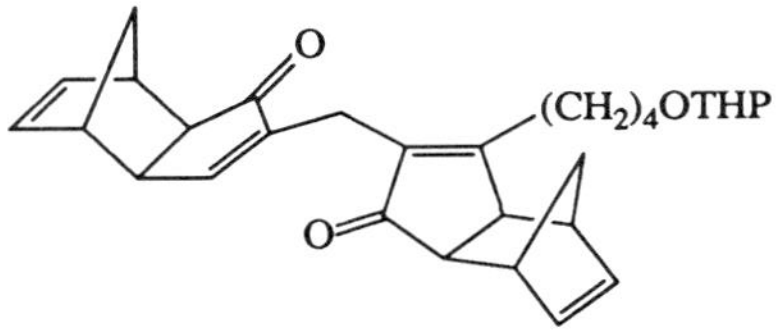

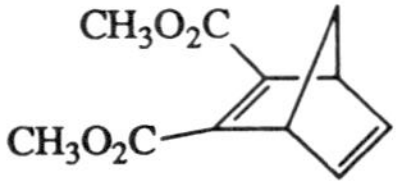

R^1	R^2	Conditions	Yield (%)	Refs.
H	H	Toluene, 60-70°, 4 h, N_2	(23)	1
CH_3	"	Toluene, 60-70°, 4 h, N_2	(38)	1
C_6H_5	"	Toluene, 60-70°, 4 h, N_2	(31)	1
C_6H_5	C_6H_5	Toluene, 60-70°, 4 h, N_2	(28)	1

TABLE VI. Alkynes with Bicyclic or Polycylic Alkenes (*Continued*)

Alkene	Alkyne[a]	Reaction Conditions	Product(s) and Yield(s) (%)	Refs.
	RC≡CH		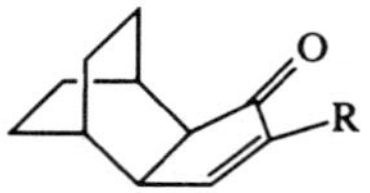 CH_3O_2C, O, R	
R				
H		Toluene, 60-80°, 4-6 h, N_2	(50)	25
CH_3		Toluene, 60-80°, 4-6 h, N_2	(19)	25
C_6H_5		Toluene, 60-80°, 4-6 h, N_2	(34)	25
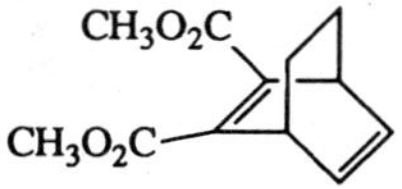 CH_3O_2C, CH_3O_2C	RC≡CH		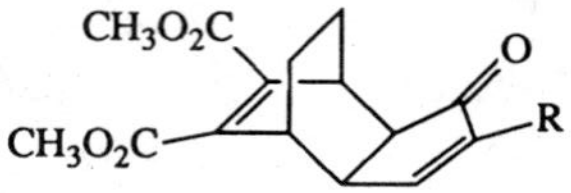 CH_3O_2C, CH_3O_2C, O, R	
R				
H		Toluene, 70-80°, 6-7 h, N_2	(80)	25
CH_3		Toluene, 70-80°, 6-7 h, N_2	(36)	25
C_6H_5		Toluene, 70-80°, 6-7 h, N_2	(85)	25

[a]This refers to the alkyne portion of the preformed alkyne·$Co_2(CO)_6$ complex.

[b]The product of this reaction was not fully characterized.

[c]This reaction was carried out in a sealed tube.

[d]The alkene was the solvent for this reaction.

[e]The alkene used in this reaction was bicyclo[3.3.0]oct-1-ene, .

[f]The product was formed in low, unstated yield.

[g]The alkyne·$Co_2(CO)_6$ complex was prepared in situ from $CoBr_2$, Zn, CO, and alkyne.

[h]The starting complex used in this reaction was a 3:2 mixture of the (+) and (-) diastereomers of $C_6H_5C{\equiv}CH{\cdot}Co_2(CO)_5{\cdot}$(+)-Glyphos.

[i]The product was formed in 36% ee with a (-) optical rotation.

[j]The starting complex used in this reaction was the pure (-) diastereomer of $C_6H_5C{\equiv}CH{\cdot}Co_2(CO)_5{\cdot}$(+)-Glyphos.

[k]The product was formed in 100% ee with a (+) optical rotation.

[l]The yield was not stated; the reaction is cited as an unpublished observation in the reference.

[m]The substrate for this reaction was prepared by esterification of polymer-bound benzoic acid (derived from 2% crosslinked Merrifield-type styrene/divinylbenzene/*p*-(chloromethyl)styrene copolymer) with 4-pentyn-1-ol.

[n]This is the yield after hydrolytic removal of the reaction product from the polymer.

[o]It is not known which structure corresponds to the major or minor isomer in this reaction.

[p]This is the total yield of a mixture of the two regioisomeric products.

[q]This product was formed in "substantial amounts".

[r]These products were not isolated when an excess of alkene was used, as in this experiment, but were formed in ≈0.1% yield each when equimolar alkene and alkyne complex were used.

[s]This reaction was carried out in an autoclave pressurized to 50 atm with CO before heating.

[t]This product was not isolated when an excess of alkene was used, but was formed in 20% yield when equimolar alkene and polymer-bound alkyne complex were used.

TABLE VII. ALKYNES WITH HETEROBICYCLIC OR HETEROPOLYCYCLIC ALKENES

Alkene	Alkyne[a]	Reaction Conditions	Product(s) and Yield(s) (%)		Refs.
Cl, CH_3O_2C (structure)	TMSC≡CH	Toluene, 90°, 12 h	Cl, CH_3O_2C, O, TMS (structure) (29) + Cl, CH_3O_2C, TMS, O (structure) (23)		64
CH_3, HO, OCH_3, O, CH_3 (structure)	RC≡CH		CH_3, HO, OCH_3, O, CH_3, R (structure) I + CH_3, HO, OCH_3, O, CH_3, R (structure) II		
R			I	II	
H		Toluene, 85°, 12 h	(24)	(24)	64
CH_3		Toluene, 85°, 12 h	(34)	(34)	64

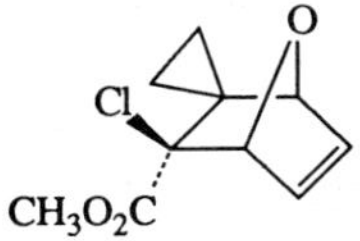

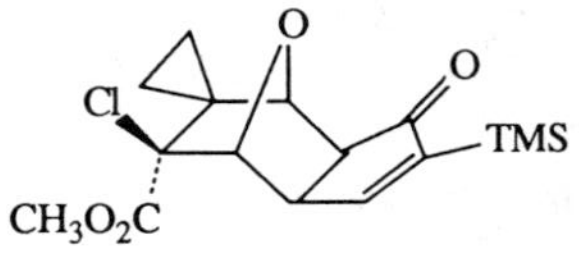

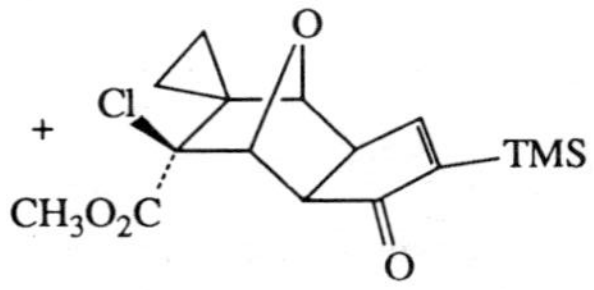

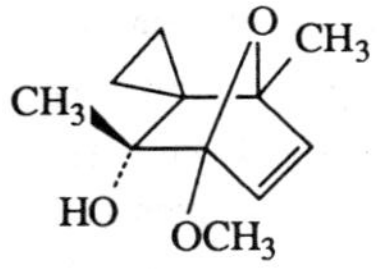

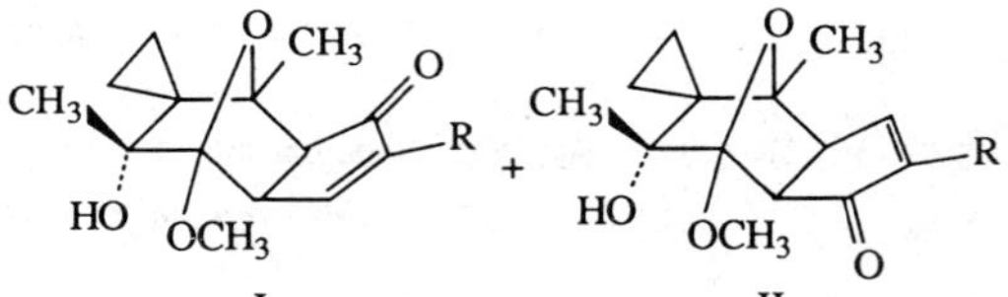

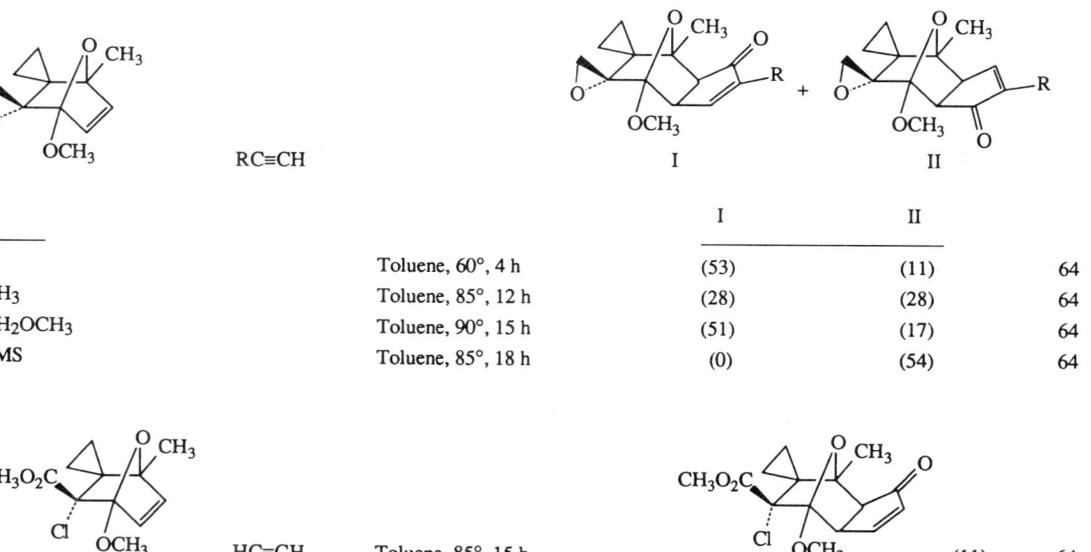

R		Conditions	I	II	Refs.
	RC≡CH				
H		Toluene, 60°, 4 h	(53)	(11)	64
CH_3		Toluene, 85°, 12 h	(28)	(28)	64
CH_2OCH_3		Toluene, 90°, 15 h	(51)	(17)	64
TMS		Toluene, 85°, 18 h	(0)	(54)	64
	HC≡CH	Toluene, 85°, 15 h	(11) + (8)		64

TABLE VII. ALKYNES WITH HETEROBICYCLIC OR HETEROPOLYCYCLIC ALKENES (*Continued*)

Alkene	Alkyne[a]	Reaction Conditions	Product(s) and Yield(s) (%)	Refs.
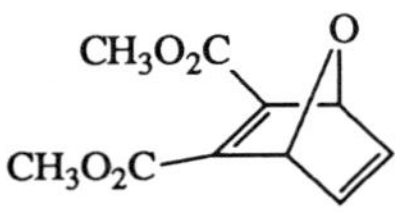 CH_3O_2C, CH_3O_2C (7-oxabicyclic diene)	$C_6H_5C{\equiv}CH$	Toluene, 70-80°, 6 h, N_2	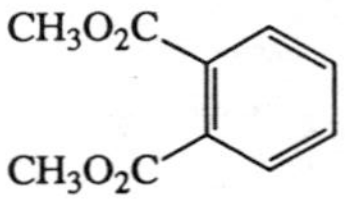 CH_3O_2C, CH_3O_2C (benzene) (39)	2
R^1O_2C, R^1O_2C—N (diazabicyclic alkene)	$R^2C{\equiv}CH$		R^1O_2C, R^1O_2C—N, O, R^2	
R^1 / R^2				
C_2H_5 / CH_3		Toluene, 80-90°, 6 h, N_2	(20)	2
" / C_6H_5		Toluene, 60-70°, 10 h, N_2	(36)	2
$TsO(CH_2)_2$ / "		Toluene, 65-70°, 6.5 h, N_2	(65)	3
RO_2C, RO_2C—N (spirocyclopropane diazabicyclic alkene)	CH_2TMS (cyclopropylethyne)		RO_2C, RO_2C—N, O, CH_2TMS	
R				
C_2H_5		Toluene, 70°, 7 h, then 20°, 14 h	(67)	29,30
CCl_3CH_2		Toluene, 80°, 44 h	(8)	29

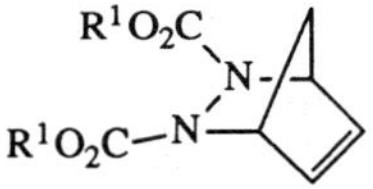

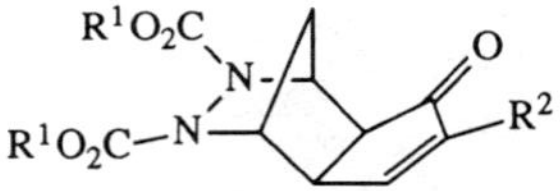

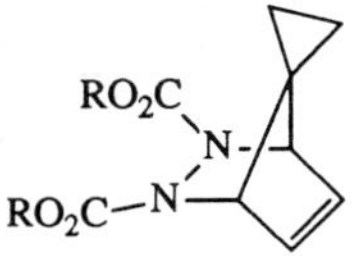

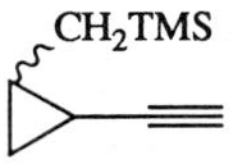

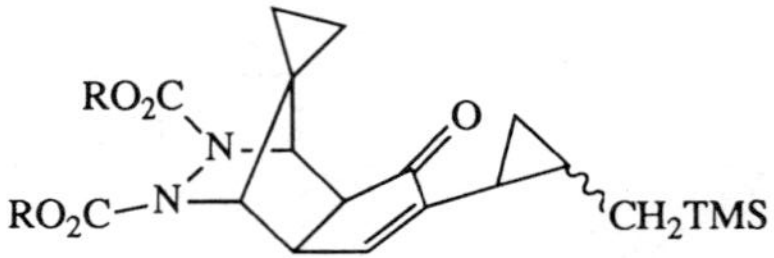

Reactant	Alkyne	Conditions	Product(s) and Yield(s) (%)	Refs.
8-Methoxycarbonyl-8-azabicyclo[3.2.1]oct-6-en-3-one (NCO_2CH_3)	HC≡CH	DME, 55-60°, 12 h	Tricyclic dione (NCO_2CH_3) (55)	38

R^1	R^2	R^3	Alkyne	Conditions	I	II	Refs.
			$R^3C≡CH$		I	II	
H	H	H		DME, 65°, 42 h, HC≡CH/CO	(45)		5
"	"	n-C_4H_9		DME, 65°, 44 h, CO	(42)		5
"	"	C_6H_5		DME, 65°, 2 d, CO	(40)		5
"	CH_3	H		DME, 65°, 64 h, HC≡CH/CO	(30)	(19)	5
"	CH_3CO	"		DME, 65°, 2 d, HC≡CH/CO	(18)	(6)	5
CH_3	H	H		DME, 65°, 64 h, HC≡CH/CO	(0)	(0)	5

TABLE VII. ALKYNES WITH HETEROBICYCLIC OR HETEROPOLYCYCLIC ALKENES (*Continued*)

Alkene			Alkyne[a]	Reaction Conditions	Product(s) and Yield(s) (%)		Refs.
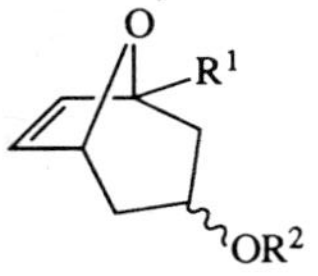			$R^3C{\equiv}CH$		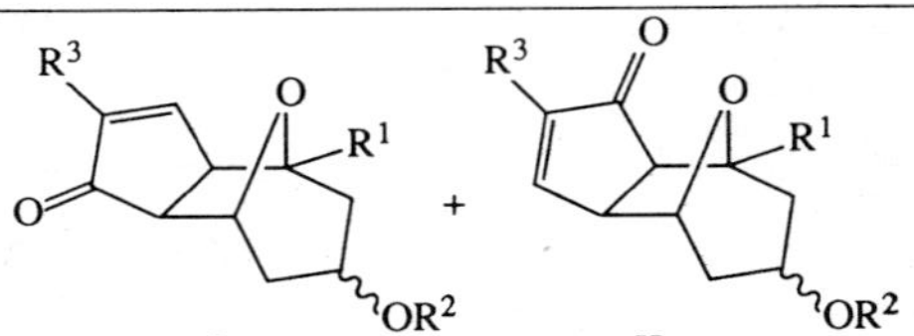 I + II		
R^1	R^2	R^3			I	II	
H	H	H		DME, 65°, 36 h, $HC{\equiv}CH$/CO	(52)		5
"	"	n-C_4H_9		DME, 65°, 1 d, CO	(43)		5
"	"	C_6H_5		DME, 65°, 1 d, CO	(57)		5
CH_3	"	CH_3		Benzene, 68°, 42 h, $CH_3C{\equiv}CH$/CO	(45)	(30)	36
CN	CH_3CO	H		DME, 65°, 19 h, $HC{\equiv}CH$/CO	(45)	(23)	5
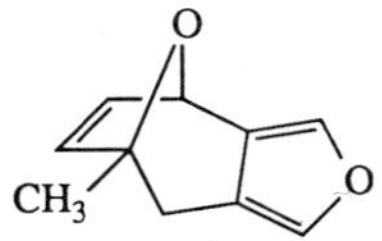			$CH_3C{\equiv}CH$	Benzene, 64°, 44 h, $CH_3C{\equiv}CH$/CO	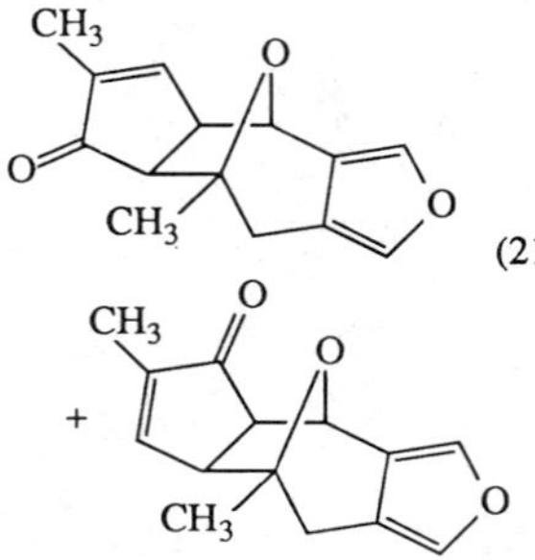 (21) + (43)		37

[a]This refers to the alkyne portion of the preformed alkyne·$Co_2(CO)_6$ complex.

TABLE VIII. INTRAMOLECULAR CYCLOADDITIONS OF ALL-CARBON ENYNES

Substrate[a]							Reaction Conditions[b]	Product(s) and Yield(s) (%)			Refs.
(hex-1-en-5-yne)							Isooctane, 95°, 4 d	No cyclic ketone products			39
R^1	R^2	R^3	R^4	R^5	R^6	R^7		I	II	III	
H	H	H	H	H	H	H	Isooctane, 95°, 4 d	(31)		(0)	39
H	H	H	H	H	H	2-(t-C_4H_9O)-c-C_3H_4	Toluene, 118°, 20 h	(18)		(0)	64
H	H	H	H	CH_3	H	H	Isooctane, 95°, 4 d[c]	(20) (I + II)		(0)	44
H	H	H	H	CH_3	H	H	Toluene, 120°, 7 d[c]	(25) (I + II)		(5)	44
H	H	H	H	CH_3	H	H	Toluene, 110°, 18 h[c]	(25) (I + II)		(15)	44
H	H	H	H	CH_3	H	TMS	Toluene, 110°, 7 d[c]	(15) (I + II)		(0)	44
H	H	H	H	TBDMSO	H	TMS	–	(14)	(0)	(0)	66
H	H	H	H	HO	CH_3	H	Cyclohexane, 65°, 4 h	(17)	(9)	(0)	47
H	H	H	H	HO	CH_3	H	SiO_2 (dry), 60°, 2 h	(33)	(19)	(0)	47

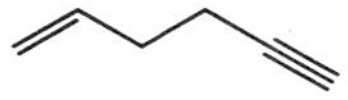

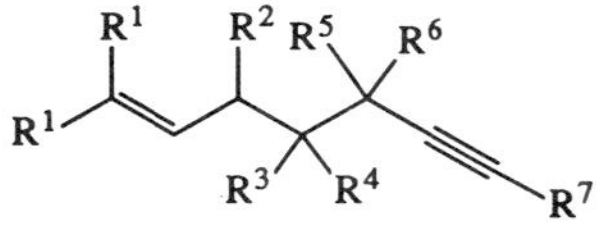

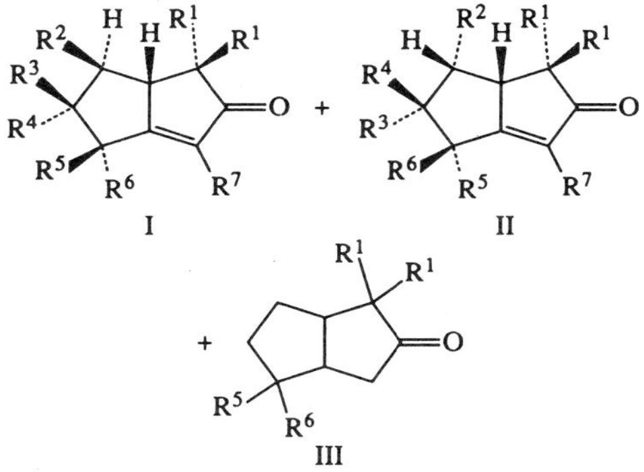

TABLE VIII. Intramolecular Cycloadditions of All-Carbon Enynes (*Continued*)

Substrate[a]							Reaction Conditions[b]	Product(s) and Yield(s) (%)			Refs.
R^1	R^2	R^3	R^4	R^5	R^6	R^7		I	II	III	
H	H	H	H	AcO	CH_3	H	SiO_2 (dry), 60-70°, 0.5-2.5 h	(29)	(29)	(0)	47
H	H	H	H	CH_3O	CH_3	H	SiO_2 (dry), 60-70°, 0.5-2.5 h	(53)	(26)	(0)	47
H	H	H	H	TBDMSO	CH_3	H	SiO_2 (dry), 60-70°, 0.5-2.5 h	(33)	(33)	(0)	47
H	H	CH_3	CH_3	H	H	H	Heptane, 120°, 3 d[c]	(58)	(0)	(0)	43
H	HO	H	H	CH_3O	CH_3	H	SiO_2 (dry), 50-100°, 1-2 h	(62) (I + II)		(0)	46
H	TBDMSO	CH_3	CH_3	H	H	TMS	Heptane, 80°, 2 d[c]	(18)[d]	(7)[d]	(0)	41
H	MOMO	CH_3	CH_3	H	H	TMS	Heptane, 90°, 36 h[c]	(68)	(0)	(0)	41
H	MOMO	CH_3	CH_3	H	H	$THPOCH(CH_3)(CH_2)_2$	Heptane, 125°, 36 h[c]	(≈20)	(0)	(0)	41
H	MOMO	CH_3	CH_3	H	H	$TBDMSOCH_2$	Heptane, 85°, 50 h[c,e]	(64)	(0)	(0)	79
H	$MOMO(CH_2)_2$	CH_3	CH_3	H	H	TMS	Heptane, 115°, 36 h[c]	(78)	(0)	(0)	41
H	$MOMO(CH_2)_2$	CH_3	CH_3	H	H	$TBDMSOCH_2$	Heptane, 86°, 30 h[e]	(45)[f]	(0)	(0)	66
H	H	CH_3	CH_3	TBDMSO	H	CH_3	Heptane, 110°, 20 h[c]	(50)	(15)	(0)	40
H	H	CH_3	CH_3	TBDMSO	H	CH_3	Heptane, 100°, 20 h	(41)	(12)	(0)	42
H	H	CH_3	CH_3	TBDMSO	H	CH_3	Heptane, 110°, 20 h[g]	(20)	(0)	(0)	42
H	H	CH_3	CH_3	TBDMSO	H	TMS	Heptane, 110°, 20 h[c]	(79)	(3)	(0)	40
H	H	CH_3	CH_3	H	H	$TBDMSOCH_2$	Heptane, 80°, 2 d[c]	(65)	(0)	(0)	69
H	H	CH_3O_2C	CH_3	H	H	TMS	Heptane, 120°, 25 h[c]	(47)	(39)	(0)	42
CH_3	H	H	H	CH_3	H	H	Benzene, 110°, 3 d[c]	(15) (I + II)			44
CH_3	H	H	H	CH_3	H	H	Toluene, 110°, 6 d[c]	(20) (I + II)		(3)	44
CH_3	H	H	H	CH_3	H	H	t-$C_4H_9C_6H_5$, 170°, 20 h	(20) (I + II)		(10)	44
CH_3	H	H	H	CH_3	H	H	Xylene, 140°, 3 h	(20) (I + II)		(15)	44
CH_3	H	H	H	CH_3	H	TMS	t-$C_4H_9C_6H_5$, 170°, 2 h	No cyclic ketone products			44
CH_3	H	H	H	TBDMSO	CH_3	H	t-$C_4H_9C_6H_5$, 170°, 2 h	(15) (I + II)		(10)	44
CH_3	H	H	H	TBDMSO	CH_3	TMS	Isooctane, 165°, 4 d	No cyclic ketone products			44

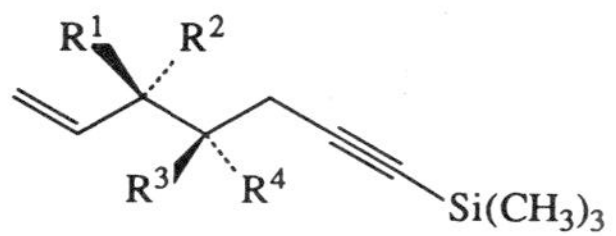

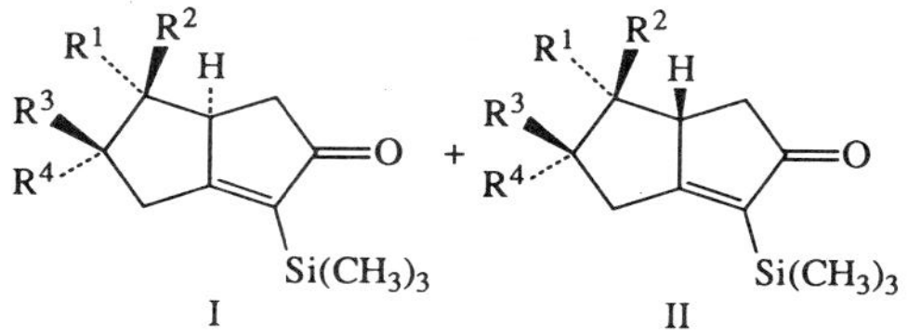

R^1	R^2	R^3	R^4	Conditions	I	II	Refs.
$BnOCH_2CH_2$	H	H	TBDMSO	Octane, 100°, 4 d	(43)[h]	(0)	68
$BnOCH_2CH_2$	H	H	TBDMSO	Hexane, 100°, 30 atm[i]	(>80)[h]	(0)	68
H	BnO	TBDMSO	H	Octane, 100°, 4 d	(8)[h]	(23)[h]	68
BnO	H	TBDMSO	H	Octane, 100°, 4 d	(18)[h]	(15)[h]	68

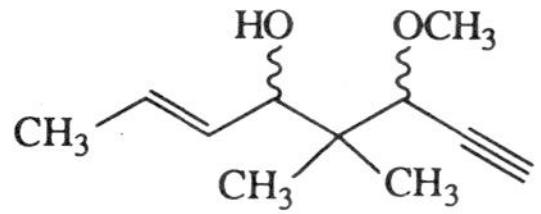

HO OCH3 CH3 CH3 CH3 — SiO2 (dry), 50-100°, 1-2 h — HO CH3 CH3 CH3 CH3O (87) 46

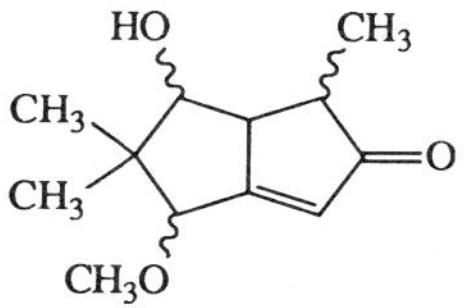

TABLE VIII. INTRAMOLECULAR CYCLOADDITIONS OF ALL-CARBON ENYNES (*Continued*)

Substrate[a]	Reaction Conditions[b]	Product(s) and Yield(s) (%)	Refs.
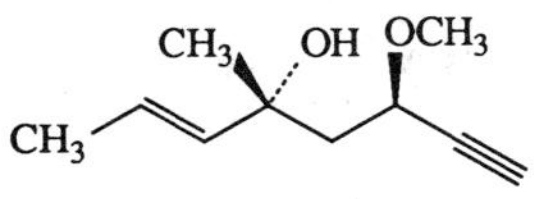 CH_3, OH, OCH_3, CH_3	SiO_2 (dry), 60°, 4 h	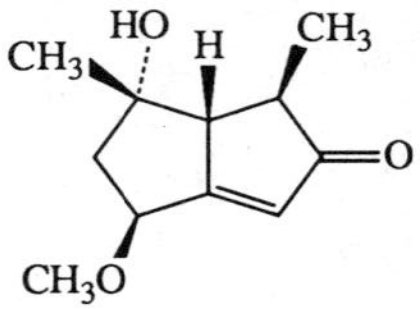 HO, H, CH_3, CH_3, O, CH_3O (74)	45
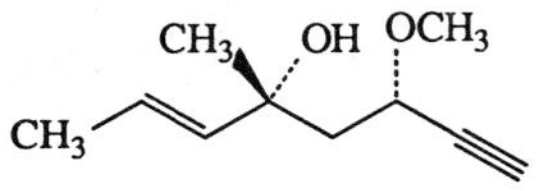 CH_3, OH, OCH_3, CH_3	SiO_2 (dry), 60°, 4 h	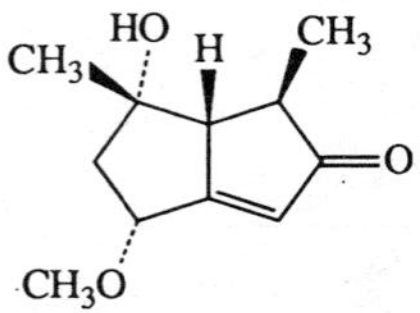 HO, H, CH_3, CH_3, O, CH_3O (40)	45
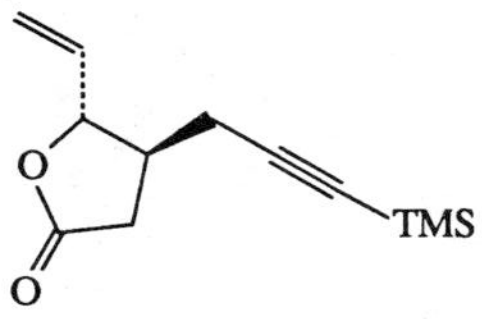 O, O, TMS	Heptane, heat	No cyclic ketone products	67

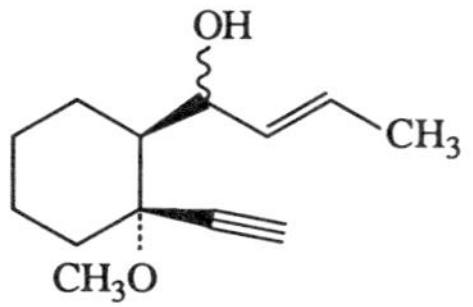	SiO_2 (dry), 50-100°, 1-2 h	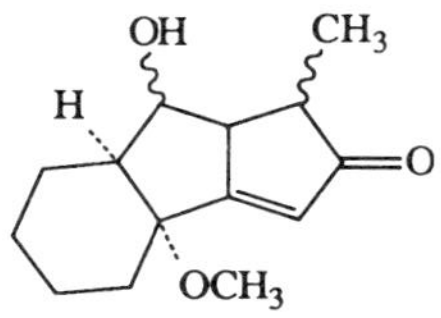	(90)	46
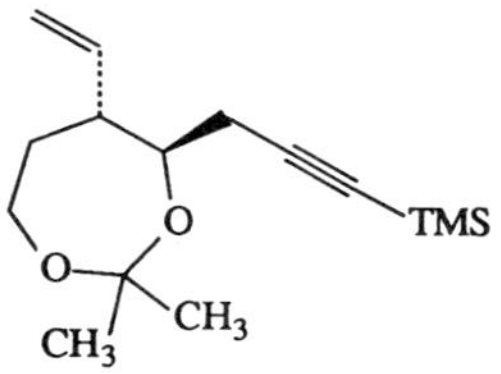	Heptane, $(n\text{-}C_4H_9)_3PO$, 85°, 3 d	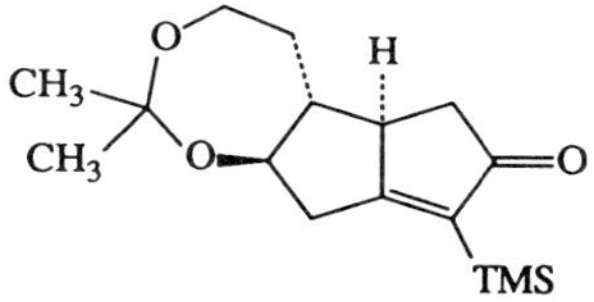	(51)[h]	67
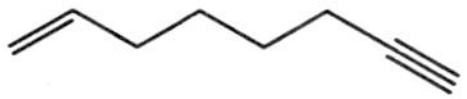	Isooctane, 95°, 4 d	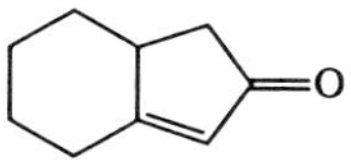	(35)	39

TABLE VIII. INTRAMOLECULAR CYCLOADDITIONS OF ALL-CARBON ENYNES (*Continued*)

Substrate[a]	Reaction Conditions[b]	Product(s) and Yield(s) (%)	Refs.
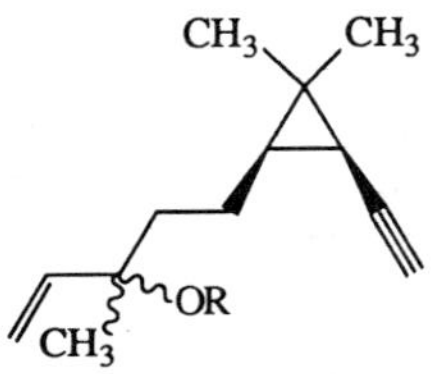 CH_3 CH_3 OR CH_3 R = H, $Si(CH_3)_3$	Toluene, 80-120°, 42-96 h	No cyclic ketone products	29

[a]This refers to the organic portion of the preformed $Co_2(CO)_6$ complex of the alkyne moiety of the substrate. All chiral substrates and products were racemic unless otherwise indicated.

[b]All reactions were carried out in the indicated solvent under an atmosphere of CO unless otherwise indicated.

[c]This reaction was carried out in a sealed tube.

[d]In this product R^2 = OH.

[e]To this reaction mixture was added 0.1 equiv 2,6-di-*tert*-butyl-4-methylpyridine.

[f]The compound with structure I, $R^7 = CH_3$, was isolated as a minor product of this reaction.

[g]This reaction was carried out in a sealed tube with the enyne in the presence of a catalytic amount of the cobalt complex.

[h]Both the substrate(s) and the product(s) were optically pure.

[i]The reaction was carried out in an autoclave under the indicated pressure of CO prior to heating.

TABLE IX. Intramolecular Cycloadditions of Heteroatom-Linked Enynes

Substrate[a]		Reaction Conditions	Product(s) and Yield(s) (%)		Refs.
N-allyl-N-(COR[1])-propargylamine with R^2 on alkyne			R^1CON bicyclic enone (I) + R^1CON bicyclic ketone (II)		
R^1	R^2		I	II	
CH_3	H	Isooctane, 100°, 1 d	(0)	(5)	61
"	"	Isooctane, 50°, $h\nu$, 20 h	(0)	(33)	61
"	"	Isooctane, ultrasound, 60°, 5 h	(0)	(36)	61
"	"	CCl_4, 50°, 18 h	(38)	(0)	61
"	"	SiO_2 (dry), 70°, 1.5 h	(0)	(67)	61
C_6H_5	"	Isooctane, ultrasound, 60°, 4.5 h	(0)	(32)	61
"	"	SiO_2 (dry), 90°, 1.5 h	(0)	(45)	61
CH_3	C_2H_5	Isooctane, 100°, 3 h	(57)		61
"	"	SiO_2 (dry), 70°, 1 h	(75)		61
"	TMS	$Cl_2C{=}CCl_2$, 110°, 20 h	(28)	(0)	61
"	"	SiO_2 (dry), 80°, 4 h	(0)	(46)	61

TABLE IX. INTRAMOLECULAR CYCLOADDITIONS OF HETEROATOM-LINKED ENYNES (*Continued*)

Substrate[a]							Reaction Conditions	Product(s) and Yield(s) (%)		Refs.
R^1	R^2	R^3	R^4	R^5	R^6	R^7		I	II	
H	H	H	H	H	H	H	Isooctane, 60°, 1 d, CO	(14)	(0)	55
H	H	H	H	H	H	H	SiO_2 (dry), 50°, 3.5 h, O_2	(58)	(0)	45,57-60
H	H	H	H	H	H	H	Al_2O_3, 70°, 1.5 h, argon	(0)	(70)	59,60
H	H	H	H	H	H	CH_3	Isooctane, 60°, 1 d, CO	(41)	(0)	55
H	H	H	H	H	H	CH_3	SiO_2 (dry), 60°, 2 h, O_2	(60)	(0)	57-60
H	H	H	H	H	H	CH_3	Al_2O_3, 80°, 1 h, argon	(0)	(57)	59,60
H	H	H	H	H	H	CH_2=CH	SiO_2 (dry), 50°, 20 min, O_2	(58)	(0)	57,58,60
H	H	H	H	H	H	$THPOCH_2CH_2$	Isooctane, 60°, 1 d, CO	(41)	(0)	55
H	H	H	H	CH_3	H	H	Isooctane, 60°, 1 d, CO	(29)	(0)	55
H	H	H	H	CH_3	H	H	SiO_2 (dry), 50°, 1 h, O_2	(43)[b]	(0)	59,60
H	H	H	H	CH_3	CH_3	H	Isooctane, 60°, 1 d, CO	(29)	(0)	55
H	H	H	H	CH_3	CH_3	H	SiO_2 (dry), 45°, 30 min, O_2	(76)	(0)	57,59,60
H	H	H	H	CH_3	CH_3	H	SiO_2 (dry), 45°, 1.5 h, argon	(15)	(40)	57
H	H	H	H	CH_3	CH_3	H	Al_2O_3, 45°, 1.5 h, argon	(0)	(40)	59,60
H	H	H	H	CH_3	CH_3	CH_2=CH	SiO_2 (dry), 50°, 30 min, O_2	(58)	(0)	59,60

							Conditions	Product(s) and Yield(s) (%)		Refs.
H	H	CH_3	H	H	H	H	SiO_2 (dry), 60°, 1 h, O_2	(48)[c]	(0)	57-60
H	H	CH_3	CH_3	H	H	H	SiO_2 (dry), 55°, 1.5 h, O_2	(59)	(0)	57,59,60
H	CH_3	H	H	H	H	H	SiO_2 (dry), 60°, 2.5 h, O_2	(56)	(0)	45,57-60
H	CH_3	H	H	H	H	H	Al_2O_3, 90°, 70 min, argon	(0)	(77)	59,60
CH_3	H	H	H	H	H	H	SiO_2 (dry), 60°, 1.5 h, O_2	(46)[d]	(0)	45,57-60
CH_3	H	H	H	H	H	H	Al_2O_3, 80°, 1 h, argon	(0)	(67)[e]	59,60
CH_3	H	H	H	CH_3	CH_3	H	SiO_2 (dry), 45°, 2 h, O_2	(92)[d]	(0)	45,57-60
CH_3	H	H	H	CH_3	CH_3	H	Al_2O_3, 80°, 45 min, argon	(0)	(73)[e]	59,60
CH_3, CH_3, O							SiO_2 (dry), 50°, 2 h, O_2	CH_3, CH_3, O, O (40)		58
O, R								O, O, R, I + HO, O, II		
R								I	II	
H							SiO_2 (dry), 45°, 45 min, O_2	(64)	(0)	45,57-60
H							Al_2O_3, 80°, 35 min, argon	(0)	(70)	59,60
C_2H_5CO							Hexane, 60°, 7-8 h	(43)	(0)	45,56
i-C_4H_9CO							Hexane, 60°, 7-8 h	(43)	(0)	45,56

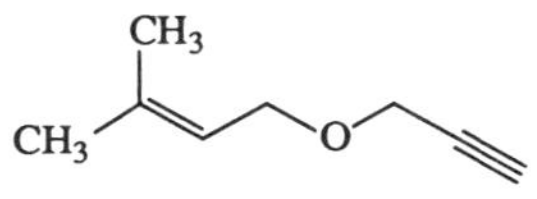

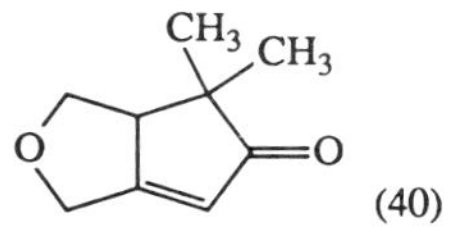

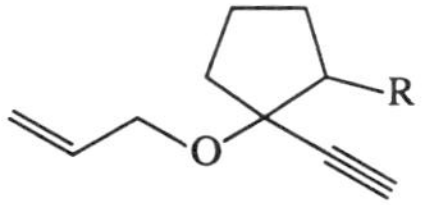

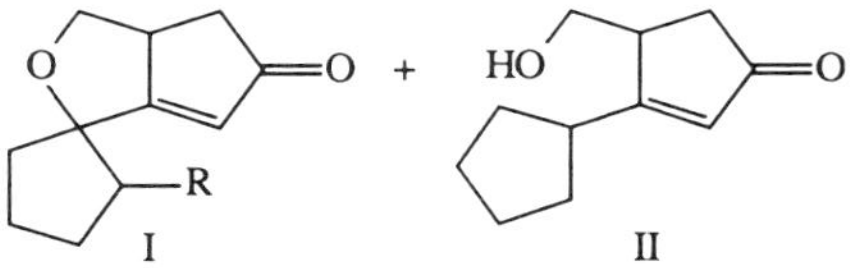

TABLE IX. INTRAMOLECULAR CYCLOADDITIONS OF HETEROATOM-LINKED ENYNES (*Continued*)

Substrate[a]	Reaction Conditions	Product(s) and Yield(s) (%)	Refs.
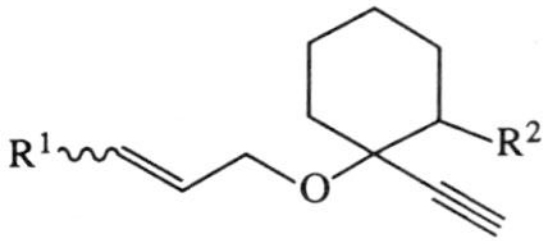		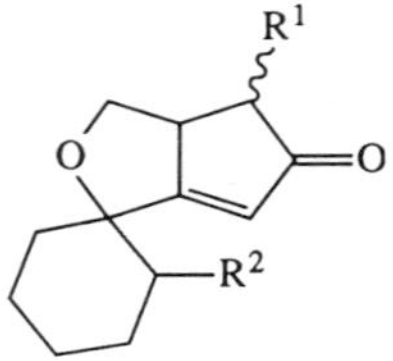	
R^1 / R^2			
H / H	SiO_2 (dry), 60°, 45 min, O_2	(80)	55,58
H / C_2H_5CO	Hexane, 60°, 4 h	(31)	45,56
H / i-C_4H_9CO	Hexane, 60°, 4 h	(45)	45,56
CH_3 / H	SiO_2 (dry), 60°, 1 h, O_2	(85)[d]	59,60
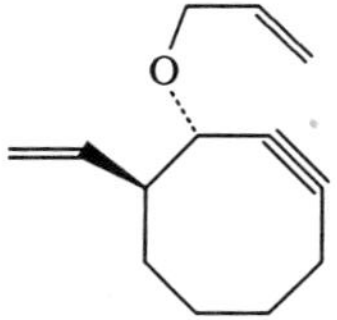	Benzene, 60°, 4 h	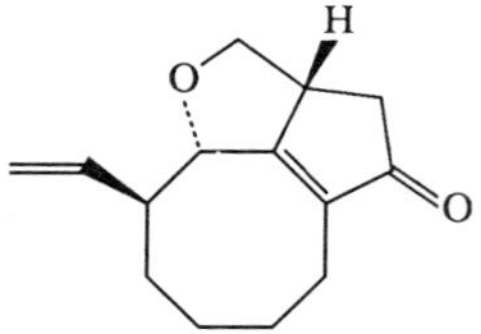 (85)	70

Substrate	Conditions	Product(s) and Yield(s) (%)	Refs.
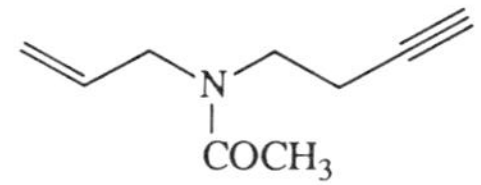 N-allyl-N-(but-3-ynyl)acetamide (N, $COCH_3$)	SiO_2 (dry), 70°, 4 h 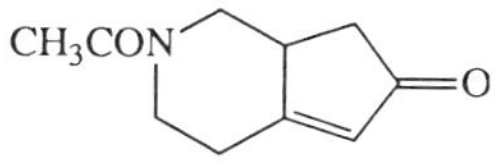	CH_3CON bicyclic enone (=O) (44)	61
vinyl ether, O–$(CH_2)_n$–C≡CH			
n = 1	Isooctane, 95°, 4 d, CO	No cyclic ketone products	20
n = 2	Isooctane, 95°, 4 d, CO	(*f*)	20
n = 3	Isooctane, 95°, 4 d, CO	(*f*)	20

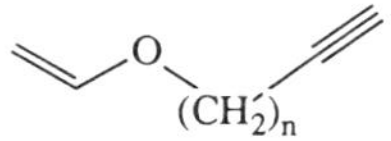

[a] This refers to the organic portion of the preformed $Co_2(CO)_6$ complex of the alkyne moiety of the substrate. All chiral substrates and products were racemic unless otherwise indicated.

[b] The product was a mixture of two stereoisomers in a 2:1 ratio. The isomer with the methyl substituent in the *exo* position was the major product.

[c] The product was a mixture of two stereoisomers in a 2.5:1 ratio. The isomer with the methyl substituent in the *exo* position was the major product.

[d] The starting material consisted of a 4:1 ratio of *E* and *Z* stereoisomers, and the product consisted of a corresponding 4:1 ratio of stereoisomers in which the isomer with the methyl substituent in the *exo* position predominated.

[e] The starting material consisted of a 4:1 ratio of *E* and *Z* stereoisomers, and the product consisted of a corresponding 4:1 ratio of stereoisomers in which the isomer with the methyl substituent *trans* to the hydroxymethyl substituent predominated.

[f] The product mixture resulting from this reaction contained a cyclopentenone which was not fully characterized as it was isolated in very low yield and could not be purified.

TABLE X. INTRAMOLECULAR CYCLOADDITIONS OF CYCLOALKENE-CONTAINING ENYNES

R^1	R^2	R^3	R^4	R^5	R^6	Reaction Conditions	I	II	III	Refs.
H	H	H	H	H	H	Benzene, 80°, 4 d, CO	(35)			48,49
H	H	HO	H	CH_3O	H	SiO_2 (dry), 50-100°, 1-2 h	(40) (I + II)			46
H	CH_3	H	CH_3	H	H	Benzene, 80°, 1 d, CO	(24)		(5)	52
H	CH_3	H	CH_3	H	H	Heptane, 110°, 20 h, CO[b]	(46)		(5)	52
H	CH_3	HO	CH_3	H	H	Benzene, 80°, 1 d, CO	No cyclic ketone products			49
H	CH_3	MOMO	CH_3	H	H	Benzene, 80°, 22 h, CO	(30) (I + II + III)			49
H	CH_3	HO	CH_3	H	TMS	Benzene, 80°, 1 d, CO	No cyclic ketone products			49
H	CH_3	MOMO	CH_3	H	TMS	Benzene, 80°, 1 d, CO	No cyclic ketone products			49
CH_3	CH_3	H	H	H	H	Benzene, 80°, 4 d, CO	([c])			48,49

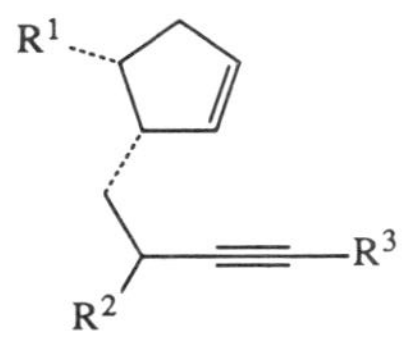

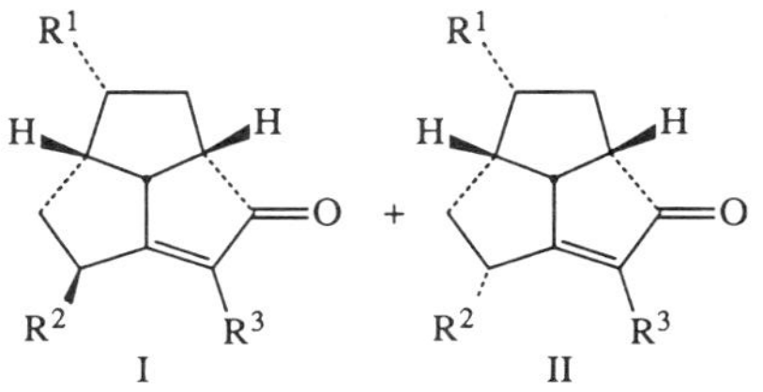

R^1	R^2	R^3		I	II	
HO	HO	H	SiO_2 (dry), 120°, 2 h	(16) (I + II)[d]		51
TBDMSO	TBDMSO	H	Isooctane, 160°, 5 d, CO[b]	([e]) (I + II)		50,53
BnO	BnO	H[f]	t-$C_4H_9C_6H_5$, 170°, 2 h	(42)[g]	(18)[g]	51,53
BnO	BnO	H[f,h]	t-$C_4H_9C_6H_5$, 170°, 2 h	(35)[g,h]	(15)[g,h]	71
BnO	BnO	H[f]	SiO_2 (dry), 150°, 6 h	(23)[i]	(7)[i]	51,53
TBDMSO	TBDMSO	TMS[f]	Isooctane, 160°, 3 d, CO[b]	(76)[i]	(0)	50,53

[a]This refers to the organic portion of the preformed $Co_2(CO)_6$ complex of the alkyne moiety of the substrate. All chiral substrates and products were racemic unless otherwise indicated.

[b]This reaction was carried out in a sealed tube.

[c]The product mixture resulting from this reaction contained a cyclopentenone which was not fully characterized as it was isolated in very low yield and could not be purified.

[d]This is the overall yield of tricyclo[5.2.1.$0^{4,10}$]decane-2,5,8-trione arising from cycloaddition, followed by catalytic hydrogenation of the carbon-carbon double bond ($Pd/C/H_2$), and finally oxidation of the two alcohols (PCC).

[e]The product of this reaction was obtained in low and somewhat variable yields, with a I:II ratio of 22:78.

[f]The starting material for this reaction was a 4:1 mixture of diastereoisomers.

[g]This is the overall yield of the saturated ketone arising from catalytic hydrogenation of the carbon-carbon double bond [$Pd/C/(C_2H_5)_3N/H_2$] of the cycloaddition product, which itself consisted of a mixture of both enone and saturated ketone. The cycloaddition mixture was obtained in a total yield of 60%.

[h]Both the substrate and the products of this reaction were optically pure.

[i]This is the overall yield of the saturated ketone arising from catalytic hydrogenation of the carbon-carbon double bond [$Pd/C/(C_2H_5)_3N/H_2$] of the cycloaddition product.

TABLE XI. INTRAMOLECULAR CYCLOADDITIONS OF OTHER ENYNES

Substrate[a]	Reaction Conditions	Product(s) and Yield(s) (%)	Refs.
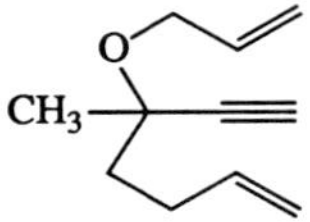	SiO_2 (dry), 65°, 1 h	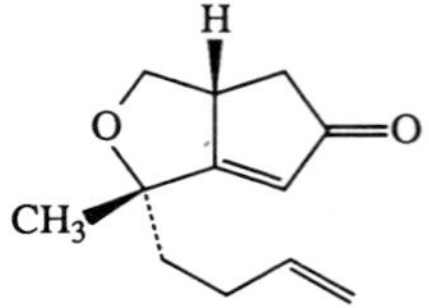 (31) + 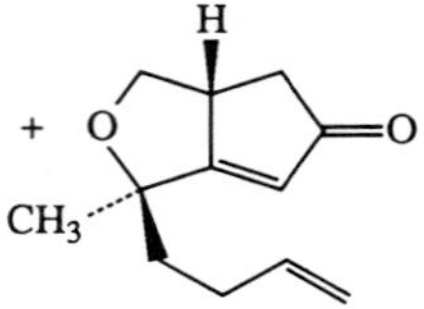 (31) + 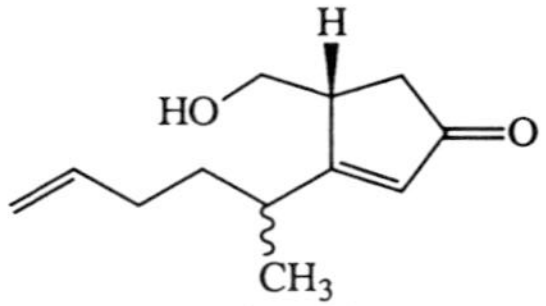 (30)[b]	47
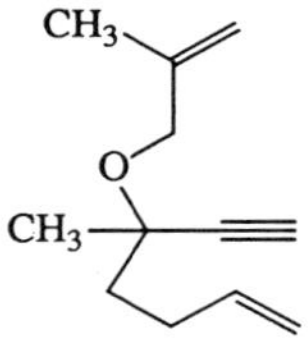	SiO_2 (dry), heat	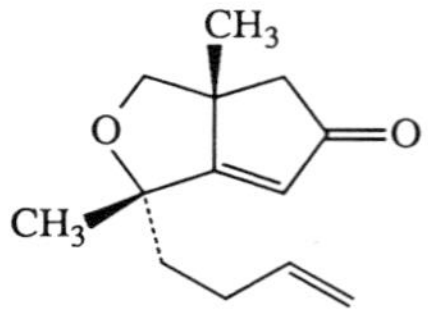 (32) + 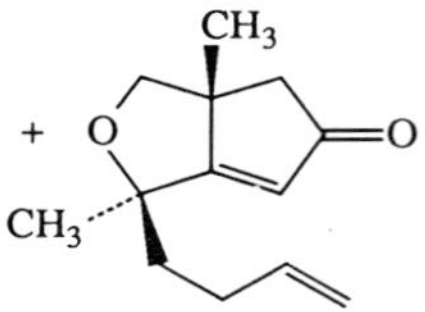 (20)	72

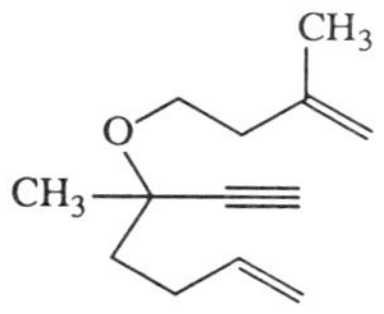 CH_3, O, CH_3	SiO_2 (dry), 60-70°, 0.5-2.5 h	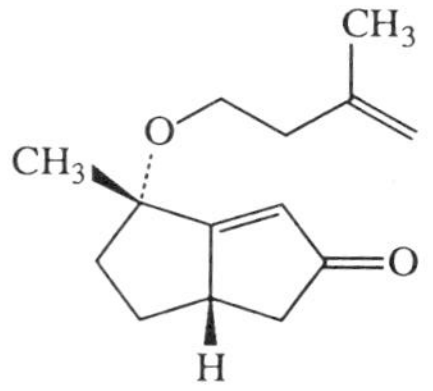 CH_3, O, CH_3, H, =O (18) 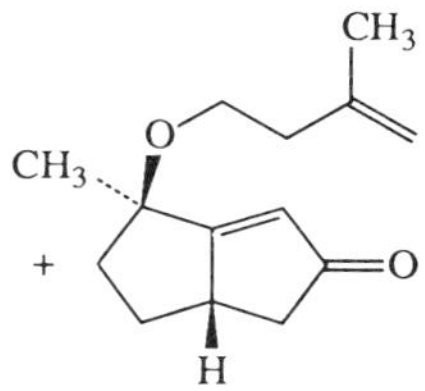 + CH_3, O, CH_3, H, =O (38)	47,72

[a]This refers to the organic portion of the preformed $Co_2(CO)_6$ complex of the alkyne moiety of the substrate. All chiral substrates and products were racemic unless otherwise indicated.

[b]This product was formed as a 2:1 mixture of diastereoisomers.

REFERENCES

[1] I. U. Khand, G. R. Knox, P. L. Pauson, and W. E. Watts, *J. Chem. Soc,. Chem. Commun.* **1971**, 36; I. U. Khand, G. R. Knox, P. L. Pauson, and W. E. Watts, *J. Chem. Soc., Perkin Trans. 1,* **1973**, 975; I. U. Khand, G. R. Knox, P. L. Pauson, W. E. Watts, and M. I. Foreman, *J. Chem. Soc., Perkin Trans. 1,* **1973**, 977.

[2] P. L. Pauson and I. U. Khand, *Ann. N.Y. Acad. Sci.,* **295**, 2 (1977).

[3] P. L. Pauson, *Tetrahedron*, **41**, 5855 (1985).

[4] P. L. Pauson, in *Organometallics in Organic Synthesis. Aspects of a Modern Interdisciplinary Field*; A. de Meijere and H. tom Dieck, Eds., Springer, Berlin, 1988, p. 233.

[5] B. E. La Belle, M. J. Knudsen, M. M. Olmstead, H. Hope, M. D. Yanuck, and N. E. Schore, *J. Org. Chem.*, **50**, 5215 (1985).

[6] M. E. Krafft, *J. Am. Chem. Soc.*, **110**, 968 (1988).

[7] M. E. Krafft, *Tetrahedron Lett.*, **29**, 999 (1988).

[8] I. U. Khand and P. L. Pauson, *J. Chem. Soc., Chem. Commun.*, **1974**, 379.

[9] I. U. Khand and P. L. Pauson, *Heterocycles*, **11**, 59 (1978).

[10] N. E. Schore, B. E. LaBelle, M. J. Knudsen, H. Hope, and X-J. Xu, *J. Organometal. Chem.*, **272**, 435 (1984). See also V. Rautenstrauch, P. Mégard, B. Gamper, B. Bourdin, E. Walther, and G. Bernadinelli, G. *Helv. Chim. Acta*, **72**, 811 (1989).

[11] I. U. Khand and P. L. Pauson, *J. Chem. Res.* (*M*), **1977**, 168.

[12] H. J. Jaffer and P. L. Pauson, *J. Chem. Res.* (*M*), **1983**, 2201.

[13] R. F. Newton, P. L. Pauson, and R. G. Taylor, *J. Chem. Res.* (*M*), **1980**, 3501.

[14] L. Daalman, R. F. Newton, P. L. Pauson, and A. Wadsworth, *J. Chem. Res.* (*M*), **1984**, 3150.

[15] D. C. Billington, I. M. Helps, P. L. Pauson, W. Thomson, and D. Willison, *J. Organometal. Chem.*, **354**, 233 (1988).

[16] D. C. Billington, P. Bladon, I. M. Helps, P. L. Pauson, W. Thomson, and D. Willison, *J. Chem. Res.* (*M*), **1988**, 2601.

[17] D. C. Billington and P. L. Pauson, *Organometallics*, **1**, 1560 (1982).

[18] I. U. Khand, E. Murphy, and P. L. Pauson, *J. Chem. Res.* (*M*), **1978**, 4434.

[19] I. U. Khand, C. A. L. Mahaffy, and P. L. Pauson, *J. Chem. Res.* (*M*), **1978**, 4454.

[20] M. C. Croudace and N. E. Schore, *J. Org. Chem.*, **46**, 5357 (1981).

[21] W. A. Smit, S. L. Kireev, and O. M. Nefedov, *Izv. Akad. Nauk SSSR, Ser. Khim.*, **1987**, 2637, *Bull. Acad. Sci. USSR, Div. Chem. Sci.*, **36**, 2452 (1987); V. A. Smit, V. A. Tarasov, E. D. Daeva, and I. I. Ibragimov, *Izv. Akad. Nauk SSSR, Ser. Khim.*, **1987**, 2870, *Bull. Acad. Sci. USSR, Div. Chem. Sci.,* **36**, 2669 (1987); W. A. Smit, S. L. Kireev, O. M. Nefedov, and V. A. Tarasov, *Tetrahedron Lett.*, **30**, 4021 (1989).

[22] D. C. Billington, W. J. Kerr, P. L. Pauson, and C. F. Farnocchi, *J. Organometal. Chem.*, **356**, 213 (1988).

[23] I. U. Khand, P. L. Pauson, and M. J. A. Habib, *J. Chem. Res.* (*M*), **1978**, 4418.

[24] D. C. Billington, W. J. Kerr, and P. L. Pauson, *J. Organometal. Chem.*, **328**, 223 (1987).

[25] I. U. Khand, P. L. Pauson, and M. J. A. Habib, *J. Chem. Res.* (*M*), **1978**, 4401.

[26] V. Sampath, E. C. Lund, M. J. Knudsen, M. M. Olmstead, and N. E. Schore, *J. Org. Chem.*, **52**, 3595 (1987).

[27] P. Bladon, I. U. Khand, and P. L. Pauson, *J. Chem. Res.* (*M*), **1977**, 153.

[28] A.-M. Montaña, A. Moyano, M. A. Pericàs, and F. Serratosa, *Tetrahedron*, **41**, 5995 (1985).

[29] T. Liese and A. de Meijere, *Chem. Ber.*, **119**, 2995 (1986); A. Weier, Ph.D. Dissertation, Hamburg, 1988; A. de Meijere, personal communication.

[30] A. de Meijere, A. Kaufmann, R. Lackmann, H.-C. Militzer, O. Reiser, S. Schömenauer, and A. Weier, in *Organometallics in Organic Synthesis 2*, H. Werner and G. Erker, Eds.; Springer, Berlin, 1989, p. 255.

[31] P. Bladon, P. L. Pauson, H. Brunner, and R. Eder, *J. Organometal. Chem.*, **355**, 449 (1988).

[32] I. U. Khand and P. L. Pauson, *J. Chem. Soc., Perkin Trans.* **1**, 30 (1976).

[33] S. E. MacWhorter, V. Sampath, M. M. Olmstead, and N. E. Schore, *J. Org. Chem.*, **53**, 203 (1988).

[34] N. E. Schore and S. D. Najdi, *J. Am. Chem. Soc.*, **112**, 441 (1990).
[35] N. E. Schore, *Synth. Commun.* **9**, 41 (1979).
[36] M. E. Price and N. E. Schore, *J. Org. Chem.*, **54**, 5662 (1989).
[37] M. E. Price and N. E. Schore, *Tetrahedron Lett.*, **30**, 5865 (1989).
[38] E. C. Lund, Ph.D. Dissertation, University of California, Davis, 1987.
[39] N. E. Schore and M. C. Croudace, *J. Org. Chem.*, **46**, 5436 (1981).
[40] C. Exon and P. Magnus, *J. Am. Chem. Soc.*, **105**, 2477 (1983).
[41] P. Magnus and L. M. Principe, *Tetrahedron Lett.*, **26**, 4851 (1985).
[42] P. Magnus, C. Exon, and P. Albaugh-Robertson, *Tetrahedron*, **41**, 5861 (1985).
[43] D. H. Hua, *J. Am. Chem. Soc.*, **108**, 3835 (1986).
[44] A.-M. Montaña, A. Moyano, M. A. Pericàs, and F. Serratosa, *An. Quím., Ser. C*, **84**, 82 (1988); *Chem. Abstr.*, **111**, 23697g (1989).
[45] W. A. Smit, A. S. Gybin, S. O. Simonyan, A. S. Shashkov, V. A. Tarasov, and I. I. Ibragimov, *Izv. Akad. Nauk SSSR, Ser. Khim.*, **1985**, 2650; *Bull. Acad. Sci. USSR, Div. Chem. Sci.*, **34**, 2455 (1985); W. A. Smit, A. S. Gybin, A. S. Shashkov, Y. T. Strychkov, L. G. Kyz'mina, G. S. Mikaelian, R. Caple, and E. D. Swanson, *Tetrahedron Lett.*, **27**, 1241 (1986).
[46] W. A. Smit, A. S. Gybin, S. O. Veretenov, and A. S. Shashkov, *Izv. Akad. Nauk SSSR, Ser. Khim.*, **1987**, 232; *Bull. Acad. Sci. USSR, Div. Chem. Sci.*, **36**, 211 (1987).
[47] W. A. Smit, S. O. Simonyan, V. A. Tarasov, A. S. Shashkov, S. S. Mamyan, A. S. Gybin, and I. I. Ibragimov, *Izv. Akad. Nauk SSSR, Ser. Khim.*, **1988**, 2796; *Bull. Acad. Sci. USSR, Div. Chem. Sci.*, **37**, 2521 (1988).
[48] M. J. Knudsen and N. E. Schore, *J. Org. Chem.*, **49**, 5025 (1984).
[49] N. E. Schore and M. J. Knudsen, *J. Org. Chem.*, **52**, 569 (1987).
[50] E. Carceller, V. Centellas, A. Moyano, M. A. Pericàs, and F. Serratosa, *Tetrahedron Lett.*, **26**, 2475 (1985); E. Carceller, M. L. García, A. Moyano, M. A. Pericàs, and F. Serratosa, *Tetrahedron*, **42**, 1831 (1986).
[51] C. Almansa, E. Carceller, E. García, A. Torrents, and F. Serratosa, *Synth. Commun.*, **18**, 381 (1988).
[52] N. E. Schore and E. G. Rowley, *J. Am. Chem. Soc.*, **110**, 5224 (1988).
[53] C. Almansa, E. Carceller, E. García, and F. Serratosa, *Synth. Commun.*, **18**, 1079 (1988).
[54] K. M. Nicholas, *Acc. Chem. Res.*, **20**, 207 (1987).
[55] D. C. Billington and D. Willison, *Tetrahedron Lett.*, **25**, 4041 (1984).
[56] G. S. Mikaelian and W. A. Smit, *Izv. Akad. Nauk SSSR, Ser. Khim.*, **1984**, 2652; *Bull. Acad. Sci. USSR, Div. Chem. Sci.*, **33**, 2434 (1984).
[57] S. O. Simonian, W. A. Smit, A. S. Gybin, A. S. Shashkov, G. S. Mikaelian, V. A. Tarasov, I. I. Ibragimov. R. Caple, and D. E. Froen, *Tetrahedron Lett.*, **27**, 1245 (1986).
[58] R. Caple, in *Organic Synthesis: Modern Trends*, O. Chizhov, Ed., Blackwell, 1987, p. 119.
[59] W. A. Smit, S. O. Simonyan, V. A. Tarasov, G. S. Mikaelian, A. S. Gybin, I. I. Ibragimov, R. Caple, D. Froen, and A. Kreager, *Synthesis*, **1989**, 472.
[60] W. A. Smit, S. O. Simonyan, A. S. Shashkov, S. S. Mamyan, V. A. Tarasov, and I. I. Ibragimov, *Izv. Akad. Nauk SSSR, Ser. Khim.*, **1987**, 234; *Bull. Acad. Sci. USSR, Div. Chem. Sci.*, **36**, 213 (1987); W. A. Smit, S. O. Simonyan, V. A. Tarasov, A. S. Gybin, G. S. Mikaélian, A. S. Shashkov, S. S. Mamyan, I. I. Ibragimov, and R. Kéipl, *Izv. Akad. Nauk SSSR, Ser. Khim.*, **1988**, 2802; *Bull. Acad. Sci. USSR, Div. Chem. Sci.*, **37**, 2526 (1988).
[61] S. W. Brown and P. L. Pauson, *J. Chem. Soc., Perkin Trans. 1*, **1990**, 1205.
[62] L. Daalman, R. F. Newton, P. L. Pauson, R. G. Taylor, and A. Wadsworth, *J. Chem. Res. (M)*, **1984**, 3131.
[63] D. C. Billington, *Tetrahedron Lett.*, **24**, 2905 (1983).
[64] A. De Meijere and L. Wessjohann, *Synlett*, **1990**, 20; H.-C. Militzer, Ph.D. Dissertation, Hamburg, 1990; A. de Meijere, personal communication.
[65] S. Keyaniyan, M. Apel, J. P. Richmond, and A. de Meijere, *Angew. Chem. Int. Ed., Engl.*, **24**, 770 (1985); De Meijere, A. *Chem. Brit.*, **1987**, 865.
[66] P. Magnus, L. M. Principe, and M. J. Slater, *J. Org. Chem.*, **52**, 1483 (1987).
[67] P. Magnus and D. P. Becker, *J. Am. Chem. Soc.*, **109**, 7495 (1987).

[68] J. Mulzer, K.-D. Graske, and B. Kirste, *Justus Liebigs Ann. Chem.*, **1988**, 891; J. Mulzer, personal communication.
[69] D. H. Hua, M. J. Coulter, and I. Badejo, *Tetrahedron Lett.*, **28**, 5465 (1987).
[70] S. L. Schreiber, T. Sammakia, and W. E. Crowe, *J. Am. Chem. Soc.*, **108**, 3128 (1986).
[71] C. Almansa, A. Moyano, and F. Serratosa, *Tetrahedron*, **44**, 2657 (1988).
[72] W. A. Smit, S. M. Bukhanyuk, A. S. Shashkov, Yu. T. Struchkov, and A. I. Yanovskii, *Izv. Akad. Nauk SSSR, Ser Khim.*, **1988**, 2876; *Bull. Acad. Sci. USSR, Div. Chem. Sci.*, **37**, 2597 (1988).
[73] R. S. Dickson and P. J. Fraser, *Adv. Organomet. Chem.*, **12**, 323 (1974).
[74] A. Devasagayarai and M. Periasamy, *Tetrahedron Lett.*, **30**, 595 (1989).
[75] W. C. Still, M. Kahn, and A. Mitra, *J. Org. Chem.*, **43**, 2923 (1978).
[76] U. Krüerke and W. Hübel, *Chem. Ber.*, **94**, 2829 (1961).
[77] T. K. Jones and S. E. Denmark, *Helv. Chim. Acta*, **66**, 2377 (1983).
[78] D. C. Billington, W. J. Kerr, and P. L. Pauson, *J. Organometal. Chem.*, **341**, 181 (1988).
[79] P. Magnus, M. J. Slater, and L. M. Principe, *J. Org. Chem.*, **54**, 5148 (1989).

CHAPTER 2

REDUCTION WITH DIIMIDE

DANIEL J. PASTO

University of Notre Dame, Notre Dame, Indiana

RICHARD T. TAYLOR

Miami University, Oxford, Ohio

CONTENTS PAGE

Organic Reactions, *Vol. 40*, Edited by Leo A. Paquette et al.
ISBN 0-471-53841-8. Published by John Wiley & Sons Inc.

INTRODUCTION

The reduction of a double bond in the presence of hydrazine appears to have been first observed in 1905 during the reaction of glyceryl oleate, which produced stearic hydrazide.[1] That hydrazine could act as a reagent for the reduction of a carbon–carbon double bond was firmly established much later,[2] at which point it was shown that oleic acid could be reduced to stearic acid by treatment with hydrazine,[2] or with hydrazine and sulfur.[3] In 1941 it was reported that vinyl groups in chlorins and porphyrins are selectively reduced to ethyl groups by hydrazine under mild conditions.[4] The synthetic potential of this type of reduction was not recognized until the early 1960s when results from several independent laboratories implicated diimide (HN=NH) as the actual reducing agent.[5–8]

Evidence for the existence of diimide* was first obtained in 1892 in the decarboxylation of dipotassium azodicarboxylate, which produced equimolar quantities of nitrogen and hydrazine by the proposed disproportionation of diimide.[9] In 1910 it was proposed that diimide was formed in the reaction of benzenesulfonylhydrazide with hot alkali.[10] Following the proposal that diimide is the reactive intermediate in these reduction reactions, numerous experimental and theoretical studies were launched to find other methods for the synthesis of diimide and to determine the structure(s) of the reactive intermediate(s) and the mechanism of the reduction reaction. An excellent review has appeared which covers the literature on the structure and molecular properties, spectral characterization, and gas-phase reactions of diimide.[11] In this chapter only highlights of such areas are covered. Two reviews covering reductions with diimide appeared in 1965,[12,13] but none since that time. Most organic texts describe diimide reductions, but not in significant detail.

There are three potential structures for diimide: *cis*- and *trans*-diimide and 1,1-diimide (aminonitrene).

*Diimide has also been referred to as diimine and diazene. As diimide is most commonly used in the current literature, this name is used in this review.

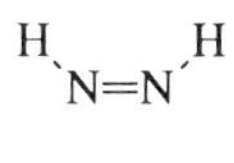

$$\underset{cis\text{-Diimide}}{H-N=N-H} \qquad \underset{trans\text{-Diimide}}{H-N=N-H} \qquad \underset{1,1\text{-Diimide}}{H_2\overset{+}{N}=\overset{-}{N}}$$

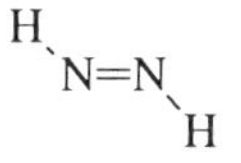

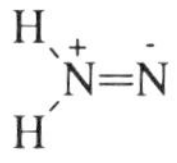

trans-Diimide can be generated and trapped at low temperature by a gas-phase electric discharge in hydrazine[14,15] and by the thermal decomposition of metal salts of *p*-toluenesulfonylhydrazide.[16] Although stable at low temperatures (−196°), diimide undergoes disproportionation to nitrogen and hydrazine at higher temperatures (~ −180°).[17] A spectral analysis of the product mixtures formed in the thermal decomposition of the metal salts of *p*-toluenesulfonylhydrazide has been interpreted in terms of *cis*- and 1,1-diimide.[16,17] A review of the experimental data, however, has led to the conclusion that these species have not been unambiguously characterized in these reaction mixtures.[11] Recently, 1,1-diimide has been generated and trapped by the low temperature photochemical decomposition of carbamoyl azide.[18] Although *cis*-diimide must be formed as a reactive intermediate in many systems, it has not yet been unambiguously characterized.

The diimide system has been subjected to several theoretical studies at many different basis set levels. *trans*-Diimide is calculated to be lowest in energy, with *cis*-diimide 4.7–7.3[19–24] and 1,1-diimide 24.5–27.4[21–23] kcal per mole higher in energy. The *trans* to *cis* inversion barrier is calculated to be 46–66.4 kcal per mole,[17–26] while the rotation barrier is calculated to be generally higher, in the range 41.2–84 kcal per mole.[22,25] The energy barrier for the isomerization of *trans*- to 1,1-diimide is calculated to be 82.6–87.6 kcal per mole.[19,20,23,24]

The results of stereochemical studies on the reduction of alkenes and alkynes have led to the suggestion that *cis*-diimide is the reactive hydrogen-transfer reagent.[5] Calculating the energy surface for the concerted transfer of hydrogen from *cis*-, *trans*-, and 1,1-diimide to ethylene gives energy barriers of 26.7, 45.3, and 45.8 kcal per mole, respectively, with the transfer of hydrogen from *cis*-diimide being very exothermic (117.8 kcal per mole when calculated at the STO-3G level).[19] These results support the suggestion that *cis*-diimide is the active hydrogen-transfer reagent. The fact that *cis*-diimide has not been observed and that the calculated inversion and rotation barriers are too large to provide a rate of isomerization of *trans*- to *cis*-diimide that would be sufficiently high to account for the observed rate of reduction provides for a mechanistic dilemma. In gas-phase reactions isomerization of *trans*- to *cis*-diimide has been proposed to be the rate-limiting step.[27] In solution, however, the isomerization in all probability occurs via a catalyzed process, probably involving a rapid protonation–deprotonation sequence. In this chapter the use of the term "diimide" implies *cis*-diimide as the reducing agent.

The energy barriers for the disproportionation of *cis*- with *cis*-, and *cis*- with *trans*-diimide are calculated to be 19.3 and 23.8 kcal per mole,[28] considerably smaller than the barriers for hydrogen transfer to a carbon–carbon double bond. From a practical point of view, this competing disproportionation requires the use of considerable excesses of the diimide precursors in the reduction reactions.

MECHANISM

The early studies indicated that symmetrical double and triple bonds are readily reduced by diimide, whereas polar bonds such as C=O and C=N are only slowly reduced, or do not react at all.[29] Cyclic transition states have been proposed for the reduction of multiple bonds by diimide,[7,29] and are supported by the observed *syn* addition of hydrogen (or deuterium) across double and triple bonds in a number of systems.[30] Theoretical calculations further support the concerted, symmetrical transfer of hydrogen from *cis*-diimide to a symmetrical alkene such as ethylene via a cyclic transition state.[19]

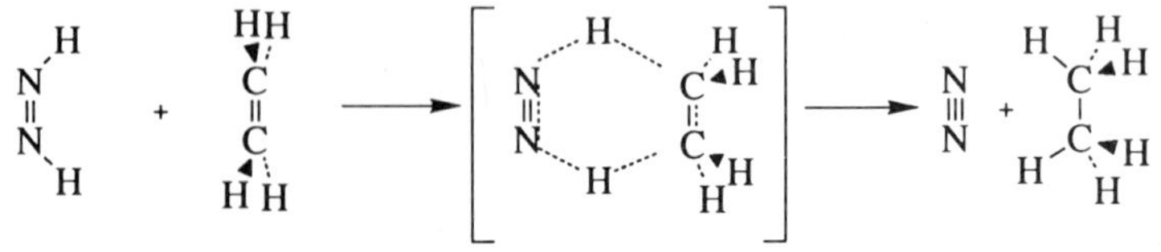

A study of the kinetics of the reduction of methyl oleate in the aprotic solvents acetonitrile, dimethylformamide, and dimethyl sulfoxide with diimide generated by the oxidation of hydrazine with oxygen indicates the reaction to be zero order in hydrazine and methyl oleate.[31] The rate of reaction appears to depend on the rate of dissolution of oxygen. The overall rate of reduction is highest in acetonitrile, followed by dimethylformamide and finally by dimethyl sulfoxide. This observation does not necessarily imply that there is a solvent effect on the rate of reduction; it could reflect the relative solubilities of oxygen in the solvents. The reaction displays autocatalytic behavior and is catalyzed by low concentrations of acetic acid, but is retarded by high concentrations of acetic acid. No rationales have been given for the autocatalytic and acid-catalyzed behaviors, although it seems obvious that water generated in the oxidation of hydrazine and trace quantities of acid should catalyze the equilibration of *trans*- and *cis*-diimide. The retardation in rate in the presence of large quantities of acetic acid must be due to the protonation of hydrazine to form the hydrazonium ion which is not oxidized by oxygen. The kinetics of the Cu(II)-catalyzed reaction of hydrazine with hydrogen peroxide to produce diimide show the reaction to be zero order in hydrazine and first order in both hydrogen peroxide and Cu(II).[32] The kinetics of the reaction in the presence of an alkene appear not to have been studied. It appears, however, that in all cases formation of diimide is rate determining, with the reaction of diimide with an alkene being fast.

Relative Reactivity. A number of relative reactivity studies have been carried out on alkyl-substituted alkenes and dienes,[33–35] and on unsaturated acids.[36] A representative selection of data is presented in Tables A and B. In general, as the degree of alkyl substitution on the double bond of an alkene or a diene increases, the relative reactivity decreases. Conjugated dienes are more reactive than monoenes, and strained double bonds are more reactive than unstrained double bonds. *Trans* double bonds are more reactive than *cis* double bonds, the only exception reported being *trans*- and *cis*-2-pentene.

The relative reactivities of alkenes have been interpreted in terms of dif-

TABLE A. RELATIVE REACTIVITIES OF ALKENES AND DIENES TOWARD REDUCTION BY DIIMIDE[a]

Substrate	k (rel.)	Refs.
Cyclohexene	1.00	33
	Acyclic Alkenes	
1-Pentene	20.2	33
trans-2-Pentene	2.59	33
cis-2-Pentene	2.65	33
2-Methyl-1-pentene	2.04	33
2-Methyl-2-butene	0.28	33
2,3-Dimethyl-2-butene	0.50	33
	Cyclic Alkenes	
Cyclopentene	15.5	33
Cycloheptene	12.1	33
Cyclooctene	17.0	33
Cyclononene	5.7	33
Cyclodecene	0.85	33
1-Methylcyclohexene	0.11	33
1,2-Dimethylcyclohexene	0.012	33
Bicyclo[2.2.1]heptene	450	33
Bicyclo[2.2.2]octene	29	33
cis-Cyclododecene[b]	1.46	35
trans-Cyclododecene[b]	8.0	35
	Dienes	
1,3-Cyclohexadiene	47[c]	34
2-Methyl-1,3-butadiene	13.6[d]	34
2,3-Dimethyl-1,3-butadiene	3.1[c]	34
2,5-Dimethyl-2,4-hexadiene	0.5[c]	34

[a] The reductions were carried out at 80° unless otherwise noted.
[b] The reductions were carried out at 25°.
[c] The relative rate constant for reduction of the first double bond.
[d] The relative rate constant for reduction of the 3,4-double bond.

TABLE B. RELATIVE REACTIVITIES OF ACIDS[36]

Acid	k (rel.)
Fumaric	100
Maleic	10
Methylfumaric	3
Methylmaleic	0.7
trans-Cinnamic	10
cis-Cinnamic	3
α-Methylcinnamic	1.4
β-Methylcinnamic	1.4

ferences in torsional and bond angle strain and α-alkyl substituent effects, and an empirical correlation has been developed to calculate rates of reduction with diimide.[33] Similar trends are observed with unsaturated acids.[36]

Stereoselectivity. The highly exothermic nature of the diimide reduction reaction and the results of theoretical calculations suggest that the transition state for hydrogen transfer occurs rather early along the reaction coordinate.[20] The results of stereochemical studies show that the approach by diimide to the double bond occurs to the less sterically hindered face of the double bond, producing, in general, the less thermodynamically stable product.[37] Numerous examples appear in Table I. The only apparent exception is the diimide reduction of 7-hydroxy-, 7-acetoxy-, and 7-*tert*-butoxynorbornadiene in which *syn–exo* addition occurs preferentially over *anti–exo* addition;[38] however, 4-*tert*-butylnorbornadiene undergoes the expected predominant *anti–exo* addition.[39] The stereoselectivities in diimide reductions have been compared with those observed in catalytic hydrogenations,[37,39,40] and it has been noted that the ratios of the stereoisomers formed in the reduction of substituted alkylidenecycloalkanes with diimide parallel those obtained on a platinum catalyst at high hydrogen pressures (irreversible addition conditions).[40]

SCOPE AND LIMITATIONS

As is evident from the Experimental Conditions section, diimide can be formed under a variety of conditions. As a result, the conditions for generation of diimide can usually be selected in such a way as to accommodate sensitive functionality in the substrate. Thus the scope and limitations of reduction with diimide can usually be attributed to factors intrinsic to this intermediate, and not to conditions of formation. On the other hand, the need to use large excesses of diimide (owing to disproportionation and other side reactions) makes selective reductions difficult (except when the reactivity of competing functionality differs widely). Even very close monitoring of such competing reductions results in only moderate yields of the desired product as part of a complex mixture.

Carbon–carbon triple bonds are, in general, the most easily reduced, but, except in special cases, cannot be selected over symmetrical double bonds such as C=C and N=N. Only when polar substituents (either electron-donating or withdrawing) or an accumulation of sterically demanding substituents have been appended can reactivity differences be exploited successfully. The reactivity of relatively nonpolar carbon–carbon double and triple bonds permits diimide reductions to be carried out under exceptionally mild conditions, which can be selected to tolerate the presence of a number of reactive functional groups that would either be reduced or would suffer hydrogenolysis under catalytic hydrogenation conditions.

Carbon–carbon double bond reductions are outlined in Table I. As can be seen, diimide reductions can be carried out in the presence of reactive allylic functional groups including halides,[41,42] esters,[41] amines,[41] and disulfides.[29] Table I also provides many examples of the reduction of allylic alcohols. More direct polarization of the double bond results in a marked decline in reactivity. Vinyl halides and vinyl ethers undergo reduction only very slowly.[41,43] Unsaturated ketones are reduced to saturated ketones by diimide generated by the hydrolysis of dipotassium azodicarboxylate.[41,44] Other functions that are rather sensitive toward reduction by other reducing agents but are not reduced by diimide are N—O,[45] O—N—N,[46] and O—O bonds of highly strained endoperoxides.[45–47] There are numerous reports of reductions of double bonds in the presence of highly strained bi- and polycyclic systems,[47–56] examples of which are given in Eqs. 1[47] and 2.[48]

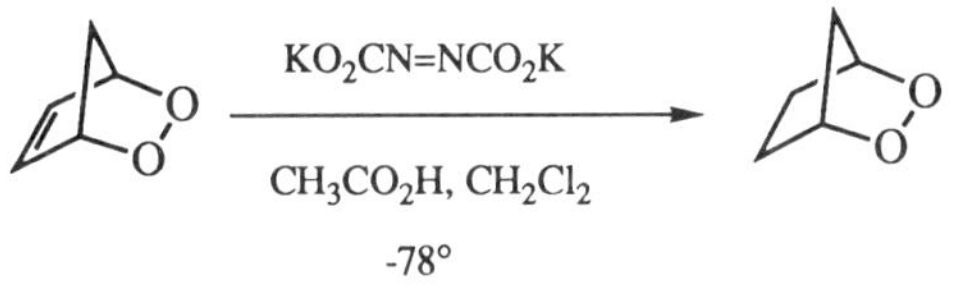

$KO_2CN{=}NCO_2K$; CH_3CO_2H, CH_2Cl_2; -78° (Eq. 1)

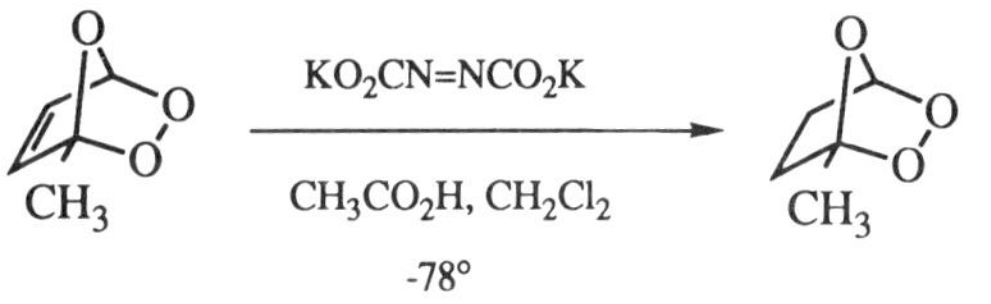

$KO_2CN{=}NCO_2K$; CH_3CO_2H, CH_2Cl_2; -78° (Eq. 2)

While steric bulk alone is normally insufficient to prevent reduction of a carbon–carbon double bond, a careful choice of conditions does permit the selective reduction of relatively unsubstituted double bonds in the presence of more highly substituted double bonds. Illustrative of this type of selectivity are the examples in Eqs. 3[57] and 4.[58] Many other examples can be seen in Table I.

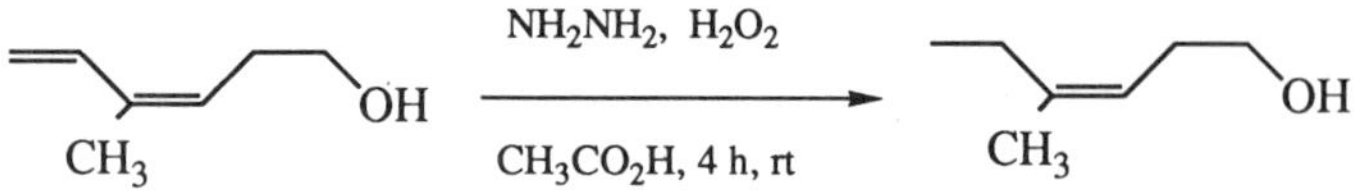

(74%) (Eq. 3)

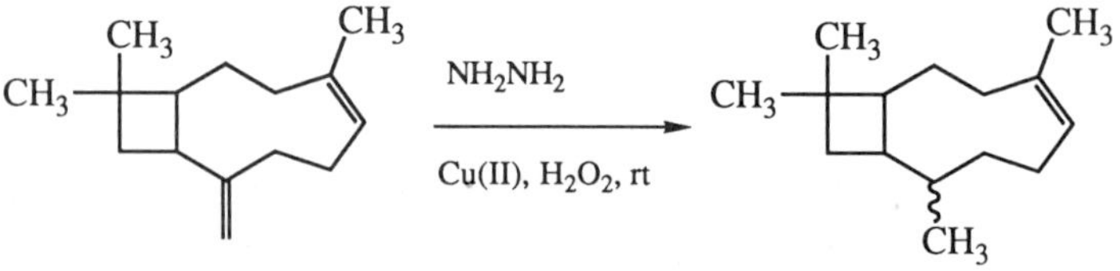

(87%) (Eq. 4)

Substituted allenes (see Table II) readily undergo reduction with diimide to produce alkenes (Eq. 5[59]) which undergo further slower reduction to alkanes. The diimide approaches the allene chromophore from the least-hindered side of the least-substituted double bond to produce the alkene having the *cis* geometry. Increasing the substitution on the allene chromophore increases the reactivity of the other double bond toward reduction, as shown in Eq. 6.[60]

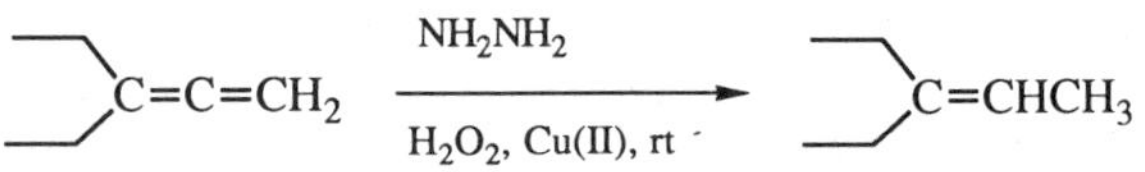

(16%) (Eq. 5)

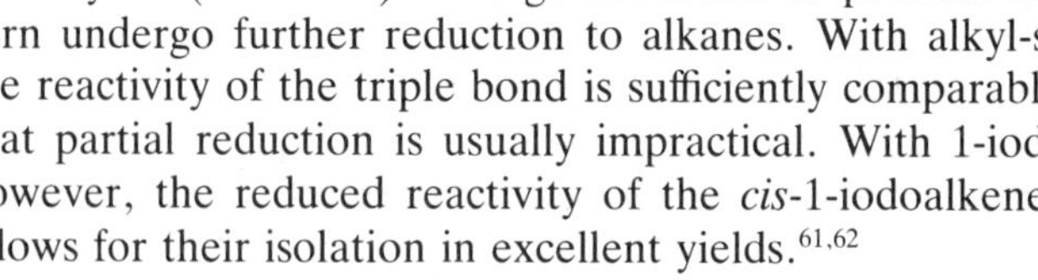

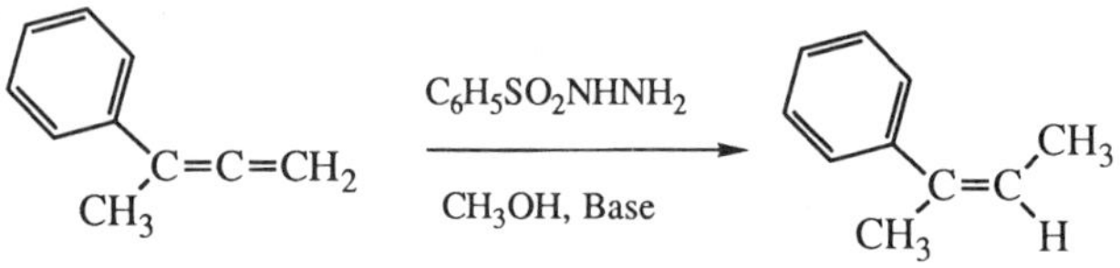

(28%) (Eq. 6)

Alkynes (Table III) undergo reduction to produce *cis*-alkenes, which in turn undergo further reduction to alkanes. With alkyl-substituted alkynes, the reactivity of the triple bond is sufficiently comparable to that of alkenes that partial reduction is usually impractical. With 1-iodoalkynes (Eq. 7),[61] however, the reduced reactivity of the *cis*-1-iodoalkenes toward reduction allows for their isolation in excellent yields.[61,62]

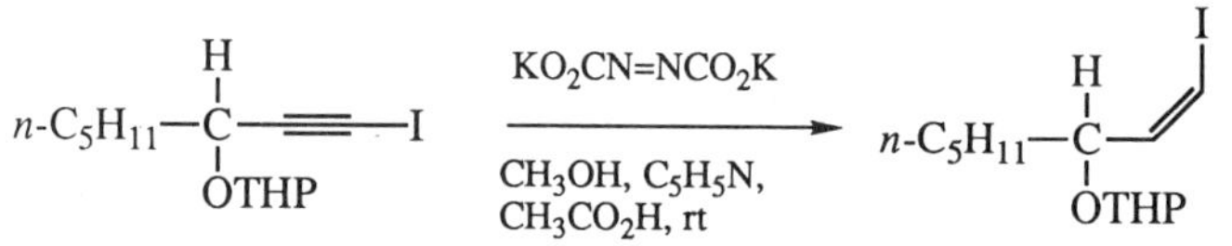

(82%) (Eq. 7)

Reduction can be conveniently carried out on thermally labile systems at −70° in methylene chloride by diimide generated from dipotassium azodicarboxylate and acetic acid.[47,48,55] Position- and stereospecific deuterium and tritium labeling can be readily accomplished in aprotic solvents using dipotassium azodicarboxylate and labeled acetic acid, or in the presence of labeled water.

Although diimide was initially thought to be useful only for the reduction of symmetrical double and triple bonds such as C═C, C≡C, and N═N, it was subsequently demonstrated that substituted aromatic aldehydes could be reduced to benzylic alcohols in excellent yields by diimide formed by hydrolysis of azodicarboxylate (the use of hydrazine as the source of diimide results in azine formation).[63] In analogy with the reactivity trend observed with substituted alkenes, aromatic ketones are reduced more slowly, while aliphatic aldehydes and ketones are reduced even more slowly.[41] The *N′*-benzoylhydrazones of aldehydes and ketones (but apparently not the corresponding hydrazones) and imines of aromatic and aliphatic aldehydes and ketones undergo reduction in good yield with diimide generated from hydrazine.[64]

COMPARISON WITH OTHER METHODS

The most commonly used method for reducing double and triple bonds involves catalytic hydrogenation. Although catalytic hydrogenation suffices very well in many instances, reduction with diimide offers certain advantages. First and foremost, diimide reductions can be carried out in simple, readily available laboratory equipment, in contrast to catalytic hydrogenations, which require the handling of hydrogen gas and often require rather expensive high-pressure equipment. Catalytic hydrogenations are often complicated by a lack of position specificity and stereospecificity owing to the reversible addition and abstraction of hydrogen to and from the organic substrate adsorbed on the surface of the catalyst. Diimide reduction followed by mass spectral analysis has been used extensively to determine the position(s) of unsaturation in naturally derived fatty acids and their derivatives.[65–67] Allylic and benzylic functions do not undergo hydrogenolysis with diimide as is often the case in catalytic hydrogenations. And finally, heteroatom bonds such as N—N, N—O, and O—O, which often suffer reductive cleavage under catalytic hydrogenation conditions, remain intact during diimide reductions.

Although there are several other less widely used methods of reducing double and triple bonds, such as hydrometallation–protonation sequences, the utility of these reactions suffers from limited applicability owing to the high reactivity of the addition reagents with other functions that might be present in the substrate molecules.

EXPERIMENTAL CONDITIONS

Although many methods have been discovered for the generation of diimide in solution, not all are synthetically useful. The following is a list of procedures that have been used to effect reductions. The numbers of the procedures correspond to those indicated under "Procedure" in Tables I–VI of the Tabular Survey.

1. Hydrazine with oxygen, generally in the presence of a catalytic quantity of Cu(II)* and/or a carboxylic acid, in a variety of protic or aprotic solvents or mixtures thereof.
2. Hydrazine and sulfur.[64]
3. Hydrazine and selenium[68] or phenylseleninic acid or anhydride.[69]
4. Hydrazine and hydrogen peroxide; generally in the presence of a catalytic quantity of Cu(II).[6]
5. Hydrazine and periodate.[70]
6. Hydrazine and ferricyanide.[7,36]
7. Hydrazine and mercuric oxide.[7]
8. Hydrazine and iodosobenzene diacetate.[71]
9. Protolytic decarboxylation of dipotassium or disodium azodicarboxylate in protic or aprotic solvents.[8]
10. The thermal and base-catalyzed[11] decomposition of (a) benzene-, (b) 4-methylbenzene-,[8] (c) 4-nitrobenzene-,[72] (d) 2,4,6-triisopropylbenzenesulfonylhydrazide,[73] and (e) polymer-bound arylsufonylhydrazide.[74] Procedure 10d is claimed to be superior to the other reagents because of the lower temperatures required to induce the elimination of diimide.
11. Base-catalyzed decomposition of benzenesulfenylhydrazide.[75]
12. Hydroxylamine in ethyl acetate.[76]
13. Pyrolysis of the anthracene–diimide adduct.[77]
14. Reaction of hydroxylamine-*O*-sulfonic acid with base.[78,79]
15. Reaction of chloramine with base.[80]
16. Oxidation of hydrazine with transition metal complexes of nickel,[80] iron,[81] vanadium,[82–84] molybdenum,[83,85,86] and iridium.[87]
17. Thermal decomposition of α,α′-dihydroxyazocyclohexane.[88]
18. Thermal decomposition of *N*-amino-2,2-diphenylaziridine.[89]
19. Reaction of sulfinate esters with hydrazine.[90]

EXPERIMENTAL PROCEDURES

The following experimental procedures have been selected to illustrate the most widely used and general procedures. Additional examples have been chosen in order to emphasize the wide variety of structural features compatible with the reaction conditions. Further, reactions which illustrate the potential selectivity of diimide reductions are included.

Diimide Reduction of 2-Chlorobenzonorbornene with Hydrazine and Oxygen.[43] 2-Chloronorbornene (4.5 g, 25.5 mmol) and 95% hydrazine (9 g, 281

*The source of Cu(II) may be *insitu* oxidation of Cu(I) or the metal.

mmol) were dissolved in 200 mL of 95% ethanol containing 0.5 g of suspended CuCl. A stream of air was bubbled through the stirred solution for 48 hours. The initially white cuprous salt turned brown and then black as the reaction proceeded. The solution was filtered and the ethanol removed by flash evaporation. The residue was dissolved in ether, washed with water, and dried. Distillation through a small Vigreux column afforded some unreacted starting material and *endo*-2-chlorobenzonorbornane as a colorless oil: 1.01 g (22%), bp 66–69° (0.45 mm), n_{D}^{9} 1.5730; IR (neat) 2934, 2840, 1453, 1314, 1282, 1181, 1136, 1126, 1098, 1012, 954, 934, 922, 870, 831, 752, 706, 664 cm^{-1}.

Diimide Reduction of $\Delta^{13(18)}$-Cholajervene (1) with Hydrazine and Oxygen.[91] Into a solution of $CuSO_4$ (0.006 g), 85% hydrazine hydrate (10 mL) and absolute ethanol (40 mL) was introduced 0.137 g (0.418 mmol) of olefin **1** in benzene (1 mL). Reduction was complete after oxygen was bubbled through the refluxing system for 25 hours. The reaction mixture was extracted with ether and the organic layer was washed successively with dilute acid, base, and water, then dried and freed of solvent. Column chromatography over silver nitrate–alumina yielded 5β,12α-cholajervane (**2**) (0.084 g, 61%) which was recrystallized from a THF–acetone mixture. Mp 38.0–39.5°; $[a]_D$ +62.2° (*c* 0.101, cyclohexane); ^1H NMR (100 MHz) (no resonance downfield from δ 3.00), 0.76 (d, 3*H*, J = 7 Hz), 0.83 (d, 3*H*, J = 7 Hz), 0.89 (s, 3*H*), among others. Anal. Calcd. for $C_{24}H_{42}$ (330.3286): C, 87.19; H, 12.81. Found: C, 87.47; H, 12.49. Mass spectrum, m/e = 330.3292.

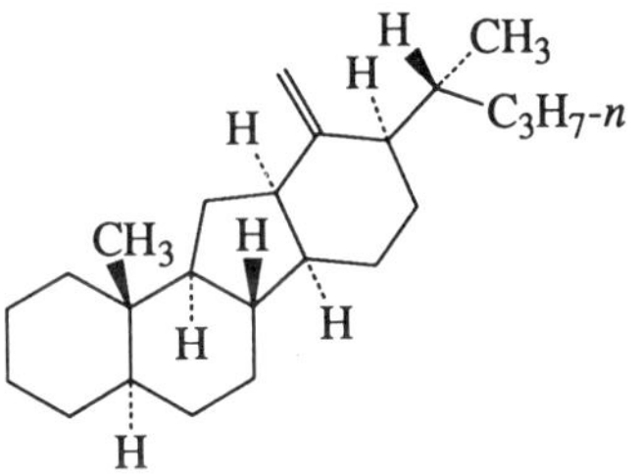

1 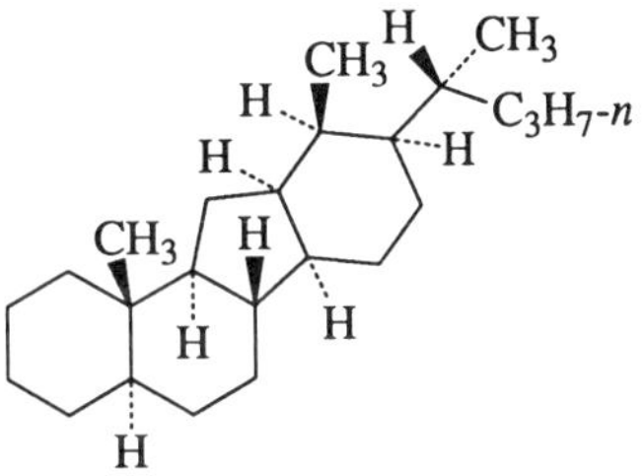 **2**

Diimide Reduction of 4-Methylhexa-3-*cis*,5-dien-1-ol (3) with Hydrazine and Hydrogen Peroxide.[57] To a stirred and ice-cooled solution of **3** (22.4 g, 0.2 mol) in 99% ethanol (300 mL) and 85% hydrazine hydrate (105 g, 2.1 mol) was added 35% hydrogen peroxide (110 mL) during 1.5 hours at a rate that kept the solution temperature below 30°. After addition was complete, the mixture was stirred for ca. 4 hours at room temperature until the IR absorption at 900 cm^{-1} disappeared. The mixture was poured into water and extracted with ether. The ethereal extract was washed with $FeSO_4$ solution, water, and brine, then dried ($MgSO_4$) and concentrated. Distillation of the residue afforded 16.9 g (74%) of 4-methylhex-3-*cis*-en-1-ol (**4**): bp 64–65° (10 mm), n_D^{18} 1.4461, IR 3300, 1640, 1045 cm^{-1}; ^{1}H NMR δ 0.90 (t, 3*H*, *J* = 7 Hz), 1.69 (d, 3*H*, *J* = 1.5 Hz), 2.03 (q, 2H, *J* = 7 Hz), 2.20 (q, 2*H*, *J* = 7 Hz), 2.98 (br s, 1*H*), 3.49 (t, 2*H*, *J* = 7 Hz), 5.04 (t, 1*H*, *J* = 7 Hz). Anal. Calcd. for $C_7H_{14}O$: C, 73.63; H, 12.36. Found: C, 73.40; H, 12.23.

3 **4**

Diimide Reduction of 1,2,3,4,5,6-Hexahydro-1,2,3:4a,5,8a-dimetheno-naphthalene (5) with Hydrazine and Hydrogen Peroxide.[92] A cooled (0°) solution of **5** (200 mg, 1.28 mmol) and 99% hydrazine (900 mg, 28 mmol) in 5 mL of 95% ethanol was treated dropwise with 1.3 mL of 30% hydrogen peroxide. The reduction mixture was allowed to warm to room temperature over 4 hours. The solution was extracted with pentane and the extract was washed with water and dried. The pentane was removed and the residue was distilled (70°, 0.5 mm) giving a 45:17 mixture of **5** and the reduced product (45% yield based on recovered starting material), which could be isolated by gas chromatography (0.2% SE-30 on glass, 135°) and was identical to the authentic material prepared in another fashion. For octahydro-1,2,3:4a,5,8a-dimethenonaphthalene: ^{1}H NMR (C_6D_6) δ 2.52–2.43 (m, 3*H*), 2.27–1.08 (m, 11*H*); ^{13}C NMR (C_6D_6) ppm 46.3, 44.5, 40.6, 40.2, 23.6, 22.2, 20.4, 20.2, 15.4, 14.9, 5.0, 4.3. Anal. Calcd. for $C_{12}H_{24}$: C, 91,08; H, 8.92. Found C, 91.09; H, 9.11.

5

Diimide Reduction of (*S*)-1-(Acetoxymethyl)-2-carbomethoxy-3-isopro-penylcyclopentene (6) with Hydrazine and Periodate.[93] To a solution of **6** (440 mg, 1.85 mmol) in 75 mL of methanol was added 9 mL of hydrazine hydrate, acetic acid (9 drops), and saturated aqueous copper sulfate (9 drops). The stirred mixture was maintained at 25° while a solution of sodium periodate (7.9 g, 20 equiv) in water (60 mL) was added dropwise during 1 hour. Upon completion of the addition, stirring was maintained for 36 hours before re-moval of most of the methanol under reduced pressure. The product was taken up in ether, washed with water and saturated salt solution, dried, and concentrated under reduced pressure to afford 330 mg (90%) of (*S*)-1-hy-droxymethyl-2-carbomethoxy-3-isopropylcyclopentene (**7**). (The acetate function of **6** suffers hydrazinolysis during the reaction). ^{1}H NMR of **7** ($CDCl_3$) δ 4.4 (m, 2*H*), 3.70 (s, 3H), 3.1–1.4 (m, 7*H*), 0.95 (d, 3H, $J = 6$ Hz), 0.70 (d, 3*H*, $J = 6$ Hz); mass spectrum, calcd (M^+) m/e 198.1256, obsd. 198.1262.

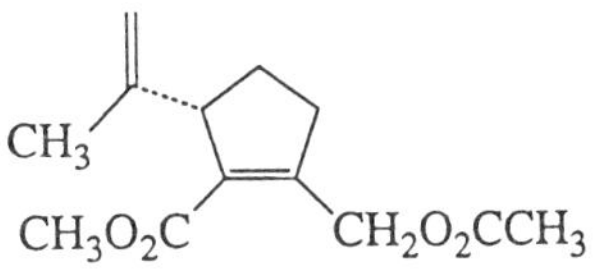

6 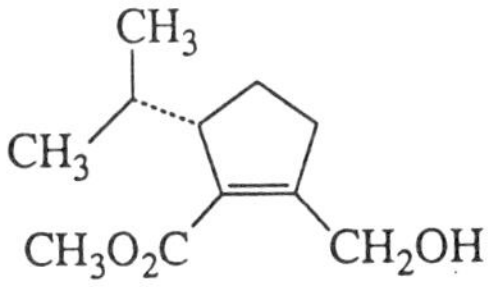 **7**

Diimide Reduction of Benzvalene with Hydrazine and Potassium Ferricyanide.[94] To a solution of 4.50 g (57.7 mmol) of benzvalene in about 200 mL of ether (containing some benzene and methylene chloride from the synthesis of benzvalene) was added 50 g (1 mol) of hydrazine hydrate, 49 g (0.8 mol) of ethanolamine, and 200 mL of 2-methoxyethanol. A solution of potassium ferricyanide (110 g, 325 mmol) in 220 mL of water was added dropwise over the course of 2 hours while maintaining the temperature of the reaction at about 20°. During the reaction the nitrogen gas which was evolved was passed through two flasks cooled to −75° to condense ether, starting benzvalene, and product. The flasks were changed from time to time during the course of the reaction. At the end of the reaction the unpleasant odor of benzvalene was no longer detectable. The condensed material was distilled under water aspirator pressure at 35° water bath temperature. The distillate was washed with water and the ether solution was dried over sodium sulfate/potassium bicarbonate. The ether solution was distilled through a spinning band column, with 5 mL of toluene used as a chaser, to give three fractions, bp 39–110°, containing a total of 3.23 g (70%) of dihydrobenzvalene contaminated by solvents. The higher boiling fractions were treated with 100 mg of lithium aluminum hydride (to remove the 2-methoxyethanol) and distilled. The distillate was redistilled through a spinning band column to give 1.50 g of dihydrobenzvalene, bp 69–69.5°, practically free of ether and methylene chloride. ^{1}H NMR (C_6D_6) δ 2.03 (m, 2*H*), 1.69 (t, 2*H*), 1.28 (m, 4*H*); ^{13}C NMR (C_6D_6) ppm 34.0, 26.1, 2.4.

Preparation of Dipotassium Azodicarboxylate.[95] To a stirred solution of potassium hydroxide (31 mL, 40% by weight) at 8° was added 5 g of azodicarboxamide (Aldrich Chemical Co.) in small portions over 2 hours. After stirring for an additional hour the bright yellow dipotassium azodicarboxylate was filtered off using a Büchner funnel, and the solid was washed 20 times with cold methanol. Yields varied from 80 to 92%.

Caution: There has been one report that dipotassium azodicarboxylate exploded violently when left exposed to bright sunlight for about 30 minutes.[41] There have been other reports of suspected nonviolent decomposition of the potassium salt in storage.[96–98] It is advisable to dry the material in vacuo, and not to store the material in a sealed container.

Diimide Reduction of 1-Iodo-3-hydroxy-7-(*tert*-butyldimethylsilyloxy)oct-1-yne with Dipotassium Azodicarboxylate and Acetic Acid.[61] To a solution of the crude iodoacetylene (18.9 g, 49.5 mmol) in 70 mL of methanol and 24.5 mL of pyridine was added 12.0 g (62 mmol) of dipotassium azodicarboxylate. Glacial acetic acid (7.5 mL) was added slowly (2 hours, room temperature) and stirring was continued overnight. An additional 18 g of dipotassium azodicarboxylate and 10.3 mL of glacial acetic acid were added over the course of 18 hours. When no starting material could be detected by GLC (5 ft, SE-30) analysis of aliquots, 200 mL of ether was added. Any remaining diimide precursor was destroyed by carefully adding 100 mL of 5% hydrochloric acid with vigorous stirring. The organic layer was separated and the aqueous layer was extracted with ether (2 × 100 mL). The combined organic layers were washed with 5% hydrochloric acid and with 5% sodium bicarbonate solution and dried over magnesium sulfate, and the solvents were removed on a rotary evaporator to give an oil that was dissolved in 50 mL of ether and stirred with 12 mL of 40% aqueous dimethylamine to remove a small amount (ca. 5%) of overreduced material. The ether solution was washed with 5% hydrochloric acid (2 × 50 mL) and dried over magnesium sulfate, and the ether was removed under reduced pressure to give crude *cis*-1-iodo-3-hydroxy-7-(*tert*-butyldimethylsilyloxy)oct-1-ene (15.63 g, 83%). Chromatography on silica gel (1.5 kg) with acetone–hexane (1.19, v/v) gave the pure material (12.9 g, 68%) as an oil: IR (film) 3300, 1245, 1090, 1045, 835, 805, 775 cm^{-1}; ^{1}H NMR ($CDCl_3$) δ 6.24 (m, 1*H*), 4.41 (m, 1*H*), 3.82 (m, 1*H*), 1.09 (d. $J = 6$ Hz, 3*H*), 0.85 (s, 9*H*), 0.02 (s, 6*H*); ^{13}C NMR ($CDCl_3$) −4.62, −4.32, 18.17, 21.20, 21.30, 23.80, 25.98, 36.05, 39.56, 68.56, 74.45, 82.28, 143.56; GC-MS (70 eV) m/e (rel intensity) 327 (9), 235 (22), 193 (46), 108 (80).

Diimide Reduction of 2,3-Dioxabicyclo[2.2.2]oct-5-ene with Dipotassium Azodicarboxylate and Acetic Acid.[50] To a 50-mL round-bottomed flask, provided with a magnetic spinbar, was charged 560 mg (5 mmol) of the substrate and 2.91 g (15 mmol) of dipotassium azodicarboxylate in 10 mL of absolute methanol. While stirring magnetically and cooling by means of an ice bath, a solution of 1.86 g (30 mmol) of acetic acid in 3 mL of absolute methanol was added dropwise within 30 minutes. After stirring for 3 hours at 30°, the solvent was evaporated and the residue taken up in 20 mL of water and extracted twice with 20 mL of dichloromethane. The organic extracts were washed once with saturated bicarbonate solution, dried, and evaporated. The endoperoxide (2,3-dioxabicyclo[2.2.2]octane) was obtained (2.74 g, 48%) by recrystallization from hexane: ^{1}H NMR (CCl_4): δ 1.4-1.9 (m, 4*H*), 1.9-2.5 (m, 4*H*), 3.9 (m, 2*H*). IR (CCl_4): 2960, 2940, 2890, 2855, 1460, 1445, 1430, 1305, 1225, 1030, 950 cm^{-1}. The pure substance deteriorates on standing within a few days. Mass spectrum m/e (relative intensity): 114 (71), 81 (100), 67 (41), 57 (85), 43(88).

Diimide Reduction–Deuteration of 3-*endo*-Phenylsulfinylbicyclo[2.2.1]-hept-5-ene-2-*endo*-carboxylic Acid (8).[89] Acetic acid-*O-d* (1.2 g) was slowly added dropwise into a solution of the carboxylic acid **8** (200 mg, 1 mmol) and dipotassium azodicarboxylate (400 mg, 2.5 mmol) in DMSO (7 mL). After stirring 4 hours at room temperature the solution was diluted with brine and extracted with pentane. The pentane layer was dried and evaporated to afford 160 mg (75%) of 5-*exo*-6-*exo*-dideuterio-3-*endo*-phenylsulfinylbicyclo[2.2.1]heptane-2-*endo*-carboxylic acid (**9**): mp 183–184°. ^{1}H NMR (DMSO-d_6): δ 7.32 (s, 5*H*) 4.37 (m, 1*H*), 4.17 (m, 1*H*), 3.68 (m, 1*H*), 3.48 (m, 1*H*), 2.60 (m, 2*H*), 1.49 (br s, 2*H*).

8 **9**

Diimide Reduction of Tetracyclo[4.1.0.0^{2,4}.0^{3,5}]deca-7,9-diene (10) with *p*-Nitrobenzenesulfonylhydrazide.[72] To a solution of 11 g (51.0 mmol) of *p*-nitrobenzenesulfonylhydrazide in 100 mL of ethanolamine–methanol (1.5 M) was added 1.20 g (9.20 mmol) of the diene **10**. The resultant solution was heated at reflux for 2 hours. Following the addition of 800 mL of water, the reaction mixture was extracted repeatedly with *n*-hexane. The combined hexane layers were dried with calcium chloride and concentrated under reduced pressure. Distillation (bp 60°/14 mm) of the residue afforded 570 mg (45%) of tetracyclo[4.1.0.0^{2,4}.0^{3,5}]decane (**11**): IR (film) 3020, 2900, 2850, 1465 (s), 1458, 1447, 1394, 1350, 1330, 1305, 1289, 1149, 1112, 1098 (s), 1025, 1010, 950, 869, 842, 819, 808, 778, 748, 690 cm^{-1}. Anal. Calcd. for $C_{10}H_{14}$: C, 89.49; H, 10.51. Found: C, 90.28; H, 10.60.

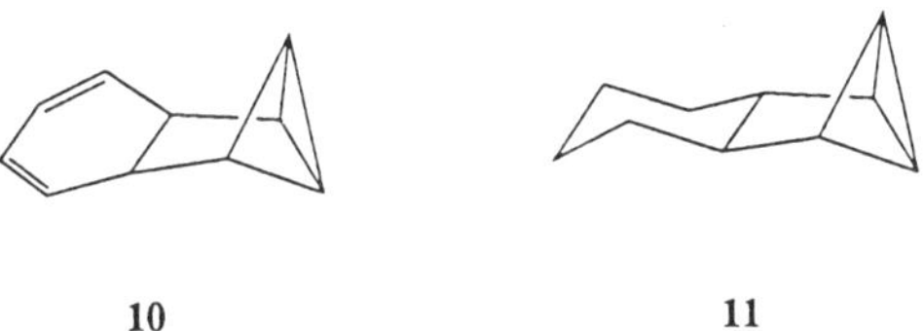

10 **11**

Diimide Reduction of Acenaphthylene with 2,4,6-Triisopropylbenzenesulfonylhydrazide.[73] 2,4,6-Triisopropylbenzenesulfonylhydrazide (4.12 g, 13.8 mmol) and 0.838 g (5.5 mmol) of acenaphthylene were dissolved in 50 mL of methanol at 20°. After 16 hours the yellow color had disappeared, indicating that the reaction was complete. Aqueous sodium hydroxide (5%, 50 mL) was added and the colorless crystalline material was collected by filtration and washed with water and dried in vacuo over phosphorus pentoxide to give 0.835 g (99%) of acenaphthene, mp 94–95°.

Diimide Reduction of 2-Decen-1-ol with Hydroxylamine and Ethyl Acetate.[76] Powdered potassium hydroxide (85%, 32.98 g) was added to a mechanically stirred solution of hydroxylamine hydrochloride (34.75 g, 0.50 mol) in 100 mL of dimethylformamide at 25–35° under a nitrogen atmosphere. The resulting mixture was stirred for 10 minutes and filtered. The filtrate (pH 8–9) was cooled in an ice bath and 19.59 g (0.22 mol) of ethyl acetate was added. This solution was then added in 25-mL portions at 45-minute intervals to 1.56 g (0.01 mol) of 2-decen-1-ol stirred at 90–100°. (Alternatively, the ethyl acetate solution could be added dropwise over 1 hour.) After addition was complete, the reaction solution was heated an additional 1 hour, cooled to room temperature and added to water. Extraction with 50:50 hexane–ether followed by routine workup of the organic extract and distillation gave 1.50 g (95%) of pure 1-decanol, bp 125–128° (25mm).

Diimide Reduction of Fumaric Acid with Hydroxylamine-*O*-sulfonic Acid.[99] Into 100 mL of water was dissolved 5.8 g (50 mmol) of fumaric acid, 12.4 g (109 mmol) of hydroxylamine-*O*-sulfonic acid (previously washed with 95% THF) and 8.2 g (50 mmol) of hydroxylamine sulfate. The solution was neutralized slowly with 24 mL of concentrated sodium hydroxide. The temperature rose to about 50° with vigorous gas evolution. After 2 hours, 30 mL of 2 N H_2SO_4 was added and the solution was extracted with ether. The ether extracts contained 5.3 g (90%) of succinic acid, mp 180–184°.

TABULAR SURVEY

The literature has been surveyed through mid-1988 using primarily a citation search for references to the early original publications on diimide reductions. A few references have been found in which diimide reductions have been used but in which no references were cited. Undoubtedly, there are many more instances of such reductions in the literature which do not cite these original publications and are not listed under "diimide" (1,2-diazene) in *Chemical Abstracts*. In many instances reductions with diimide have been reported without a specific procedure, only a reference. In such cases, the procedures and conditions are not listed in the Tabular Survey and are thus indicated by a dash (—). In several studies of the relative reactivity of diimide reductions, products have not been isolated nor have yields been determined. In such cases, product structures have been assumed, but no yields are given. Finally, there have been several studies on the selectivity of diimide reductions in which only product ratios were reported. As noted previously, the maximum yield of product is often limited by the amount of excess diimide precursor used.

In the following tables the procedures used are indicated by number corresponding to the procedures given in the Experimental Conditions section.

The following abbreviations have been used in Tables I–VI.

Bn	benzyl
DMF	dimethylformamide
Diglyme	ethylene glycol dimethyl ether
DMSO	dimethyl sulfoxide
EET	2-ethoxyethanol
ETA	ethanolamine
MET	2-methoxyethanol
TEA	triethylamine
THF	tetrahydrofuran
THP	tetrahydropyranyl
TMED	tetramethylethylenediamine

TABLE I. Diimide Reductions of Substituted Alkenes

Reactant	Procedure	Conditions	Product(s) and Yield(s) (%)	Refs.
C_2				
$CH_2{=}CH_2$	14	NaOH, 40°	C_2H_6 (—)	100
C_3				
$CH_2{=}CHCN$	8	CH_2Cl_2, rt	C_2H_5CN (97)	71
$CH_2{=}CHCH_2Br$	9	CH_3OH, CH_3CO_2H, 25°, 4.5 h	n-C_3H_7Br (—)	63
$CH_2{=}CHCH_2OH$	1	C_2H_5OH, $C_9H_{19}CO_2H$	n-C_3H_7OH (—)	101
	9	CH_3OH, CH_3CO_2H, 25°	" (99)	63
	10	Diglyme, heat	" (99)	102
	10	EET, $C_7H_{15}CO_2H$	" (78)	7
	10	MET, 100°[a]	" (—)	103
C_4				
Maleic Anhydride	5	C_6H_6, CH_3CO_2H or Cu(II)	Succinic anhydride (—)	70
	8	CH_2Cl_2, rt	" (83)	71
Maleic Acid	1,4	CH_3OH, 25°	Succinic acid (82)	5
	1,8 (N_2D_2)	D_2O, 25°	*meso*-2,3-d_2-Succinic acid (100)	30
	13	C_2H_5OH, reflux	Succinic acid (87)	77
Diethyl maleate	8	CH_2Cl_2, rt	Diethyl succinate (—)	95
Fumaric acid	1,4	CH_3OH	Succinic acid (76)	5
	1,8 (N_2D_2)	D_2O, 25°	*dl*-2,3-d_2-Succinic acid (100)	30
	6	—	Succinic Acid (—)	37
	13	C_2H_5OH, reflux	" (80)	77
	1	C_2H_5OH, $C_3H_7CO_2H$	" (—)	102
	14	50°, 2 h	" (90)	99
	14	NaOH, 40°	" (—)	100
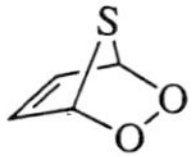	9	CH_2Cl_2, CH_3CO_2H, −78°	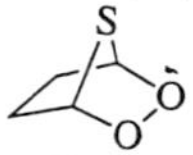 (—)	52
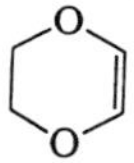	9	CH_3OH, RCO_2H[b]	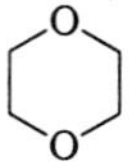 (—)	34
$CH_3CH{=}CHCH_2OH$	1	C_2H_5OH, RCO_2H[b]	n-C_4H_9OH (—)	101, 102
cis-$HOCH_2CH{=}CHCH_2OH$	1,4 (N_2D_2)	D_2O	*meso*-$HOCH_2CHDCHDCH_2OH$ (—)	30

C_5				
CH_2=$CHCO_2C_2H_5$	10	MET, 100°[a]	$C_2H_5CO_2C_2H_5$ (—)	103
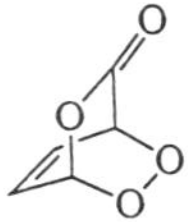	9	CH_2Cl_2, CH_3CO_2H, −78°	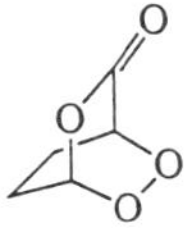 (—)	55
Cyclopentadiene	9	CH_3OH, CH_3CO_2H, 25°	Cyclopentane (—)	35
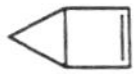	9	C_2H_5OH, CH_3CO_2H, 5 min, rt	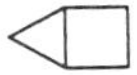 (—)	104
	9	CH_2Cl_2, CH_3CO_2H, −78°	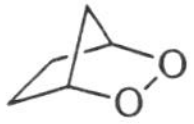 (—)	47, 105
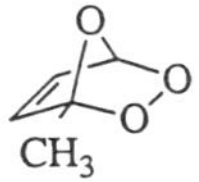	9	CH_2Cl_2, CH_3CO_2H, −78°	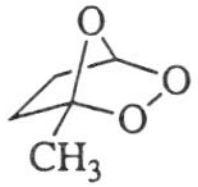 (—)	48
(*E*)-$HO_2CC(CH_3)$=$CHCO_2H$	6	—	$HO_2CCH(CH_3)CH_2CO_2H$ (—)	36
(*Z*)-$HO_2CC(CH_3)$=$CHCO_2H$	6	—	$HO_2CCH(CH_3)CH_2CO_2H$ (—)	36
CH_2=$C(CH_3)CH$=CH_2	9	CH_3OH, CH_3CO_2H	*i*-C_5H_{12} (—)	34
	10	Diglyme, ETA, 80°	" (—)	34
Cyclopentene	9	CH_3OH, CH_3CO_2H	Cyclopentane (—)	35
	10	Diglyme, 80°	" (—)	32
	9	CH_3OH, CH_3CO_2H	 (0)	63
CH_2=$CHCH_2O_2CCH_3$	1	CH_3OH, *n*-$C_3H_7CO_2H$	*n*-$C_3H_7O_2CCH_3$ (—)	101
CH_3CH=$C(CH_3)CO_2H$	1	C_2H_5OH, *n*-$C_3H_7CO_2H$	*s*-$C_4H_9CO_2H$ (—)	102
$(CH_3)_2C$=$CHCO_2H$	1	C_2H_5OH, *n*-$C_3H_7CO_2H$	*i*-$C_4H_9CO_2H$ (—)	102
n-C_3H_7CH=CH_2	10	Diglyme, 80°	*n*-C_5H_{12} (—)	32, 34
cis-C_2H_5CH=$CHCH_3$	10	Diglyme, 80°	*n*-C_5H_{12} (—)	32
trans-C_2H_5CH=$CHCH_3$	10	Diglyme, 80°	*n*-C_5H_{12} (—)	32
C_6				
1,4-Benzoquinone	17	Alkaline, rt	Hydroquinone (49)	88
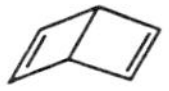	9	H_2O, C_5H_5N, CH_3CO_2H	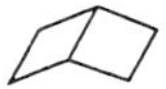 (—)	106

TABLE I. DIIMIDE REDUCTIONS OF SUBSTITUTED ALKENES (*Continued*)

Reactant	Procedure	Conditions	Product(s) and Yield(s) (%)	Refs.
 [structure]	6	MET, H_2O, ETA, 20°, 2 h	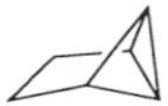 [structure] (—)	94
1,3-Cyclohexadiene	9	CH_3OH, CH_3CO_2H, 25°	Cyclohexane (—)	34, 35
	10	Diglyme, 80°	" (—)	34
1,4-Cyclohexadiene	10	Diglyme, 80°	" (—)	34
Cyclohexen-3-one	9	CH_3OH, RCO_2H[b]	Cyclohexanone (6)	63
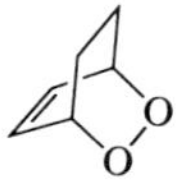 [structure]	9 (N_2D_2)	CH_3OD, CH_3CO_2D, 0°	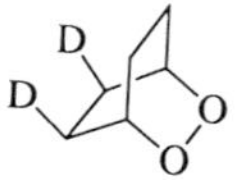 [structure, D, D] (40–48)	50, 105
$CH_3CH{=}CHCH{=}CHCO_2H$	1	C_2H_5OH, $C_3H_7CO_2H$	n-$C_5H_{11}CO_2H$ (—)	102
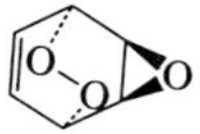 [structure]	9	CH_2Cl_2, CH_3CO_2H, 0°	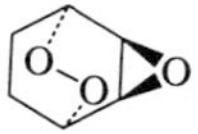 [structure] (45)	107
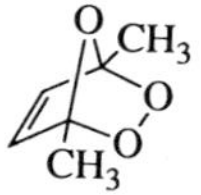 [structure]	9	CH_2Cl_2, CH_3CO_2H, −60°	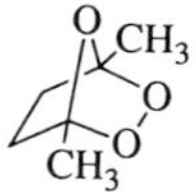 [structure] (—)	48
2,3-Dimethylmaleic acid	1,4 (N_2D_2)	CH_3OD, D_2O	*meso*-$HO_2CCD(CH_3)CD(CH_3)CO_2H$ (—)	30
Cyclohexene	10	MET, 100°[a]	Cyclohexane (—)	103
	9	CH_3OH, CH_3CO_2H, 25°	" (—)	35
	10	Diglyme, heat	" (98)	102, 32, 34
	1	C_2H_5OH, n-$C_3H_7CO_2H$	" (—)	102
	14	NaOH, 40°	" (—)	100
$CH_2{=}C(CH_3)C(CH_3){=}CH_2$	9	CH_3OH, CH_3CO_2H, 25°	i-$C_3H_7C_3H_7$-i (—)	34
1-Methylcyclopentene	9	CH_3OH, CH_3CO_2H, 25°	1-Methylcyclopentane (—)	32
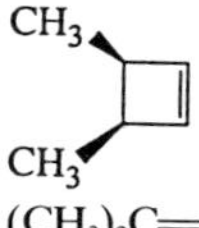 [structure]	—(N_2D_2)		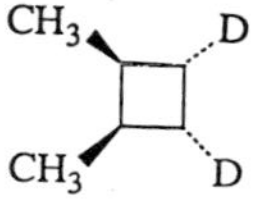 [structure] (—)	108
$(CH_3)_2C{=}CHCOCH_3$	9	CH_3OH, RCO_2H[b]	i-$C_4H_9COCH_3$ (—)	63

$CH_3OCH_2CH{=}CHCH{=}CH_2$	—	—	$CH_3OC_5H_{11}$-*n* (—)	109
$(CH_2{=}CHCH_2S)_2$	10	$HOCH_2CH_2OH$, 3 h	$(n\text{-}C_3H_7S)_2$ (93–100)	29
n-$C_3H_7C(CH_3){=}CH_2$	10	Diglyme, 80°	*n*-$C_3H_7C_3H_7$-*i* (—)	32, 34
$C_2H_5CH{=}C(CH_3)_2$	10	Diglyme, 80°	" (—)	34
$(CH_3)_2C{=}C(CH_3)_2$	9	CH_3OH, RCO_2H[b]	*i*-$C_3H_7C_3H_7$-*i* (—)	41
	10	Diglyme, 80°	" (—)	32
	1	CH_3OH, RCO_2H[b]	" (—)	101
C_7				
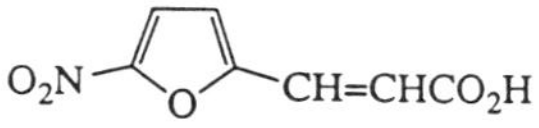 O_2N–(furan)–$CH{=}CHCO_2H$	9	CH_3OH, CH_3CO_2H, 10 h, rt	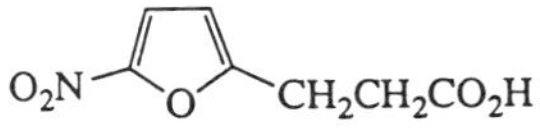 O_2N–(furan)–$CH_2CH_2CO_2H$ (89)	110
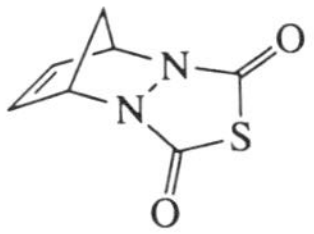	9	CH_3OH, CH_3CO_2H	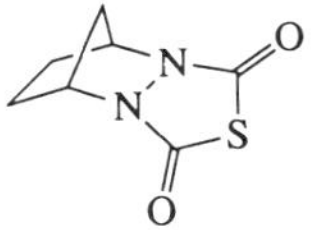 (—)	111
Norbornadiene	9	Diglyme, 80°	Norbornane (—)	32
	9 (N_2D_2)	CH_3OD, CH_3CO_2D, 2 h, rt	 (24) +  (47)	38
7-Hydroxynorbornadiene	9 (N_2H_2, N_2D_2)	$CH_3OH(D)$, $CH_3CO_2H(D)$[b]	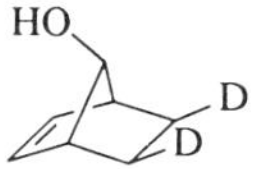 (27) + 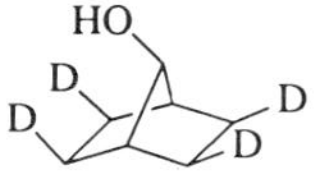 (18)	38
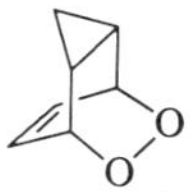	9	CH_3OH, CH_3CO_2H, 0°	 (—)	53, 112
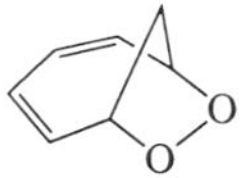	9	CH_3OH, CH_3CO_2H, 0°	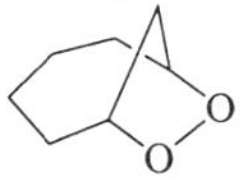 (—)	53, 112
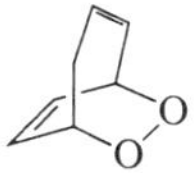	9	CH_3OH, CH_3CO_2H, 0°	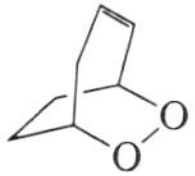 (—)	53, 112

TABLE I. DIIMIDE REDUCTIONS OF SUBSTITUTED ALKENES (*Continued*)

Reactant	Procedure	Conditions	Product(s) and Yield(s) (%)	Refs.
	9	CH_2Cl_2, CH_3CO_2H, −78°	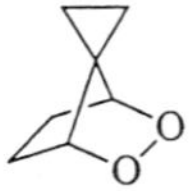 (—)	54
1,3-Cycloheptadiene	9	CH_3OH, CH_3CO_2H	Cycloheptene (—) + Cycloheptane (—)	35
Norbornene	9 (N_2H_2, N_2D_2)	$CH_3OH(D)$, $CH_3CO_2H(D)$ 0.5 h, rt	 (—)	113
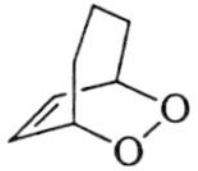	9	CH_2Cl_2, CH_3CO_2H, 0°	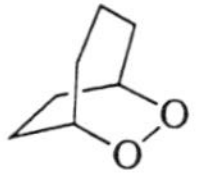 (72)	53, 105, 112
—CH=CH$(CH_2)_3OH$	1	13.5 h, rt	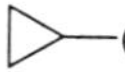—$(CH_2)_5OH$ (88)	114
1,5-Dimethylcyclopentene	10	Diglyme, 80°	*cis* + *trans*-1,2-Dimethylcyclopentane(—) *cis*/*trans* = 31/69	40
Methylenecyclohexane	9	Diglyme, 80°	Methylcyclohexane (—)	32
	10	Diglyme, 80°	" (—)	35
	18	CH_2Cl_2, rt, 24 h	" (—)	87
1-Methylcyclohexene	10	Diglyme 80°	" (—)	32
Cycloheptene	9	CH_3OH, CH_3CO_2H, 25°	Cycloheptane (—)	35
	10	Diglyme, ETA, 80°	" (—)	32
(*Z*)-CH_2=CHC(CH_3)=$CHCH_2CH_2OH$	4	C_2H_5OH, 30–40°, 7 h	(*Z*)-$C_2H_5C(CH_3)$=$CHCH_2CH_2OH$ (74)	57
C_8				
4-ClC_6H_4CH=CH_2	9	CH_3OH, CH_3CO_2H, 0°	4-$ClC_6H_4C_2H_5$ (98)	115
3-ClC_6H_4CH=CH_2	9	CH_3OH, CH_3CO_2H, 0°	3-$ClC_6H_4C_2H_5$ (98)	115
4-BrC_6H_4CH=CH_2	9	CH_3OH, CH_3CO_2H, 0°	4-$BrC_6H_4C_2H_5$ (96)	115
4-$O_2NC_6H_4CH$=CH_2	9	CH_3OH, CH_3CO_2H, 0°	4-$O_2NC_6H_4C_2H_5$ (98)	115
Styrene	9	CH_3OH, CH_3CO_2H, 0°	Ethylbenzene (98)	115
C_6H_5SCH=CH_2	8	CH_2Cl_2, rt	$C_6H_5SC_2H_5$ (85)	71
cis-C_6H_5CH=CHBr	9	Dioxane, CH_3CO_2H, 15°	$C_6H_5CH_2CH_2Br$ (22)	97

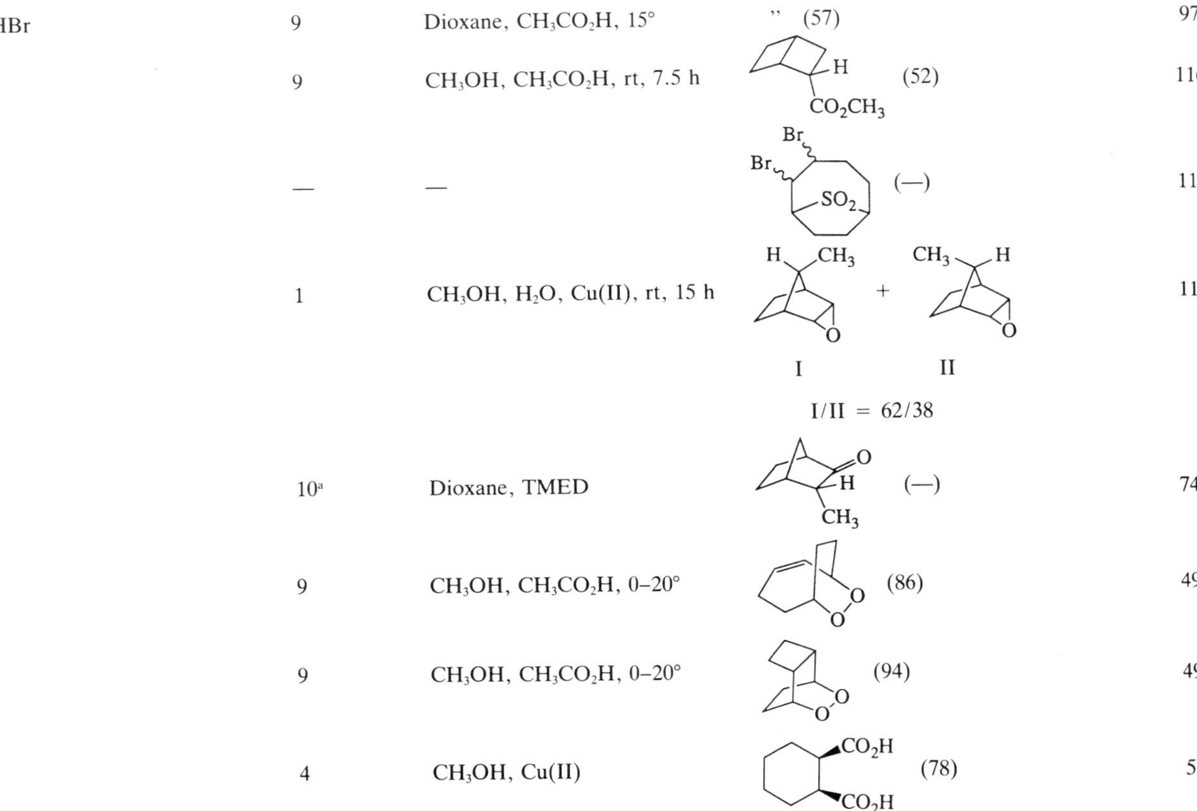

trans-$C_6H_5CH{=}CHBr$	9	Dioxane, CH_3CO_2H, 15°	" (57)	97
	9	CH_3OH, CH_3CO_2H, rt, 7.5 h	(52)	116
	—	—	(—)	117
	1	CH_3OH, H_2O, Cu(II), rt, 15 h	I + II; I/II = 62/38	118
	10[a]	Dioxane, TMED	(—)	74
	9	CH_3OH, CH_3CO_2H, 0–20°	(86)	49
	9	CH_3OH, CH_3CO_2H, 0–20°	(94)	49
	4	CH_3OH, Cu(II)	(78)	5
	13	C_2H_5OH, reflux	" (45)	77

TABLE I. Diimide Reductions of Substituted Alkenes (*Continued*)

Reactant	Procedure	Conditions	Product(s) and Yield(s) (%)	Refs.
$(CH_2{=}CHCH_2)_2CHCO_2Na$	1 (N_2D_2)	C_2H_5OD, D_2O	$(CH_2DCHDCH_2)_2CHCO_2Na$ (—)	119
1,3-Cyclooctadiene	9	CH_3OH, CH_3CO_2H, 25°	Cyclooctane (—)	35
1,4-Cyclooctadiene	9	CH_3OH, CH_3CO_2H, 25°	" (—)	35
1,5-Cyclooctadiene	9 (N_2D_2)	$(C_2H_5)_2O$, CH_3CO_2D	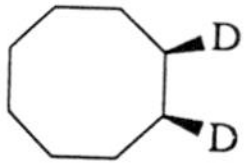 D, D (—)	98
	10	Diglyme, 80°	 (—)	32
$=CH_2$	10	Diglyme, 80°	H, CH_3 (—)	32
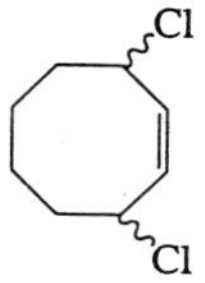 Cl, Cl	4	CH_3OH, reflux, 1 h	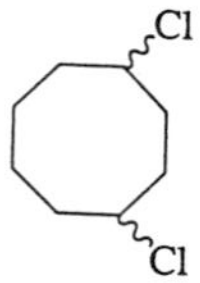 Cl, Cl (92)	42
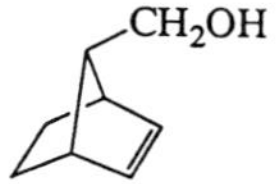 CH_2OH	9 (N_2D_2)	CH_3OD, CH_3CO_2D	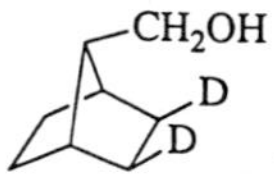 CH_2OH, D, D (79)	96
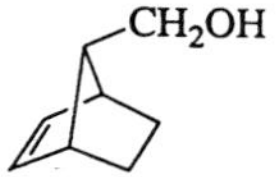 CH_2OH	9 (N_2D_2)	CH_3OD, CH_3CO_2D	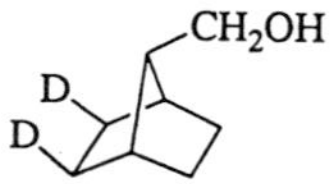 CH_2OH, D, D (—)	96
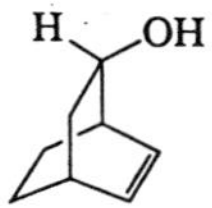 H, OH	9 (N_2D_2)	CH_3OD, CH_3CO_2D	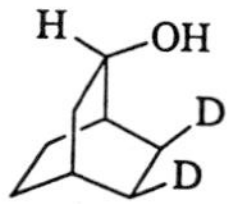 H, OH, D, D (—)	96

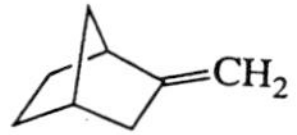

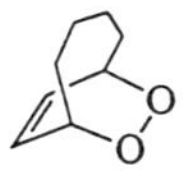 O, O	9	CH_3OH, CH_3CO_2H, 0°	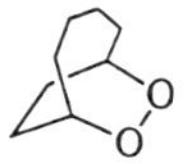 O, O (—)	50, 105
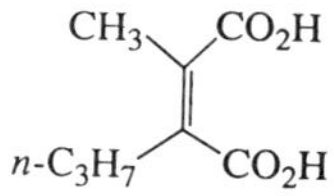 CH_3, CO_2H, n-C_3H_7, CO_2H	—	—	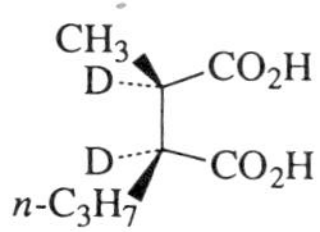 CH_3, D, CO_2H, D, CO_2H, n-C_3H_7 (trace)	120
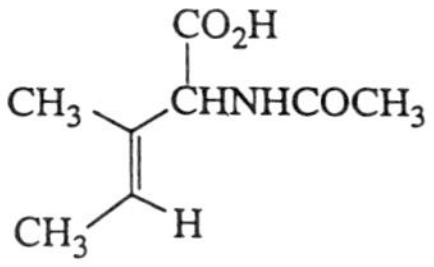 CO_2H, CH_3, $CHNHCOCH_3$, CH_3, H	9 (N_2T_2)	CH_3CO_2T, C_5H_5N, rt, 5 h	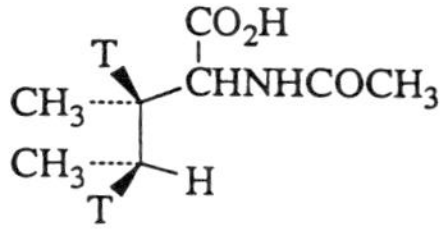 CO_2H, T, $CHNHCOCH_3$, CH_3, CH_3, H, T (—)	121
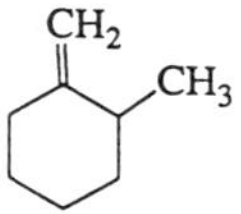 CH_2, CH_3	10	Diglyme, TEA, 80°	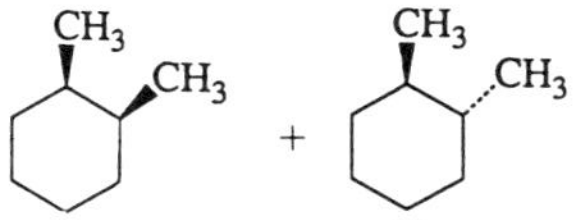 CH_3, CH_3 (I) + CH_3, CH_3 (II) I/II = 61/39	40
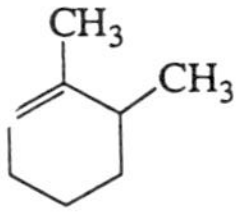 CH_3, CH_3	1	C_2H_5OH, 55°	" I/II = 63/37	37
	10	Diglyme, TEA, 80°	" I/II = 29/71	40
	1	C_2H_5OH, 55°	" I/II = 24/76	37
1,4-Dimethylcyclohexene	10	Diglyme, TEA, 80°	*cis* + *trans*-1,4-Dimethylcyclohexane *cis*/*trans* = 45/55	40
cis-Cyclooctene	9	CH_3OH, CH_3CO_2H, 25°	Cyclooctane (—)	35
	10	Diglyme, 80°	" (—)	32
trans-Cyclooctene	10	CH_3OH, CH_3CO_2H, 25°	" (—)	35
1-Methylcycloheptene	10	Diglyme, 80°	Methylcycloheptane (—)	32
1,2-Dimethylcyclohexene	10	Diglyme, 80°	*cis*-1,2-Dimethylcyclohexane (—)	32
	3	C_2H_5OH, 0°, 24 h	" (—)	68
n-$C_6H_{13}CH{=}CH_2$	1	C_2H_5OH, n-$C_3H_7CO_2H$	n-C_8H_{18} (—)	102

TABLE I. Diimide Reductions of Substituted Alkenes (*Continued*)

	Reactant	Procedure	Conditions	Product(s) and Yield(s) (%)	Refs.
	$n\text{-}C_5H_{11}CH{=}CHCH_3$	1	C_2H_5OH, $n\text{-}C_3H_7CO_2H$	" (—)	102
	$(CH_3)_2C{=}CHCH{=}C(CH_3)_2$	10	Diglyme, 80°	$i\text{-}C_4H_9C_4H_9\text{-}i$ (—)	34
	$(CH_3)_2C{=}CHCH_2CH_2COCH_3$	10[a]	—	$i\text{-}C_3H_7(CH_2)_3COCH_3$ (—)	74
	cyclopropyl—$C(CH_3){=}CHCH_2CH_2OH$	1	13.5 h, rt	cyclopropyl—$CH(CH_3)(CH_2)_3OH$ (78)	114
C_9	$4\text{-}BrC_6H_4CH{=}CHCO_2H$	1	3 h, 50°	$4\text{-}BrC_6H_4CH_2CH_2CO_2H$ (60)	122
	$4\text{-}ClC_6H_4CH{=}CHCO_2H$	6	—	$4\text{-}ClC_6H_4CH_2CH_2CO_2H$ (—)	36
	$2\text{-}ClC_6H_4CH{=}CHCO_2H$	6	—	$2\text{-}ClC_6H_4CH_2CH_2CO_2H$ (—)	36
	$4\text{-}O_2NC_6H_4CH{=}CHCO_2H$	6	—	$4\text{-}O_2NC_6H_4CH_2CH_2CO_2H$ (—)	36
		13	NaOH, 2 h, 50°	" (80)	99
	$cis\text{-}C_6H_5CH{=}CHCO_2H$	6	—	$C_6H_5CH_2CH_2CO_2H$ (—)	36
	$trans\text{-}C_6H_5CH{=}CHCO_2H$	1	CH_3OH, $n\text{-}C_9H_{19}CO_2H$	" (—)	101
		5	C_6H_6, CH_3CO_2H, Cu(II)	" (—)	70
		6	—	" (—)	36
		12	DMF	" (67)	76
		13	C_2H_5OH, reflux	" (81)	77
		14	—	" (70)	123
	$2\text{-}HOC_6H_4CH{=}CHCO_2H$	1	CH_3OH, $n\text{-}C_{10}H_{21}CO_2H$	$2\text{-}HOC_6H_4CH_2CH_2CO_2H$ (—)	101
	[structure]	6	MET, ETA	[structure] (—)	124
	$4\text{-}CH_3C_6H_4CH{=}CH_2$	9	CH_3OH, CH_3CO_2H, 0°	$4\text{-}CH_3C_6H_4C_2H_5$ (98)	117
	$3\text{-}CH_3C_6H_4CH{=}CH_2$	9	CH_3OH, CH_3CO_2H, 0°	$3\text{-}CH_3C_6H_4C_2H_5$ (98)	117
	$2\text{-}CH_3C_6H_4CH{=}CH_2$	9	CH_3OH, CH_3CO_2H, 0°	$2\text{-}CH_3C_6H_4C_2H_5$ (99)	117
	$C_6H_5C(CH_3){=}CH_2$	9	CH_3OH, CH_3CO_2H, 0°	$C_6H_5C_3H_7\text{-}i$ (99)	117
	$trans\text{-}C_6H_5CH{=}CHCH_3$	9	CH_3OH, CH_3CO_2H, 0°	$C_6H_5C_3H_7\text{-}n$ (99)	117
	$4\text{-}CH_3OC_6H_4CH{=}CH_2$	9	CH_3OH, CH_3CO_2H, 0°	$4\text{-}CH_3OC_6H_4C_2H_5$ (99)	117
	$3\text{-}CH_3OC_6H_4CH{=}CH_2$	9	CH_3OH, CH_3CO_2H, 0°	$3\text{-}CH_3OC_6H_4C_2H_5$ (98)	117
	$C_6H_5CH{=}CHCH_2OH$	12	CH_3OH, RCO_2H[b]	$C_6H_5(CH_2)_3OH$ (69)	76
	$C_6H_5OCH_2CH{=}CH_2$	9	CH_3OH, RCO_2H[b]	$C_6H_5OC_3H_7\text{-}n$ (87)	63

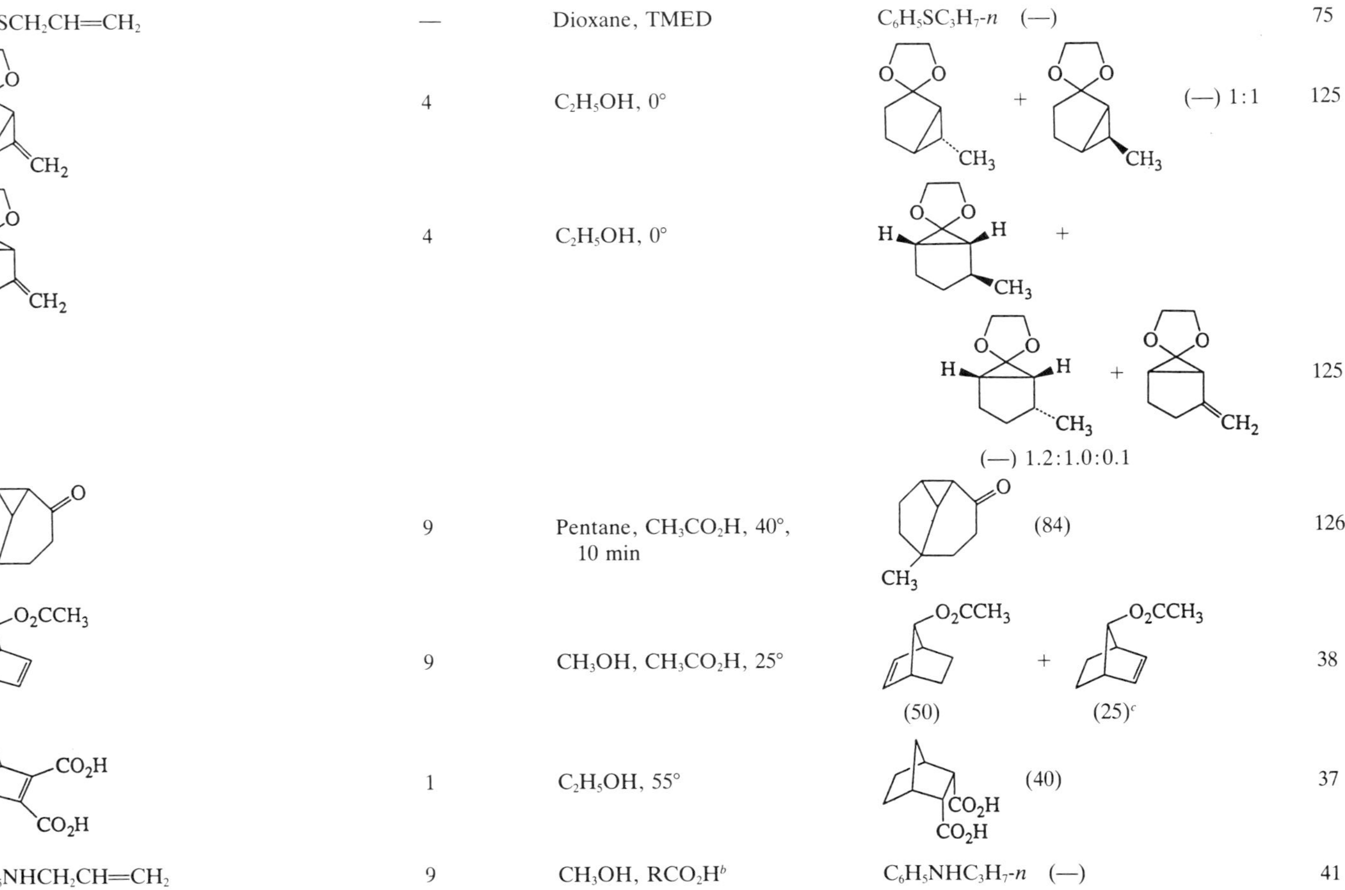

$C_6H_5SCH_2CH{=}CH_2$	—	Dioxane, TMED	$C_6H_5SC_3H_7$-*n* (—)	75
	4	C_2H_5OH, 0°	+ (—) 1:1	125
	4	C_2H_5OH, 0°	+ + (—) 1.2:1.0:0.1	125
	9	Pentane, CH_3CO_2H, 40°, 10 min	(84)	126
	9	CH_3OH, CH_3CO_2H, 25°	(50) + (25)[c]	38
	1	C_2H_5OH, 55°	(40)	37
$C_6H_5NHCH_2CH{=}CH_2$	9	CH_3OH, RCO_2H[b]	$C_6H_5NHC_3H_7$-*n* (—)	41

TABLE I. DIIMIDE REDUCTIONS OF SUBSTITUTED ALKENES (*Continued*)

Reactant	Procedure	Conditions	Product(s) and Yield(s) (%)	Refs.
$COCH_3$	10	Dioxane, TMED[a]	$COCH_3$ (—)	74
O, O	9 (N_2D_2)	CH_3OD, CH_3CO_2D, rt	O, O, H, D, D, H (92)	127
7,7-Dimethylnorbornene	9	CH_3OH, CH_3CO_2H, rt 20 min	7,7-Dimethylnorbornane (10)	115
Cyclononene	10	Diglyme, 80°	Cyclononane (—)	32
1-Methylcyclooctene	10	Diglyme, 80°	Methylcyclooctane (—)	32
1,4,4-Trimethylcyclohexene	10	Diglyme, 80°	1,4,4-Trimethylcyclohexane (—)	32
1-*tert*-Butylcyclopentene	10	Diglyme, 80°	*tert*-Butylcyclopentane (—)	32
C_{10}				
O	9 (N_2D_2)	C_2H_5OD, CH_3CO_2D, THF	O, D, D (96)	128
	10	CH_3OH, ETA, reflux, 2 h	(45)	72
	1	CH_3OH, Cu(II), rt, 12 h	(97)	129
$C_6H_5CH{=}CHCO_2CH_3$	9	CH_3OH, Dioxane, 23°, 14 h	$C_6H_5CH_2CH_2CO_2CH_3$ (96)	97
$C_6H_5CH{=}CHO_2CCH_3$	9	CH_3OH, Dioxane, 23°, 18 h	$C_6H_5CH_2CH_2O_2CCH_3$ (50)	97
$C_6H_5C(CH_3){=}CHCO_2H$	6	—	$C_6H_5CH(CH_3)CH_2CO_2H$ (—)	36

$C_6H_5CH{=}C(CH_3)CO_2H$	6	—	$C_6H_5CH_2CH(CH_3)CO_2H$ (—)	36
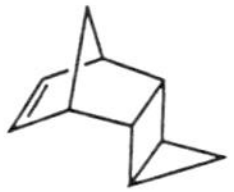			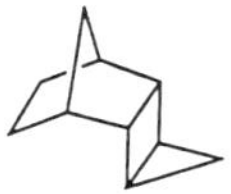 (—)	130
	9	CH_3OH, CH_3CO_2H	 (40)	131
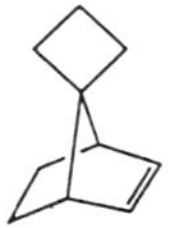	—	—	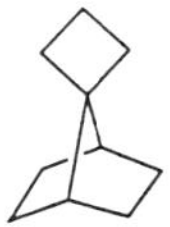 (13)	132
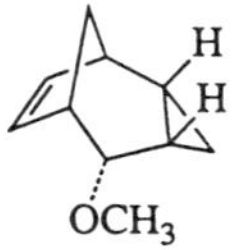 OCH_3	1	C_2H_5OH, Cu(II)	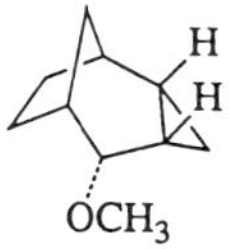 OCH_3 (77)	133
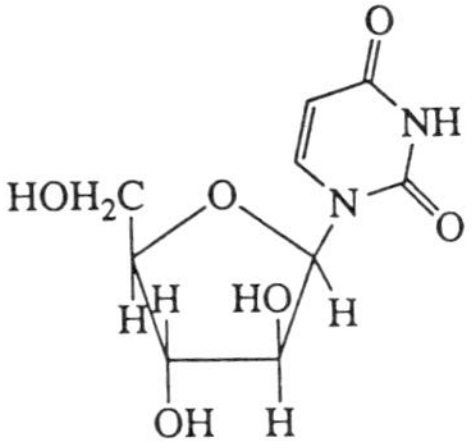	1	H_2O, Cu, pH 8.0, 21°	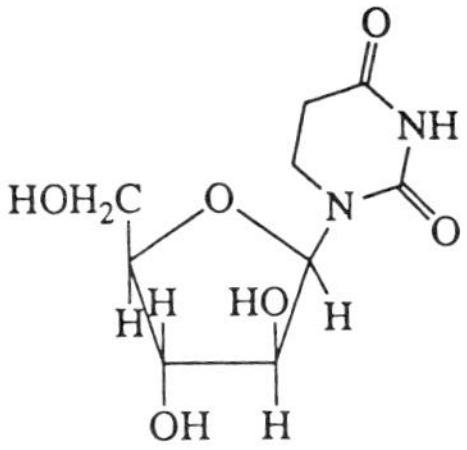 (—)	134
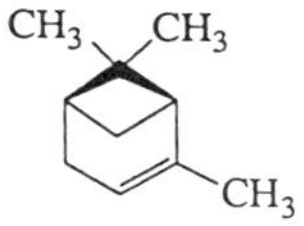	1	C_2H_5OH, 55°	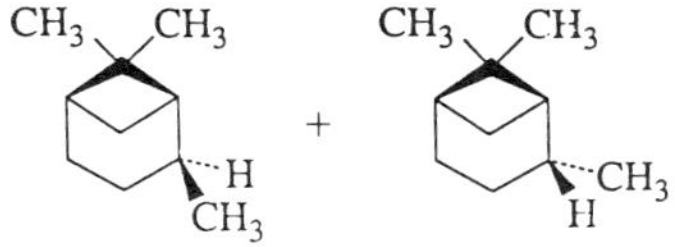 I + II I/II = 99/1 (—)	37

TABLE I. DIIMIDE REDUCTIONS OF SUBSTITUTED ALKENES (*Continued*)

Reactant	Procedure	Conditions	Product(s) and Yield(s) (%)	Refs.
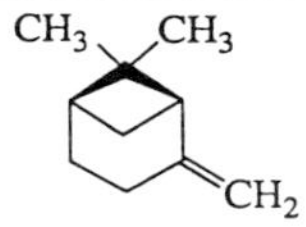 CH_3 CH_3 CH_2	1	C_2H_5OH, 55°	" I + II I/II = 96/4 (—)	37
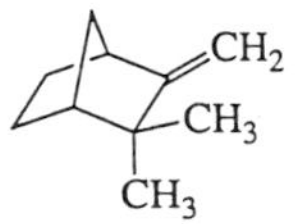 CH_2 CH_3 CH_3	1	C_2H_5OH, 55°	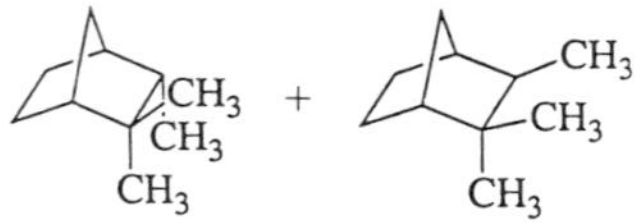 CH_3 CH_3 CH_3 I + CH_3 CH_3 CH_3 II (—) I/II = 92/8	37
	5	C_2H_5OH, CH_3CO_2H, Cu(II)	" I/II = 88/12 (—)	89
	10	Diglyme, ETA, 80°	" I/II = 96/4 (—)	40
	18	CH_2Cl_2, rt, 24 h	" I/II = 92/8 (—)	89
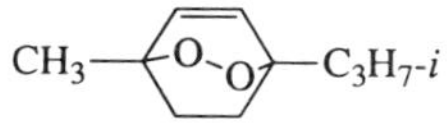 CH_3—O-O—C_3H_7-*i*	9	CH_3OH, CH_3CO_2H, 0°, 3 h	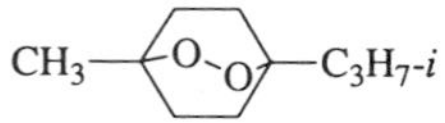 CH_3—O-O—C_3H_7-*i* (—)	50
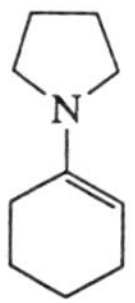 N	9	CH_3OH, RCO_2H[b]	— (0)	41
cis-Cyclodecene	10	Diglyme, 80°	Cyclodecane (—)	32
1-*tert*-Butylcyclohexene	10	Diglyme, 80°	1-*tert*-Butylcyclohexane (—)	32
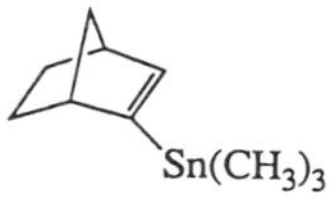 $Sn(CH_3)_3$	4	C_2H_5OH, Cu(II), 0°	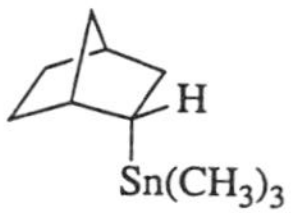 H $Sn(CH_3)_3$ (74)	135

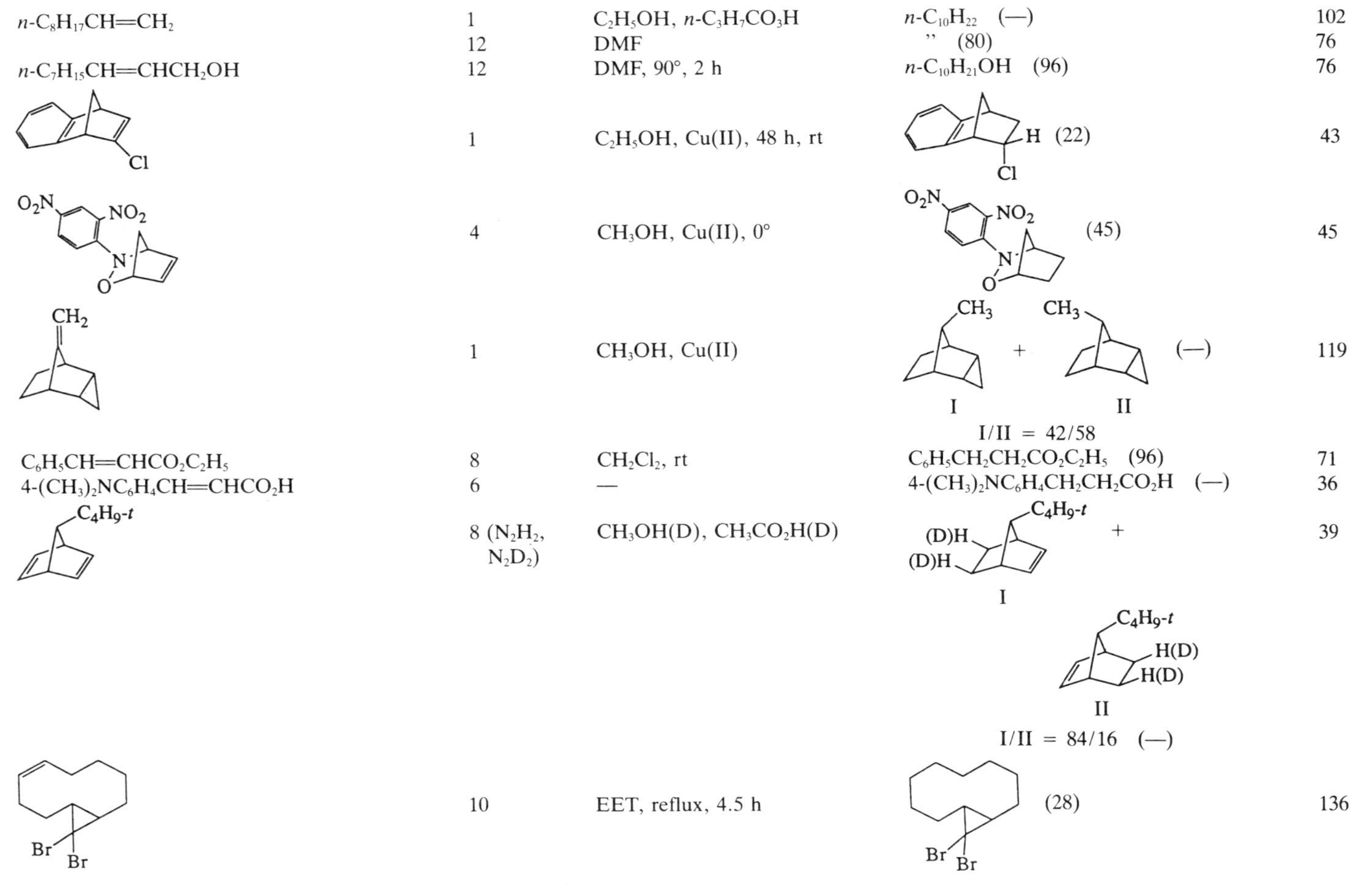

n-$C_8H_{17}CH{=}CH_2$	1	C_2H_5OH, n-$C_3H_7CO_3H$	n-$C_{10}H_{22}$ (—)	102
	12	DMF	" (80)	76
n-$C_7H_{15}CH{=}CHCH_2OH$	12	DMF, 90°, 2 h	n-$C_{10}H_{21}OH$ (96)	76
C_{11}				
Cl	1	C_2H_5OH, Cu(II), 48 h, rt	H, Cl (22)	43
O_2N, NO_2, N, O	4	CH_3OH, Cu(II), 0°	O_2N, NO_2, N, O (45)	45
CH_2	1	CH_3OH, Cu(II)	CH_3 (I) + CH_3 (II) (—) I/II = 42/58	119
$C_6H_5CH{=}CHCO_2C_2H_5$	8	CH_2Cl_2, rt	$C_6H_5CH_2CH_2CO_2C_2H_5$ (96)	71
4-$(CH_3)_2NC_6H_4CH{=}CHCO_2H$	6	—	4-$(CH_3)_2NC_6H_4CH_2CH_2CO_2H$ (—)	36
C_4H_9-t	8 (N_2H_2, N_2D_2)	$CH_3OH(D)$, $CH_3CO_2H(D)$	(D)H, (D)H, C_4H_9-t (I) + C_4H_9-t, H(D), H(D) (II) I/II = 84/16 (—)	39
Br, Br	10	EET, reflux, 4.5 h	Br, Br (28)	136

TABLE I. DIIMIDE REDUCTIONS OF SUBSTITUTED ALKENES (Continued)

Reactant	Procedure	Conditions	Product(s) and Yield(s) (%)	Refs.
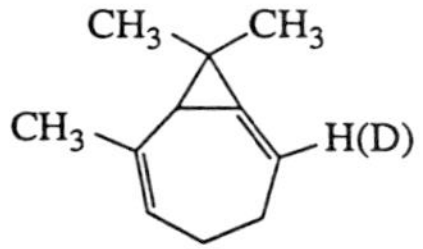 [structure: 3,7,7-trimethylbicyclo[5.1.0]octa-2,?-diene, H(D)]	10	—	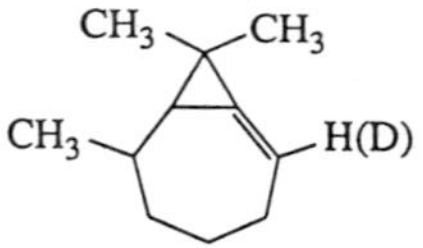 [structure: CH$_3$, CH$_3$, CH$_3$, H(D) bicyclic reduced product] (—)	137
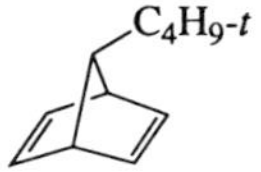 [structure: 7-C_4H_9-t norbornadiene]	9 (N_2H_2, N_2D_2)	$CH_3OH(D)$, $CH_3CO_2H(D)$	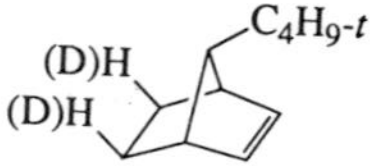 [structure: (D)H, (D)H, C_4H_9-t norbornene] (63)[c]	38
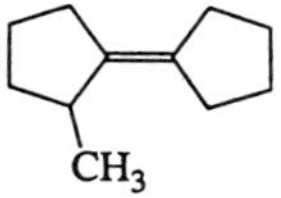 [structure: 2-methyl-1-cyclopentylidenecyclopentane, CH_3]	10	Diglyme, ETA, 80°	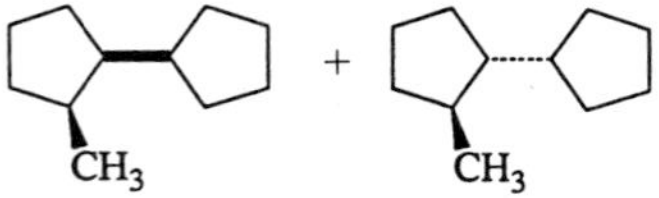 [structure: cis CH_3] + [structure: trans CH_3] (*cis*/*trans* = 28/72) (—)	40
4-*tert*-Butyl-1-methylene-cyclohexane	1	C_2H_5OH	*cis* + *trans*-1-*tert*-Butyl-4-methylcyclohexane (—) *cis*/*trans* = 49/51	40
	1	C_2H_5OH, 55°	" (—) *cis*/*trans* = 51/49	34
	5	C_2H_5OH, CH_3CO_2H, Cu(II)	" (—) *cis*/*trans* = 51/49	89
	9	CH_3OH, CH_3CO_2H, 25°	" (—)	34
	9	CH_3OH, CH_3CO_2H	" (—) *cis*/*trans* = 50/50	89
	10	Diglyme, ETA, 80°	" (—) *cis*/*trans* = 49/51	40
	18	CH_2Cl_2, rt, 24 h	" (—) *cis*/*trans* = 48/52	89
1-Methyl-4-*tert*-butyl-cyclohexene	10	Diglyme, ETA, 80°	" (—) *cis*/*trans* = 30/70	40
1-*tert*-Butyl-4-methyl-cyclohexene	9	CH_3OH, CH_3CO_2H, 25°	" (—) *cis*/*trans* = 49/51	40
	10	Diglyme, 80°	" (—)	32
1-*tert*-Butylcycloheptene	10	Diglyme, 80°	*tert*-Butylcycloheptane (—)	32
n-$C_8H_{17}CH{=}CHCO_2H$	6	—	*n*-$C_{10}H_{21}CO_2H$ (—)	36

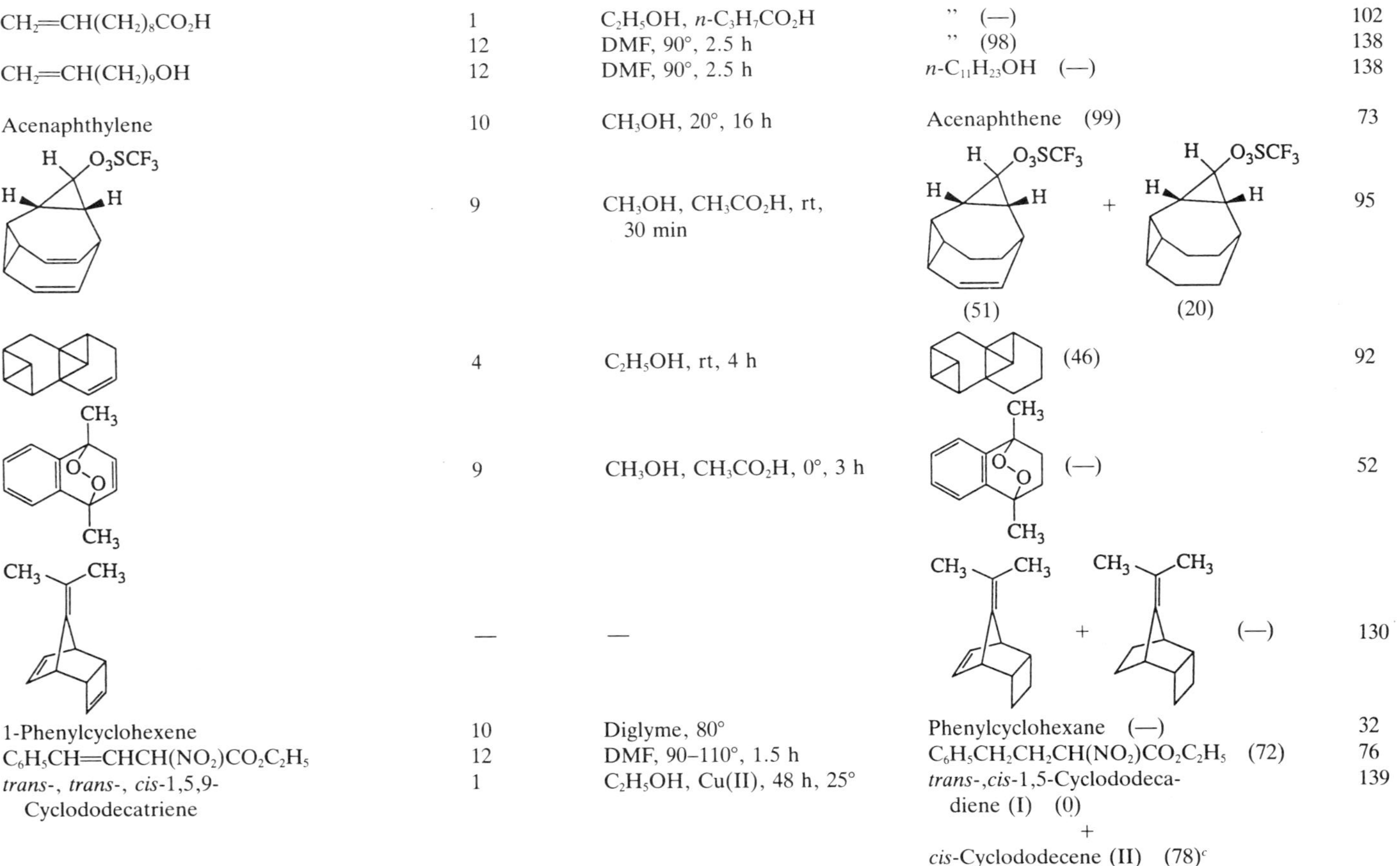

	Substrate	Catalyst (%)	Conditions	Product(s) (%)	Refs.
	$CH_2{=}CH(CH_2)_8CO_2H$	1	C_2H_5OH, n-$C_3H_7CO_2H$	" (—)	102
		12	DMF, 90°, 2.5 h	" (98)	138
	$CH_2{=}CH(CH_2)_9OH$	12	DMF, 90°, 2.5 h	n-$C_{11}H_{23}OH$ (—)	138
C_{12}	Acenaphthylene	10	CH_3OH, 20°, 16 h	Acenaphthene (99)	73
	[structure]	9	CH_3OH, CH_3CO_2H, rt, 30 min	[structure] (51) + [structure] (20)	95
	[structure]	4	C_2H_5OH, rt, 4 h	[structure] (46)	92
	[structure]	9	CH_3OH, CH_3CO_2H, 0°, 3 h	[structure] (—)	52
	[structure]	—	—	[structure] + [structure] (—)	130
	1-Phenylcyclohexene	10	Diglyme, 80°	Phenylcyclohexane (—)	32
	$C_6H_5CH{=}CHCH(NO_2)CO_2C_2H_5$	12	DMF, 90–110°, 1.5 h	$C_6H_5CH_2CH_2CH(NO_2)CO_2C_2H_5$ (72)	76
	trans-, *trans*-, *cis*-1,5,9-Cyclododecatriene	1	C_2H_5OH, Cu(II), 48 h, 25°	*trans*-,*cis*-1,5-Cyclododecadiene (I) (0) + *cis*-Cyclododecene (II) (78)[c]	139

TABLE I. DIIMIDE REDUCTIONS OF SUBSTITUTED ALKENES (*Continued*)

Reactant	Procedure	Conditions	Product(s) and Yield(s) (%)	Refs.
	1	C_2H_5OH, Cu(II), 35°–50°, 50 h	" I(0) + II(79)[c]	140
	4	C_2H_5OH, Cu(II), 0°, 1 h	" I(25) + II(44)[c]	140
	10	EET, 95–100°, 7 h	" I(0) + II(43)	140
	—	C_2H_5OH, NaOH, rt, 7 h	" I(29) + II(11)	140
*trans-,trans-,trans-*1,5,9-Cyclododecatriene	4	C_2H_5OH, Cu(II), 0°, 1 h	*trans-,trans-*1,5-Cyclododecadiene (I) (24) + *trans*-cyclododecene (II) (20)[c]	140
	10	EET, 95–100°, 7 h	" I(25) + II(11)[c]	140
*trans-,cis-,cis-*1,5,9 Cyclododecatriene	4	C_2H_5OH, Cu(II), 0°, 1 h	*trans-,cis-*1,5-Cyclododecadiene (I) (5) + *cis-,cis-*1,5-Cyclododecadiene (II) (41) + *cis*-Cyclododecene (III) (19)[c]	140
	10	EET, 95–100°, 7 h	" I(8) + II(31) + III(5)[c]	140
*cis-,cis-,cis-*1,5,9-Cyclododecatriene	10	EET, 95–100°, 7 h	" II(41) + III(11)[c]	140
4-*t*-$C_4H_9C_6H_4CH{=}CH_2$	10	CH_3OH, CH_3CO_2H, 0°	4-*t*-$C_4H_9C_6H_4C_2H_5$ (96)	117
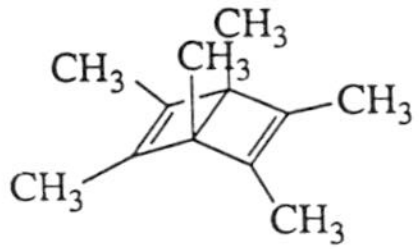	9 (N_2D_2)	CH_3OD, CH_3CO_2D[b]	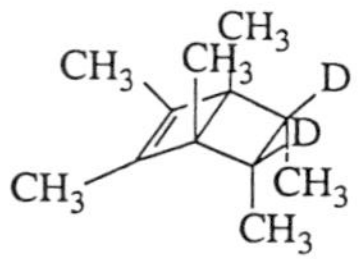 (95)	141
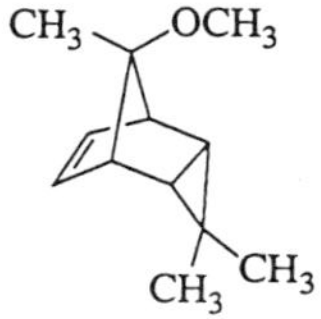	4	CH_3OH, Cu(II), rt, 24 h	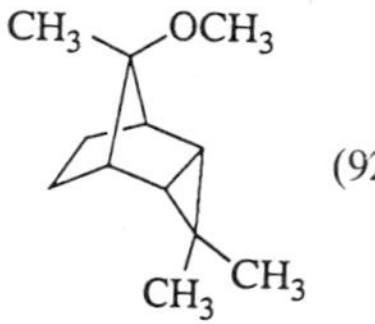 (92)	142

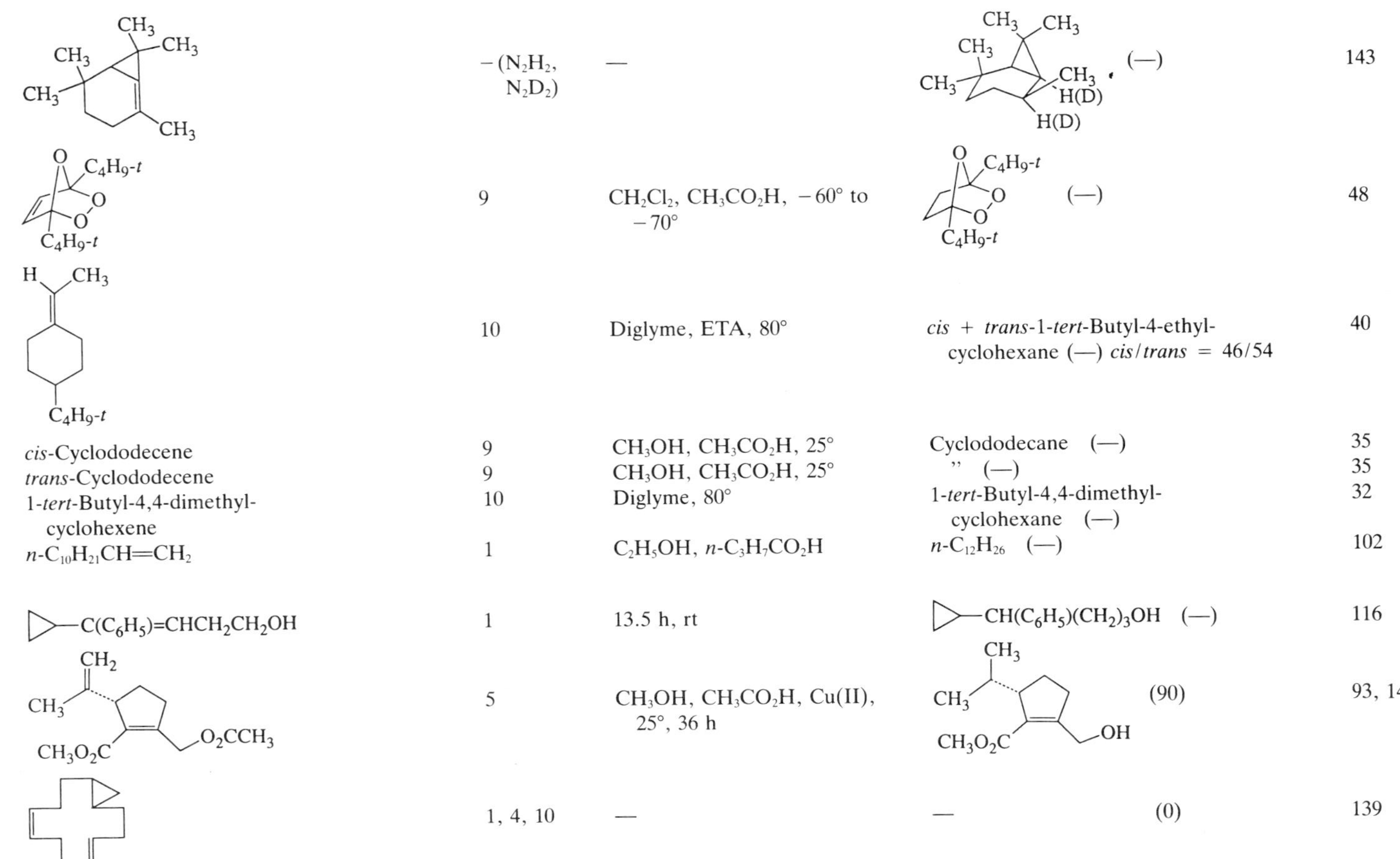

	Substrate	Method	Conditions	Product(s) (% Yield)	Refs.
		$-(N_2H_2, N_2D_2)$	—	(—)	143
		9	CH_2Cl_2, CH_3CO_2H, $-60°$ to $-70°$	(—)	48
		10	Diglyme, ETA, 80°	*cis* + *trans*-1-*tert*-Butyl-4-ethylcyclohexane (—) *cis*/*trans* = 46/54	40
	cis-Cyclododecene	9	CH_3OH, CH_3CO_2H, 25°	Cyclododecane (—)	35
	trans-Cyclododecene	9	CH_3OH, CH_3CO_2H, 25°	" (—)	35
	1-*tert*-Butyl-4,4-dimethylcyclohexene	10	Diglyme, 80°	1-*tert*-Butyl-4,4-dimethylcyclohexane (—)	32
	n-$C_{10}H_{21}CH{=}CH_2$	1	C_2H_5OH, n-$C_3H_7CO_2H$	n-$C_{12}H_{26}$ (—)	102
C_{13}					
	$\triangleright$—$C(C_6H_5){=}CHCH_2CH_2OH$	1	13.5 h, rt	$\triangleright$—$CH(C_6H_5)(CH_2)_3OH$ (—)	116
		5	CH_3OH, CH_3CO_2H, Cu(II), 25°, 36 h	(90)	93, 144
		1, 4, 10	—	— (0)	139

TABLE I. DIIMIDE REDUCTIONS OF SUBSTITUTED ALKENES (*Continued*)

Reactant	Procedure	Conditions	Product(s) and Yield(s) (%)	Refs.
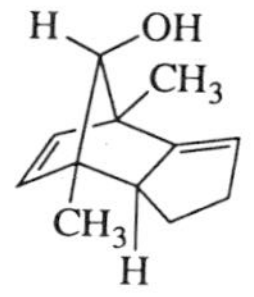 H, OH, CH_3, CH_3, H	—	—	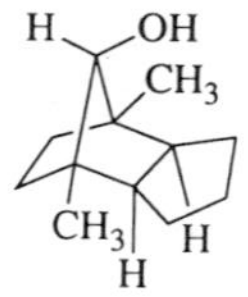 H, OH, CH_3, CH_3, H, H (—)	145
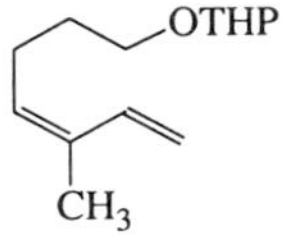 OTHP, CH_3	4	C_2H_5OH, Cu(II), 0°	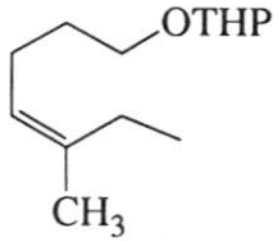 OTHP, CH_3 (—)	146
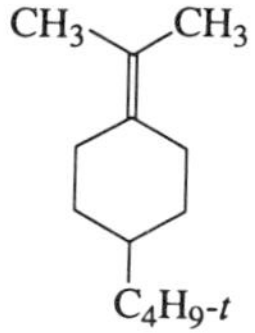 CH_3, CH_3, C_4H_9-*t*	10	Diglyme, ETA, 80°	*cis* + *trans*-1-*tert*-Butyl-4-isopropylcyclohexane (—) *cis*/*trans*-30/70	40
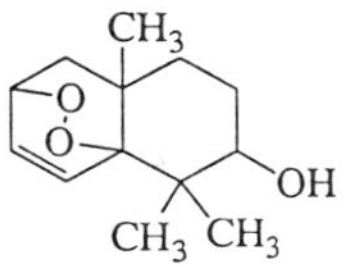 CH_3, O, O, OH, CH_3 CH_3	9	CH_2Cl_2, CH_3CO_2H, 25°	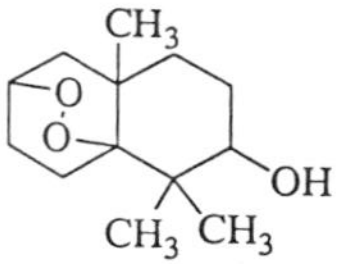 CH_3, O, O, OH, CH_3 CH_3 (84)	147
trans-$C_6H_5CH{=}CHC_6H_5$	1, 4	CH_3OH, Cu(II), 25°	$C_6H_5CH_2CH_2C_6H_5$ (88)	5
	10	CH_3OH, CH_3CO_2H, 0°	" (99)	117
	1,9 (N_2D_2)	CH_3OD, Cu(II), 25°	*dl*-$C_6H_5CHDCHDC_6H_5$ (—)	30
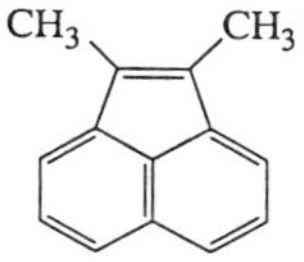 CH_3, CH_3	—	—	— (0)	148

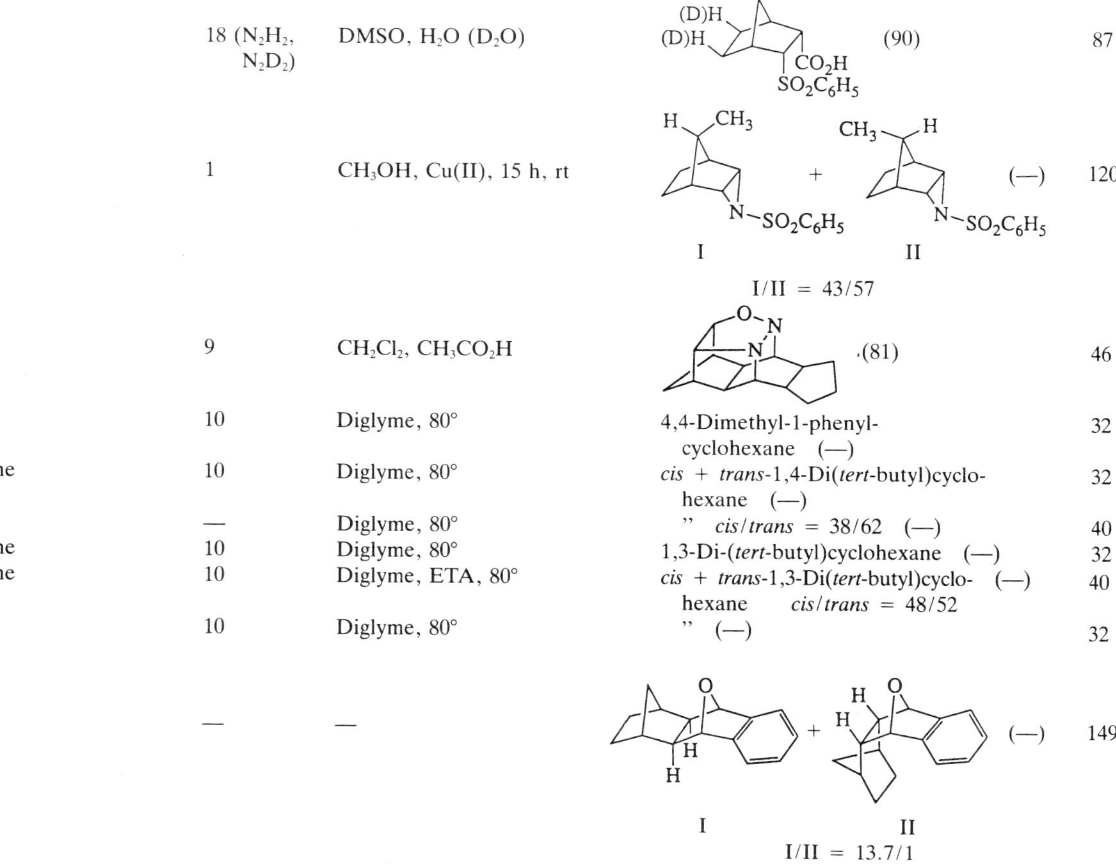

Reactant	Equiv.	Conditions	Product(s) (yield %)	Refs.
(structure)	18 (N_2H_2, N_2D_2)	DMSO, H_2O (D_2O)	(structure) (90)	87
(structure)	1	CH_3OH, Cu(II), 15 h, rt	I + II (—); I/II = 43/57	120
(structure)	9	CH_2Cl_2, CH_3CO_2H	(structure) (81)	46
4,4-Dimethyl-1-phenyl-cyclohexene	10	Diglyme, 80°	4,4-Dimethyl-1-phenyl-cyclohexane (—)	32
1,4-Di-(*tert*-butyl)cyclohexene	10	Diglyme, 80°	*cis* + *trans*-1,4-Di(*tert*-butyl)cyclo-hexane (—)	32
	—	Diglyme, 80°	" *cis*/*trans* = 38/62 (—)	40
3,5-Di-(*tert*-butyl)cyclohexene	10	Diglyme, 80°	1,3-Di-(*tert*-butyl)cyclohexane (—)	32
1,5-Di-(*tert*-butyl)cyclohexene	10	Diglyme, ETA, 80°	*cis* + *trans*-1,3-Di(*tert*-butyl)cyclo-hexane *cis*/*trans* = 48/52 (—)	40
	10	Diglyme, 80°	" (—)	32
C_{15}				
(structure)	—	—	I + II (—); I/II = 13.7/1	149

TABLE I. DIIMIDE REDUCTIONS OF SUBSTITUTED ALKENES (*Continued*)

Reactant	Procedure	Conditions	Product(s) and Yield(s) (%)	Refs.
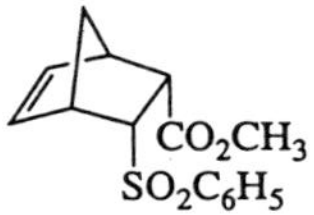 CO_2CH_3, $SO_2C_6H_5$	18 (N_2D_2)	$CDCl_3$, D_2O, rt	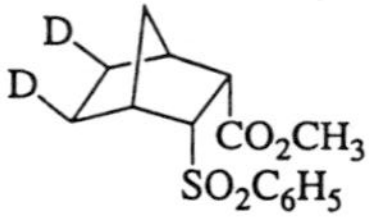 D, D, CO_2CH_3, $SO_2C_6H_5$ (83)	89
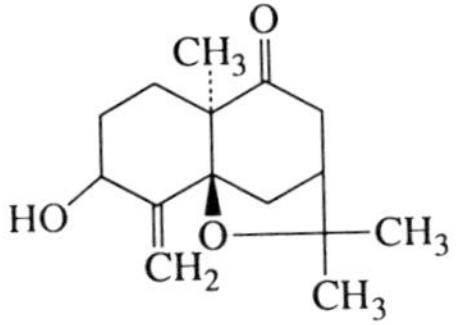 CH_3, O, HO, CH_2, O, CH_3, CH_3	1	Cu(II), CH_3CO_2H, 6 h	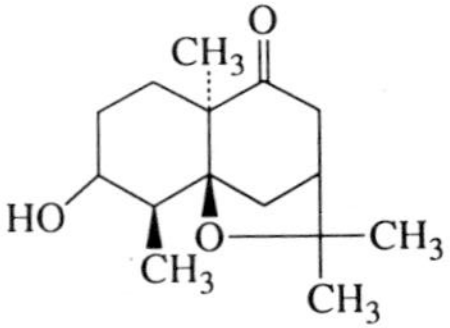 CH_3, O, HO, CH_3, O, CH_3, CH_3 (90)	150
CH_3, O, O, O_2CCH_3, CH_3 CH_3	4	C_2H_5OH, rt, 16 h	" (85)	150
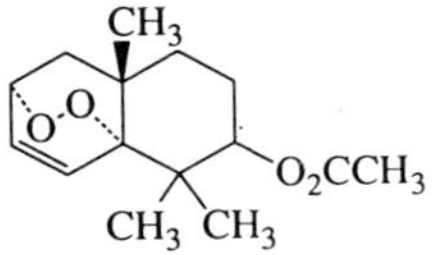	9	CH_2Cl_2, CH_3CO_2H, 25°	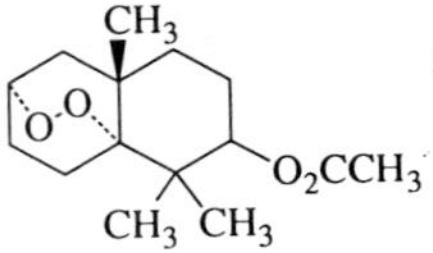 CH_3, O, O, O_2CCH_3, CH_3 CH_3 (80)	147
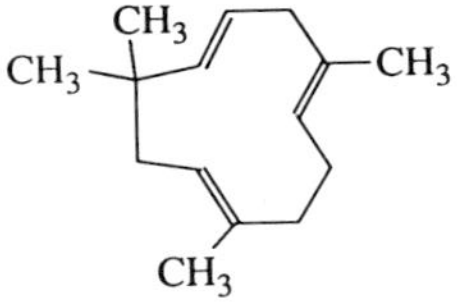 CH_3, CH_3, CH_3, CH_3	12	—	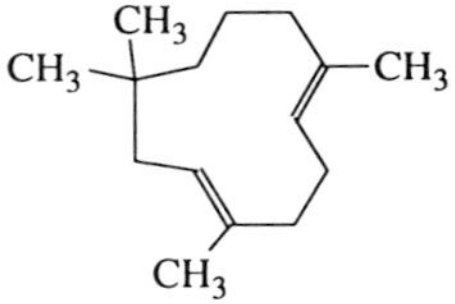 CH_3, CH_3, CH_3, CH_3 (—)	151
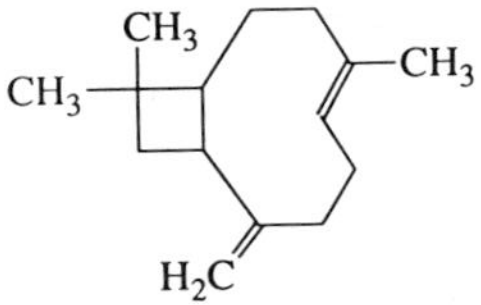 CH_3, CH_3, CH_3, H_2C	4	C_2H_5OH, Cu(II), rt	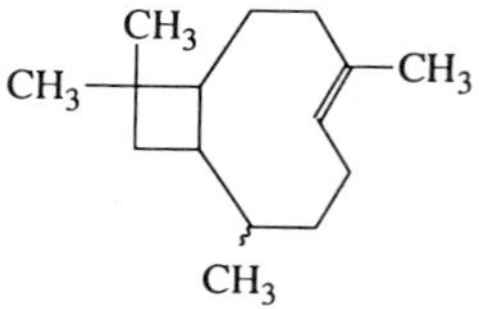 CH_3, CH_3, CH_3, CH_3 (87)	58

Reactant		Conditions	Product(s) and Yield(s) (%)	Refs.
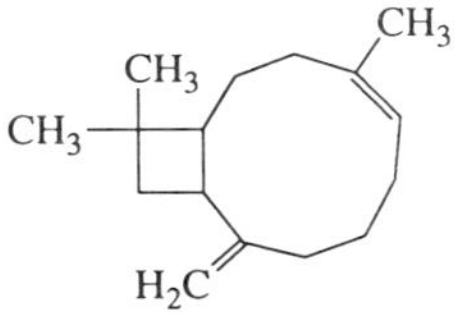 [structure]	4	C_2H_5OH, Cu(II), rt	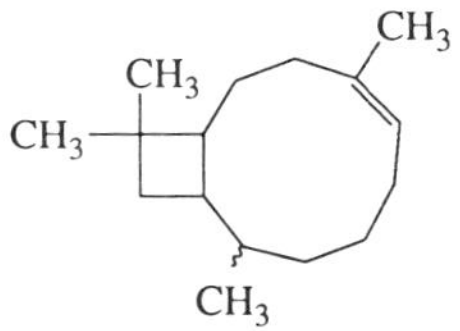 [structure] (98)	58
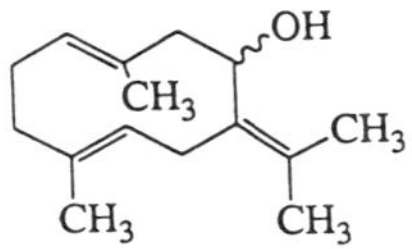 [structure]	12	DMF, 60–70°	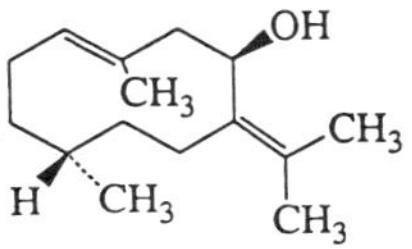 [structure] (I) 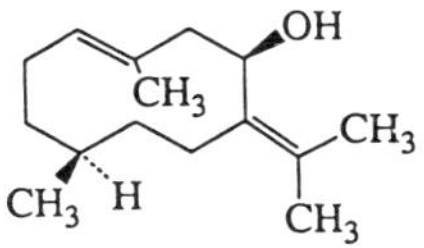 + [structure] (II) I/II = 60/40 (—)	151
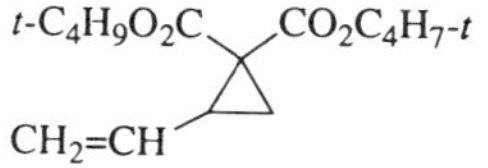 $t\text{-}C_4H_9O_2C$, $CO_2C_4H_7\text{-}t$, $CH_2{=}CH$ (cyclopropane)	9	CH_3OH, CH_3CO_2H	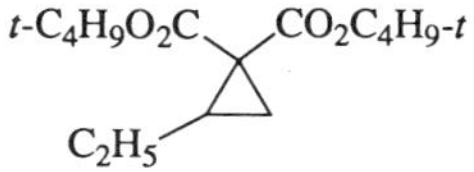 $t\text{-}C_4H_9O_2C$, $CO_2C_4H_9\text{-}t$, C_2H_5 (cyclopropane) (100)	152
$(CH_3)_2C{=}CHCH_2C(CH_3){=}CH$ / $HOCH_2CH{=}C(CH_3)CH_2CH_2$ (2-*trans*-6-*trans*-)	1	C_2H_5OH, 80°, 5 h	*trans*- $(CH_3)_2C{=}CHCH_2C(CH_3){=}CH$ / $HOCH_2CH_2CH(CH_3)CH_2CH_2$ (—)	153
(2-*cis*-6-*trans*-)	1	C_2H_5OH, 80°, 5 h	" (—)	153
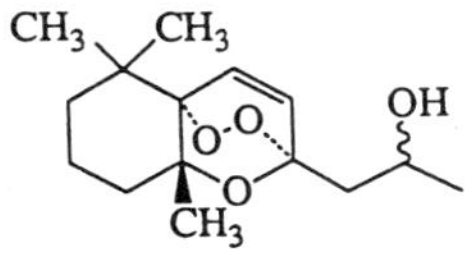 [structure]	9	CH_2Cl_2, CH_3CO_2H, 36 h	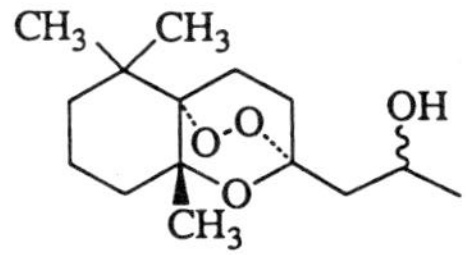 [structure] (99)	154
$s\text{-}C_4H_9CH{=}CHC_9H_{19}\text{-}n$	1	C_2H_5OH, Cu(II)	$s\text{-}C_4H_9C_{11}H_{23}\text{-}n$ (95)	155

TABLE I. DIIMIDE REDUCTIONS OF SUBSTITUTED ALKENES (*Continued*)

Reactant	Procedure	Conditions	Product(s) and Yield(s) (%)	Refs.
C_{16}				
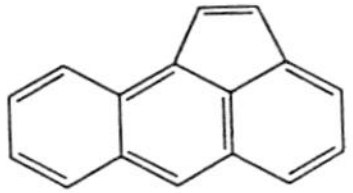	1	C_2H_5OH, Pd/C	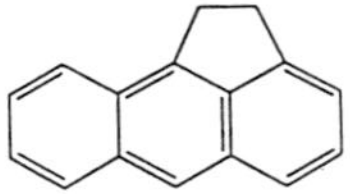 (90)	156
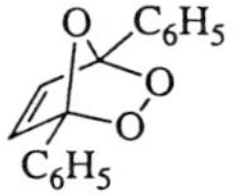 O, C_6H_5, O, O, C_6H_5	9	CH_2Cl_2, CH_3CO_2H, −60 to −70°	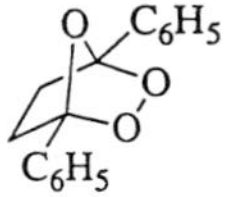 O, C_6H_5, O, O, C_6H_5 (—)	48
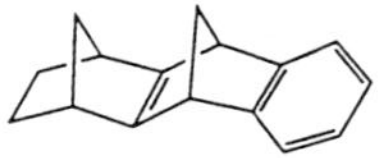	9	CH_3OH, CH_3CO_2H	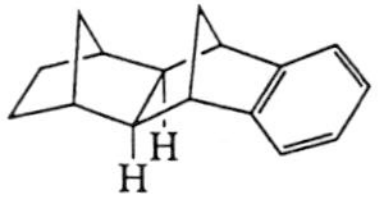 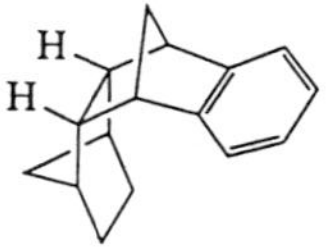 H, H (I) + H, H (II) I/II = 1/5 (—)	157
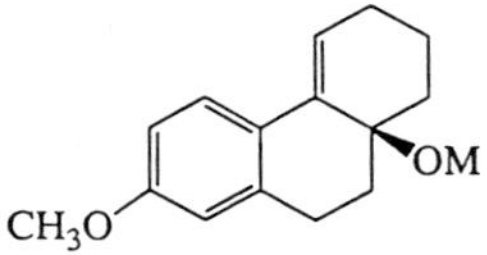 CH_3O, OM	10	$(n\text{-}C_4H_9)_2O$, diglyme, reflux	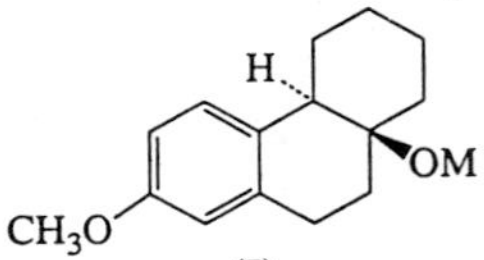 H, CH_3O, OM (I) 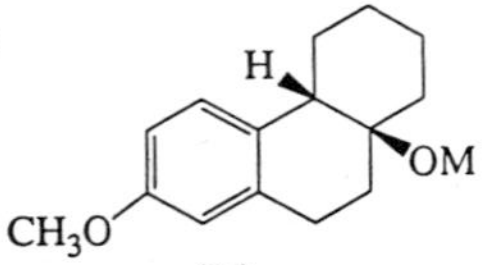 + H, CH_3O, OM (II)	158

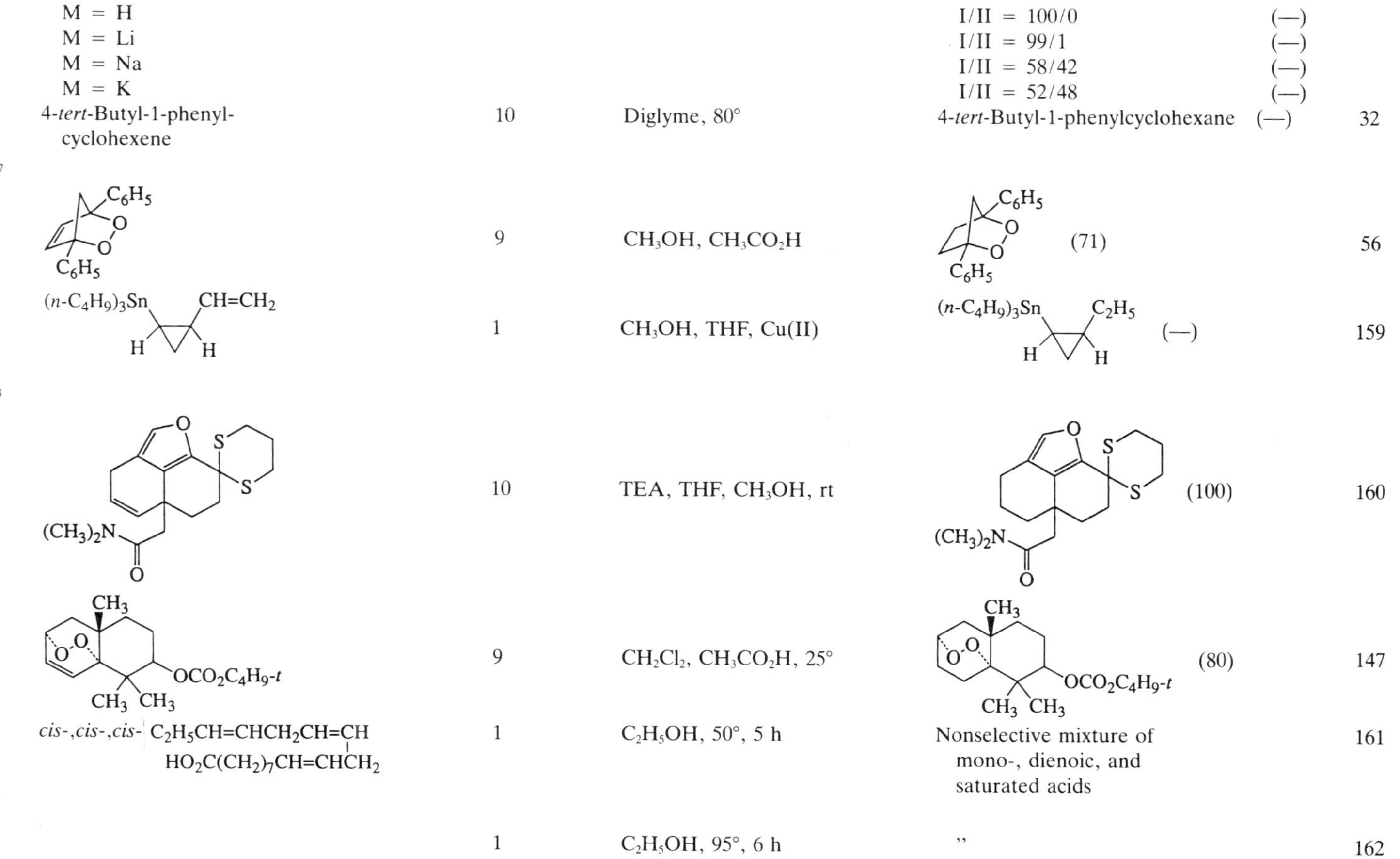

	M = H			I/II = 100/0 (—)	
	M = Li			I/II = 99/1 (—)	
	M = Na			I/II = 58/42 (—)	
	M = K			I/II = 52/48 (—)	
	4-*tert*-Butyl-1-phenyl-cyclohexene	10	Diglyme, 80°	4-*tert*-Butyl-1-phenylcyclohexane (—)	32
C_{17}					
	C_6H_5, O, O, C_6H_5	9	CH_3OH, CH_3CO_2H	C_6H_5, O, O, C_6H_5 (71)	56
	$(n\text{-}C_4H_9)_3Sn$, $CH{=}CH_2$, H, H	1	CH_3OH, THF, Cu(II)	$(n\text{-}C_4H_9)_3Sn$, C_2H_5, H, H (—)	159
C_{18}					
	O, S, S, $(CH_3)_2N$, O	10	TEA, THF, CH_3OH, rt	O, S, S, $(CH_3)_2N$, O (100)	160
	CH_3, O, O, $OCO_2C_4H_9$-*t*, CH_3 CH_3	9	CH_2Cl_2, CH_3CO_2H, 25°	CH_3, O, O, $OCO_2C_4H_9$-*t*, CH_3 CH_3 (80)	147
	cis-,*cis*-,*cis*- $C_2H_5CH{=}CHCH_2CH{=}CH$ / $HO_2C(CH_2)_7CH{=}CHCH_2$	1	C_2H_5OH, 50°, 5 h	Nonselective mixture of mono-, dienoic, and saturated acids	161
		1	C_2H_5OH, 95°, 6 h	"	162

TABLE I. DIIMIDE REDUCTIONS OF SUBSTITUTED ALKENES (*Continued*)

Reactant	Procedure	Conditions	Product(s) and Yield(s) (%)	Refs.
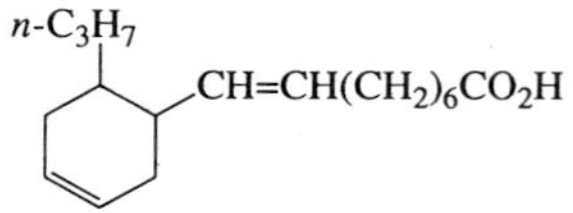 n-C_3H_7, $CH{=}CH(CH_2)_6CO_2H$	—	—	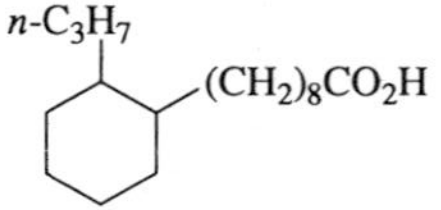 n-C_3H_7, $(CH_2)_8CO_2H$ (—)	163
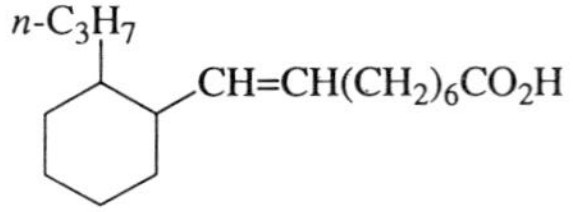 n-C_3H_7, $CH{=}CH(CH_2)_6CO_2H$	—	—	" (—)	163
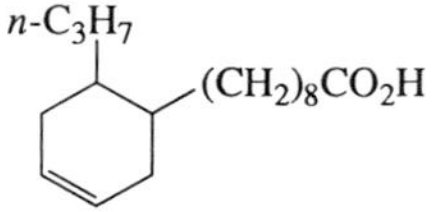 n-C_3H_7, $(CH_2)_8CO_2H$	—	—	" (—)	163
 n-C_7H_{15}, $(CH_2)_7CO_2H$	1	C_2H_5OH, 1 h, 35°	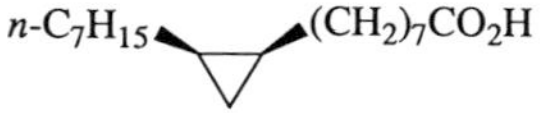 n-C_7H_{15}, $(CH_2)_7CO_2H$ (—)	164
Elaidic Acid	1	—	Stearic acid (—)	101
	1 (N_2D_2)	CH_3OD, 50°	Stearic acid-d_2 (—)	165
	9	CH_3OH, CH_3CO_2H, rt	Stearic acid (62)	7
	10	Diglyme, heat	" (70)	7
Linoleic Acid	12	DMF, 90–95°	Stearic acid (61) + Oleic acid (34)	138
Oleic Acid	1	—	Stearic acid (—)	101
	1 (N_2D_2)	Dioxane, D_2O, 52°	Stearic acid-d_2 (—)	66
	1 (N_2D_2)	CH_3OD, 50°	Stearic acid-d_2 (—)	165
	9	CH_3OH, CH_3CO_2H, rt	Stearic acid (51)	7
	10	Diglyme, heat	" (73)	7
	10	MET, 100°[a]	" (—)	103
	12	DMF, 90–95°, 1.5 h	" (85)	138
	13	C_2H_5OH, 85°, 1.5 h	" (—)	166
cis-6-Octadecenoic Acid	1	C_2H_5OH, n-$C_3H_7CO_2H$	" (—)	102
trans-11-Octadecenoic Acid	1	C_2H_5OH, n-$C_3H_7CO_2H$	" (—)	102

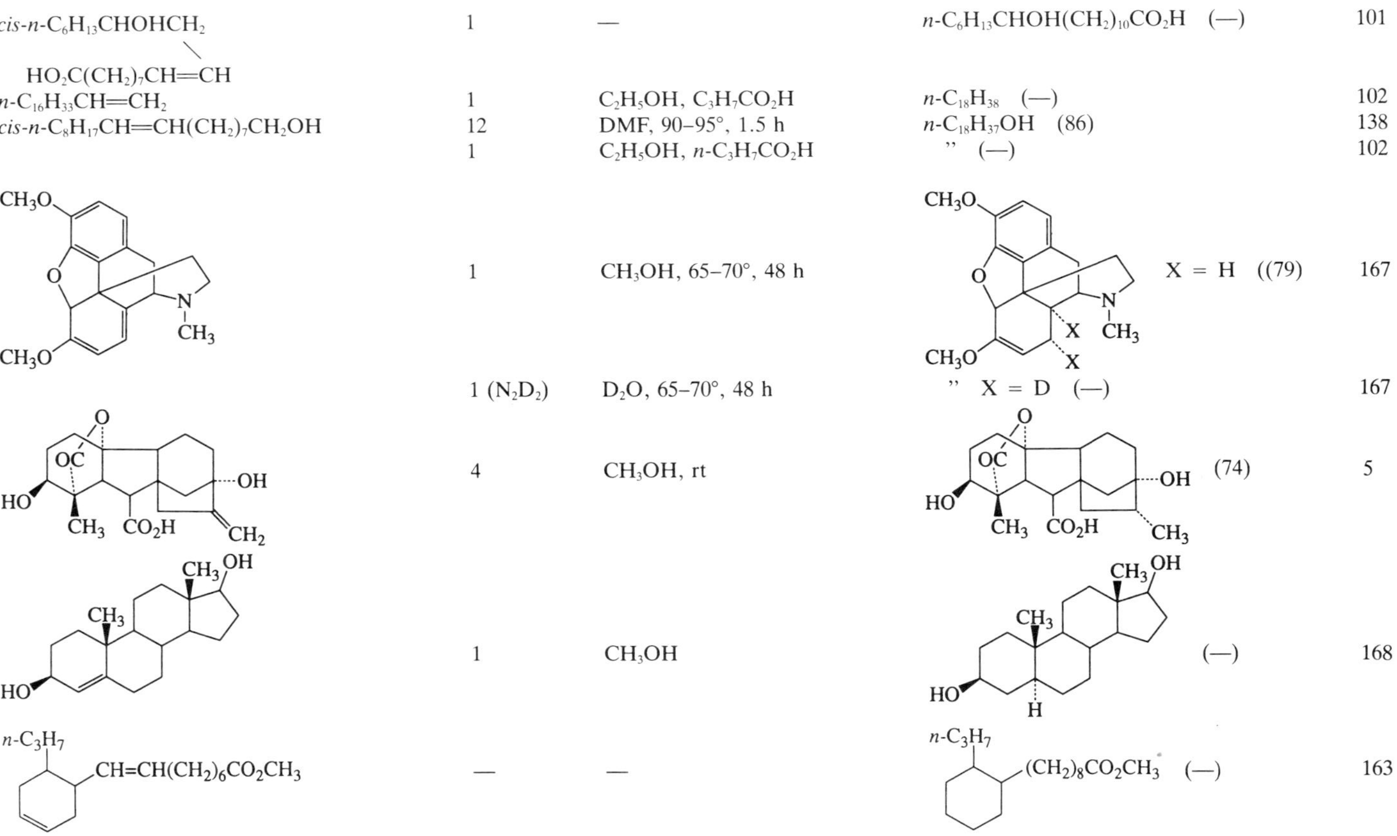

	cis-*n*-$C_6H_{13}CHOHCH_2$ \ $HO_2C(CH_2)_7CH{=}CH$	1	—	*n*-$C_6H_{13}CHOH(CH_2)_{10}CO_2H$ (—)	101
	n-$C_{16}H_{33}CH{=}CH_2$	1	C_2H_5OH, $C_3H_7CO_2H$	*n*-$C_{18}H_{38}$ (—)	102
	cis-*n*-$C_8H_{17}CH{=}CH(CH_2)_7CH_2OH$	12	DMF, 90–95°, 1.5 h	*n*-$C_{18}H_{37}OH$ (86)	138
		1	C_2H_5OH, *n*-$C_3H_7CO_2H$	" (—)	102
C_{19}		1	CH_3OH, 65–70°, 48 h	X = H ((79)	167
		1 (N_2D_2)	D_2O, 65–70°, 48 h	" X = D (—)	167
		4	CH_3OH, rt	(74)	5
		1	CH_3OH	(—)	168
	n-C_3H_7 … $CH{=}CH(CH_2)_6CO_2CH_3$	—	—	*n*-C_3H_7 … $(CH_2)_8CO_2CH_3$ (—)	163

TABLE I. DIIMIDE REDUCTIONS OF SUBSTITUTED ALKENES (*Continued*)

Reactant	Procedure	Conditions	Product(s) and Yield(s) (%)	Refs.
2-n-C_3H_7-cyclohexyl-$CH{=}CH(CH_2)_6CO_2CH_3$	—	—	" (—)	163
6-n-C_3H_7-cyclohex-3-enyl-$(CH_2)_8CO_2CH_3$	—	—	" (—)	163
9-*cis*-11-*trans*-n-C_6H_{13} / $CH_3O_2C(CH_2)_7(CH{=}CH)_2$	—	—	n-$C_{17}H_{35}CO_2CH_3$ (—)	31
cis-n-$C_8H_{17}CH{=}CH(CH_2)_7CO_2CH_3$	—	C_2H_5OH, CH_3CN, DMF, or DMSO	" (—)	31
10-*trans*-12-*cis*-n-$C_5H_{11}CH{=}CH$ / $CH_3O_2C(CH_2)_7CHOHCH{=}CH$	9	CH_3OH, CH_3CO_2H, 2 h, rt	*cis*-n-$C_5H_{11}CH{=}CHCH_2CH_2$ / $CH_3O_2C(CH_2)_7CHOH$ (18) *trans*-n-$C_7H_{15}CH{=}CHCHOH$ / $(CH_3O_2C(CH_2)_7$ (9) + n-$C_9H_{19}CHOH(CH_2)_7CO_2CH_3$ (34)	169
9-*cis*-11-*trans*-n-$C_5H_{11}CHOH$ / $CH_3O_2C(CH_2)_7CH{=}CHCH{=}CH$	9	CH_3OH, CH_3CO_2H, 2 h, rt	*cis*-n-$C_5H_{11}CHOHCH_2CH_2CH{=}CH$ / $CH_3O_2C(CH_2)_7$ (19) + *trans*-n-$C_5H_{11}CHOHCH{=}CH(CH_2)_9CO_2CH_3$ (9) + n-$C_5H_{11}CHOH(CH_2)_{11}CO_2CH_3$ (30)	169
n-C_8H_{17}-cyclopropene-$(CH_2)_7CO_2H$	1	C_2H_5OH, 1 h, 35°	*cis*-n-C_8H_{17}-cyclopropane-$(CH_2)_7CO_2H$ (—)	164

C_{20}

Substrate		Conditions	Product(s) and Yield(s) (%)	Refs.
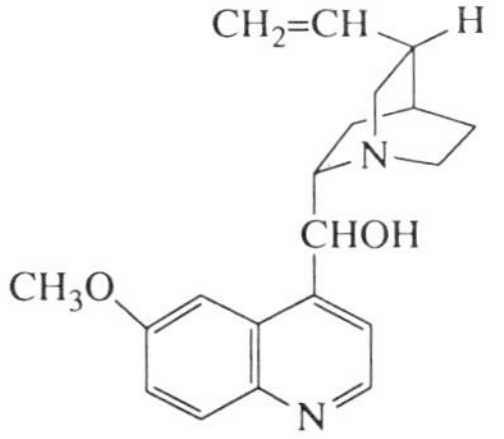 $CH_2{=}CH$, H, N, CHOH, CH_3O, N	9	CH_3OH, RCO_2H^{b}	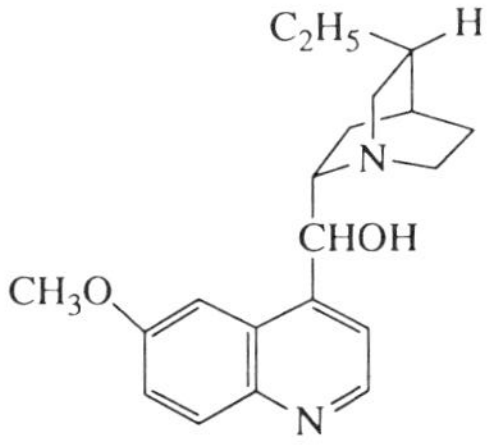 C_2H_5, H, N, CHOH, CH_3O, N (78)	7
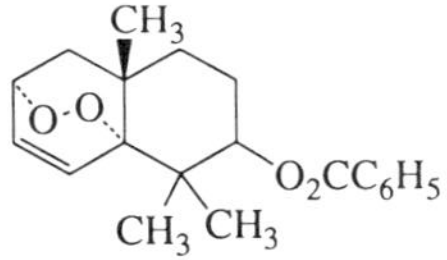 CH_3, O-O, $O_2CC_6H_5$, CH_3 CH_3	9	CH_2Cl_2, CH_3CO_2H, 25°	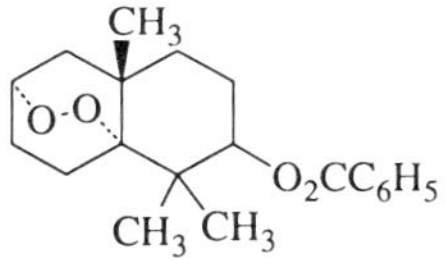 CH_3, O-O, $O_2CC_6H_5$, CH_3 CH_3 (84)	147
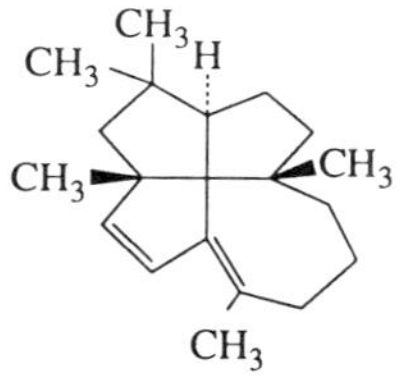 CH_3, CH_3, H, CH_3, CH_3, CH_3	10	$CHCl_3$, reflux, 72 h	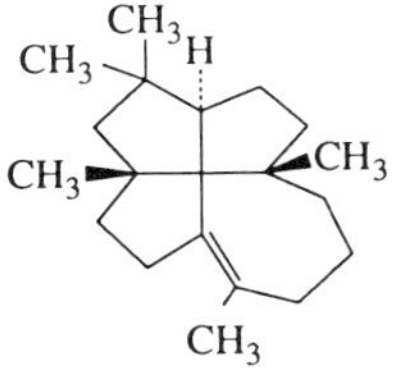 CH_3, CH_3, H, CH_3, CH_3, CH_3 (—)	170
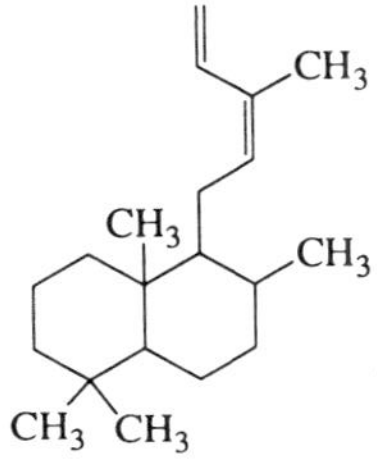 CH_3, CH_3, CH_3, CH_3 CH_3	1	CH_3OH, Cu(II), 25°, 4 d	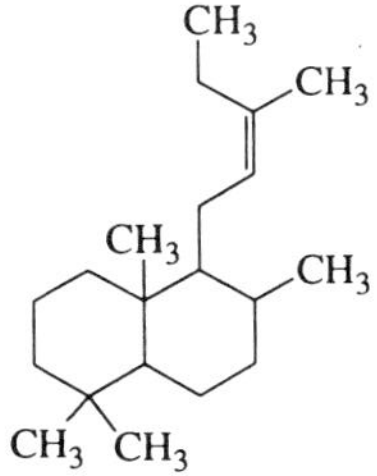 CH_3, CH_3, CH_3, CH_3, CH_3 CH_3 (—)	171

TABLE I. DIIMIDE REDUCTIONS OF SUBSTITUTED ALKENES (*Continued*)

	Reactant	Procedure	Conditions	Product(s) and Yield(s) (%)	Refs.
C_{21}	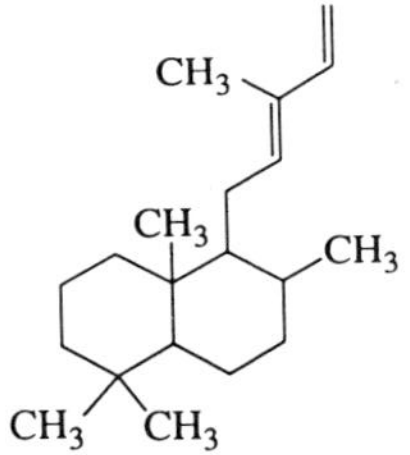 (structure)	1	CH_3OH, Cu(II), 25°, 22 h	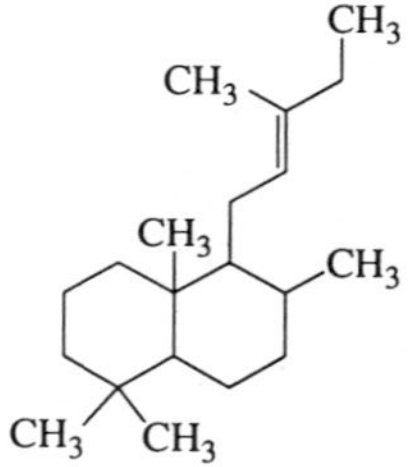 (structure) (100)	171
	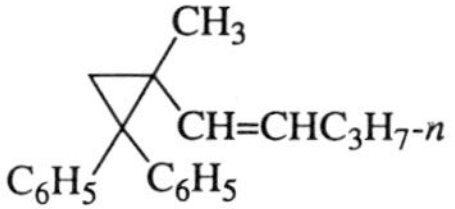 (structure) $CH=CHC_3H_7$-*n*, C_6H_5, C_6H_5, CH_3	9	C_2H_5OH, CH_3CO_2H, rt, 6 h	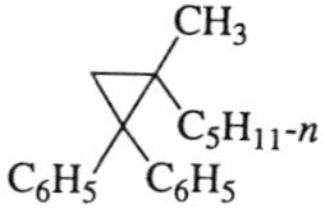 (structure) C_5H_{11}-*n*, C_6H_5, C_6H_5, CH_3 (75)	172
	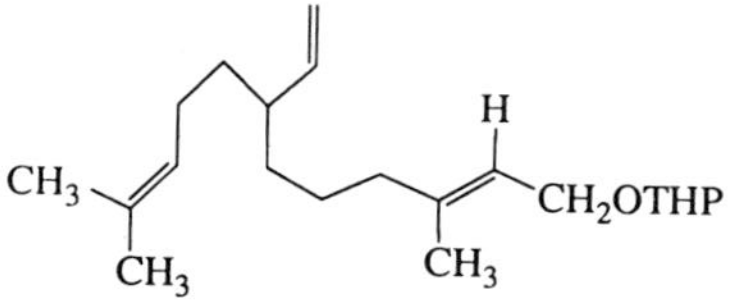 (structure) CH_2OTHP	4	C_2H_5OH, Cu(II)	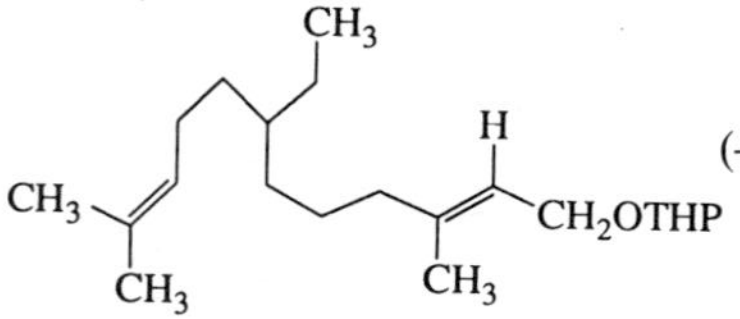 (structure) CH_2OTHP (—)	173
C_{22}	(structure)	1[d]	—	(structure) (—)	174
	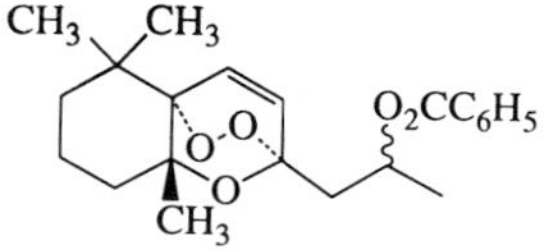 (structure) $O_2CC_6H_5$	9	CH_2Cl_2, CH_3CO_2H, 36 h	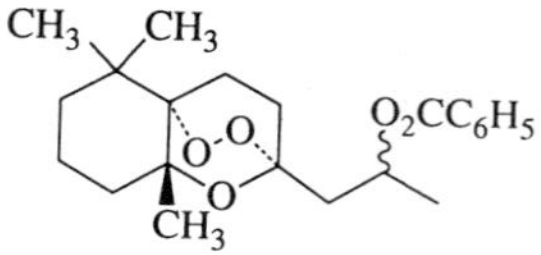 (structure) $O_2CC_6H_5$ (—)	154

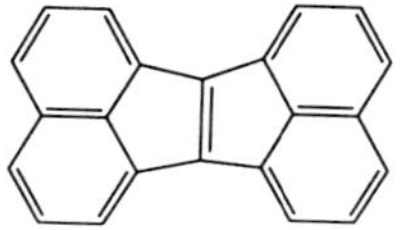

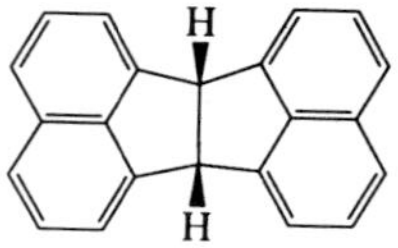

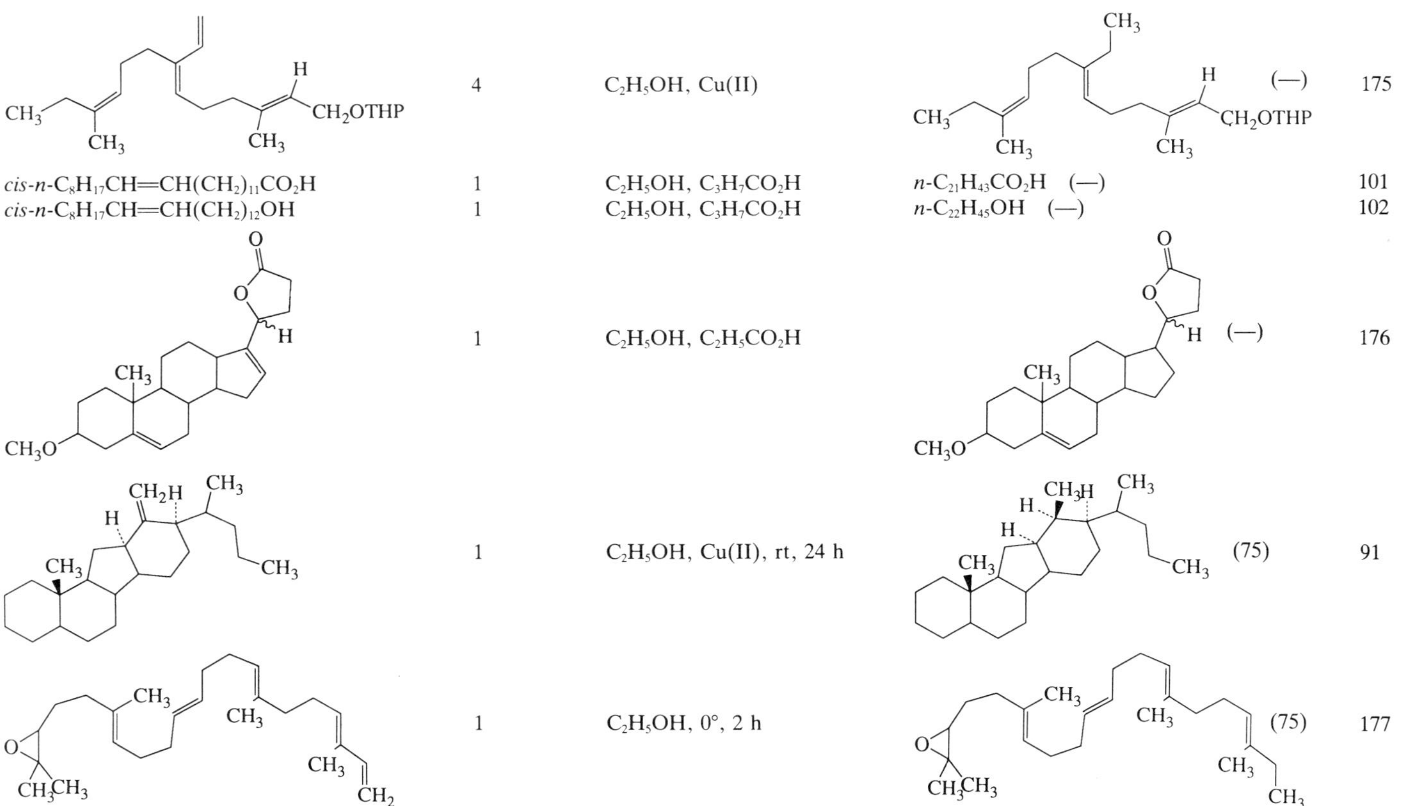

Substrate		Conditions	Product(s) (Yield, %)	Refs.
(structure)	4	C_2H_5OH, Cu(II)	(structure) (—)	175
cis-*n*-$C_8H_{17}CH{=}CH(CH_2)_{11}CO_2H$	1	C_2H_5OH, $C_3H_7CO_2H$	*n*-$C_{21}H_{43}CO_2H$ (—)	101
cis-*n*-$C_8H_{17}CH{=}CH(CH_2)_{12}OH$	1	C_2H_5OH, $C_3H_7CO_2H$	*n*-$C_{22}H_{45}OH$ (—)	102
(structure)	1	C_2H_5OH, $C_2H_5CO_2H$	(structure) (—)	176
(structure)	1	C_2H_5OH, Cu(II), rt, 24 h	(structure) (75)	91
(structure)	1	C_2H_5OH, 0°, 2 h	(structure) (75)	177

TABLE I. DIIMIDE REDUCTIONS OF SUBSTITUTED ALKENES (*Continued*)

Reactant	Procedure	Conditions	Product(s) and Yield(s) (%)	Refs.
C_{23}				
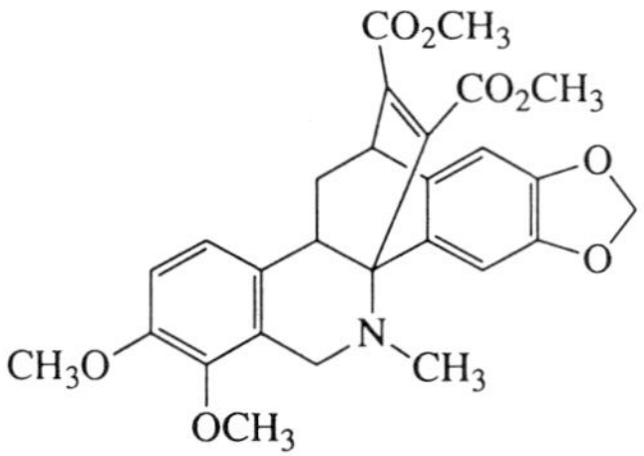 CO_2CH_3, CO_2CH_3, CH_3O, OCH_3, $N-CH_3$	1	CH_3OH, reflux, 48 h	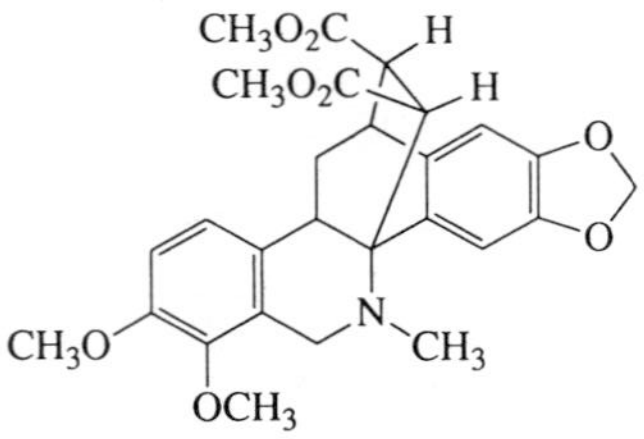 CH_3O_2C, H, CH_3O_2C, H, CH_3O, OCH_3, $N-CH_3$ (—)	178
	9	CH_3OH, CH_3CO_2H, rt, 30 min	" (26)	178
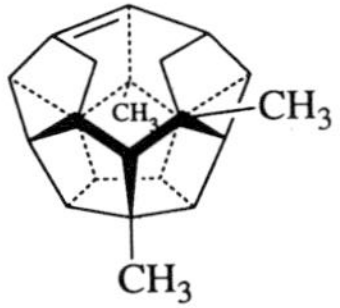 CH_3, CH_3, CH_3	4	C_2H_5OH, rt, 6 h	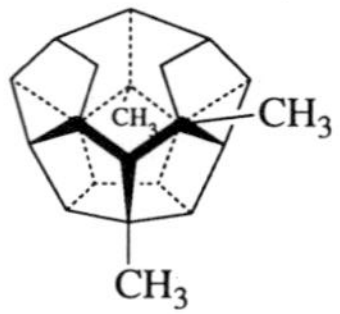 CH_3, CH_3, CH_3 (86)	179
C_{24}				
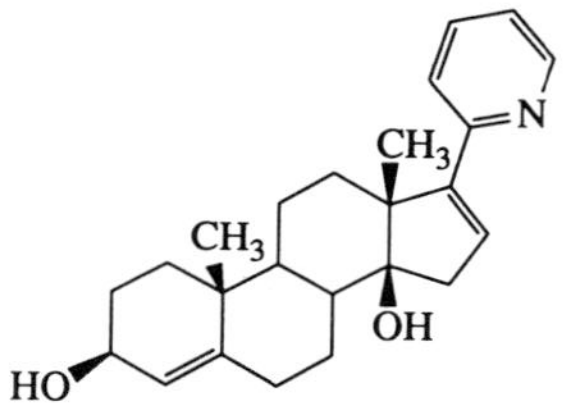 CH_3, CH_3, N, OH, HO	—	—	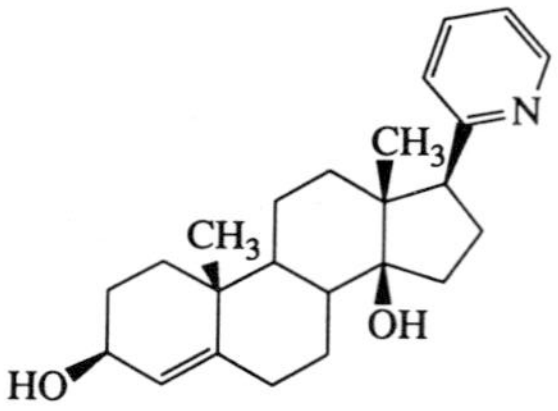 CH_3, CH_3, N, OH, HO (—)	180

Substrate	Method	Conditions	Product(s) and Yield(s) (%)	Refs.
C_{25}				
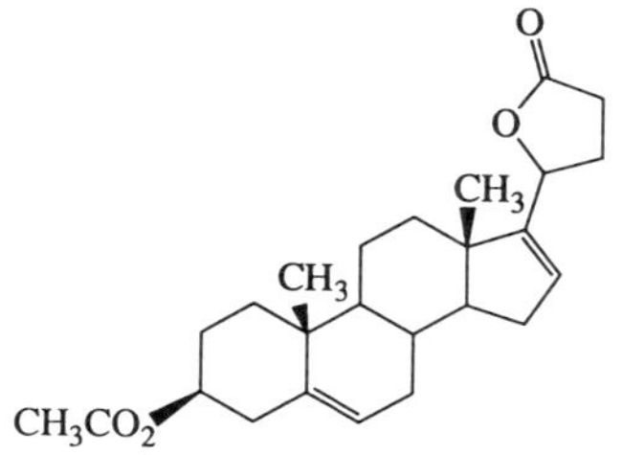 CH_3CO_2, CH_3, CH_3, O, O	1	C_2H_5OH, $C_2H_5CO_2H$	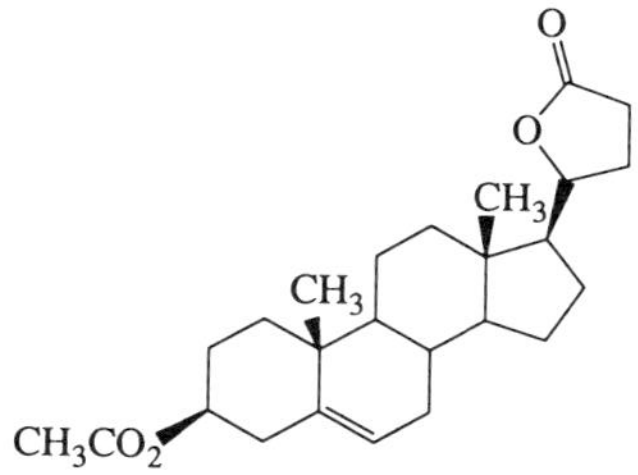 CH_3CO_2, CH_3, CH_3, O, O (58)	181
C_{26}				
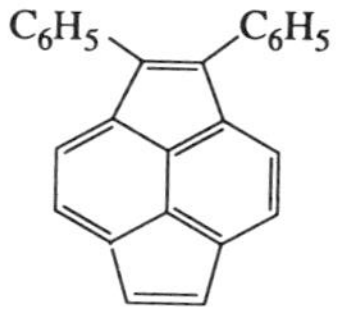 C_6H_5, C_6H_5	9 (N_2D_2)	CH_3OD, CH_3CO_2D	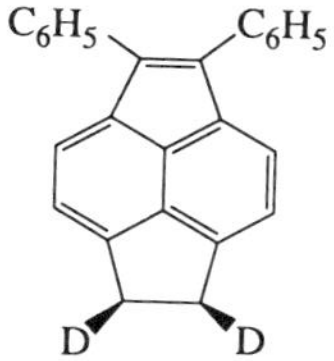 C_6H_5, C_6H_5, D, D (73)	182
C_{27}				
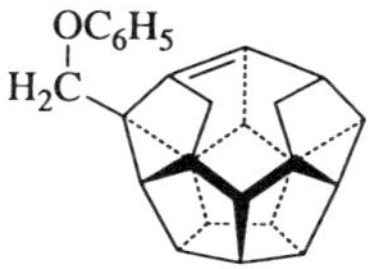 OC_6H_5, H_2C	5	THF, CH_3CO_2H, Cu(II), 2 h	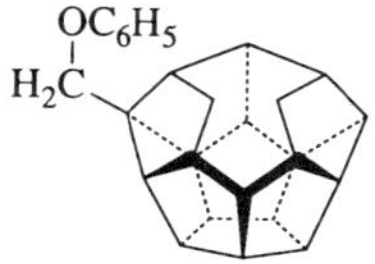 OC_6H_5, H_2C (90)	183
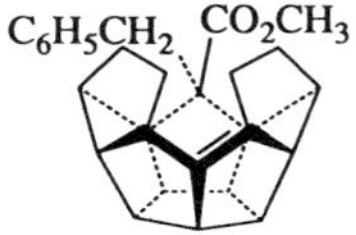 $C_6H_5CH_2$, CO_2CH_3	4	C_2H_5OH, rt	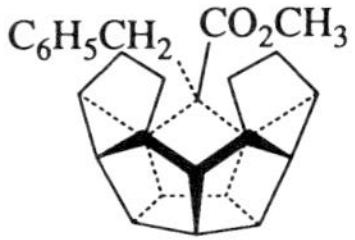 $C_6H_5CH_2$, CO_2CH_3 (95)	179

TABLE I. DIIMIDE REDUCTIONS OF SUBSTITUTED ALKENES (*Continued*)

Reactant	Procedure	Conditions	Product(s) and Yield(s) (%)	Refs.
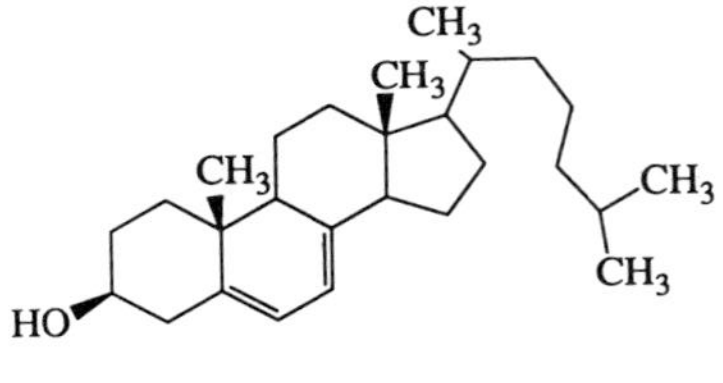	1 (N_2H_2, N_2D_2)	$CH_3OH(D)$, THF, Cu(II) rt, 6 h	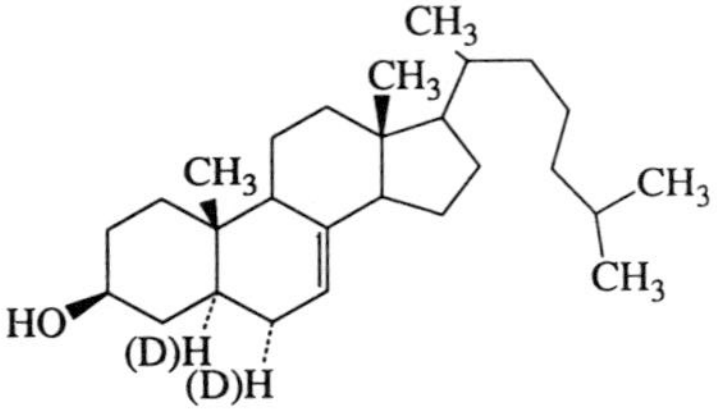 (—)	184, 101
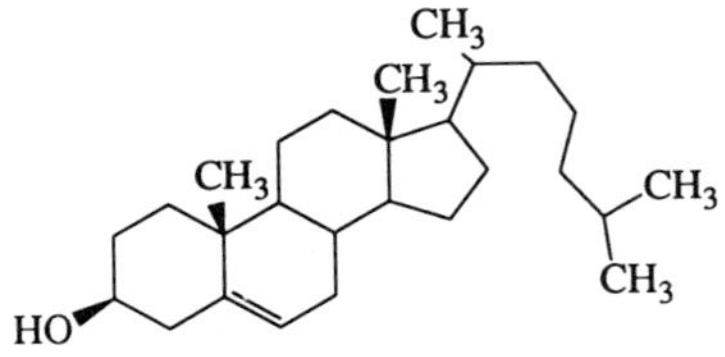	1	CH_3OH, Cu(II)	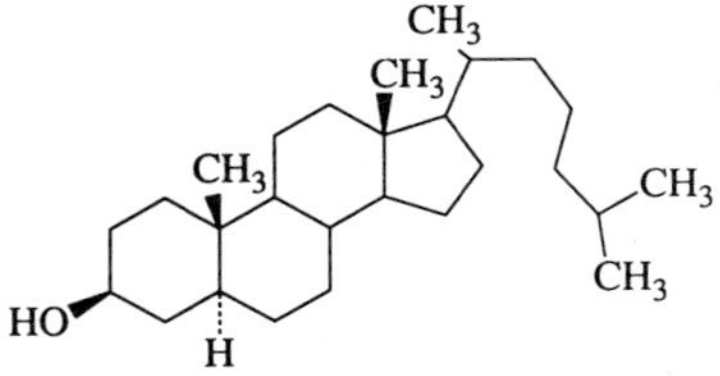 (20)	5, 101
C_{28} 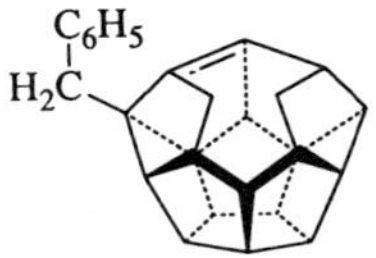	4	C_2H_5OH, rt	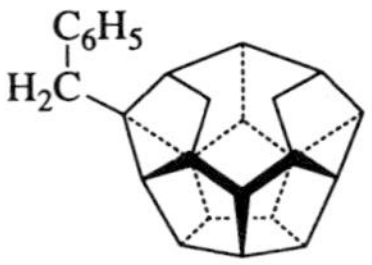 (55)	179
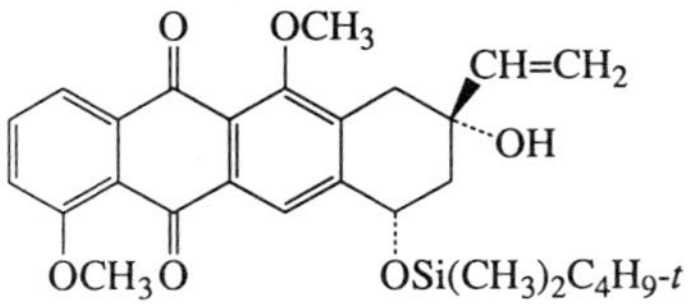	4	C_2H_5OH, Cu(II), rt, 16 h	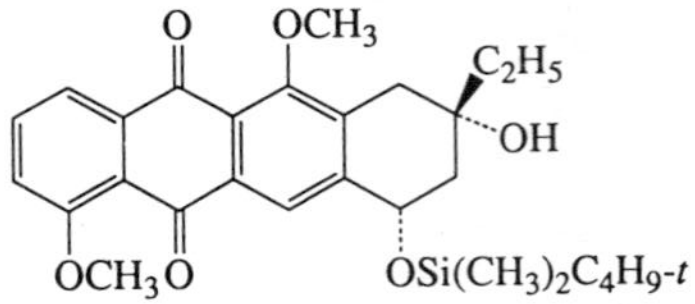 (95)	185

Formula	Substrate	Catalyst	Conditions	Product(s) and Yield(s) (%)	Refs.
C_{29}	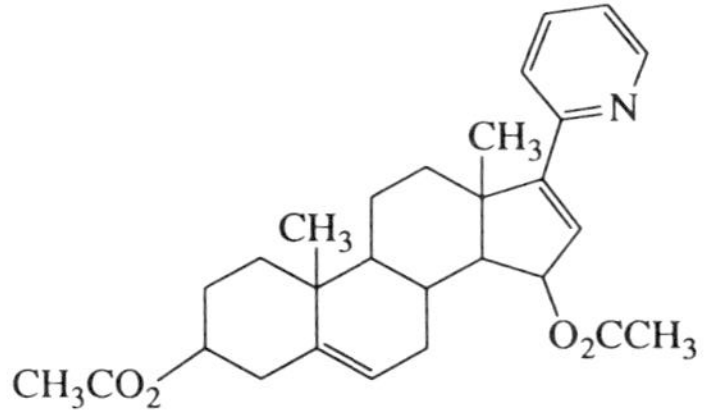 [structure: 3-acetoxy, 15-acetoxy, 17-(2-pyridyl) steroid diene]	1	CH_3OH, $C_3H_7CO_2H$ 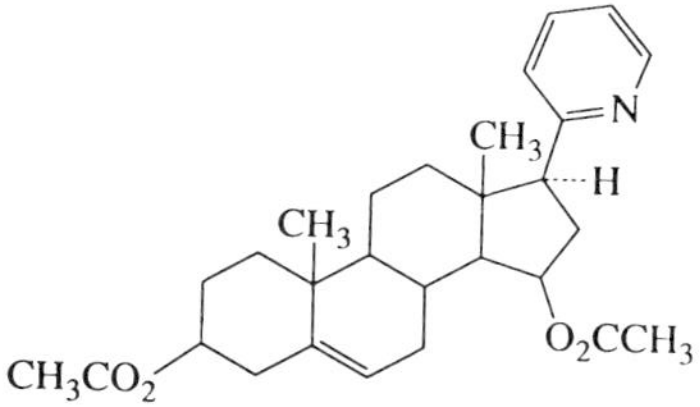	[structure] (75)	186
	[structure: HO-steroid with unsaturated side chain]	1, 10	—	[structure] (—)	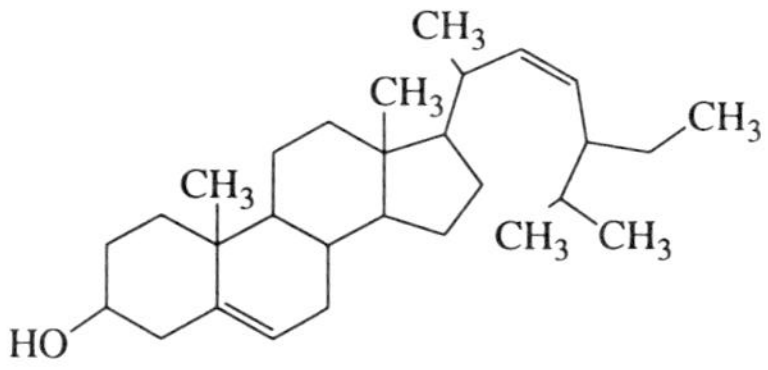 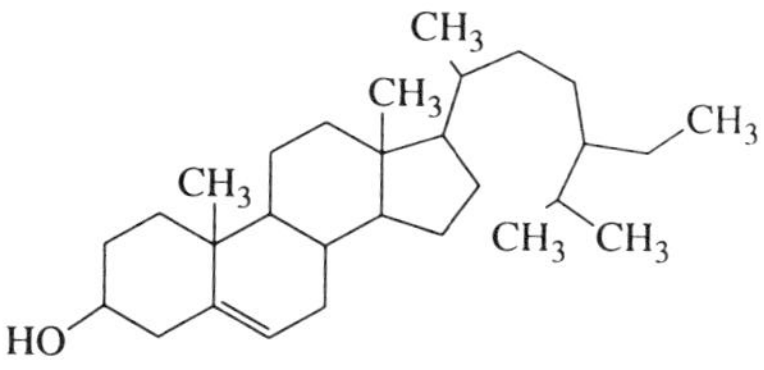 187
C_{30}	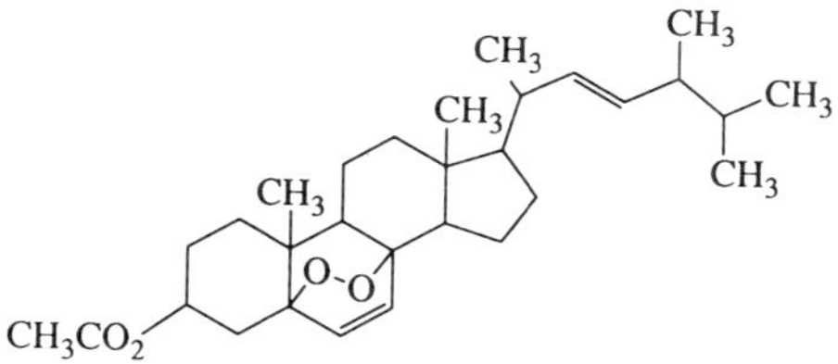 [structure: 3-acetoxy 5,8-epidioxy steroid]	9	CH_3OH, CH_3CO_2H, 0°, 3 h 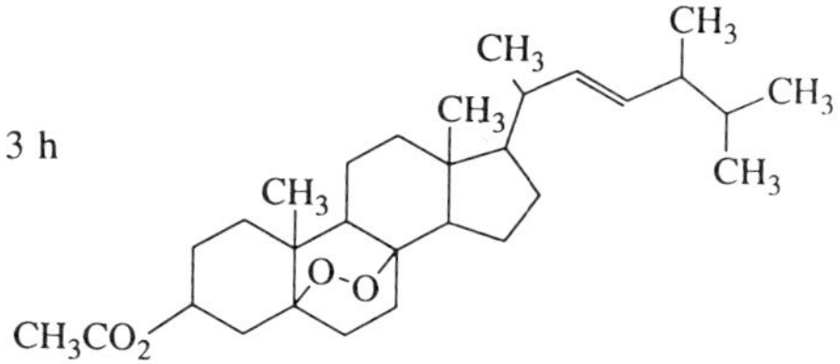	[structure] (—)	50
	[structure: 4,4-dimethyl HO-steroid]	1	—	— (—)	101

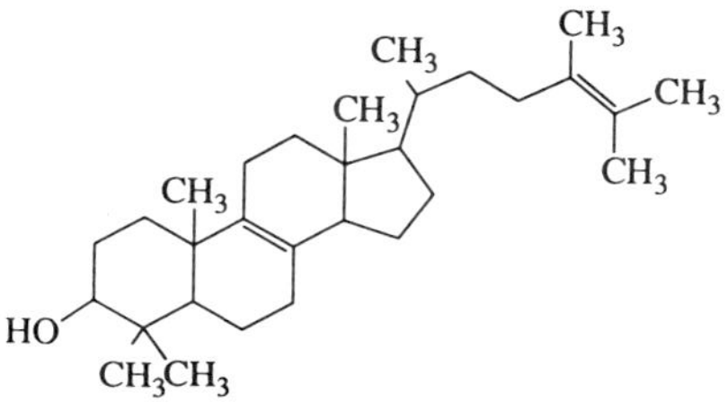

TABLE I. DIIMIDE REDUCTIONS OF SUBSTITUTED ALKENES (*Continued*)

Reactant	Procedure	Conditions	Product(s) and Yield(s) (%)	Refs.
C_{35}				
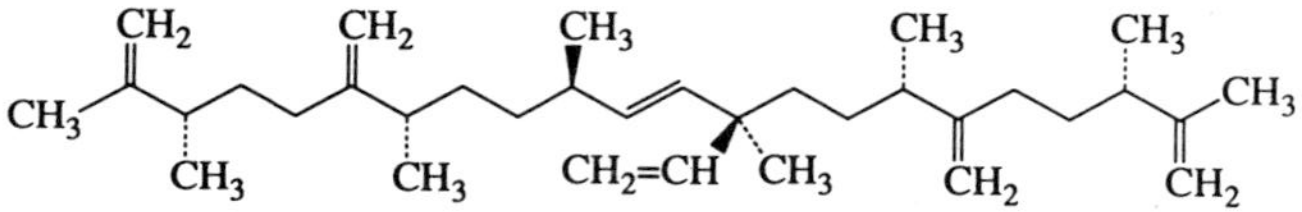	—		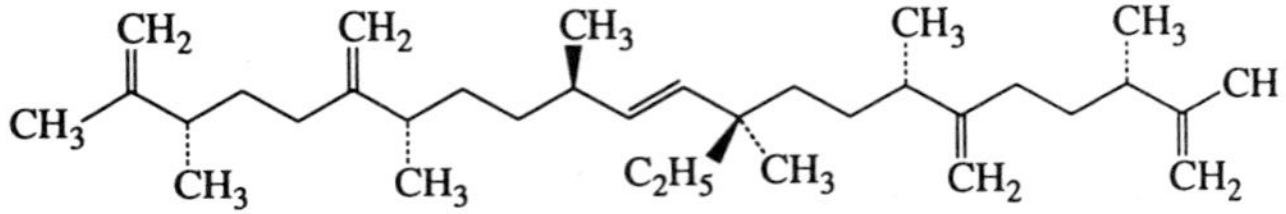 (—)	188
C_{36}				
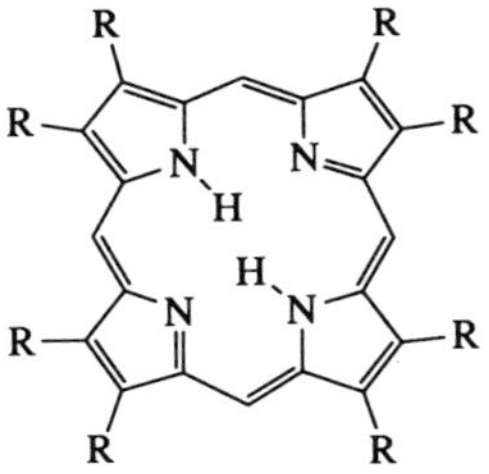 R = C_2H_5	10	C_5H_5N, K_2CO_3, 105°, 6.5 h	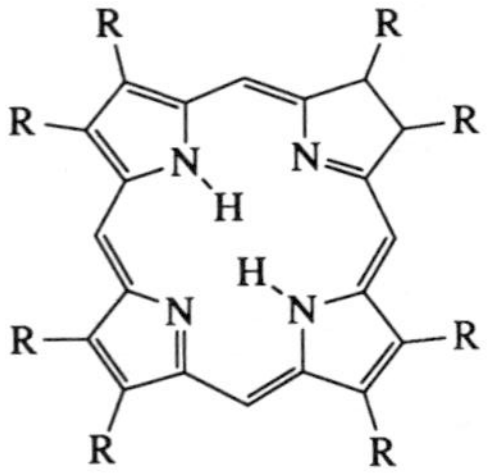 (11)	189
C_{44}				
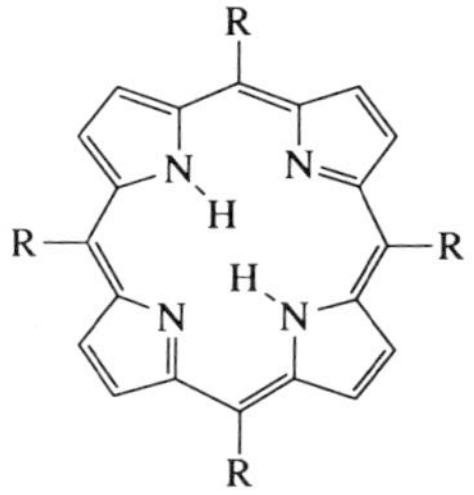 R = C_6H_5	10	C_5H_5N, K_2CO_3, 105°, 6.5 h	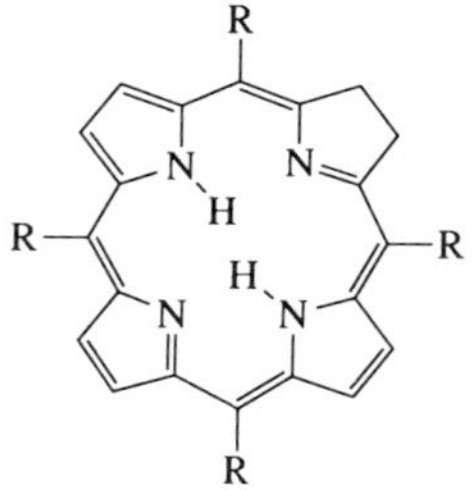 (68) +	189

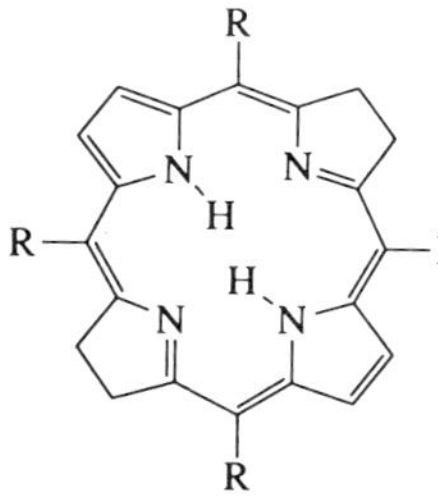

(32)

C_n (polymers)				
cis-1,4-Polyisoprene	10	Aromatic solvents, 100–140°	Partial reduction (—)	190
	10	Xylene, DMF, or diglyme 135–140°	" (—)	191
Polyisoprene	10	—	" (—)	191, 192
cis- and *trans*-1,4-Polybutadiene	10	—	" (—)	191, 192
Butadiene polymers	10	—	" (—)	191, 192
Polycyclohexadiene	10	—	" (—)	192
Poly(styrene-co-butadiene)	10	—	" (—)	192
Lignan	10	MET, pH 8, 12 h	" (—)	193

[a] An arylsulfonylhydrazide bound to a stationary polymer was used.
[b] The carboxylic acid was not specified.
[c] Overreduction also occurs.
[d] Reduction occurred under Wolff–Kishner conditions.

TABLE II. DIIMIDE REDUCTIONS OF SUBSTITUTED ALLENES

	Reactant	Procedure	Conditions	Product(s) and Yield(s) (%)	Refs.
C_4					
	$CH_2{=}C{=}CHCO_2H$	1	C_2H_5OH, Cu(II)	*cis*-$CH_3CH{=}CHCO_2H$ (20) + *trans*-$CH_3CH{=}CHCO_2H$ (6) + $CH_2{=}CHCH_2CO_2H$ (31) + *n*-$C_3H_7CO_2H$ (22)	194
C_7					
	$(C_2H_5)_2C{=}C{=}CH_2$	1, 4	C_2H_5OH	$(C_2H_5)_2C{=}CHCH_3$ (16)	59
		4	C_2H_5OH, Cu(II), 0°, 0.5 h	" (16)	195
	$(CH_3)_2C{=}C{=}C(CH_3)_2$	1, 4	C_2H_5OH	$(CH_3)_2C{=}CHC_3H_7$-*i* (4)	59
C_9					
	$C_6H_5CH{=}C{=}CH_2$	1, 4	C_2H_5OH	*cis*-$C_6H_5CH{=}CHCH_3$ (I) (26) + *trans*-$C_6H_5CH{=}CHCH_3$ (II) (5) + $C_6H_5CH_2CH{=}CH_2$ (III) (2)	59
		10	CH_3OH, ETA	" I(79) + II(3) + III(18)	60
	1,2,6-Cyclononatriene	1, 4	C_2H_5OH	1,5-Cyclononadiene (81)	59
		4		" (100)	195
	n-$C_6H_{13}CH{=}C{=}CH_2$	1, 4	C_2H_5OH	*cis*-(2)-Nonene (32) + *trans*-(2)-Nonene (1)	59
		4	—	*cis*-2-Nonene (17)	195

C_{10}					
	$C_6H_5C(CH_3){=}C{=}CH_2$	1, 4	C_2H_5OH	$C_6H_5C(CH_3){=}CHCH_3$ (20) + $C_6H_5CH(CH_3)CH{=}CH_2$ (72)	59
		10	CH_3OH, ETA	(Z)-$C_6H_5C(CH_3){=}CHCH_3$ (28) + (E)-$C_6H_5C(CH_3){=}CHCH_3$ (0.5).	60
	1,2,6-Cyclodecatriene	1, 4	C_2H_5OH	1,5-Cyclodecadiene (23) + 1,6-Cyclodecadiene (27)	59
	1,2-Cyclodecadiene	1, 4	C_2H_5OH	*cis*-Cyclodecene (49)	59
		4	C_2H_5OH, Cu(II), 0°, 0.5 h	" (24)	195
C_{11}					
	$C_6H_5CH{=}C{=}CHC_2H_5$	4	C_2H_5OH, Cu(II), 0°, 0.5 h	*cis*-$C_6H_5CH{=}CHC_3H_7$-*n* (28) + *trans*-$C_6H_5CH{=}CHC_3H_7$-*n* (0.5) + *cis*-$C_6H_5CH_2CH{=}CHC_2H_5$ (56) + *trans*-$C_6H_5CH_2CH{=}CHC_2H_5$ (16)	60
C_{13}					
	C_6H_5, H (cyclopropyl) $C{=}C{=}C(CH_3)_2$	4	C_2H_5OH, rt	C_6H_5, H, H (cyclopropyl) $CH{=}C(CH_3)_2$ (—)	196
	1,2-Cyclotridecadiene	4	C_2H_5OH, Cu(II), 0°, 0.5 h	*cis*-Cyclotridecene (—)	195

TABLE III. DIIMIDE REDUCTIONS OF SUBSTITUTED ALKYNES

	Reactants	Procedure	Conditions	Product(s) and Yield(s) (%)	Refs.
C_4					
	$HO_2CC{\equiv}CCO_2H$	1, 4	CH_3OH, 25°	Succinic acid (40)	5
		14	—	Maleic (—) + Succinic (—) acid	124
	$CH_3C{\equiv}CCO_2H$	1	C_2H_5OH, Cu(II), rt, 2.7 h	*cis*-$CH_3CH{=}CHCO_2H$ (20) + *n*-$C_3H_7CO_2H$ (40)	194
C_6					
	n-$C_4H_9C{\equiv}CI$	—	—	*cis-n*-$C_4H_9CH{=}CHI$ (—)	197
	(*Z*)-$HC{\equiv}CC(CH_3){=}CHCH_2OH$	4	C_2H_5OH, 30–40°, 7 h	(*Z*)-$C_2H_5C(CH_3){=}CHCH_2OH$ (72)	57
C_8					
	c-C_3H_5–C≡C–c-C_3H_5	9	C_2H_5OH, CH_3CO_2H, 0°, 4 h	*cis*-c-C_3H_5CH=CH-c-C_3H_5 (30) + c-C_3H_5–$(CH_2)_2$–c-C_3H_5 (70)	198
C_9					
	$C_6H_5C{\equiv}CCO_2H$	14	NaOH, 40°	*cis*-Cinnamic Acid (—) + 3-Phenylpropionic Acid (—)	78, 100
C_{13}					
	n-$C_5H_{11}CH(OTHP)C{\equiv}CI$	9	CH_3OH, C_5H_5N, CH_3CO_2H	*cis-n*-$C_5H_{11}CH(OTHP)CH{=}CHI$ (82) + *n*-$C_5H_{11}CH(OTHP)CH_2CH_2I$ (—)	62
	t-$C_4H_9(CH_3)_2SiOCH(CH_3)$ | $IC{\equiv}CCHOH(CH_2)_2$	9	CH_3OH, C_5H_5N, CH_3CO_2H	*cis-t*-$C_4H_9(CH_3)_2SiOCHCH_3$ | $ICH{=}CHCHOH(CH_2)_2$ (82)	61
C_{14}					
	$C_6H_5C{\equiv}CC_6H_5$	1	CH_3OH, Cu(II)	$C_6H_5CH_2CH_2C_6H_5$ (80)	5
		8	CH_2Cl_2, rt	*cis*-Stilbene (80)	71
		9 (N_2D_2)	—	*cis*-$C_6H_5CD{=}CDC_6H_5$ (—) + $C_6H_5CD_2CD_2C_6H_5$ (—)	30

C_{18}				
Stearolic Acid	12	DMF, 90–95°, 1.5 h	Stearic acid (40) + Oleic acid (22)	137
C_{19}				
n-$C_8H_{17}C{\equiv}C(CH_2)_7CO_2CH_3$	—	—	n-$C_{17}H_{35}CO_2CH_3$ (—)	31
C_{22}				
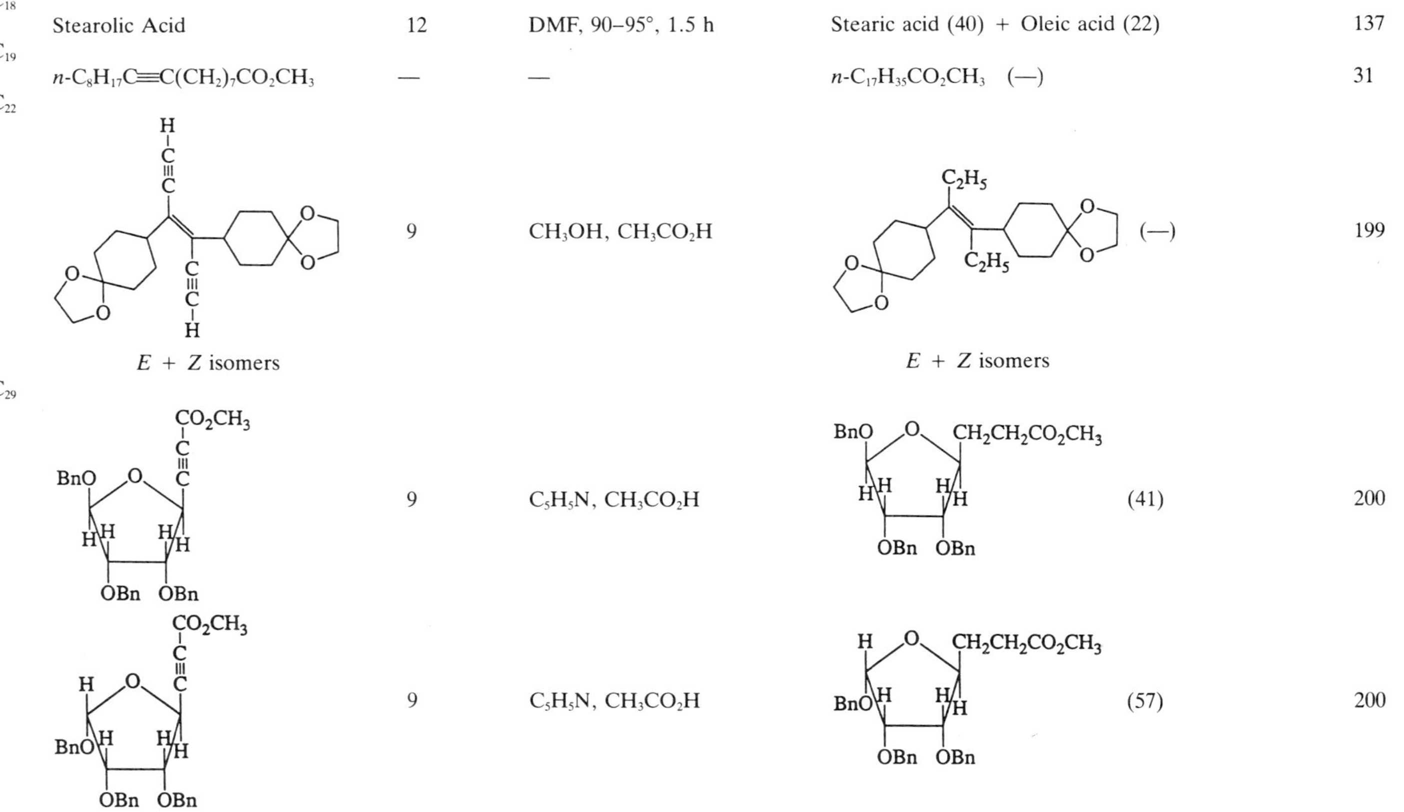 $E + Z$ isomers	9	CH_3OH, CH_3CO_2H	$E + Z$ isomers (—)	199
C_{29}				
—	9	C_5H_5N, CH_3CO_2H	(41)	200
—	9	C_5H_5N, CH_3CO_2H	(57)	200

TABLE IV. DIIMIDE REDUCTIONS OF SUBSTITUTED CARBONYL COMPOUNDS

	Reactants	Procedure	Conditions	Product(s) and Yield(s) (%)	Refs.
C_3					
	Acetone	9	CH_3OH, RCO_2H, rt, 4.5 h	i-C_3H_7OH (14)	41
C_5					
	Furfural	9	CH_3OH, CH_3CO_2H, rt, 4.5 h	Furfuryl alcohol (84)	63
	3-Pentanone	9	CH_3OH, CH_3CO_2H	3-Pentanol (7)	41
C_6					
	Cyclohexanone	9	CH_3OH, CH_3CO_2H	Cyclohexanol (9)	41
C_7					
	4-ClC_6H_4CHO	9	CH_3OH, CH_3CO_2H, rt, 4.5 h	4-$ClC_6H_4CH_2OH$ (69)	63
	4-$O_2NC_6H_4CHO$	9	CH_3OH, CH_3CO_2H, rt, 4.5 h	4-$O_2NC_6H_4CH_2OH$ (5)	63
	Benzaldehyde	9	CH_3OH, CH_3CO_2H, rt, 4.5 h	Benzyl alcohol (62)	41, 63
	n-$C_6H_{13}CHO$	9	CH_3OH, RCO_2H	n-$C_7H_{15}OH$ (—)	41
C_8					
	3-$CH_3C_6H_4CHO$	9	CH_3OH, CH_3CO_2H, rt, 4.5 h	3-$CH_3C_6H_4CH_2OH$ (72)	63
	4-$CH_3C_6H_4CHO$	9	CH_3OH, CH_3CO_2H, rt, 4.5 h	4-$CH_3C_6H_4CH_2OH$ (75)	63
	Acetophenone	9	CH_3OH, RCO_2H	1-Phenylethanol (31)	41
	4-$CH_3OC_6H_4CHO$	9	CH_3OH, CH_3CO_2H, rt, 4.5 h	4-$CH_3OC_6H_4CH_2OH$ (77)	63
C_{13}					
	Benzophenone	9	CH_3OH, RCO_2H	Benzhydrol (26)	41

TABLE V. Diimide Reductions of Substituted Imines and Hydrazones

	Reactants	Procedure	Conditions	Product(s) and Yield(s) (%)	Refs.
C_9	$CH_3CH{=}NNHCOC_6H_5$	2	C_2H_5OH, reflux, 3 h	$C_2H_5NHNHCOC_6H_5$ (64)	64
C_{10}	$(CH_3)_2C{=}NNHCOC_6H_5$	2	"	i-$C_3H_7NHNHCOC_6H_5$ (75)	64
C_{12}	2-furyl-$CH{=}NNHCOC_6H_5$	2	"	2-furyl-$CH_2NHNHCOC_6H_5$ (55)	64
C_{13}	$C_6H_5N{=}CHC_6H_5$	2	"	$C_6H_5NHCH_2C_6H_5$ (67)	64
	$C_6H_5N{=}CHC_6H_4OH$-4	2	"	$C_6H_5NHCH_2C_6H_4OH$-4 (65)	64
	cyclohexylidene=$NNHCOC_6H_5$	2	"	cyclohexyl-$NHNHCOC_6H_5$ (65)	64
C_{14}	4-$CH_3OC_6H_4CH{=}NC_6H_5$	2	"	4-$CH_3OC_6H_4CH_2NHC_6H_5$ (70)	64
	4-$CH_3OC_6H_4N{=}CHC_6H_5$	2	"	4-$CH_3OC_6H_4NHCH_2C_6H_5$ (60)	64
	4-$CH_3C_6H_4N{=}CHC_6H_4OH$-4	2	"	4-$CH_3C_6H_4NHCH_2C_6H_4OH$-4 (66)	64
C_{15}	4-$CH_3C_6H_4N{=}CHC_6H_4OCH_3$-4	2	"	4-$CH_3C_6H_4NHCH_2C_6H_4OCH_3$-4 (64)	64

TABLE VI. DIIMIDE REDUCTIONS OF SUBSTITUTED AZO COMPOUNDS

	Reactants	Procedure	Conditions	Product(s) and Yield(s) (%)	Refs.
C_6					
	Diethyl azodicarboxylate	8	CH_2Cl_2, rt	$C_2H_5O_2CNHNHCO_2C_2H_5$ (90)	71
C_9					
	$CCl_3CCl{=}CHN{=}NC_6H_3Cl_2$-2,4	9	$CHCl_3$, CH_3CO_2H, rt, 15 min	$CCl_3CCl{=}CHNHNHC_6H_3Cl_2$-2,4 (78)	201
C_{12}					
	Azobenzene	1, 4	CH_3OH	1,2-Diphenylhydrazine (95)	5
		5	C_6H_6, CH_3CO_2H or Cu(II)	" (—)	70
		9	CH_3OH, CH_3CO_2H	" (99)	7
		9	CH_3OH or DMSO, 23°, 1.5 h	" (100)	96
		17	Alkaline solution, rt	" (55)	88
		18	CH_2Cl_2, rt, 3 h	" (—)	89
C_{26}					
	$C_6H_5CH(N{=}NC_6H_5)CH(N{=}NC_6H_5)C_6H_5$	9	$CHCl_3$, CH_3CO_2H	$C_6H_5CH(NHNHC_6H_5)CH(NHNHC_6H_5)C_6H_5$ (—)	202

REFERENCES

[1] J. Hanuš, reference 2, footnote 2.
[2] J. Hanuš and J. Voříšek, *Collect. Czech. Chem. Commun.*, **1**, 223 (1929).
[3] J. Voříšek, *Chem. Listy*, **26**, 286 (1932) [*C. A.*, **26**, 5864 (1932)].
[4] H. Fischer and H. Gibian, *Justus Liebigs Ann. Chem.*, **548**, 143 (1941); *ibid.*, **550**, 183 (1942).
[5] E. J. Corey, W. L. Mock, and D. J. Pasto, *Tetrahedron Lett.*, **1961**, 347.
[6] S. Hünig, R. Müller, and W. Thier, *Tetrahedron Lett.*, **1961**, 353.
[7] E. E. van Tamelen, R. S. Dewey, and R. J. Timmons, *J. Am. Chem. Soc.*, **83**, 3725 (1961).
[8] F. Aylward and M. Sawistowska, *Chem. Ind. (London)*, **1962**, 484.
[9] J. Thiele, *Justus Liebigs Ann. Chem.*, **271**, 127 (1892).
[10] F. Raschig, *Angew. Chem.*, **23**, 972 (1910).
[11] R. A. Back, *Reviews of Chemical Intermediates*, **5**, 293 (1984).
[12] S. Hünig, H. R. Müller, and W. Thier, *Angew. Chem. Int. Ed. Engl.*, **4**, 271 (1965).
[13] C. E. Miller, *J. Chem. Ed.*, **42**, 254 (1965).
[14] S. F. Foner and R. C. Hudson, *J. Chem. Phys.*, **28**, 719 (1958).
[15] N. Wiberg, G. Fischer, and H. Bachhuber, *Chem. Ber.*, **107**, 1456 (1974).
[16] N. Wiberg, G. Fischer, and H. Bachhuber, *Angew. Chem. Int. Ed. Engl.*, **15**, 385 (1976).
[17] N. Wiberg, G. Fischer, and H. Bachhuber, *Angew. Chem. Int. Ed. Engl.*, **16**, 780 (1977).
[18] A. P. Sylvester and P. B. Dervan, *J. Am. Chem. Soc.*, **106**, 4648 (1984).
[19] L. G. Spears, Jr. and J. S. Hutchinson, *J. Chem. Phys.*, **88**, 240 (1988); *ibid.*, 250; K. Ito and S. Nagase, *Chem. Phys. Lett.*, **126**, 531 (1986).
[20] D. J. Pasto and D. M. Chipman, *J. Am. Chem. Soc.*, **101**, 2290 (1979).
[21] C. J. Casewit and W. A. Goddard, III, *J. Am. Chem. Soc.*, **102**, 4057 (1980).
[22] R. Ahlrichs and V. Staemmler, *Chem. Phys. Lett.*, **37**, 77 (1976).
[23] C. A. Parsons and C. E. Dykstra, *J. Chem. Phys.*, **71**, 3025 (1979).
[24] J. M. Howell and J. L. Kirschenbaum, *J. Am. Chem. Soc.*, **98**, 877 (1976).
[25] N. C. Baird and J. R. Swenson, *Can. J. Chem.*, **51**, 3097 (1973); R. Cimiraglia, J. M. Riera, and J. Tomasi, *Theor. Chim. Acta*, **46**, 222 (1977); G. Merenyi and G. Wettermark, *Chem. Phys.*, **1**, 340 (1973).
[26] E. Flood and P. N. Skanche, *Chem. Phys. Lett.*, **54**, 53 (1978).
[27] C. Willis, R. A. Bach, J. M. Parsons, and J. G. Purdon, *J. Am. Chem. Soc.*, **99**, 4451 (1977); S. K. Vidyarthi, C. Willis, R. A. Back, and R. M. McKitrick, *ibid.*, **96**, 7647 (1974).
[28] D. J. Pasto, *J. Am Chem. Soc.*, **101**, 6852 (1979).
[29] E. E. van Tamelen, R. S. Dewey, M. F. Lease, and W. H. Pirkle, *J. Am. Chem. Soc.*, **83**, 4302 (1961).
[30] E. J. Corey, D. J. Pasto, and W. L. Mock, *J. Am. Chem. Soc.*, **83**, 2957 (1961).
[31] G. K. Koch, *J. Labelled Comp. Radiopharm.*, **5**, 99 (1969).
[32] C. R. Wellman, J. R. Ward, and L. P. Kuhn, *J. Am. Chem. Soc.*, **98**, 1683 (1976).
[33] E. W. Garbisch, Jr., S. M. Schildcrout, D. B. Patterson, and C. M. Spreecher, *J. Am. Chem. Soc.*, **87**, 2932 (1965).
[34] S. Siegel, M. Foreman, R. P. Fisher, and S. E. Johnson, *J. Org. Chem.*, **40**, 3599 (1975).
[35] N. Garti and S. Siegel, *J. Org. Chem.*, **41**, 3922 (1976).
[36] S. Hünig and H. R. Müller, *Angew. Chem. Int. Ed. Engl.*, **1**, 213 (1962).
[37] E. E. van Tamelen and R. J. Timmons, *J. Am. Chem. Soc.*, **84**, 1067 (1962).
[38] W. C. Baird, Jr., B. Franzus, and J. H. Surridge, *J. Am. Chem. Soc.*, **89**, 410 (1967).
[39] W. C. Baird, Jr. and J. H. Surridge, *J. Org. Chem.*, **37**, 304 (1972).
[40] S. Siegel, G. M. Foreman, and D. Johnson, *J. Org. Chem.*, **40**, 3589 (1975).
[41] E. E. van Tamelen, M. Davis, and M. F. Deem, *Chem. Commun.*, **1965**, 71.
[42] S. Uemura, A. Onoe, H. Akazaki, M. Okano, and K. Ichikawa, *Bull. Chem. Soc. Jpn.*, **49**, 1437 (1976).
[43] J. W. Wilt, G. Gutman, W. J. Ranus, Jr., and A. R. Zigman, *J. Org. Chem.*, **32**, 893 (1967).
[44] W. G. Dauben and C. H. Schallhorn, *J. Am. Chem. Soc.*, **93**, 2254 (1971).
[45] G. Just and L. Cutrone, *Can J. Chem.*, **54**, 867 (1976).
[46] S. Hünig and M. Schmitt, *Tetrahedron Lett.*, **25**, 1725 (1984).

[47] W. Adam and H. J. Eggelte, *J. Org. Chem.*, **42**, 3987 (1977).
[48] W. Adam. H. J. Eggelte, and A. Rodriguez, *Synthesis*, **1979**, 383.
[49] W. Adam and I. Erden, *Tetrahedron Lett.*, **1979**, 2781.
[50] W. Adam and H. J. Eggelte, *Angew. Chem. Int. Ed. Engl.*, **16**, 713 (1977).
[51] W. Adam, A. J. Bloodworth, H. J. Eggelte, and M. E. Loveitt, *Angew. Chem. Int. Ed. Engl.*, **17**, 209 (1978).
[52] W. Adam and H. J. Eggelte, *Angew. Chem. Int. Ed. Engl.*, **17**, 765 (1978).
[53] W. Adam and M. Balci, *Angew. Chem. Int. Ed. Engl.*, **17**, 954 (1978).
[54] W. Adam and I. Erden, *J. Org. Chem.*, **43**, 2737 (1978).
[55] W. Adam and I. Erden, *Angew. Chem. Int. Ed. Engl.*, **17**, 210 (1978).
[56] D. J. Coughlin and R. G. Salomon, *J. Am. Chem. Soc.*, **99**, 655 (1977).
[57] K. Mori, M. Ohki, A. Sato, and M. Matsui, *Tetrahedron*, **28**, 3739 (1972).
[58] V. V. R. Rao and D. Devaprabhakara, *Tetrahedron*, **39**, 2981 (1978).
[59] G. Nagendrappa, S. N. Moorthy, and D. Devaprabhakara, *Ind. J. Chem.*, **14**, 81 (1976).
[60] T. Okuyama, K. Toyoshima, and T. Fueno, *J. Org. Chem.*, **45**, 1604 (1980).
[61] A. F. Kluge, K. G. Untch, and J. H. Fried, *J. Am. Chem. Soc.*, **94**, 9256 (1972).
[62] C. Lüthy, P. Konstantin, and J. G. Untch, *J. Am. Chem. Soc.*, **100**, 6211 (1978).
[63] D. C. Curry, B. C. Uff, and N. D. Ward, *J. Chem. Soc. C*, **1967**, 1120.
[64] M. O. Abdel-Rahman, M. N. El-Enien, M. A. Kira, and A. H. Fayed, *J. Chem. U. A. R.*, **9**, 87 (1966). [*C. A.*, **1967**, *67*, 32413m].
[65] A. Kawaguchi, Y. Kobayashi, Y. Ogawa, and S. Okuda, *Chem. Pharm. Bull.*, **31**, 3228 (1983).
[66] H. Rakoff, *Prog. in Lipid Res.*, **21**, 225 (1982).
[67] B. A. Anderson, F. Dinger, and N. Dinh-Nguyen, *Chemica Scripta*, **18**, 197 (1981).
[68] K. Kondo, S. Murai, and N. Sonoda, *Tetrahedron Lett.*, **1977**, 3727.
[69] T. G. Back, S. Collins, and R. G. Kerr, *J. Org. Chem.*, **46,** 1564 (1981).
[70] J. M. Hoffman, Jr. and R. H. Schlessinger, *Chem. Commun.*, **1971**, 1245.
[71] R. M. Moriarty, R. K. Vaid, and M. P. Duncan, *Synth. Commun.*, **17**, 703 (1987).
[72] R. Herbert and M. Cristl, *Chem. Ber.*, **112**, 2012 (1979).
[73] N. J. Cusack, C. B. Reese, A. C. Risius, and B. Roozpeikar, *Tetrahedron*, **32**, 2157 (1976).
[74] D. W. Emerson, R. R. Emerson, S. C. Joshi, E. M. Sorenson, and J. E. Turek, *J. Org. Chem.*, **44**, 4634 (1979).
[75] J. L. Kice, T. E. Rogers, and A. C. Warheit, *J. Am. Chem. Soc.*, **96**, 8020 (1974).
[76] P. A. Wade and N. A. Amin, *Synth. Commun.*, **12**, 287 (1982).
[77] E. J. Corey and W. L. Mock, *J. Am. Chem. Soc.*, **84**, 685 (1962).
[78] R. Appel and W. Büchner, *Angew. Chem.*, **73**, 807 (1961).
[79] E. Schmitz, R. Ohme, and G. Kozakiewicz, *Z. Anorg. Allg. Chem.*, **339**, 44 (1965).
[80] S. Acharya, G. Neogi, R. K. Panda, and D. Ramaswamy, *J. Chem. Soc. Dalt. Trans.*, **1984**, 1477.
[81] Y. Imasaka, K. Tanaka, and T. Tanaka, *Chem. Lett.*, **1983**, 1477.
[82] S. I. Zones, T. M. Vickrey, J. G. Palmer, and G. N. Schrauzer, *J. Am. Chem. Soc.*, **98**, 7289 (1976); S. I. Zones, M. R. Palmer, J. G. Palmer, J. M. Doemeny, and G. Schrauzer, *ibid.*, **100**, 2113 (1978); G. N. Schrauzer and M. R. Palmer, *ibid.*, **103**, 2659 (1981).
[83] G. N. Schrauzer, N. Strampach, M. R. Palmer, and S. I. Jones, *Nouv. J. Chim.*, **5**, 5 (1981).
[84] G. N. Schrauzer, *Angew. Chem. Int. Ed. Engl.*, **14**, 514 (1975).
[85] T. Huang and J. T. Spence, *J. Phys. Chem.*, **72**, 4198 (1968).
[86] G. N. Schrauzer, G. W. Kiefer, K. Tano, and P. A. Doemeny, *J. Am. Chem. Soc.*, **96**, 641 (1974).
[87] D. F. C. Morris and T. J. Ritter, *J. Chem. Soc. Dalt. Trans.*, **1980**, 216.
[88] E. Schmitz, R. Ohme, and S. Schramm, *Angew. Chem. Int. Ed. Engl.*, **2**, 157 (1963); E. Schmitz, ibid., **3**, 333 (1964).
[89] R. Annunziata, R. Fornasier, and F. Montanari, *J. Org. Chem.*, **39**, 3195 (1974).
[90] M. Kobayashi and A. Yamamoto, *Bull. Chem. Soc. Jpn.*, **39**, 2736 (1966).
[91] R. C. Ebersole and F. C. Chang, *J. Org. Chem.*, **38**, 2579 (1973).
[92] L. A. Paquette, A. R. Browne, E. Chamot, and J. F. Blount, *J. Am. Chem. Soc.*, **102**, 643 (1980).

93 R. A. Roberts, V. Schüll, and L. A. Paquette, *J. Org. Chem.*, **48**, 2076 (1983).
94 M. Christl and G. Brüntrup, *Chem. Ber.*, **107**, 3908 (1974).
95 J. T. Groves and K. W. Ma, *J. Am. Chem. Soc.*, **99**, 4076 (1977).
96 J. A. Berson and M. S. Poonian, *J. Am. Chem. Soc.*, **88**, 170 (1966); J. A. Berson, M. S. Poonian, and W. J. Libbey, *ibid.*, **91**, 5567 (1969).
97 J. W. Hamersma and E. I. Snyder, *J. Org. Chem.*, **30**, 3985 (1965).
98 A. J. Bellamy, *J. Chem. Soc., Perkin Trans. 1*, **1972**, 342.
99 W. Dürckheimer, *Justus Liebigs Ann. Chem.*, **721**, 240 (1969).
100 R. Appel and W. Büchner, *Justus Liebigs Ann. Chem.*, **654**, 1 (1962).
101 F. Aylward and M. Sawistowska, *Chem. Ind. (London)*, **1962**, 484.
102 F. Aylward and M. Sawistowska, *Chem. Ind. (London)*, **1961**, 404.
103 F. Gavino, S. V. Luis, and A. M. Costero, *Reactive Polymers*, **6**, 291 (1987).
104 E. E. van Tamelen, J. I. Brauman, and L. E. Ellis, *J. Am. Chem. Soc.*, **93**, 6145 (1971).
105 E. Bascetta, F. D. Gunstone, and C. M. Scrimgeour, *J. Chem. Soc., Perkin Trans. 1*, 2199 (1984).
106 E. E. van Tamelen and S. P. Pappas, *J. Am. Chem. Soc.*, **85**, 3297 (1963).
107 N. Akbulut and M. Balci, *J. Org. Chem.*, **53**, 3338 (1988).
108 R. Srinivasan and J. N. C. Hsu, *J. Chem. Soc., Chem. Commun.*, **1972**, 1213.
109 W. Kirmse, P. V. Chiem, and P.-G. Henning, *Tetrahedron*, **41**, 1441 (1985).
110 P. B. Hulbert, E. Bueding, and C. H. Robinson, *J. Med. Chem.*, **16**, 72 (1973).
111 S. W. Mojé and P. Beak, *J. Org. Chem.*, **39**, 2951 (1974).
112 W. Adam and M. Balci, *J. Am. Chem. Soc.*, **101**, 7537 (1979).
113 H. C. Brown, J. H. Kawakami, and K. T. Liu, *J. Am. Chem. Soc.*, **95**, 2209 (1973).
114 S. Nishida, K. Fushimi, and T. Tsuji, *J. Chem. Soc., Chem. Commun.*, **1973**, 525.
115 T. Negoro and Y. Ikeda, *Org. Prep. Proceed. Int.*, **19**, 71 (1987).
116 R. N. McDonald and G. E. Davis, *J. Org. Chem.*, **34**, 1916 (1969).
117 W. L. Mock, *J. Am. Chem. Soc.*, **92**, 3807 (1970).
118 R. W. Hoffmann, N. Hauel, and B. Landmann, *Chem. Ber.*, **116**, 389 (1983).
119 V. V. Ranade and A. Alter, *J. Labelled Comp. Radiopharm.*, **17**, 733 (1980).
120 B. Akermark and N.-G. Johansson, *Acta Chem. Scand.*, **21**, 583 (1967).
121 R. Cahill, D. H. G. Crout, M. V. M. Gregorio, M. B. Mitchell and U. S. Muller, *J. Chem. Soc., Perkin Trans. 1*, 173 (1983).
122 J. M. Bobbitt, A. R. Katritzky, P. D. Kennewell, and M. Snarey, *J. Chem. Soc. B*, **1968**, 550.
123 E. Schmitz and R. Ohme, *Angew. Chem.*, **73**, 807 (1961).
124 U. Burger and B. Bianco, *Helv. Chim. Acta*, **66**, 60 (1983).
125 S. Pikulin and J. A. Berson, *J. Am. Chem. Soc.*, **110**, 8500 (1988).
126 U. Burger and D. Zellweger, *Helv. Chim. Acta*, **69**, 676 (1986).
127 R. L. Cargill, A. B. Sears, J. Boehm, and M. R. Wilcott, *J. Am. Chem. Soc.*, **95**, 4346 (1973).
128 H. Hart and Y. Takehira, *J. Org. Chem.*, **47**, 4370 (1982).
129 G. Schroeder, *Chem. Ber.*, **97**, 3140 (1964).
130 H. D. Martin, S. Kagabu, and R. Schwesinger, *Chem. Ber.*, **107**, 3130 (1974).
131 R. Kiwus, W. Schwarz, I. Rossnagel, and H. Musso, *Chem. Ber.*, **120**, 435 (1987).
132 A. de Meijere and L.-U. Meyer, *Chem. Ber.*, **110**, 2561 (1977).
133 A. F. Diaz and R. D. Miller, *J. Am. Chem. Soc.*, **100**, 5905 (1978).
134 D. M. Brown, A. D. McNaught, and P. Schell, *Biochem. Biophys. Res. Commun.*, **24**, 967 (1966).
135 J. D. Kennedy, H. G. Kuivila, F. L. Pelzar, and R. Y. Tien, *J. Organomet. Chem.*, **61**, 167 (1973).
136 E. V. Dehmlow and R. Kramer, *Z. Naturforsch.*, **42b**, 489 (1987).
137 M. Baumann and G. Köbrich, *Tetrahedron Lett.*, **1974**, 1217.
138 A. Gangadhar, R. Subbarao, and G. Lakshminarayano, *J. Am. Oil Chem. Soc.*, **61**, 1239 (1984).
139 M. Ohno and M. Okamoto, *Tetrahedron Lett.*, **1964**, 2423.
140 H. Nozaki, Y. Simokawa, T. Mori, and R. Noyori, *Can. J. Chem.*, **44**, 2921 (1966).

[141] H. C. Volger and H. Hogeveen, *Recl. Trav. Chim. Pays-Bas,* **86**, 1356 (1967).
[142] R. W. Hoffmann, H. R. Kurz, J. Becherer, and M. T. Reetz, *Chem. Ber.*, **111**, 1264 (1978).
[143] G. Köbrich and M. Baumann, *Angew. Chem Int. Ed. Engl.*, **11**, 52 (1972).
[144] J. Wright, G. J. Drtina, R. A. Roberts, and L. A. Paquette, *J. Am. Chem. Soc.*, **110**, 5806 (1988).
[145] A. Nickon, R. Ferguson, A. Bosch, and T. Iwadare, *J. Am. Chem. Soc.*, **99**, 4518 (1977).
[146] E. J. Corey, H. Yamamoto, D. K. Herron, and K. Achiwa, *J. Am. Chem. Soc.*, **92**, 6635 (1970).
[147] J. A. Kepler, A. Philip, Y. W. Lee, H. A. Musallam, and F. I. Carroll, *J. Med. Chem.*, **30**, 1505 (1987).
[148] A. Bosch and R. K. Brown, *Can. J. Chem.*, **46**, 715 (1968).
[149] P. D. Bartlett and G. L. Combs, *J. Org. Chem.*, **49**, 625 (1984).
[150] J. W. Huffman and P. C. Raveendranath, *Tetrahedron,* **43**, 5557 (1987).
[151] H. R. Fransen, G. J. M. Dormas, and H. M. Buck, *Tetrahedron,* **39**, 2981 (1983).
[152] L. A. Blanchard and J. A. Schneider, *J. Org. Chem.*, **51**, 1372 (1986).
[153] G. Bergstrom, B. Kullenberg, S. Stallburg-Stenhagen, and E. Stenhagen, *Ark. Kemi*, **28**, 1372 (1967) [*C. A.*, **69**, 9787 (1968)].
[154] J. A. Kepler, A. Philip, Y. W. Lee, M. C. Morey, and F. I. Carroll, *J. Med. Chem.*, **31**, 713 (1988).
[155] C. A. Brown, M. C. Desai, and P. K. Jadhav, *J. Org. Chem.*, **51**, 162 (1986).
[156] B. F. Plummer, Z. Y. Alsaigh, and M. Arfan, *J. Org. Chem.*, **49**, 2069 (1984).
[157] L. A. Paquette, M. C. Bohm, R. V. C. Carr, and R. Gleiter, *J. Am. Chem. Soc.*, **102**, 7218 (1980).
[158] H. W. Thompson and E. McPherson, *J. Org. Chem.*, **42**, 3350 (1977).
[159] E. J. Corey and T. M. Eckrich, *Tetrahedron Lett.,* **25**, 2415 (1984).
[160] Y. Yamaguchi, K. Hayakawa, and K. Kanematsu, *J. Chem. Soc., Chem. Commun.*, **1987**, 515.
[161] H. J. Dutton, C. R. Scholfield, E. P. Jones, E. H. Pryde, and J. C. Cowan, *J. Am. Oil Chem. Soc.*, **40**, 175 (1963).
[162] G. Mallet, O. Morin, and E. Ucciani, *Rev. Fran. Coprs. Gras,* **30**, 478 (1983) [*C. A.*, **100**, 190375e (1984)].
[163] R. A. Awl, E. N. Franke., and L. W. Tjarks, *Chem. Phys. Lipids,* **34**, 25 (1983).
[164] J. Conway, W. M. N. Ratnayake, and R. G. Ackman, *J. Am Oil Chem. Soc.*, **62**, 1340 (1985).
[165] L. J. Morris, R. V. Harris, W. Kelly, and A. T. James, *Biochem. J.*, **109**, 673 (1968).
[166] B. R. Talamo and K. Bloch, *Anal. Biochem.*, **29**, 300 (1969).
[167] U. Eppenberger, M. E. Warren, and H. Rapoport, *Helv. Chim. Acta*, **51**, 381 (1968).
[168] Y. J. Abul-Hajj, *J. Org. Chem.*, **36**, 2730 (1971).
[169] R. G. Powell, C. R. Smith, Jr., and I. A. Wolff, *J. Org. Chem.*, **32**, 1442 (1967).
[170] N. B. Berry and R. T. Weavers, *Aust. J. Chem.*, **41**, 81 (1988).
[171] J. S. Mills, *J. Chem. Soc., C,* **1967**, 2514.
[172] J. B. Pierce and H. M. Walborsky, *J. Org. Chem.*, **33**, 1962 (1968).
[173] E. J. Corey and H. Yamamoto, *J. Am. Chem. Soc.*, **92**, 6637 (1970).
[174] W. C. Agosta, *J. Am. Chem. Soc.*, **89**, 3505 (1967).
[175] E. J. Corey and H. Yamamoto, *J. Am. Chem. Soc.*, **92**, 6636 (1970).
[176] M. Kočor and B. Bersz, *Tetrahedron*, **41**, 197 (1985).
[177] M. Herin, P. Delbar, J. Remion, P. Sandra, and A. Krief, *Tetrahedron Lett.*, **1979**, 3107.
[178] M. Onda, K. Yonezawa, K. Abe, H. Tayama, and T. Suzuki, *Chem. Pharm, Bull.*, **19**, 317 (1971).
[179] L. A. Paquette and Y. Miyahara, *J. Org. Chem.*, **52**, 1265 (1987).
[180] J. Wicha and M. Masnyk, *Bull. Acad. Pol. Sci., Ser. Sci. Chim.*, **33**, 19 (1985) [*C. A.* **103**, 160756z (1985).
[181] M. Kočor and B. Bersz, *Tetrahedron,* **41**, 197 (1985).
[182] B. M. Trost, G. M. Bright, C. Frihart, and D. Brittelli, *J. Am. Chem. Soc.*, **93**, 737 (1971)

183 L. A. Paquette, J. C. Weber, T. Kobayashi, and Y. Miyahara, *J. Am. Chem. Soc.*, **110**, 8591 (1988).
184 A. M. Paliokas and G. J. Schroepfer, Jr., *J. Biol. Chem.*, **243**, 453 (1968).
185 F. M. Hauser, P. Hewawasam, and D. Mal, *J. Am. Chem. Soc.*, **110**, 2919 (1988).
186 J. Wicha, M. Masnyk, W. Schonfeld, and K. R. H. Repke, *Heterocycles*, **20**, 231 (1983).
187 H. W. Kirchner, *Lipids*, **8**, 101 (1973) [*C. A.*, **78**, 124807y (1973)].
188 J. D. White, T. C. Somers, and G. N. Reddy, *J. Am. Chem. Soc.*, **108**, 5352 (1986).
189 H. W. Whitlock, Jr., R. Hanauer, M. Y. Oester, and B. K. Bower, *J. Am. Chem. Soc.*, **91**, 7485 (1969).
190 T. D. Nang, Y. Katabe, and Y. Minoura, *Polymer*, **17**, 117 (1976).
191 H. J. Harwood, D. B. Russell, J. J. A. Verthe, and J. Zymonas, *Makromol. Chem.*, **163**, 1 (1973).
192 L. A. Mango and R. W. Lenz, *Makromol. Chem.*, **163**, 13 (1973).
193 G. S. Furman and W. F. W. Lonsky, *J. Wood Chem.*, **8**, 191 (1988).
194 L. Crombie, P. A. Jenkins, and J. Roblin, *J. Chem. Soc. Perkin Trans. 1*, **1975**, 1099.
195 N. Nagendrappa and D. Devaprabhakara, *Tetrahedron Lett.*, **1970**, 4243.
196 D. J. Pasto and J. K. Borchardt, *Tetrahedron Lett.*, **1973**, 2517.
197 J. K. Stille and J. H. Simpson, *J. Am. Chem. Soc.*, **109**, 2138 (1987).
198 G. Köbrich, D. Merkel, and K.-W. Thiem, *Chem. Ber.*, **105**, 1683 (1972).
199 H. H. Inhoffen, K. Radscheit, and H. Dettmer, *Justus Liebigs Ann. Chem.*, **692**, 66 (1966).
200 C. M. Gupta, G. H. Jones, and J. G. Moffat, *J. Org. Chem.*, **41**, 3000 (1976).
201 A. Roedig and W. Wenzel, *Chem. Ber.*, **102**, 3135 (1969).
202 E. Fahr and H.-D. Rupp, *Justus Liebigs Ann. Chem.*, **712**, 93 (1968).

CHAPTER 3

THE PUMMERER REACTION OF SULFINYL COMPOUNDS

Ottorino De Lucchi

Dipartimento di Chimica dell'Università, Sassari, Italy

Umberto Miotti and Giorgio Modena

Centro Studi Meccanismi di Reazioni Organiche del C.N.R., Dipartimento di Chimica Organica dell'Università, Padova, Italy

CONTENTS

Organic Reactions, Vol. 40, Edited by Leo A. Paquette et al.
ISBN 0-471-53841-8 Published by John Wiley & Sons, Inc.

INTRODUCTION

The Pummerer reaction involves the formation of an α-functionalized sulfide from a sulfoxide bearing at least one α-hydrogen atom.[1–14] The reaction can also be described as an internal redox process where the S=X group is reduced and the α carbon is oxidized (Eq. 1).

$$\underset{\displaystyle \text{RSCHR'R''}}{\overset{\displaystyle \text{X}}{\|}} \longrightarrow \text{RSCYR'R''} \qquad \text{(Eq. 1)}$$

X = O, NR
Y = OH, O_2CR, Halogen, OR, SR, NR_2, etc.

The first report by Pummerer on the reaction which now bears his name appeared in 1909 and described the formation of thiophenol and glyoxylic acid on heating phenylsulfinylacetic acid with mineral acids.[15,16] The products Pummerer observed resulted from hydrolysis of the initially formed α-substituted sulfides, which are the typical products of the reaction. The term "Pummerer reaction" was later extended to the reaction of sulfoxides with acid anhydrides.[17]

Selenium and nitrogen analogs undergo similar reactions. The former is known as the seleno–Pummerer reaction, and the latter is usually referred to as the Polonovski reaction.[6,6a,18] The sila–Pummerer reaction, which is also discussed in this chapter, is the rearrangement of sulfoxides bearing a silyl group on the α carbon.

From a mechanistic point of view there are many other reactions, sometimes given specific names, such as the Sommelet–Hauser, Stevens, and Vilsmeier rearrangements, that appear to resemble the Pummerer reaction. Reactions in which the sulfoxide group acts as an oxidant in an intermolecular redox process have characteristics similar to the typical Pummerer reaction. The α-halogenation of sulfides, in which the sulfide sulfur may first be oxidized to a halosulfonium salt that rearranges to the final product, is formally similar to the Pummerer reaction.

$$\text{RSCHR'R''} + \text{X}^+ \longrightarrow \text{R}\overset{+}{\text{S}}\text{(X)CHR'R''} \longrightarrow \text{RSCXR'R''} + \text{H}^+ \quad \text{(Eq. 2)}$$

For the sake of clarity and to be as exhaustive as possible, we have limited the scope of this chapter to the restrictive definition of Eq. 1.

MECHANISM

A generalized mechanism for the Pummerer reaction is illustrated in Eq. 3.

$$\text{R}\overset{\text{O}}{\overset{\|}{\text{S}}}\text{CHR'R''} \underset{}{\overset{\text{E}^+}{\rightleftharpoons}} \text{R}\overset{\text{OE}}{\overset{|}{\underset{+}{\text{S}}}}\text{CHR'R''} \overset{-\text{H}^+}{\rightleftharpoons} \left[\text{R}\overset{\text{OE}}{\overset{|}{\text{S}}}\text{=CR'R''} \leftrightarrow \text{R}\overset{\text{OE}}{\overset{|}{\underset{+}{\text{S}}}}\text{-}\underset{-}{\text{C}}\text{R'R''}\right] \overset{-\text{EO}^-}{\rightleftharpoons} \left[\text{R}\overset{+}{\text{S}}\text{=CR'R''} \leftrightarrow \text{RS-}\overset{+}{\text{C}}\text{R'R''}\right] \overset{\text{Y}^-}{\longrightarrow} \text{RS}\overset{\text{Y}}{\overset{|}{\text{C}}}\text{R'R''} \quad \text{(Eq. 3)}$$

The detailed mechanism has been investigated by several authors and is the subject of a recent review.[1]

In general, the Pummerer reaction requires (1) an electrophile (E^+) to activate the sulfoxide and to transform the oxygen into a good leaving group, (2) a general base to remove the proton, and (3) a nucleophile (Y^-) to be incorporated into the final product.

From the synthetic point of view, much of the potential of the Pummerer reaction depends upon the group to be introduced α to the sulfide, which can be hydroxy, carboxy, alkoxy, alkylthio, sulfido, alkylamino, alkenyl, and even an aromatic nucleus. Under some conditions, the process can be catalyzed by bases, most commonly tertiary amines, pyridine, triethylamine, lutidine, collidine, or proton sponges.

With suitable substrates and reaction conditions, the Pummerer reaction leads to vinyl sulfides either by direct β elimination from the cationic intermediates or by elimination of HY from the Pummerer products.

$$\text{R}\overset{+}{\text{S}}\text{C-}\underset{\text{H}}{\overset{|}{\underset{|}{\text{C}}}}\text{—} + \text{Y}^- \longrightarrow \overset{\text{RS}}{\diagdown}\text{C=C}\overset{\diagup}{\diagdown} + \text{HY} \quad \text{(Eq. 4)}$$

REGIO- AND STEREOSELECTIVITY

The regioselectivity of the Pummerer reaction depends on the relative kinetic acidity of the α protons. Thus the methyl group is substituted in a sulfoxide containing a methyl group and an alkyl chain, and an electron-withdrawing group (EWG) directs the nucleophile to the carbon bearing it.

$$RCH_2S(O)CH_3 + Ac_2O \longrightarrow RCH_2SCH_2OAc$$

$$RCH_2S(O)CH_2EWG + Ac_2O \longrightarrow RCH_2SCH(OAc)EWG$$

However, the usual regioselectivity can be overcome by steric factors. As an example, the different behavior of the isomeric sulfoxides **1** and **2** is attributed to steric factors.[19]

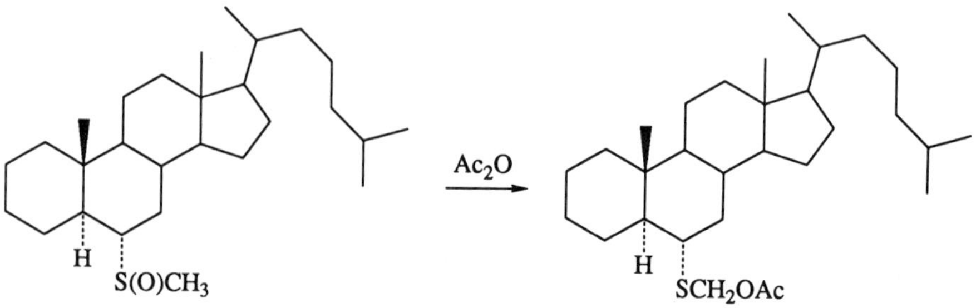

H S(O)CH$_3$ **1** $\xrightarrow{Ac_2O}$ H SCH$_2$OAc

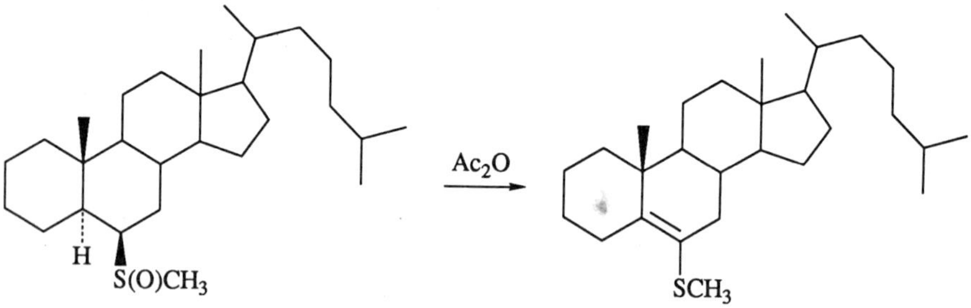

H S(O)CH$_3$ **2** $\xrightarrow{Ac_2O}$ SCH$_3$

There are a few examples of remote Pummerer reactions. For example, the Pummerer reaction of Eq. 5[20] is preceded by transfer of the "oxidation state" from the diaryl-substituted sulfur to the dialkyl-substituted sulfur. A dication intermediate[20] or nucleophilic attack of the counterion at the distal sulfur[21] is suggested for the intramolecular rearrangement.

Chirality transfer from the sulfoxide sulfur to the α carbon via the Pummerer reaction is reported for only a few reactions, and generally with rather low efficiency in both inter- and intramolecular processes.[22–27] Chiral sulfoxides undergo racemization under most of the reaction conditions used.

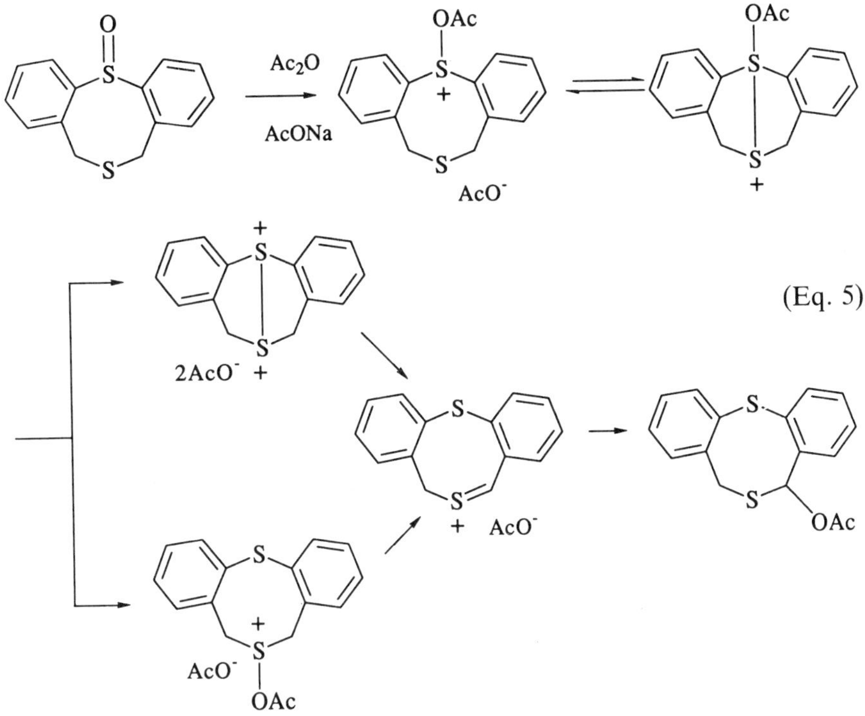

(Eq. 5)

The prerequisites for a stereoselective process appear to be (1) that the configuration of the sulfoxide not change under the reaction conditions, (2) that the leaving group EO in Eq. 3 not depart before attack of the nucleophile Y, and (3) that the addition of the nucleophile to the thionium ion not be reversible. The possibility that the nucleophile is intramolecularly and concertedly transferred from sulfur to the neighboring carbon has been studied by using both the chirality of sulfur and ^{18}O labeling.[26] Addition of dicyclohexylcarbodiimide (DCC) increases stereoselectivity.[26]

Stereoselection derived from stereocenters other than sulfoxide is usually more effective. For example, single stereoisomers are formed in the reactions of Eqs. 6 and 7[28,29] and Eqs. 8 and 9.[25] Additional examples are given in the tables.

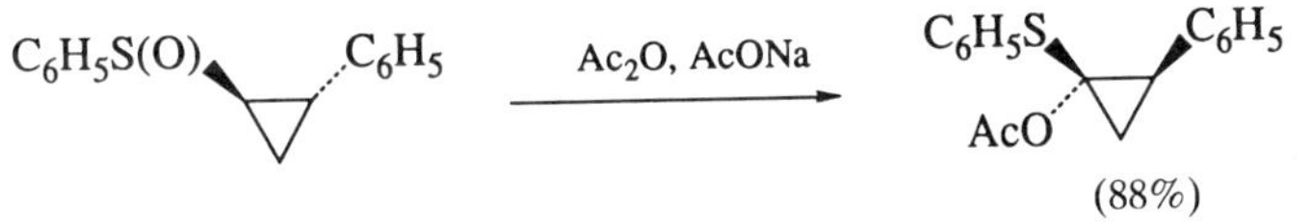

(Eq. 6)

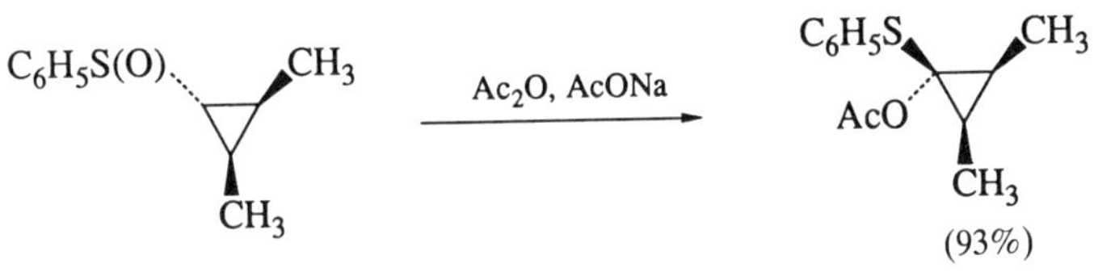

(Eq. 7)

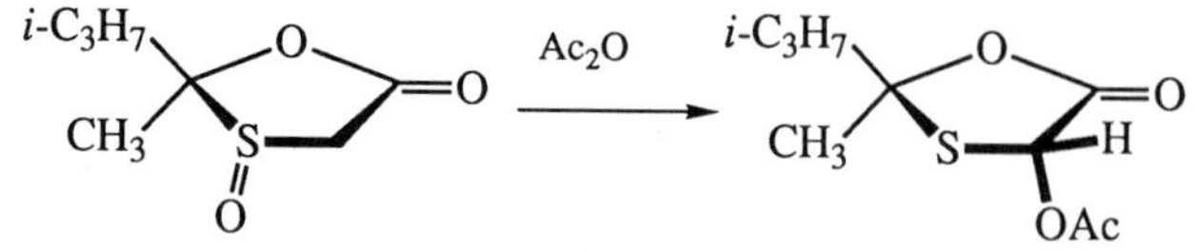

$\xrightarrow{Ac_2O}$ (Eq. 8)

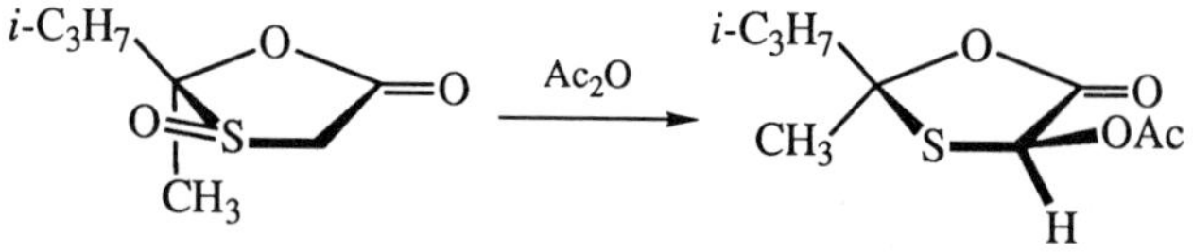

$\xrightarrow{Ac_2O}$ (Eq. 9)

REAGENTS

Mineral Acids. Hydrochloric and sulfuric acids were the first reagents used by Pummerer, but they are rarely used now unless the hydrolysis products are desired. Moreover, with these acids, decomposition may occur and condensation products may be formed.[30] With α-keto sulfides or α-carboxy sulfides it is possible to drive the reaction to an isolable α-hydroxy sulfide, but the isolation of such a product appears to be limited to these classes of compounds.[31–33]

$$CH_3S(O)CH_2COC_6H_5 \xrightarrow{\text{aq. HCl}} CH_3SCHOHCOC_6H_5$$

$$C_6H_5S(O)CH_2CO_2H \xrightarrow{\text{spontaneous}} C_6H_5SCHOHCO_2H$$

Pummerer-type reactions also occur when a sulfoxide is treated with hydrogen chloride gas, either in the absence of solvent[15,16] or in anhydrous diethyl ether or ethanol.[34,35] Concentrated sulfuric acid in benzene promotes phenylation reactions.[36]

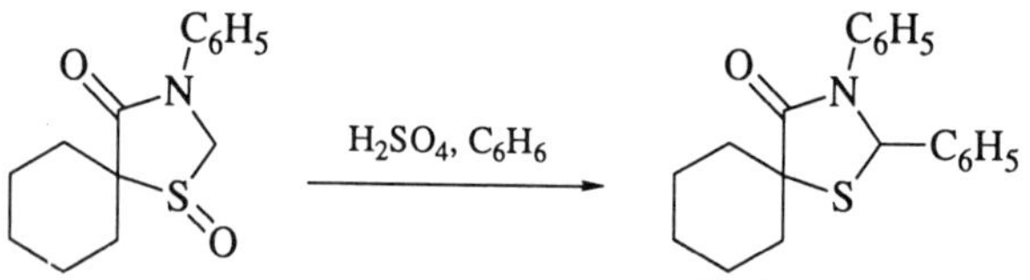

$\xrightarrow{H_2SO_4,\ C_6H_6}$ (93%)

***p*-Toluenesulfonic Acid (TsOH).** Suitable sulfoxides are converted into vinyl sulfides on treating with *p*-toluenesulfonic acid in refluxing benzene in a Dean–Stark apparatus for azeotropic water removal. The reaction corresponds to a Pummerer reaction carried out in the absence of a nucleophile so that the intermediate carbocation loses a proton to form a double bond.

p-Toluenesulfonic acid is the reagent of choice for the synthesis of naphthalene and phenanthrene derivatives,[37] as well as condensed heterocycles, carbazoles, indoles, and benzothiophenes[38] by cyclization of β-keto sulfoxides.[39]

$$\text{2-(}C_6H_5COCH_2S(O)CH_2CH_2\text{)benzene} \xrightarrow[\text{reflux}]{\text{TsOH, } C_6H_6} \text{1-benzoylisothiochroman (86\%)}$$

$$\text{pyrrole-}CH_2CH_2COCH_2S(O)CH_3 \text{ (5-}C_6H_5\text{, NH)} \xrightarrow[\text{reflux}]{\text{TsOH, THF}} \text{4-}SCH_3\text{-tetrahydroindol-5-one, 2-}C_6H_5 \text{ (63\%)}$$

If the sulfoxide bears at the α position an electron-withdrawing group such as ethoxycarbonyl, treatment with *p*-toluenesulfonic acid under conditions of continuous water removal brings about electrophilic aromatic substitution even on simple aromatic hydrocarbons such as benzene.[40]

$$CH_3S(O)CH_2CO_2C_2H_5 \xrightarrow[\text{reflux}]{C_6H_6,\ \text{TsOH}} CH_3SCH(C_6H_5)CO_2C_2H_5$$

Acetic Anhydride. Acetic anhydride is by far the most commonly used reagent for the Pummerer reaction. It is generally used in large excess as the solvent or in such solvents as benzene, carbon tetrachloride, or ethyl acetate. The addition of a cocatalyst such as acetic acid,[41] *p*-toluenesulfonic acid,[41] or trifluoroacetic anhydride[42,43] is often recommended because the cocatalyst minimizes side reactions and increases product yields. Whatever the activating species, the product is usually the acetoxy sulfide. Likewise, base catalysis is also beneficial to the reaction. The most common base used is sodium acetate because it probably acts both as a base and as an additional nucleophile.

Trifluoroacetic Anhydride (TFAA). The more electrophilic trifluoroacetic anhydride promotes Pummerer reactions under mild conditions.[44] It is suitable for introducing aryloxy, arylamino, and alkylthio groups α to sulfur.[45,46] A mixture of trifluoroacetic acid and trifluoroacetic anhydride can be used to generate the Pummerer intermediate from α-methylsulfinylacetamide,[47] and also effects the reaction with alkenes (Eq. 10).

$$CH_3S(O)CH_2CONH_2 \xrightarrow[CF_3CO_2H]{\text{TFAA}} [CH_3\overset{+}{S}{=}CHCONH_2]\ CF_3CO_2^- \xrightarrow{n\text{-}C_6H_{13}CH{=}CH_2}$$

$$n\text{-}C_5H_{11}CH{=}CHCH_2CH(SCH_3)CONH_2 \qquad \text{(Eq. 10)}$$

Trifluoroacetic anhydride in combination with the Lewis acid tin tetrachloride is very effective, and permits the preparation of benzylic thioethers from unactivated substrates (Eq. 11).[48,49]

$$CH_3S(O)CH_3 + C_6H_6 \xrightarrow[\text{2. } SnCl_4\text{, rt}]{\text{1. TFAA}} CH_3SCH_2C_6H_5 \quad \text{(Eq. 11)}$$

On the other hand, when a nonnucleophilic base such as pyridine is added, the reaction leads to the α-trifluoroacetoxy sulfide.[50]

$$C_6H_5S(O)CH_2\text{-(2,2-dimethyl-1,3-dioxolan-4-yl)} \xrightarrow[C_6H_6\text{, Pyr, rt}]{\text{TFAA}} C_6H_5S\text{-}CH(O_2CCF_3)\text{-(2,2-dimethyl-1,3-dioxolan-4-yl)} \quad (84\%)$$

Isopropenyl Acetate. Despite its potential, this acetyl transfer agent is rarely employed in the Pummerer reaction. The product of acyl transfer is the acetone anion, which, once protonated, has little nucleophilicity. An example is a vinylogous Pummerer reaction of vinyl sulfoxides, which cannot be effected with other reagents (see below).[51]

***tert*-Butyl Bromide.** The activation of a sulfoxide, particularly dimethyl sulfoxide, by *tert*-butyl bromide is an efficient and mild method for carrying out a Pummerer reaction. This halide is especially effective when admixed with dimethyl sulfoxide in the presence of sodium bicarbonate or triethylamine at room temperature. Under these conditions, carboxylic acids, *N*-protected amino acids, and phenols are converted into the corresponding methylthiomethyl derivatives in quantitative yields.[52,53]

$$CH_3S(O)CH_3 + (CH_3)_3CBr + C_6H_5CO_2H \xrightarrow{NaHCO_3} CH_3SCH_2O_2CC_6H_5 \quad (99\%)$$

Trimethylsilyl Halides. When a sulfoxide is treated with trimethylsilyl chloride, a Pummerer reaction occurs as the result of electrophilic attack by the silyl group. The final product is predominantly the vinyl sulfide, with minor amounts of the sulfide (Eq. 12).[54]

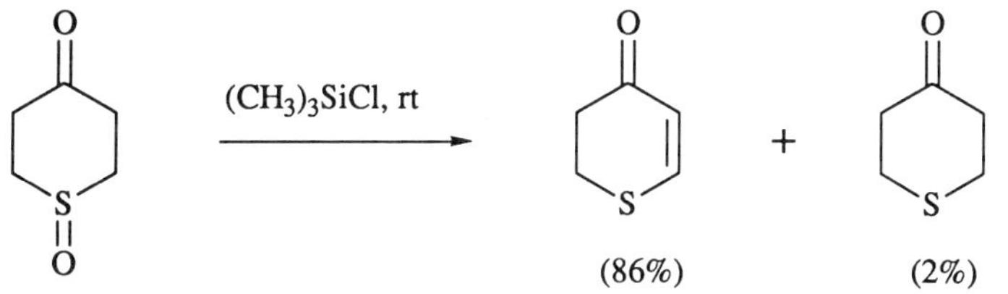

(Eq. 12)

Attack by the trimethylsilyloxy group has not been observed in this reaction, but with more complex reagents it is possible to prepare α-silyloxy sulfides. (Eq. 13).[55]

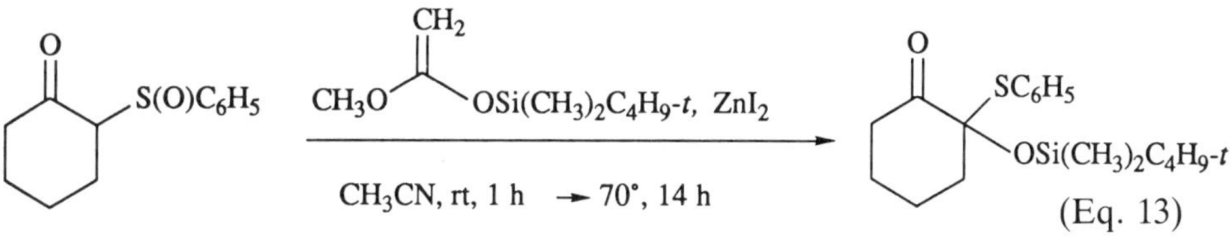

(Eq. 13)

The Pummerer reaction of thiazolidine *S*-oxides takes different paths depending on the counterion of the silyl group (iodide vs. trifluoromethanesulfonyl) (Eq. 14).[56]

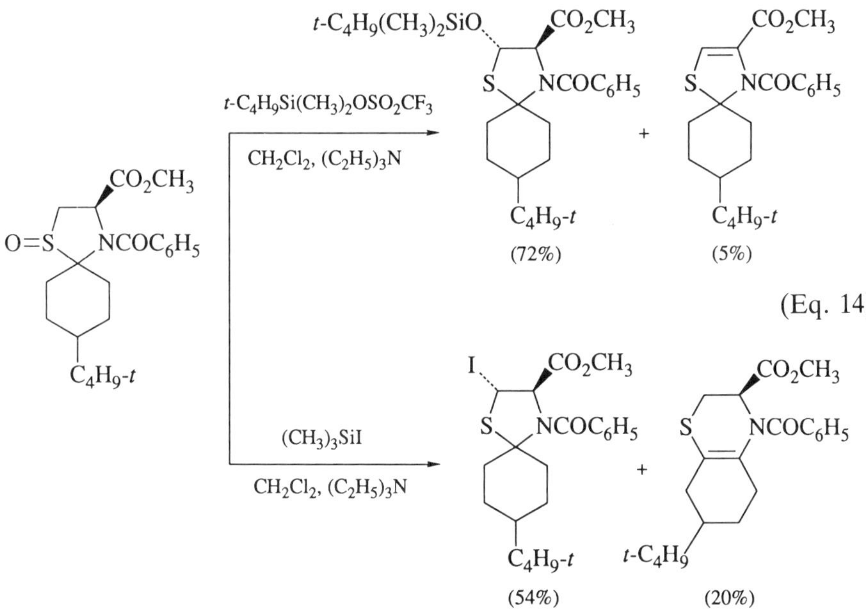

(Eq. 14)

Trimethylsilyl iodide is particularly effective for the preparation of vinyl sulfides from sulfoxides and of dienic sulfides from α,β- and β,γ-unsaturated sulfoxides.[57]

$$C_6H_5S(O)\text{-}CH_2CH{=}CH\text{-}C_3H_7\text{-}n \xrightarrow[(i\text{-}C_3H_7)_2NC_2H_5,\ 2\ h,\ 25^\circ]{(CH_3)_3SiI,\ CH_2Cl_2} C_6H_5S\text{-diene}$$

(mixture of isomers, 91%)

$$C_6H_5S(O)\text{-}CH_2CH{=}CH\text{-} \xrightarrow[(i\text{-}C_3H_7)_2NC_2H_5,\ 2.5\ h,\ 25^\circ]{(CH_3)_3SiI,\ CH_2Cl_2} C_6H_5S\text{-}CH{=}CHCH{=}CH_2$$

(>95% E, 85%)

More complex reagents such as polyphosphoric acid trimethylsilyl ester can also be used.[58]

Lewis Acids. Lewis acids are often used as catalysts in standard Pummerer reactions,[48] and are only rarely used in stoichiometric quantities. Tin(II) trifluoromethanesulfonate is an example (Eq. 15).[59]

$$C_6H_5COCH_2S(O)C_6H_4(C_4H_9\text{-}t)\text{-}p + C_6H_5C(=CH_2)OSi(CH_3)_3 \xrightarrow[CH_2Cl_2,\ 0^\circ]{Sn(OSO_2CF_3)_2,\ CH_3C(OSi(CH_3)_3)=NSi(CH_3)_3} C_6H_5COCH_2CH(SC_6H_4(C_4H_9\text{-}t)\text{-}p)COC_6H_5 \quad (82\%)$$

(Eq. 15)

Phosphorus trichloride and phosphorus pentachloride are other useful Pummerer reagents in this category.

Iodine in Alcohols. This reagent is especially useful for the transformation of a sulfoxide into an acetal under mild conditions (Eq. 16).[60–62] Iodine serves both as a source of acid to catalyze the rearrangement and as a thiol scavenger, oxidizing the product thiol and removing it from the equilibrium.

$$C_6H_5COCH_2S(O)CH_3 \xrightarrow[[HI]]{I_2,\ CH_3OH} C_6H_5COCH(OCH_3)_2 + 1/2\ CH_3SSCH_3 + 2\ HI$$

(88%)

(Eq. 16)

Diethylaminosulfur Trifluoride (DAST). There is one report on the use of this reagent for converting sulfoxides into α-fluorosulfides, which are convenient precursors for other fluoro compounds.[63] Xenon difluoride can also be used to prepare α-fluorosulfides.[64]

$$C_6H_5S(O)CH_3 + F_3SN(C_2H_5)_2 \xrightarrow{ZnI_2} C_6H_5SCH_2F \quad (85\%)$$

Photolysis. There are a few examples of Pummerer-type reactions initiated by ultraviolet light.[65,66] The aromatic group and the tertiary carbon adjacent to the sulfur atom are probably indispensable requisites.

$h\nu$; C_6H_6, Vycor; product: 2-(C_3H_7-*i*)benzothiophene

Acyl Halides. Although the Pummerer reaction can be carried out with acyl halides, these reagents are generally of little synthetic value because of the presence in the reaction mixture of both halide ion and acyloxy anion formed in the reaction. The usual reaction product is the dithioacetal derived from decomposition of the initially formed α-acyloxysulfide.[67]

$$C_6H_5S(O)CH_2C_6H_5 \xrightarrow{CHCl_2COCl} [C_6H_5SCH(C_6H_5)O_2CCHCl_2] \longrightarrow C_6H_5CH(SC_6H_5)_2 \quad (38\%)$$

Other Reagents. Other reagents that can effect the Pummerer reaction are described in the tables. Among these, the reaction of sulfoxides with a solution of a mercury(II) salt in water or an alcohol merits attention.[68,69] Although this reaction has not been studied extensively, it is potentially a useful synthetic method because the carbonyl compound is produced directly in a single step.

$$CH_3S(O)CH(CH_3)CO_2H \xrightarrow{HgCl_2} CH_3SHgCl + CH_3COCO_2H + HCl$$

RELATED REACTIONS

Several reactions related to the Pummerer reaction are postulated to involve similar intermediates. Among these are the Sommelet–Hauser, Stevens, and Vilsmeier rearrangements, which involve treatment of a sulfonium[70] or heterosulfonium salt[2,71,72] with a base. These reactions may be considered variations of the same theme applied to different substrates. A common, although oversimplified, picture is shown in Scheme 1.

$$\underset{\mathbf{4}}{\overset{\overset{X}{|}}{\underset{+}{R S}}CHR'R''} \xrightarrow{\text{base}} \underset{\mathbf{3}}{\overset{\overset{X}{|}}{\underset{+}{R S}}\text{-}\underset{-}{C}R'R''}$$

A X= O-Aryl, Sommelet-Hauser → o-(CR'R''SR)phenol (OH)

B X = Alkyl, Stevens → RSCR'R''Alkyl

C X = -NH–C(=O)–, Vilsmeier → RSCR'R''O–C(=N–)–

Scheme 1

Path **A** is an extension by Hauser of the Sommelet rearrangement originally observed with ammonium salts, which results in regiospecific *ortho* alkylation. The Stevens rearrangement of path **B** produces α-alkylated products, and the Vilsmeier rearrangement of path **C** occurs with azasulfonium salts.[2,73] All of these reactions proceed through the same ylide precursor **3**. Reactions in which the sulfonium or heterosulfonium salt **4** is derived from a sulfoxide are included in the Tabular Survey.

A reaction that is often compared to the Pummerer reaction is the α-halogenation of a sulfide.[74,75] Since sulfides are precursors of sulfoxides, this procedure may avoid one reaction step. When the α-halogenation of a sulfide is carried out with chlorine[76–78] or bromine,[79] it is often difficult to limit the

$$RSCH_2R' \xrightarrow[\text{Halogenating agent}]{\text{Halogen or}} RSCHXR'$$

reaction to monosubstitution and to avoid the formation of byproducts. Accordingly, such halogenations are usually carried out with milder reagents such as *N*-halosuccinimides,[80,81] thionyl chloride,[82] Chloreal® (trichloroisocyanuric acid),[83] and cyclic phosphorus chlorides.[84]

There are several other reagents or methods for introducing a functional group α to a sulfide via thionium ions. Singlet oxygen reacts with certain sulfides to give α-hydroperoxides rather than the expected sulfur oxidation products. These hydroperoxides can be readily reduced to α-hydroxysulfides (Eq. 17).[85,86]

$$\text{thiazolidine}(CO_2CH_3, S, NAc) \xrightarrow{^1O_2} \text{HO}_2\text{-thiazolidine}(CO_2CH_3) \xrightarrow{CH_3SCH_3} \text{HO-thiazolidine}(CO_2CH_3) \quad \text{(85-95\% overall)} \qquad \text{(Eq. 17)}$$

Sulfides react with 4-phenyl-1,2,4-triazoline-3,5-dione to afford thioaminals.[87]

+ N-C_6H_5 —($Cl_2CHCHCl_2$, 75°, 24 h)→ NH, N, =O, N-C_6H_5, O

(29%)

Pummerer-type products are produced by treatment of α-oxosulfides with thallium(III) nitrate.[88,89]

$$C_6H_5COCH(CH_3)SC_2H_5 \xrightarrow[CH_3OH,\ 30\ min]{Tl(NO_3)_3} C_6H_5COCOCH_3 \quad (59\%)$$

Sulfides that bear electron-withdrawing functionality at the α carbon (especially β-ketosulfides) can be acetoxylated in high yields by lead tetraacetate (LTA).[90] This procedure leads to the Pummerer product in a single step without the need for oxidizing the sulfide and reacting the sulfoxide with acetic anhydride.

O, SC_6H_5, C_4H_9-*t* —(LTA, C_6H_6, reflux, 3 h)→ O, SC_6H_5, OAc, C_4H_9-*t* (95%)

A transformation closely related to the Pummerer reaction is the reaction of sulfides with dibenzoyl peroxide[91,92] or other acyl peroxides[93–96] to form an α-thioester.

$$RSCH_2R' + R''\overset{O}{\overset{\|}{C}}OO\overset{O}{\overset{\|}{C}}R'' \longrightarrow RSCH(OCOR'')R'$$

The Pummerer reaction can also be brought about by anodic oxidation of alkyl aryl sulfides in boiling acetic acid containing acetic anhydride and sodium acetate.[97,98] This reaction was used in a synthesis of pellitorine.[99]

$$n\text{-}C_8H_{17}\overset{SC_6H_5}{\overset{|}{C}}HCO_2C_4H_9\text{-}t \xrightarrow[AcOH]{2.5\ F/mol} n\text{-}C_8H_{17}\overset{SC_6H_5}{\overset{|}{C}}(OAc)CO_2C_4H_9\text{-}t \xrightarrow[steps]{several}$$

n-C_5H_{11} ... $CONHC_4H_9$-*t*

pellitorine

Electrolysis of α-phenylthiocarboxylic acids leads directly to aldehydes or acetals in high yields by concomitant oxidative decarboxylation and desulfurization.[100]

$$n\text{-}C_4H_9CH(SC_6H_5)CO_2H \xrightarrow[\text{NaOH (1.5 eq)}]{H_2O} n\text{-}C_4H_9CHO + (C_6H_5S)_2 \quad (48\%)$$

$$n\text{-}C_4H_9CH(SC_6H_5)CO_2H \xrightarrow[\text{LiClO}_4\text{ (cat.)}]{CH_3OH} n\text{-}C_4H_9CH(OCH_3)_2 + C_6H_5SO_2CH_3 \quad (72\%)$$

The reaction of sulfoxides with Grignard reagents[101–104] is not considered a Pummerer reaction because a mechanism different from that of Eq. 3 is involved. The overall reaction, however, corresponds to a Pummerer process in which a carbon nucleophile is introduced α to a sulfoxide.

$$CH_3S(O)CH_3 \xrightarrow[\text{ether, reflux, 3 h}]{C_6H_5MgBr} CH_3SCH_2C_6H_5 \quad (31\%)$$

SYNTHETIC APPLICATIONS

The value of the Pummerer reaction lies mainly in the variety of synthetically useful transformations of the products.[105–107] The Pummerer reaction can be considered as a mild method for generating α-sulfur-substituted carbocations, which can be trapped by nucleophiles or can lose a proton to give the vinyl sulfide (Scheme 2).

$$RS(O)CHR'R'' \longrightarrow R\overset{+}{S}CR'R''$$

$$R\overset{+}{S}CR'R'' \xrightarrow{Y^-} \alpha\text{-functionalized sulfide}$$

$$R\overset{+}{S}CR'R'' \xrightarrow{H_2O,\ H^+} \text{carbonyl compound and thiol}$$

$$R\overset{+}{S}CR'R'' \xrightarrow{-H^+} \text{vinyl sulfide}$$

Scheme 2

Synthesis of α-Functionalized Sulfides

Carbon atoms functionalized with a sulfur and another heteroatom, such as thioketals and their oxides, are well known as one-carbon-homologizing reagents. Closely related to the Pummerer reaction products or derived therefrom are synthons containing sulfur–sulfur,[108] sulfur–oxygen,[109] sulfur–silicon,[110] or sulfur–halogen[74] functionalities.

Alcoholic hydroxy groups can be protected as their methylthiomethyl ethers by carrying out their Pummerer reaction with dimethyl sulfoxide.[111–113] Although these ethers are often byproducts of the oxidation of alcohols by dimethyl sulfoxide-acetic anhydride, they are the principal products if acetic acid is added to the reagent. Deprotection can be effected by any of several reagents, including aqueous sodium bicarbonate, copper(II) or mercury(II) chloride, or wet silica gel. The procedure can be applied to primary, secondary, and tertiary alcohols.

$$(CH_3)_2SO + ROH \xrightarrow{AcOH,\ Ac_2O} CH_3SCH_2OR \longrightarrow ROH + CH_3SH + CH_2O$$

Methylthiomethyl ethers can also be reduced to the corresponding methyl ethers by Raney nickel (Eq. 18). This reaction is carried out under mildly acidic or neutral conditions and is complementary to standard methylation procedures, which require basic media.[113]

$$ROCH_2SCH_3 \xrightarrow{\text{Raney Nickel}} ROCH_3 \qquad \text{(Eq. 18)}$$

Synthesis of Carbonyl Compounds and Thiols

The Pummerer reaction followed by hydrolysis of the product affords a carbonyl compound and a thiol. The reactions can be carried out sequentially in a one-pot operation, or directly by using hydrolytic conditions in the Pummerer step. This strategy has synthetic importance when only one of the two products is desired and the other can be easily removed from the reaction mixtures (Eqs. 19[114] and 20[115]).

$$R\text{-}C_6H_4S(O)CH_3 \xrightarrow[40^\circ]{TFAA} R\text{-}C_6H_4SCH_2OTFA \xrightarrow[2.\ NH_4Cl]{1.\ CH_3OH,\ (C_2H_5)_3N} R\text{-}C_6H_4SH \ (80\text{-}100\%) \qquad \text{(Eq. 19)}$$

$$C_6H_5S(O)CH_2R \xrightarrow[0^\circ]{TFAA} C_6H_5SCH(OTFA)R \longrightarrow RCHO \ (43\text{-}86\%) \qquad \text{(Eq. 20)}$$

Although the reaction can be used to synthesize virtually any carbonyl compound, in practice it is largely limited to the preparation of aldehydes. The reason may be the fact that α-acetoxysulfides are easily converted into vinyl sulfides instead of undergoing hydrolysis to ketones. However, β-dicarbonyl compounds, unlike simple ketones, are produced in high yields by the Pummerer reaction.[116]

$CH_3S(O)$, $C_2H_5O_2C$ — AcOH, H_2O, reflux, 4 h → $C_2H_5O_2C$ (quant.)

A very mild method that leads directly from sulfoxides to acetals employs iodine in alcohols such as methanol or ethanol (Eq. 12).[117,118] Standard Pummerer products are also converted into acetals by treatment with iodine in alcoholic solvents.[119]

Synthesis of Vinyl Sulfides

Vinyl sufides are formed by β elimination of the nucleophile from the Pummerer product or by loss of a proton from the ylide intermediate (Eq. 4). Usually, the elimination occurs directly under the Pummerer reaction conditions. The overall reaction corresponds to dehydration of a sulfoxide, and can be performed with a variety of dehydrating agents such as alumina[120] or phosphorus pentoxide. This procedure is widely used (see Table VII) because of the variety of synthetic applications of vinyl sulfides and their oxides. It is especially effective for the preparation of α-keto vinyl sulfides,[121] which are good dienophiles and Michael acceptors. The same transformation can be brought about by tin(II) trifluoromethanesulfonate.[122]

$S(O)C_6H_5$ — Ac_2O, CH_3SO_3H, CH_2Cl_2, rt, 16 h → SC_6H_5 (86%)

As the Pummerer reaction affords vinyl sulfides, the vinylogous Pummerer reaction gives dienyl sulfides. The latter reaction can be used for the synthesis of unstable molecules and for the in situ generation of reactive unsaturated sulfur heterocycles. For example, there is evidence for the transient existence of tetravalent sulfur species (Eq. 21),[120,123–125] and the three thiophthenes **5–7** have been synthesized.[126,127]

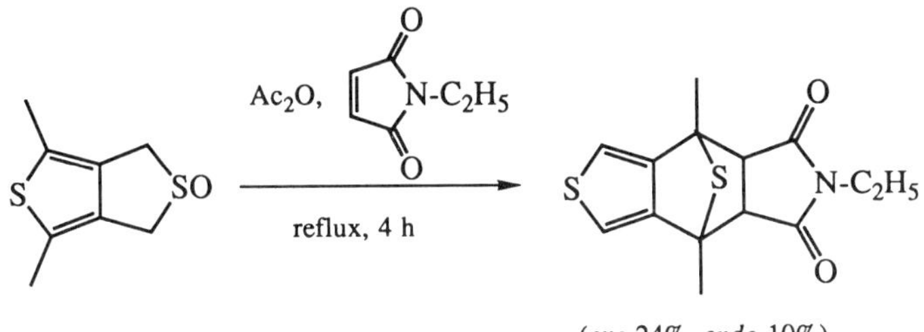

Ac_2O, N-ethylmaleimide; reflux, 4 h (Eq. 21)

(*exo* 24%, *endo* 10%)

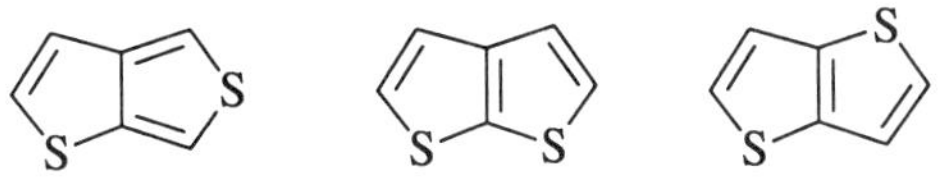

5 **6** **7**

Similar sulfoxide dehydrations can be brought about by a base-catalyzed process.[128]

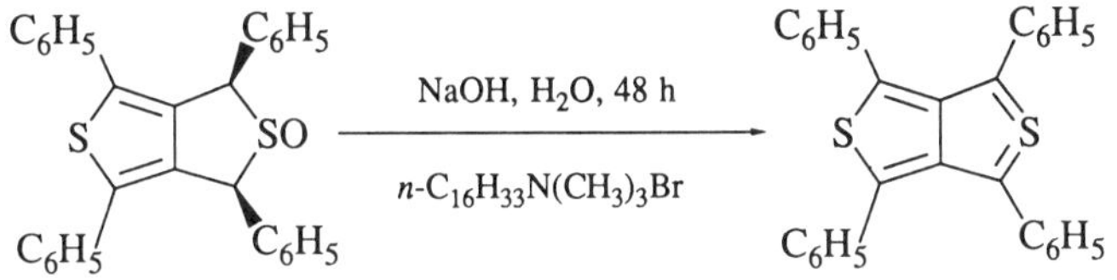

NaOH, H_2O, 48 h; n-$C_{16}H_{33}N(CH_3)_3Br$

(45%)

A related dehydration of sulfoxide is used in a convenient preparation of dehydro-1,4-dithiins.[129,130]

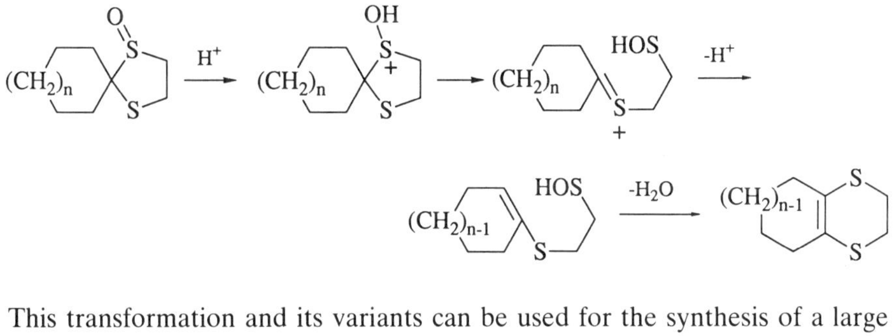

H^+; $-H^+$; $-H_2O$

This transformation and its variants can be used for the synthesis of a large number of similar compounds.[131–133] In the example of Eq. 22, the Pummerer reaction is used repeatedly to prepare the completely unsaturated product.[134]

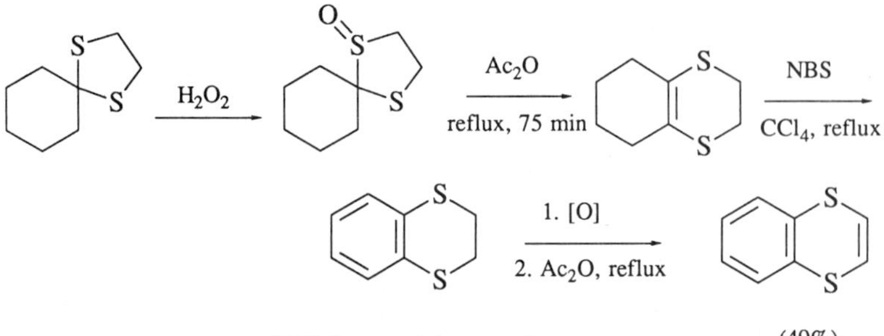

H_2O_2 ; Ac_2O, reflux, 75 min ; NBS, CCl_4, reflux

1. [O]
2. Ac_2O, reflux

(Eq. 22)

(78% from cyclohexanone) (49%)

It is also possible to convert sulfinamides into sulfenimines, with concomitant formation of acetamides.[135,136]

$$p\text{-}CH_3C_6H_4S(O)NHCH(C_6H_4Cl\text{-}p)_2 \xrightarrow{Ac_2O,\ rt,\ 70h} p\text{-}CH_3C_6H_4SN{=}C(C_6H_4Cl\text{-}p)_2 + AcNHCH(C_6H_4Cl\text{-}p)_2$$

(48%) (43%)

Application to Penicillin and Cephalosporin Chemistry[137,138]

The Pummerer reaction has been widely used in the development of the chemistry of β-lactam antibiotics. In this class of compounds, β and γ substitution occur rather than the normal α substitution. The rearrangement depends on the presence of a tertiary carbon at one side of the sulfoxide. When the sulfur atom becomes positive because of electrophilic attack, ring opening occurs with formation of a sulfenic acid derivative and the stable carbocation. Subsequent ring closure leads to product.

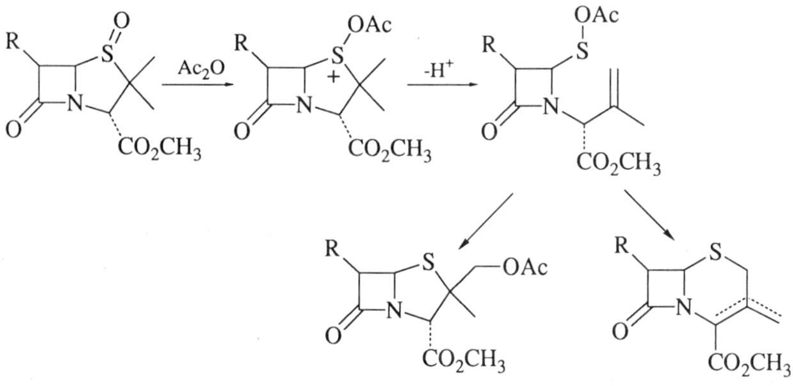

Ac_2O ; $-H^+$

R, O, S, N, CO_2CH_3, OAc

Breaking of the S—C bond in tertiary or benzyl sulfides occurs in many reactions. The reaction may take any of several routes, depending on the nature of the carbocation, but the most common is formation of a double bond and subsequent addition of the sulfenic acid derivative. This operation can transform a penicillin into a cephalosporin and is therefore important in pharmaceutical research.[139–141]

The entry to β-lactams via the Pummerer reaction (Eq. 23) does not succeed with standard reagents,[142] but can be carried out with trimethylsilyl trifluoromethanesulfonate.[143] This reagent converts the optically active sulfoxide **8** into β-lactam **9** with 67% ee, which is high for a stereoselective Pummerer reaction.[144]

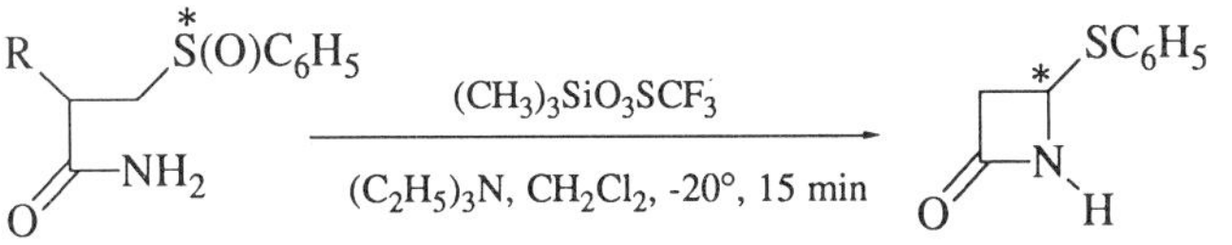

$(CH_3)_3SiO_3SCF_3$; $(C_2H_5)_3N$, CH_2Cl_2, -20°, 15 min (Eq. 23)

8 → **9** (74%)

Natural Products Synthesis

The Pummerer reaction is particularly useful in natural product chemistry because of its mild conditions and its compatibility with other functional groups in the molecule. Representative are the syntheses of berberastine,[145] leukotrienes,[146] illudin,[147,148] olivacine and ellipticine,[149] pseudo-guaianes,[150] and several saccharides.[151] The Pummerer reaction involved in the preparation of *dl*-illudin M is illustrated.

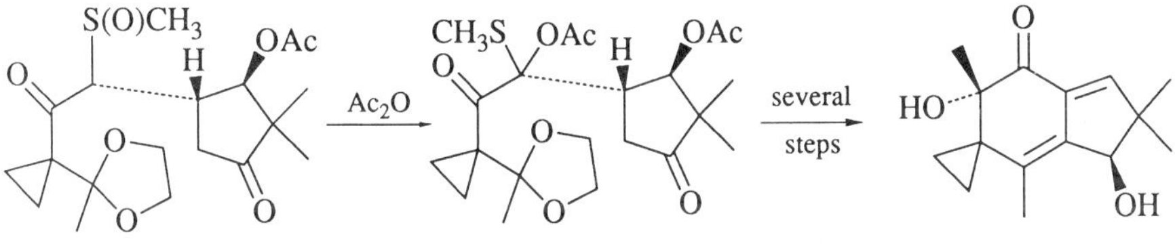

Ac_2O; several steps

illudin M

Pummerer conditions are effective in delicate cyclizations involved during the synthesis of several indole alkaloids.[152]

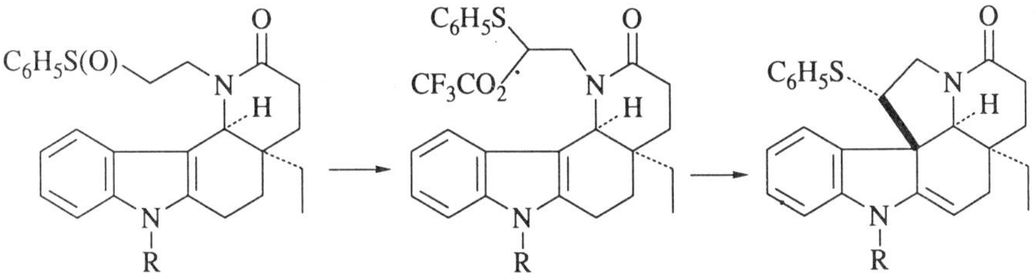

(81%)

A synthesis of enantiomerically pure polyhydroxylated natural products using an iterative two-carbon extension cycle has been reported. One of the four steps is a Pummerer reaction. The generality and effectiveness of this methodology are demonstrated by the total synthesis of all eight L-hexoses through the sequence of Scheme 3.[153]

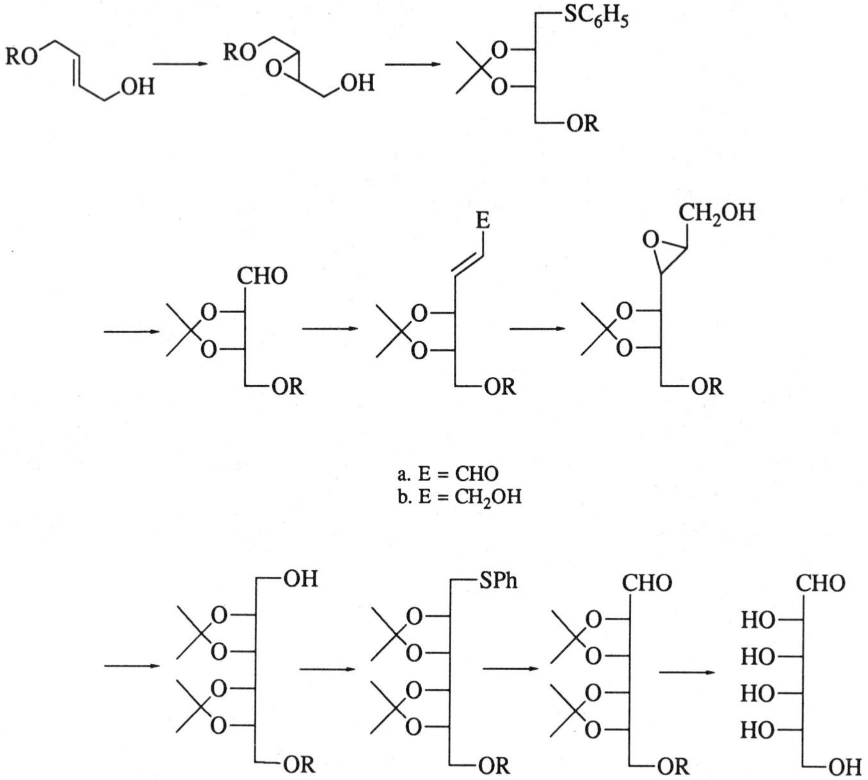

a. E = CHO
b. E = CH_2OH

Scheme 3

THE SELENO–PUMMERER REACTION

The Pummerer reaction applied to a selenoxide is called the seleno–Pummerer reaction.[154,155] However, because selenoxides bearing a β hydrogen atom readily lose RSeOH to give olefins,[156] this reaction has limited scope. Nevertheless, in a few examples, the desired seleno–Pummerer reaction can be accomplished if elimination is not possible or if the α proton is suitably activated.[157–159] As with most Pummerer reactions, the secondary products are vinyl selenides.

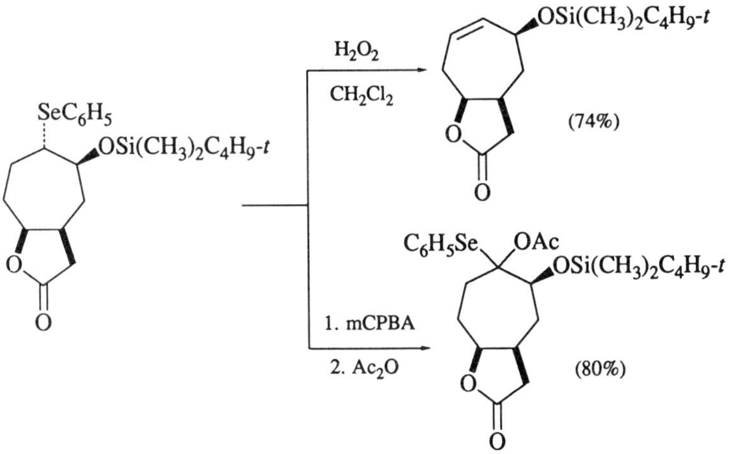
H2O2
CH2Cl2
SeC6H5
OSi(CH3)2C4H9-t
(74%)
C6H5Se
OAc
1. mCPBA
2. Ac2O
(80%)

THE SILA-PUMMERER REACTION

Sulfoxides bearing a trimethylsilyl group on the α carbon are thermally unstable and spontaneously rearrange to the Pummerer products: α-silyloxy sulfides.[160] The reaction is of interest because of the mildness of the conditions

$$R'S(O)CH_2SiR_3 \longrightarrow R'SCH_2OSiR_3$$

and because it occurs without added reagents.[161,162] It can be used for the synthesis of ketones (Eq. 24),[163,164] thioesters (Eq. 25),[165] and unsaturated aldehydes (Eq. 26),[166–168] as well as for the formylation of alkyl halides (Eq. 27).[161,169,170] The secondary products most often observed are vinyl sulfides.

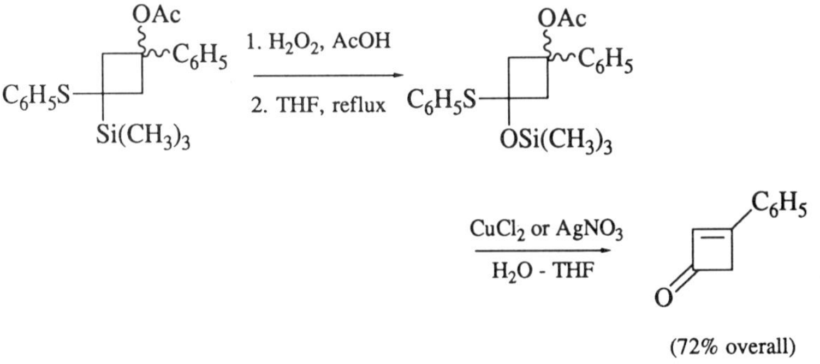
OAc
C6H5
C6H5S
Si(CH3)3
1. H2O2, AcOH
2. THF, reflux
OSi(CH3)3
CuCl2 or AgNO3
H2O - THF
(72% overall)

(Eq. 24)

$$C_6H_5CHClS(O)C_6H_5 \xrightarrow[\text{2. }(CH_3)_3SiCl,\ -78°]{\text{1. LDA, THF}} \left[C_6H_5C(Si(CH_3)_3)(Cl){-}S(O)C_6H_5 \right] \xrightarrow{0°}$$

$$\left[C_6H_5C(OSi(CH_3)_3)(Cl){-}SC_6H_5 \right] \longrightarrow \underset{(74\%)}{C_6H_5COSC_6H_5} + (CH_3)_3SiCl \quad \text{(Eq. 25)}$$

$$\xrightarrow[CH_2Cl_2,\ -48°]{\text{mCPBA,}} \xrightarrow[20°,\ 18\ h]{THF,\ H_2O,\ (CO_2H)_2} \underset{(40\text{-}60\%)}{} \quad \text{(Eq. 26)}$$

(Eq. 26: 1-vinylcyclohexyl-$CH(Si(CH_3)_3)SCH_3$ → 1-vinylcyclohexyl-$CH(OSi(CH_3)_3)SCH_3$ → 1-vinylcyclohexyl-CHO (40-60%))

$$C_6H_5SCH_2Si(CH_3)_3 \xrightarrow[\text{2. RX}]{\text{1. base}} C_6H_5SCHRSi(CH_3)_3 \xrightarrow[\text{2. }C_6H_6,\ \text{reflux}]{\text{1. mCPBA}}$$

$$C_6H_5SCHROSi(CH_3)_3 \xrightarrow[H_2O]{HCl} RCHO \quad \text{(Eq. 27)}$$

The sila–Pummerer reaction is rather sensitive to steric and electronic effects, and in some instances requires more vigorous conditions than the Pummerer reaction. For example, the diastereomeric silylsulfoxides **10** and **11** rearrange at much different rates, which has been viewed as evidence for a concerted four-center mechanism that is unfavored for **11** because it requires more severe repulsive interactions in the transition state.[171]

(Structures **10** and **11**: $(CH_3)_3Si$, C_6H_5, H on carbon bonded to S(=O)(:)C_4H_9-*t*; diastereomers)

10 **11**

Application of the sila–Pummerer reaction to bis(trimethylsilylmethyl)sulfoxide is a general method for the generation of thiocarbonyl ylides (Eq. 28).[172]

$$(CH_3)_3SiCH_2S(O)CH_2Si(CH_3)_3 \longrightarrow [(CH_3)_3SiCH_2\overset{+}{S}{=}CH_2]\ OSi(CH_3)_3^- \xrightarrow{-[(CH_3)_3Si]_2O} CH_2{\cdots}\overset{+}{S}{\cdots}CH_2$$

(Eq. 28)

If the sulfoxide and trimethylsilyl groups are attached to an sp^2 carbon, the sila–Pummerer reaction is not prevented, but the primary product is not stable and rearranges to the products shown.[173]

$$\mathrm{RCH{=}C(Si(CH_3)_3)S(O)C_6H_5} \xrightarrow[\text{20 h}]{\mathrm{C_6H_6},\ \text{reflux}} \mathrm{RCH{=}CHS(O)C_6H_5} + \mathrm{RC{\equiv}CSC_6H_5} + \mathrm{RCH{=}C(SC_6H_5)(OSi(CH_3)_3)}$$

Examples of the sila–seleno–Pummerer reaction have also been reported (Eq. 29).[174,175] As expected, the reaction occurs at low temperatures, and isolation of the α-silylselenoxides is often unfeasible.

$$\mathrm{C_5H_6Se(O)\text{-}C(Si(CH_3)_3)(CH_3)\text{-}C_6H_5} \xrightarrow[\mathrm{(i\text{-}C_3H_7)_2NH}]{\text{THF, }25^\circ} \underset{(22\%)}{\mathrm{C_5H_6Se\text{-}C(OSi(CH_3)_3)(CH_3)\text{-}C_6H_5}} + \mathrm{C_6H_5COCH_3} + \underset{(24\%)}{\mathrm{(C_6H_5)(C_6H_5Se)C{=}CH_2}} + \underset{(30\%)}{\mathrm{(C_6H_5)((CH_3)_3Si)C{=}CH_2}} \quad \text{(Eq. 29)}$$

The competition between sila–seleno–Pummerer and selenoxide *syn* elimination in Eq. 29 has been studied.[175] The ratio of the two processes can be varied to some extent by changing reaction conditions, or by changing the electron demand of the selenide substituent. With an electron-withdrawing group such as trifluoromethyl in the *meta* position of the phenyl ring, *syn* elimination is somewhat favored.[175]

THE VINYLOGOUS AND ADDITIVE PUMMERER REACTIONS

The substrates for vinylogous and additive Pummerer reactions are vinyl sulfoxides; both processes usually occur simultaneously in the same reaction. As shown in Eq. 30, in the vinylogous Pummerer reaction the nucleophile adds to the allylic carbon atom in a process that resembles the standard Pummerer reaction. Because the mechanism involves an allylic carbocation, there are two sites for attack by the nucleophile with possible formation of two products (Eq. 30).

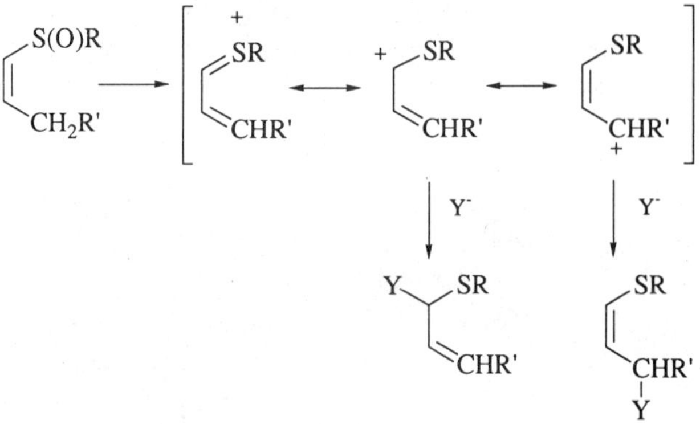

(Eq. 30)

The vinylogous Pummerer reaction cannot occur when the double bond does not have allylic hydrogens, in which event the normal Pummerer products are formed.[176]

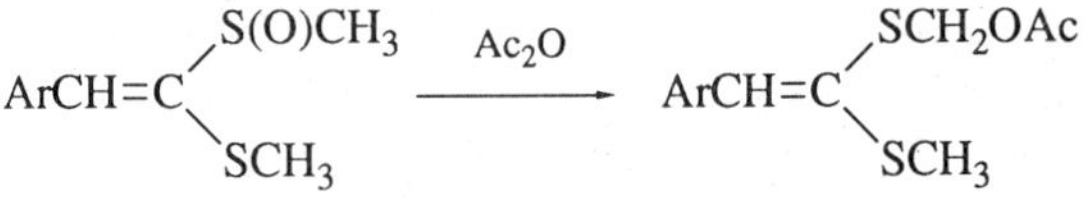

$$\text{ArCH=C(S(O)CH}_3\text{)(SCH}_3\text{)} \xrightarrow{\text{Ac}_2\text{O}} \text{ArCH=C(SCH}_2\text{OAc)(SCH}_3\text{)}$$

The additive Pummerer reaction involves addition of two molecules of the nucleophile to the double bond (Eq. 31).[177–180]

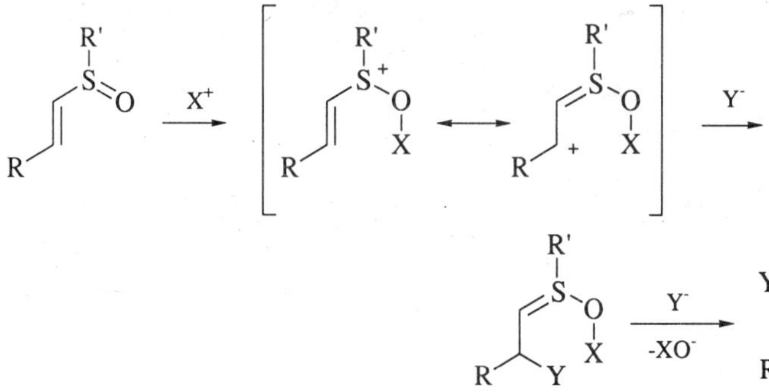

(Eq. 31)

Alternatively, a [3,3] sigmatropic rearrangement may occur with some reagents, for example thionyl chloride.[177]

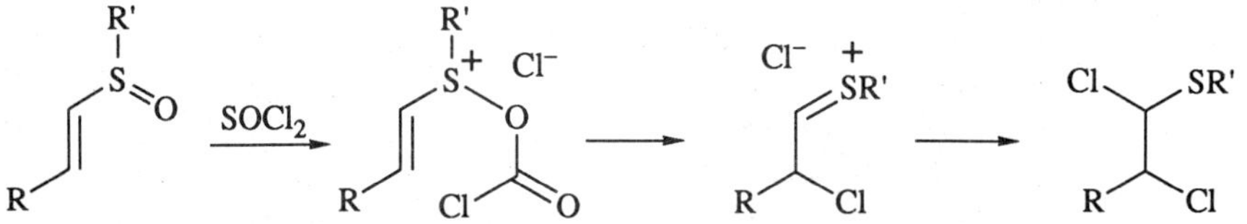

(Eq. 32)

The additive mechanism can account for formation of the interesting cycloadducts derived from addition of dichloroketene to vinyl sulfoxides (Eq. 33).[177]

(Eq. 33)

Particularly attractive from a synthetic point of view is the process carried out on chiral sulfoxides, where it is possible to transfer the chirality from sulfur to several carbon centers.[181]

Although the reaction of standard Pummerer reagents with vinyl sulfoxides fails to give the vinyl acetate or products derived from it,[182] the reaction does occur with isopropenyl acetate[51] to afford β-keto sulfides. The absence of nucleophiles is probably responsible for this anomalous Pummerer reaction, which can be rationalized as shown in Eq. 34.

(Eq. 34)

A more complex transformation takes place in the reaction of the unsaturated sulfinyl substrates **12**.[183,184] This transformation may be accounted for by a mechanism involving acetylation of the amino group and Pummerer rearrangement of the sulfinyl group with concomitant migration of the methylthio group. This reaction is of especial utility in the synthesis of amino acids from nitriles.[183]

12

SCOPE AND LIMITATIONS

The Pummerer reaction can be performed on almost any sulfide that bears α-hydrogen atoms.[182] Limitations arise from other functional groups in the molecule that may also react with the reagent. For example, hydroxy or amino groups may be acetylated if the reaction is carried out in acetic anhydride. Other possible side reactions are summarized in Scheme 4.

$$\mathrm{RS(O)CHR'R''} \xrightarrow{\text{elimination}} \mathrm{RSOH} + \;>\!C{=}C\!<$$

$$\mathrm{RS(O)CHR'R''} \underset{}{\overset{E^+}{\rightleftharpoons}} \mathrm{R\overset{+}{S}(X)CHR'R''}$$

$$\mathrm{R\overset{+}{S}(X)CHR'R''} \xrightarrow{\text{fragmentation}} \mathrm{RSX} + \mathrm{R'R''\overset{+}{C}H}$$

$$\mathrm{R\overset{+}{S}(X)CHR'R''} \xrightarrow[\text{Nu}^-]{\text{substitution}} \mathrm{R\overset{+}{S}(Nu)CHR'R''} + X^-$$

Scheme 4

Concerted elimination of sulfenic or seleninic acid[156] becomes important when the low reactivity of the substrate requires more vigorous conditions.

Fragmentation reactions occur when relatively stable cations (allylic, benzylic, or tertiary carbocations) can be formed by cleavage of the C—S bond (see section on penicillins and cephalosporins). The reaction is useful in the protection–deprotection of thiol groups.[185]

Nucleophilic substitution reactions on the heterosulfonium cation appear to be fast but reversible; they are responsible for the racemization of optically active sulfoxides under acid catalysis.[186] They do not usually interfere with the Pummerer reaction except when the substitution products lead to stable derivatives. For example, the reaction with primary amines may lead to sulfilimines by elimination of hydrogen chloride (Scheme 5, path A).[187]

$$\mathrm{R\overset{+}{S}(X)CHR'R''} \xrightarrow[\mathrm{RNH_2,\ -H^+}]{\text{formation of sulfilimines}} \mathrm{RS({=}NR)CHR'R''} + \mathrm{HX}$$

$$\mathrm{R\overset{+}{S}(X)CHR'R''} \xrightarrow{\text{sulfoxide reduction}} \mathrm{RSCHR'R''}$$

$$\mathrm{R\overset{+}{S}(X)CHR'R''} \xrightarrow[\text{oxidation},\ -\mathrm{H^+}]{>\mathrm{CHOH}} \mathrm{R\overset{+}{S}(\overset{-}{C}R'R'')\text{–}O\text{–}C(H)<} \longrightarrow \;>\!C{=}O + \mathrm{RSCHR'R''}$$

Scheme 5

Alternatively, reduction of the sulfoxide to the sulfide may occur when the nucleophile is, for example, iodide ion.[188,189] When the nucleophile is a primary or secondary alcohol, reduction of the sulfoxide is accompanied by oxidation to the corresponding aldehyde or ketone (Kornblum, Moffat, Swern oxidations, and variants).[10,81,190–192] Such a conversion is usually carried out with dimethyl sulfoxide and proceeds under very mild conditions. The byproduct dimethyl sulfide is low-boiling and easy to remove.

Finally, as Pummerer observed, when the α carbon bears a carboxy group, the acetic anhydride reaction induces decarboxylation, which can be avoided by esterifying the carboxy group.[16]

EXPERIMENTAL PROCEDURES

4,5-Di-*O*-isopropylidene-β-D-fructopyranose Methylthiomethyl Ether [Oxygen Nucleophile].[112] A mixture of 4,5-di-*O*-isopropylidene-β-D-fructopyranose (8 g) in dimethyl sulfoxide (100 mL), acetic acid (20 mL), and acetic anhydride (66 mL) was stored at room temperature for two days, then poured into a cold solution of sodium carbonate (100 g) in water (1 L). The alkaline solution was extracted with chloroform (3 × 200 mL), the combined extracts were washed with water (5 × 200 mL) and evaporated, finally under high vacuum, to yield the product as a syrup (8.1 g, 82%) which was pure as judged by TLC. Passage through a silica gel column afforded the title compound (7.5 g, 76%), mp 82–83°. Similarly prepared were the methylthiomethyl ethers of 1,2:5,6-di-*O*-isopropylidene-α-D-glucofuranose, *cis* and *trans*-4-*tert*-butylcyclohexanol, *n*-butyl alcohol, *tert*-butyl alcohol, and 1-methylcyclohexanol.

1,2-Diacetoxy-2-phenethyl *p*-Tolyl Sulfide [Oxygen Nucleophile].[106] A stirred mixture of 2-hydroxy-2-phenethyl *p*-tolyl sulfoxide (1.79 g, 6.86 mmol) and sodium acetate (1.79 g, 22 mmol) in acetic anhydride (20 mL) was heated from room temperature to reflux during 0.5 hour and then refluxed for 3 hours. Excess acetic anhydride and acetic acid were removed under reduced pressure, and the residue was suspended in benzene and passed through silica gel. Evaporation of the solvent followed by drying under vacuum gave an oil (2.25 g, 95%). The product thus obtained was almost homogeneous in TLC with benzene as eluant. IR (film): 700, 1021, 1210, 1235, 1370, 1493, 1750 cm^{-1}; ^1H NMR ($CDCl_3$) δ 1.97 (s, 3*H*, CH_3CO_2), 2.05 (s, 3*H*, CH_3CO_2), 2.08 (s, 3*H*, CH_3CO_2), 2.14 (s, 3*H*, CH_3CO_2), 2.35 (s, 6*H*, 2 × aryl-CH_3), 5.97, 6.05, 6.30, 6.37 (AB, q, J = 7 Hz, 2H), 5.99, 6.05, 6.29, 6.34 (AB, q, J = 5.5 Hz, 2*H*), 7.0-7.5 (18*H*, aryl); mass spectrum, m/z (rel intensity): 344 (M^+, 8), 221 (34), 124 (80), 119(39), 91 (40), 43 (100).

4-Phenylthio-4-butanolide [Intramolecular Oxygen Nucleophile].[41] A mixture of 4-(phenylsulfinyl)butyric acid (0.69 g, 3 mmol), 1.53 g (15 mmol) of acetic anhydride and a catalytic amount of *p*-toluenesulfonic acid in 20 mL of toluene was heated under reflux for 1 hour. The solvent and excess acetic anhydride were removed under reduced pressure. The residue was chromatographed on silica gel using benzene to give 0.44 g (75%) of the title compound. IR (NaCl) 1780 (CO) cm^{-1}; ^1H NMR ($CDCl_3$) δ 1.88–2.52 (m, 4*H*), 5.44 (m, 1*H*), 6.75–7.22 (m, 5*H*); mass spectrum, m/z: 194 (M^+).

4-(Phenylthio)-2-azetidinone [Intramolecular Nitrogen Nucleophile].[143] To a solution of 3-(phenylsulfinyl)propionamide (99 mg, 0.5 mmol) in 20 mL of dichloromethane at −20° were added triethylamine (251 μL, 1.8 mmol) and trimethylsilyl trifluoromethanesulfonate (348 μL, 1.8 mmol). The solution was stirred at −20° for 15 minutes and then quenched by addition of 5% sodium bicarbonate solution and brine. Drying over anhydrous sodium sulfate and removal of the solvent gave a colorless oil. A preparative silica gel TLC (5% CH_3OH-CH_2Cl_2) of this material yielded, beside the starting material (18%) and trans-3-(phenylthio)acrylamide (8%), 37 mg (41%) of 4-phenylthio-2-azetidinone, which was recrystallized from diethyl ether: mp 72–73°; IR (film) 1740 cm^{-1}; ^1H NMR ($CDCl_3$) δ 2.90 (ddd, J = 15.0, 2.26 and 1.3 Hz, 1*H*), 3.33 (ddd, J = 15.0, 5.0 and 2.1 Hz, 1*H*), 5.06 (dd, J = 5.0 and 2.6 Hz, 1*H*), 6.49 (br. s, 1H), 7.47 (m, 5H).

***N*-Isobutyl-2-methylthiodec-4-enamide [Carbon Nucleophile].**[47] Trifluoroacetic anhydride (21 mmol of TFAA ≈ 3 mL) was added to a stirred solution of α-methylsulfinyl-(*N*-isobutyl)acetamide (3.72 g, 21 mmol) in trifluoroacetic acid (2 mL) at 0°, and 1-octene (21 mmol, ≈3.32 mL) was added to the mixture. Stirring was continued for 1 hour at the same temperature, the solvent was removed in vacuo, and the residue was chromatographed on silica gel using diethyl ether as eluant. The title compound was obtained as a mixture of isomers (81%, 88/12 *E*/*Z* ratio). IR ($CDCl_3$) 3360 (NH), 1655 (C=O), 970 (C=C) cm^{-1}; ^1H NMR ($CDCl_3$) δ 0.7–1.1 (m, 9*H*), 1.1–1.6 (m, 6*H*), 1.5–2.2 (m, 3*H*); 2.09 (s, 3*H*), 2.3–2.7 (m, 2*H*), 3.11 (br t, 1*H*), 5.1–5.7 (m, 2*H*), 6.6–6.9 (br. s, 1*H*)

1,1-Dimethyl-3-methylthio-2-oxo-2,4,5,6,7,7a-hexahydroindene [Intramolecular Carbon Nucleophile].[39] *p*-Toluenesulfonic acid monohydrate (3.35 g, 17.6 mmol) was added to benzene (30 mL) and the mixture was heated under reflux with azeotropic removal of water for 2 hours, then cooled to room temperature under nitrogen. To this benzene solution containing anhydrous *p*-toluenesulfonic acid, a solution of methyl 3-(1-cyclohexenyl)-3-methyl-2-oxobutyl sulfoxide (2.0 g, 8.8 mmol) in dry benzene (5 mL) was added by syringe in one portion and the mixture was again heated under reflux with azeotropic removal of water for 2 hours. After cooling to room temperature, the mixture was washed with water (2 × 5 mL) and dried over magnesium sulfate. The solvent was removed in vacuo and the residue was chromatographed on silica gel using benzene as eluent to give an oil: 1.04 g (56%). IR (neat) 1705 (CO), 1600 (C=C) cm^{-1}; ^1H NMR ($CDCl_3$) δ 0.7–2.4 (m, 8*H*), 0.98 (s, 3*H*), 1.10 (s, 3*H*), 2.30 (s, 3*H*), 2.95–3.50 (m, 1*H*).

Octanal from Octyl Phenyl Sulfoxide [Preparation of Aldehydes].[115]

Method A: To an acetonitrile solution (60 mL) of octyl phenyl sulfoxide (2.38 g, 10 mmol) and 2,6-lutidine (2.14 g, 20 mmol) was added an acetonitrile solution (20 mL) of trifluoroacetic anhydride (2.82 mL, 20 mmol) at 0° under nitrogen. After the reaction mixture was stirred at 0° for 10 minutes, an aqueous solution (100 mL) of sodium bicarbonate (60 mmol) was added. The mixture was stirred at room temperature for 2 hours. The resultant octanal was extracted with diethyl ether, the ether extract was washed with dilute hydrochloric acid and aqueous sodium bicarbonate solution. The extract was dried over magnesium sulfate and the solvent was evaporated. The residual oil was purified by column chromatography on silica gel with *n*-hexane as eluant and distilled under reduced pressure to give 0.92 g (72%) of the title compound (bp 80°/32 torr); the purity was more than 90% as estimated from the ^{1}H NMR spectrum. To the undistilled crude product was added 2,4-dinitrophenylhydrazine solution (60 mL). Precipitation occurred immediately. The hydrazone was filtered and the solid was washed with 50% aqueous ethanol and recrystallized from ethanol to give octanal 2,4-dinitrophenylhydrazone, mp 105°.

Method B: Using the first part of method A, an aqueous solution (60 mL) of copper(II) chloride (60 mmol) was added instead of a solution of sodium bicarbonate. The mixture was stirred at room temperature for 2 hours. The resultant aldehyde was extracted with diethyl ether and treated in a similar manner to give 0.94 g (74%) of octanal.

Method C: After the Pummerer reaction, an aqueous solution (60 mL) of mercury(II) chloride (3.8 g, 14 mmol) was added. The mixture was stirred at room temperature for 2 hours. The resultant aldehyde was extracted with diethyl ether and treated in the similar manner to give 1.10 g (86%) of octanal.

2-Formylchromone [Preparation of Aldehydes].[193]

2-[Acetoxy(methylthio)methyl]chromone. A solution of 2-(methanesulfinylmethyl)chromone (2.22 g, 0.01 mol) in acetic anhydride (15 mL) was heated under reflux under nitrogen for 5 hours. The solvent was removed under reduced pressure to give a brown gum which crystallized on standing. Recrystallization from methanol gave white crystals: 2.36 g (90%); mp 123–125°; UV (C_2H_5OH) nm max (ϵ): 222 (18800), 295 (6800); IR (Nujol) 1754 (CO), 1652 (CO) cm^{-1}; ^{1}H NMR ($CDCl_3$) δ 2.25 (s, 3*H*, SCH_3), 6.50 (s, 1*H*, HC-3), 6.70 [s, 1*H*, CH(OAc)-SCH_3], 7.20-7.90 (m, $3H_{arom}$), 8.20 (m, $1H_{arom}$).

2-(Dimethoxymethyl)chromone. A mixture of 2-[acetoxy(methylthio)methyl]chromone (2.64 g, 0.01 mol) and iodine (1.39 g, 0.011 mol) in methanol (50 mL) was heated under reflux for 6 hours. The solvent was evaporated under reduced pressure and the residue was dissolved in chloroform. The chloroform solution was washed three times with saturated sodium thiosulfate solution, dried over magnesium sulfate, and evaporated to give a solid product. Recrystallization from ethyl acetate–hexane gave yellow crystals: 1.90 g (86%); mp 64–68°; UV (C_2H_5OH) nm max (ε) 221 (19000), 298 (6500); IR (Nujol) 1655 (CO) cm^{-1}; ^{1}H NMR ($CDCl_3$) δ 3.48 (s, 6*H*, 2 × H_3CO-), 5.24 [s, 1*H*, -CH(OCH_3)$_2$], 6.58 (s, 1*H*, HC-3), 7.00-8.40 (m, 4H_{arom}).

2-Formylchromone. A mixture of 2-(dimethoxymethyl)chromone (1.1 g, 5 mmol) in 5 N hydrochloric acid (15 mL) was heated at 100° for 3 hours. The mixture was cooled and extracted with chloroform. The extracts were dried over magnesium sulfate and evaporated to give a crystalline product. Recrystallization from ethyl acetate gave yellow crystals: 710 mg (81%); mp 159–161°; UV (C_2H_5OH) nm max (ε) 222 (17600), 299 (6700); IR (Nujol) 1760 (CO), 1745 (CO) cm^{-1}; ^{1}H NMR (dimethyl sulfoxide-d_6) δ 7.10 (s, 1*H*, HC-3), 7.25-8.25 (m, 4H_{arom}), 9.81 (s, 1*H*, CHO).

2-(Phenylthio)cyclohexen-2-one [Preparation of Vinyl Sulfides].[121] Acetic anhydride (0.5 mL, 5.3 mmol) and methanesulfonic acid (0.04 mL, 0.6 mmol) were added to a solution of 2-phenylsulfinylcyclohexanone (1.0 g, 4.5 mmol) in dichloromethane (25 mL) under nitrogen at room temperature. The solution was left standing for 16 hours, the solvent and the resulting acetic acid were removed under vacuum, and the crude product was chromatographed on a dry Florisil® column (petroleum ether 35–60°). After discarding the early fractions, which contained small amounts of diphenyl disulfide, 2-(phenylthio)cyclohexen-2-one was recovered by evaporation of the solvent and crystallization from isopropyl alcohol: 0.79 g (86%); mp 50–51° (from isopropanol); UV (*n*-hexane) nm max (ε) 235 (6500), 256 (4100), 272 (4100); IR (neat) 1673 cm^{-1}; ^{1}H NMR (CCl_4) δ 6.32 (t, J = 6 Hz, 1*H*).

2-(6β-*tert*-Butyldimethylsilyloxy-2β-hydroxy-5α-phenylseleno-5β-acetoxycyclohept-1β-yl)acetic Acid Lactone [Seleno–Pummerer].[158] A 1-L three-necked flask equipped with a reflux condenser, two glass stoppers and a magnetic stirring bar was charged with 150 mL of tetrahydrofuran and 4.12 g (22.8 mmol) of 95% *m*-chloroperbenzoic acid. After the acid had dissolved, the solution was cooled to −78° and a solution of 10.0 g (22.8 mmol) of 2-[6β-*tert*-butyldimethylsilyloxy-2β-hydroxy-5α-(phenylseleno)-cyclohept-1β-yl]acetic acid lactone in 25 mL of tetrahydrofuran was added dropwise via syringe with stirring. After 20 minutes, 10 mL (0.105 mol) of acetic anhydride and 5.0 g (0.060 mol) of anhydrous sodium acetate were added, and the solution was allowed to warm slowly to 20°. The reaction mixture was heated to reflux for 3 hours. The mixture was cooled to 30°, 50 mL of methanol was added, and stirring was continued for 30 minutes. The solution was diluted with 300 mL of ethyl acetate and washed with four 75-mL portions of 10% sodium hydroxide solution and once with 100 mL of brine. The solution was dried over magnesium sulfate, and the solvents were removed by rotary evaporation, affording 10.3 g (96%) of crude product. Purification by medium-pressure liquid chromatography afforded 8.40 g (80%) of the title compound as a yellow oil. IR (film) 3.26, 3.38, 3.40, 3.50, 5.60, 5.78, 6.33, 7.30, 7.94, 9.76, 11.83 μm; ^1H NMR ($CDCl_3$) δ 0.09 (s, methyls), 0.92 (s, *tert*-butyl), 2.04 (s, CH_3CO), 1.7–3.1 (m, 9*H*), 4.51 (m, H-7 and H-3), 7.20 (m, 3*H*, aromatic), 7.52 (m, 2*H*, aromatic).

S-Phenyl Thiolbenzoate [Sila-Pummerer].[165] α-Chlorobenzyl phenyl sulfoxide (1.25 g, 5.0 mmol) in 2 mL of tetrahydrofuran was added dropwise with stirring over a 3-minute period to lithium diisopropylamide (5.0 mmol, prepared from 0.50 g of diisopropylamine and 3.2 mL of 1.58 M *n*-butyllithium in hexane at 0°) in 10 mL of tetrahydrofuran under a nitrogen atmosphere at −78°. It was stirred at −78° for 30 minutes and then the anion solution was transferred dropwise to excess chlorotrimethylsilane (1.62 g, 15 mmol) in tetrahydrofuran (10 mL) at −78° while stirring over a 5-minute period. The reaction mixture was allowed to warm to 0°. After 1 hour of stirring at 0°, 2% hydrochloric acid (5 mL) was added dropwise and the mixture was extracted with dichloromethane (3 × 40 mL), dried over sodium sulfate, and concentrated under reduced pressure to give the product as a yellow solid. Crystallization from hexane–ether gave white crystals (0.79 g, 74%): mp 54-55°; IR (KBr) 1685 cm^{-1}; ^1H NMR ($CDCl_3$) δ 7.20-7.60 (m, 8*H*), 7.85-8.50 (m, 2*H*).

5,6,7,7a-Tetrahydro-7a-hydroxy-3-(phenylthio)benzofuran-2(4*H*)-one [Vinylogous Pummerer Reaction].[179] A solution of 3-(phenylsulfinyl)-furanone (0.32 g, 1.5 mmol) in 6 N sulfuric acid (1 mL) and dioxane (4 mL) was heated under reflux for 4 hours. The reaction mixture was diluted with water and extracted with ether. The combined extracts were washed with water and brine and evaporated to dryness. Preparative TLC of the residue (0.34 g) on silica gel [elution with a mixture of light petroleum–diethyl ether (1:1)] gave the thermally very unstable (polymerized on standing) 5,6-dihydro-3-(phenylthio)benzofuran-2-(4*H*)-one (0.053 g, 18%); IR ($CHCl_3$) 1775, 1655, 1590, 980 cm^{-1}; ^1H NMR (CCl_4) δ 1.30–2.60 (m, 6H), 5.68 (t, $J = 4$ Hz, 1H), 7.00–7.60 (m, 5H) and the title compound (0.075 g, 24%); mp 135–136° from carbon tetrachloride. IR ($CHCl_3$) 3300, 1755, 1640, 975 cm^{-1}; ^1H NMR (CCl_4) δ 0.70–3.00 (m, 8*H*), 4.10–5.70 (br s, OH), 6.90–7.70 (m, 5*H*).

2,3-Bis(acetoxy)-2,3-dihydro-8-methoxy-3-methylthio-4*H*-1-benzopyran-4-one [Additive Pummerer Reaction].[194] A mixture of 8-methoxy-3-methylsulfinyl-4*H*-1-benzopyran-4-one (5.0 g, 0.021 mol) in acetic anhydride (25 mL) was refluxed under nitrogen for 10 hours, cooled, and poured into ice-water. The precipitate was filtered off and recrystallized from ethyl acetate to give crystals (4.47 g, 63%), mp 181–182°; UV nm max(ε) 264 (9000), 330 (2000); IR (Nujol) 1775, 1755, 1700 cm^{-1}; ^1H NMR ($CDCl_3$) δ 2.09 (s, 3*H*,

$COCH_3$), 2.20 (s, 3*H*, $COCH_3$), 2.32 (s, 3H, SCH_3), 3.90 (s, 3*H*, OCH_3), 7.70–7.80 (m, 4*H*, ArH and CH-2).

***trans*-2,2-Dichloro-3-phenyl-4-phenylthio-γ-butyrolactone [Additive Pummerer Reaction].**[195] (*E*)-β-Styryl phenyl sulfoxide (134 mg, 0.6 mmol) was dissolved in 30 mL of ether, and 0.8 g (12 mg-atom) of zinc was added. The resulting suspension was heated to reflux under nitrogen. A solution of 0.35 mL (3 mmol) of freshly distilled trichloroacetyl chloride in 20 mL of ether was added dropwise to the refluxing zinc suspension over a period of 15 minutes. The reaction mixture was cooled to room temperature, filtered through Celite, and poured into 50 mL of cold sodium bicarbonate solution. The two-phase mixture was stirred for 15 minutes at room temperature while a white precipitate formed. The aqueous layer was separated and extracted once with ether. The organic portions were combined, dried over magnesium sulfate, and evaporated. The solid residue was crystallized from cyclohexane to give 108 mg of the title compound (55%): mp 114–115°: Another 10% yield of product was isolated from flash chromatography of the mother liquors with 9:1 petroleum ether/ether on silica.

TABULAR SURVEY

The tabular survey covers as exhaustively as possible only those reactions in which the oxidation state of the products is the same as that of the starting material, with the exception of carbon nucleophiles.

The literature is covered through 1986. Papers published in primary journals in 1987 and beginning 1988 are also included.

Dimethyl sulfoxide is a special case. This reagent, as mentioned, is well known to give a number of reactions related to the Pummerer reaction. It usually necessitates activating electrophilic species which are often the same reagents that promote the Pummerer reaction. It is therefore obvious that Pummerer products may be formed in traces during these processes[196–199] and which for an exhaustive coverage of the literature should also be cited. Nonetheless, because these examples are repetitive we have not included them in the tabular survey.

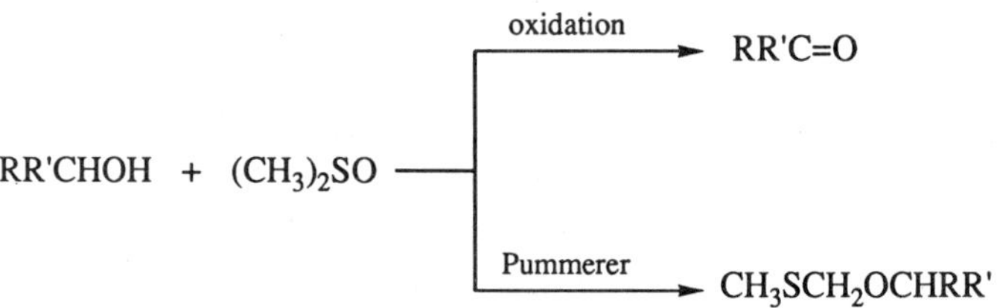

Only such reactions in which the Pummerer products predominated have been included. When the expected Pummerer product was not isolated but directly transformed, only the final product is shown. When more than one type of product was formed, the entry is located in the table for the more abundant product.

The following abbreviations are used in the tables:

Ac	acetyl
anh	anhydrous
aq	aqueous
BnOC	benzyloxycarbonyl
BOC	*tert*-butoxycarbonyl
C_4H_3S	thienyl
C_6H_{11}	cyclohexyl
$C_{10}H_7$	naphthyl
DABCO	1,4-diazabicyclo[2.2.2]octane
DAST	diethylaminosulfur trifluoride
DBN	1,5-diazabicyclo[4.3.0]non-5-ene
DBU	1,8-diazabicyclo[5.4.0]undec-7-ene
DCC	dicyclohexylcarbodiimide
DIBAH	diisobutylaluminum hydride
DMF	dimethylformamide
DNPH	2,4-dinitrophenylhydrazine
ee	enantiomeric excess
ether	diethyl ether
LDA	lithium diisopropylamide
mCPBA	*m*-chloroperbenzoic acid
NCS	*N*-chlorosuccinimide
NPS	*p*-nitrophenylthio
Pet. Et	petroleum ether
Pyr	pyridine
PPSE	polyphosphoric acid trimethylsilyl ester
rt	room temperature
TBDMS	*tert*-butyldimethylsilyl
TFAA	trifluoroacetic anhydride
THF	tetrahydrofuran
THP	tetrahydropyranyl
TMS	trimethylsilyl
Ts	*p*-toluenesulfonyl
*	a nonracemic carbon or sulfur atom

Usual acronyms are used for α-amino acids.

TABLE I. SULFOXIDES WITH AN OXYGEN NUCLEOPHILE

Sulfoxide	Nucleophile	Reaction Conditions	Product(s) and Yield(s) (%)	Refs.
C_2				
$(CH_2)_2SO$	CH_3OH	90°, 15 h	$CH_3OCH_2CH(OCH_3)SS(CH_2)_2OCH_3$ I (85)	200
	CH_3OH	Reflux, 50 h	$CH_3O(CH_2)_2S(O)S(CH_2)_2OCH_3$ (13)[a] + I (9)[a]	200
$(CH_3)_2SO$	Ac_2O	C_6H_6, 80°, 6 h	CH_3SCH_2OAc (85)	201
	Ac_2O	Ether, reflux 1.5 d[b]	" (20)	202
	$C_6H_5OP(O)OH(OAc)$	Pyr, 40°, 1 d	" (10)	203
	AcOH	t-C_4H_9Br, $NaHCO_3$, 24 h	" (95)	52, 53
	TFAA	CH_2Cl_2, −30° to rt	$CH_3SCH_2O_2CCF_3$ (65)	44, 204, 205
	$(C_6H_5CO)_2O$	C_6H_6, 80°, 6 h	$CH_3SCH_2O_2CC_6H_5$ (79)	92, 201
	$(C_6H_5CO)_2O$	Dioxane, 48 h	" (71)	206
	$C_6H_5CO_2H$	C_6H_5NCO, C_6H_6, reflux overnight	" (50)	207
	$C_6H_5CO_2H$	P_2O_5, 65–70°	" (60)	208
	$C_6H_5CO_2H$	t-C_4H_9Br, $NaHCO_3$, 24 h	" (98)	52, 53
	$(p$-$O_2NC_6H_4CO)_2O$	Dioxane, 6 h	$CH_3SCH_2O_2CC_6H_4NO_2$-p (71)	206
	p-$O_2NC_6H_4CO_2H$	P_2O_5, 65–70°	" (51)	208
	p-$O_2NC_6H_4CO_2H$	DCC, H_3PO_4, C_6H_6, 30 min	" (42)[c]	209
	p-$O_2NC_6H_4CO_2H$	Ac_2O, 23°, 3 d	" (33)	209
	p-$O_2NC_6H_4CONHOH$	DCC, H_3PO_4, C_6H_6, 1 h	" (31)[c]	209
	$(p$-$ClC_6H_4CO)_2O$	Dioxane, 24 h	$CH_3SCH_2O_2CC_6H_4Cl$-p (68)	206
	$(p$-$CH_3OC_6H_4CO)_2O$	Dioxane, 48 h	$CH_3SCH_2O_2CC_6H_4OCH_3$-p (36)	206

$(C_6H_5)_2CHCO_2H$	DCC, H_3PO_4, C_6H_6, overnight	$CH_3SCH_2O_2CCH(C_6H_5)_2$ (69)[c]	209
$(C_6H_5)_2CHCONHOH$	"	" (50)[c]	209
p-$O_2NC_6H_4CONHOCH_3$	DCC, H_3PO_4, C_6H_6, 18 h	$CH_3SCH_2OC(C_6H_4NO_2\text{-}p)$=$NOCH_3$ (Z, 5; E, 10)	209
C_6H_5OH	t-C_4H_9Br, base, 35°, 24 h	$CH_3SCH_2OC_6H_5$ (49–84) + o-$(CH_3SCH_2)C_6H_4OH$ (0–43) + o-$(CH_3SCH_2)C_6H_4OCH_2SCH_3$ (0–37)	52, 210
p-$O_2NC_6H_4OH$	t-C_4H_9Br, $(C_2H_5)_3N$, 35°, 24 h	p-$(CH_3SCH_2O)C_6H_4NO_2$ (100)	210
2-$CH_3C_6H_4OH$	t-C_4H_9Br, $NaHCO_3$	2-$CH_3C_6H_4OCH_2SCH_3$ I (56)[a] + 2-CH_3-6-$(CH_3SCH_2)C_6H_3OH$ II (40)[a] + 2-CH_3-6-$(CH_3SCH_2)C_6H_3OCH_2SCH_3$ III (4)[a]	52
2-$CH_3C_6H_4OH$	t-C_4H_9Br, $(C_2H_5)_3N$, 35°, 24 h	I (66) + II (20) + III (14)	210
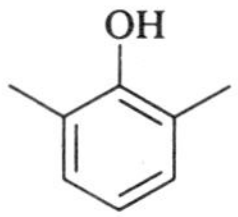 OH	t-C_4H_9Br, $(C_2H_5)_3N$, 35°, 24 h	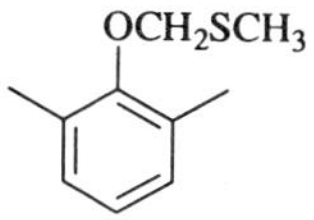 OCH_2SCH_3 (49) 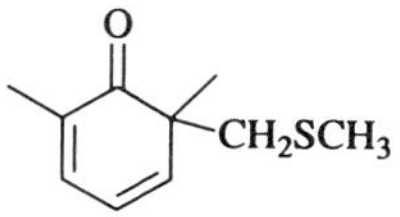 + O CH_2SCH_3 (51)	210

TABLE I. Sulfoxides with an Oxygen Nucleophile (*Continued*)

Sulfoxide	Nucleophile	Reaction Conditions	Product(s) and Yield(s) (%)	Refs.
	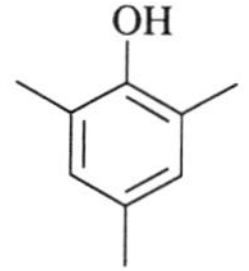 OH	t-C_4H_9Br, $(C_2H_5)_3N$, 35°, 24 h	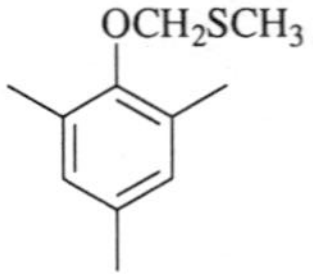 OCH_2SCH_3 (48) 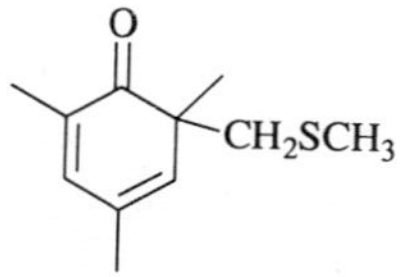 + O CH_2SCH_3 (52)	210
	C_6Cl_5OH	DCl, H_3PO_4, C_6H_6, overnight	$CH_3SCH_2OC_6Cl_5$ (60)	211, 212
	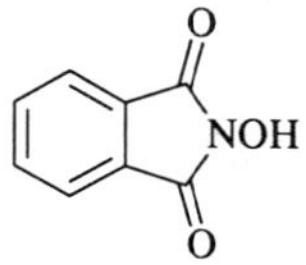 O NOH O	DCC, H_3PO_4, C_6H_6, 23°, 3 h	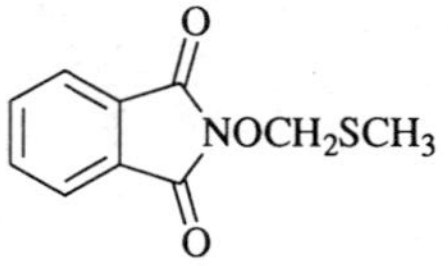 O $NOCH_2SCH_3$ O (58)	209
	(E)-p-$BrC_6H_4CH{=}NOH$	DCC, CF_3CO_2H, C_6H_6, 2.5 h	$CH_3SCH_2ON(O){=}CHC_6H_4Br$-$p$ (—)[c]	213
	s-$C_4H_9CO_2H$	t-C_4H_9Br, $NaHCO_3$, 24 h	$CH_3SCH_2O_2CC_4H_9$-s (95)	52, 53
	$(CH_3)_2C{=}CHCO_2H$	"	$CH_3SCH_2O_2CCH{=}C(CH_3)_2$ (95)	52, 53
	$[(CH_2)_2CO_2H]_2$	"	$[CH_3SCH_2O_2C(CH_2)_2]_2$ (98)	52, 53
	$C_6H_5CH{=}CHCO_2H$	"	$CH_3SCH_2O_2CCH{=}CHC_6H_5$ (95)	52, 53
	$CH_2{=}CH(CH_2)_8CO_2H$	"	$CH_3SCH_2O_2C(CH_2)_8CH{=}CH_2$ (90)	52, 53
	n-$C_{17}H_{35}CO_2H$	P_2O_5, 65–70°	$CH_3SCH_2O_2CC_{17}H_{35}$-$n$ (40)	208
	$C_6H_5CHOHCO_2H$	t-C_4H_9Br, $NaHCO_3$, 24 h	$CH_3SCH_2O_2CCHOHC_6H_5$ (80)	52, 53
	N-BnOC-L-Trp-OH	"	CH_3SCH_2OTrp-L-BnOC-N (94)	52, 53, 214
	N-BnOC-L-Phe-OH	"	CH_3SCH_2OPhe-L-BnOC-N (95)	52, 53, 214

N-BnOC-L-Ser-OH	"	CH_3SCH_2OSer-L-BnOC-*N* (88)	53, 214
N-BnOC-L-Ala-OH	"	CH_3SCH_2OAla-L-BnOC-*N* (98)	53, 214
N-BnOC-L-Met-OH	*t*-C_4H_9Br,$NaHCO_3$, 30°, 5 h	CH_2SCH_2OMet-L-BnOC-*N* (82)	214
N-BnOC-L-Asp-OH	"	$(CH_3SCH_2O)_2$Asp-L-BnOC-*N* (95)	214
N-BnOC-L-Glu-OH	"	$(CH_3SCH_2O)_2$Glu-L-BnOC-*N* (94)	214
N,*O*-$(BnOC)_2$-L-Tyr-OH	"	CH_3SCH_2OTyr-L-$(BnOC)_2$-*N*,*O* (89)	214
N-BOC-L-Phe-OH	*t*-C_4H_9Br, $NaHCO_3$, 24 h	CH_3SCH_2OPhe-L-BOC-*N* (90)	53, 214
N-BOC-L-Tyr-OH	*t*-C_4H_9Br, $NaHCO_3$, 30°, 5 h	CH_3SCH_2OTyr-L-BOC-*N* (85)	214
N-NPS-L-Phe-OH	*t*-C_4H_9Br, $NaHCO_3$, 24 h	CH_3SCH_2OPhe-L-NPS-*N* (85)	53, 214
N-NPS-Gly-OH	*t*-C_4H_9Br, $NaHCO_3$, 30°, 5 h	CH_3SCH_2OGly-NPS-*N* (91)	214
N-NPS-L-Met-OH	"	CH_3SCH_2OMet-L-NPS-*N* (62)	214
N-NPS-L-Pro-OH	"	CH_3SCH_2OPro-L-NPS-*N* (87)	214
N-NPS-L-Trp-OH	"	CH_3SCH_2OTrp-L-NPS-*N* (82)	214
N-HOC-L-Phe-OH	*t*-C_4H_9Br, $NaHCO_3$, 24 h	CH_3SCH_2OPhe-L-OHC-*N* (80)	53, 214
N-PHT-L-Phe-OH	"	CH_3SCH_2OPhe-L-PHT-*N* (90)	53, 214
N-CF_3CO-L-Phe-OH	"	CH_3SCH_2OPhe-L-$OCCF_3$-*N* (80)	53, 214
N-Trityl-L-Phe-OH	*t*-C_4H_9Br, $NaHCO_3$, 30°, 5 h	CH_3SCH_2OPhe-L-Trityl-*N* (82)	214
$C_6H_5SO_3Na$	1. Ac_2O, 80°, 24 h 2. AcOH, AcONa, 100°, 26 h	$CH_3SCH_2O_3SC_6H_5$ (38)	108
TsOH	1. Ac_2O, 80°, 24 h 2. AcOH, AcONa, 100°, 26 h	CH_3SCH_2OTs (71)	108
ROH	1. TFAA, CH_2Cl_2, −55 to −60°, 15 min 2. ROH, CH_2Cl_2, BF_3·ether, $< -55°$, 30 min	CH_2SCH_2OR	215

TABLE I. SULFOXIDES WITH AN OXYGEN NUCLEOPHILE (*Continued*)

Sulfoxide	Nucleophile	Reaction Conditions	Product(s) and Yield(s) (%)	Refs.
		3. $(C_2H_5)_3N$, −55° to rt		
			R = C_6H_{11} (50)	
			R = $CH(CH_3)C_6H_{13}$-*n* (50)	
			R = $C_{10}H_{21}$-*n* (40)[c]	
			R = $(CH_2)_2C_6H_5$ (50)	
			R = (*E*)-2-Hexenyl (40)	
	n-C_4H_9OH	Ac_2O, AcOH, 2 d	$CH_3SCH_2OC_4H_9$-*n* (97)	112, 113
	t-C_4H_9OH	"	$CH_3SCH_2OC_4H_9$-*t* (—)	112, 113
	t-C_4H_9OH	Ac_2O, 6 d	" (46) + CH_3SCH_2OH (—)	216
	$C_6H_{11}OH$	Ac_2O, AcOH, H_2O	$CH_3SCH_2OC_6H_{11}$ (64–90)	217
	OH	"	OCH_2SCH_3 (64–90)	217
	OH	Ac_2O, AcOH, 2 d	OCH_2SCH_3 (—)	112, 113
	OH C_4H_9-*t*	"	OCH_2SCH_3 C_4H_9-*t* (—)	112, 113
	CH_2OH	Ac_2O, AcOH, H_2O	$CH_2OCH_2SCH_3$ (64)	217
	n-$C_7H_{15}OH$	"	$CH_3SCH_2OC_7H_{15}$-*n* (64–90)	217

	n-$C_6H_{13}CH(CH_3)OH$	"	$CH_3SCH_2OCH(CH_3)C_6H_{13}$-$n$ (64–90)	217
	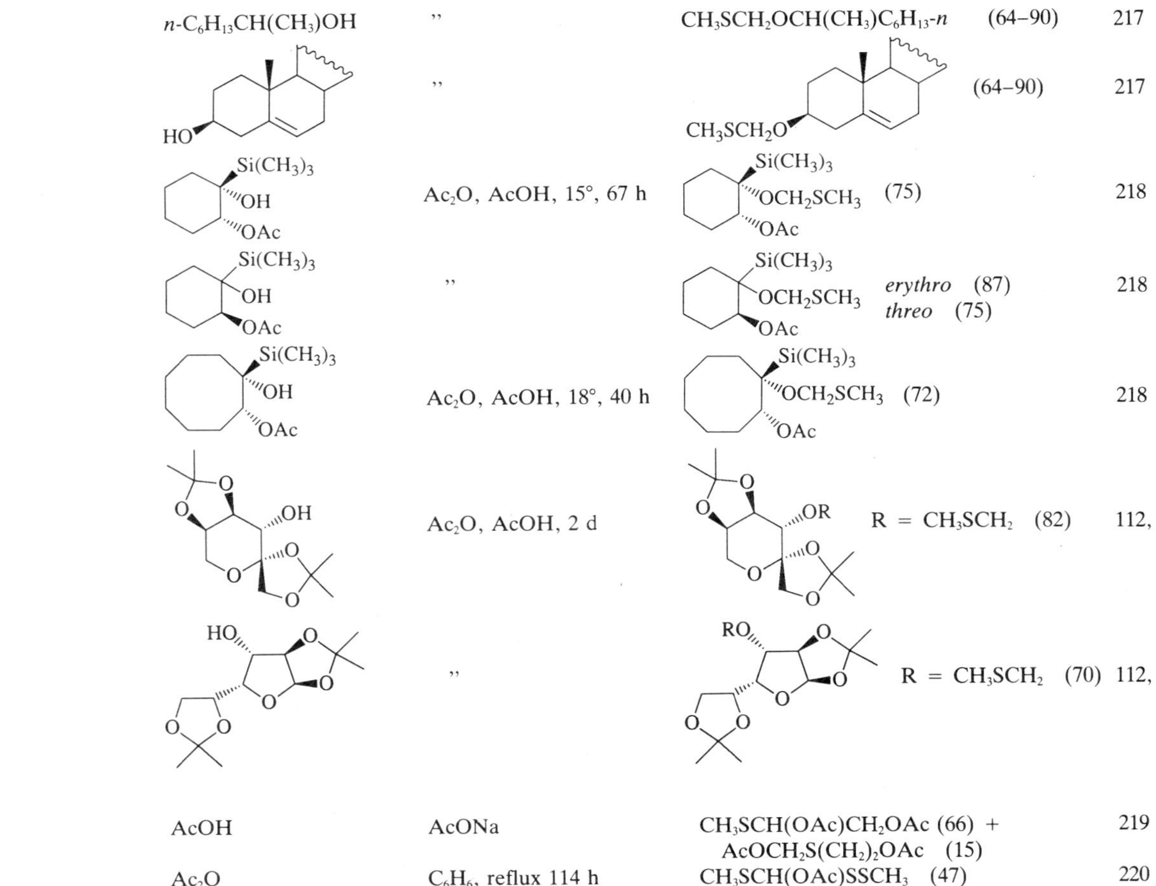	"	(64–90)	217
		Ac_2O, AcOH, 15°, 67 h	(75)	218
		"	*erythro* (87) *threo* (75)	218
		Ac_2O, AcOH, 18°, 40 h	(72)	218
		Ac_2O, AcOH, 2 d	R = CH_3SCH_2 (82)	112, 113
		"	R = CH_3SCH_2 (70)	112, 113
C_3				
$CH_3S(O)(CH_2)_2OH$	AcOH	AcONa	$CH_3SCH(OAc)CH_2OAc$ (66) + $AcOCH_2S(CH_2)_2OAc$ (15)	219
$CH_3S(O)CH_2SSCH_3$	Ac_2O	C_6H_6, reflux 114 h	$CH_3SCH(OAc)SSCH_3$ (47)	220

TABLE I. SULFOXIDES WITH AN OXYGEN NUCLEOPHILE (*Continued*)

Sulfoxide	Nucleophile	Reaction Conditions	Product(s) and Yield(s) (%)	Refs.
C_4				
$(CH_2)_4S$	Ac_2O	$CHCl_3$, 25°, 3 d	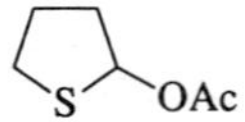 S OAc (85)	201
	Ac_2O	C_6H_6, 80°, 3 h	" (84)	201
	p-$O_2NC_6H_4OH$	DCl, H_3PO_4, C_6H_6, 3 h	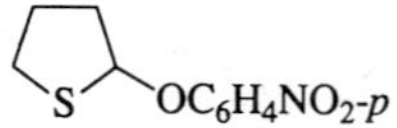 S $OC_6H_4NO_2$-p (59)	221
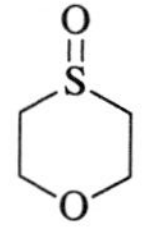 O=S, O	Ac_2O	TsOH, C_6H_6, reflux 3.5 h	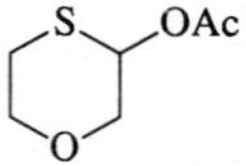 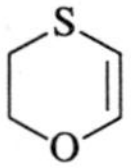 S, OAc, O (~60) + S, O (tr)	222
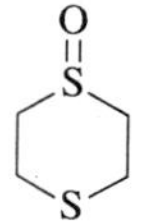 O=S, S	Ac_2O	"	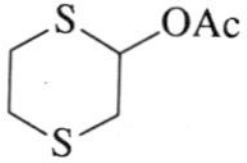 S, OAc, S (18)[a]	222
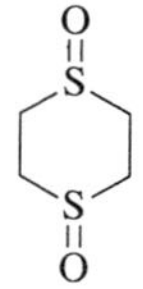 O=S, S=O	Ac_2O	100°, 71 h	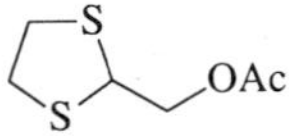 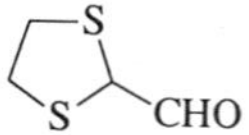 " (17) 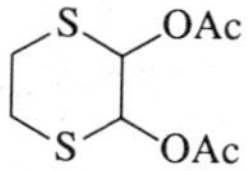 + S, OAc, S, OAc I + S, S, OAc II + S, S, CHO (5) 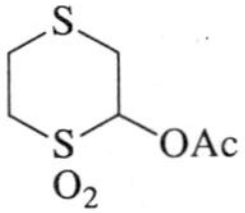 + S, S, O_2, OAc (tr) I + II (53)	223

$(C_2H_5)_2SO$	Ac_2O	$CHCl_3$, 25°, 3 d	$C_2H_5SCH(OAc)CH_3$ (93)	201
	Ac_2O	C_6H_6, 80°, 4 h	" (69)	201
$[HO(CH_2)_2]_2SO$	AcONa	Ac_2O, 140°	$AcO(CH_2)_2SCH(OAc)CH_2OAc$ (92)	219
$C_2H_5SS(O)C_2H_5$	CH_3OH	90°, 9 h	$C_2H_5SSCH(OCH_3)CH_3$ (97)	200
n-$C_3H_7S(O)CH_3$	Ac_2O	C_6H_6, reflux	*n*-$C_3H_7SCH_2OAc$ (73)	224
i-$C_3H_7S(O)CH_3$	Ac_2O	"	*i*-$C_3H_7SCH_2OAc$ (69)	224
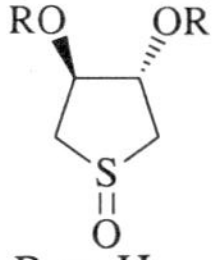 RO OR S=O			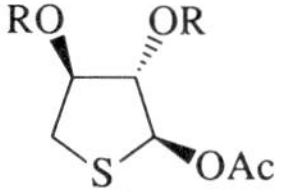 RO OR S OAc	225
R = H	Ac_2O	AcONa, reflux 3.5 h	R = Ac (21)	
R = Ac	Ac_2O	C_6H_6, reflux	(20)	
R = O_2SCH_3	Ac_2O	C_6H_6, reflux 15 min	(15)	
R = COC_6H_5	AcONa	Ac_2O, reflux 3 h	Mixture of isomers (81)	
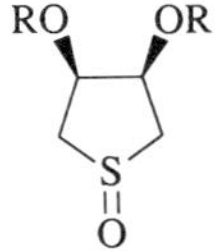 RO OR S=O			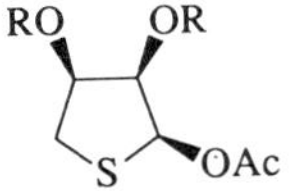 RO OR S OAc	225
R = H	AcONa	Ac_2O, reflux 3.5 h	(61)	
R = Ac	Ac_2O	C_6H_6, reflux	(16)	
R = O_2SCH_3	Ac_2O	40°, 5 h	(44)	
R = COC_6H_5	AcONa	Ac_2O, reflux 3 h	(48)	
R-R = C_6H_5B	AcONa	"	R = Ac (27)	
R-R = CO	AcONa	Ac_2O, reflux	(9–53)	
$(CH_3O)_2P(O)CH_2S(O)CH_3$	AcONa	120°, 2 h	$(CH_3O)_2P(O)CH(OAc)SCH_3$ (90)	226
	TFAA	−78°, 15 min	$(CH_3O)_2P(O)CH(O_2CF_3)SCH_3$ (74)	226
	C_6H_5COCl	5 h	$(CH_3O)_2P(O)CHClSCH_3$ (84)	226
	SO_2Cl_2	CH_2Cl_2, 0°, 2 h	$(CH_3O)_2P(O)CCl_2SCH_3$ (90)	226
	CH_3OH	I_2, reflux 2 h	$(CH_3O)_2P(O)CH(OCH_3)SCH_3$ (70)	226

TABLE I. SULFOXIDES WITH AN OXYGEN NUCLEOPHILE (*Continued*)

Sulfoxide	Nucleophile	Reaction Conditions	Product(s) and Yield(s) (%)	Refs.
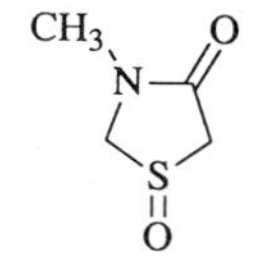 CH_3 N O S O	TFAA	CF_3CO_2H, C_6H_6, 0°, 3–4 h	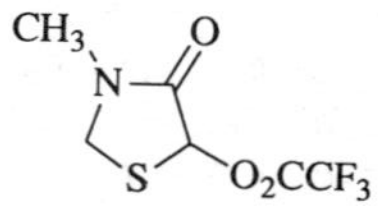 CH_3 N O S O_2CCF_3 (100)	36
C_5				
n-$C_4H_9S(O)CH_3$	Ac_2O	C_6H_6, reflux	n-$C_4H_9SCH_2OAc$ (64)	214
	Ac_2O	Reflux, 6 h	" (61)	203
	Ac_2O	100°, 4.5 h	" (50) + $C_2H_5CH{=}CHSCH_3$ (4)	182
$CH_3S(O)CH_2CO_2C_2H_5$	TFAA	CH_2Cl_2	$CH_3SCH(O_2CCF_3)CO_2C_2H_5$ (—)	227
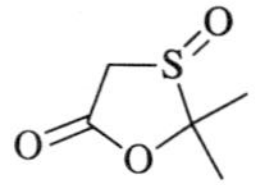 O S O O	TFAA	CF_3CO_2H, C_6H_6, 0°, 3–4 h	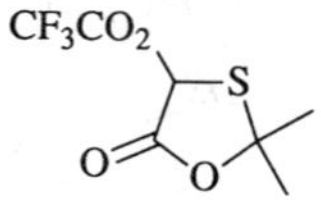 CF_3CO_2 S O O (45)	36
 O O S O	Ac_2O	TsOH, C_6H_6, reflux 3–12 h	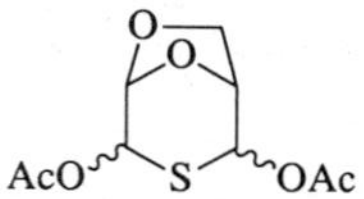 O O AcO S OAc mixture of 4 isomers (—)	228
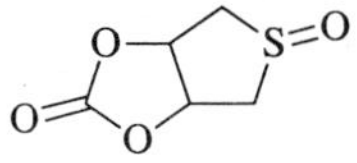 O O O S=O	Ac_2O	100°, 18 h	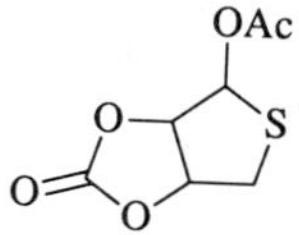 OAc O O O S (42)	225
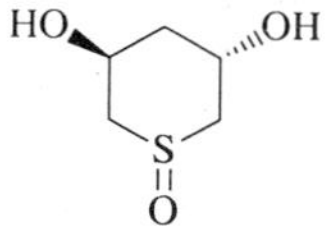 HO OH S O	AcONa	Ac_2O, 140°, 3 h	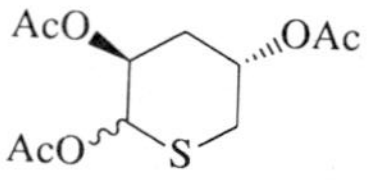 AcO OAc AcO S (21) 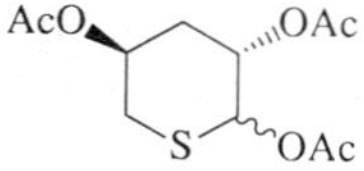 + AcO OAc S OAc (5)	229

	Sulfoxide	Reagent	Conditions	Product(s) (Yield, %)	Refs.
C_6	$(i\text{-}C_3H_7)_2SO$	Ac_2O	$CHCl_3$, 25°, 4 d	$i\text{-}C_3H_7SC(OAc)(CH_3)_2$ (56)	201
		Ac_2O	C_6H_6, 80°, 4 h	" (89)	201
		AcOH	Ac_2O, C_6H_6	" + (thiirane) 2:1 (—)	230
	$n\text{-}C_3H_7COCH_2S(O)CH_3$	AcONa	Ac_2O, toluene, reflux 30 min	$n\text{-}C_3H_7CH(OAc)COSCH_3$ (89)	107
	$CH_3O(CH_2)_2S(O)S(CH_2)_2OCH_3$	CH_3OH	90°, 20 h	$CH_3OCH_2CH(OCH_3)SS(CH_2)_2OCH_3$ (94)	200
	$(CH_3O)_2P(O)S(CH_2)_2S(O)C_2H_5$	TFAA	30°, 60 min or 100°, 15 min	$(CH_3O)_2P(O)SCH_2CH(O_2CCF_3)SC_2H_5$ (41–85) + $(CH_3O)_2P(O)SCH{=}CHSC_2H_5$ (—)	231
	$(C_2H_5O)_2P(O)CH_2S(O)CH_3$	Ac_2O	120°, 2 h	$(C_2H_5O)_2P(O)CH(OAc)SCH_3$ (86)	226
		Ac_2O	CH_3SO_3H, CH_2Cl_2, reflux 3 h	" (81)	232
		$(C_2H_5CO)_2O$	"	$(C_2H_5O)_2P(O)CH(O_2CC_2H_5)SCH_3$ (73)	232
		$(n\text{-}C_3H_7CO)_2O$	"	$(C_2H_5O)_2P(O)CH(O_2CC_3H_7\text{-}n)SCH_3$ (74)	232
		$(t\text{-}C_4H_9CO)_2O$	"	$(C_2H_5O)_2P(O)CH(O_2CC_4H_9\text{-}t)SCH_3$ (66)	232
		TFAA	−78°, 15 min	$(C_2H_5O)_2P(O)CH(O_2CCF_3)SCH_3$ (76)	226, 233
		C_6H_5COCl	5 h	$(C_2H_5O)_2P(O)CHClSCH_3$ (90)	226
		SO_2Cl_2	CH_2Cl_2, 0°, 2 h	$(C_2H_5O)_2P(O)CCl_2SCH_3$ (92)	226
		CH_3OH	I_2, reflux 1.5 h	$(C_2H_5O)_2P(O)CH(OCH_3)SCH_3$ (82)	226, 234
		C_2H_5OH	I_2, reflux, 15 min	$(C_2H_5O)_2P(O)CH(OC_2H_5)SCH_3$ (73)	226, 234
		AcONa	Ac_2O, C_6H_6, reflux 24 h	OAc (75)	235
		"	"	OAc + OAc (76)[b]	235

TABLE I. SULFOXIDES WITH AN OXYGEN NUCLEOPHILE (*Continued*)

Sulfoxide	Nucleophile	Reaction Conditions	Product(s) and Yield(s) (%)	Refs.
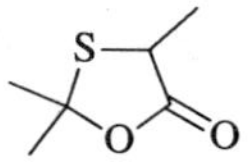 [structure]	H_2O	AcOH, H_2O_2 (30%), 0° to rt, overnight	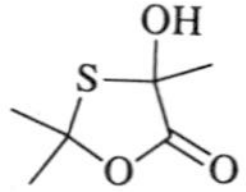 [structure] (25)	36
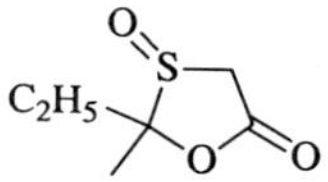 [structure, C_2H_5]	Ac_2O	TsOH, CH_2Cl_2, 48 h	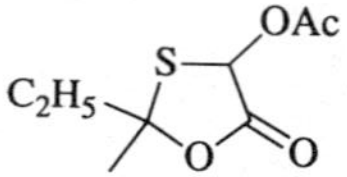 [structure, C_2H_5, OAc] (80–90)[d]	25
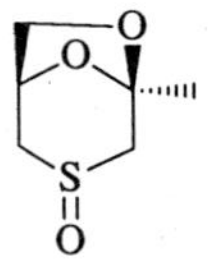 [structure]	Ac_2O	TsOH, C_6H_6, reflux 3–12 h	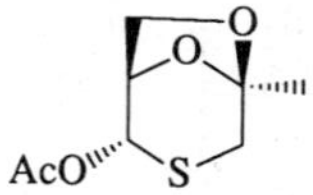 [structure, AcO] (75) + other isomers (—)	228
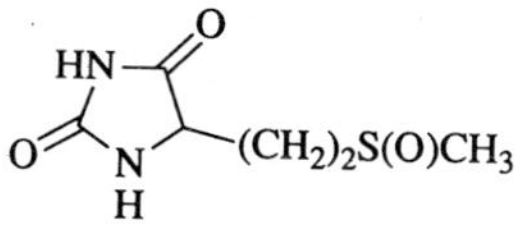 [structure]$(CH_2)_2S(O)CH_3$	Ac_2O	Reflux 1 h	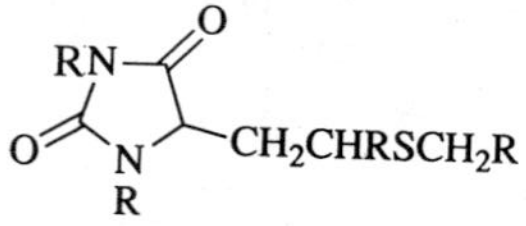 [structure]$CH_2CHRSCH_2R$ (—) R = H or OAc	236
C_7				
$C_6H_5S(O)CH_3$	Ac_2O	120°, 8 h	$C_6H_5SCH_2OAc$ (92)	201, 237[b]
	$(C_6H_5CO)_2O$	120°, 8 h	$C_6H_5SCH_2O_2CC_6H_5$ (81)	201
p-$ClC_6H_4S(O)CH_3$	TFAA	CH_2Cl_2, 3 h	p-$C_6H_4SCH_2O_2CCF_3$ (quant)	238
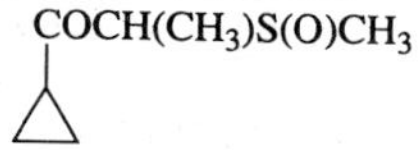 (cyclopropyl)$COCH(CH_3)S(O)CH_3$	AcCl	Pyr, CH_2Cl_2, reflux 3 h, rt 48 h	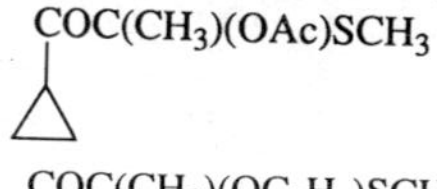 (cyclopropyl)$COC(CH_3)(OAc)SCH_3$ (84)	239
	C_2H_5OH	AcOH, Zn, reflux 5 h	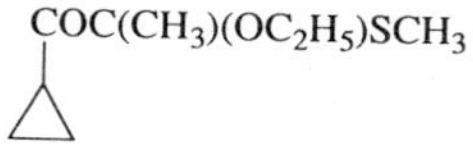 (cyclopropyl)$COC(CH_3)(OC_2H_5)SCH_3$ (36)[c]	239

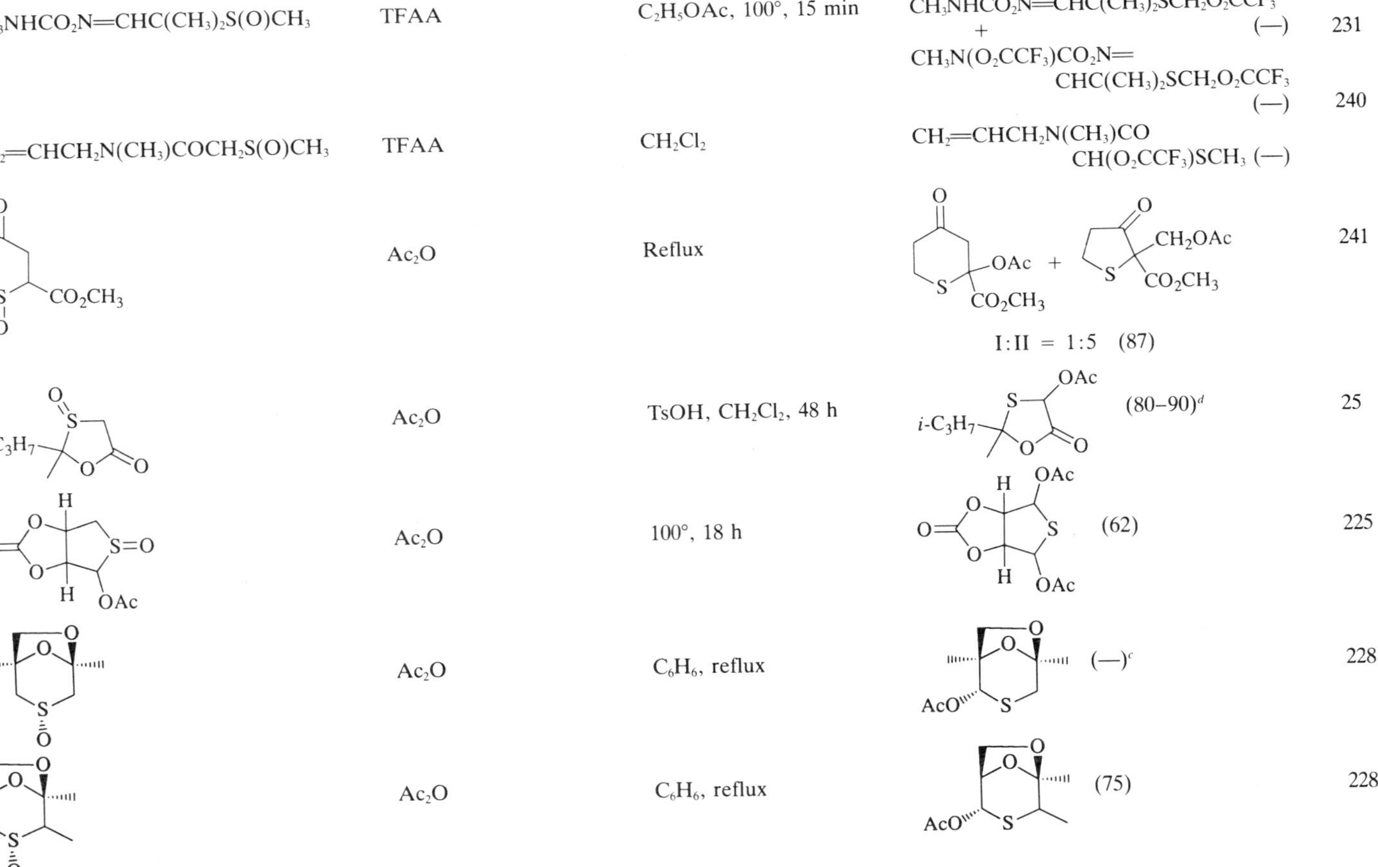

$CH_3NHCO_2N{=}CHC(CH_3)_2S(O)CH_3$	TFAA	C_2H_5OAc, 100°, 15 min	$CH_3NHCO_2N{=}CHC(CH_3)_2SCH_2O_2CCF_3$ (—) + $CH_3N(O_2CCF_3)CO_2N{=}CHC(CH_3)_2SCH_2O_2CCF_3$ (—)	231
$CH_2{=}CHCH_2N(CH_3)COCH_2S(O)CH_3$	TFAA	CH_2Cl_2	$CH_2{=}CHCH_2N(CH_3)COCH(O_2CCF_3)SCH_3$ (—)	240
(structure)	Ac_2O	Reflux	(structures) I:II = 1:5 (87)	241
(structure)	Ac_2O	TsOH, CH_2Cl_2, 48 h	(structure) (80–90)[d]	25
(structure)	Ac_2O	100°, 18 h	(structure) (62)	225
(structure)	Ac_2O	C_6H_6, reflux	(structure) (—)[c]	228
(structure)	Ac_2O	C_6H_6, reflux	(structure) (75)	228

TABLE I. SULFOXIDES WITH AN OXYGEN NUCLEOPHILE (*Continued*)

Sulfoxide	Nucleophile	Reaction Conditions	Product(s) and Yield(s) (%)	Refs.
C_8				
$C_6H_5CH_2S(O)CH_3$	Ac_2O	130°	$C_6H_5CH(OAc)SCH_3$ + $C_6H_5CH_2SCH_2OAc$ 45:55 (39)	242
	$(Cl_2CHCO)_2O$	$CHCl_3$, 25°	$C_6H_5CH(O_2CCHCl_2)SCH_3$ + $C_6H_5CH_2SCH_2O_2CCHCl_2$ 53:47 (100)	242
	TFAA	$CHCl_3$, 0°	$C_6H_5CH(O_2CCF_3)SCH_3$ + $C_6H_5CH_2SCH_2O_2CCF_3$ 53:47 (100)	242
$C_6H_5S(O)CH_2CF_3$	Ac_2O	120°, 24 h	$C_6H_5SCH(OAc)CF_3$ (42)	243
$CH_3S(O)CHDC_6H_5$	Ac_2O	130°, 2 h	$CH_3SCH(OAc)C_6H_5$ (—) + $CH_3SCD(OAc)C_6H_5$ (—)	244
	TFAA	CCl_4, 25°	$CH_3SCHD(O_2CCF_3)C_6H_5$ (—)	244
$C_6H_5CH_2S(O)CH_3$	Ac_2O	85°, 4 h[b]	$C_6H_5CH_2SCH_2OAc$ (23)[a] + $C_6H_5CH(SCH_3)_2$ (54)[a]	237[b]
p-$ClC_6H_4CH_2S(O)CH_3$	TFAA	$CHCl_3$, 25°	p-$ClC_6H_4CH(O_2CCF_3)SCH_3$ + p-$ClC_6H_4CH_2SCH_2O_2CCF_3$ 61:39 (—)	242
p-$O_2NC_6H_4CH_2S(O)CH_3$	TFAA	$CHCl_3$, 25°	p-$O_2NC_6H_4CH(O_2CCF_3)SCH_3$ + p-$O_2NC_6H_4CH_2SCH_2O_2CCF_3$ 87:13 (—)	242
$C_6H_5S(O)CH_2CO_2H$	Ac_2O		$C_6H_5SCH_2OAc$ (20)	16
p-$XC_6H_4S(O)CH_2CN$	Ac_2O	120°, 3–5 h	p-$XC_6H_4SCH(OAc)CN$ (85–90)	24, 245[d]
$AcO(CH_2)_2S(O)S(CH_2)_2OAc$	CH_3OH	90°, 13 h	$AcOCH_2CH(OCH_3)SS(CH_2)_2OAc$ (90)	200
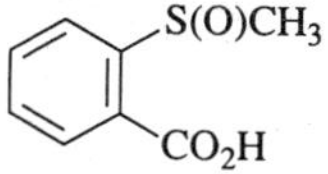 [structure: $S(O)CH_3$, CO_2H]	"	Ac_2O, 100°, 1–2 h	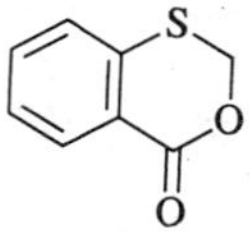 [structure: S, O, O] (97)	246–248

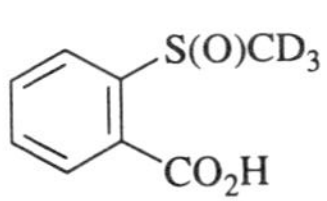 $S(O)CD_3$, CO_2H	,,	Ac_2O, 100°, 1 h	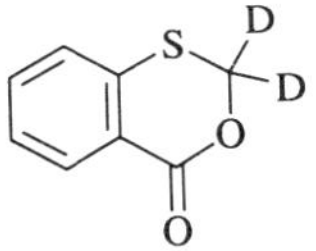 S, D, D, O, O (95)	246–248
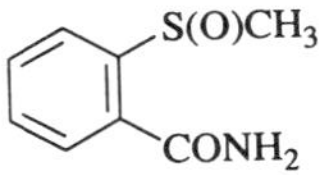 $S(O)CH_3$, $CONH_2$	,,	Ac_2O, 100°, 5 h	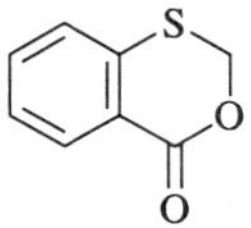 S, O, O I (44) 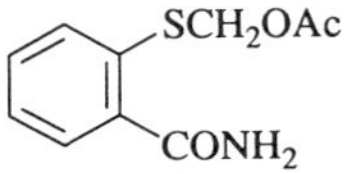 + SCH_2OAc, $CONH_2$ II (rest)	246
	Ac_2O	140°, 2 h	I (5) + II (90)	246
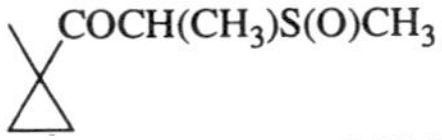 $COCH(CH_3)S(O)CH_3$	C_2H_5OH	AcOH, Zn, reflux 4 h	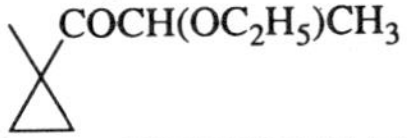 $COCH(OC_2H_5)CH_3$ (8)	239
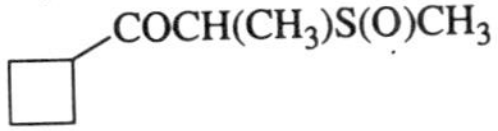 $COCH(CH_3)S(O)CH_3$	C_2H_5OH	AcOH, Zn, reflux 4–5 h	 $COCH(OC_2H_5)CH_3$ (37)	239
2-$(C_4H_3S)CH{=}C(SCH_3)S(O)CH_3$	Ac_2O	110°, 11 h	2-$(C_4H_3S)CH{=}C(SCH_3)SCH_2OAc$ (64)	176
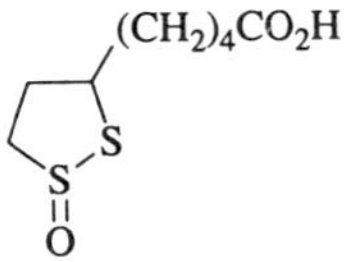 $(CH_2)_4CO_2H$, S, S, O	Ac_2O	$CHCl_3$, ether, 45°, 2 d	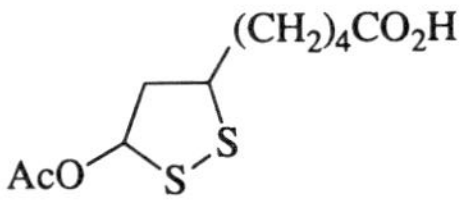 $(CH_2)_4CO_2H$, AcO, S, S (6)	249
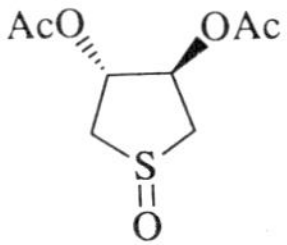 AcO, OAc, S, O	Ac_2O	C_6H_6, reflux	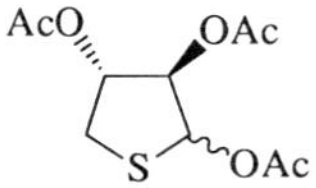 AcO, OAc, S, OAc (—)	250
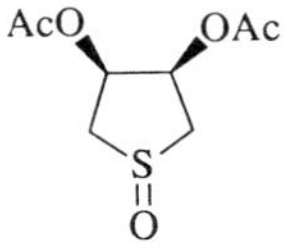 AcO, OAc, S, O	Ac_2O	C_6H_6, reflux	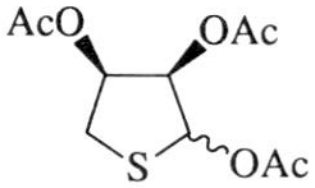 AcO, OAc, S, OAc (—)	250

TABLE I. SULFOXIDES WITH AN OXYGEN NUCLEOPHILE (*Continued*)

	Sulfoxide	Nucleophile	Reaction Conditions	Product(s) and Yield(s) (%)	Refs.
	[structure: bicyclic sulfoxide, CH_3SO_2, O_2SCH_3, O, S=O]	Ac_2O	50°, 3 h	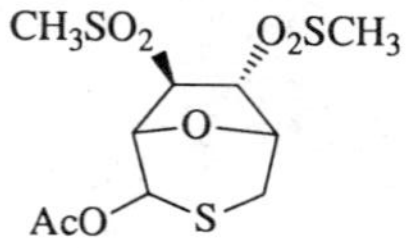 [structure: CH_3SO_2, O_2SCH_3, O, AcO, S] (73)	251
	[structure: OH, HO, OH, CH_3O, O, $CH_2S(O)CH_3$]	Ac_2O	1. 20 h 2. Pyr	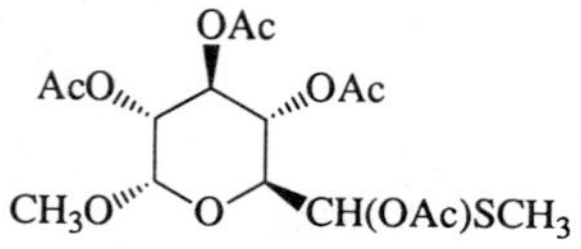 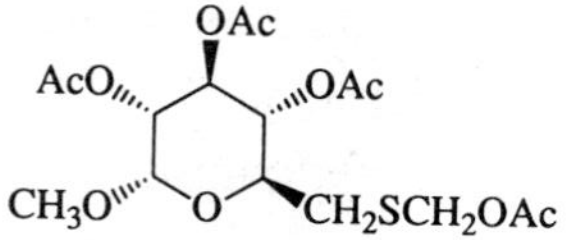 [structure: OAc, AcO, OAc, CH_3O, O, $CH(OAc)SCH_3$] (—) [structure: OAc, AcO, OAc, CH_3O, O, CH_2SCH_2OAc] (—)	252
	[structure: $t\text{-}C_4H_9$, O=S, O, =O]	Ac_2O	TsOH, CH_2Cl_2, 48 h	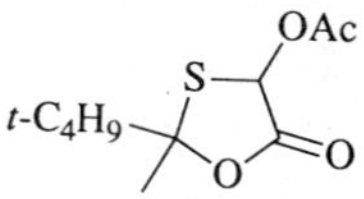 [structure: OAc, S, $t\text{-}C_4H_9$, O, =O] (80–90)[d]	25
C_9	$p\text{-}CH_3C_6H_4CH_2S(O)CH_3$	TFAA	$CHCl_3$	$p\text{-}CH_3C_6H_4CH(O_2CCF_3)SCH_3$ + $p\text{-}CH_3C_6H_4CH_2SCH_2O_2CCF_3$ 34:66 (—)	242
	$CH_3COCH_2S(O)C_6H_5$	Ac_2O	1. Pyr, 120°, 1.75 h 2. $(C_2H_5)_3N$, C_6H_6, reflux 1 h	$CH_3CH(OAc)COSC_6H_5$ (73)	253
	$C_6H_5COCH_2S(O)CH_3$	Ac_2O	Pyr, 125°, 1.25 h, then 140°, 5 min	$C_6H_5COCH(OAc)SCH_3$ (98)	253
		Ac_2O	NaH, THF (before Ac_2O)	" (75)	254

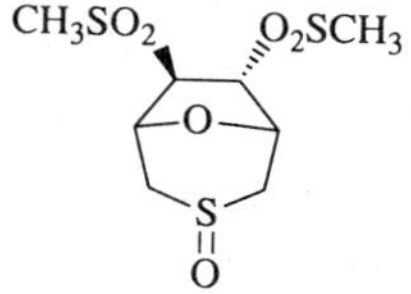

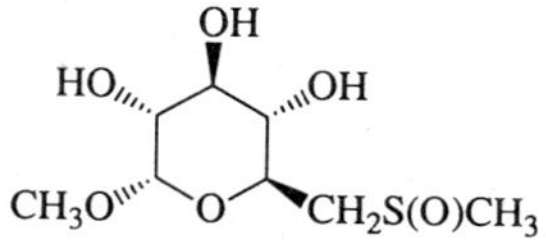

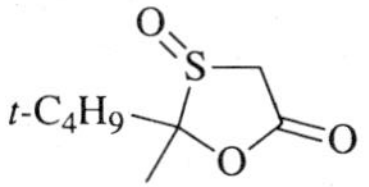

	Ac_2O	Reflux, 2 h	” (>95)	255
	Ac_2O	Pyr, 1 week	” (>95)	255
	Ac_2O	$HgCl_2$, 3 h	” (95)	255
	Ac_2O	1. Pyr, reflux 30 min 2. mCPBA, CH_2Cl_2, 0°, 45 min 3. $C_6H_{11}NH_2$, CH_3CN, 1 h	$C_6H_5CH(OAc)CONHC_6H_{11}$ (53)	253
	AcONa	Ac_2O, toluene, reflux 30 min	$C_6H_5CH(OAc)COSCH_3$ (98)	107
	HCl, H_2O		$C_6H_5COCHOHSCH_3$ (95)	32, 33
$XC_6H_4COCH_2S(O)CH_3$ X = H, *p*-Cl, *m*-Cl, *p*-CH_3, *m*-CH_3, *p*-CH_3O	HCl, H_2O[e]		$XC_6H_4COCHOHSCH_3$ (—)[e]	82[e]
p-$BrC_6H_4COCH_2S(O)CH_3$	HCl, H_2O		*p*-$BrC_6H_4COCHOHSCH_3$ (77)	33
p-$CH_3C_6H_4S^*(O)CH_2CN$	AcOH	Ac_2O, 120°, 3.5 h	*p*-$CH_3C_6H_4SC^*H(OAc)CN$ 29% ee (81)	245[a,b,d]
	Ac_2O	120°, 3–5 h	” (85–90)	24
p-$CH_3C_6H_4S(O)CD_2CN$	AcOH	Ac_2O, 20°[e]	*p*-$CH_3C_6H_4SCD(OAc)CN$ (—)[e]	245[e]
p-$CH_3OC_6H_4S(O)CH_2CN$	Ac_2O	120°, 3–6 h	*p*-$CH_3OC_6H_4SCH(OAc)CN$ (85–90)	24, 245
	AcOH	Ac_2O, 120°[e]	” (—)[e]	245[e]
$C_6H_5CH{=}CHS(O)CH_3$	Ac_2O	100°, 12 h	$C_6H_5CH{=}CHSCH_2OAc$ (92)	254
$S(O)C_6H_5$ (cyclopropane)	AcONa	Ac_2O, 170°, 3 h	AcO, SC_6H_5 (cyclopropane) (95)	29, 256, 257
	CH_3ONa	1. $(CH_3)_3O^+BF_4^-$, CH_2Cl_2 2. CH_3OH, 25°	CH_3O, SC_6H_5 (cyclopropane) (54)[c]	258
(cyclobutane)–$COCH(CH_3)S(O)CH_3$	C_2H_5OH	AcOH, Zn, reflux 4–5 h	(cyclobutane)–$COCH(OC_2H_5)CH_3$ (41)	239
$C_6H_{11}COCH_2S(O)CH_3$	Ac_2O	NaH, THF (before Ac_2O)	$C_6H_{11}COCH(OAc)SCH_3$ (71)	254

TABLE I. SULFOXIDES WITH AN OXYGEN NUCLEOPHILE (*Continued*)

Sulfoxide	Nucleophile	Reaction Conditions	Product(s) and Yield(s) (%)	Refs.
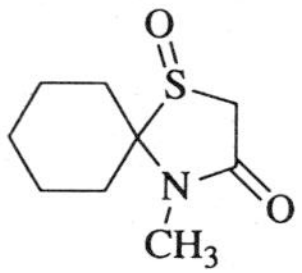 1-oxo-4-methyl-1-thia-4-azaspiro[4.5]decan-3-one (S=O, N–CH_3, C=O)	H_2O	1. TFAA, CF_3CO_2H, C_6H_6, 0°, 3–4 h 2. H_2O	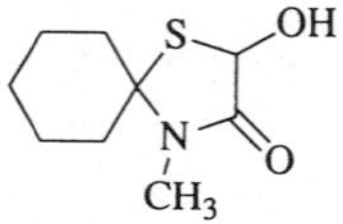 2-hydroxy-4-methyl-1-thia-4-azaspiro[4.5]decan-3-one (S, OH, N–CH_3, C=O) (70)	36
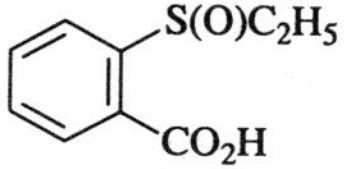 2-$S(O)C_2H_5$-benzoic acid (CO_2H)	Ac_2O	100°, 1–2 h	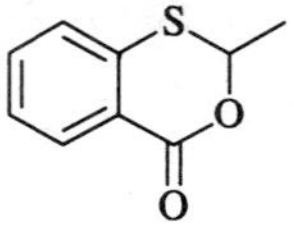 2-methyl-3,1-benzoxathiin-4-one (S, O, C=O) (95)	246, 247
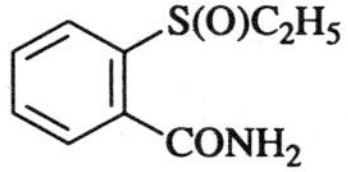 2-$S(O)C_2H_5$-benzamide ($CONH_2$)	Ac_2O	130°, 3 h	" (29)	246
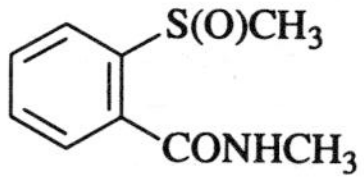 2-$S(O)CH_3$-N-methylbenzamide ($CONHCH_3$)	Ac_2O	120°, 3 h	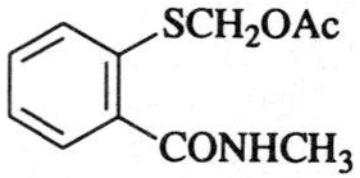 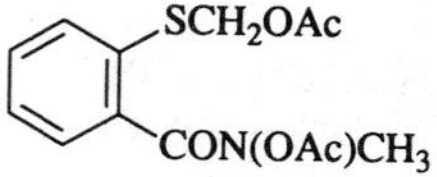 2-SCH_2OAc-C6H4-$CONHCH_3$ I 2-SCH_2OAc-C6H4-$CON(OAc)CH_3$ II I:II = 53:47 (100)	246, 259
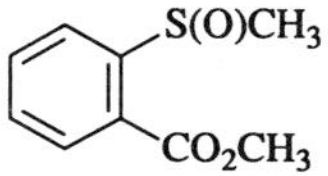 2-$S(O)CH_3$-C6H4-CO_2CH_3	Ac_2O	100°, 3 h	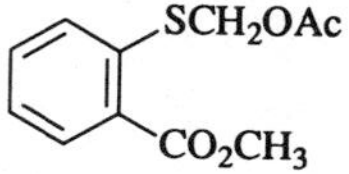 2-SCH_2OAc-C6H4-CO_2CH_3 (27)[a]	246
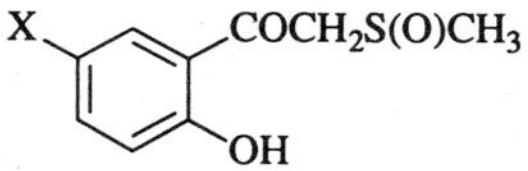 5-X-2-OH-C6H3-$COCH_2S(O)CH_3$		CF_3CO_2H, C_6H_6, reflux 40–90 min	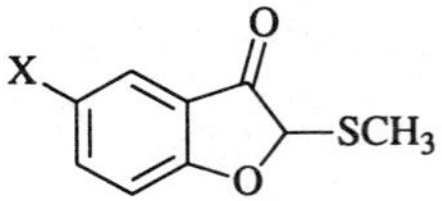 5-X-2-SCH_3-benzofuran-3(2H)-one X = H (45) X = Cl (54)	260

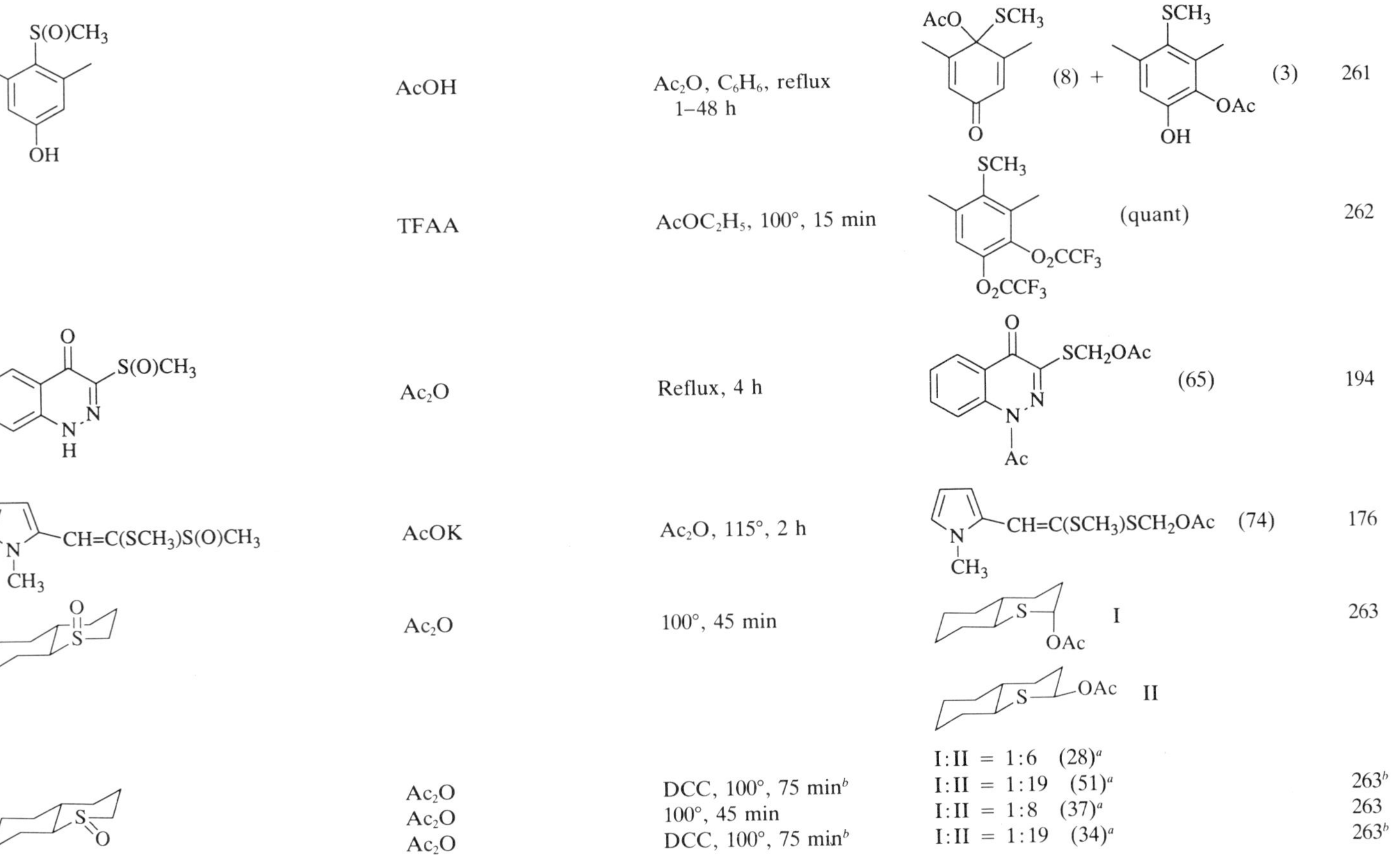

Reagent	Conditions	Product(s) and Yield(s) (%)	Refs.
AcOH	Ac_2O, C_6H_6, reflux 1–48 h	(8) + (3)	261
TFAA	$AcOC_2H_5$, 100°, 15 min	(quant)	262
Ac_2O	Reflux, 4 h	(65)	194
AcOK	Ac_2O, 115°, 2 h	$CH{=}C(SCH_3)SCH_2OAc$ (74)	176
Ac_2O	100°, 45 min	I, II; I:II = 1:6 (28)[a]	263
Ac_2O	DCC, 100°, 75 min[b]	I:II = 1:19 (51)[a]	263[b]
Ac_2O	100°, 45 min	I:II = 1:8 (37)[a]	263
Ac_2O	DCC, 100°, 75 min[b]	I:II = 1:19 (34)[a]	263[b]

TABLE I. SULFOXIDES WITH AN OXYGEN NUCLEOPHILE (*Continued*)

	Sulfoxide	Nucleophile	Reaction Conditions	Product(s) and Yield(s) (%)	Refs.
	[structure]	Ac_2O	Reflux, 30 min	[structure] I	264
				[structure] II	
				[structure] III	
				I:II = 2:1 (40) + III (30)	
			TsOH, xylene, reflux 1 h	III (55)	264
C_{10}	p-$CH_3C_6H_4S(O)CH_2C{\equiv}CH$	Ac_2O	60°, 10 h[b]	p-$CH_3C_6H_4SCH(OAc)C{\equiv}CH$ I (83) +	265
				[structure] II (tr)	
		Ac_2O	90°, 2 h	I (57) + II (22)	265
	$C_6H_5S(O)CH_2C_3H_7$-i	AcONa	Ac_2O, 90°, 20 min	$C_6H_5SCH(OAc)C_3H_7$-i (9)[a]	237
	$C_6H_5S(O)(CH_2)_2OC_2H_5$	AcONa	Ac_2O, reflux 7 h	$C_6H_5SCH(OAc)CH_2OC_2H_5$ (81)	42
	$C_6H_5S(O)CH_2CO_2C_2H_5$	Ac_2O	70°, 12 h, then reflux 30 min	$C_6H_5SCH(OAc)CO_2C_2H_5$ (80)	16

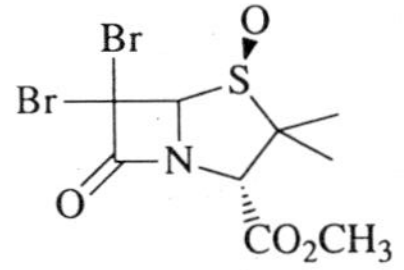

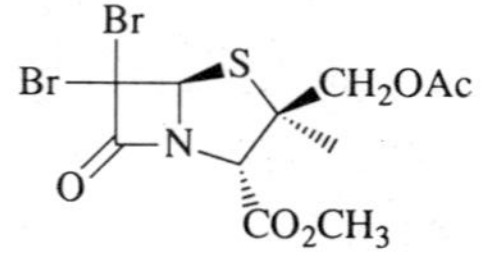

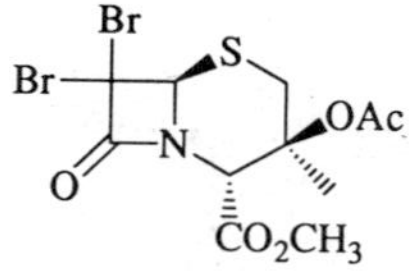

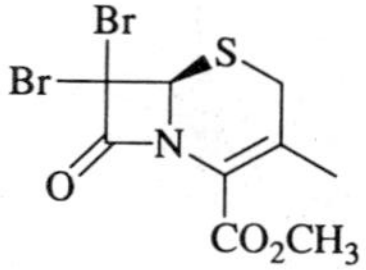

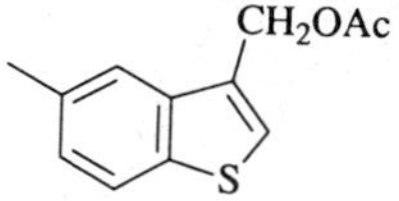

$C_6H_5S(O)(CH_2)_3CO_2H$	Ac_2O	TsOH, toluene, reflux 1 h	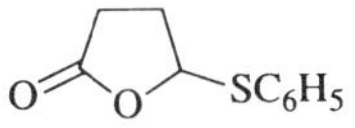 O O SC_6H_5 (75)	41
	Ac_2O	C_6H_6, reflux 10 h	" (—)	41
	Ac_2O	TsOH, C_6H_6, reflux 3 h	" (56)	41
	Ac_2O	$ClCH_2CO_2H$, C_6H_6, reflux 14 h	" (10)	41
	Ac_2O	H_3PO_4, C_6H_6, reflux 6 h	" (15)	41
	Ac_2O	TsOH, xylene, reflux 30 min	" (50)	41
p-$CH_3C_6H_4COCH_2S(O)CH_3$	HCl, H_2O	24 h	p-$CH_3C_6H_4COCHOHSCH_3$ (96)	32, 33
	HCl, H_2O	$(CH_3)_2SO$, 75 min	" (74)	32
p-$CH_3OC_6H_4COCH_2S(O)CH_3$	HCl, H_2O	$(CH_3)_2SO$, 12–24 h	p-$CH_3OC_6H_4COCHOHSCH_3$ (87)	33
	HCl, H_2O	$(CH_3)_2SO$, 30 min	" (82)	32
$(CH_3O)_2P(O)CH_2S(O)C_6H_4CH_3$-$p$	Ac_2O	120°, 3 h	$(CH_3O)_2P(O)CH(OAc)SC_6H_4CH_3$-$p$ 24% ee (84)	226
	TFAA	−78°, 15 min	$(CH_3O)_2P(O)CH(O_2CCF_3)SC_6H_4CH_3$-$p$ (72)	226
	CH_3OH	I_2, reflux 7 h	$(CH_3O)_2P(O)CH(OCH_3)SC_6H_4CH_3$-$p$ (65)	226
$CH_3CHOHCH_2S(O)C_6H_4CH_3$-$p$	AcONa	Ac_2O, reflux 3 h	$CH_3CH(OAc)CH(OAc)SC_6H_4CH_3$-$p$ (90)	106
p-$XC_6H_4CH{=}C(SCH_3)S(O)CH_3$	Ac_2O	110°, 11 h	p-$XC_6H_4CH{=}C(SCH_3)SCH_2OAc$ X = H (93); X = Cl (86)	176
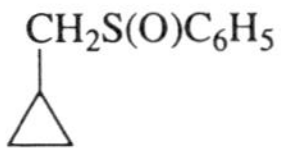 $CH_2S(O)C_6H_5$	AcONa	Ac_2O, 170°, 3 h	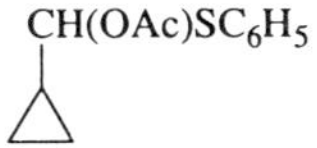 $CH(OAc)SC_6H_5$ (96)	29, 257
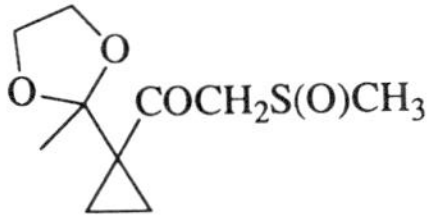 O O $COCH_2S(O)CH_3$	Ac_2O	Reflux, 2 h	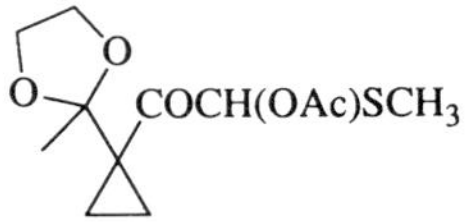 O O $COCH(OAc)SCH_3$ (79)	147

TABLE I. SULFOXIDES WITH AN OXYGEN NUCLEOPHILE (*Continued*)

Sulfoxide	Nucleophile	Reaction Conditions	Product(s) and Yield(s) (%)	Refs.
	Ac_2O	Pyr, 1 week	" (85)	147
	Ac_2O	$HgCl_2$, 3 h	" (72)	147
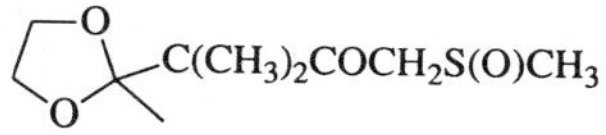 $C(CH_3)_2COCH_2S(O)CH_3$ (O, O dioxolane ring)	Ac_2O	Reflux, 2 h	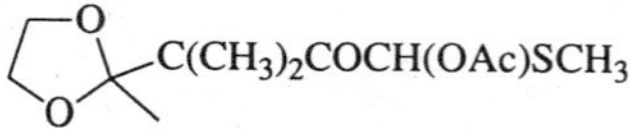 $C(CH_3)_2COCH(OAc)SCH_3$ (O, O dioxolane ring) (95)	147
	Ac_2O	Pyr, 1 week	" (90)	147
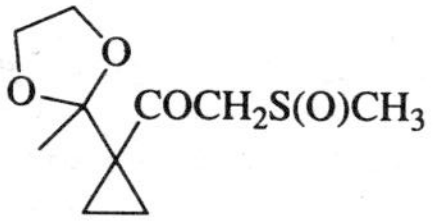 $COCH_2S(O)CH_3$ (O, O dioxolane ring; cyclopropane)	Ac_2O	Reflux, 2 h	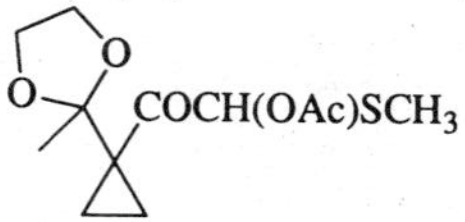 $COCH(OAc)SCH_3$ (O, O dioxolane ring; cyclopropane) (79)	147
	Ac_2O	Pyr, 1 week	" (85)	147
	Ac_2O	$HgCl_2$, 3 h	" (75)	147
$C_6H_{11}COCH(CH_3)S(O)CH_3$	H_2O	$(CH_3)_2SO$, HCl, 25°, 24 h	$C_6H_{11}COCHOHSCH_3$ (94)	239
	C_2H_5OH	AcOH, Zn, reflux 4–5 h	$C_6H_{11}COCH(OC_2H_5)CH_3$ (40)	239
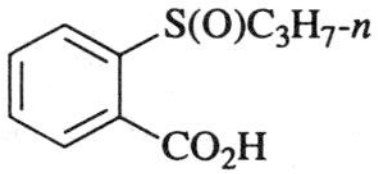 $S(O)C_3H_7$-*n*, CO_2H (benzene ring)		Ac_2O, 100°, 1–2 h	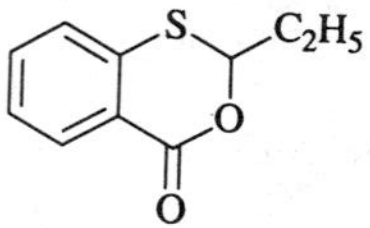 S, C_2H_5, O, O (ring structure) (93)	246–248
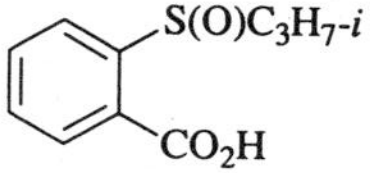 $S(O)C_3H_7$-*i*, CO_2H (benzene ring)		Ac_2O, 100°, 1–2 h	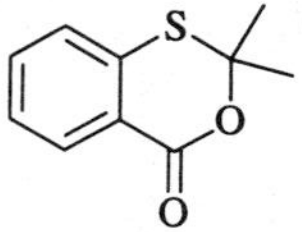 S, O, O (ring structure) (95)	246–248
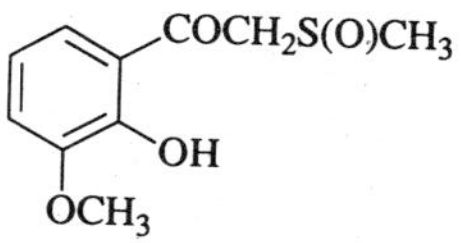 $COCH_2S(O)CH_3$, OH, OCH_3 (benzene ring)		$COCl_2$, pyr, C_6H_6, 80°, 4 h	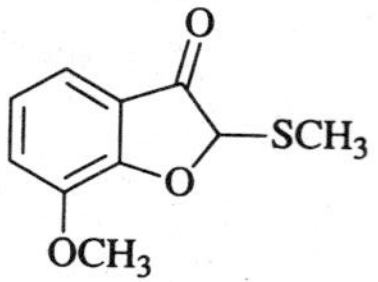 O, SCH_3, O, OCH_3 (ring structure) (48)[c]	266

Substrate	Reagent	Conditions	Product(s) (Yield %)	Ref.
$S(O)C_2H_5$, $CONHCH_3$	Ac_2O	80°, 10 h	$SCH(OAc)CH_3$, $CONRCH_3$ R = H (77); R = Ac (6)	267
O, $S(O)CH_3$, N, N, CH_3	HCl, H_2O	Reflux 6 h	O, SCH_3, N, N, CH_3 (11) + O, N, N, CH_3 (4)	268
	Ac_2O		O, SCH_2OAc, N, N, CH_3 (—)	194
$CH_2S(O)CH_3$, N, O, O	Ac_2O	Reflux	$CH(OAc)SCH_3$, N, O, O (63)	119
$COCH_2S(O)CH_3$, OH, OCH_3		H^+	$COCHOHSCH_3$, OH, OCH_3 (—)	269
$COCH_2S(O)CH_3$, OCH_3, OH		H^+	$COCHOHSCH_3$, OCH_3, OH (—)	269

TABLE I. SULFOXIDES WITH AN OXYGEN NUCLEOPHILE (*Continued*)

Sulfoxide	Nucleophile	Reaction Conditions	Product(s) and Yield(s) (%)	Refs.
2-($S(O)CH_3$)-indane-1,3-dione	HCl, H_2O	15 min	2-SCH_3-2-OH-indane-1,3-dione (90)	270
	AcOH	50°, 20 h	2-SCH_3-2-OAc-indane-1,3-dione (71)	270
	Ac_2O	50°, 15 h	" (75)	270
	C_2H_5OH	55°, 12 h	2-SCH_3-2-OC_2H_5-indane-1,3-dione (52)	270
C_6H_5-B(O,O)-fused tetrahydrothiophene S=O	Ac_2O	C_6H_6, reflux 20 h	C_6H_5-B(O,O)-fused tetrahydrothiophene with OAc α to S (60–70)	225
3,3-Br,Br-4-$S(O)CH_3$-azetidin-2-one, N-$C(CO_2CH_3)=C(CH_3)_2$	Ac_2O	Reflux, 4 h	3,3-Br,Br-4-SCH_2OAc-azetidin-2-one, N-$C(CO_2CH_3)=C(CH_3)_2$ (30)[c]	106
C_{11}				
$C_6H_5S(O)CH_2C_4H_9$-*t*	Ac_2O	100°, 66 h	$C_6H_5SCH(OAc)C_4H_9$-*t* (31)	182
$C_6H_5(CH_2)_2COCH_2S(O)CH_3$	Cl_3CCO_2H	C_6H_6, reflux 1.5 h	$C_6H_5(CH_2)_2COCH(O_2CCCl_3)SCH_3$ (65) + $C_6H_5(CH_2)_2COCH(SCH_3)_2$ (3)	37

$C_6H_5S(O)(CH_2)_2C(CH_3)_2NO_2$	Ac_2O	TFAA, 2,6-lutidine, 3 h	$C_6H_5SCH(OAc)CH_2C(CH_3)_2NO_2$ (64)	42
	AcONa	Ac_2O, reflux 11 h	" (38)	42
$C_6H_5S(O)(CH_2)_2CH(CH_3)CO_2H$		Ac_2O, TsOH (cat.), toluene, reflux 1 h	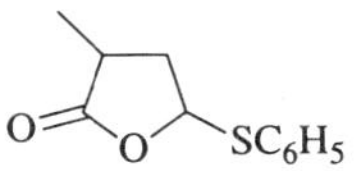 SC_6H_5 (51)	41
$C_6H_5S(O)CH_2CH(CH_3)CH_2CO_2H$		"	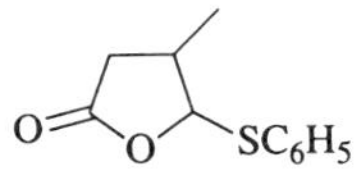 SC_6H_5 (52)	41
p-$CH_3C_6H_4S^*(O)CH_2CO_2C_2H_5$	Ac_2O	110°, 4 h[e]	p-$CH_3C_6H_4SC^*H(OAc)CO_2C_2H_5$ 29% ee (26)[e]	26, 245
	Ac_2O	DCC (2 eq), 110°, 6 h	" 70% ee (10)	26, 245
	Ac_2O	DCC (4 eq), 120°, 8 h	" 50% ee (43)[e]	26, 245
p-$CH_3C_6H_4S^*(O)CH_2CON(CH_3)_2$	Ac_2O	120°, 4 h	p-$CH_3C_6H_4SC^*H(OAc)CON(CH_3)_2$ 29% ee (51)	26, 245
	Ac_2O	DCC (2 eq), 110°, 6 h[e]	" 65% ee (35)[e]	26, 245
	Ac_2O	DCC (4 eq), 120°, 8 h	" 57% ee (57)	26, 245
$(C_2H_5O)_2P(O)CH_2S(O)C_6H_5$			$(C_2H_5O)_2P(O)CH(OR)SC_6H_5$	232
	Ac_2O	CH_3SO_3H, CH_2Cl_2, reflux 3 h	R = Ac (72)	
	$(C_2H_5CO)_2O$	CH_3SO_3H, CH_2Cl_2, reflux 3 h	R = COC_2H_5 (58)	
	$(n$-$C_3H_7CO)_2O$	"	R = COC_3H_7-n (55)	
	$(t$-$C_4H_9CO)_2O$	"	R = COC_4H_9-t (62)	
	$(C_6H_5CO)_2O$	"	R = COC_6H_5 (38)	
	Ac_2O	120°, 3 h	R = Ac (88)	226
	CH_3OH	I_2, reflux 4 h	R = CH_3 (68)	226
	C_2H_5OH	I_2, reflux 50 min	R = C_2H_5 (78)	226, 234
p-$CH_3OC_6H_4CH{=}C(SCH_3)S(O)CH_3$	Ac_2O	110°, 11 h	p-$CH_3OC_6H_4CH{=}C(SCH_3)SCH_2OAc$ (80)	176
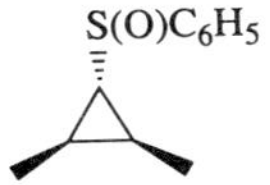 $S(O)C_6H_5$	AcONa	Ac_2O, 170°, 3 h	 AcO SC_6H_5 (93)	28, 29 261

TABLE I. SULFOXIDES WITH AN OXYGEN NUCLEOPHILE (*Continued*)

Sulfoxide	Nucleophile	Reaction Conditions	Product(s) and Yield(s) (%)	Refs.
 $S(O)C_6H_5$ (cyclopropane structure)	"	"	 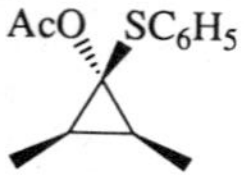 C_6H_5S, OAc (cyclopropane structure) I + AcO, SC_6H_5 (cyclopropane structure) II; I:II = 3:1 (92)	28
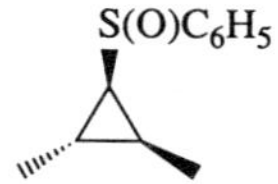 $S(O)C_6H_5$ (cyclopropane structure)	"	"	 AcO SC_6H_5 (cyclopropane structure) (90)	29, 261
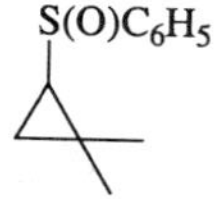 $S(O)C_6H_5$ (cyclopropane structure)	"	"	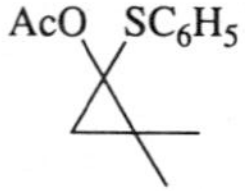 AcO SC_6H_5 (cyclopropane structure) (92)	29, 261
$C_6H_5S(O)CH_2C_4H_9$	AcONa	Ac_2O, 170°, 3 h	$C_6H_5SCH(OAc)C_4H_9$ (84)	29
p-$[(C_2H_5O)_2P(X)O]C_6H_4S(O)CH_3$	TFAA	15 min	p-$[(C_2H_5O)_2P(X)O]C_6H_4SCH_2O_2CCF_3$ X = O, S (—)	231
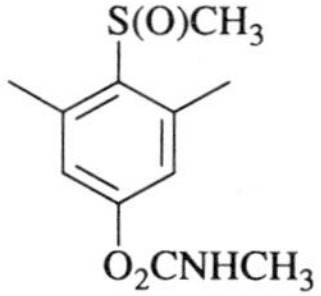 $S(O)CH_3$, O_2CNHCH_3 (aryl structure)	TFAA		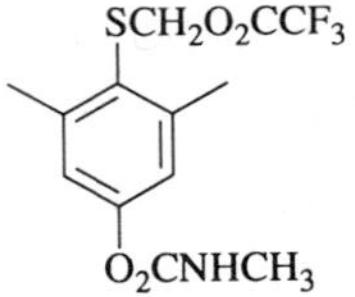 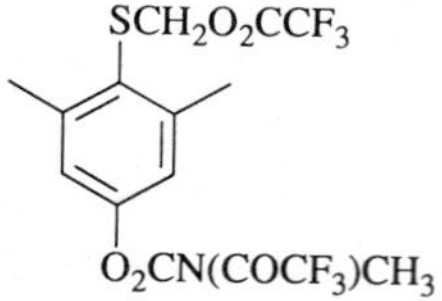 $SCH_2O_2CCF_3$, O_2CNHCH_3 (aryl structure) I (—) + $SCH_2O_2CCF_3$, $O_2CN(COCF_3)CH_3$ (aryl structure) II (—)	231
	TFAA	$AcOC_2H_5$, 100°	II (—)	262
	TFAA	$AcOC_2H_5$, 15°	I (—)	262

Substrate	Reagent	Conditions	Product(s) and Yield(s) (%)	Refs.
N-$CH_2C_6H_5$ thiomorpholine S-oxide	Ac_2O	C_6H_6, TsOH (cat.), 60°, 3.5 h	3-OAc, N-$CH_2C_6H_5$ thiomorpholine (80) + N-$CH_2C_6H_5$ 2,3-dihydro-1,4-thiazine (20)	271
2,2-dimethylthiochroman S-oxide	Ac_2O		2-CH_2OAc, 2-$CH(OAc)_2$ thiochroman (—)	272
4-$CH_2S(O)CH_3$ coumarin	Ac_2O	Reflux	4-$CH(OAc)SCH_3$ coumarin (83)	119
6-Cl, 4-$CH_2S(O)CH_3$ quinolin-2(1H)-one (N–H)	Ac_2O	Reflux	6-Cl, 4-$CH(OAc)SCH_3$, N-Ac quinolin-2-one (39) + 6-Cl, 4-$CH(OAc)SCH_3$ quinolin-2(1H)-one (N–H) (32)	119

TABLE I. SULFOXIDES WITH AN OXYGEN NUCLEOPHILE (*Continued*)

Sulfoxide	Nucleophile	Reaction Conditions	Product(s) and Yield(s) (%)	Refs.
2-(chromon-2-yl) $CH_2S(O)CH_3$	Ac_2O	Reflux, 5 h	2-(chromon-2-yl) $CH(OAc)SCH_3$ (90)	193
1-methyl-4-quinolon-3-yl $S(O)CH_3$	Ac_2O	Reflux, 5 h	1-methyl-4-quinolon-3-yl SCH_2OAc (92)	194
phthalimido $N(CH_2)_2S(O)CH_3$	AcONa	Ac_2O, reflux 3 h	phthalimido $NCH_2CH(OAc)SCH_3$ (62) +	273
			phthalimido $N(CH_2)_2SCH_2OAc$ (33)	274
pyrimido-thiazine sulfoxide (C_2H_5, n-C_3H_7)		Neat, 11 months	2-OR pyrimido-thiazine (C_2H_5, n-C_3H_7), R = H (~80)	274
	H_2O	24 h	R = H (~85)	
	CH_3OH	Reflux, 5 min	R = CH_3 (76)	
	C_2H_5OH	Reflux, 5 min	R = C_2H_5 (~70)	

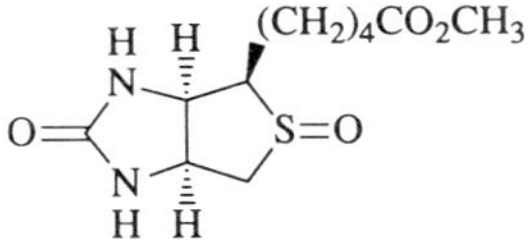	AcOH	Reflux, 3 min	R = Ac (—)	
	TFAA	$CHCl_3$, −60° to rt	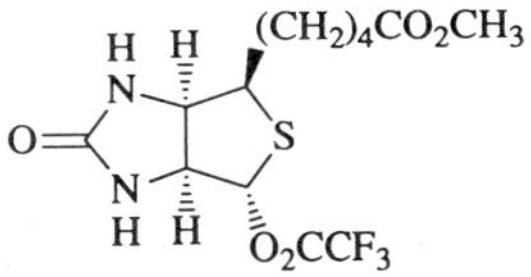 (—)	275
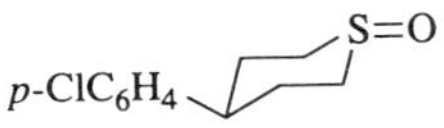			 I + 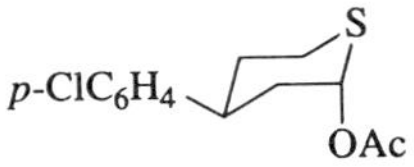 II + 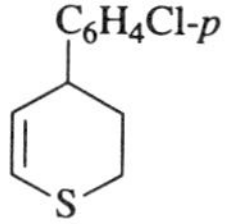 III	263, 276[b]
	Ac_2O	100°, 45 min	I:II = 1: 1 (40) + III (—)	
	Ac_2O	100°, 3 h	I:II = 1:15 (70) + III (23)	
	Ac_2O	DCC (1 eq), 100°, 4 h	I:II = 2: 1 (78) + III (20)	
	Ac_2O	DCC (3 eq), 100°, 4 h	I:II = 9:1 (85) + III (11)	
	Ac_2O	DCC (5 eq), 100°, 4 h	I:II = 13:1 (83) + III (9)	263, 276[b]
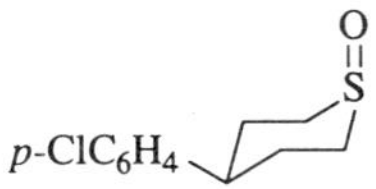	Ac_2O	100°, 45 min	I:II = 1:1 (40) + III (—)	
	Ac_2O	100°, 3 h	I:II = 1:15 (71) + III (21)	
	Ac_2O	DCC (1 eq), 100°, 4 h	I:II = 2:1 (77) + III (20)	
	Ac_2O	DCC (5 eq), 100°, 4 h	I:II = 14:1 (83) + III (12)	
	Ac_2O	2,6-Lutidine (5 eq), 100°, 4.5 h	I:II = 5:2 (81) + III (15)	
	Ac_2O	2,6-Lutidine (10 eq), 100°, 4.5 h	I:II = 5:1 (86) + III (12)	

TABLE I. SULFOXIDES WITH AN OXYGEN NUCLEOPHILE (*Continued*)

Sulfoxide	Nucleophile	Reaction Conditions	Product(s) and Yield(s) (%)	Refs.
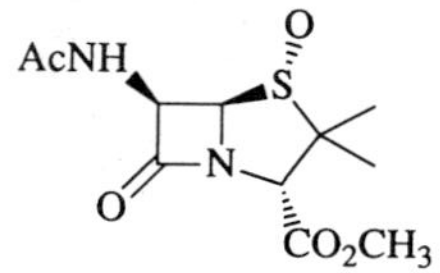 [penam sulfoxide: AcNH, S→O, CO_2CH_3]	Ac_2O		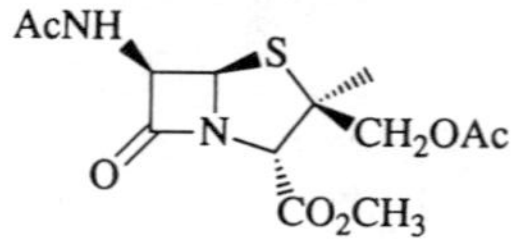 [AcNH penam, CH_2OAc, CO_2CH_3] (28)	277
C_{12}				
$C_6H_5S(O)C_6H_{13}$-*n*	Ac_2O	TFAA, 2,6-lutidine, 30 min	$C_6H_5SCH(OAc)C_5H_{11}$-*n* (84)	42
	AcONa	Ac_2O, reflux 7 h	" (62)	42
n-$C_3H_7COCH_2S(O)C_6H_4CH_3$-*p*	AcONa	Ac_2O, toluene, reflux 2 h	*n*-$C_3H_7COCH(OAc)SC_6H_4CH_3$-*p* (77)	107
n-$C_3H_7CHOHCH_2S(O)C_6H_4CH_3$-*p*	AcONa	Ac_2O, reflux 2 h	*n*-$C_3H_7CH(OAc)CH(OAc)SC_6H_4CH_3$-*p* (71)	106
$AcOCH_2CHOHCH_2S(O)C_6H_4CH_3$-*p*	AcONa	Ac_2O, reflux 3 h	$AcOCH_2CH(OAc)CH(OAc)SC_6H_4CH_3$-*p* (90)	106
$C_6H_5S(O)CH_2CH(C_2H_5)CH_2CO_2H$	Ac_2O	TsOH (cat.), toluene, reflux 1 h	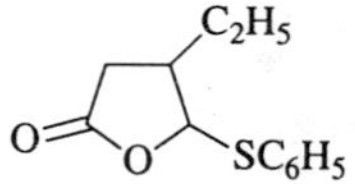 [γ-lactone: 4-C_2H_5, 5-SC_6H_5] (22)	41
2,4-$(CH_3O)_2C_6H_3CH{=}C(SCH_3)S(O)CH_3$	Ac_2O	110°, 11 h	2,4-$(CH_3O)_2C_6H_3CH{=}C(SCH_3)SCH_2OAc$ (94)	176
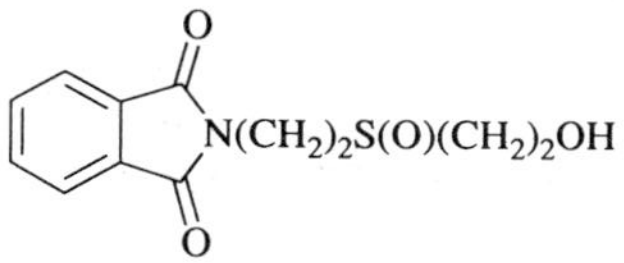 [phthalimido]$N(CH_2)_2S(O)(CH_2)_2OH$	AcONa	Ac_2O, reflux 3 h	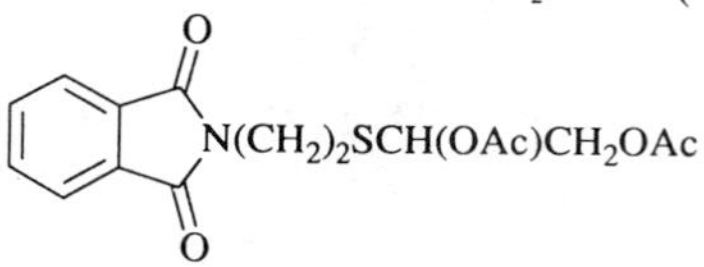 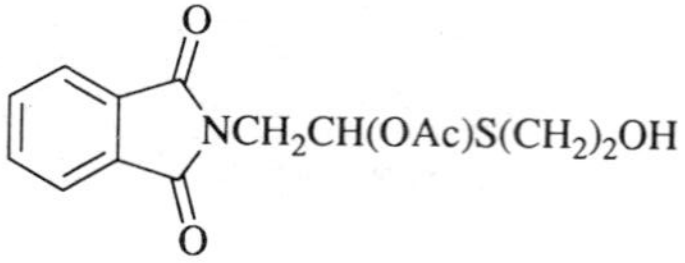 [phthalimido]$N(CH_2)_2SCH(OAc)CH_2OAc$ I + [phthalimido]$NCH_2CH(OAc)S(CH_2)_2OH$ II; I:II = 3:1 (—)	273

p-[(i-$C_3H_7NH(C_2H_5O)P(O)O]C_6H_4S(O)CH_3$	TFAA	100°, 1 h	p-[(i-$C_3H_7NR(C_2H_5O)P(O)O]C_6H_4SCH_2O_2CCF_3$ R = O_2CCF_3 (12)	231
	TFAA	15 min	" R = H (—)	231
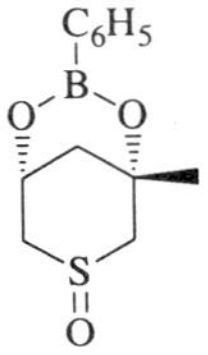	Ac_2O	Heat	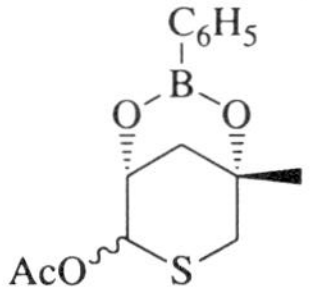 (10) +	229
			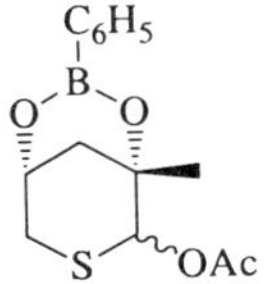 (10)	
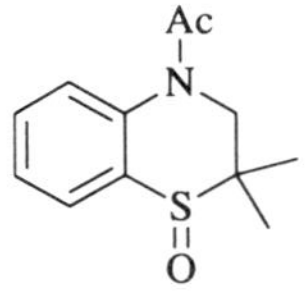	Ac_2O	Reflux, 4 h	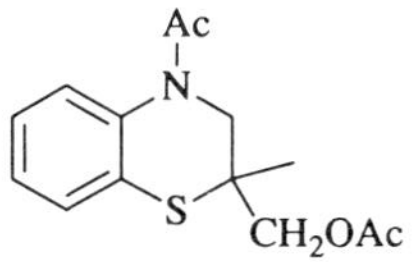 (67)	278
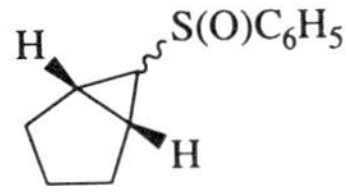	CH_3OH	1. $(CH_3)_3O^+BF_4^-$, CH_2Cl_2 (before CH_3OH) 2. CH_3ONa, 25°	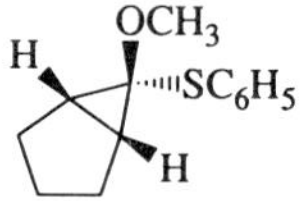 (14)[c]	279
			(28)[c]	
C_{13}				
$C_6H_5S(O)CH_2C_6H_5$			$C_6H_5SCH(OAc)C_6H_5$ I + C_6H_5SAc II + C_6H_5SH III + C_6H_5CHO IV	201
	Ac_2O (1 eq)	140°, 6 h	I (22) + II (12) + III (—) + IV (—)	
	Ac_2O (6 eq)	140°, 2 h	I (42) + II (14) + III (27) + IV (12)	
	TFAA	$CDCl_3$, 0°	$C_6H_5SCH(O_2CCF_3)C_6H_5$ (100)	242
p-$CH_3C_6H_4S(O)CH_2CHOHCH(OAc)CH_3$	AcONa	Ac_2O, reflux, 3 h	p-$CH_3C_6H_4S(CHOAc)_3CH_3$ (87)	106
$CH_2{=}CHCCH_3(OAc)CH_2S(O)C_6H_5$	AcONa	Ac_2O, reflux, 130°, 8 h	$CH_2{=}CHCCH_3(OAc)CH(OAc)SC_6H_5$ (82)	280

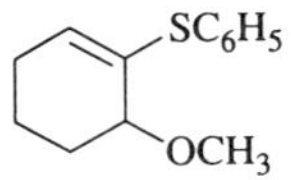

TABLE I. SULFOXIDES WITH AN OXYGEN NUCLEOPHILE (*Continued*)

Sulfoxide	Nucleophile	Reaction Conditions	Product(s) and Yield(s) (%)	Refs.
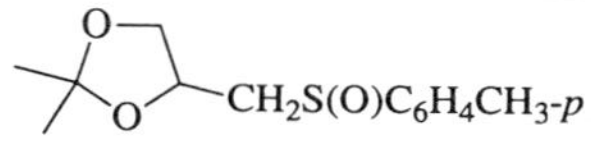 (2,2-dimethyl-1,3-dioxolan-4-yl)$CH_2S(O)C_6H_4CH_3$-*p*	AcONa	Ac_2O, reflux 3.5 h	 (2,2-dimethyl-1,3-dioxolan-4-yl)$CH(OAc)SC_6H_4CH_3$-*p* ~1:1 (100)	253
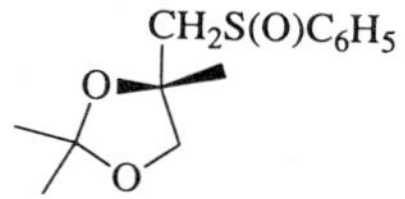 $CH_2S(O)C_6H_5$ (dioxolane)	AcONa	Ac_2O, reflux 11 h	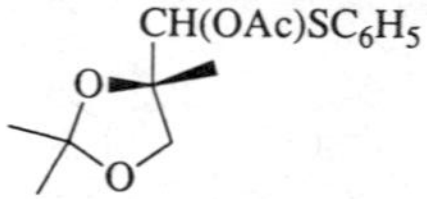 $CH(OAc)SC_6H_5$ (dioxolane) (88)	281
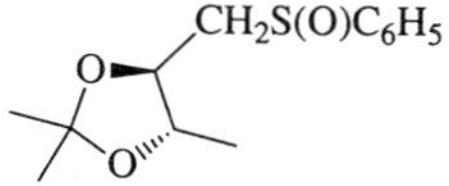 $CH_2S(O)C_6H_5$ (dioxolane)	TFAA	Pyr, C_6H_6, 5 min	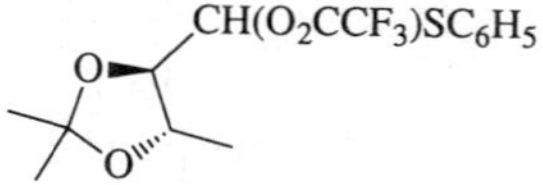 $CH(O_2CCF_3)SC_6H_5$ (dioxolane) (84)	50
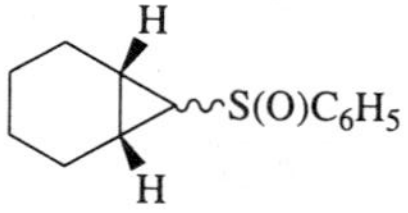 (bicyclo, H, H)$S(O)C_6H_5$	CH_3ONa	1. $(CH_3)_3O^+BF_4^-$, CH_2Cl_2 (before CH_3OH) 2. CH_3ONa, 25°	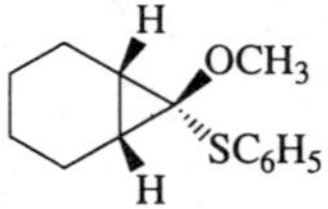 (bicyclo, H, H) OCH_3, SC_6H_5 (44)[c]	279
1-$C_{10}H_7COCH_2S(O)CH_3$	HCl, H_2O	$(CH_3)_2SO$, 12–24 h	1-$C_{10}H_7COCHOHSCH_3$ (63)	33
2-$C_{10}H_7COCH_2S(O)CH_3$	HCl, H_2O	"	2-$C_{10}H_7COCHOHSCH_3$ (90)	33
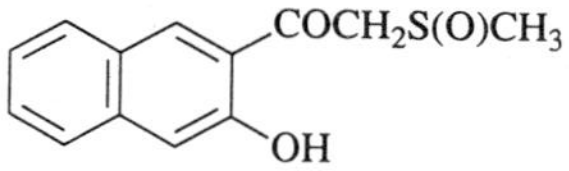 (naphthalene)$COCH_2S(O)CH_3$, OH		CF_3CO_2H, C_6H_6, reflux 40–90 min	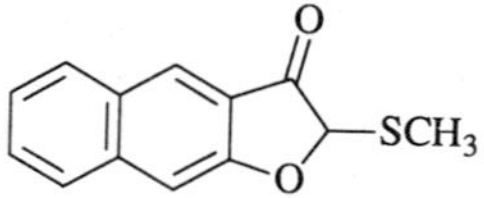 (O, SCH_3, O) I (81)	260
		$COCl_2$, pyr, C_6H_6, 80°, 4 h	I (24) 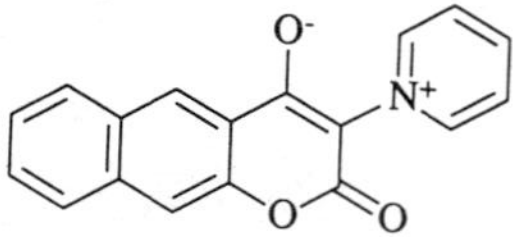 + (O^-, N^+, O, O) (14)	266
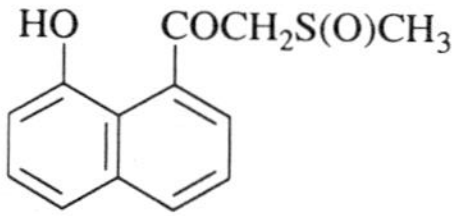 HO, $COCH_2S(O)CH_3$ (naphthalene)		CF_3CO_2H, C_6H_6, reflux 40–90 min	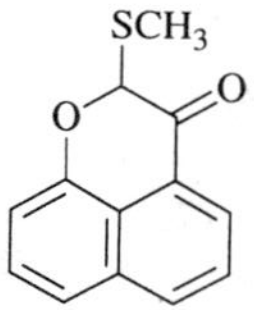 (SCH_3, O, O) (64)	260

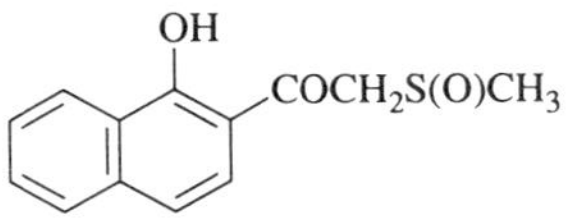

Reactant	Additive	Conditions	Product(s) and Yield(s) (%)	Refs.
1-hydroxy-2-naphthyl $COCH_2S(O)CH_3$		$COCl_2$, pyr, C_6H_6, 80°, 5 h	2-SCH_3 naphthofuranone (17) + pyridinium naphthopyranone olate (10)	266
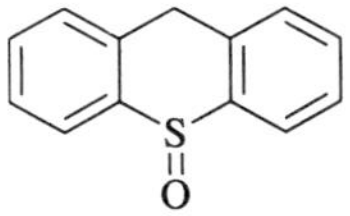 thioxanthene S-oxide		H_2SO_4 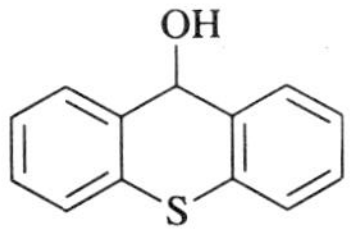	9-hydroxythioxanthene (—)	282
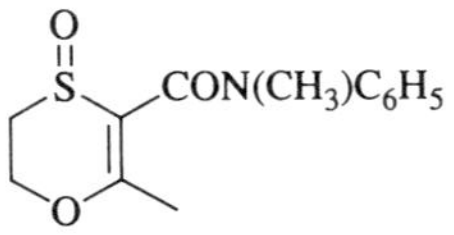 1-oxide, $CON(CH_3)C_6H_5$	AcOH	Ac_2O, C_6H_6, reflux 1.5 h 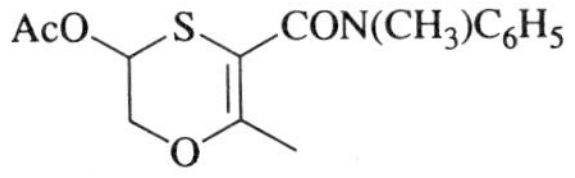	AcO, $CON(CH_3)C_6H_5$ (~70)	178
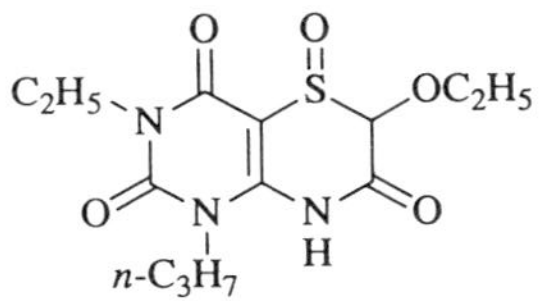 C_2H_5, OC_2H_5, n-C_3H_7	C_2H_5OH	Reflux 4 h 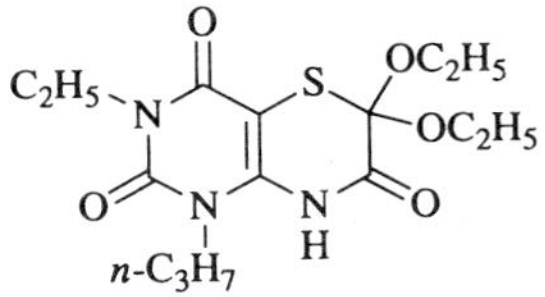	C_2H_5, OC_2H_5, OC_2H_5, n-C_3H_7 (60)	274
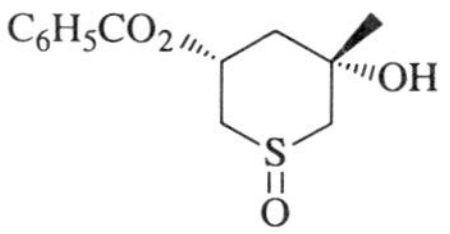 $C_6H_5CO_2$, OH, S=O	AcONa	Ac_2O, 140°, 3 h	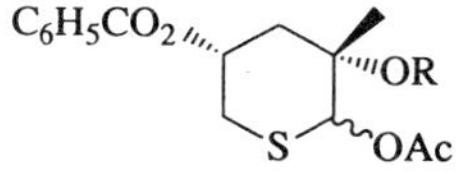 $C_6H_5CO_2$, OR, OAc; R = Ac (9)	229
	Ac_2O	100°, 18 h	R = H (5)	229

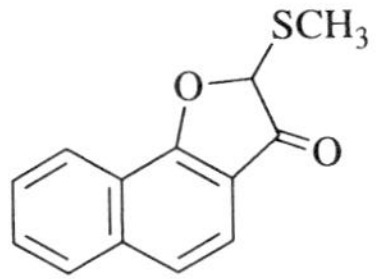

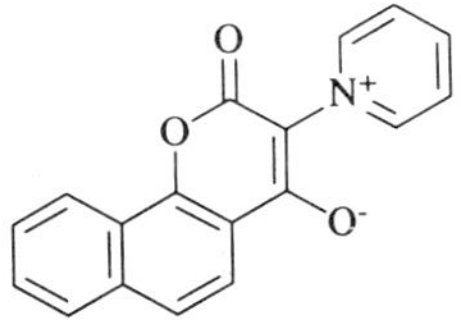

TABLE I. SULFOXIDES WITH AN OXYGEN NUCLEOPHILE (*Continued*)

Sulfoxide	Nucleophile	Reaction Conditions	Product(s) and Yield(s) (%)	Refs.
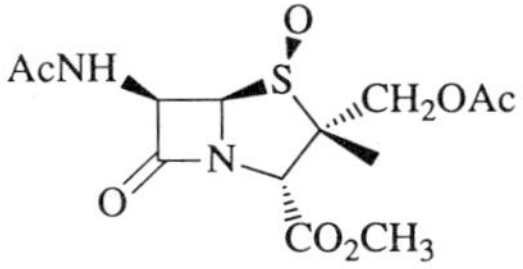	Ac_2O		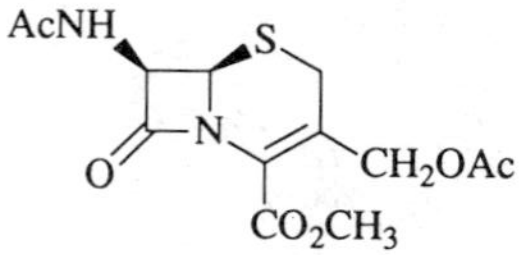 (7) + 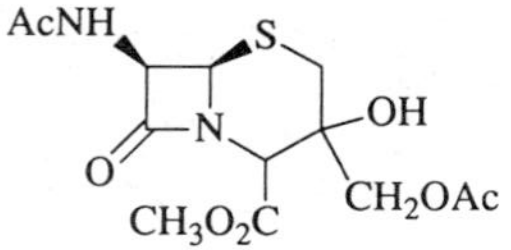 (41)	277
C_{14}				
$(C_6H_5CH_2)_2SO$	Ac_2O	$CDCl_3$, 76°	$C_6H_5CH_2SCH(OAc)C_6H_5$ (—) + $(C_6H_5CH_2S)_2$ (—) + $(C_6H_5CH_2)_2S$ (—) + C_6H_5CHO (—) + $C_6H_5CH(SCH_2C_6H_5)_2$ (—)	283
$C_6H_5CH_2S(O)SCH_2C_6H_5$	AcOH	Ac_2O, 60°, 2 h[b]	$C_6H_5CH_2S(O)CH(SAc)C_6H_5$ (37)	284, 285
$C_6H_5S(O)(CH_2)_2OC_6H_5$	AcONa	Ac_2O, reflux 7 h	$C_6H_5SCH(OAc)CH_2C_6H_5$ (77)	42
	Ac_2O	TFAA, 2,6-lutidine, 30 min	" (76)	42
	Ac_2O	100°, 23.5 h	" (76)[a] + $C_6H_5SO_2(CH_2)_2OC_6H_5$ (5)[a]	182
$C_6H_5S(O)(CH_2)_2SC_6H_5$	AcONa	Ac_2O, reflux 7 h	$C_6H_5SCH(OAc)CH_2C_6H_5$ (45)	42
	Ac_2O	TFAA, 2,6-lutidine, 30 min	" (56)	42
p-$ClC_6H_4S(O)(CH_2)_2OC_6H_5$	AcONa	Ac_2O, reflux 9 h	p-$ClC_6H_4SCH(OAc)CH_2OC_6H_5$ (60)	42
	Ac_2O	TFAA, 2,6-lutidine, 30 min	" (63)	42
$C_6H_5S(O)CH_2COC_6H_5$	Ac_2O	110°, 1 h[b]	$C_6H_5SCH(OAc)COC_6H_5$ (74)	245[b]
p-$CH_3C_6H_4S^*(O)CH_2C_6H_5$	Ac_2O	DCC, (5 eq), 100°, 10 h	p-$CH_3C_6H_4SC^*H(OAc)C_6H_5$(~7)[a]	27
p-$CH_3C_6H_4S^*(O)CH_2C_6H_4Cl$-$m$	Ac_2O	DCC (5 eq), 120°, 4 h	p-$CH_3C_6H_4SC^*H(OAc)C_6H_4Cl$-$m$ (9)[a]	27
p-$CH_3C_6H_4S^*(O)CH_2C_6H_3Cl_2$-$m,m'$	Ac_2O	DCC (5 eq), 130°, 4 h	p-$CH_3C_6H_4SC^*H(OAc)C_6H_3Cl_2$-$m,m'$ (7)[a]	27

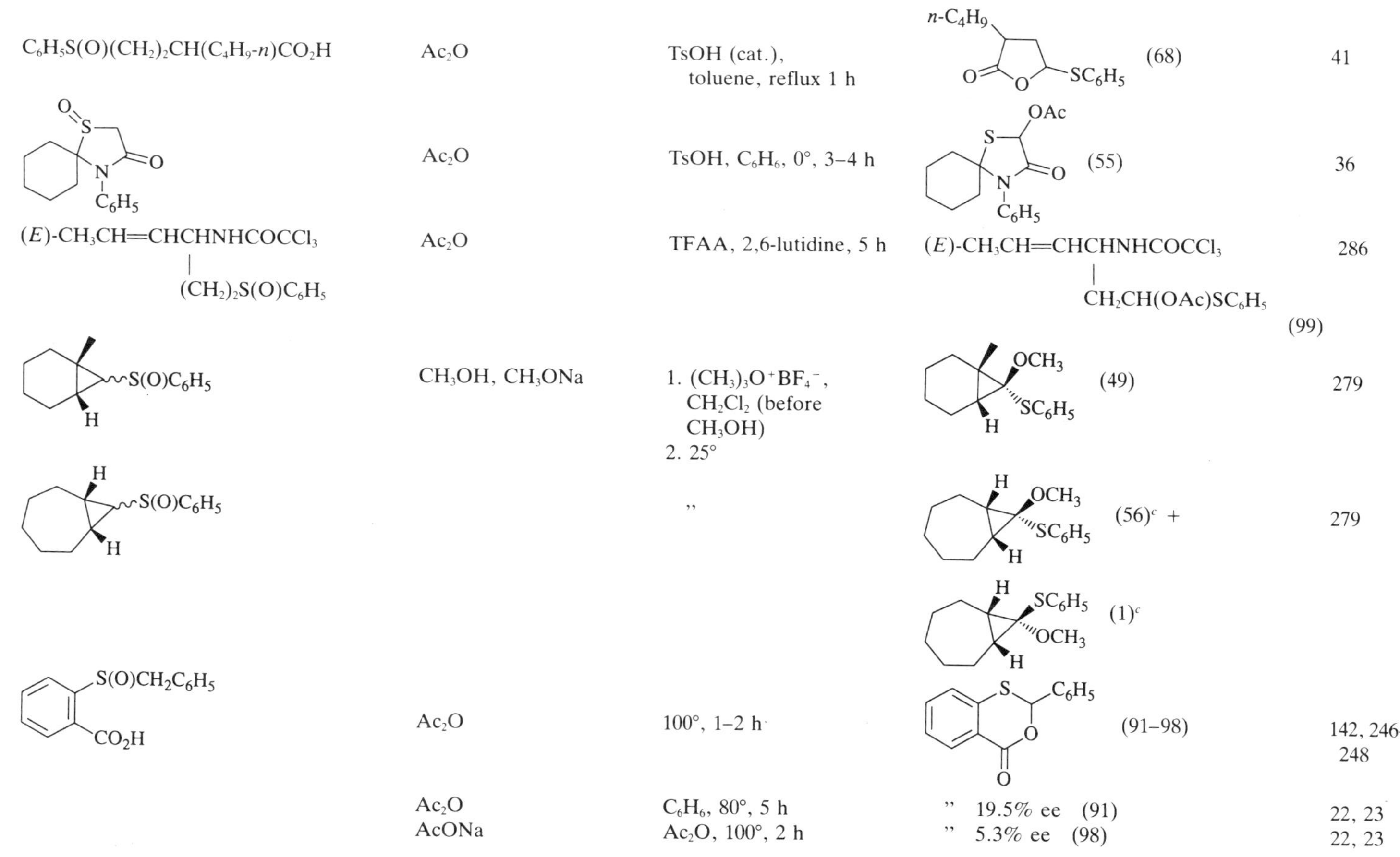

Substrate	Activator	Conditions	Product(s) and Yield(s) (%)	Refs.
$C_6H_5S(O)(CH_2)_2CH(C_4H_9\text{-}n)CO_2H$	Ac_2O	TsOH (cat.), toluene, reflux 1 h	(68)	41
	Ac_2O	TsOH, C_6H_6, 0°, 3–4 h	(55)	36
(E)-$CH_3CH{=}CHCH(NHCOCCl_3)(CH_2)_2S(O)C_6H_5$	Ac_2O	TFAA, 2,6-lutidine, 5 h	(E)-$CH_3CH{=}CHCH(NHCOCCl_3)CH_2CH(OAc)SC_6H_5$ (99)	286
	CH_3OH, CH_3ONa	1. $(CH_3)_3O^+BF_4^-$, CH_2Cl_2 (before CH_3OH) 2. 25°	(49)	279
		"	(56)[c] + (1)[c]	279
	Ac_2O	100°, 1–2 h	(91–98)	142, 246–248
	Ac_2O	C_6H_6, 80°, 5 h	" 19.5% ee (91)	22, 23
	AcONa	Ac_2O, 100°, 2 h	" 5.3% ee (98)	22, 23

TABLE I. Sulfoxides with an Oxygen Nucleophile (*Continued*)

Sulfoxide	Nucleophile	Reaction Conditions	Product(s) and Yield(s) (%)	Refs.
		TsOH, C_6H_6, 80°, 2 h	" (96)	22, 23
	CF_3CO_2OAc	CH_2Cl_2, 4°, 30 min	" (73)[a]	142
		DCC, $(CH_2Cl)_2$, 25°, 15 h	" 29.9% ee (91)	22, 23
	H_3PO_4 (1.4 eq)	DCC, $(CH_2Cl)_2$, 25°, 2.5 h	" 20.4% ee (73)	22, 23
	TFAA	CH_2Cl_2, 0°, 3–4 min	" (41) + $(o\text{-}HO_2CC_6H_4S)_2$ (58)	142
o-($\overset{*}{S}(O)CHDC_6H_5$)$C_6H_4CO_2H$	Ac_2O	114°, 2 h[b]	2-C_6H_5-2-(H,D)-4H-3,1-benzoxathiin-4-one (—)	287[b]
o-($S(O)CH_2C_6H_4Cl$-p)$C_6H_4CO_2H$	Ac_2O	100°, 1–2 h	2-(C_6H_4Cl-p)-4H-3,1-benzoxathiin-4-one (99)	246, 247
o-($CH_2S(O)C_6H_5$)$C_6H_4CO_2H$	Ac_2O	100°, 3.5 h	3-(SC_6H_5)phthalide (89)	23
		TsOH, C_6H_6, 80°, 11 h	" (81)	23
		DCC, $(CH_2Cl)_2$, 84°, 16 h	" (32)	23
o-($S(O)C_6H_5$)$C_6H_4CH_2SCH_3$	Ac_2O	100°, 13 h	o-(SC_6H_5)$C_6H_4CH_2SCH_2OAc$ (23)[a]	20

Reactant	Reagent	Conditions	Product(s) and Yield(s) (%)	Refs.
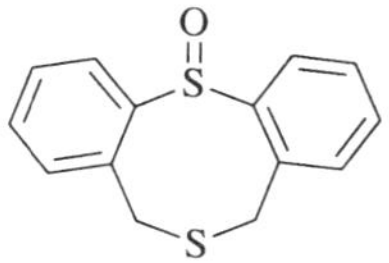 O, S, S	AcONa	Ac_2O, 135°, 8 h	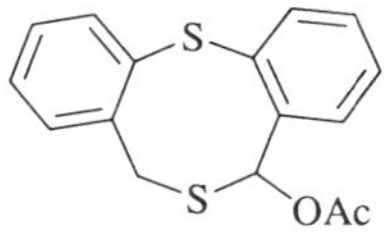 S, S, OAc (45)[a]	20
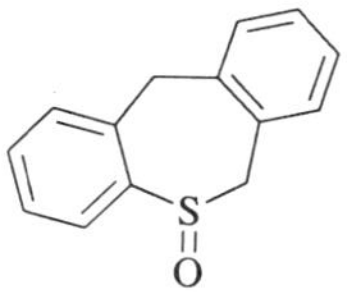 S, O	Ac_2O	Reflux	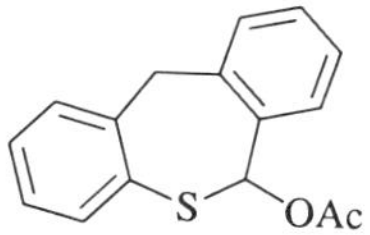 S, OAc (—)	288
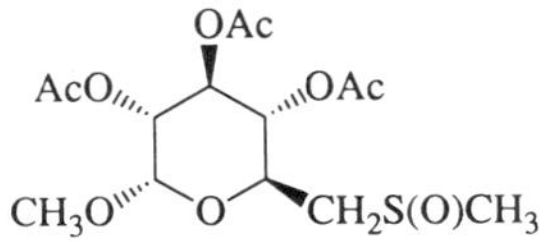 OAc, AcO, OAc, CH_3O, O, $CH_2S(O)CH_3$	Ac_2O	100°, 10 h	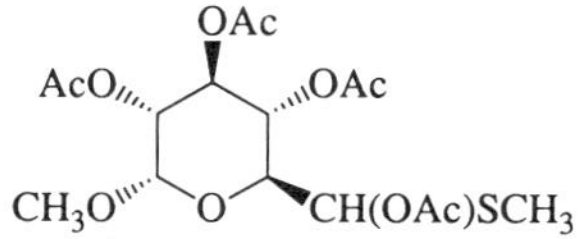 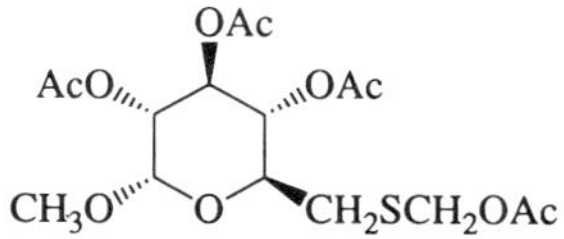 OAc, AcO, OAc, CH_3O, O, $CH(OAc)SCH_3$ I + OAc, AcO, OAc, CH_3O, O, CH_2SCH_2OAc II I:II = 2:1 (43)	252
C_{15}				
n-$C_{12}H_{25}S(O)CH_2CO_2CH_3$	HCl, H_2O	Heat	n-$C_{12}H_{25}SCHOHCO_2H$ (—) + $OHCCO_2H$ (—) + $(n$-$C_{12}H_{25}S)_2CHCO_2H$ (—)	289
p-$CH_3C_6H_4S(O)CH_2COC_6H_5$	AcONa	Ac_2O, toluene, reflux overnight	p-$CH_3C_6H_4SCOCH(OAc)C_6H_5$ (74)	107
p-$CH_3C_6H_4S^*(O)CH_2CO_2C_6H_5$	Ac_2O	110°, 1 h	p-$CH_3C_6H_4SC^*H(OAc)CO_2C_6H_5$ 0.5% ee (80)	26, 245

TABLE I. SULFOXIDES WITH AN OXYGEN NUCLEOPHILE (*Continued*)

Sulfoxide	Nucleophile	Reaction Conditions	Product(s) and Yield(s) (%)	Refs.
	Ac_2O	DCC (2 eq), 110°, 2 h	" 6% ee (58)	26, 245
	Ac_2O	DCC (4 eq), 110°, 1.25 h	" 38% ee (32)	26, 245
	Ac_2O	DCC (4 eq), 130°, 1 h	" 10% ee (88)	26, 245
p-$CH_3C_6H_4S(O)CH_2CHOHC_6H_5$	AcONa	Ac_2O, reflux 3 h	p-$CH_3C_6H_4S[CH(OAc)]_2C_6H_5$ (95)	106
$CH(OH)CH_2S(O)C_6H_4CH_3$-p 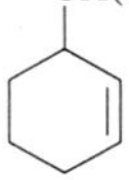 (on cyclohexenyl ring)	AcONa	Ac_2O, reflux 4 h	$CH(OAc)CH(OAc)SC_6H_4CH_3$-$p$ 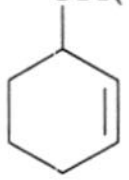 (on cyclohexenyl ring) (92)	106
	Ac_2O	AcONa, reflux 4 h		
p-$CH_3C_6H_4S(O)CH_2CHOHC_6H_4Cl$-$p$	AcONa	Ac_2O, reflux 2 h	p-$CH_3C_6H_4S[CH(OAc)]_2C_6H_4Cl$-$p$ (76)	106
(E)-$CH_3CH{=}CHC^*(CH_3)NHCOCCl_3$ with $(CH_2)_2S(O)C_6H_5$	Ac_2O	TFAA, lutidine, 5 h	(E)-$CH_3CH{=}CHC^*(CH_3)NHCOCCl_3$ with $CH_2CH(OAc)SC_6H_5$ (89)	286
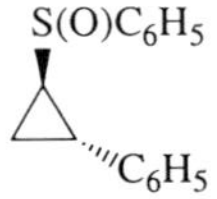 cyclopropane: $S(O)C_6H_5$, C_6H_5	AcONa	Ac_2O, 170°, 3 h	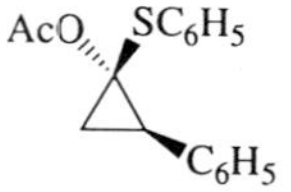 cyclopropane: AcO, SC_6H_5, C_6H_5 (88)	29, 257
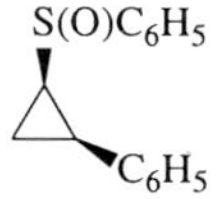 cyclopropane: $S(O)C_6H_5$, C_6H_5	AcONa	Ac_2O, 170°, 3 h	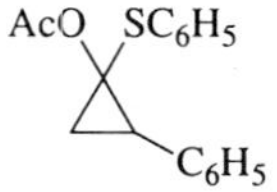 cyclopropane: AcO, SC_6H_5, C_6H_5 31:69 (84)	28, 29
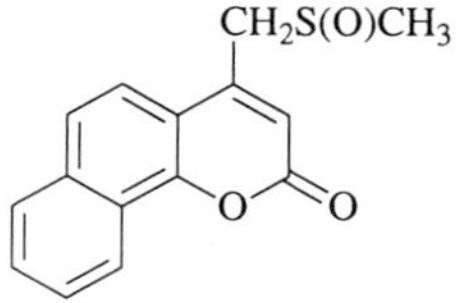 benzocoumarin: $CH_2S(O)CH_3$, O, O	Ac_2O	Reflux	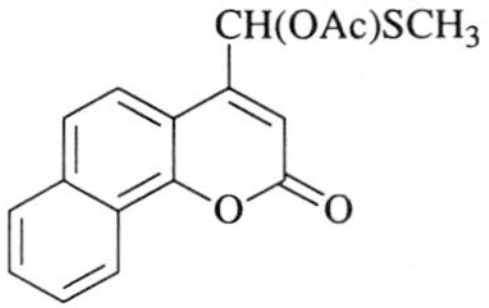 benzocoumarin: $CH(OAc)SCH_3$, O, O (85)	119

Reactant	Activator	Conditions	Product(s) and Yield(s) (%)	Refs.
CF_3, N, CHO, S=O (dibenzothiazepine sulfoxide structure)	C_2H_5OH	1. TFAA, C_6H_6, 48 h (before C_2H_5OH) 2. Ether	CF_3, S, O, N, C_2H_5O (structure) (43)	290
CH_3O, O, CH_3O, $CH_2S(O)CH_3$ (furochromone structure)	Ac_2O	TsOH, 40°, 5 min	CH_3O, O, CH_3O, $CH(OAc)SCH_3$ (structure) (71)	291
C_{16}				
$C_6H_5S(O)CH_2C(CH_3)_2C_6H_5$	Ac_2O	95°, 21 h	$C_6H_5SCH(OAc)C(CH_3)_2C_6H_5$ (54)	182
p-$CH_3C_6H_4S(O)CH_2CHOHCH_2C_6H_5$	AcONa	Ac_2O, reflux 3 h	p-$CH_3C_6H_4S[CH(OAc)]_2CH_2C_6H_5$ (74)	106
p-$CH_3C_6H_4S(O)CH_2CHOH$-(3,4-methylenedioxyphenyl)	AcONa	Ac_2O, reflux 6 h	p-$CH_3C_6H_4S[CH(OAc)]_2$-(3,4-methylenedioxyphenyl) (99)	106
p-$CH_3C_6H_4S(O)CH_2CH(OCH_3)C_6H_5$	AcONa	Ac_2O, reflux 3 h	p-$CH_3C_6H_4SCH(OAc)CH(OCH_3)C_6H_5$ (99)	106
$C_6H_5S(O)CH_2CH(C_6H_5)CH_2CO_2H$	Ac_2O	TsOH (cat.), toluene, reflux 1 h	C_6H_5, O, O, SC_6H_5 (lactone structure) (56)	41
$C_6H_5S(O)(CH_2)_2CH(C_6H_5)CO_2H$	Ac_2O	TsOH	C_6H_5, O, O, SC_6H_5 (lactone structure) (—)	292
$C_2H_5O_2CCH_2NHCOCH(NHAc)CH_2S(O)CH_2C_6H_5$	Ac_2O	75–80°, 3.75 h	$C_2H_5O_2CCH_2NHCO$, N, S, O, C_6H_5 (structure) (59)	142

TABLE I. SULFOXIDES WITH AN OXYGEN NUCLEOPHILE (*Continued*)

Sulfoxide	Nucleophile	Reaction Conditions	Product(s) and Yield(s) (%)	Refs.
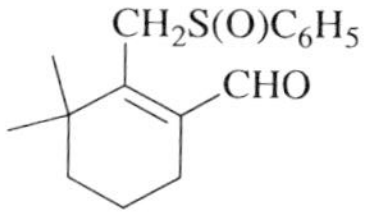 $CH_2S(O)C_6H_5$ CHO	Ac_2O	110°, 1.54 h	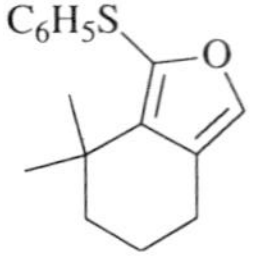 C_6H_5S O (70)	293
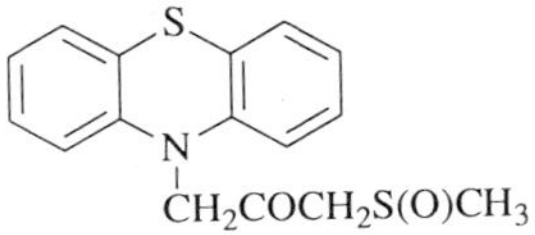 S N $CH_2COCH_2S(O)CH_3$	TsOH	THF, 65°, 15 min	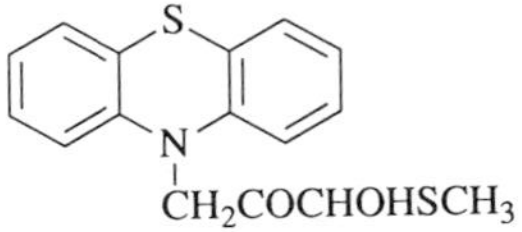 S N $CH_2COCHOHSCH_3$ (90–95)	294
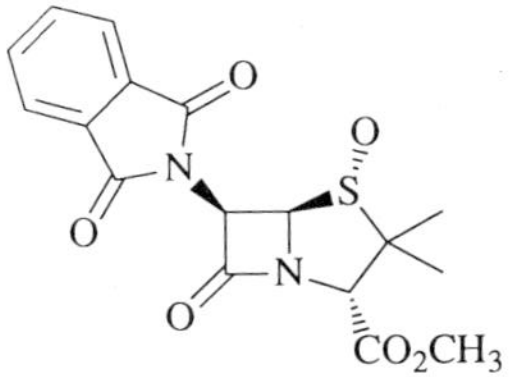 O N O S O N CO_2CH_3	Ac_2O	Oxidation	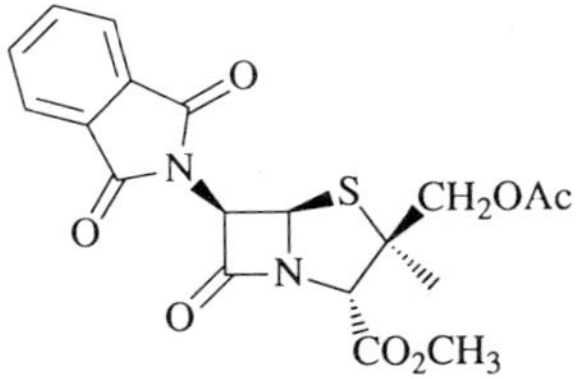 O N O S CH_2OAc N O CO_2CH_3 (30–40) 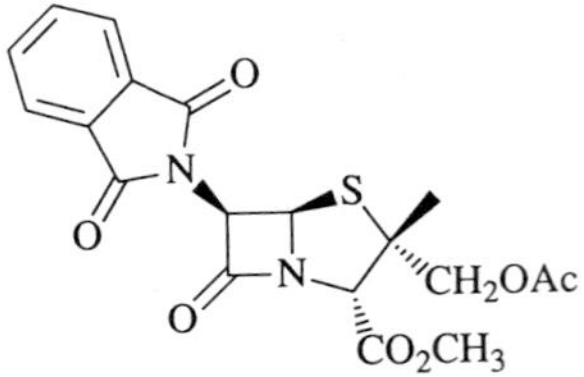 O N O S N CH_2OAc O CO_2CH_3 (10)	272, 277
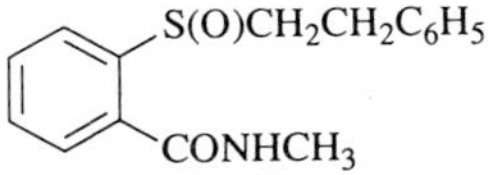 $S(O)CH_2CH_2C_6H_5$ $CONHCH_3$	Ac_2O	80°, 15 h	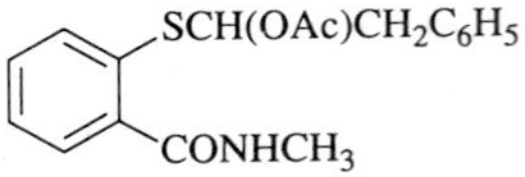 $SCH(OAc)CH_2C_6H_5$ $CONHCH_3$ (65)	267

C_{17}				
$C_6H_5S(O)(CH_2)_2C(CO_2C_2H_5)_2C_2H_5$	AcONa	Ac_2O, reflux 7 h	$C_6H_5SCH(OAc)CH_2C(CO_2C_2H_5)_2C_2H_5$ (40)	42
	Ac_2O	TFAA, 2,6-lutidine, 3 h	" (60)	42
p-$CH_3C_6H_4S(O)CH_2CHOHCH_2OCH_2C_6H_5$	AcONa	Ac_2O, reflux 3 h	p-$CH_3C_6H_4S[CH)OAc)]_2CH_2OCH_2C_6H_5$ (97)	106
$C_6H_5S(O)(CH_2)_3CONHCH_2C_6H_5$	Ac_2O	Reflux 3.5 h	$C_6H_5SCH(OAc)(CH_2)_2CONRCH_2C_6H_5$ R = H (18), R = Ac (62)	259
$C_6H_5S(O)(CH_2)_2CH(CH_2C_6H_5)CO_2H$	Ac_2O	TsOH (cat.), toluene, reflux 1 h	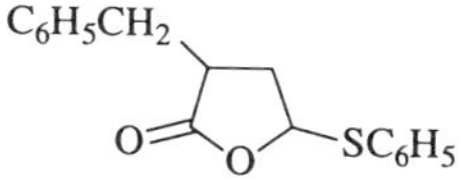 $C_6H_5CH_2$ … SC_6H_5 (63)	41, 292
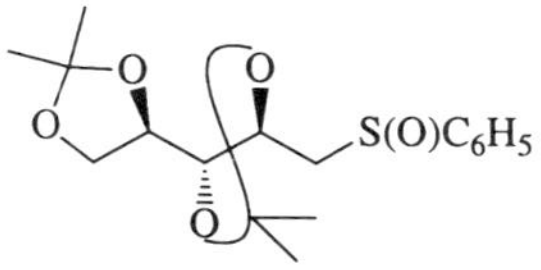 $S(O)C_6H_5$	AcONa	Ac_2O, reflux 8 h	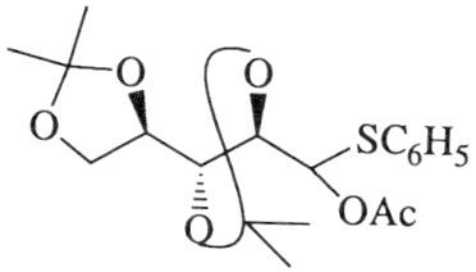 SC_6H_5, OAc (—)	295, 151
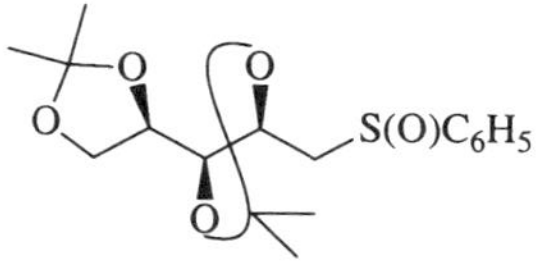 $S(O)C_6H_5$	AcONa	Ac_2O, reflux 8 h	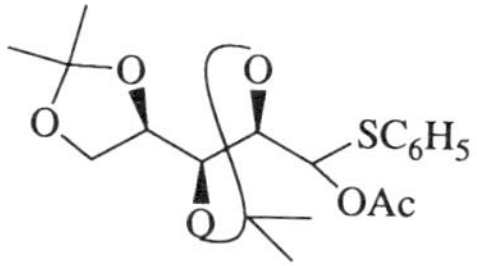 SC_6H_5, OAc (—)	295, 151
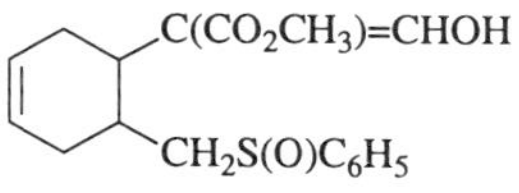 $C(CO_2CH_3)=CHOH$, $CH_2S(O)C_6H_5$	TFAA	2,6-Lutidine, CH_3CN, −50 to 0°, 1 h	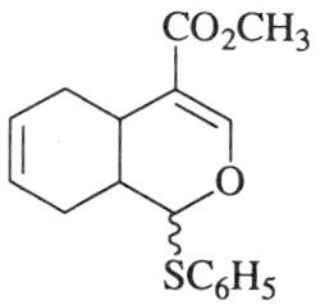 CO_2CH_3, SC_6H_5 (15) +	
			CO_2CH_3, OH (56)	296

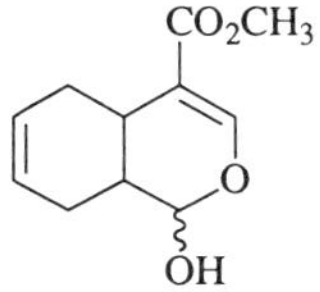

TABLE I. SULFOXIDES WITH AN OXYGEN NUCLEOPHILE (*Continued*)

Sulfoxide	Nucleophile	Reaction Conditions	Product(s) and Yield(s) (%)	Refs.
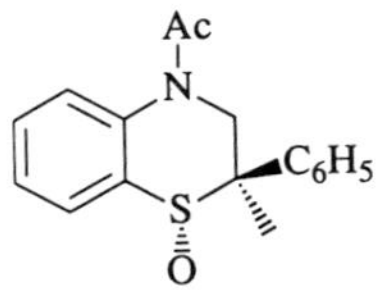 Ac, N, S, O, C_6H_5	Ac_2O	Reflux 1.5 h[f]	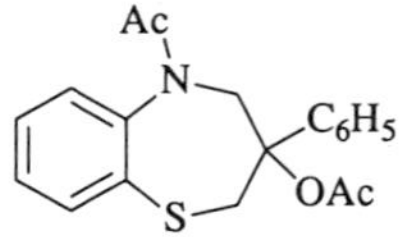 Ac, N, S, C_6H_5, OAc (26) 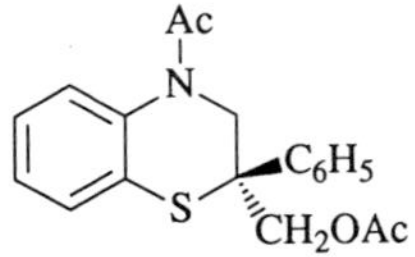 + Ac, N, S, C_6H_5, CH_2OAc (52)	278
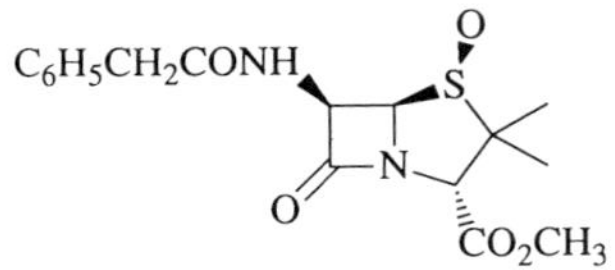 $C_6H_5CH_2CONH$, S, O, N, O, CO_2CH_3	Ac_2O	C_6H_6, reflux 21 h	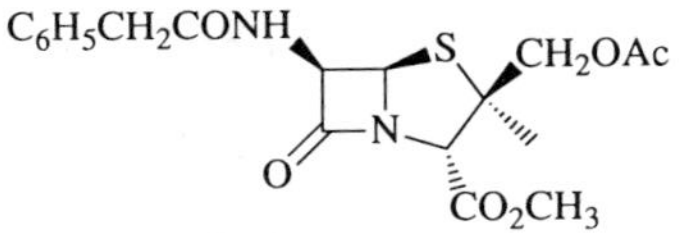 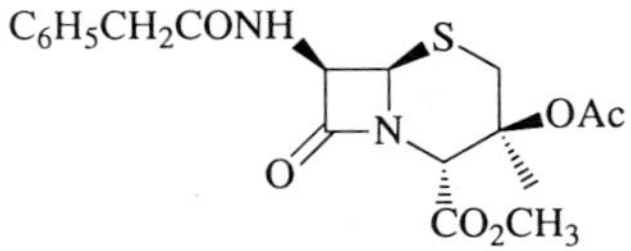 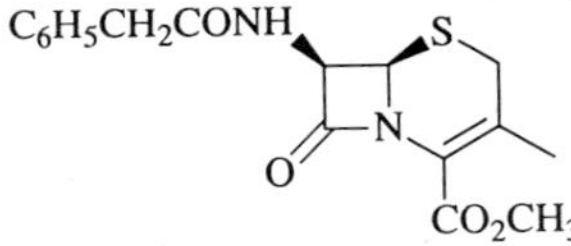 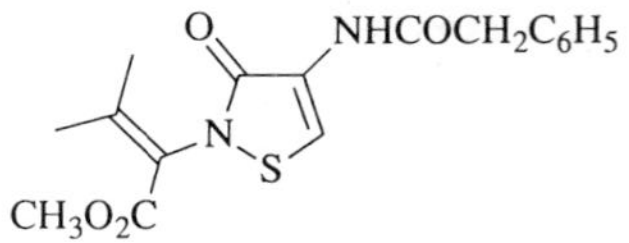 $C_6H_5CH_2CONH$, S, CH_2OAc, N, O, CO_2CH_3 I $C_6H_5CH_2CONH$, S, OAc, N, O, CO_2CH_3 (—) $C_6H_5CH_2CONH$, S, N, O, CO_2CH_3 II O, $NHCOCH_2C_6H_5$, N, S, CH_3O_2C (—) (I + II ~55)	297

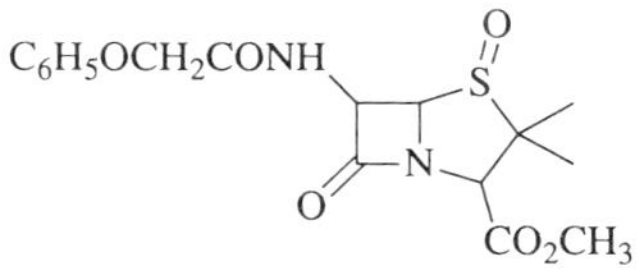

	Substrate	Reagent	Conditions	Product(s) (Yield, %)	Refs.
	$C_6H_5OCH_2CONH$ (penam sulfoxide, CO_2CH_3)	Ac_2O	Reflux 30 min	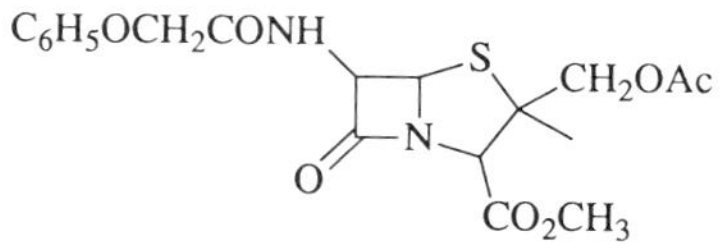 $C_6H_5OCH_2CONH$ (penam, CH_2OAc, CO_2CH_3) I (~40) + $C_6H_5OCH_2CONH$ (cepham, OAc, CO_2CH_3) II (—)	140
		AcONa	Ac_2O, reflux 10 min	I (—) + II 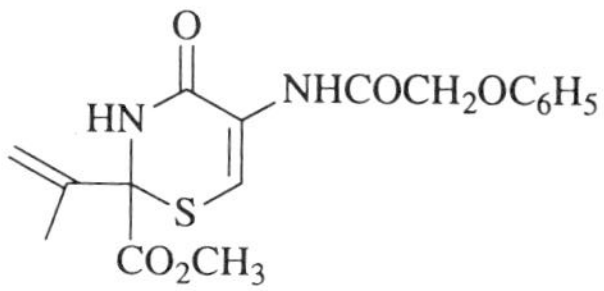 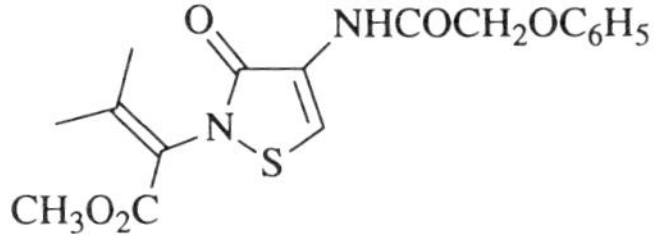 (—) + (HN, S ring, $NHCOCH_2OC_6H_5$, CO_2CH_3) (—) + (isothiazolone, $NHCOCH_2OC_6H_5$, CH_3O_2C) (—)	140
C_{18}	$C_6H_5S(O)CH_2CO(CH_2)_2NHCO_2CH_2C_6H_5$	Ac_2O	Pyr, toluene, reflux 3 h	$C_6H_5SCOCH(OAc)(CH_2)_2NHCO_2CH_2C_6H_5$ (63)	253
		AcONa	Ac_2O, toluene, reflux 5.5 h	" (61)	253

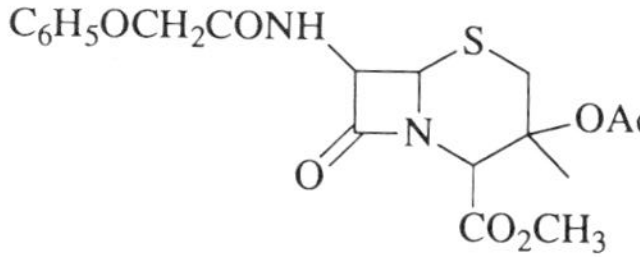

TABLE I. SULFOXIDES WITH AN OXYGEN NUCLEOPHILE (*Continued*)

Sulfoxide	Nucleophile	Reaction Conditions	Product(s) and Yield(s) (%)	Refs.
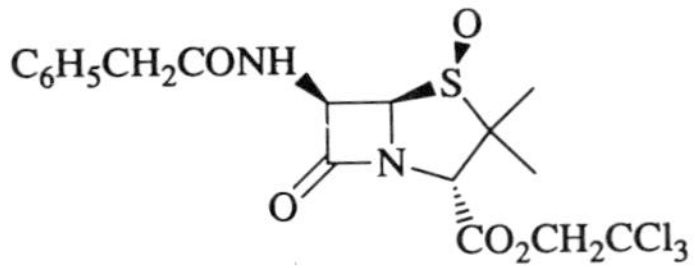 $C_6H_5CH_2CONH$ … $CO_2CH_2CCl_3$ (penicillin sulfoxide structure)	AcONa	Ac_2O, toluene, reflux 2.5 h	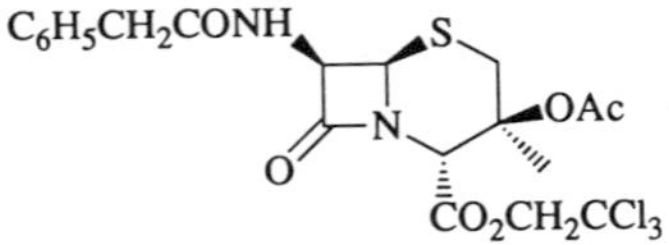 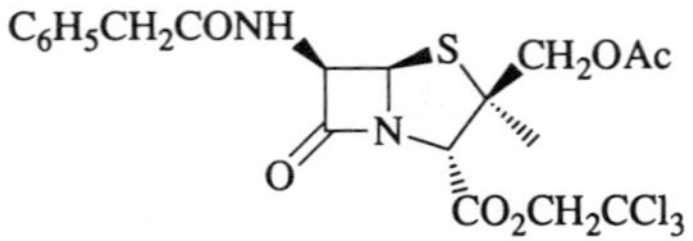 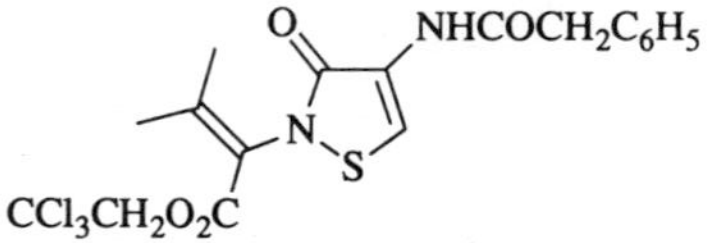 $C_6H_5CH_2CONH$ … OAc, $CO_2CH_2CCl_3$ (—) + $C_6H_5CH_2CONH$ … CH_2OAc, $CO_2CH_2CCl_3$ (—) + $CCl_3CH_2O_2C$ … $NHCOCH_2C_6H_5$ (—)	297
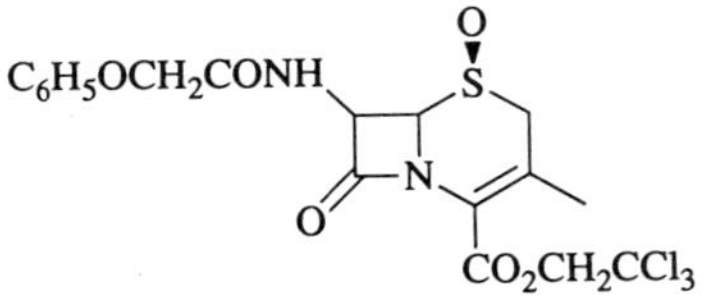 $C_6H_5OCH_2CONH$ … $CO_2CH_2CCl_3$ (cephem sulfoxide structure)	$ClCO_2C_2H_5$	$(C_2H_5)_3N$[g]	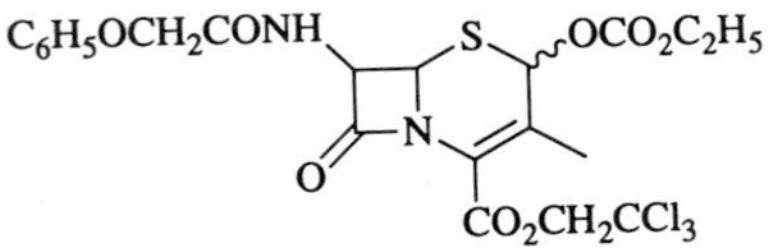 $C_6H_5OCH_2CONH$ … $OCO_2C_2H_5$, $CO_2CH_2CCl_3$ (68 + 4)[h]	298, 299[g]
C_{19}				
p-$CH_3C_6H_4S(O)CH_2CHOHCH_2C_6H_3(OCH_3$-$m)OAc$-$p$	AcONa	Ac_2O, reflux 5 h	p-$CH_3C_6H_4S[CH(OAc)]_2CH_2C_6H_3(OCH_3$-$m)OAc$-$p$ (—)	106
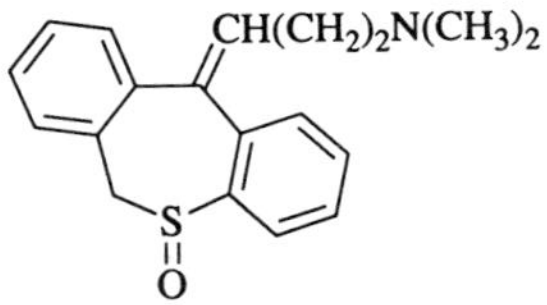 $CH(CH_2)_2N(CH_3)_2$ … S=O (dibenzothiepin sulfoxide structure)	Ac_2O	100°	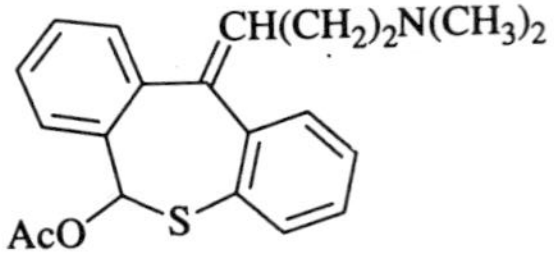 $CH(CH_2)_2N(CH_3)_2$ … AcO … S (—)	288, 300

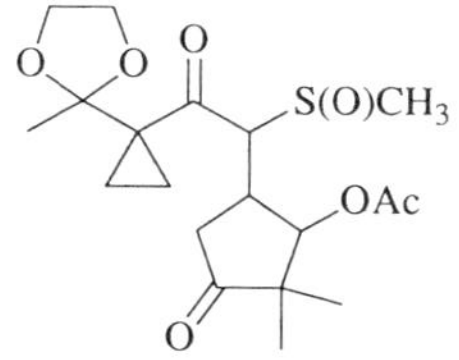 Substrate (O, O, S(O)CH$_3$, OAc)	Ac_2O	Pyr, 1 week	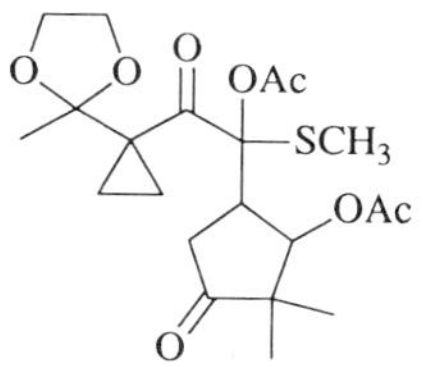 Product (O, O, O, OAc, SCH$_3$, OAc) (70)	147
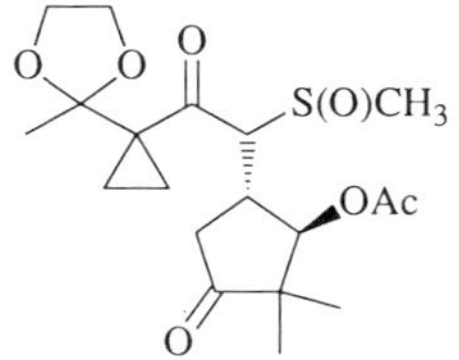 Substrate (O, O, S(O)CH$_3$, OAc)	Ac_2O	Pyr, 1 week	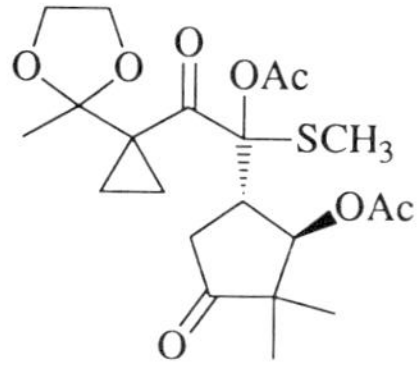 Product (O, O, O, OAc, SCH$_3$, OAc) (—)	148, 301
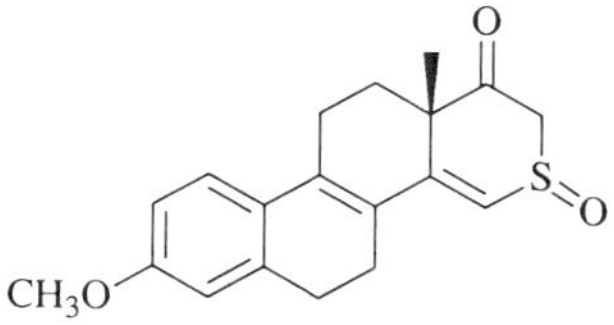 Substrate (CH_3O, O, S=O)	AcOH	Ac_2O, heat overnight	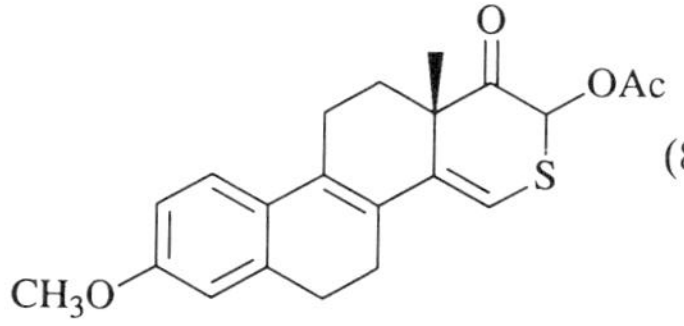 Product (CH_3O, O, OAc, S) (80–90)	302
	Ac_2O	83°, 2 h	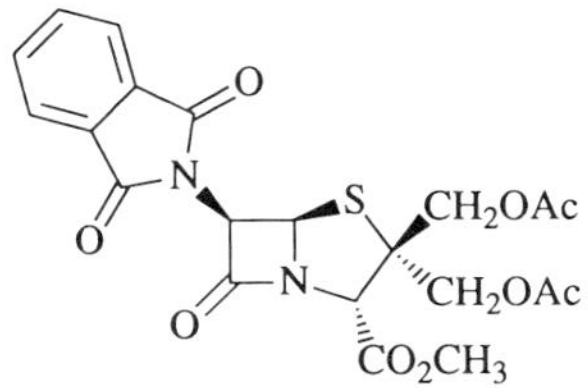 Product (O, N, O, O, S, N, CH_2OAc, CH_2OAc, CO_2CH_3) (19)	272

TABLE I. SULFOXIDES WITH AN OXYGEN NUCLEOPHILE (*Continued*)

	Sulfoxide	Nucleophile	Reaction Conditions	Product(s) and Yield(s) (%)	Refs.
	[structure]: CH_2OAc, CO_2CH_3	Ac_2O	84°, 3 h	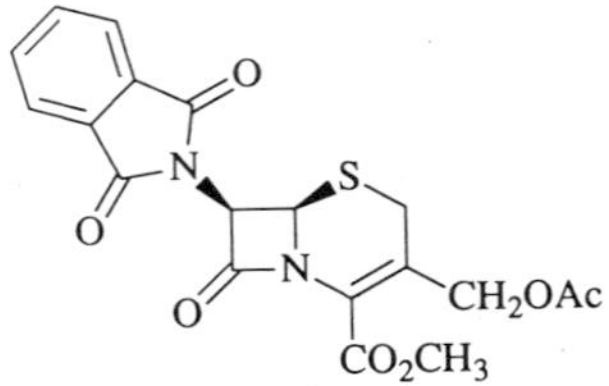 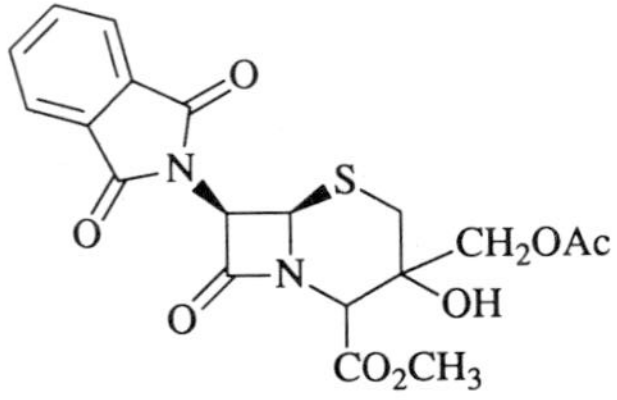 [structure]: CH_2OAc, CO_2CH_3 (6–7) +	277
				[structure]: CH_2OAc, OH, CO_2CH_3 (19–22)	
C_{20}	$C_6H_5CH_2OCH_2$ [dioxolane] $CH_2S(O)C_6H_5$	AcONa	Ac_2O, reflux 6 h	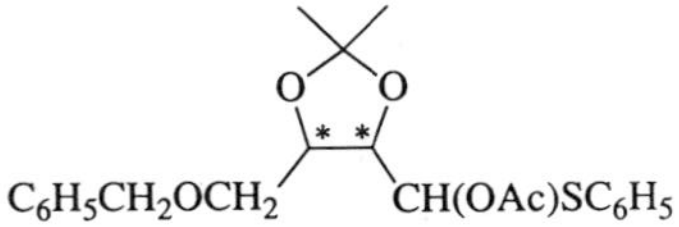 $C_6H_5CH_2OCH_2$ [dioxolane] $CH(OAc)SC_6H_5$ (>81)	295
	[structure]: $CH_2S(O)C_6H_5$	Ac_2O		[structure]: $CH(OAc)SC_6H_5$ (—) +	303
				[structure]: SC_6H_5 (—)	

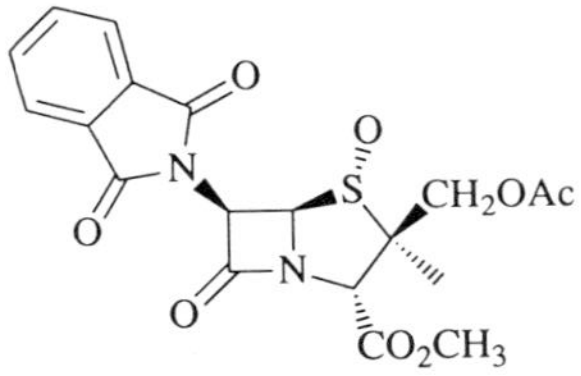

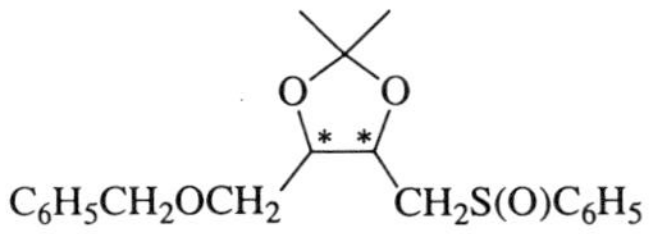

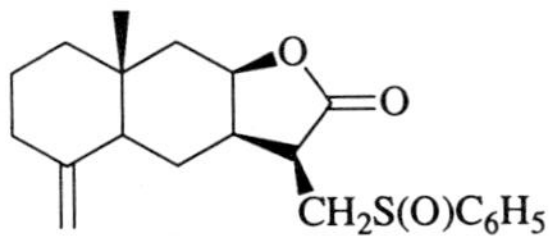

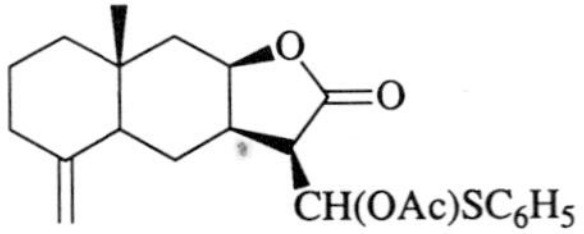

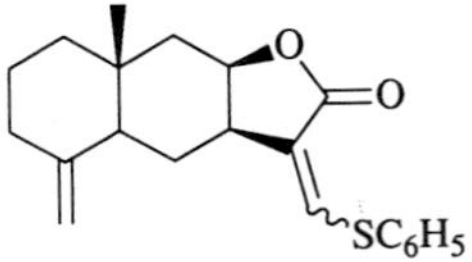

Reactant	Reagent	Conditions	Product(s) and Yield(s) (%)	Refs.
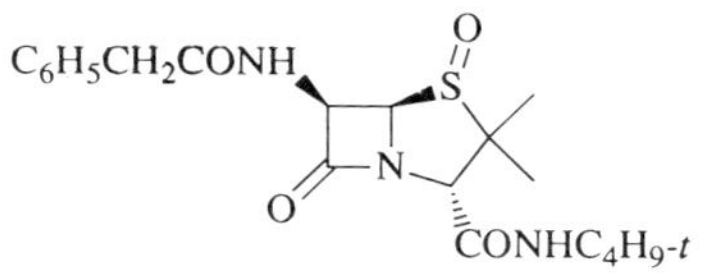 $C_6H_5CH_2CONH$ … $CONHC_4H_9$-*t* (penicillin sulfoxide)	Ac_2O	Toluene, reflux 7 h 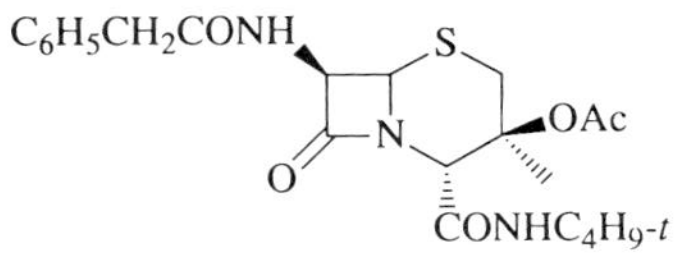	$C_6H_5CH_2CONH$ … OAc, $CONHC_4H_9$-*t* (—) +	297
			$NHCOCH_2C_6H_5$, *t*-C_4H_9NHCO (isothiazolone) (—)	
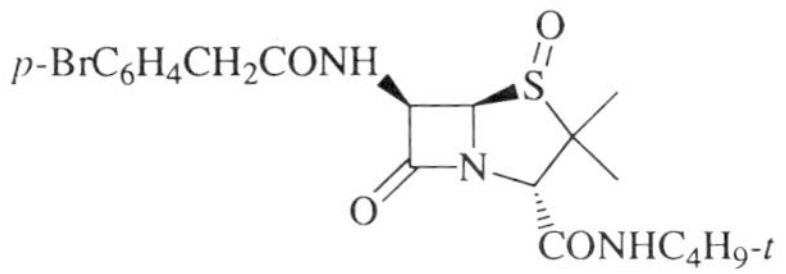 *p*-$BrC_6H_4CH_2CONH$ … $CONHC_4H_9$-*t* (penicillin sulfoxide)	Ac_2O	Toluene, reflux 7 h 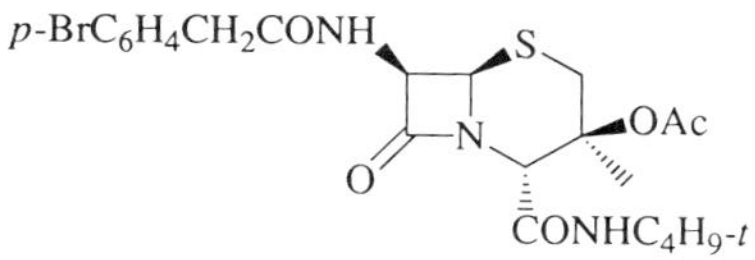	*p*-$BrC_6H_4CH_2CONH$ … OAc, $CONHC_4H_9$-*t* (—)	297
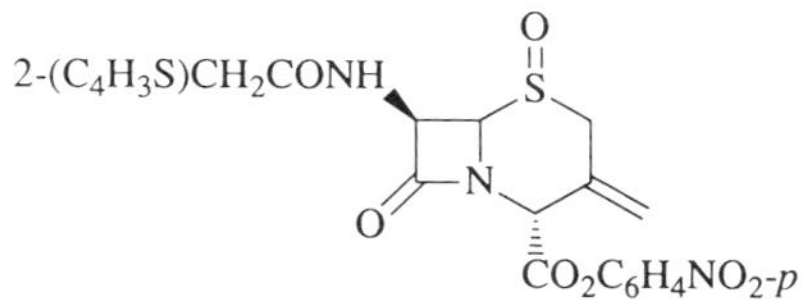 2-$(C_4H_3S)CH_2CONH$ … $CO_2C_6H_4NO_2$-*p* (sulfoxide)	Ac_2O, AcOH (2:1)	Reflux 2 h 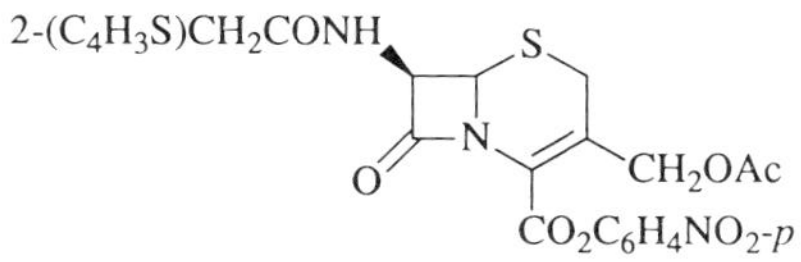	2-$(C_4H_3S)CH_2CONH$ … CH_2OAc, $CO_2C_6H_4NO_2$-*p* (—) +	304
			2-$(C_4H_3S)CH_2CONH$ … CH_2OAc, $CO_2C_6H_4NO_2$-*p* (—)	
C_{21}				
$C_2H_5O_2CCH_2NHCOCHNHCOC_6H_5$ with $CH_2S(O)CH_2C_6H_5$	Ac_2O	75–80°, 3 h 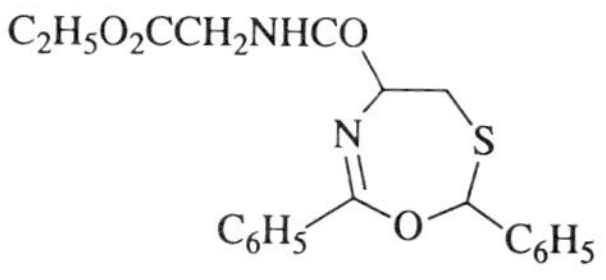	$C_2H_5O_2CCH_2NHCO$ … N, S, O, C_6H_5, C_6H_5 (42)	142

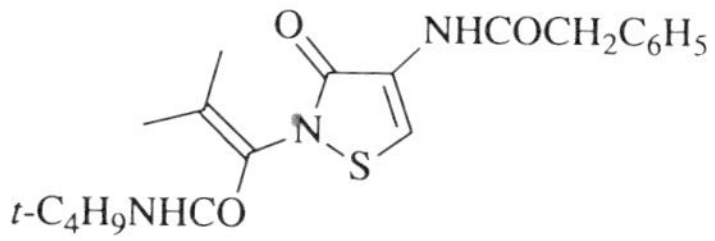

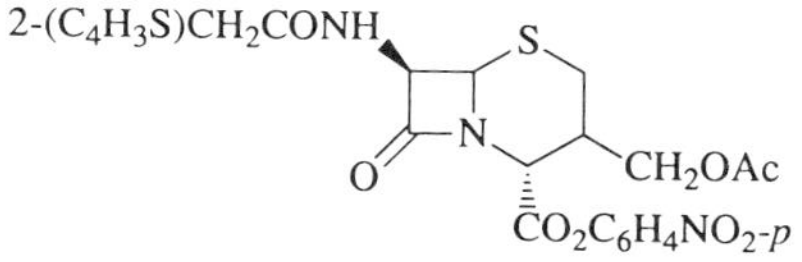

TABLE I. SULFOXIDES WITH AN OXYGEN NUCLEOPHILE (*Continued*)

Sulfoxide	Nucleophile	Reaction Conditions	Product(s) and Yield(s) (%)	Refs.
(*E*)-$(CH_3)_2C{=}CH(CH_2)_2C(CH_3){=}CH$–$CH_2{=}C[(CH_2)_2S(O)C_6H_5](CH_2)_2$	Ac_2O	TFAA (cat.), 72 h	(*E*)-$(CH_3)_2C{=}CH(CH_2)_2C(CH_3){=}CH(CH_2)_2$–$C[CH_2CH(OAc)SC_6H_5]{=}CH_2$ (70)	305
[structure: S(O)CH₃, AcO, AcOCH₂]	Ac_2O	Pyr, 1 week	[structure: SCH₃, OAc, AcO, AcOCH₂] (65)	147
[structure: CH₂S(O)C₆H₅, CHO, R, R′, R′]	Ac_2O	110°, 1.5 h	[structure: C₆H₅S, R, R′, R′]	
			R = H, R′ = CH_3 (77)	293
			R = CH_3, R′ = $=CH_2$ (61)	293
[structure: CH₃O, R, R′, COCH₂S(O)CH₃, N, H]	AcOH	100°, 2 h	[structure: CH₃O, R, R′, COCH(OAc)SCH₃, N, H]	
			R = H, R′ = H (–)	306
			R = H, R′ = OH (–)	306
			R = OH, R′ = H (–)	306
			R = OH, R′ = OH (–)	306

Sulfoxide	Reagent	Conditions	Product(s) and Yield(s) (%)	Refs.
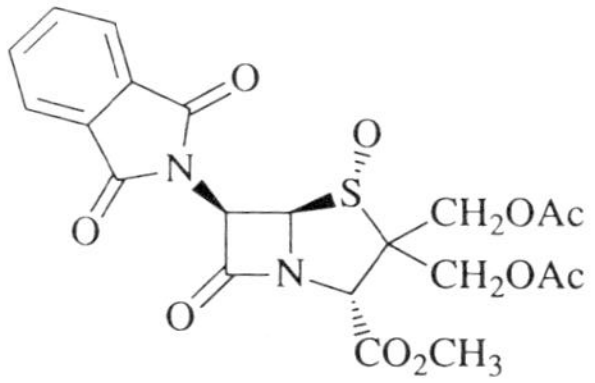 (structure) CH_2OAc, CH_2OAc, CO_2CH_3	Ac_2O		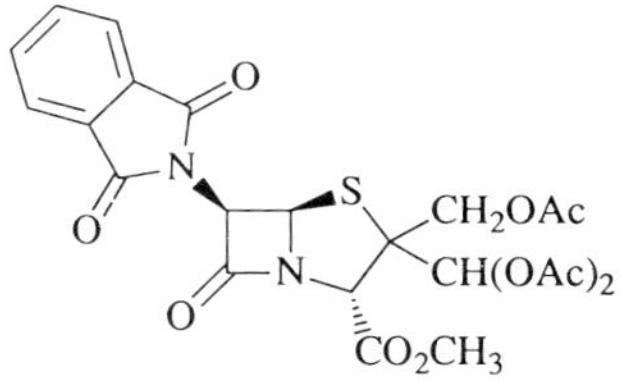 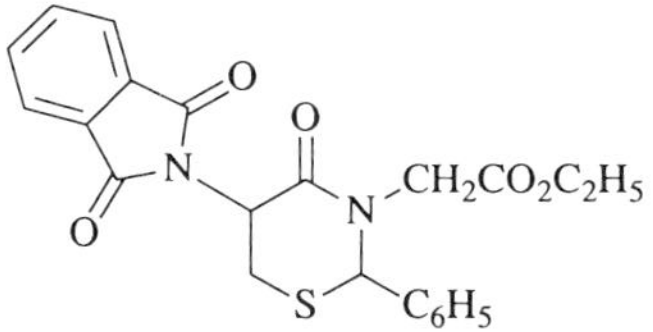 (structure) CH_2OAc, $CH(OAc)_2$, CO_2CH_3; (structure) $CH_2CO_2C_2H_5$, C_6H_5; 2:1 mixture (21)	272
C_{22} $C_2H_5O_2CCH_2NHCOCH(PHT)CH_2S(O)CH_2C_6H_5$	Ac_2O	75–80°, 20 min	(31)[a]	142
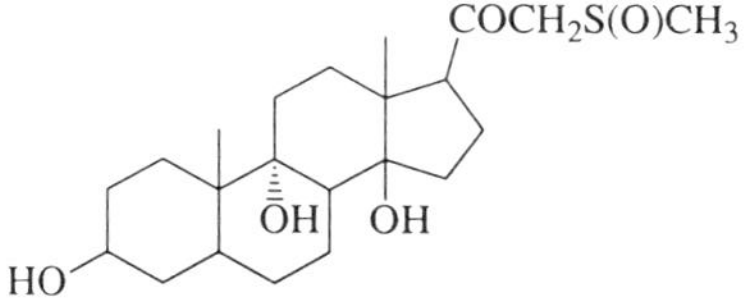 (structure) $COCH_2S(O)CH_3$, HO, OH, OH	Ac_2O	Pyr, 60 h	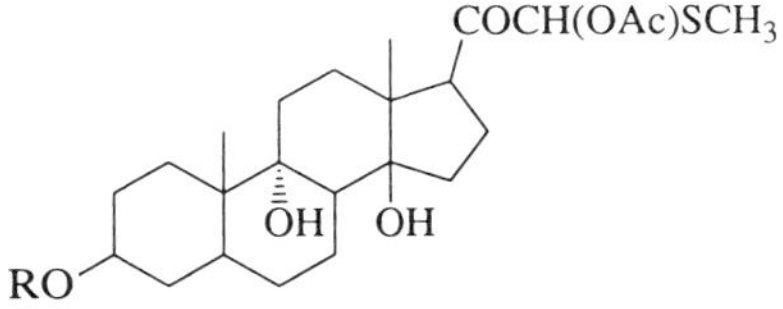 (structure) $COCH(OAc)SCH_3$, RO, OH, OH; R = H (55) + R = OAc (33)	307
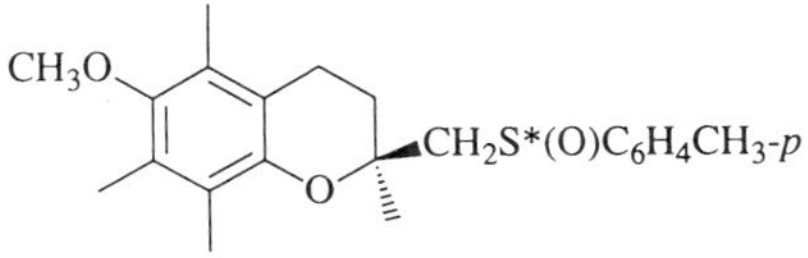 (structure) CH_3O, $CH_2S^*(O)C_6H_4CH_3$-*p*	AcONa	Ac_2O, 135°	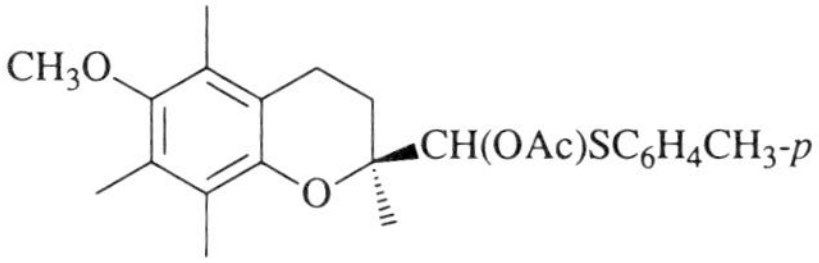 (structure) CH_3O, $CH(OAc)SC_6H_4CH_3$-*p* (83)	308

TABLE I. SULFOXIDES WITH AN OXYGEN NUCLEOPHILE (*Continued*)

Sulfoxide	Nucleophile	Reaction Conditions	Product(s) and Yield(s) (%)	Refs.
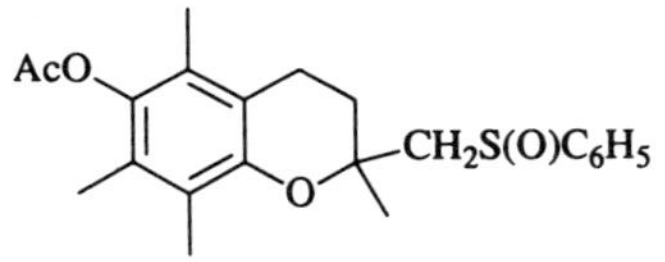 AcO-chroman-$CH_2S(O)C_6H_5$ (structure)	AcONa	Ac_2O, 135°, 7 h	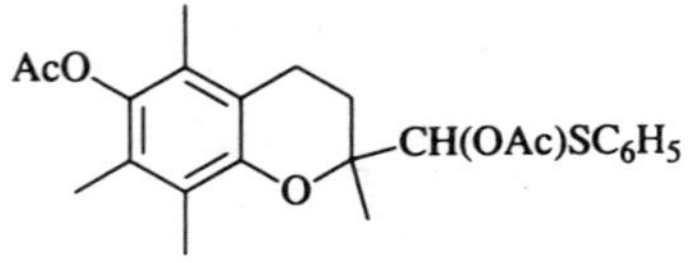 AcO-chroman-$CH(OAc)SC_6H_5$ (structure) (92)	309
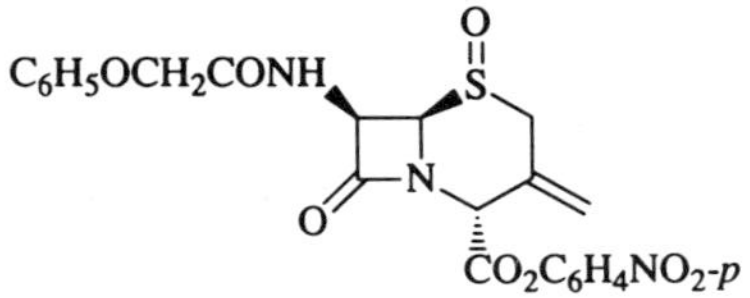 $C_6H_5OCH_2CONH$ … $CO_2C_6H_4NO_2$-*p* (structure)	AcOH	Ac_2O (2:1), reflux 2 h	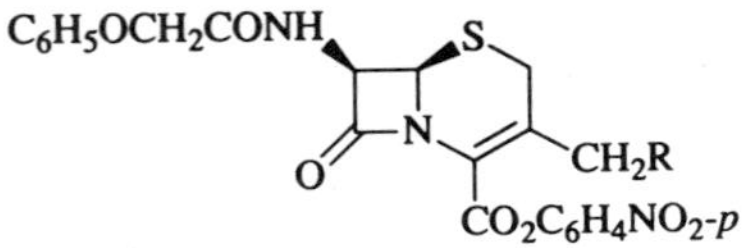 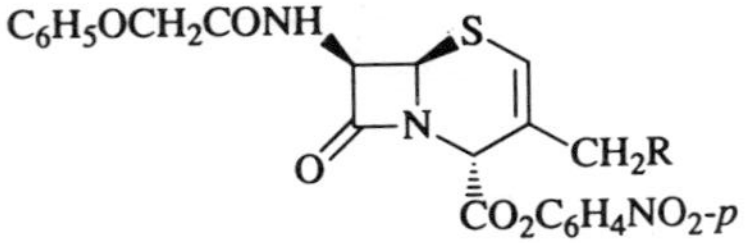 $C_6H_5OCH_2CONH$ … CH_2R, $CO_2C_6H_4NO_2$-*p* (structure) + $C_6H_5OCH_2CONH$ … CH_2R, $CO_2C_6H_4NO_2$-*p* (structure) R = OAc (—) R = OAc (—)	304
	C_2H_5OH	$(C_2H_5CO)_2O$, 120°	R = $O_2CC_2H_5$ (—) R = $O_2CC_2H_5$ (—)	304
C_{23}				
$CH_3S(O)CH_2COC_6H_3(OCH_2C_6H_5)_2$-3,4		H^+	$CH_3SCHOHCOC_6H_3(OCH_2C_6H_5)_2$-3,4 (—)	269
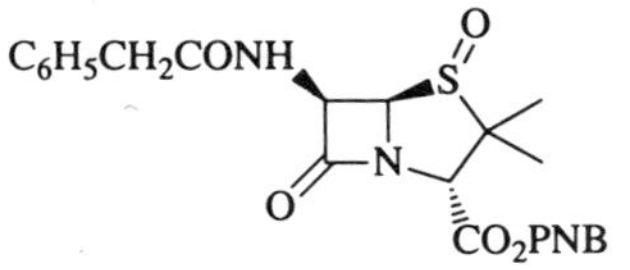 $C_6H_5CH_2CONH$ … CO_2PNB (structure)	Ac_2O	Reflux 12 min	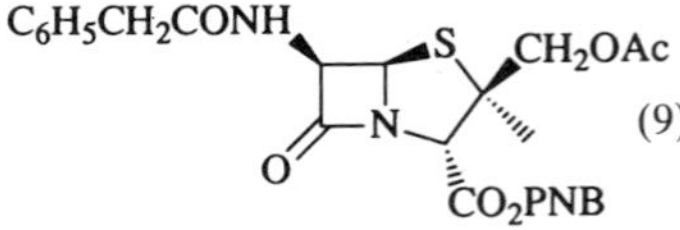 $C_6H_5CH_2CONH$ … CH_2OAc, CO_2PNB (structure) (9) +	297

			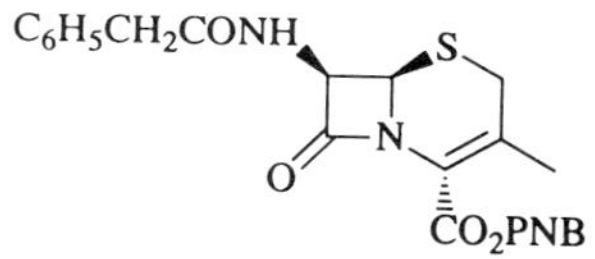 $C_6H_5CH_2CONH$, S, N, O, CO_2PNB (5)	
			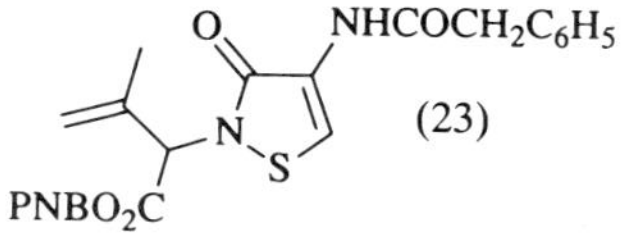 + $PNBO_2C$, N, S, O, $NHCOCH_2C_6H_5$ (23)	
	$(CH_2ClCO)_2O$	Toluene, reflux 2.5 h	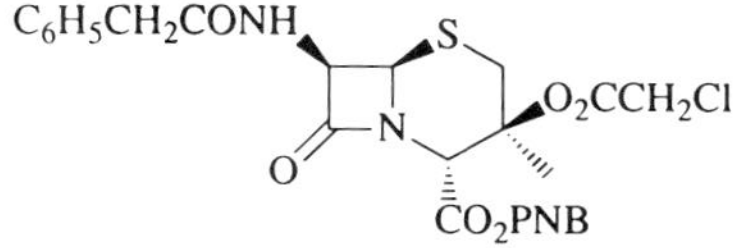 $C_6H_5CH_2CONH$, S, N, O, O_2CCH_2Cl, CO_2PNB (~15)	297
C_{24}				
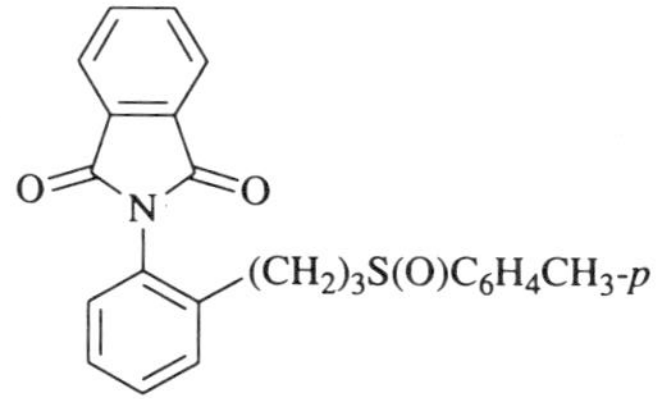 O, N, O, $(CH_2)_3S(O)C_6H_4CH_3$-*p*	AcOH	Ac_2O, reflux 3 h	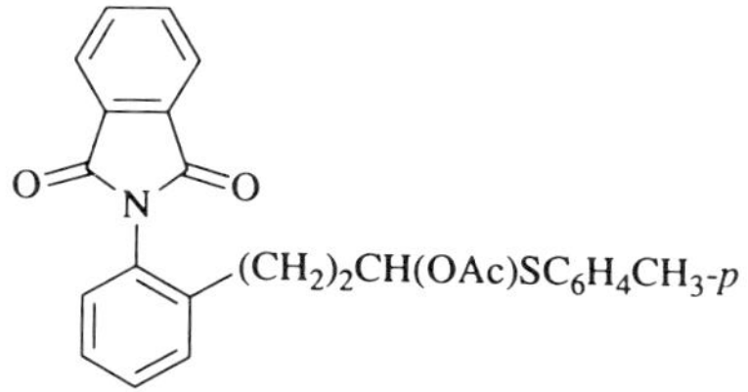 O, N, O, $(CH_2)_2CH(OAc)SC_6H_4CH_3$-*p* (90)	310
C_{25}				
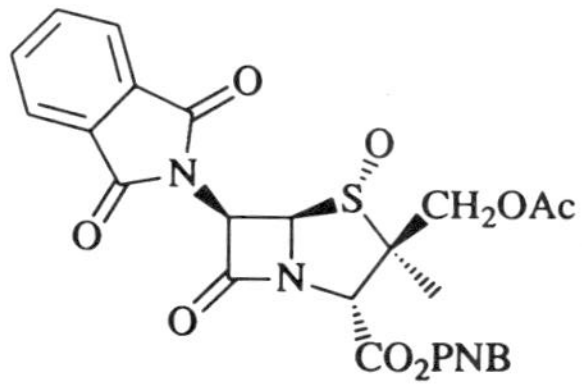 O, N, O, S, O, CH_2OAc, N, O, CO_2PNB	Ac_2O	1. "Strong base" 2. N_2H_4	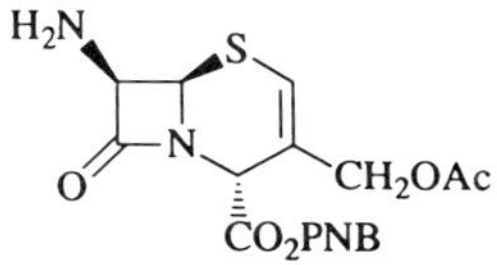 H_2N, S, N, O, CH_2OAc, CO_2PNB (30)[a]	277

TABLE I. SULFOXIDES WITH AN OXYGEN NUCLEOPHILE (*Continued*)

Sulfoxide	Nucleophile	Reaction Conditions	Product(s) and Yield(s) (%)	Refs.
C_{32}				
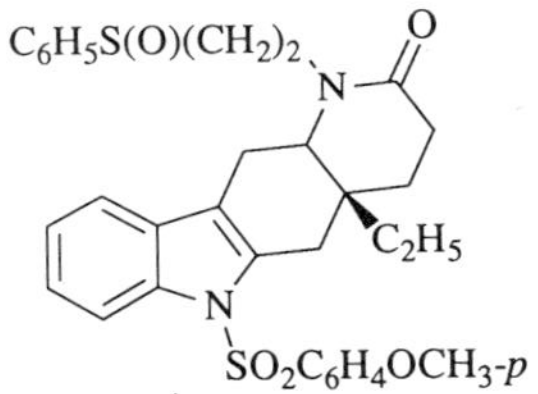 $C_6H_5S(O)(CH_2)_2$-N (tetracyclic indole lactam, C_2H_5, N-$SO_2C_6H_4OCH_3$-*p*)	TFAA	CH_2Cl_2, 0°, 10 min	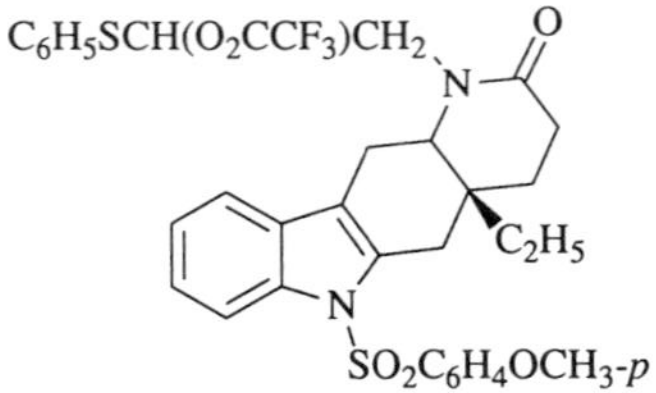 $C_6H_5SCH(O_2CCF_3)CH_2$-N (tetracyclic indole lactam, C_2H_5, N-$SO_2C_6H_4OCH_3$-*p*) (—)	311
C_{36}				
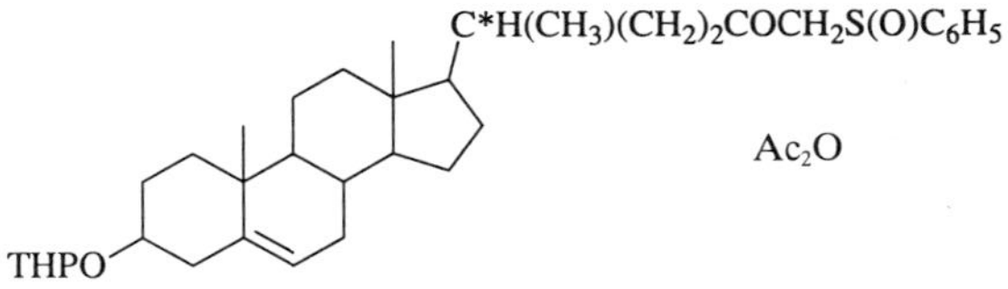 THPO-steroid-$C^*H(CH_3)(CH_2)_2COCH_2S(O)C_6H_5$	Ac_2O	Pyr, toluene, 125–130°, 6 h (partial)	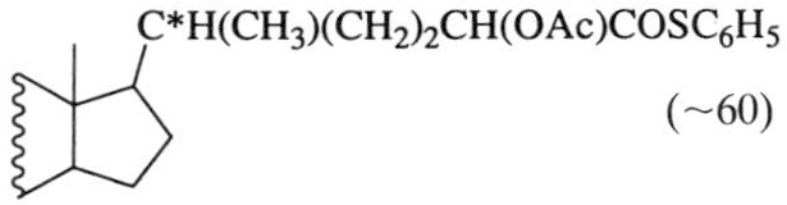 steroid-$C^*H(CH_3)(CH_2)_2CH(OAc)COSC_6H_5$ (~60)	312
C_{38}				
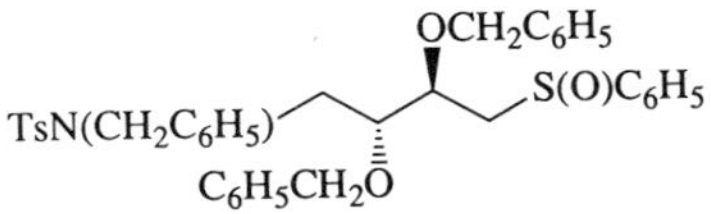 $TsN(CH_2C_6H_5)CH_2CH(OCH_2C_6H_5)CH(OCH_2C_6H_5)CH_2S(O)C_6H_5$	Ac_2O	TFAA, 2,6-lutidine, 3 h	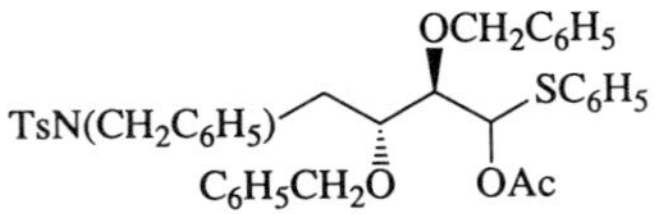 $TsN(CH_2C_6H_5)CH_2CH(OCH_2C_6H_5)CH(OCH_2C_6H_5)CH(OAc)SC_6H_5$ (71)	43

[a] The reaction was carried out to partial conversion.
[b] The sulfoxide oxygen was ^{18}O.
[c] Other non-Pummerer products were also formed.
[d] The reaction was 85–90% stereoselective, with the OAc group in the products having the same stereochemistry as the starting sulfoxide oxygen.
[e] The reaction has been studied from a kinetic–mechanistic point of view.
[f] The epimeric sulfoxide did not react under similar conditions.
[g] The (*R*)-1-oxide did not react under the same conditions.
[h] The yields represent the two diastereoisomers formed in the reaction.

TABLE II. SULFOXIDES WITH A SULFUR NUCLEOPHILE

Sulfoxide	Nucleophile	Reaction Conditions	Product(s) and Yield(s) (%)	Refs.
C_2				
$(CH_3)_2SO$	$(CH_3)_2S$[a]	TFAA, CH_2Cl_2, 0°	$CH_3SCH_2S^+(CH_3)_2\ ^-O_2CCF_3$ (64)	45
	$CH_3SC_8H_{17}$-*n*[a]	"	$CH_3SCH_2S^+(CH_3)C_8H_{17}$-*n* $^-O_2CCF_3$ (65)	45
	$CH_3SC_4H_9$-*t*[a]	"	$CH_3SCH_2S^+(CH_3)C_4H_9$-*t* $^-O_2CCF_3$ (54)	45
	$CH_3SCH_2C_6H_5$[a]	"	$CH_3SCH_2S^+(CH_3)CH_2C_6H_5\ ^-O_2CCF_3$ (67)	45
	$(C_2H_5)_2S$[a]	"	$CH_3SCH_2S^+(C_2H_5)_2\ ^-O_2CCF_3$ (70)	45
	$(CH_2)_4S$[a]	"	$CH_3SCH_2S^+(CH_2)_4\ ^-O_2CCF_3$ (73)	45
	C_6H_5SH[a]	"	$CH_3SCH_2SC_6H_5$ (94)	45
	$C_6H_5CH_2SH$[a]	"	$CH_3SCH_2SCH_2C_6H_5$ (89)	45
	C_6H_5SH	TFAA, CH_3CN	$CH_3SCH_2SC_6H_5$ (59)	46
	p-ClC_6H_4SH	TFAA, CH_3CN	$CH_3SCH_2SC_6H_4Cl$-*p* (55)	46
	p-$CH_3C_6H_4SH$	TFAA, CH_3CN	$CH_3SCH_2SC_6H_4CH_3$-*p* (59)	46
$CH_3S(O)SCH_3$		C_6H_6, H_2O, reflux 52 h	$CH_3SSCH_2S(O)CH_3$ I (84)[b,c]	220, 313
		C_6H_6, 96°, 6 h	I (14)[c] + $CH_3SSCH_2SCH_3$ (8)[c] + $CH_3SSCH_2SO_2CH_3$ (3)[c]	220
C_3				
$C_2H_5S(O)SCH_3$		H_2O (1 eq), C_6H_6, reflux 23 h	$CH_3SSCH_2S(O)C_2H_5$ (27) + $C_2H_5SSCH_2S(O)C_2H_5$ (9)	220, 313
C_4				
$C_2H_5S(O)C_2H_5$		C_6H_6, 96°, 4 h	$C_2H_5SSCH(CH_3)S(O)C_2H_5$ (15)	220
i-$C_3H_7S(O)SCH_3$		Neat, 96°, 9.3 h	$CH_3SSCH_2S(O)C_3H_7$-*i* (12)[c]	220
C_5				
t-$C_4H_9S(O)SCH_3$		H_2O, 96°, 5.5 h	$CH_3SSCH_2S(O)C_4H_9$-*t* (27)[c]	220, 313
C_7				
$CH_2{=}CH(CH_2)_2COCH_2S(O)CH_3$		TFAA, CH_2Cl_2, 0°, 1 h	$CH_2{=}CH(CH_2)_2COCH(SCH_3)_2$ (14)	314
$C_6H_5S(O)CH_3$	$(CH_3)_2S$[a]	TFAA, CH_2Cl_2, 0°	$C_6H_5SCH_2S^+(CH_3)_2\ ^-O_2CCF_3$ (40)	45
	C_6H_5SH[a]	"	$C_6H_5SCH_2SC_6H_5$ (69)	45
C_8				
$C_6H_5S(O)C_2H_5$	C_6H_5SH[a]	"	$C_6H_5SCH(CH_3)SC_6H_5$ (76)	45
C_9				
$C_6H_5COCH_2S(O)CH_3$	CH_3SOCl[a]	NaH, THF	$C_6H_5COCH(SO_2CH_3)SCH_3$ (75)	254

[a] The nucleophile was added after other reagents.
[b] The reaction was carried out to partial conversion.
[c] Other non-Pummerer products were also formed.

TABLE III. SULFOXIDES WITH A NITROGEN NUCLEOPHILE

Sulfoxide	Nucleophile	Reaction Conditions	Product(s) and Yield(s) (%)	Refs.
C_2				
$(CH_3)_2SO$	CH_3CN	TFAA, CF_3CO_2H, 24 h	$CH_3SCH_2NHCOCH_3$ (32) + $CH_3SCH_2O_2CCF_3$ I (—)	315
	$CH_2{=}CHCN$	"	$CH_3SCH_2NHCOCH{=}CH_2$ (30) + I (—)	315
	C_6H_5CN	"	$CH_3SCH_2NHCOC_6H_5$ (53) + I (—)	315
	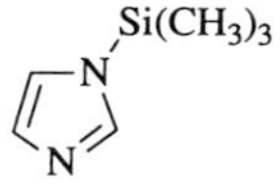 $Si(CH_3)_3$, N, N	180°, 6 h	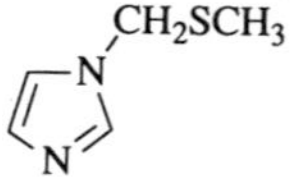 CH_2SCH_3, N, N (60)	316
	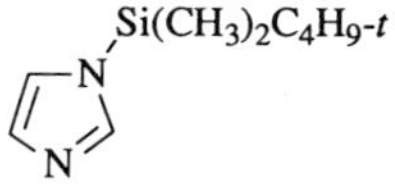 $Si(CH_3)_2C_4H_9$-*t*, N, N	150°, 3 h	" (40)	316
	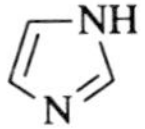 NH, N	$[(CH_3)_3Si]_2NH$, 180°, 6 h	" (70)	316
	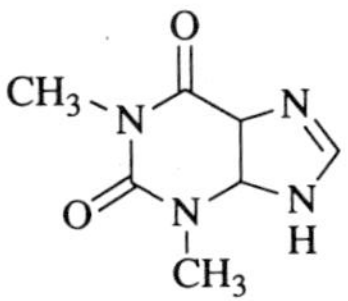 O, CH_3, N, N, N, O, N, H, CH_3	P_2O_5, 70–75°, 15 h	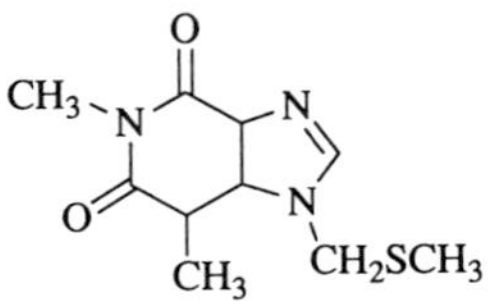 O, CH_3, N, N, O, N, CH_3, CH_2SCH_3 (62)	208
	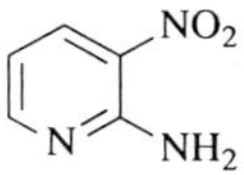 NO_2, N, NH_2	$C_2H_5COCOCl$, 18 h	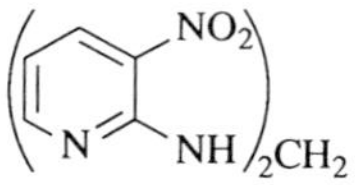 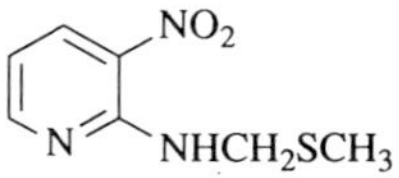 (NO_2, N, NH)$_2CH_2$ (98) +	317
			NO_2, N, $NHCH_2SCH_3$ (—)	
	O_2N, N, NH_2	$C_2H_5COCOCl$, C_6H_6, reflux	(O_2N, N, NH)$_2CH_2$ (70)	317

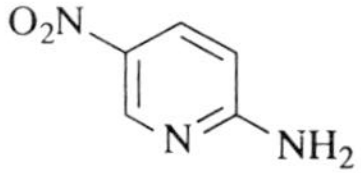

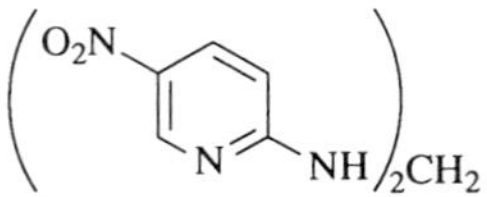

Substrate	Conditions	Product	(Yield)	Refs.
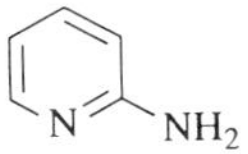 N, NH_2	$C_2H_5COCOCl$, C_6H_6, $(C_2H_5)_3N$, reflux 3 h	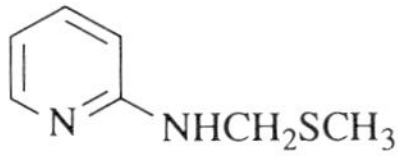 N, $NHCH_2SCH_3$	(~32)	317
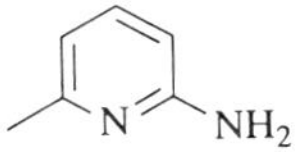 N, NH_2	"	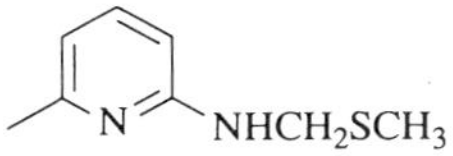 N, $NHCH_2SCH_3$	(—)	317
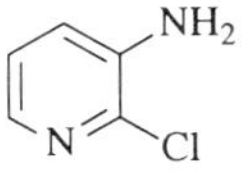 NH_2, N, Cl	AcCl, $(C_2H_5)_3N$, THF, reflux 18 h	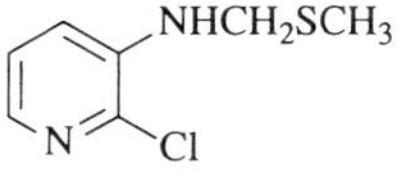 $NHCH_2SCH_3$, N, Cl	(93)	317
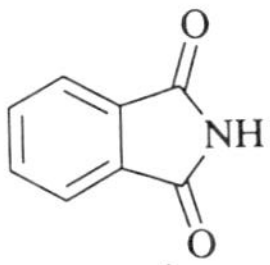 O, NH, O	DCC, H_3PO_4, C_6H_6, 2 d	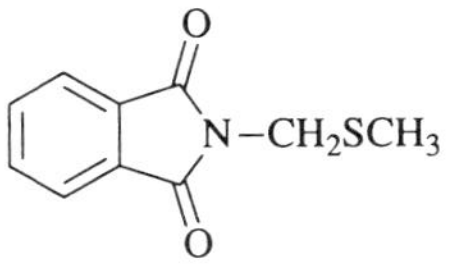 O, $N-CH_2SCH_3$, O	(38)[a,b]	209
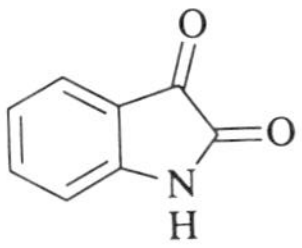 O, =O, N, H	DCC, H_3PO_4, C_6H_6, 5 d	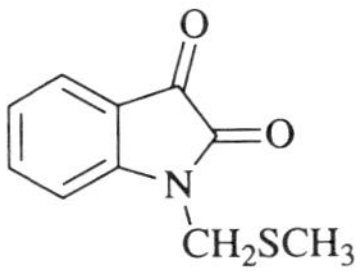 O, =O, N, CH_2SCH_3	(1)[a,b]	209
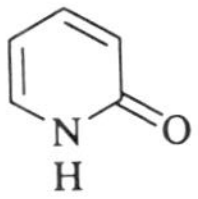 N, O, H	Reflux 35 h	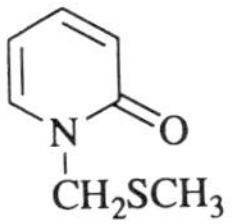 N, O, CH_2SCH_3	(84)	318
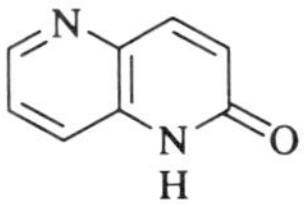 N, N, O, H	Reflux 8 h	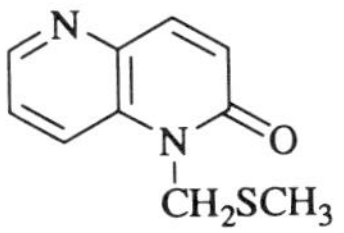 N, N, O, CH_2SCH_3	(46)	318
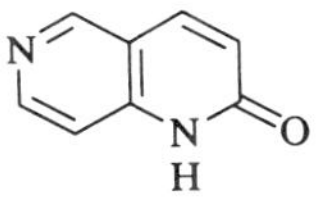 N, N, O, H	Reflux 8 h	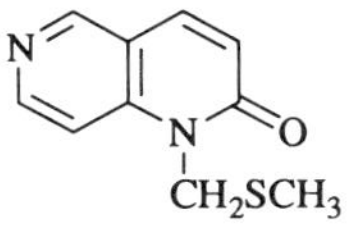 N, N, O, CH_2SCH_3	(83)	318

TABLE III. Sulfoxides with a Nitrogen Nucleophile (*Continued*)

Sulfoxide	Nucleophile	Reaction Conditions	Product(s) and Yield(s) (%)	Refs.
	N, N H, O	Reflux 8 h	N, N–CH_2SCH_3, O (35)	318
	N, N H, O	Reflux 8 h	N, N–CH_2SCH_3, O (42)	318
	N, N, N H, O	Reflux 14 h	N, N, N–CH_2SCH_3, O (66)	318
	OH, F, N, F, N, F	1. TFAA, CH_2Cl_2, $\leq -50°$, 2 h 2. $(C_2H_5)_3N$	O, CH_3SCH_2–N, F, F, N, F (71)	319
	OCH_3, F, N, HO, N, OCH_3	"	OCH_3, F, N, O, N–CH_2SCH_3, OCH_3 (59)	319
	O, HN, O, N H	DCC, H_3PO_4, C_6H_6, 7 d	O, HN, O, N–CH_2SCH_3 (16)[a,b] +	209

				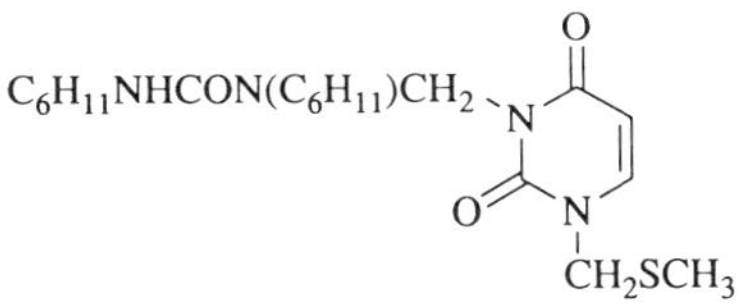 $C_6H_{11}NHCON(C_6H_{11})CH_2$–N (uracil, N-$CH_2SCH_3$) (2)[a,b]	
		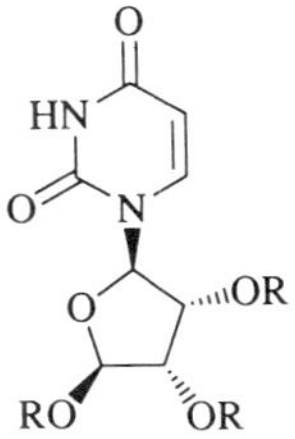 (uracil riboside; HN, O, N, OR, RO, OR)	DCC, H_3PO_4, C_6H_6, 5 d	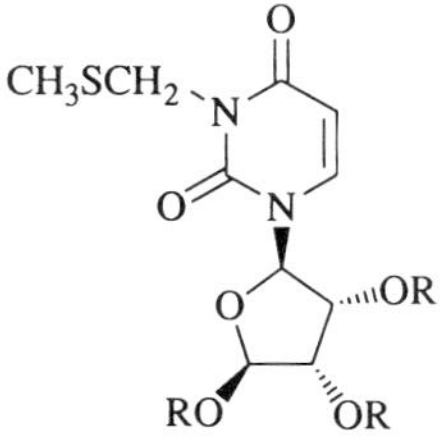 CH_3SCH_2–N (uracil riboside; O, N, OR, RO, OR) (10)[a,b] R = COC_6H_5	209
		$(C_6H_5)_2C{=}NOH$	DCC, CF_3CO_2H, C_6H_6, 25°, 2 h	$CH_3SCH_2N(O){=}C(C_6H_5)_2$ (74)[a,b] + $CH_3SCH_2ON{=}C(C_6H_5)_2$ (3)[a,b]	213
		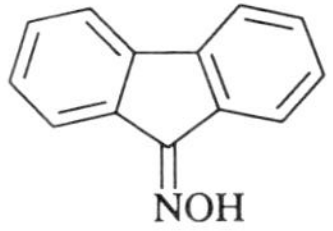 (fluorenone oxime) NOH	DCC, CF_3CO_2H, C_6H_6, 25°, 24 h	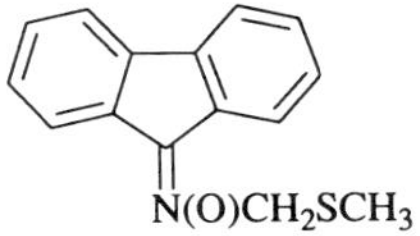 (fluorenylidene) $N(O)CH_2SCH_3$ (71) 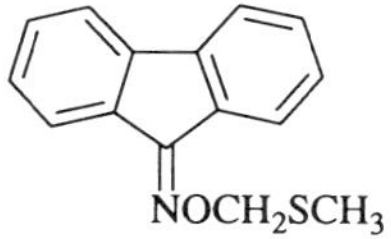 + (fluorenylidene) $NOCH_2SCH_3$ (5)	213
C_7					
	$C_6H_5S(O)CH_3$	CH_3CN	TFAA, CF_3CO_2H, 48 h	$C_6H_5SCH_2NHCOCH_3$ (45) + $C_6H_5SCH_2O_2CCF_3$ I (36)	315
		$CH_2{=}CHCN$	"	$C_6H_5SCH_2NHCOCH{=}CH_2$ (57) + I (38)	315
		C_6H_5CN	TFAA, CF_3CO_2H, 72 h	$C_6H_5SCH_2NHCOC_6H_5$ (46) + I (38)	315

TABLE III. SULFOXIDES WITH A NITROGEN NUCLEOPHILE (*Continued*)

Sulfoxide	Nucleophile	Reaction Conditions	Product(s) and Yield(s) (%)	Refs.
C_8				
p-$CH_3C_6H_4S(O)CH_3$	CH_3CN	"	p-$CH_3C_6H_4SCH_2NHCOCH_3$ (40) + p-$CH_3C_6H_4SCH_2O_2CCF_3$ I (36)	315
	$CH_2{=}CHCN$	TFAA, CF_3CO_2H, 48 h	p-$CH_3C_6H_4SCH_2NHCOCH{=}CH_2$ (43) + I (48)	315
	C_6H_5CN	"	p-$CH_3C_6H_4SCH_2NHCOC_6H_5$ (46) + I (35)	315
C_9				
$C_6H_5COCH_2S(O)CH_3$	TsNSO	C_6H_6, 80°, 1.5 h	$C_6H_5COCH(NHTs)SCH_3$ (5) I + $C_6H_5COCH(NHTs)_2$ (71) II + $C_6H_5COCH_2SCH_3$ (11)	320
	TsNSO	Ether, 35°, 14 h	I (61) + II (15)	320
	MsNSO	C_6H_6, 80°, 9 h	$C_6H_5COCH(NHMs)SCH_3$ (15) + $C_6H_5COCH(NHMs)_2$ (50)	320
	C_6H_5CONSO	C_6H_6, 45–50°, 5.5 h	$C_6H_5COCH(NHCOC_6H_5)SCH_3$ (21) + $C_6H_5COCH_2SCH_3$ (19)	320
$C_6H_{11}COCH_2S(O)CH_3$	TsNSO	C_6H_6, 80°, 1.5 h	$C_6H_{11}COCH(NHTs)SCH_3$ (29)	320
$C_6H_5S^*(O)(CH_2)_2CONH_2$		$(CH_3)_3SiOSO_2CF_3$ (3.6 eq), $(C_2H_5)_3N$, CH_2Cl_2, −20°, 15 min	4-(C_6H_5S)-azetidin-2-one (*) (76), 67% ee	143, 144
2-$CH_3S(O)C_6H_4CONHCH_3$	CH_3CN	$CH_3O(C{=}CH_2)OTBDMS$, ZnI_2, rt, 5 h	3-methyl-2,3-dihydro-4H-1,3-benzothiazin-4-one (85)	259
C_{10}				
p-$CH_3C_6H_4COCH_2S(O)CH_3$	TsNSO	C_6H_6, 80°, 7 h	p-$CH_3C_6H_4COCH(NHTs)SCH_3$ I (32) + p-$CH_3C_6H_4COCH(NHTs)_2$ (23)	320
	TsNSO	Ether, 35°, 14 h	I (54)	320
p-$CH_3OC_6H_4COCH_2S(O)CH_3$	C_6H_5CONSO	C_6H_6, 40–50°, 8 h	p-$CH_3OC_6H_4COCH(NHCOC_6H_5)SCH_3$ (49)	320

	Reactant	Conditions	Product(s) and Yield(s) (%)	Refs.
	$C_6H_5S(O)CH_2CH(CH_3)CONH_2$	$(CH_3)_3SiOSO_2CF_3$ (5 eq), $(C_2H_5)_3N$, CH_2Cl_2, 20°	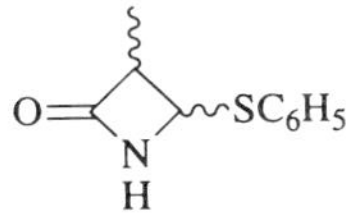 O, N, H, SC_6H_5 (41) *cis*:*trans* = 2.7:1	143
	$C_6H_5CH_2S(O)(CH_2)_2CONH_2$	Ac_2O, 110°, 30 min	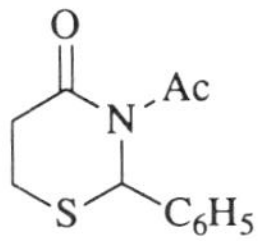 O, N, Ac, S, C_6H_5 (24)	142
C_{11}	$C_6H_5CH_2S(O)(CH_2)_2CONHCH_3$	Ac_2O, 114–120°, 30 min	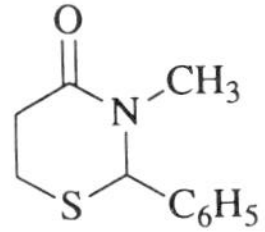 O, N, CH_3, S, C_6H_5 (62)	142
	p-$XC_6H_4CH_2S(O)(CH_2)_2CONHCH_3$	Ac_2O, C_6H_6, reflux 1 h	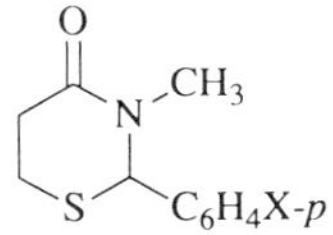 O, N, CH_3, S, C_6H_4X-*p* X = H (82) X = Cl (50)	321
	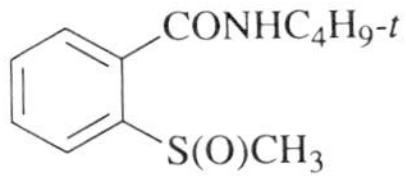 $CONHC_4H_9$-*t*, $S(O)CH_3$	AcCl, CH_2Cl. overnight	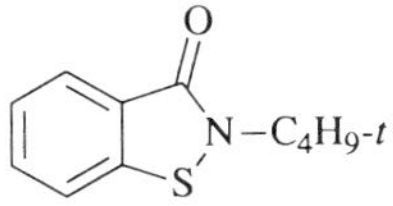 O, N–C_4H_9-*t*, S (77)	322
		$SOCl_2$, CH_2Cl_2, 1 h	" (77)	322
		Ac_2O, reflux	" (tr) 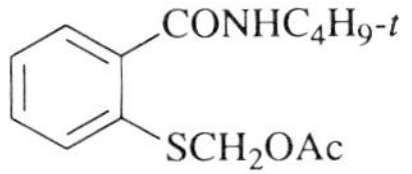 + $CONHC_4H_9$-*t*, SCH_2OAc (60)	322
		$(C_6H_5CO)_2$, *o*-$Cl_2C_6H_4$, 180°, 10 h	" (tr) 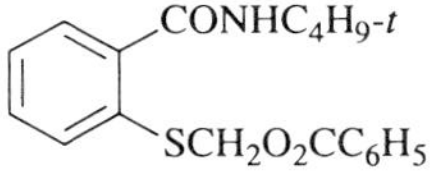 + $CONHC_4H_9$-*t*, $SCH_2O_2CC_6H_5$ (55)	322

TABLE III. SULFOXIDES WITH A NITROGEN NUCLEOPHILE (*Continued*)

Sulfoxide	Nucleophile	Reaction Conditions	Product(s) and Yield(s) (%)	Refs.
C_{12}				
p-$ClC_6H_4CH_2S(O)(CH_2)_2CONHC_2H_5$	Ac_2O, C_6H_6, reflux 1 h		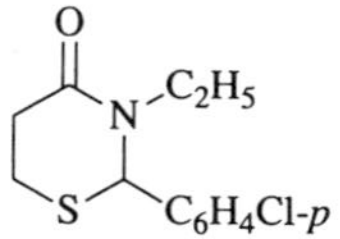 O, N, C_2H_5, S, C_6H_4Cl-p (71)	321
$C_6H_5S(O)(CH_2)_3CONHAc$	CH_3O OTBDMS, ZnI_2, CH_3CN, rt, 25 h		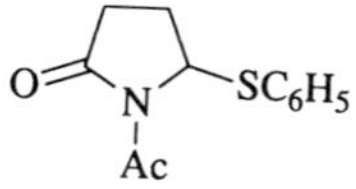 O, N, Ac, SC_6H_5 (57)	259
C_{14}				
$C_6H_5CH_2S(O)(CH_2)_2CONHR$	Ac_2O, 110°, 30 min		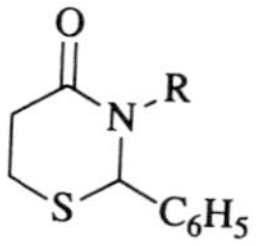 O, N, R, S, C_6H_5; R = n-C_4H_9 (28); R = $CH_2CO_2C_2H_5$ (53)	142
$C_6H_5S(O)(CH_2)_3CONHCH_2CO_2C_2H_5$	CH_3O OTBDMS, ZnI_2, CH_3CN, rt, 14 h		O, N, $CH_2CO_2C_2H_5$, SC_6H_5 (88)	259
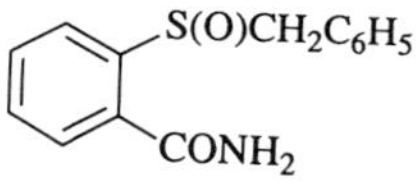 $S(O)CH_2C_6H_5$, $CONH_2$	Ac_2O, 130°, 3 h		S, N−Ac, O (90)	246

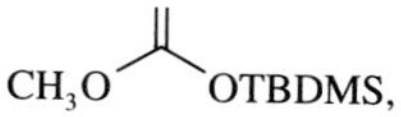

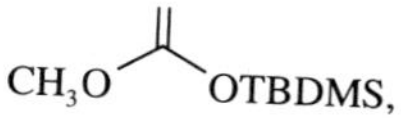

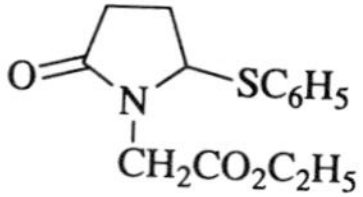

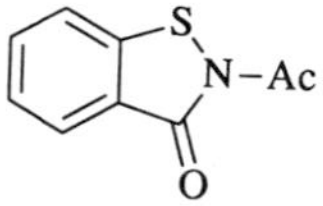

C_{15}	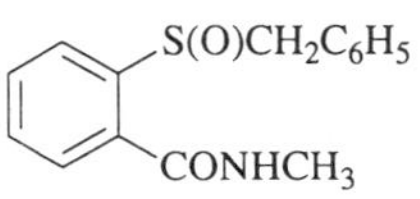	Ac_2O, 100°, 15 h	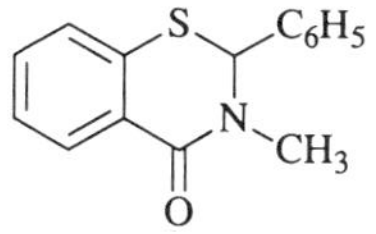 I (87) + 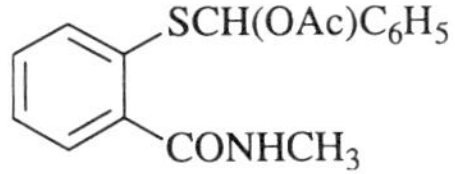 II (—)	246, 267
		Ac_2O, 70°, 10 h	(26) + II (70)	267
C_{16}	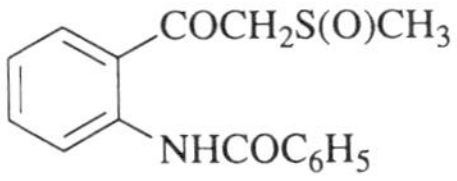	CF_3CO_2H, C_6H_6, reflux 40–90 min	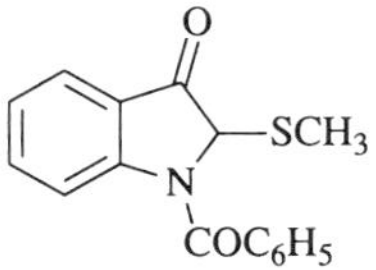 (67)	260
	$C_6H_5S(O)(CH_2)_2CONHCH_2C_6H_5$	$(CH_3)_3SiOSO_2CF_3$, $(C_2H_5)_3N$, CH_2Cl_2, −20°, 30 min	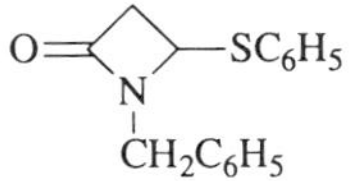 (14)	143
C_{17}	*p*-$ClC_6H_4CH_2S(O)(CH_2)_2CONHCH_2C_6H_5$	Ac_2O, C_6H_6, reflux 1 h	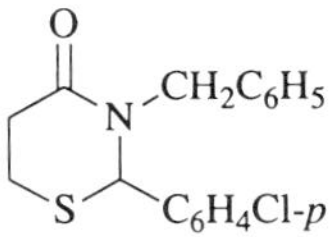 (42)	321
	$C_6H_5S(O)(CH_2)_3CONHCH_2C_6H_5$	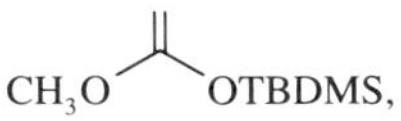 ZnI_2, CH_3CN, rt, 1 h	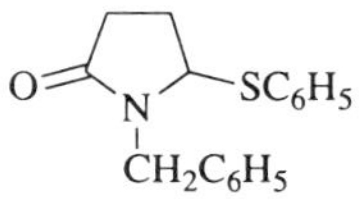 (quant)	259

TABLE III. Sulfoxides with a Nitrogen Nucleophile (*Continued*)

Sulfoxide	Nucleophile	Reaction Conditions	Product(s) and Yield(s) (%)	Refs.
C_{18}				
$C_6H_5S(O)CH_2C(CH_3)_2CONHCH_2C_6H_5$		$(CH_3)_3SiOSO_2CF_3$ (2.2 eq), $(C_2H_5)_3N$, CH_2Cl_2, −20°, 30 min	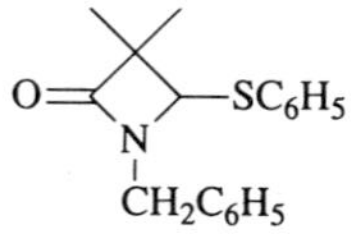 O=, N, $CH_2C_6H_5$, SC_6H_5 (51)	143
$C_6H_5S(O)(CH_2)_4CONHCH_2C_6H_5$	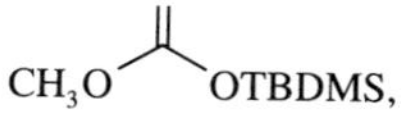	CH_3O / OTBDMS, ZnI_2, CH_3CN, rt, 4 h	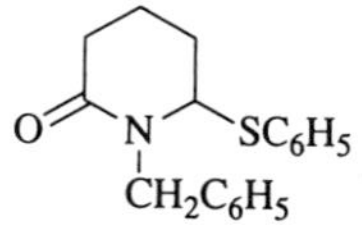 O, N, $CH_2C_6H_5$, SC_6H_5 (54)	259
C_{19}				
C_{19}				
$C_6H_5S(O)(CH_2)_5CONHCH_2C_6H_5$		CH_3O / OTBDMS, ZnI_2, CH_3CN, rt, 18 h	O, N, $CH_2C_6H_5$, SC_6H_5 (57)	
C_{20}				
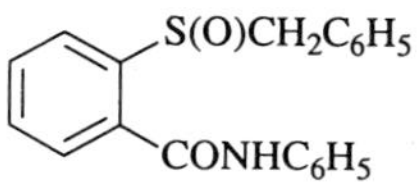 $S(O)CH_2C_6H_5$, $CONHC_6H_5$		Ac_2O, CH_3CN, 130°, 3 h	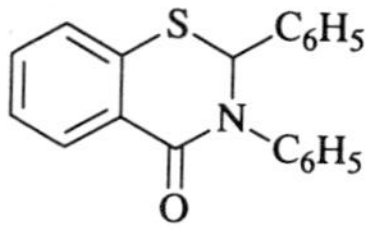 S, C_6H_5, N, C_6H_5, O (80)	246

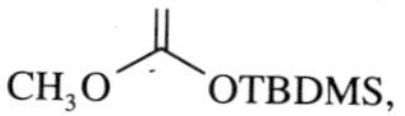

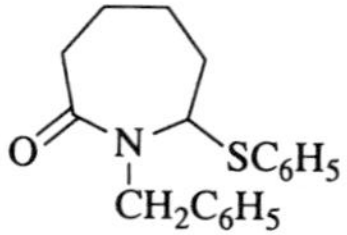

[a] The reaction was carried out to partial conversion.
[b] Other non-Pummerer products were formed.

TABLE IV. SULFOXIDES WITH A CARBON NUCLEOPHILE

Sulfoxide	Nucleophile	Reaction Conditions	Product(s) and Yield(s) (%)	Refs.
C_2				
$(CH_3)_2SO$	C_6H_6	1. TFAA, 0°, 1 h 2. $SnCl_4$, 10.5 h	$CH_3SCH_2C_6H_5$ (62)	48
	$C_6H_5CH_3$	1. TFAA, 0°, 1 h 2. $SnCl_4$, 8 h	$CH_3SCH_2C_6H_4CH_3$-*p* (62)	48
	$C_6H_5C_2H_5$	1. TFAA, 0°, 1 h 2. $SnCl_4$, 11 h	$CH_3SCH_2C_6H_4C_2H_5$-*p* (67)	48
	p-Xylene	1. TFAA, CH_2Cl_2, 0°, 1 h 2. $SnCl_4$, 9.5 h	$CH_3SCH_2C_6H_3(CH_3)_2$-*o,m* (60)	48
	$C_6H_5C_4H_9$-*s*	1. TFAA, CH_2Cl_2, 0°, 1 h 2. $SnCl_4$, 12 h	$CH_3SCH_2C_6H_4(C_4H_9$-*s*)-*p* (64)	48
	$C_6H_5C_4H_9$-*t*	1. TFAA, CH_2Cl_2, 0°, 1 h 2. $SnCl_4$, 4 h	$CH_3SCH_2C_6H_4(C_4H_9$-*t*)-*p* (62)	48
	$C_6H_4(C_3H_7$-*i*$)_2$-*m*	1. TFAA, CH_2Cl_2, 1 h 2. $SnCl_4$, 12 h	$CH_3SCH_2C_6H_3(C_3H_7$-*i*$)_2$-*m,m′* (58)	48
	$C_{10}H_8$	1. TFAA, CH_2Cl_2, 1 h 2. $SnCl_4$, 10 h	$CH_3SCH_2C_{10}H_7$-1 (74)	48
	Phenanthrene	1. TFAA, CH_2Cl_2, 0°, 1 h 2. $SnCl_4$, 10.5 h	CH_3SCH_2-phenanthryl-9 (80)	48
	$(CH_3)_2C{=}CHOCH_3$	Ac_2O, BF_3 · ether, reflux 5 h	$CH_3SCH_2C(CH_3)_2CH(OAc)OCH_3$ (82)	323
	n-$C_4H_9CHCH(OCH_3)_2$ with C_2H_5 on the CH	1. Ac_2O, BF_3 · ether 100°, 5 h 2. H_2SO_4, H_2O, 72 h	$CH_3SCH_2C(C_2H_5)(C_4H_9$-*n*)CHO (45)	323
	$(C_2H_5)_2CHCH(OCH_3)_2$	1. Ac_2O, BF_3 · ether, 2. H_2SO_4, H_2O, 72 h	$CH_3SCH_2C(C_2H_5)_2CHO$ (48)	323

TABLE IV. SULFOXIDES WITH A CARBON NUCLEOPHILE (Continued)

Sulfoxide	Nucleophile	Reaction Conditions	Product(s) and Yield(s) (%)	Refs.
	$C_6H_5OCH_3$	Ac_2O, $BF_3 \cdot$ ether, heat	$CH_3SCH_2C_6H_4OCH_3$-*p* (—)	323
	C_6H_5OH	Ac_2O, 24 h	*o*-$CH_3SCH_2C_6H_4OH$ I (31)[a] + *o,o'*-$(CH_3SCH_2)_2C_6H_3OH$ II (20)[a]	324
	C_6H_5OH	Pyr · SO_3, $(C_2H_5)_3N$, overnight	I (37) + II (27)	325
	C_6H_5OH (−50°)	1. $C_6H_5OS(O)Cl$, CH_2Cl_2, −55°, 50 min 2. −50° 3. $(C_2H_5)_3N$, −50°	I (57)	326
	C_6H_5OH[b]	1. $SOCl_2$, CH_2Cl_2, −60°, 18 min 2. CH_2Cl_2, −55°, 70 min 3. $(C_2H_5)_3N$, −50°	I (78)	326
	$C_6H_5OS(O)Cl$	1. CH_2Cl_2, −50°, 130 min 2. $(C_2H_5)_3N$	I (39)[a] + II (17)[a]	327
	C_6H_5OH	Ac_2O, P_2O_5	I (2) + *o*-HOC_6H_4CHO (0.5) 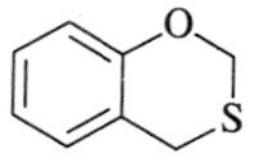 + [4H-1,3-benzoxathiin structure] III (14)	324
	C_6H_5OH	DCC, pyrH⁺ · $CF_3CO_2^-$, C_6H_6, 0° 1 h, rt 3 h	I (30) + II (20) + III (4)	328, 329
	C_6H_5OH	DCC, H_3PO_4, C_6H_6	I (27) + II (17) + III (4) 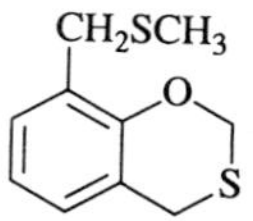 + [8-(CH_2SCH_3)-4H-1,3-benzoxathiin structure] (4)	212, 221
	o-RC_6H_4OH[b]	1. $SOCl_2$, CH_2Cl_2, −60°, 20 min	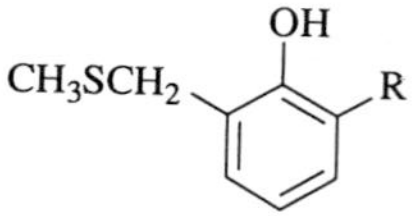 [2-OH, 3-R, 1-CH_3SCH_2 benzene structure] I +	327

Reactant	Conditions	Products (Yield, %)	Refs.
	2. CH_2Cl_2, $-50°$, 40 min 3. $(C_2H_5)_3N$, CH_2Cl_2, -40 to $-50°$	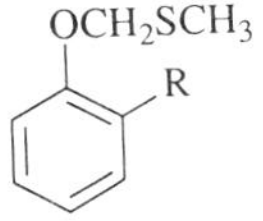 OCH_2SCH_3 / R (II) R = CH_3 I (77)[c] R = Cl I (75)[c] R = OCH_3 I (74)[c] R = NO_2 I (80)[c] + II (11)[c] R = CO_2CH_3 I (80)[c]	
p-RC_6H_4OH[b]	1. $SOCl_2$, CH_2Cl_2, $-60°$, 20 min 2. CH_2Cl_2, $-50°$, 40 min 3. $(C_2H_5)_3N$, CH_2Cl_2, -40 to $-50°$	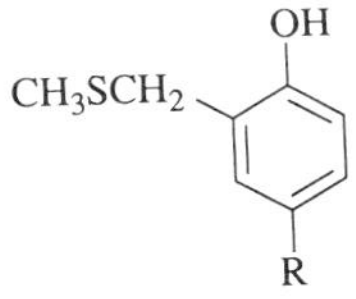 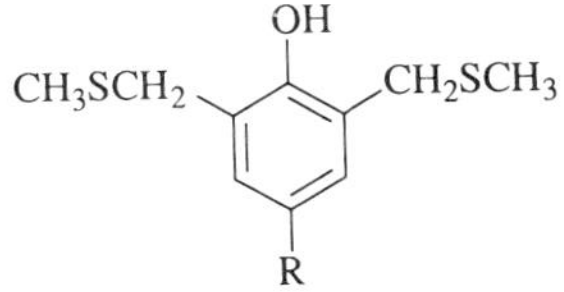 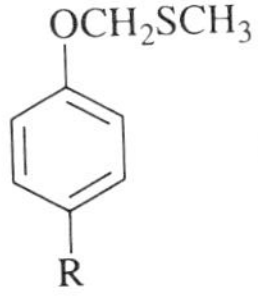 OH / CH_3SCH_2 / R (I) + OH / CH_3SCH_2 / CH_2SCH_3 / R (II) + OCH_2SCH_3 / R (III) R = CH_3 I (81)[c] + II (7)[c] R = Cl I (75)[c] + II (3)[c] R = OCH_3 I (60)[c] + II (3)[c] R = NO_2 I (45)[c] + II (5)[c] R = CO_2CH_3 I (62)[c] + II (4)[c] + III (10)[c]	327

TABLE IV. SULFOXIDES WITH A CARBON NUCLEOPHILE (*Continued*)

Sulfoxide	Nucleophile	Reaction Conditions	Product(s) and Yield(s) (%)	Refs.
	m-RC_6H_4OH[b]	1. $SOCl_2$, CH_2Cl_2, $-60°$, 20 min 2. CH_2Cl_2, $-50°$, 40 min 3. $(C_2H_5)_3N$, CH_2Cl_2, -40 to $-50°$	2-CH_3SCH_2-3-R-phenol (OH, CH_3SCH_2, R) I + 2-CH_3SCH_2-5-R-phenol (OH, CH_3SCH_2, R) II + 2,6-bis(CH_3SCH_2)-3-R-phenol (OH, CH_3SCH_2, CH_2SCH_3, R) III R = CH_3 I (39)[c] + II (39)[c] + III (1)[c] R = Cl I (36)[c] + II (29)[c] + III (4)[c] R = NO_2 I (36)[c] + II (23)[c] R = CO_2CH_3 I (32)[c] + II (21)[c] + III (2)[c]	327
	o-$O_2NC_6H_4OH$	DCC, H_3PO_4, C_6H_6, 2 h	2-CH_3SCH_2-6-NO_2-phenol (OH, CH_3SCH_2, NO_2) I (38) + CH_3SCH_2, OCH_2SCH_3, NO_2-benzene II (7) + o-$CH_3SCH_2OC_6H_4NO_2$ (10)	212, 221

o-$O_2NC_6H_4OH$	DCC, $pyrH^+CF_3CO_2^-$, C_6H_6, 0°, 15 min, rt 3 h	I (40) + II (5)	329
p-$O_2NC_6H_4OH$	DCC, H_3PO_4, C_6H_6, 1 h	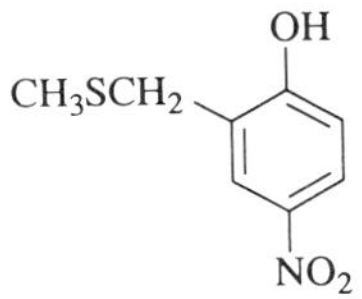 OH, CH_3SCH_2, NO_2 (26) 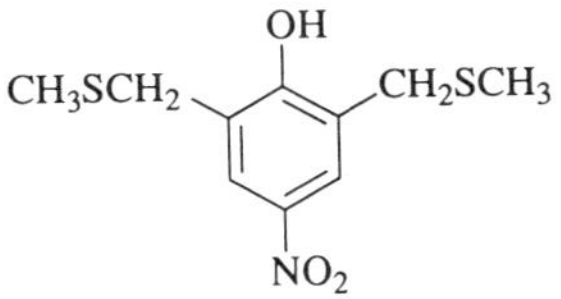 + OH, CH_3SCH_2, CH_2SCH_3, NO_2 (11) 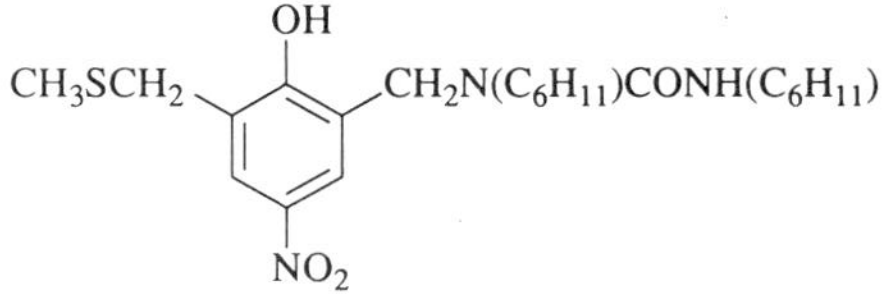 + OH, CH_3SCH_2, $CH_2N(C_6H_{11})CONH(C_6H_{11})$, NO_2 (14) + p-$CH_3SCH_2OC_6H_4NO_2$ (3) 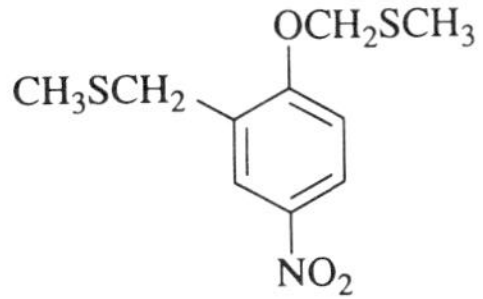 + OCH_2SCH_3, CH_3SCH_2, NO_2 (14) 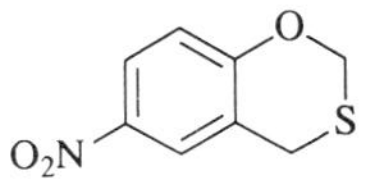 + O, S, O_2N I (3)	212, 221, 329
p-$O_2NC_6H_4OH$	Ac_2O, P_2O_5	I (21)	324

TABLE IV. SULFOXIDES WITH A CARBON NUCLEOPHILE (*Continued*)

Sulfoxide	Nucleophile	Reaction Conditions	Product(s) and Yield(s) (%)	Refs.
	HO (estrone structure, O)	DCC, H_3PO_4	CH_3SCH_2, HO (—) + HO, CH_2SCH_3 (—)	212, 221
	o-$HOCC_6H_4OH$	DCC, $pyrH^+ \cdot CF_3CO_2^-$, C_6H_6, 0° 15 min, rt 3 h	CH_3SCH_2, OH, CHO (18)[c] + CH_3SCH_2, OH (16)[c]	328, 329
	o-ClC_6H_4OH	$Pyr \cdot SO_3$, $(C_2H_5)_3N$, overnight	CH_3SCH_2, OH, Cl (4)	325
	p-ClC_6H_4OH	"	CH_3SCH_2, OH, Cl (33) + CH_3SCH_2, OH, CH_2SCH_3, Cl (10)	325

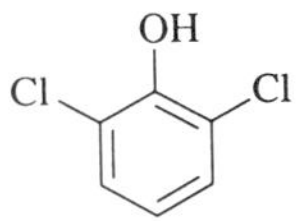 2,6-Dichlorophenol (structure: OH, Cl, Cl)	DCC, H_3PO_4, C_6H_6, overnight	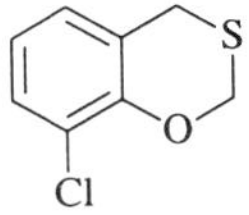 8-Chloro-4H-1,3-benzoxathiin (structure) (25)[c]	211
o,p-$Cl_2C_6H_3OH$	DCC, pyrH$^+$ $CF_3CO_2^-$, C_6H_6, 0° 15 min, rt 3 h	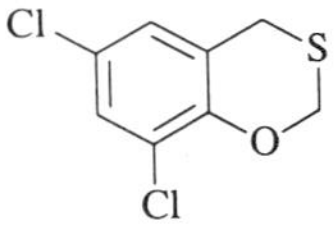 6,8-Dichloro-4H-1,3-benzoxathiin (structure) I (10)[c] 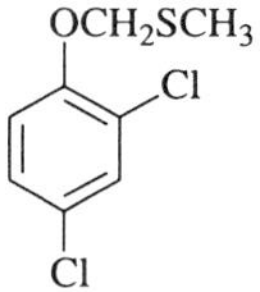 + 2,4-Dichlorophenyl OCH_2SCH_3 (structure) (7)[c]	328, 329
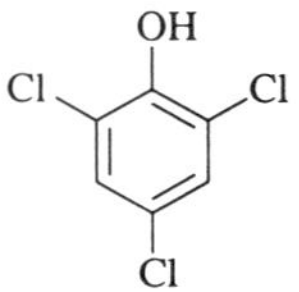 2,4,6-Trichlorophenol (structure: OH, Cl, Cl, Cl)	DCC, H_3PO_4, C_6H_6, overnight	I (42)	211
C_6F_5OH	DCC, H_3PO_4, <12°, 1 h, rt, 16 h	$CH_3SCH_2OC_6F_5$ I (3) 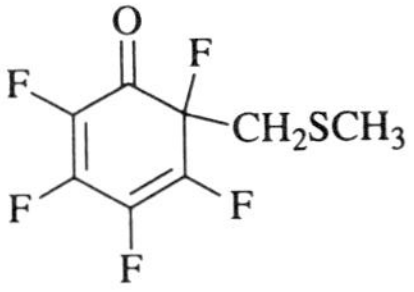 + 6-(CH_2SCH_3)-2,3,4,5,6-pentafluorocyclohexa-2,4-dienone (structure) II (13) 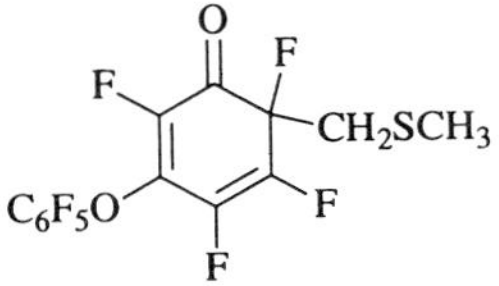 + 6-(CH_2SCH_3)-3-(C_6F_5O)-2,4,5,6-tetrafluorocyclohexa-2,4-dienone (structure) III (36)	319

TABLE IV. SULFOXIDES WITH A CARBON NUCLEOPHILE (*Continued*)

Sulfoxide	Nucleophile	Reaction Conditions	Product(s) and Yield(s) (%)	Refs.
	C_6F_5OH	1. TFAA, CH_2Cl_2, −60°, 2 h 2. $(C_2H_5)_3N$, −60° to rt, 18 h	II (71) + III (17)	319
	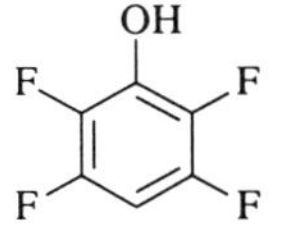 [structure] OH, F, F, F, F	DCC, H_3PO_4, <12° 1 h, rt 16 h	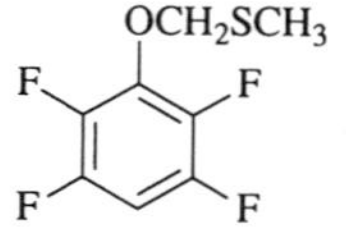 [structure] OCH_2SCH_3, F, F, F, F (8) 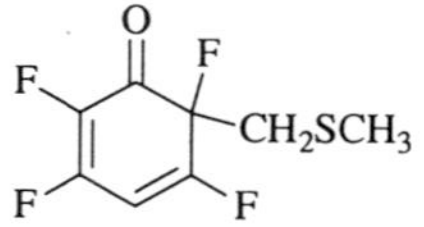 + [structure] O, F, CH_2SCH_3, F, F, F (69)	319
	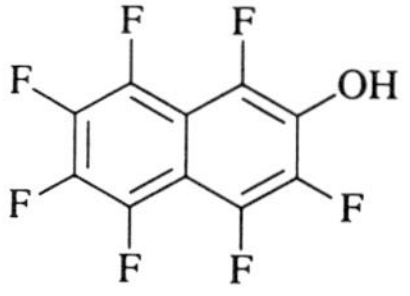 [structure] F, F, F, OH, F, F, F	DCC, H_3PO_4, <12° 1 h, rt 21 h	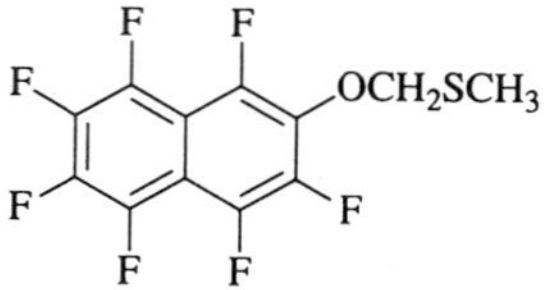 [structure] F, F, F, OCH_2SCH_3, F, F, F I (10) 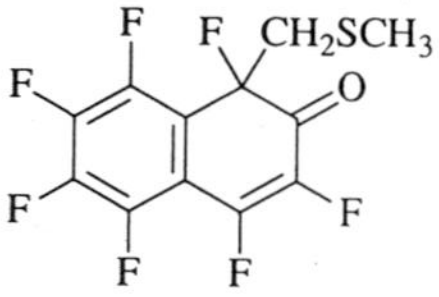 + [structure] F, F, CH_2SCH_3, F, O, F, F, F, F II (72)	319
		1. TFAA, CH_2Cl_2, −60°, 2 h 2. $(C_2H_5)_3N$, −60° to rt, 18 h	II (78)[c]	319

Substrate	Conditions	Product(s) and Yield(s) (%)	Refs.
4-Br-3,5,6-F_3-pyridin-2-ol (Br, F, F, F, OH on pyridine)	TFAA, CH_2Cl_2, −50°, 2 h	4-Br-3,5,6-F_3-2-(OCH_2SCH_3)pyridine (1.5) +	319
		4-Br-3,5-F_2-3-CH_2SCH_3-3-F-piperidine-2,6-dione (N–H) (8) +	
		4-Br-3,5,6-F_3-1-(CH_2SCH_3)pyridin-2-one (44)	
2,4,5,6-F_4-pyridin-3-ol (F, F, F, F, OH on pyridine)	"	4,5,6-F_3-3-OH-2-CHO-pyridine (6) +	319
		4,5,6-F_3-3-OH-2-$CH(SCH_3)_2$-pyridine (6)	
o-$CH_3C_6H_4OH$	Ac_2O, 15–25°, 2 d	2-OH-3-CH_3-1-(CH_3SCH_2)benzene I (25) +	324
		4-OH-3-CH_3-1-(CH_2SCH_3)benzene (36)	

TABLE IV. SULFOXIDES WITH A CARBON NUCLEOPHILE (*Continued*)

Sulfoxide	Nucleophile	Reaction Conditions	Product(s) and Yield(s) (%)	Refs.
	o-$CH_3C_6H_4OH$	DCC, H_3PO_4, C_6H_6, overnight	I (28) 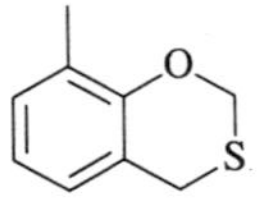 + [structure: O, S] (7)	212, 221
	o-$CH_3C_6H_4OH$	DCC, $pyrH^+$, $CF_3CO_2^-$, C_6H_6, 0° 15 min, rt 3 h	I (59–65)	328, 329
	p-$CH_3C_6H_4OH$	Ac_2O, 15–25°, 2 d	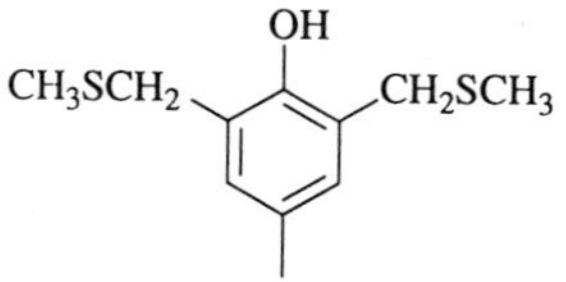 [structure: OH, CH_3SCH_2, CH_2SCH_3] I (9)	324
	p-$CH_3C_6H_4OH$	DCC, $pyrH^+$, $CF_3CO_2^-$, C_6H_6, 0° 15 min, rt 3 h	I (20) 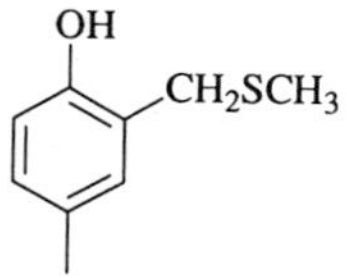 + [structure: OH, CH_2SCH_3] (18)	328, 329
	m-$CH_3C_6H_4OH$	Ac_2O, 15–25°, 2 d	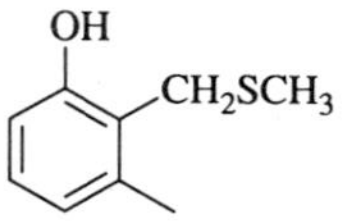 [structure: OH, CH_2SCH_3] (16) 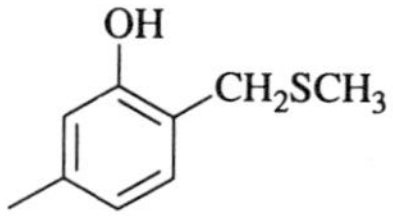 + [structure: OH, CH_2SCH_3] (18)	324
	o,*p*-$(CH_3)_2C_6H_3OH$	"	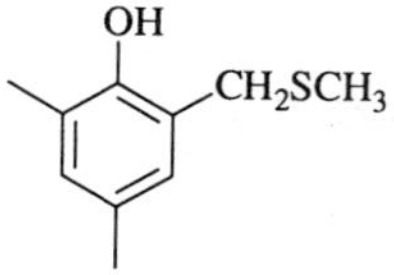 [structure: OH, CH_2SCH_3] (61)	324

o,o'-$(CH_3)_2C_6H_3OH$	"	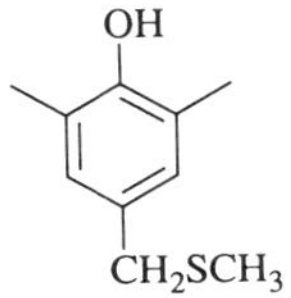 OH, CH_2SCH_3 (26)	324
o,o'-$(CH_3)_2C_6H_3OH$	DCC, H^+	" (—)	329
o,o'-$(CH_3)_2C_6H_3OH$	1. DCC, H_3PO_4, C_6H_6, 2. Silica gel	" (66)	211, 212
o,o'-$(CH_3)_2C_6H_3OH$	DCC, H_3PO_4, C_6H_6, 45 min	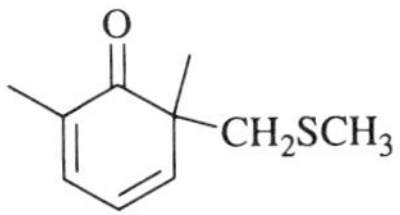 O, CH_2SCH_3 (66)	211
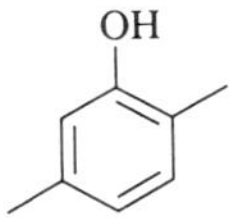 OH	DCC, H_3PO_4, C_6H_6, overnight	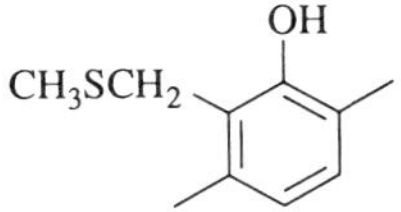 OH, CH_3SCH_2 (35)	212, 221
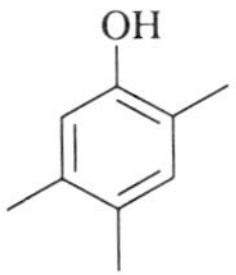 OH	"	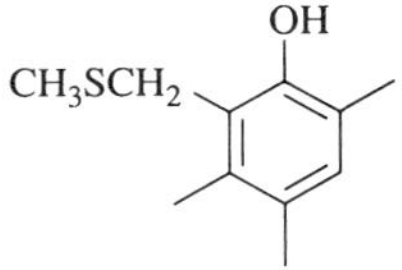 OH, CH_3SCH_2 (23)	221
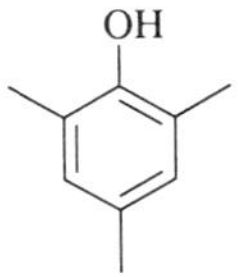 OH	1. DCC, H_3PO_4, C_6H_6, 2 h 2. Silica gel	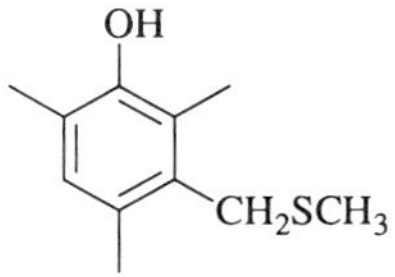 OH, CH_2SCH_3 (88)	211, 212
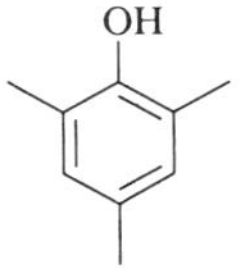 OH	DCC, H_3PO_4, C_6H_6, 2 h	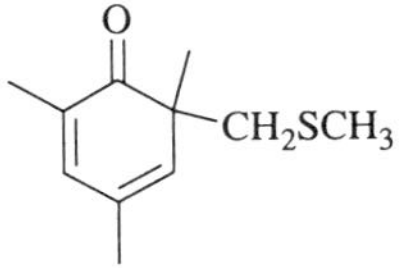 O, CH_2SCH_3 (93)	211

TABLE IV. SULFOXIDES WITH A CARBON NUCLEOPHILE (*Continued*)

Sulfoxide	Nucleophile	Reaction Conditions	Product(s) and Yield(s) (%)	Refs.
	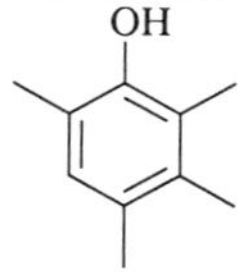 OH	1. DCC, H_3PO_4, C_6H_6, overnight 2. TFAA, CH_2Cl_2, 1 h	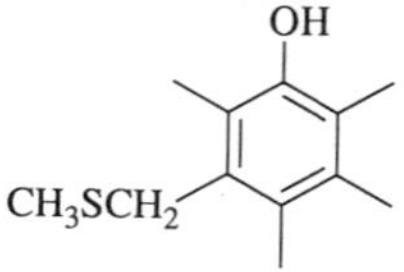 OH, CH_3SCH_2 (57)	211
	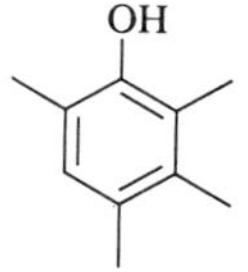 OH	DCC, H_3PO_4, C_6H_6, overnight	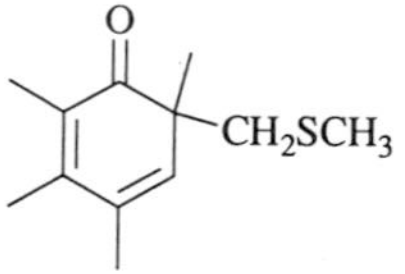 O, CH_2SCH_3 (57) +	211
			O, CH_2SCH_3 (38)	
	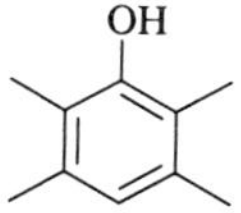 OH	1. DCC, H_3PO_4, C_6H_6, 2 h 2. CF_3CO_2H, CH_2Cl_2, 10 min	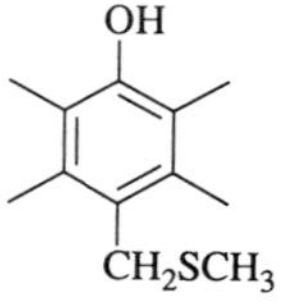 OH, CH_2SCH_3 (80)	211
	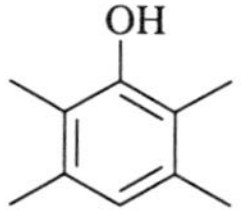 OH	DCC, H_3PO_4, C_6H_6, 2 h	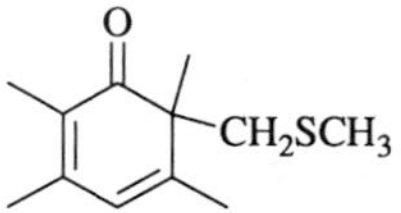 O, CH_2SCH_3 (86)	211, 212
	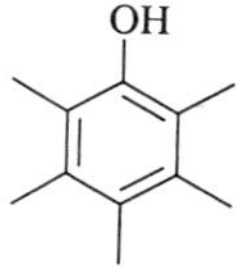 OH	DCC, H_3PO_4, C_6H_6	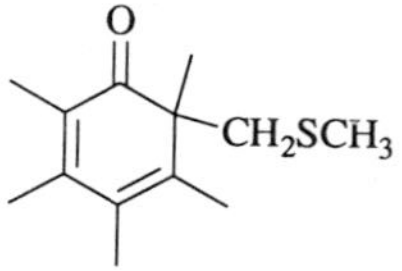 O, CH_2SCH_3 (85)	211, 212

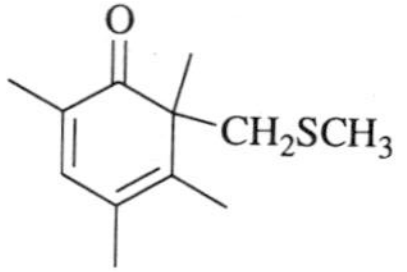

$o\text{-}CH_3OC_6H_4OH$	Ac_2O, 4 d	[ring: OH, CH_3SCH_2, OCH_3] I (40) + $o\text{-}CH_3OC_6H_4OCH_2SCH_3$ (4) + [ring: OCH_2SCH_3, CH_3SCH_2, OCH_3] (4)	330
$o\text{-}CH_3OC_6H_4OH$	Pyr · SO_3, $(C_2H_5)_3N$, overnight	I (64)[d]	325
$p\text{-}CH_3OC_6H_4OH$	"	[ring: OH, CH_2SCH_3, OCH_3] (38) + [ring: OH, CH_3SCH_2, CH_2SCH_3, OCH_3] (21)	325
$p\text{-}(n\text{-}C_9H_{19})C_6H_4OH$	"	[ring: OH, CH_2SCH_3, $C_9H_{19}\text{-}n$] (33) + [ring: OH, CH_3SCH_2, CH_2SCH_3, $C_9H_{19}\text{-}n$] (36)	325

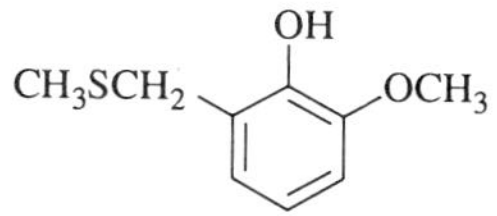

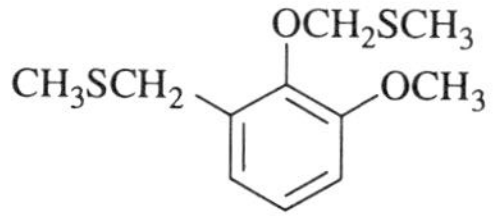

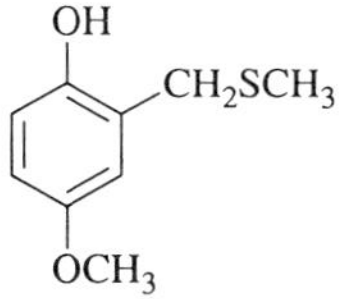

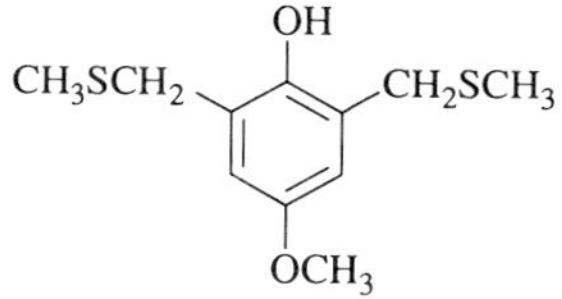

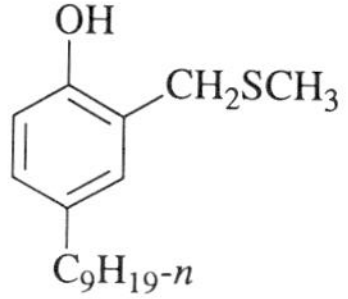

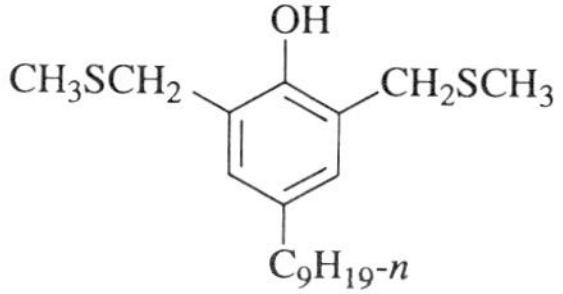

TABLE IV. SULFOXIDES WITH A CARBON NUCLEOPHILE (*Continued*)

Sulfoxide	Nucleophile	Reaction Conditions	Product(s) and Yield(s) (%)	Refs.
	p-$CH_3SC_6H_4OH$	"	OH, CH_2SCH_3, SCH_3 (24) +	325
			CH_3SCH_2, OH, CH_2SCH_3, SCH_3 (20)	
	CH_3O, OH	"	CH_3O, OH, CH_2SCH_3 (40)[d]	325
	CH_3O, OH	"	CH_3O, OH, CH_2SCH_3 I (67)[d]	325
	CH_3O, OH	Ac_2O, 4 d	I (18) + CH_3O, O, CH_2SCH_3, CH_2SCH_3 (40)	330
	CH_3O, OH	"	CH_3O, OH, CH_2SCH_3 (25)	330

Reactant	Conditions	Product(s) (Yield, %)	Refs.
2-Hydroxy-3-methoxy-5-methylbenzyl alcohol (CH_3O, OH, CH_2OH, CH_3)	"	2-hydroxy-3-methoxy-5-methylbenzyl methyl sulfide (CH_2SCH_3) (39)[a]	330
		OCH_2SCH_3 / CH_2SCH_3 / CH_3O / CH_3 derivative (1)[a]	
Vanillin (CH_3O, OH, CHO)	Pyr · SO_3, $(C_2H_5)_3N$, overnight	CH_3O, OH, CH_2SCH_3, CHO (8)	325
Vanillin (CH_3O, OH, CHO)	Ac_2O, 4 d	" (15) + CH_3O, OCH_2SCH_3, CHO (5) +	330
		CH_3O, OCH_2SCH_3, CH_2SCH_3, CHO (3) +	
		CH_3O, OH, CH_2SCH_3, CH_2SCH_3 (4)	

TABLE IV. SULFOXIDES WITH A CARBON NUCLEOPHILE (*Continued*)

Sulfoxide	Nucleophile	Reaction Conditions	Product(s) and Yield(s) (%)	Refs.
	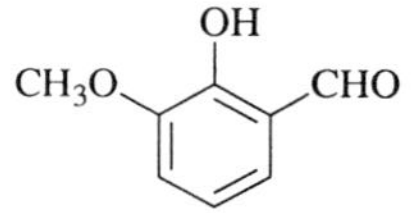 OH, CH_3O, CHO	DCC, H_3PO_4, C_6H_6	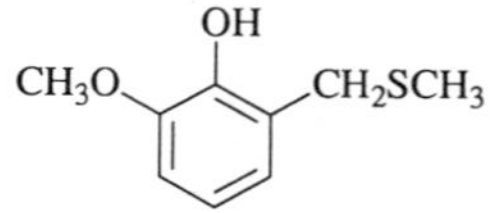 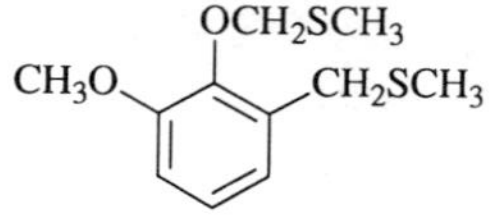 OH, CH_3O, CH_2SCH_3 (11) +	330
			OCH_2SCH_3, CH_3O, CH_2SCH_3 (3)	
	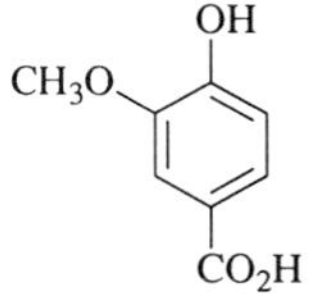 OH, CH_3O, CO_2H	Pyr · SO_3, $(C_2H_5)_3N$, overnight	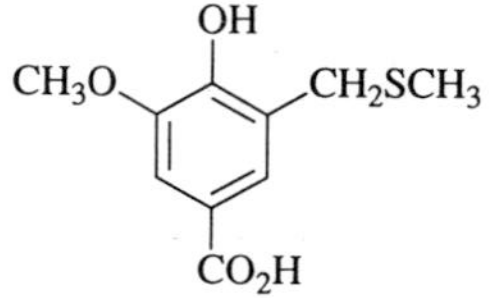 OH, CH_3O, CH_2SCH_3, CO_2H (15)	325
	OH, t-C_4H_9, OCH_3	"	OH, t-C_4H_9, CH_2SCH_3, OCH_3 (75)	325
	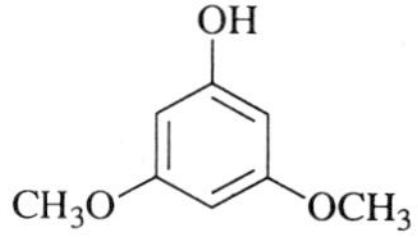 OH, CH_3O, OCH_3	Ac_2O, 4 d	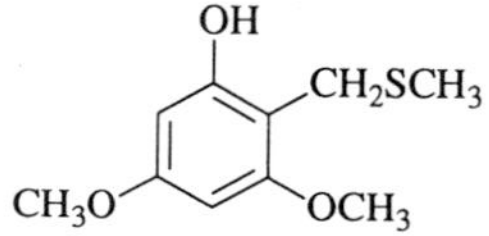 OH, CH_2SCH_3, CH_3O, OCH_3 (4)[a]	330
	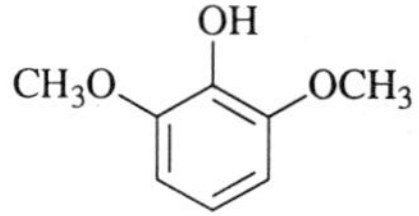 OH, CH_3O, OCH_3	Pyr · SO_3, $(C_2H_5)_3N$, overnight	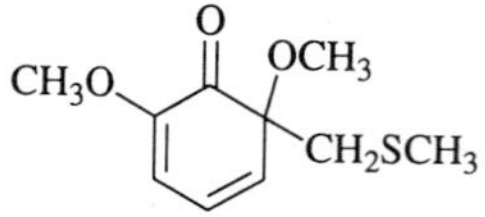 O, CH_3O, OCH_3, CH_2SCH_3 (—)	325

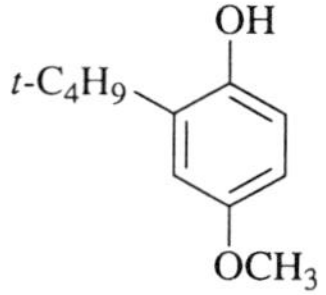

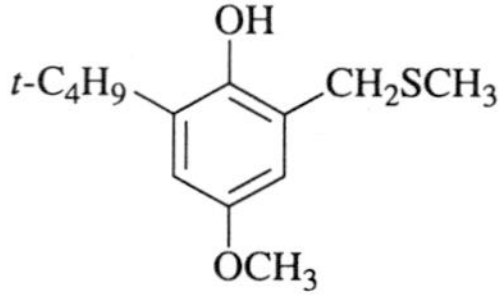

Substrate		Products	Ref.
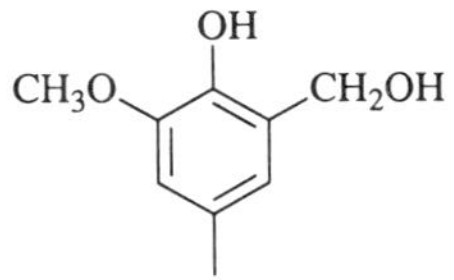 OH, CH_3O, CH_2OH	”	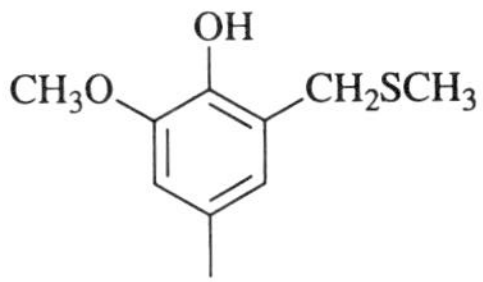 OH, CH_3O, CH_2SCH_3 (—) +	325
		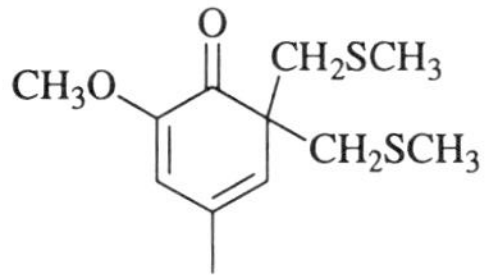 O, CH_3O, CH_2SCH_3, CH_2SCH_3 (—)	
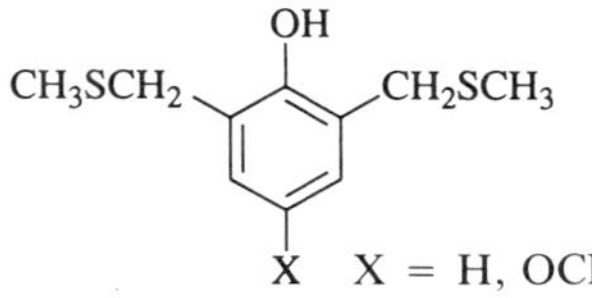 OH, CH_3SCH_2, CH_2SCH_3, X X = H, OCH_3	”	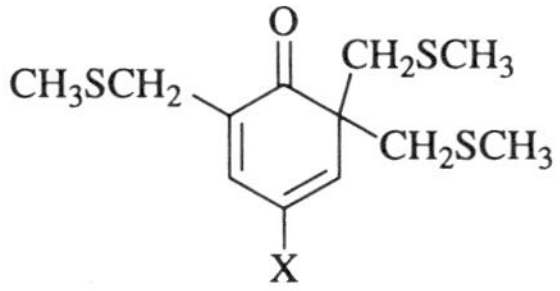 O, CH_3SCH_2, CH_2SCH_3, CH_2SCH_3, X (—)	
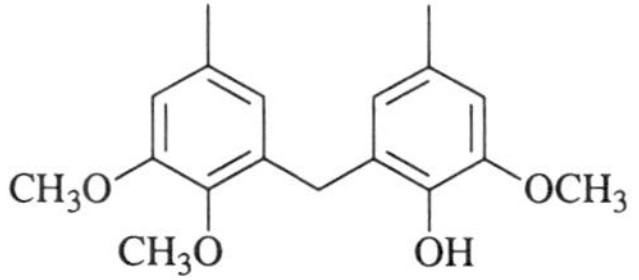 CH_3O, CH_3O, OCH_3, OH	”	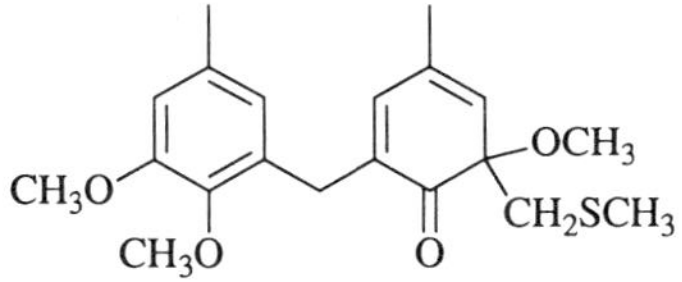 CH_3O, CH_3O, OCH_3, CH_2SCH_3, O (—) +	325
		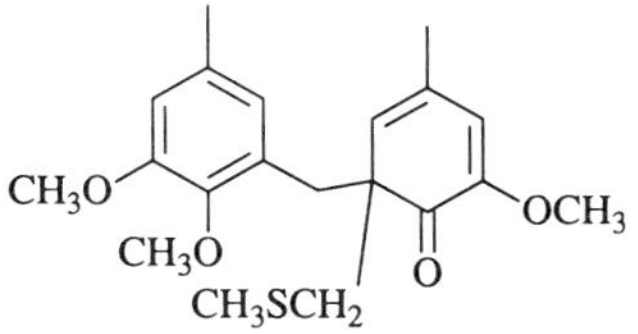 CH_3O, CH_3O, OCH_3, O, CH_3SCH_2 (—)	
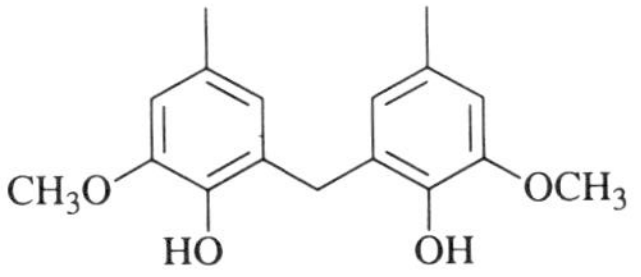 CH_3O, HO, OCH_3, OH	”	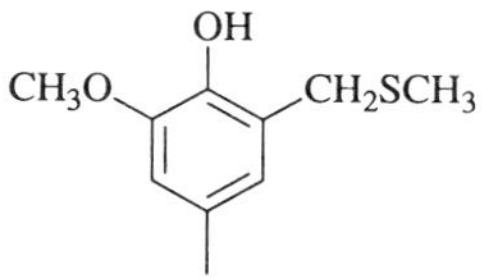 OH, CH_3O, CH_2SCH_3 (31)	325

TABLE IV. SULFOXIDES WITH A CARBON NUCLEOPHILE (*Continued*)

Sulfoxide	Nucleophile	Reaction Conditions	Product(s) and Yield(s) (%)	Refs.
	1-$C_{10}H_7OH$	Ac_2O, 20 h	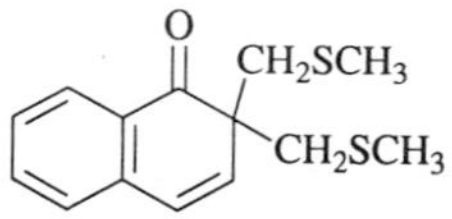 [structure: 2,2-bis(CH_2SCH_3)-1(2H)-naphthalenone] I (21) 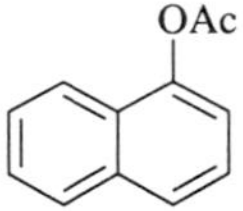 + [structure: 1-naphthyl acetate, OAc] (30)	324, 331
	1-$C_{10}H_7OH$	DCC, H_3PO_4, C_6H_6, overnight	I (36) 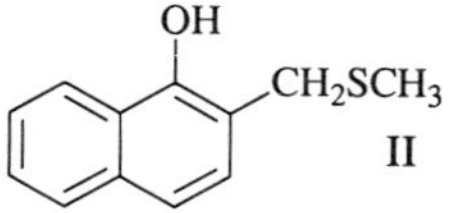 + [structure: 2-(CH_2SCH_3)-1-naphthol, OH] II (12) 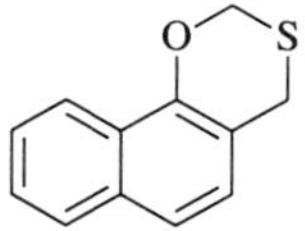 + [structure: naphtho-oxathiin, O, S] III (3)	212, 221
	1-$C_{10}H_7OH$	DCC, pyrH$^+$, $CF_3CO_2^-$, C_6H_6, 0° 15 min, rt 3 h	I (32)[c] + II (20)[c] + III (~5)	329, 328
	1-$C_{10}H_7OH$	Pyr · SO_3, $(C_2H_5)_3N$, overnight	II (16)	325
	2-$C_{10}H_7OH$	Ac_2O, 20 h	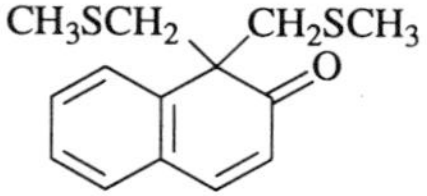 [structure: 1,1-bis(CH_2SCH_3)-2(1H)-naphthalenone; CH_3SCH_2, CH_2SCH_3, O] I (11) 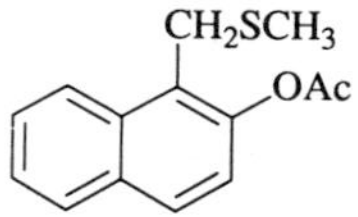 + [structure: 1-(CH_2SCH_3)-2-naphthyl acetate, OAc] (18) 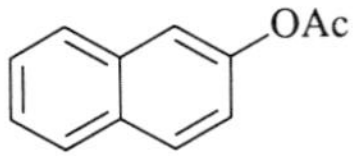 + [structure: 2-naphthyl acetate, OAc] (16)	331

		2-$C_{10}H_7OH$	DCC, H_3PO_4, C_6H_6,	I (36) 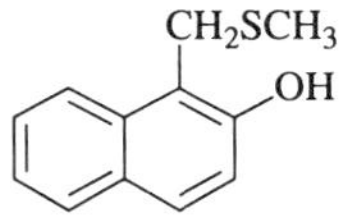 + [structure: CH_2SCH_3, OH] II (15) 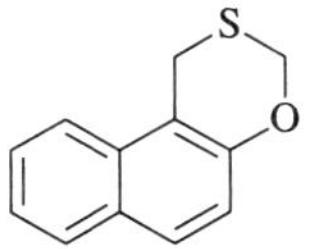 + [structure: S, O] III (3)	221
		2-$C_{10}H_7OH$	DCC, pyrH^+, $CF_3CO_2^-$, C_6H_6, 0°, 15 min, rt, 3 h	I (22)[c] + II (19)[c] + III (4)[c]	329, 328
	$CH_3S(O)CH_2Cl$	C_6H_5OH[b]	1. $C_6H_5OS(O)Cl$, CH_2Cl_2, −55°, 50 min 2. −50° 3. $(C_2H_5)_3N$, −50°	*o*-$OHCC_6H_4OH$ I (29)	326
		C_6H_5OH[b]	1. $SOCl_2$, CH_2Cl_2, −60°, 18 min 2. CH_2Cl_2, −55°, 70 min 3. $(C_2H_5)_3N$, −50°	I (13)	326
C_3	$CH_3S(O)CH_2SCH_3$	C_6H_5OH[b]	1. $C_6H_5OS(O)Cl$, CH_2Cl_2, 2. −50° 3. $(C_2H_5)_3N$, −50°	*o*-$[(CH_3S)_2CH]C_6H_4OH$ (26) + I (27)	326
	$CH_3S(O)CH_2CN$	$C_{10}H_8$	$TiCl_4$, $CHCl_3$, 0°, 2 h	$CH_3SCH(CN)C_{10}H_7$-1 (91)	332
		1,3,5-$(CH_3)_3C_6H_3$	"	$CH_3SCH(CN)C_6H_2(CH_3)_3$-2,4,6	332
	$CH_3S(O)CH_2CONH_2$	*n*-$C_6H_{13}CH{=}CH_2$[b]	1. TFAA, CF_3CO_2H, 0° 2. 0°, 1 h	$CH_3SCH(CONH_2)CH_2CH{=}CHC_5H_{11}$-*n* (65) $E:Z$ = 86:14	322
C_4	$(C_2H_5)_2SO$	C_6H_5OH[b]	1. $C_6H_5O_2SOCl$, CH_2Cl_2, −55°, 50 min 2. −50° 3. $(C_2H_5)_3N$, −50°	*o*-$[C_2H_5SCH(CH_3)]C_6H_4OH$ (67)	326

TABLE IV. SULFOXIDES WITH A CARBON NUCLEOPHILE (*Continued*)

Sulfoxide	Nucleophile	Reaction Conditions	Product(s) and Yield(s) (%)	Refs.
	$C_6H_5OH^b$	1. $SOCl_2$, CH_2Cl_2, −60°, 18 min 2. CH_2Cl_2, −55°, 70 min 3. $(C_2H_5)_3N$, −50°	" (69)	326
$(CH_2)_4SO$	*o*-$CH_3C_6H_4OH$	DCC, H_3PO_4, C_6H_6, overnight	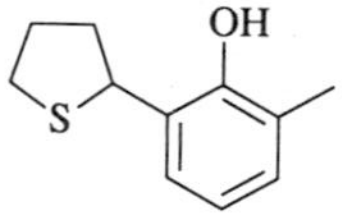 OH, S (45) 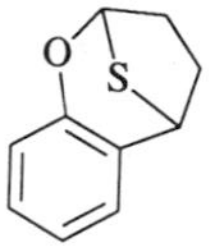 + O, S (9)	221
$CH_3S(O)CH_2PO(OCH_3)_2$	$C_6H_5C_2H_5$	1. TFAA, CH_2Cl_2, 0°, 4–5 min 2. $SnCl_4$, 0°, 1 h	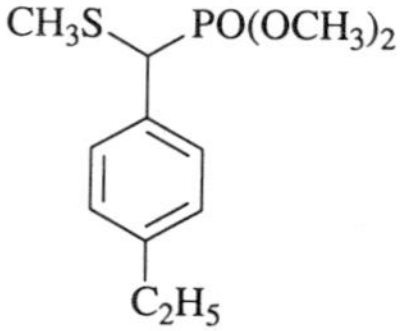 CH_3S, $PO(OCH_3)_2$, C_2H_5 (80)	333
	$C_6H_5C_4H_9$-*n*	1. TFAA, CH_2Cl_2, 0°, –5 min 2. $SnCl_4$, 0°, 1 h	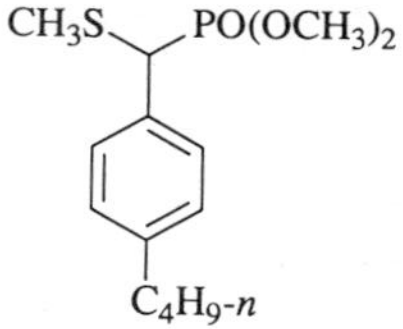 CH_3S, $PO(OCH_3)_2$, C_4H_9-*n* (91)	333

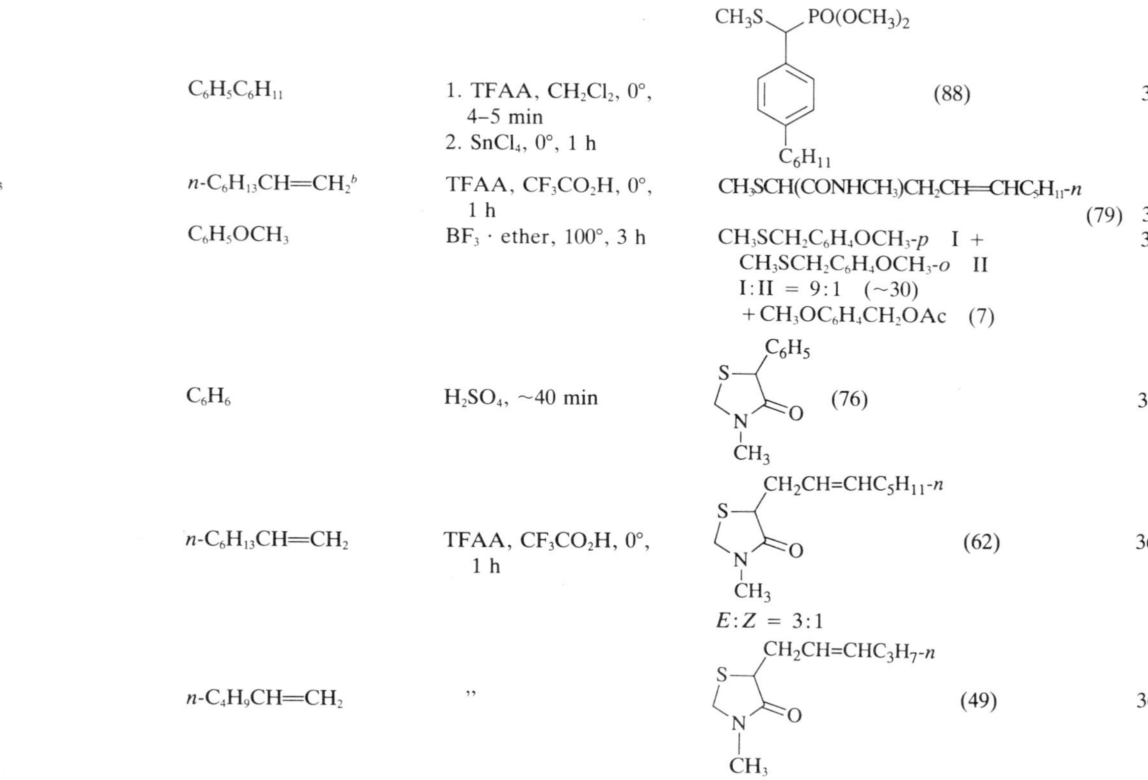

Sulfoxide	Substrate	Conditions	Product(s) (Yield, %)	Refs.
	$C_6H_5C_6H_{11}$	1. TFAA, CH_2Cl_2, 0°, 4–5 min 2. $SnCl_4$, 0°, 1 h	CH_3S–CH($PO(OCH_3)_2$)–C_6H_4–C_6H_{11} (88)	333
$CH_3S(O)CH_2CONHCH_3$	n-$C_6H_{13}CH{=}CH_2$[b]	TFAA, CF_3CO_2H, 0°, 1 h	$CH_3SCH(CONHCH_3)CH_2CH{=}CHC_5H_{11}$-$n$ (79)	322
CH_3SCH_2OAc	$C_6H_5OCH_3$	BF_3 · ether, 100°, 3 h	$CH_3SCH_2C_6H_4OCH_3$-p I + $CH_3SCH_2C_6H_4OCH_3$-o II I:II = 9:1 (~30) + $CH_3OC_6H_4CH_2OAc$ (7)	323
3-methyl-1-oxothiazolidin-4-one (CH_3 on N)	C_6H_6	H_2SO_4, ~40 min	5-C_6H_5-3-CH_3-thiazolidin-4-one (76)	36
	n-$C_6H_{13}CH{=}CH_2$	TFAA, CF_3CO_2H, 0°, 1 h	5-($CH_2CH{=}CHC_5H_{11}$-n)-3-CH_3-thiazolidin-4-one (62) $E:Z$ = 3:1	36
	n-$C_4H_9CH{=}CH_2$	"	5-($CH_2CH{=}CHC_3H_7$-n)-3-CH_3-thiazolidin-4-one (49) $E:Z$ = 2:1	36

TABLE IV. SULFOXIDES WITH A CARBON NUCLEOPHILE (*Continued*)

Sulfoxide	Nucleophile	Reaction Conditions	Product(s) and Yield(s) (%)	Refs.
C_5				
$CH_3S(O)C_4H_9$-*t*	*o*-$CH_3C_6H_4OH$	DCC, H_3PO_4, ether, overnight	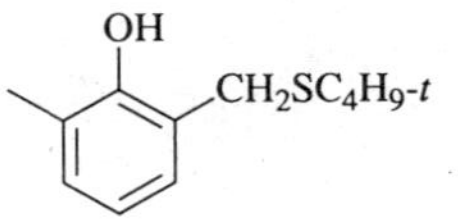 OH, $CH_2SC_4H_9$-*t* (27)	211
	o,o'-$(CH_3)_2C_6H_3OH$	DCC, H_3PO_4, C_6H_6, overnight	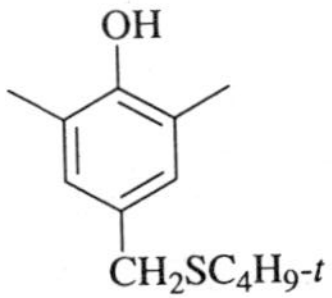 OH, $CH_2SC_4H_9$-*t* (2)	211
	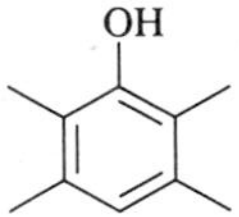 OH	DCC, Cl_2CHCO_2H, C_6H_6, overnight	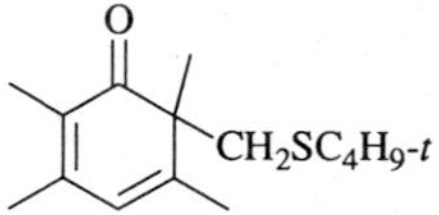 O, $CH_2SC_4H_9$-*t* (—)[a]	211
$CH_3S(O)CH_2CON(CH_3)_2$	*n*-$C_6H_{13}CH{=}CH_2$[b]	TFAA, CF_3CO_2H, 0°, 1 h	$CH_3SCH(CH_2CH{=}CHC_5H_{11}$-*n*$)_2$ $CON(CH_3)_2$ $E{:}Z$ = 82:18 (70)	47
$CH_3S(O)CH_2CO_2C_2H_5$	*n*-$C_nH_{2n+1}CH_2CH{=}CH_2$[b]	TFAA, 0°, 1 h	$CH_3SCH(CO_2C_2H_5)CH_2CH{=}CHC_nH_{2n+1}$-*n* n = 2–7 (72–79)	227
	C_6H_6	$SnCl_4$, 0° 30 min, rt 1 h	$CH_3SCH(CO_2C_2H_5)C_6H_5$ I (quant)	332
	C_6H_6	TsOH, reflux with H_2O removal, 1 h	I (88) + $(CH_3S)_2CHCO_2C_2H_5$ (—)	40
	$C_6H_5CH_3$	$SnCl_4$, 0° 20 min, rt 20 min	$CH_3SCH(CO_2C_2H_5)C_6H_4CH_3$-*p* (99)	332
	$C_6H_4CH_3$	TsOH, reflux with H_2O removal, 1 h	" (89)	40

		C_6H_5Cl	$SnCl_4$, 2 h	$CH_3SCH(CO_2C_2H_5)C_6H_4Cl$-*p* (83)	332
		C_6H_5Cl	TsOH, reflux with H_2O removal, 1 h	" (60)	40
		p-$CH_3C_6H_4CH_3$	$SnCl_4$, CH_2Cl_2, 0°, 30 min, rt 30 min	$CH_3SCH(CO_2C_2H_5)C_6H_3(CH_3)_2$-*o,m* (98)	332
		p-$CH_3C_6H_4CH_3$	TsOH, $(CH_2Cl)_2$, reflux with H_2O removal, 1 h	" (55)	40
		Thiophene	TsOH, 70°, 30 min	$CH_3SCH(CO_2C_2H_5)C_4H_3S$-2 (56)	40
		$C_{10}H_8$	$SnCl_4$, CH_2Cl_2, 0°, 10 min, rt, 20 min	$CH_3SCH(CO_2C_2H_5)C_{10}H_7$-1 (95)	332
		$C_{10}H_8$	TsOH, $(CH_2Cl)_2$, reflux with H_2O removal, 1 h	" (62)	40
		o-$ClC_6H_4OCH_2CH{=}CH_2$	TsOH, $(CH_2Cl)_2$, reflux with H_2O removal, 40 min	CH_3S, $CO_2C_2H_5$ on CH attached to benzene ring bearing Cl and $OCH_2CH{=}CH_2$ (74)	40
		$C_6H_5C_4H_9$-*i*	TsOH, $(CH_2Cl)_2$, reflux with H_2O removal, 1 h	CH_3S, $CO_2C_2H_5$ on CH attached to benzene ring bearing C_4H_9-*i* (58) + $(CH_3S)_2CHCO_2C_2H_5$ (9)	40
C_6					
	$[Cl(CH_2)_3]_2SO$	C_6H_5OH[b]	1. $SOCl_2$, CH_2Cl_2, −60°, 18 min 2. CH_2Cl_2, −55°, 70 min 3. $(C_2H_5)_3N$, −50°	$Cl(CH_2)_3SCH[(CH_2)_2Cl]C_6H_6OH$-*o* (61)	327

TABLE IV. SULFOXIDES WITH A CARBON NUCLEOPHILE (*Continued*)

Sulfoxide	Nucleophile	Reaction Conditions	Product(s) and Yield(s) (%)	Refs.
$CH_3S(O)CH_2PO(OC_2H_5)_2$	ArH	1. TFAA, CH_2Cl_2, 0°, 7 min 2. $SnCl_4$, 0°, 30 min	$CH_3SCHArPO(OC_2H_5)_2$	
	Ar = C_6H_6		Ar = C_6H_5 (82)	333, 334
	$CH_3C_6H_5$		p-$CH_3C_6H_4$ (86)	333, 334
	i-$C_3H_7C_6H_5$		p-(i-C_3H_7)C_6H_4 (92)	334
	1,3,5-$(CH_3)_3C_6H_3$		2,4,6-$(CH_3)_3C_6H_2$ (75)	334
	n-$C_4H_9C_6H_5$		p-(n-C_4H_9)C_6H_4 (91)	333
	i-$C_4H_9C_6H_5$		p-(i-C_4H_9)C_6H_4 (94)	334
	t-$C_4H_9C_6H_5$		p-(t-C_4H_9)C_6H_4 (94)	334
	p-$CH_3C_6H_4CH_3$		2,5-$(CH_3)_2C_6H_3$ (88)	334
	$CH_3OC_6H_5$		p-$CH_3OC_6H_4$ (85)	334
	1,2,4,5-$(CH_3)_4C_6H_2$		2,3,5,6-$(CH_3)_4C_6H$ (75)	334
	1,2,3,4-$(CH_3)_4C_6H_2$		2,3,4,5-$(CH_3)_4C_6H$ (72)	334
	p-$(C_2H_5)_2C_6H_4$		2,5-$(C_2H_5)_2C_6H_3$ (92)	334
C_7				
$C_6H_5S(O)CH_3$	o-$CH_3C_6H_4OH$	DCC, H_3PO_4, ether, reflux	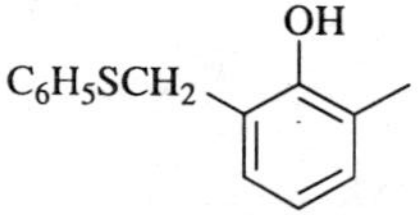 (10)	211, 221
	C_6H_5OH[b]	1. $C_6H_5OS(O)Cl$, CH_2Cl_2, −55°, 50 min 2. −50° 3. $(C_2H_5)_3N$, −50°	o-$(C_6H_5SCH_2)C_6H_4OH$ (76)	327
	C_6H_5OH[b]	1. $SOCl_2$, CH_2Cl_2, −60°, 18 min 2. CH_2Cl_2, −55°, 70 min 3. $(C_2H_5)_3N$, −50°	" (65)	327

	Sulfoxide	Substrate	Conditions	Product(s) (% Yield)	Refs.
	p-$ClC_6H_4S(O)CH_3$	$RCH_2CH{=}CH_2$[b]	1. TFAA, CH_2Cl_2 2. CF_3CO_2H, 0°	p-$ClC_6H_4S(CH_2)_2CH{=}CHR$ R = C_2H_5 $E:Z$ = 85:15 (86) C_4H_9-n (77–83) C_5H_{11}-n (77–83) C_7H_{15}-n (77–83) $(CH_2)_7CO_2CH_3$ (77–83)	335
	$CH_3S(O)CH_2CONHC_4H_9$-s	n-$C_6H_{13}CH{=}CH_2$[b]	TFAA, CF_3CO_2H, 0°, 1 h	$CH_3SCH(CONHC_4H_9$-s)-$CH_2CH{=}CHC_5H_{11}$-n $E:Z$ = 88:12 (81)	47
	$CH_3S(O)CH_2CON(CH_3)CH_2CH{=}CH_2$		1. TFAA, CH_2Cl_2 2. CF_3CO_2H, neat	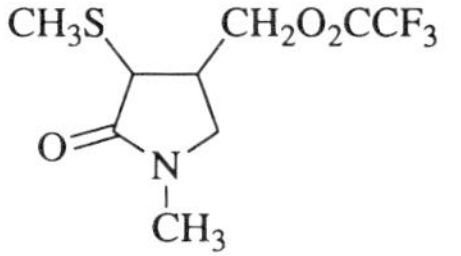 3-CH_3S-4-($CH_2O_2CCF_3$)-1-CH_3-pyrrolidin-2-one (9) 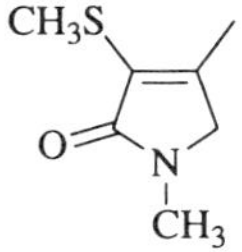 + 3-CH_3S-4-CH_3-1-CH_3-3-pyrrolin-2-one (39)	240
C_8	$C_6H_5CH_2S(O)CH_3$	o-$CH_3C_6H_4OH$	DCC, H_3PO_4, ether, overnight	2-[$C_6H_5CH(SCH_3)$]-6-$CH_3C_6H_3OH$ (34)	211
		2,6-$(CH_3)_2C_6H_3OH$	DCC, H_3PO_4, ether, DMF, overnight	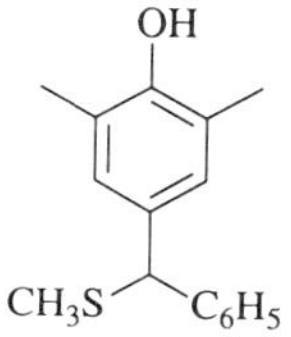 4-[$C_6H_5CH(SCH_3)$]-2,6-$(CH_3)_2C_6H_2OH$ (11)[a,c]	211

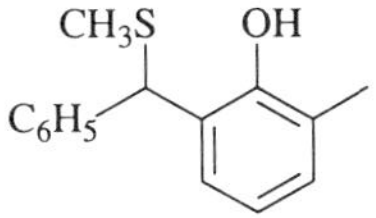

TABLE IV. SULFOXIDES WITH A CARBON NUCLEOPHILE (*Continued*)

Sulfoxide	Nucleophile	Reaction Conditions	Product(s) and Yield(s) (%)	Refs.
$[CH_3O_2C(CH_2)_2]_2SO$	RC_6H_4OH[b]	1. $SOCl_2$, CH_2Cl_2, $-60°$, 18 min 2. CH_2Cl_2, $-55°$, 70 min 3. $(C_2H_5)_3N$, $-50°$	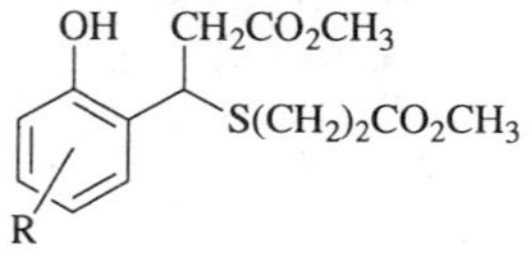 $2\text{-HO-}(R)C_6H_3CH(CH_2CO_2CH_3)S(CH_2)_2CO_2CH_3$ (structure)	326
			R = H (84)	
			o-CH_3 (53)	
			o-OCH_3 (42)	
			o-Cl (19)[a]	
			o-OAc (60)	
			o-CO_2CH_3 (41)	
			o-NO_2 (53)	
			p-CH_3 (70)	
			p-OCH_3 (39)	
			p-Cl (30)	
			p-OAc (77)	
			p-CO_2CH_3 (13)	

Reactant	Conditions	Product(s) (Yield, %)	Refs.
		p-NO_2 (45)	
		[2-hydroxy-4-R-phenyl]$CH(CH_2CO_2CH_3)S(CH_2)_2CO_2CH_3$ I +	326
		[2-hydroxy-6-R-phenyl]$CH(CH_2CO_2CH_3)S(CH_2)_2CO_2CH_3$ II	
		R = m-Cl I (3) + II (3)	
		m-OAc I (23) + II (23)	
		m-CO_2CH_3 I (20) + II (22)	
		m-NO_2 I (7)	
$CH_2{=}C(CH_3)CH_2N(CH_3)COCH_2S(O)CH_3$	TFAA, CH_2Cl_2	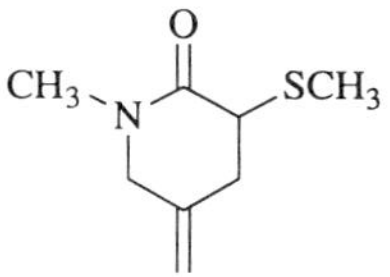 1-methyl-3-(methylthio)-5-methylene-2-piperidinone (43) +	240
		1-methyl-3-(methylthio)-5-methyl-3,4-dihydro-2-pyridinone (35)	
$CH_3CH{=}CHCH_2N(CH_3)COCH_2S(O)CH_3$	TFAA, CH_2Cl_2	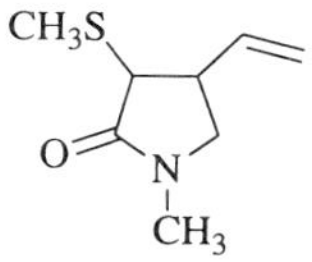 1-methyl-3-(methylthio)-4-vinyl-2-pyrrolidinone (92)	240

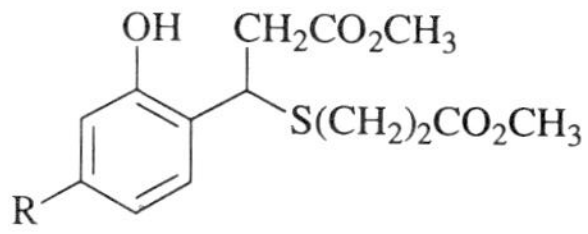

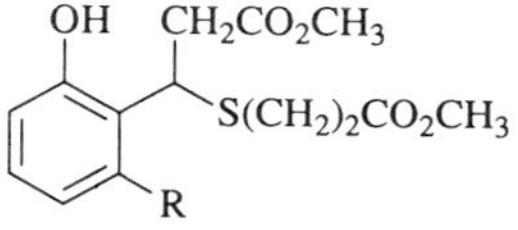

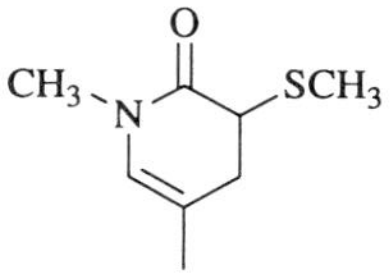

TABLE IV. SULFOXIDES WITH A CARBON NUCLEOPHILE (*Continued*)

	Sulfoxide	Nucleophile	Reaction Conditions	Product(s) and Yield(s) (%)	Refs.
	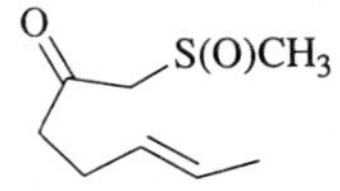		TFAA; CH_2Cl_2, 0° 1 h, rt 2 h	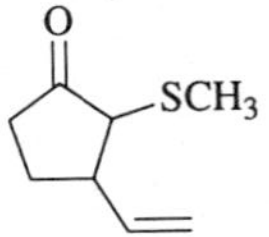 (78)	314
	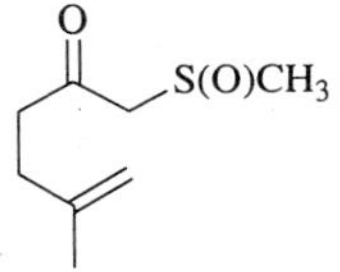		TFAA, CH_2Cl_2, 0°	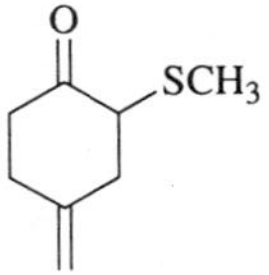 (33) + 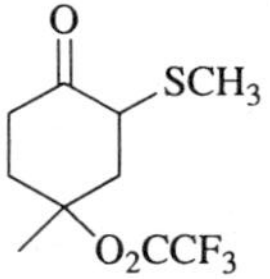 (24)	314
C_9	$CH_3S(O)CH_2COC_6H_5$	C_6H_6	$SnCl_4$, CH_2Cl_2, 0°, 30 min	$CH_3SCH(COC_6H_5)C_6H_5$ (quant)	332
		$CH_3C_6H_5$	"	$CH_3SCH(COC_6H_5)C_6H_4CH_3$-*p* (92)	332
		p-$CH_3C_6H_4CH_3$	"	$CH_3SCH(COC_6H_5)C_6H_3(CH_3)_2$-2,3 (quant)	332
		1,3,5-$(CH_3)_3C_6H_3$	"	$CH_3SCH(COC_6H_5)C_6H_2(CH_3)_3$-2,4,6 (98)	332
		C_6H_5Cl	"	$CH_3SCH(COC_6H_5)C_6H_4Cl$ *o*:*p* = 1:3 (77)	332
		$C_{10}H_8$	"	$CH_3SCH(COC_6H_5)C_{10}H_7$-1 (quant)	332
		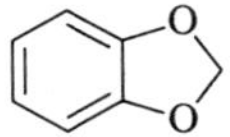	"	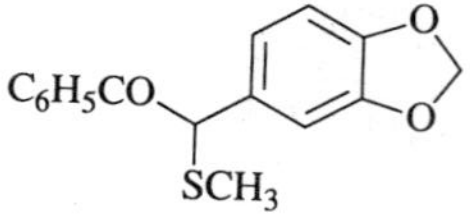 (90)	332

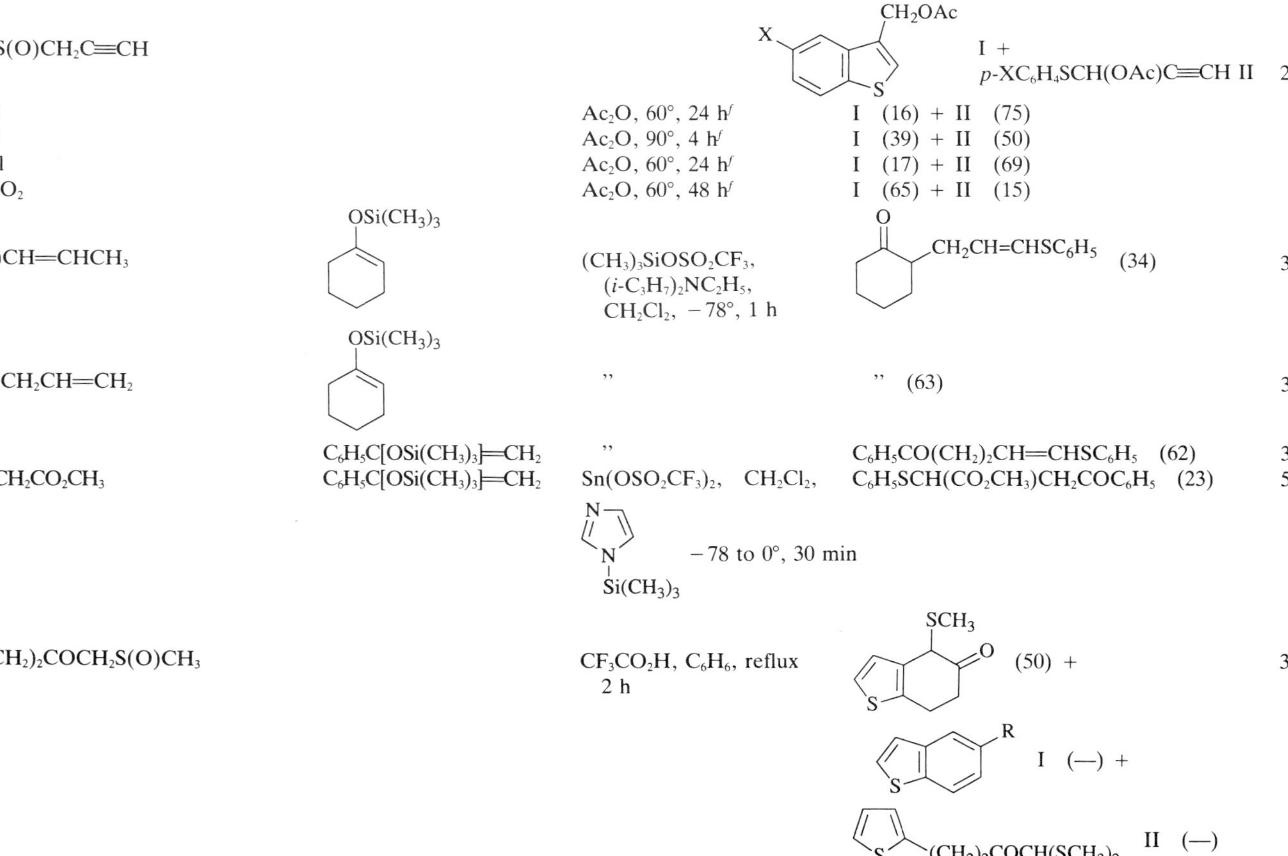

p-$XC_6H_4S(O)CH_2C{\equiv}CH$			[structure: 5-X-3-(CH_2OAc)benzo[b]thiophene] I + p-$XC_6H_4SCH(OAc)C{\equiv}CH$ II	265[f]
X = H		Ac_2O, 60°, 24 h[f]	I (16) + II (75)	
X = H		Ac_2O, 90°, 4 h[f]	I (39) + II (50)	
X = Cl		Ac_2O, 60°, 24 h[f]	I (17) + II (69)	
X = NO_2		Ac_2O, 60°, 48 h[f]	I (65) + II (15)	
$C_6H_5S(O)CH{=}CHCH_3$	[structure: 1-$OSi(CH_3)_3$-cyclohexene]	$(CH_3)_3SiOSO_2CF_3$, $(i\text{-}C_3H_7)_2NC_2H_5$, CH_2Cl_2, −78°, 1 h	[structure: 2-($CH_2CH{=}CHSC_6H_5$)cyclohexanone] (34)	336
$C_6H_5S(O)CH_2CH{=}CH_2$	[structure: 1-$OSi(CH_3)_3$-cyclohexene]	"	" (63)	336
	$C_6H_5C[OSi(CH_3)_3]{=}CH_2$	"	$C_6H_5CO(CH_2)_2CH{=}CHSC_6H_5$ (62)	336
$C_6H_5S(O)CH_2CO_2CH_3$	$C_6H_5C[OSi(CH_3)_3]{=}CH_2$	$Sn(OSO_2CF_3)_2$, CH_2Cl_2, [structure: 1-$Si(CH_3)_3$-imidazole], −78 to 0°, 30 min	$C_6H_5SCH(CO_2CH_3)CH_2COC_6H_5$ (23)	59
2-$C_4H_3S(CH_2)_2COCH_2S(O)CH_3$		CF_3CO_2H, C_6H_6, reflux 2 h	[structure: 4-SCH_3-6,7-dihydrobenzo[b]thiophen-5(4H)-one] (50) + [structure: 5-R-benzo[b]thiophene] I (—) + [structure: 2-thienyl]$(CH_2)_2COCH(SCH_3)_2$ II (—)	38, 337

TABLE IV. SULFOXIDES WITH A CARBON NUCLEOPHILE (*Continued*)

Sulfoxide	Nucleophile	Reaction Conditions	Product(s) and Yield(s) (%)	Refs.
		TsOH, CH_3CN, reflux 2 h	I, R = OH (56), R = SCH_3 (14) + II (7)	38, 337
		TsOH, CH_3OH, reflux	I, R = OCH_3 (14) + II (5)	38, 337
		CCl_3CO_2H, CH_3OH, CH_3CN, reflux 4 h	I, R = OCH_3 (22)	38, 337
$CH_2{=}C(CH_3)C(CH_3)_2COCH_2S(O)CH_3$		TsOH, C_6H_6, reflux with H_2O removal, 2 h	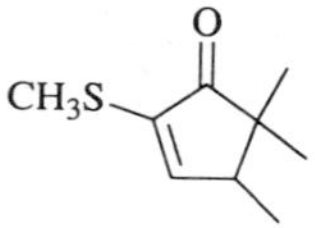 CH_3S [structure] (58)	39
$AcOCH(OCH_3)C(CH_3)_2CH_2S(O)CH_3$	$(CH_3)_2C{=}CHOCH_3$	1. Ac_2O, $BF_3 \cdot$ ether, 100°, 5 min 2. H_2SO_4, H_2O, 2 d	$[OHCC(CH_3)_2CH_2]_2S$ (—)	323
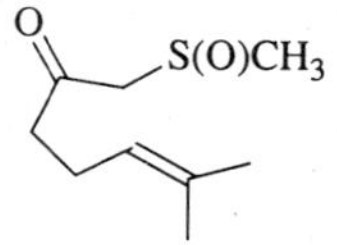 [structure] $S(O)CH_3$		1. TFAA, CH_2Cl_3, 0°, 1 h, rt 4 h 2. K_2CO_3, CH_3OH, H_2O, 1 h	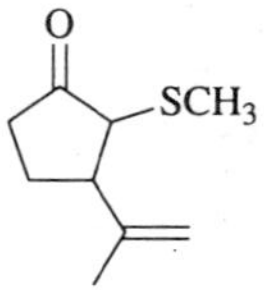 [structure] SCH_3 (52) 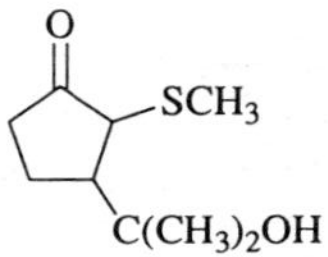 + [structure] SCH_3, $C(CH_3)_2OH$ (>12)	314
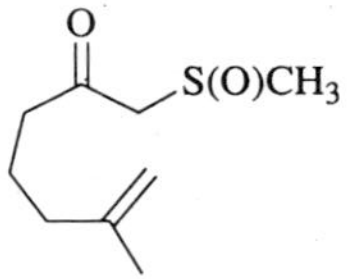 [structure] $S(O)CH_3$		TFAA, CH_2Cl_2, 0°	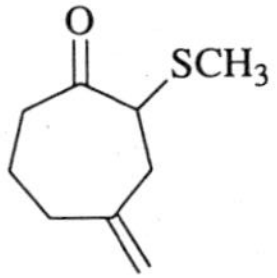 [structure] SCH_3 (54)	314

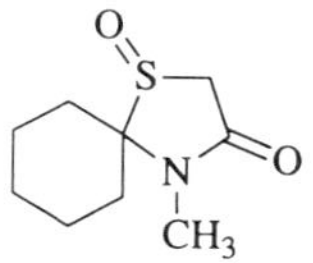 C_{10} (structure: O=S, N, O, CH_3)	C_6H_6	H_2SO_4, ~40 min	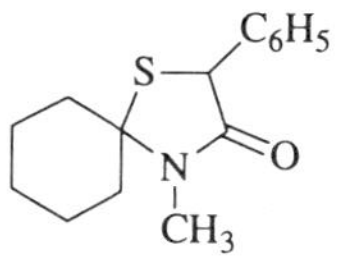 (structure: S, C_6H_5, N, O, CH_3) (95)	36
$C_6H_5S(O)CH_2C(CH_3){=}CH_2$	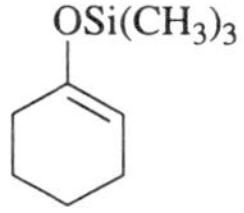 (structure: $OSi(CH_3)_3$)	$(i\text{-}C_3H_7)_2NC_2H_5$, CH_2Cl_2, $-78°$, 1 h	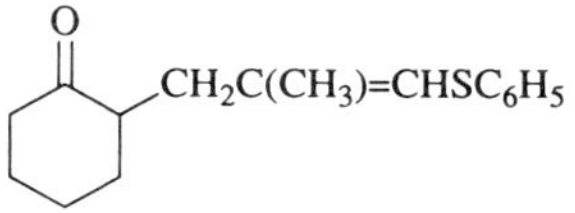 (structure: O, $CH_2C(CH_3){=}CHSC_6H_5$) (64)	336
	$C_6H_5C[OSi(CH_3)_3]{=}CH_2$	"	$C_6H_5SCH{=}C(CH_3)(CH_2)_2COC_6H_5$ (72)	336
$C_6H_5S(O)CH(CH_3)CH{=}CH_2$	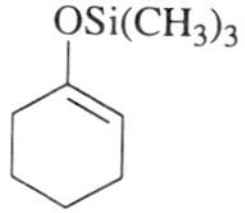 (structure: $OSi(CH_3)_3$)	"	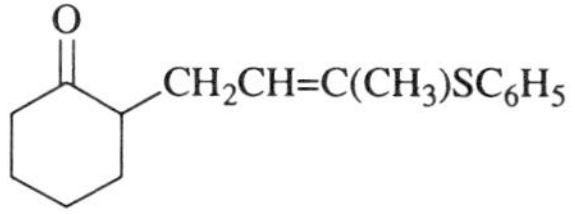 (structure: O, $CH_2CH{=}C(CH_3)SC_6H_5$) (24)	336
$C_6H_5S(O)CH_2CH{=}CHCH_3$	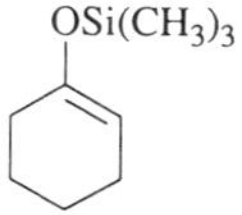 (structure: $OSi(CH_3)_3$)	"	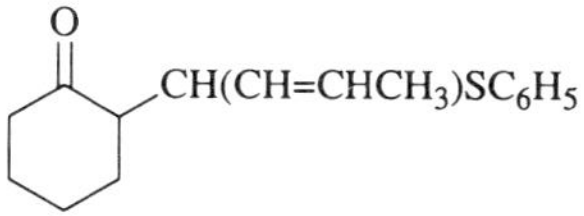 (structure: O, $CH(CH{=}CHCH_3)SC_6H_5$) (23) + $C_6H_5SCH{=}CHCH{=}CH_2$ (36)	336
(structure: $N(CH_3)COCH_2S(O)CH_3$)		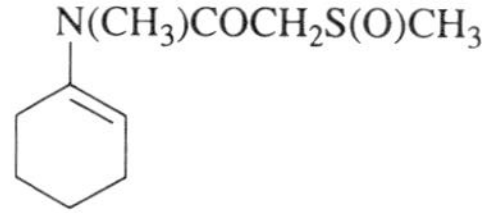 TFAA, CH_2Cl_2, 0°	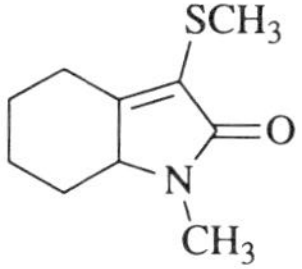 (structure: SCH_3, O, N, CH_3) (22)	338
(structure: $COCH_2S(O)CH_3$)		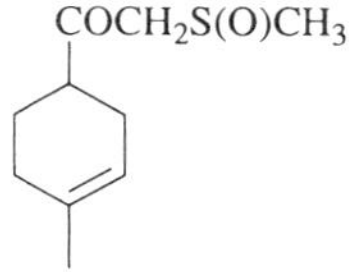 1. TFAA, C_6H_6, 5°, 30 min 2. H_2O, $NaHCO_3$	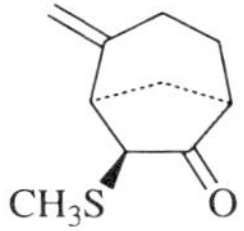 (structure: CH_3S, O) (—)	339

TABLE IV. SULFOXIDES WITH A CARBON NUCLEOPHILE (*Continued*)

	Sulfoxide	Nucleophile	Reaction Conditions	Product(s) and Yield(s) (%)	Refs.
	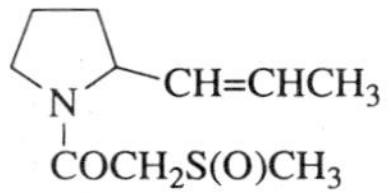 CH=CHCH$_3$ N COCH$_2$S(O)CH$_3$		TFAA, CH_2Cl_2	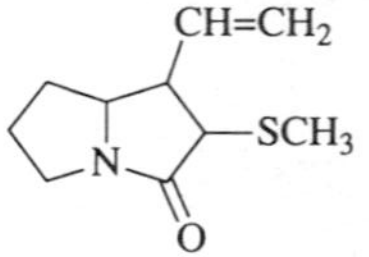 CH=CH$_2$ N —SCH$_3$ (87) O	340
C_{11}	—C(CH$_3$)$_2$COCH$_2$S(O)CH$_3$		TsOH, C_6H_6, reflux with H_2O removal, 2 h	SCH$_3$ =O (40)	39
	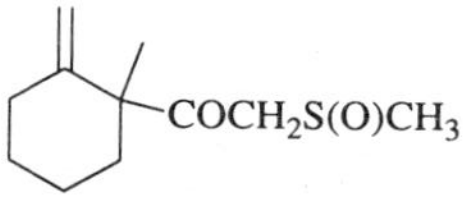 —COCH$_2$S(O)CH$_3$		"	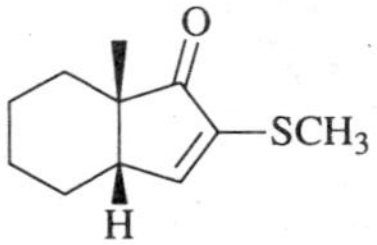 O —SCH$_3$ (34) H	39
	2-(C_4H_3S)CH_2CH(NHAc)$COCH_2S$(O)CH_3		TsOH, CH_3CN, reflux 2 h	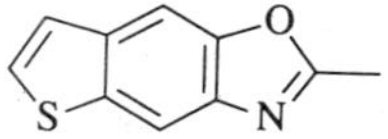 O S N (53)	38, 337
	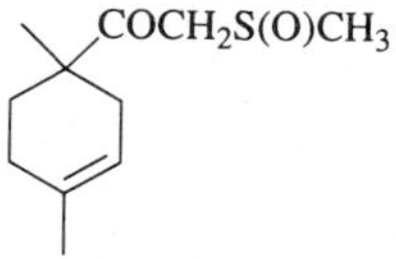 COCH$_2$S(O)CH$_3$		1. TsOH, C_6H_6, 5°, 30 min 2. H_2O, $NaHCO_3$	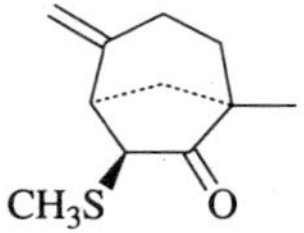 CH$_3$S O (—)	339
	C_6H_5S(O)CH_2PO(OC_2H_5)$_2$	C_6H_6	1. TFAA, CH_2Cl_2, 0°, 4–5 min 2. $SnCl_4$, 0°, 1 h	C_6H_5SCH(C_6H_5)PO(OC_2H_5)$_2$ (87)	333
		p-$CH_3C_6H_4CH_3$	"	C_6H_5SCH[C_6H_3(CH_3)$_2$-2,3]PO(OC_2H_5)$_2$ (84)	333
		1,3,5-(CH_3)$_3C_6H_3$	"	C_6H_5SCH[C_6H_2(CH_3)$_3$-2,4,6]PO(OC_2H_5)$_2$ (79)	333
		$C_6H_5C_3H_7$-i	"	C_6H_5SCH[C_6H_4(C_3H_7-i)-p]PO(OC_2H_5)$_2$ (85)	333

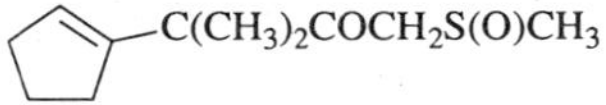

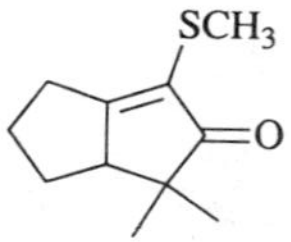

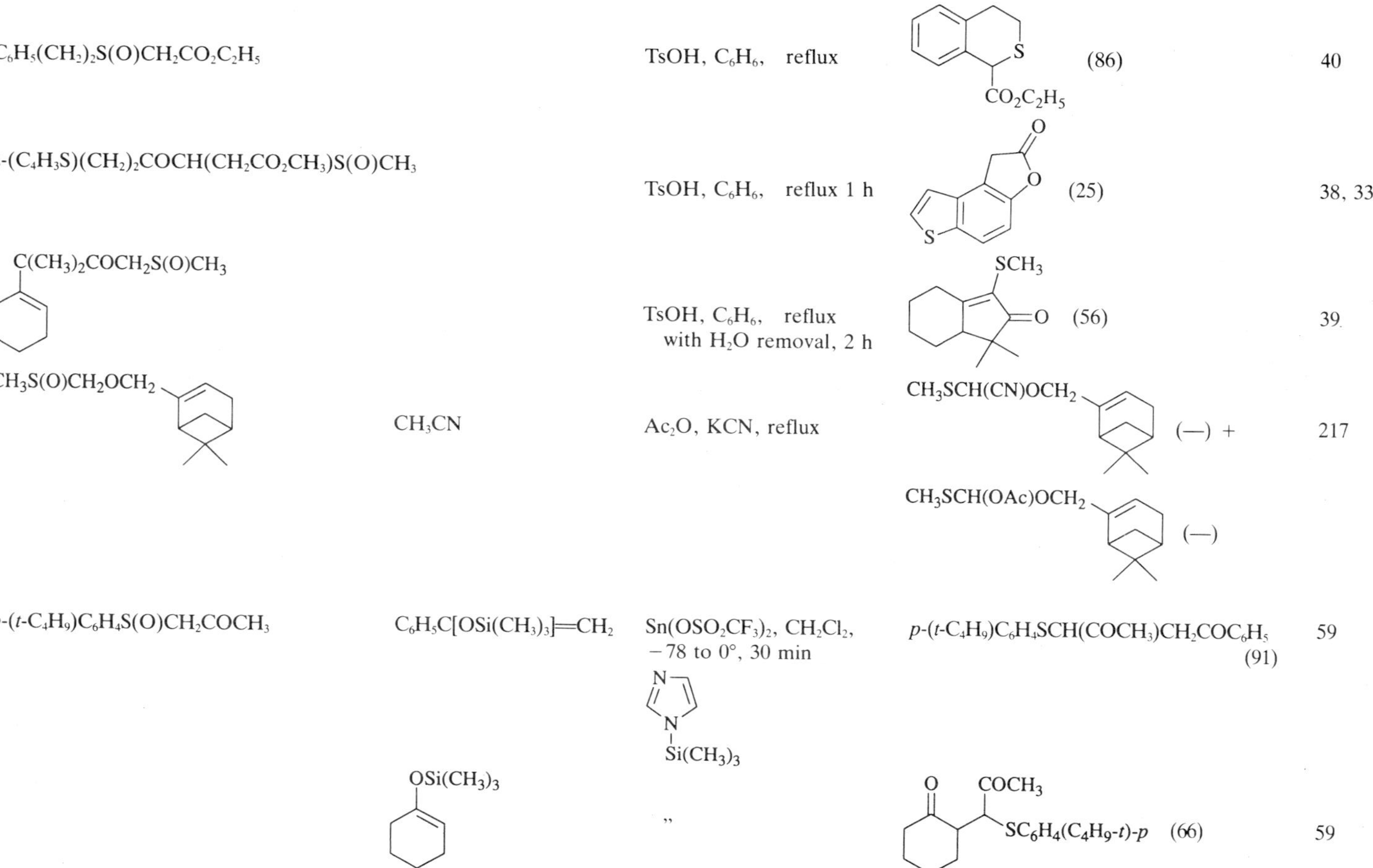

	Reactant	Co-reactant	Conditions	Product(s) and Yield(s) (%)	Refs.
C_{12}	$C_6H_5(CH_2)_2S(O)CH_2CO_2C_2H_5$		TsOH, C_6H_6, reflux	$CO_2C_2H_5$ (86)	40
	2-$(C_4H_3S)(CH_2)_2COCH(CH_2CO_2CH_3)S(O)CH_3$		TsOH, C_6H_6, reflux 1 h	(25)	38, 337
	$C(CH_3)_2COCH_2S(O)CH_3$		TsOH, C_6H_6, reflux with H_2O removal, 2 h	SCH_3 (56)	39
	$CH_3S(O)CH_2OCH_2$	CH_3CN	Ac_2O, KCN, reflux	$CH_3SCH(CN)OCH_2$ (—) + $CH_3SCH(OAc)OCH_2$ (—)	217
C_{13}	p-(t-C_4H_9)$C_6H_4S(O)CH_2COCH_3$	$C_6H_5C[OSi(CH_3)_3]{=}CH_2$	$Sn(OSO_2CF_3)_2$, CH_2Cl_2, −78 to 0°, 30 min, N-$Si(CH_3)_3$-imidazole	p-(t-C_4H_9)$C_6H_4SCH(COCH_3)CH_2COC_6H_5$ (91)	59
		$OSi(CH_3)_3$	"	$COCH_3$, $SC_6H_4(C_4H_9$-$t)$-p (66)	59

TABLE IV. SULFOXIDES WITH A CARBON NUCLEOPHILE (*Continued*)

Sulfoxide	Nucleophile	Reaction Conditions	Product(s) and Yield(s) (%)	Refs.
	$C_2H_5C[OSi(CH_3)_3]{=}CHCH_3$	"	p-$(t$-$C_4H_9)C_6H_4SCH(COCH_3)CH(CH_3)COC_2H_5$ (52)	59
$(CH_2)_2COCH_2S(O)CH_3$, OCH_3, OCH_3 (benzene ring)		CF_3CO_2H, C_6H_6, reflux 1 h	CH_3O, CH_3O, O, SCH_3 (tetralone) (72)	37, 341, 342
		CH_3CN, CCl_3CO_2H, reflux 1 h	" (70)	341, 342
		CCl_4, $CHCl_2CO_2H$, 60°, 2.5 h	" (30) + 3,4-$(CH_3O)_2C_6H_3(CH_2)_2COCH(SCH_3)O_2CCHCl_2$ (22)	
			CH_3O, CH_3O, R (naphthalene); R	
		TsOH, CH_3CN, reflux 1 h	SCH_3 (3), OH (50)	341, 342
		TsOH, CH_3CN, Ac_2O, 17 h	SCH_3 (30), OH (37)	341, 342
		TsOH, CH_3OH, CH_3CN, reflux 1 h	SCH_3 (4), OH (29), OCH_3 (40)	341, 342
		TsOH, C_2H_5SH, C_6H_6, reflux 1 h	$SCH_3 : SC_2H_5$ = 1:1.83	341, 342
		CCl_3CO_2H, CH_3OH, CH_3CN, reflux 4 h	OCH_3 (55), OH (6)	341, 342
		CCl_3CO_2H, C_2H_5OH, CH_3CN, reflux 4 h	OC_2H_5 (51), OH (6)	341, 342

		TsOH, $AcOC_2H_5$, 36 h	SCH_3 (22), OC_2H_5 (40), OH (4)	343
		TsOH, $AcOCH_3$, 18 h	SCH_3 (30), OCH_3 (20), OH (5)	343
		TsOH, i-C_3H_7OH, 20 h	SCH_3 (17), OC_3H_7-i (20)	343
		CH_3SO_3H, C_2H_5OH, 20 h	SCH_3 (24), OC_2H_5 (36), OH (5)	343
		TsOH, $TsOCH_3$, CH_2Cl_2, 20 h	SCH_3 (56), OCH_3 (23), H (14)	343
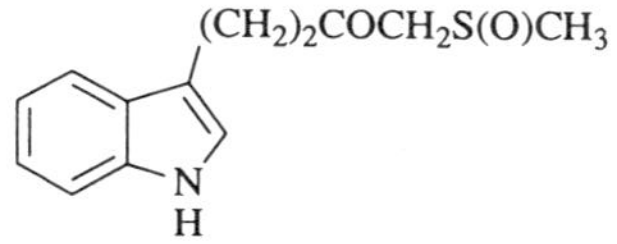 $(CH_2)_2COCH_2S(O)CH_3$ N H		CH_2Cl_2, CCl_3CO_2H (0.5 eq), reflux 2.5 h	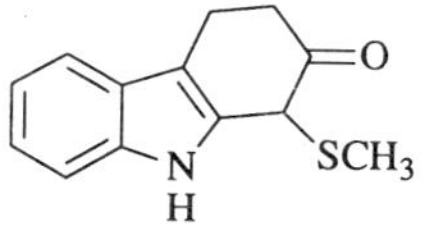 O N H SCH_3 (68)	38, 341
		TsOH, dioxane, reflux	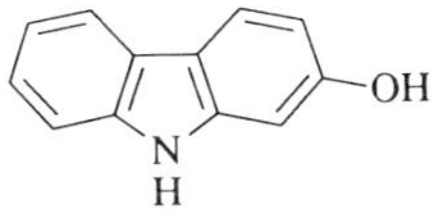 OH N H (55)	38, 337, 341
C_{14}				
$(C_6H_5CH_2)_2SO$	o-$CH_3C_6H_4OH$	DCC, H_3PO_4, ether, overnight	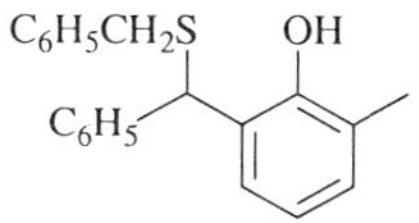 $C_6H_5CH_2S$ OH C_6H_5 (26)	212, 221
	C_6H_5OH[b]	1. $C_6H_5OS(O)Cl$, CH_2Cl_2, 2. $-50°$ 3. $(C_2H_5)_3N$, $-50°$	o-$C_6H_5CH_2SCH(C_6H_5)C_6H_4OH$ (75)	326
	C_6H_5OH[b]	1. $SOCl_2$, CH_2Cl_2, $-60°$, 18 min 2. CH_2Cl_2, $-55°$, 70 min 3. $(C_2H_5)_3N$, $-50°$	" (68)	326
$C_6H_5S(O)(CH_2)_2C_6H_5$	$CH_3OC[OSi(CH_3)_3]{=}C(CH_3)_2$	ZnI_2, CH_3CN, 15 min	$C_6H_5SCH(CH_2C_6H_5)C(CH_3)_2CO_2CH_3$ (55)	344
	$CH_3OC[OSi(CH_3)_3]{=}CHCH_3$	"	$C_6H_5SCH(CH_2C_6H_5)CH(CH_3)CO_2CH_3$ (56)	344

TABLE IV. SULFOXIDES WITH A CARBON NUCLEOPHILE (*Continued*)

Sulfoxide	Nucleophile	Reaction Conditions	Product(s) and Yield(s) (%)	Refs.
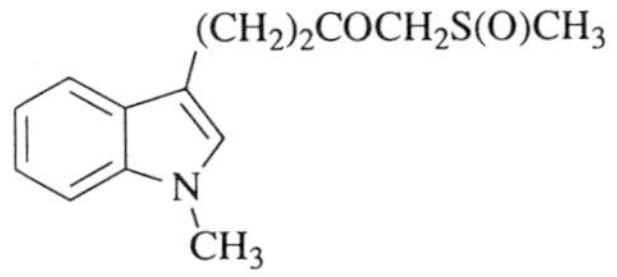 1-CH_3-indol-3-yl-$(CH_2)_2COCH_2S(O)CH_3$		CCl_3CO_2H, C_6H_6, reflux	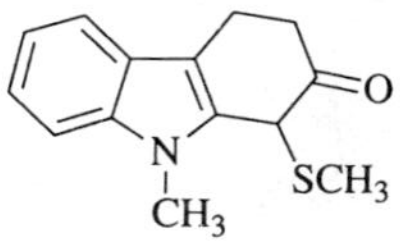 N-CH_3, SCH_3, =O (60)	38, 341
		CF_3CO_2H, C_6H_6, reflux	" (74)	38, 341
		TsOH, THF, reflux 2 h	" (82)	38, 337, 341
			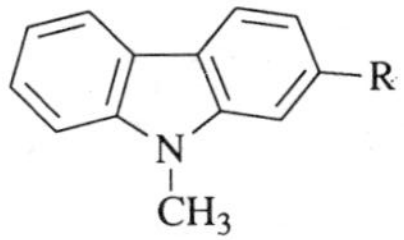 N-CH_3, R	
		TsOH, CH_3CN, 50°, 1.5 h	R = OH (34), R = SCH_3 (22)	38, 341
		TsOH, dioxane, reflux 3 h	R = OH (54)	38, 337, 341
		TsOH, CH_3OH, acetone, reflux 5 h	R = OCH_3 (47)	38, 337, 341
		TsOH, C_2H_5OH, acetone, reflux 5 h	R = OC_2H_5 (40)	38, 337, 341
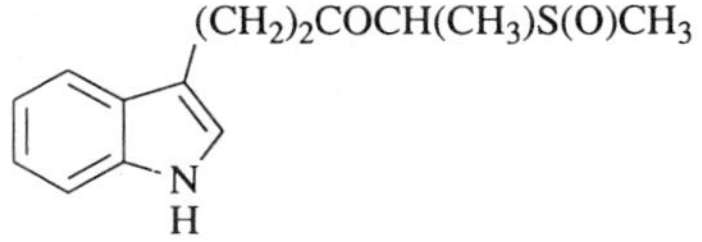 1H-indol-3-yl-$(CH_2)_2COCH(CH_3)S(O)CH_3$		$(CH_2Cl)_2$, CCl_3CO_2H, reflux 2 h	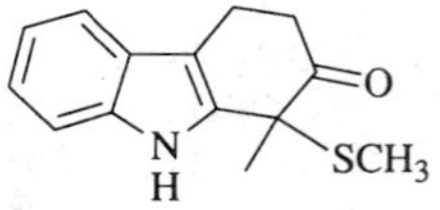 NH, SCH_3, =O (11)	38

Reactant	Reagent	Conditions	Product(s) and Yield(s) (%)	Refs.
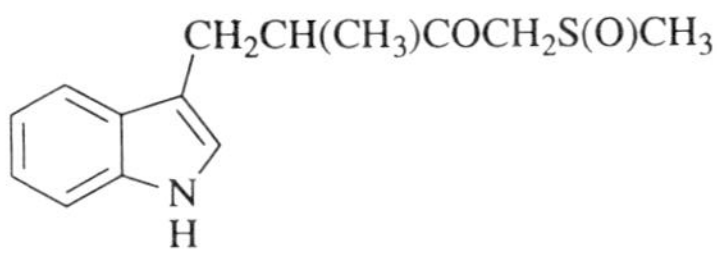 3-[$CH_2CH(CH_3)COCH_2S(O)CH_3$]indole (N–H)		"	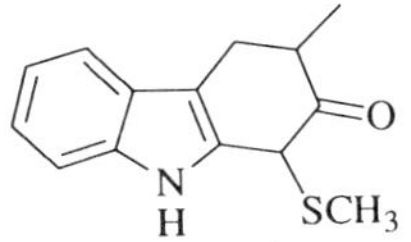 (45) SCH_3, O, N–H	38
		TsOH, CH_3CN, reflux 2 h	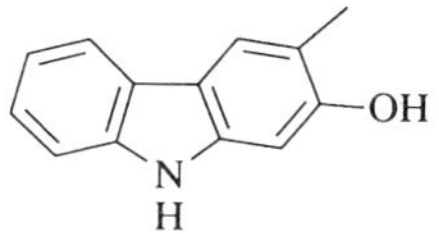 OH (50), N–H	38
p-$XC_6H_4S(O)CH_2COC_6H_5$	$C_6H_5C[OSi(CH_3)_3]{=}CH_2$	$Sn(OSO_2CF_3)_2$, CH_2Cl_2, 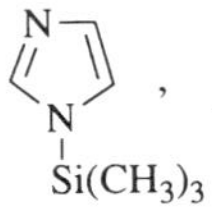 1-(trimethylsilyl)imidazole [N, N–$Si(CH_3)_3$], −78 to 0°, 30 min	p-$XC_6H_4SCH(COC_6H_5)CH_2COC_6H_5$ X = H (78), X = Cl (65)	59
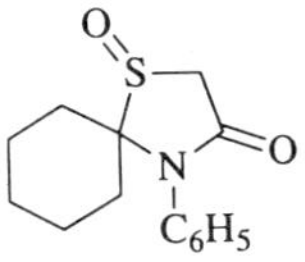 S-oxide spiro thiazolidinone (O=S, N–C_6H_5, =O)	C_6H_6	H_2SO_4, ~40 min	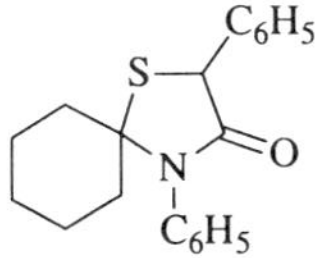 C_6H_5, S, N–C_6H_5, =O (93)	36
C_{15}				
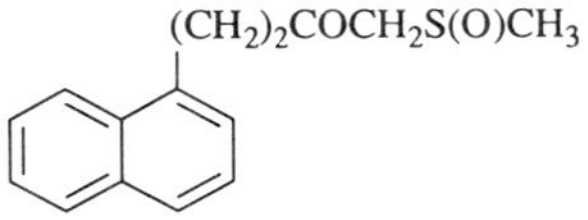 1-[$(CH_2)_2COCH_2S(O)CH_3$]naphthalene		CF_3CO_2H, CH_3CN, reflux 1 h	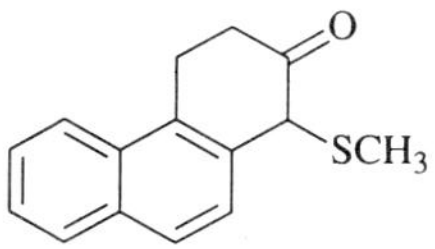 O, SCH_3 (27)	37

TABLE IV. SULFOXIDES WITH A CARBON NUCLEOPHILE (*Continued*)

Sulfoxide	Nucleophile	Reaction Conditions	Product(s) and Yield(s) (%)	Refs.
		TFAA, CH_3CN, reflux 1 h	" (58)	37
		TFAA	" (41)	341
		TsOH, CH_3CN, reflux 1.5 h	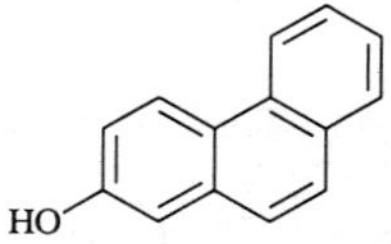 HO-phenanthrene (30) 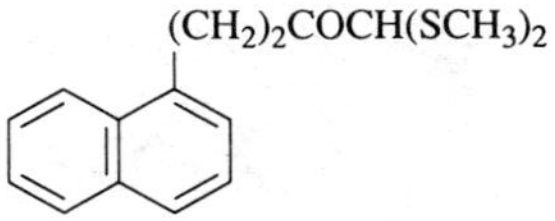 + 1-naphthyl-$(CH_2)_2COCH(SCH_3)_2$ (62)	37
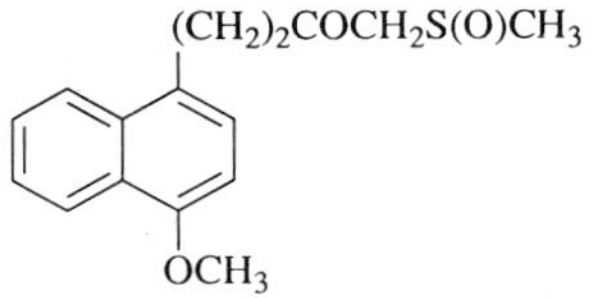 4-methoxy-1-naphthyl-$(CH_2)_2COCH_2S(O)CH_3$		TFAA, C_6H_6, 14 h	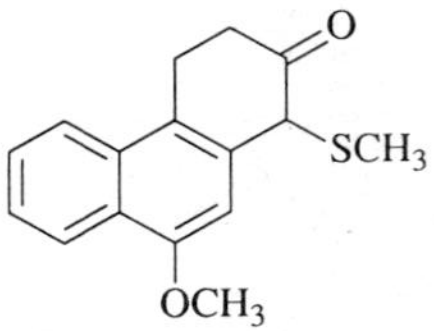 SCH_3, O, OCH_3 I (85)	345

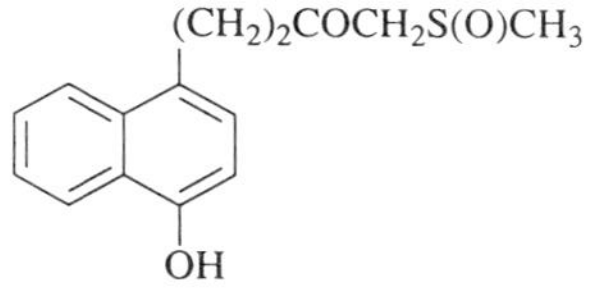	1. TFAA, C_6D_6, 1 min 2. H_2O, 2.5 h	I (20) + 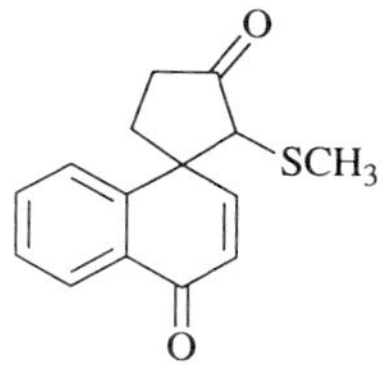 II (75)	345
	TFAA, C_6H_6, 40°, 5 h	II (80)	345
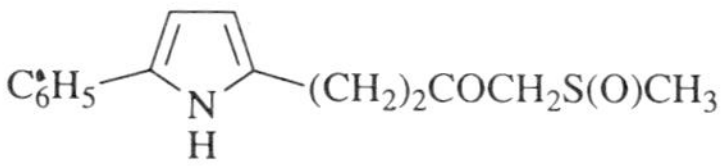	TsOH, THF, reflux 1 h	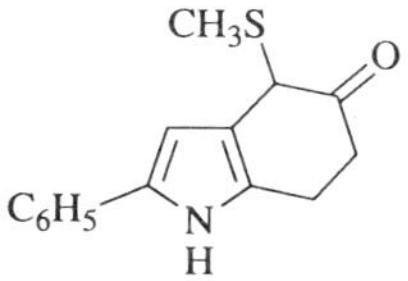 (63)	38, 337
	TsOH, CH_3OH, reflux 40 min	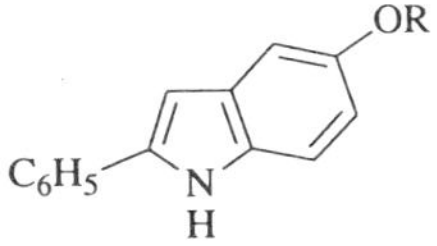 R = CH_3 (80)	38, 337
	TsOH, C_2H_5OH, reflux 40 min	" R = C_2H_5 (72)	38, 337
C_{16} 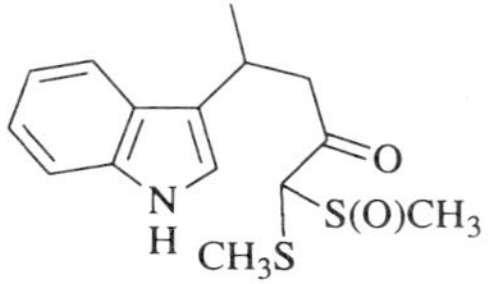	TsOH (anh), C_6H_6-THF (4:1), 60°	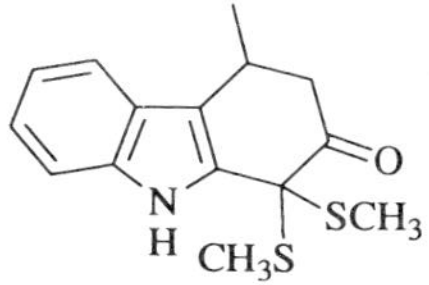 (81)	346

TABLE IV. Sulfoxides with a Carbon Nucleophile (*Continued*)

Sulfoxide	Nucleophile	Reaction Conditions	Product(s) and Yield(s) (%)	Refs.
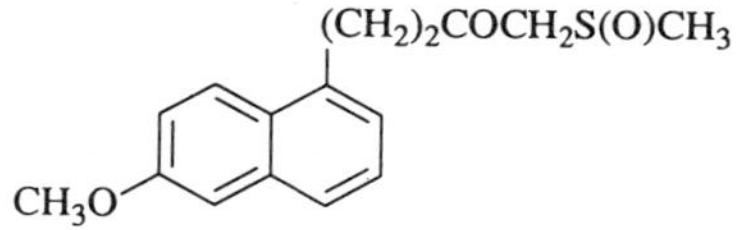 [structure: CH_3O-naphthalene bearing $(CH_2)_2COCH_2S(O)CH_3$]		TsOH, C_6H_6, 14 h	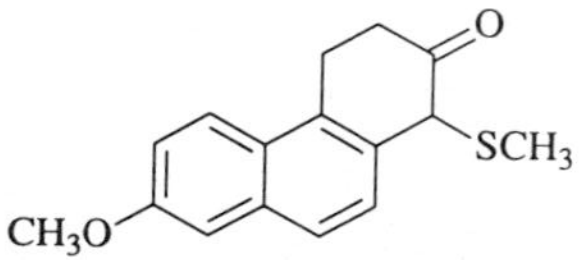 [structure: CH_3O, SCH_3, O] (80)	345
$C_6H_5(CH_2)_2S(O)CH_2COC_6H_5$		TsOH, C_6H_6, reflux	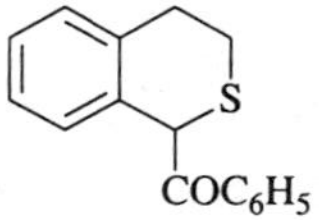 [structure: S, COC_6H_5] (78)	40
p-(t-C_4H_9)$C_6H_4S(O)CH_2COC_4H_9$-t		CH_2Cl_2, TFAA, 0°	" (82)	347, 348
	$C_6H_5[OSi(CH_3)_3]{=}CH_2$	$Sn(OTf)_2$, CH_2Cl_2, 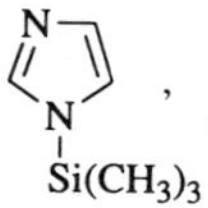 [structure: N-$Si(CH_3)_3$ imidazole], −78 to 0°, 30 min	p-(t-C_4H_9)$C_6H_4SCH(COC_4H_9$-t)-$CH_2COC_6H_5$ (65)	59
2-(C_4H_3S)$(CH_2)_2COCH(CONHC_6H_5)S(O)CH_3$		TsOH, C_6H_6, reflux 1 h	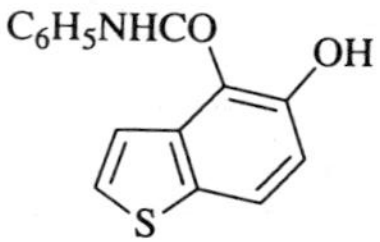 [structure: C_6H_5NHCO, OH, S] (55)	38, 337
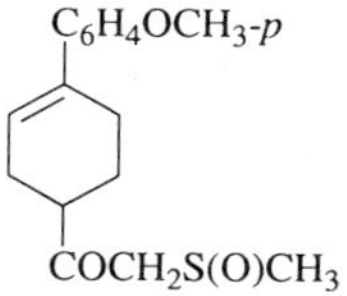 [structure: cyclohexene bearing $C_6H_4OCH_3$-p and $COCH_2S(O)CH_3$]		1. TFAA, C_6H_6, 5°, 8 h 2. H_2O, $NaHCO_3$	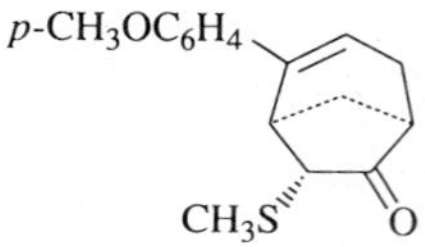 [structure: p-$CH_3OC_6H_4$, CH_3S, O] (70)	339

Reactant	Conditions	Product(s) and Yield(s) (%)	Refs.
3,4-$(CH_3O)_2C_6H_3(CH_2)_2COCH(CH_2CO_2CH_3)S(O)CH_3$	TsOH, CH_3CN, reflux 1 h	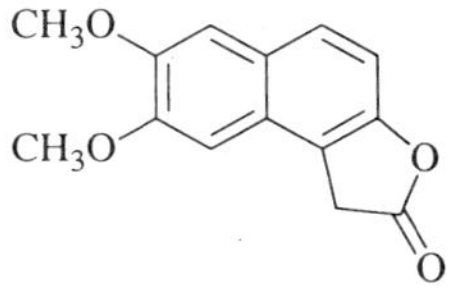 CH_3O, CH_3O (structure) (60)	37
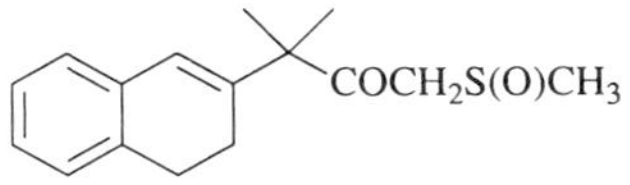 $COCH_2S(O)CH_3$ (structure)	TsOH, C_6H_6, reflux with H_2O removal, 2 h	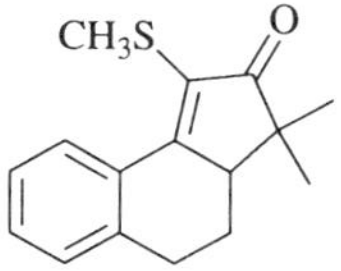 CH_3S (structure) (43)	39
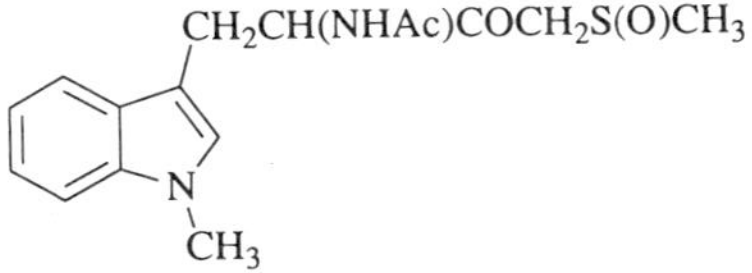 $CH_2CH(NHAc)COCH_2S(O)CH_3$, N–$CH_3$ (structure)	CF_3CO_2H, C_6H_6, reflux 1 h	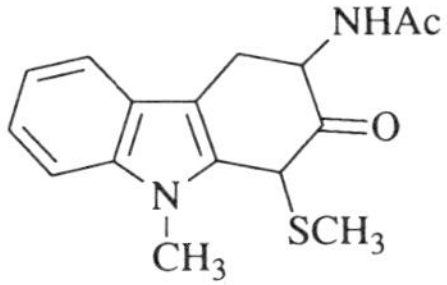 NHAc, SCH_3, N–CH_3 (structure) (66)	38, 341
	TsOH, C_6H_6, reflux	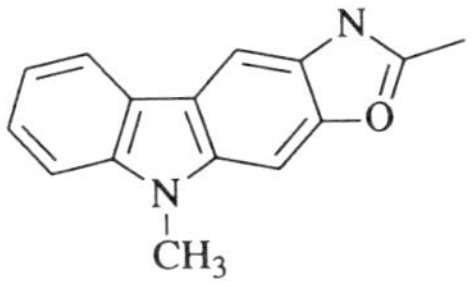 N–CH_3 (structure) (60)	38, 341
	TsOH, CH_3CN, reflux 3.5 h	" (80)	38, 337 341
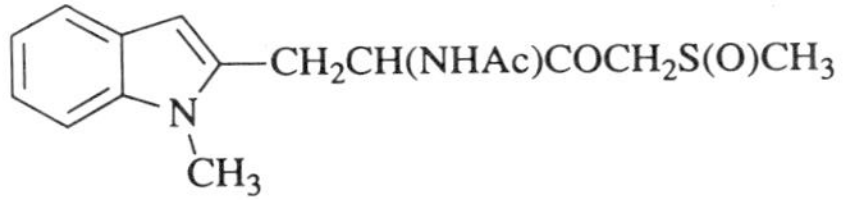 N–CH_3, $CH_2CH(NHAc)COCH_2S(O)CH_3$ (structure)	TsOH, CH_3CN, reflux 1 h	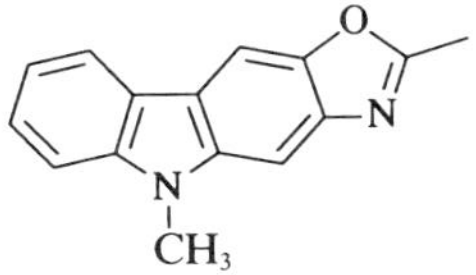 N–CH_3 (structure) (3)	38, 337
C_{17}			
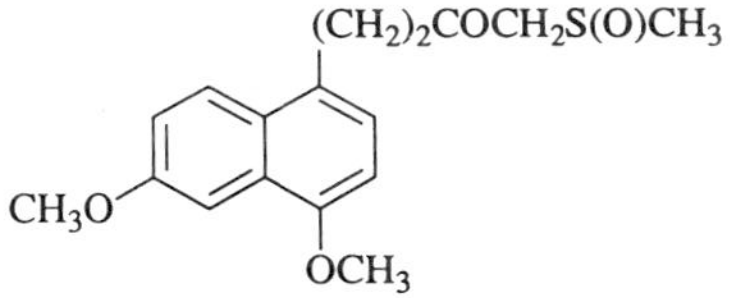 $(CH_2)_2COCH_2S(O)CH_3$, CH_3O, OCH_3 (structure)	TFAA, C_6H_6, 7 h	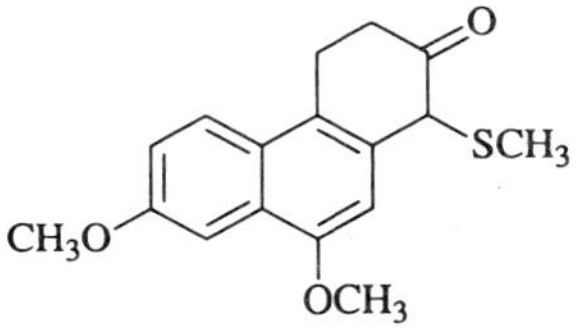 SCH_3, CH_3O, OCH_3 (structure) (80)	345

TABLE IV. SULFOXIDES WITH A CARBON NUCLEOPHILE (*Continued*)

	Sulfoxide	Nucleophile	Reaction Conditions	Product(s) and Yield(s) (%)	Refs.
	$C_6H_5(CH_2)_2S(O)CH_2COC_6H_4OCH_3$-*p*		TFAA, CH_2Cl_2, overnight	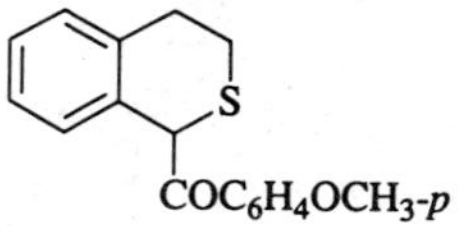 S; $COC_6H_4OCH_3$-*p* (47)	347, 348
C_{18}	$COCH_2S(O)CH_3$; CH_3O		1. TFAA, C_6H_6, 5°, 9 h 2. H_2O, $NaHCO_3$	O; CH_3O; SCH_3 (65)	339
	N; $CH_2S(O)CH_3$; CH_3O; OCH_3		TsOH, $(CH_2Cl)_2$, reflux with H_2O removal	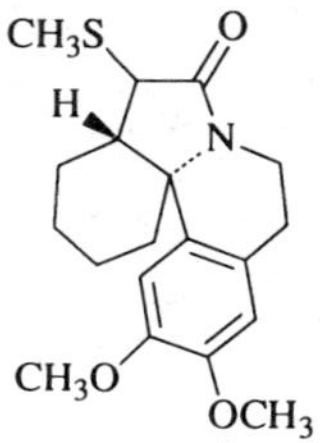 CH_3S; O; H; N; CH_3O; OCH_3 (60) + CH_3O; CH_3O; NH; O; CH_3S (8)	338

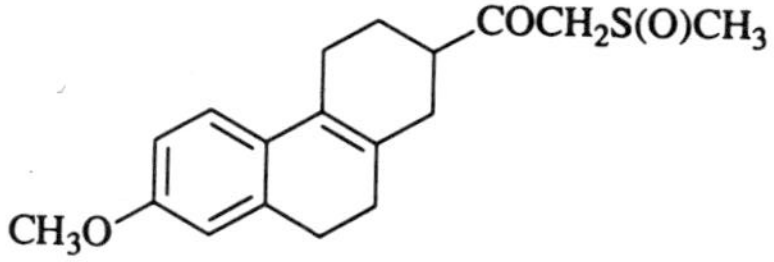

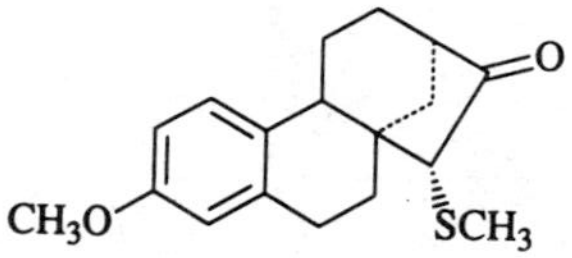

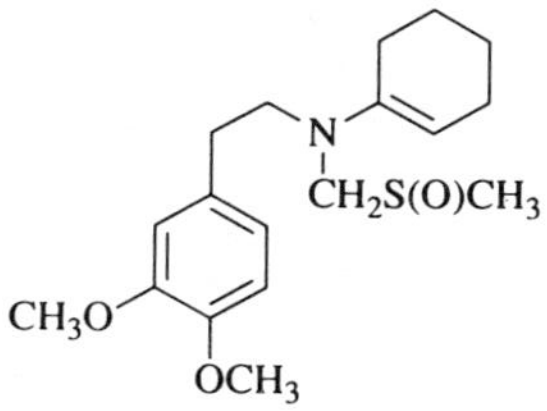

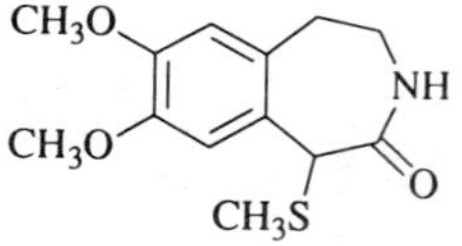

$p\text{-}(t\text{-}C_4H_9)C_6H_4S(O)CH_2COC_6H_5$	$C_6H_5C[OSi(CH_3)_3]{=}CH_2$	$Sn(OSO_2CF_3)_2$, base, CH_2Cl_2, 0°	$p\text{-}(t\text{-}C_4H_9)C_6H_4SCH(COC_6H_5)CH_2COC_6H_5$ (0–82) + $C_6H_5COCHOHCH_2COC_6H_5$ (0–16)	59
	$t\text{-}C_4H_9[OSi(CH_3)_3]{=}CH_2$	$Sn(OTf)_2$, CH_2Cl_2, 1-(trimethylsilyl)imidazole [N-Si(CH$_3$)$_3$], −78 to 0°, 30 min	$p\text{-}(t\text{-}C_4H_9)C_6H_4SCH(COC_6H_5)CH_2COC_4H_9\text{-}t$ (73)	59
	1-[$OSi(CH_3)_3$]cyclohexene	"	2-[$CH(COC_6H_5)SC_6H_4(C_4H_9\text{-}t)\text{-}p$]cyclohexanone (66)	59
	$i\text{-}C_3H_7[OSi(CH_3)_3]{=}CH_2$	"	$p\text{-}(t\text{-}C_4H_9)C_6H_4SCH(COC_6H_5)CH_2COC_3H_7\text{-}i$ (58)	59
	$C_2H_5C[OSi(CH_3)_3]{=}CHCH_3$	"	$p\text{-}(t\text{-}C_4H_9)C_6H_4SCH(COC_6H_5)CH(CH_3)COC_2H_5$ 1:1 diasteromeric mixture (74)	59
C_{19} 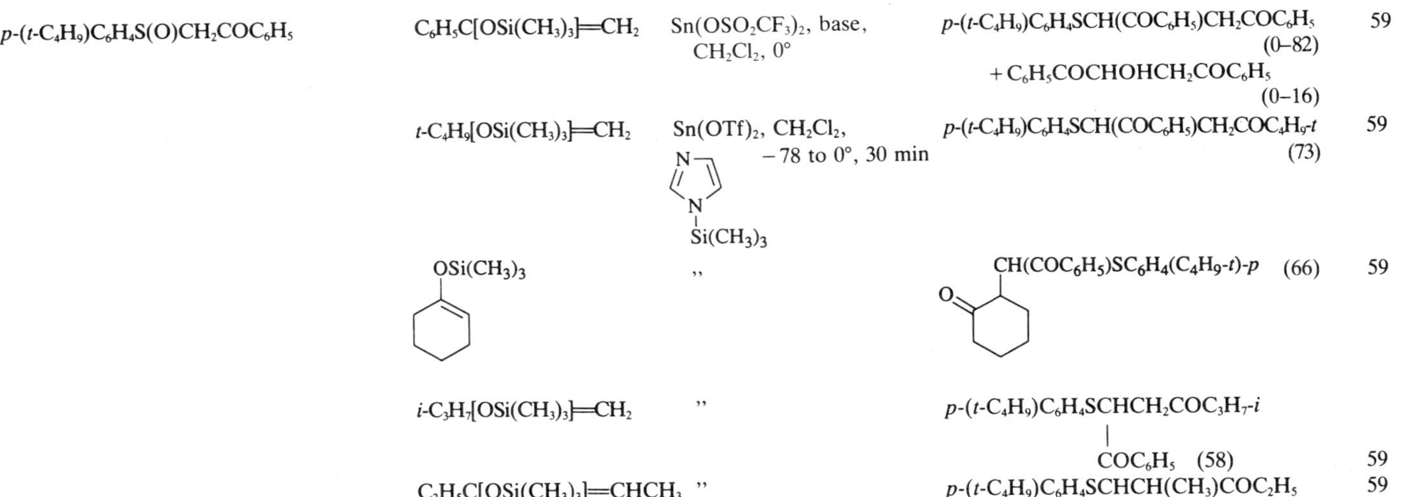		1. TFAA, C_6H_6, 5°, 2 h 2. H_2O, $NaHCO_3$	(76)	339

TABLE IV. SULFOXIDES WITH A CARBON NUCLEOPHILE (*Continued*)

	Sulfoxide	Nucleophile	Reaction Conditions	Product(s) and Yield(s) (%)	Refs.
	3-indolyl–$CH_2CH(CH_3)COCH(CH_2CH=C(CH_3)_2)S(O)CH_3$ (indole N–H)		TsOH, CH_3CN, reflux 3 h	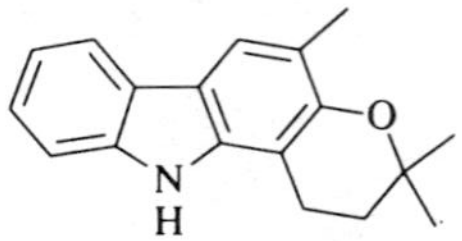 (23)	349
	1-methyl-3-indolyl–$(CH_2)_2COCH(CONHC_2H_5)S(O)CH_3$		TsOH, CH_3CN, reflux 2 h	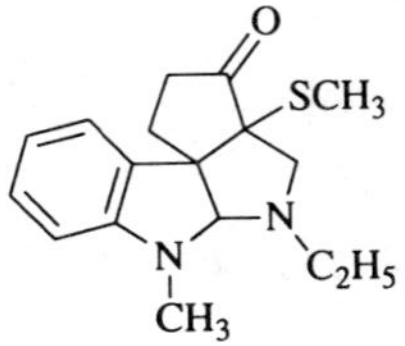 I (33) 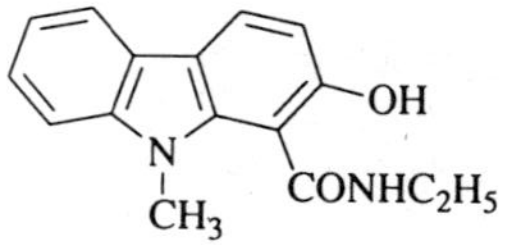 + (SCH_3, C_2H_5, CH_3, O) (41) (OH, $CONHC_2H_5$, CH_3)	38
			TsOH, dioxane, 50°, 5 h	I (40) 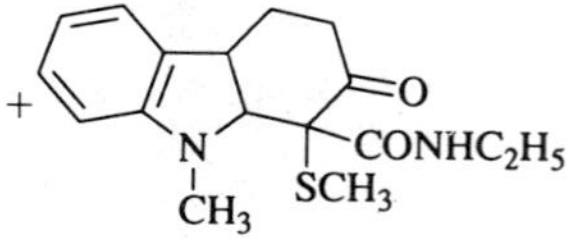 + ($CONHC_2H_5$, SCH_3, CH_3, O) (44)	38
C_{20} 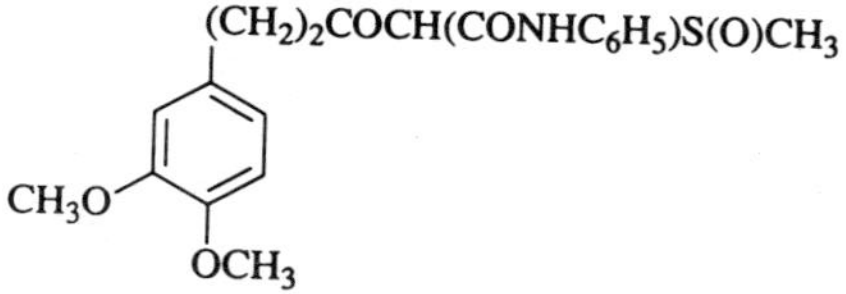	3,4-$(CH_3O)_2C_6H_3$–$(CH_2)_2COCH(CONHC_6H_5)S(O)CH_3$		TsOH, CH_3CN, reflux 1 h	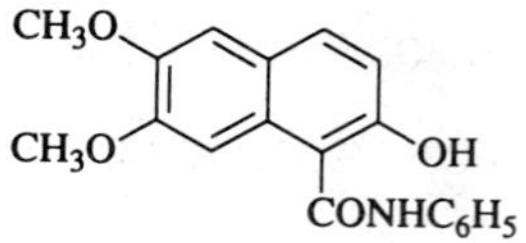 (CH_3O, CH_3O, OH, $CONHC_6H_5$) (100)	37

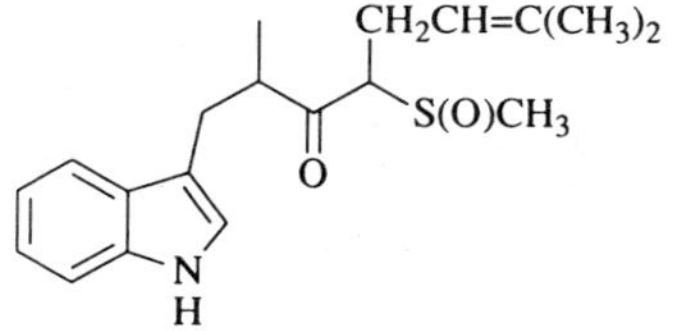

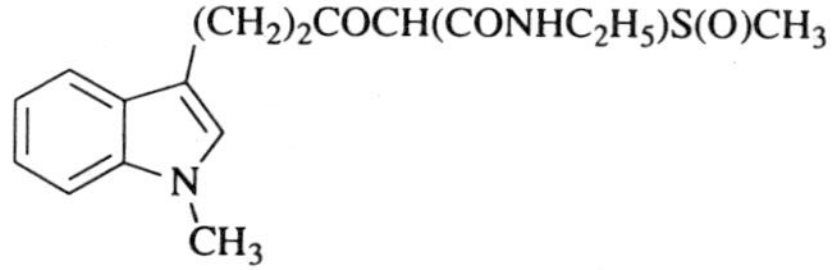

	Reactant	Conditions	Product(s) (Yield, %)	Refs.
C_{21}	$(CH_2)_2COCH_2S(O)CH_3$; N–$CH_2C_6H_5$	TsOH, THF, reflux 3 h	$C_6H_5CH_2$; SCH_3; O (75)	38, 337
	$(CH_2)_2COCH(CONHC_6H_5)S(O)CH_3$; N–$CH_3$	TsOH, CH_3CN, THF, reflux 1 h	OH; $CONHC_6H_5$; N–CH_3 (30) +	38
			O; SCH_3; N–C_6H_5; N–CH_3 (45)	
	$CH(CH_3)CH_2COCH_2S(O)CH_3$; N–$CH_2C_6H_5$	TsOH, THF, reflux 3 h	O; SCH_3; $C_6H_5CH_2$ (42)	38
		TsOH, CH_3OH, THF, acetone, reflux 6 h	OCH_3; N–$CH_2C_6H_5$ (23)	38
C_{22}	C_6H_5; N H; $(CH_2)_2COCH(CONHC_6H_5)S(O)CH_3$	TsOH, *i*-C_3H_7OH, THF, reflux 1 h	$CONHC_6H_5$; OH; C_6H_5; N H (91)	38, 337

TABLE IV. SULFOXIDES WITH A CARBON NUCLEOPHILE (*Continued*)

	Sulfoxide	Nucleophile	Reaction Conditions	Product(s) and Yield(s) (%)	Refs
C_{23}	1-benzylindol-3-yl–$(CH_2)_2COCH(CH_3)S(O)CH_3$; N–$CH_2C_6H_5$		CF_3CO_2H, CH_3CN, THF, reflux 25 h	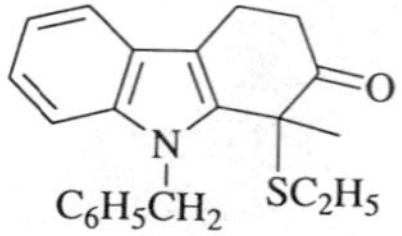 tetrahydrocarbazolone, N–$C_6H_5CH_2$, SC_2H_5 (51)	149
	1-benzylindol-3-yl–$CH(CH_3)CH_2COCH(CH_3)S(O)C_2H_5$; N–$CH_2C_6H_5$		"	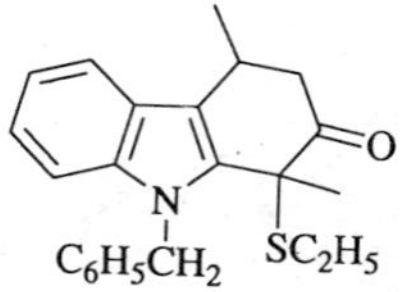 tetrahydrocarbazolone, N–$C_6H_5CH_2$, SC_2H_5 (53)	149
C_{25}	$C_6H_5S(O)(CH_2)_2-N$ (pyridinone-fused carbazole, H; N–CO_2CH_3)		1. TFAA, CH_2Cl_2, 0° to rt, 1 h 2. Reflux $CH_3C_6H_5$, 1 h	C_6H_5S, O, H, N–CO_2CH_3 (70, from the sulfide)	350
C_{27}	methylenedioxy tetrahydroisoquinoline, N–CH₂-aryl (OCH_3, OCH_3); p-$CH_3C_6H_4S^*(O)CH_2$		TFAA, $CH_3C_6H_5$, 0°, 10 min, 90°, 3 h	H, p-$CH_3C_6H_4S$, OCH_3, OCH_3 (62) 1:1 mixture	351

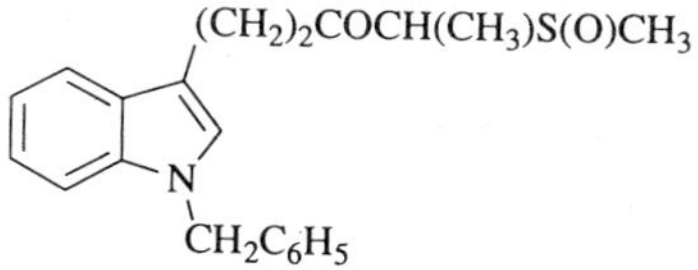

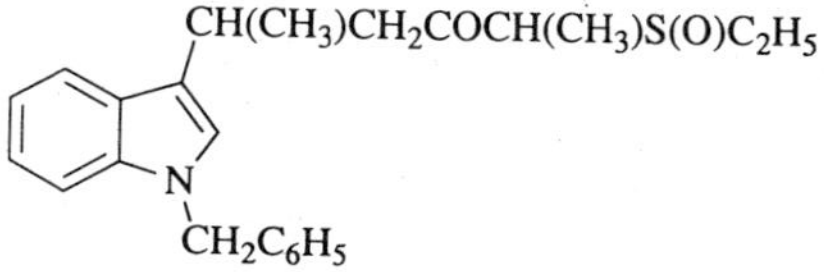

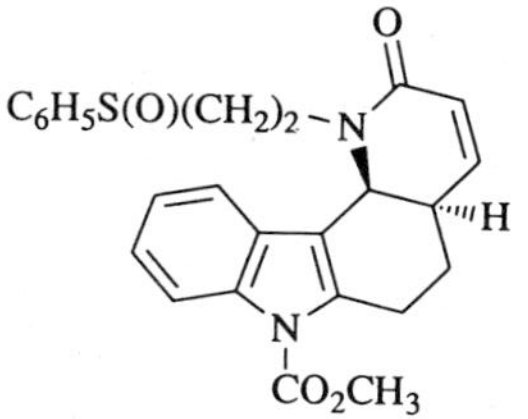

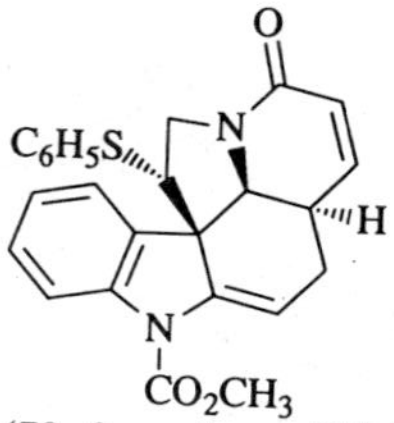

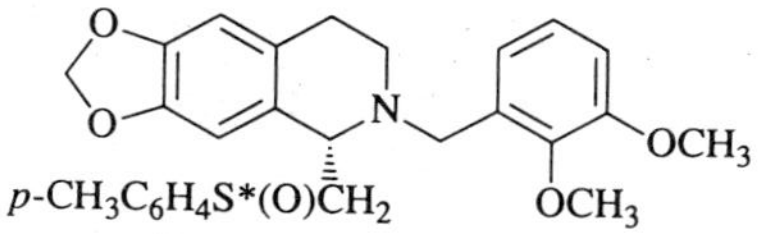

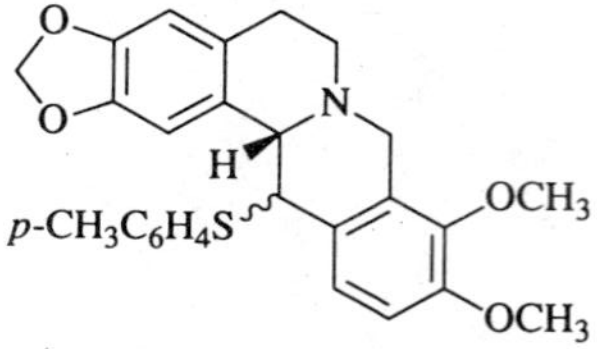

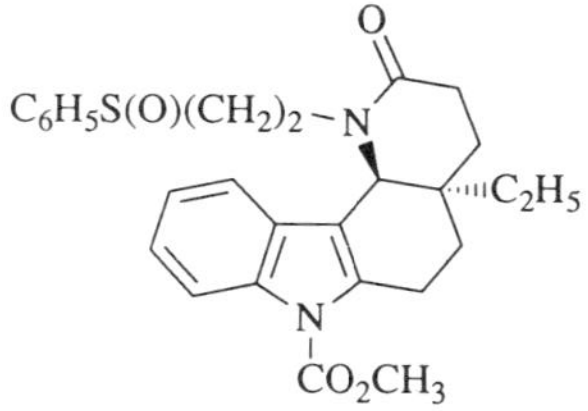

	$C_6H_5S(O)(CH_2)_2$–N, O, C_2H_5, N, CO_2CH_3	TFAA, heat	O, C_6H_5S, N, C_2H_5, N, CO_2CH_3	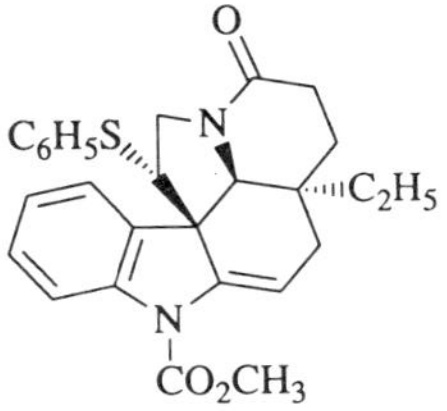 (<82)	352
C_{29}	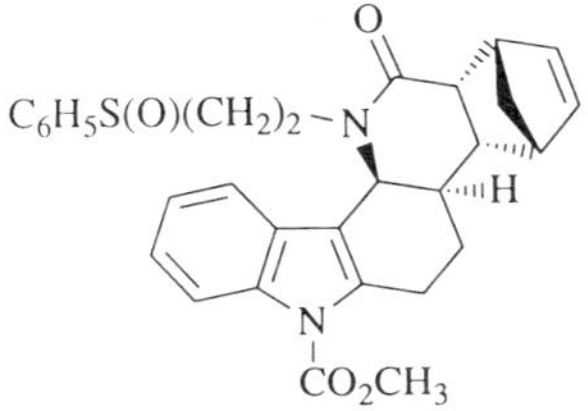 $C_6H_5S(O)(CH_2)_2$–N, O, H, N, CO_2CH_3	1. TFAA, CH_2Cl_2, 0° to rt, 1 h 2. $CH_3C_6H_5$, reflux 1 h	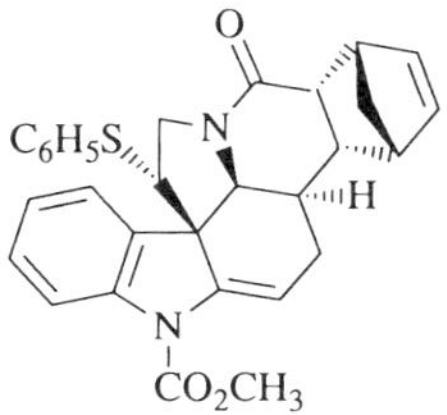 O, C_6H_5S, N, H, N, CO_2CH_3 (80 from the sulfide)		350
C_{30}	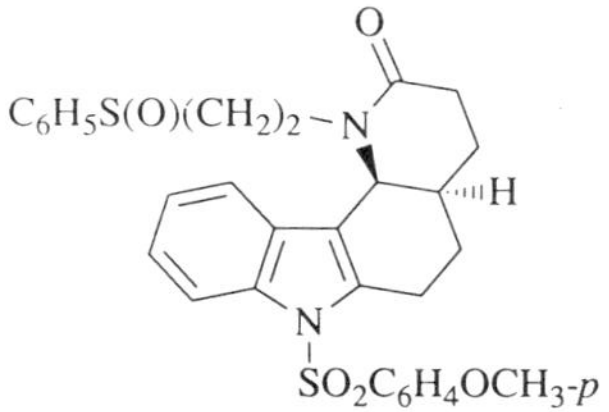 $C_6H_5S(O)(CH_2)_2$–N, O, H, N, $SO_2C_6H_4OCH_3$-*p*	1. TFAA, CH_2Cl_2, 0°, 1 h 2. C_6H_5Cl, 135°, 1.5 h	O, C_6H_5S, N, H, N, $SO_2C_6H_4OCH_3$-*p*	(84)	353
	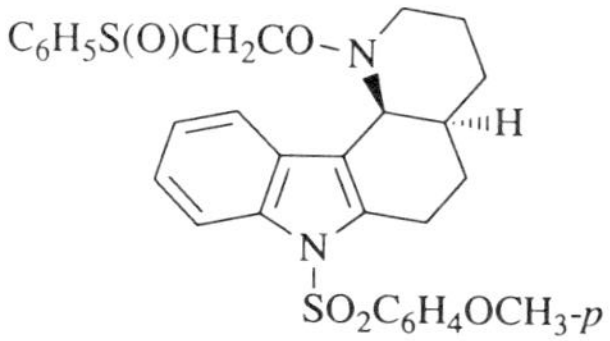 $C_6H_5S(O)CH_2CO$–N, H, N, $SO_2C_6H_4OCH_3$-*p*	1. TFAA, CH_2Cl_2, 0°, 10 min 2. C_6H_6Cl, 140°	O, C_6H_5S, N, H, N, $SO_2C_6H_4OCH_3$-*p*	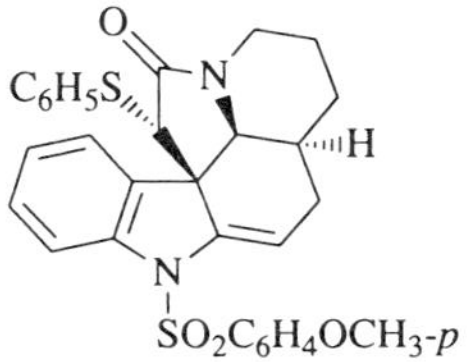 (55)	353, 354

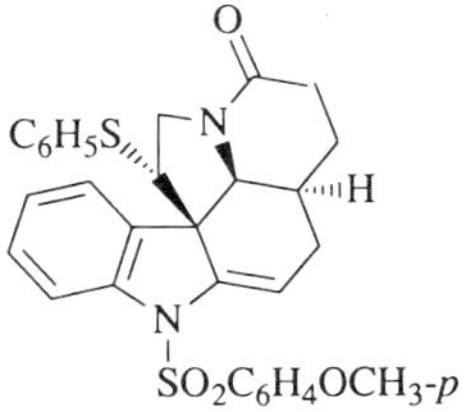

TABLE IV. SULFOXIDES WITH A CARBON NUCLEOPHILE (*Continued*)

	Sulfoxide	Nucleophile	Reaction Conditions	Product(s) and Yield(s) (%)	Refs.
	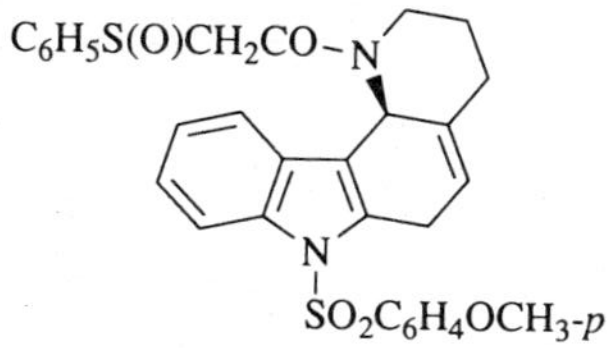 $C_6H_5S(O)CH_2CO-N$; $SO_2C_6H_4OCH_3$-*p*		1. TFAA, CH_2Cl_2, 0°, 15 min 2. C_6H_5Cl, 135°, 30 min	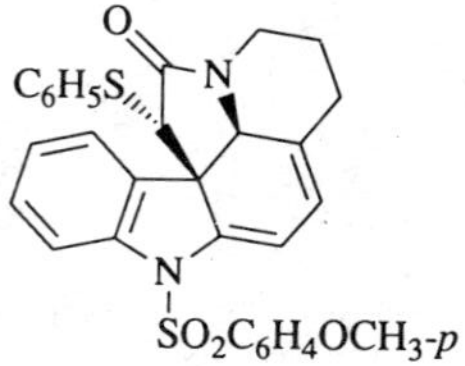 C_6H_5S ; O ; N ; $SO_2C_6H_4OCH_3$-*p* (78)	355
C_{31}	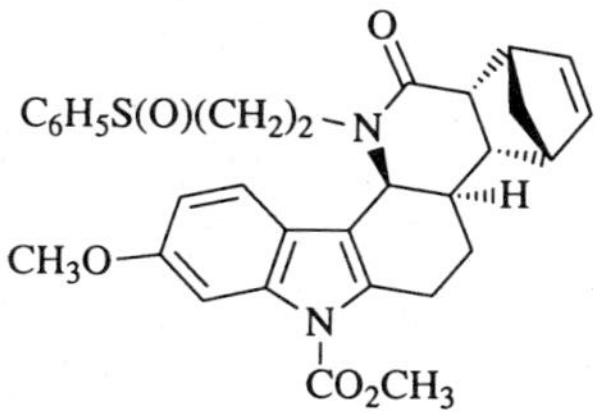 $C_6H_5S(O)(CH_2)_2-N$; O ; H ; CH_3O ; N ; CO_2CH_3		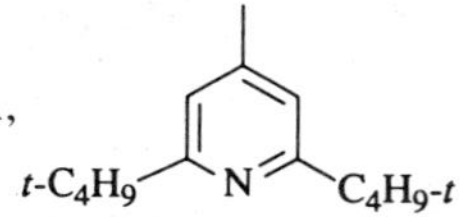 TFAA, *t*-C_4H_9, N, C_4H_9-*t* (pyridine) $CH_3C_6H_5$, 0–110°	(1:1), 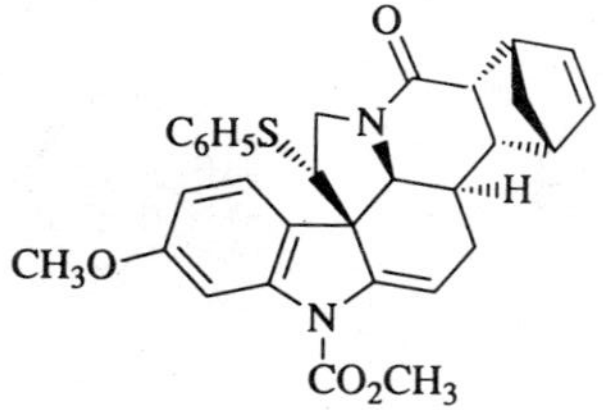 C_6H_5S ; O ; N ; H ; CH_3O ; CO_2CH_3 (65)	356
	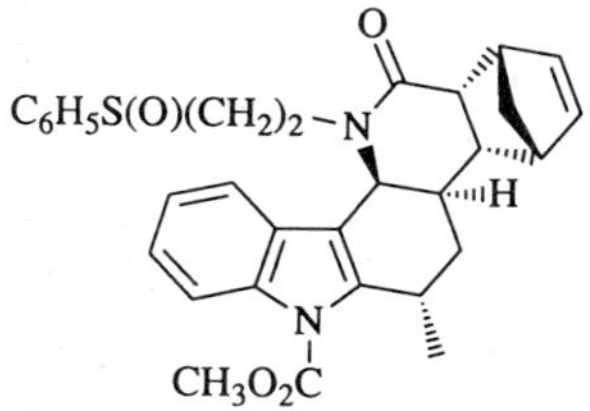 $C_6H_5S(O)(CH_2)_2-N$; O ; H ; N ; CH_3O_2C		1. TFAA, CH_2Cl_2, 0°, 1 h 2. $CH_3C_6H_5$, reflux 1.5 h	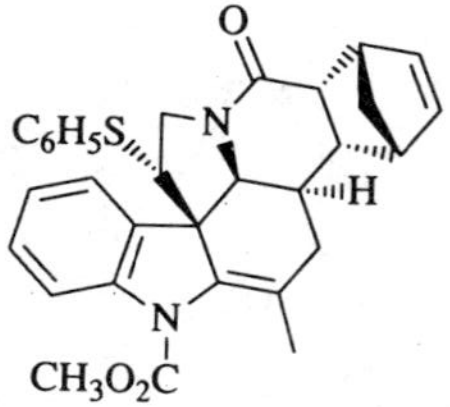 C_6H_5S ; O ; N ; H ; CH_3O_2C (87)	357

C_{32}

Reactant	Conditions	Product(s) (Yield, %)	Refs.
$C_6H_5S(O)(CH_2)_2-N$, O, C_2H_5, N, $SO_2C_6H_4OCH_3$-*p*	TFAA, C_6H_5Cl, 135°	C_6H_5S, N, O, C_2H_5, N, $SO_2C_6H_4OCH_3$-*p* (65)	358
$C_6H_5S(O)(CH_2)_2-N$, O, C_2H_5, N, $SO_2C_6H_4OCH_3$-*p*	1. TFAA, CH_2Cl_2, 0° 2. C_6H_5Cl, 130°, 2.5 h	C_6H_5S, N, O, C_2H_5, N, $SO_2C_6H_4OCH_3$-*p* (81)	353
$C_6H_5S(O)CH_2CO-N$, C_2H_5, N, $SO_2C_6H_4OCH_3$-*p*	1. TFAA, CH_2Cl_2, 0° 2. C_6H_5Cl, 135°, 2.5 h	C_6H_5S, O, N, C_2H_5, N, $SO_2C_6H_4OCH_3$-*p* (91)	353

[a] Other non-Pummerer products were also formed.
[b] The nucleophile was added in the second step.
[c] The reaction was carried out to partial conversion.
[d] The product was formed by rearrangement of a cyclohexadienone intermediate.
[e] The product was a mixture of isomers.
[f] The marked sulfoxide oxygen was ^{18}O.

TABLE V. SULFOXIDES WITH A HALOGEN NUCLEOPHILE

Sulfoxide	Nucleophile	Reaction Conditions	Product(s) and Yield(s) (%)	Refs.
C_2				
$(CH_3)_2SO$	CH_3COCl (2 eq)	CH_2Cl_2	CH_3SCH_2Cl (98)	359
	C_6H_5COCl (0.5 eq)	CH_2Cl_2 or ether	" (98)	360
	$(C_6H_5)_2PCl$ (0.5 eq)	"	" (83)	360
	$C_6H_5PCl_2$ (0.3 eq)	"	" (96)	360
	PCl_3 (0.25 eq)	"	" (90)	360
	$N_3C_3Cl_3$ (0.3 eq)	"	" (73)	360
	$(C_6H_5)_2P(O)Cl$ (0.5 eq)	"	" (95)	360
	$C_6H_5SO_2Cl$ (0.5 eq)	"	" (97)	360
	5-$N(CH_3)_2$-naphthalene-1-SO_2Cl	Ether, 24 h	" (quant)	361
	$SiCl_4$	0°	" (77)	362
	BCl_3	1. CH_2Cl_2, 0° 2. Quinoline, 0°	" (70)	362
	BCl_3	CH_2Cl_2, 0°	$CH_3SCH_2Cl\cdot BOCl$ (85–97) + $CH_3SCH_2Cl\cdot BCl_3$ (—)	362
	spiro bis(O,O-P(O)Cl)	0°	$[CH_3SCH_2Cl]^a$ (—)	83, 363
	$(NPCl_2)_3$		" (—)	363
	SO_2Cl_2, Cl_2	CCl_4	$ClCH_2SCCl_3$ (62)	364

		HCl[b]	1. 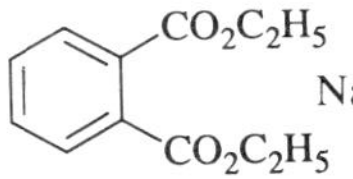 [structure: $CO_2C_2H_5$, $CO_2C_2H_5$], NaH, 2. 0°	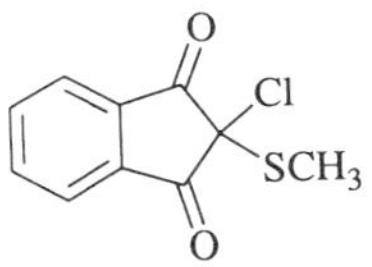 [structure: O, Cl, SCH_3, O] (80)	365, 366
		HCl[b]	1. 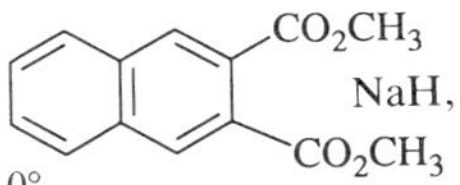 [structure: CO_2CH_3, CO_2CH_3], NaH, 2. 0°	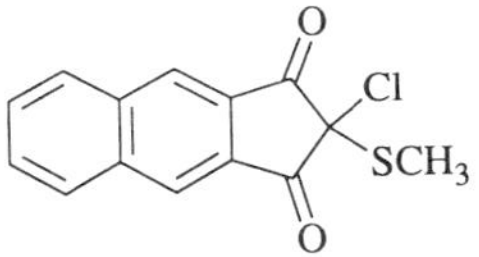 [structure: O, Cl, SCH_3, O] (59)	365
C_3	$CH_3S(O)C_2H_5$	DAST	1. $CHCl_3$, 16 h 2. mCPBA	$CH_3SO_2CHFCH_3$ (83)	63
C_4	$CH_3S(O)C_3H_7$-*n*	DAST	1. $CHCl_3$, 16 h 2. mCPBA	$CH_3SO_2CHFC_2H_5$ (69)	63
	$CH_3SCH_2S(O)C_2H_5$	SO_2Cl_2	CH_2Cl_2, reflux 1 h	CH_3SCH_2Cl (75) + $(C_2H_5S)_2$ (39)	367
C_5	$C_2H_5SCH_2S(O)C_2H_5$	SO_2Cl_2	"	$C_2H_5SCH_2Cl$ (70) + $(C_2H_5S)_2$ (35)	367
		C_6H_5COCl	"	" (53) + $OHCSC_2H_5$ (20)	367
	$CH_3S(O)CH_2CO_2C_2H_5$	HCl	$CHCl_3$	$CH_3SCHClCO_2C_2H_5$ (80)	368
C_7	$C_6H_5S(O)CH_3$	DAST	$CHCl_3$, 50°, 3 h	$C_6H_5SCH_2F$ (85)	63
	$C_6H_5S(O)CH_2F$	DAST	"	$C_6H_5SCHF_2$ (23)	63
C_8	*p*-$CH_3OC_6H_4S(O)CH_3$	DAST	"	*p*-$CH_3OC_6H_4SCH_2F$ (95)	63
	p-$CH_3OC_6H_4S(O)CH_2F$	DAST	"	*p*-$CH_3OC_6H_4SCHF_2$ (68)	63
	$C_6H_5S(O)CH_2CO_2H$	HCl (gas)	5–8 h	$C_6H_5SCHClCO_2H$ (—)	15
		HCl (gas)	C_2H_5OH, 0°, 2 d	$C_6H_5SCHClCO_2C_2H_5$ (—)	15
	p-$CH_3C_6H_4S(O)CH_3$	$(Cl_3Si)_2$	C_6H_6, 25°, 1 h	*p*-$CH_3C_6H_4SCH_2Cl$ I + *p*-$CH_3C_6H_4SCH_3$ II 1:1.7 (83)	189
		$(Cl_3Si)_2$	"	I + II 1:0.4 (100)	189
		$(Cl_3Si)_2$	$CHCl_3$, 25°, 1 h	I + II 1:3.6 (86)	189

TABLE V. SULFOXIDES WITH A HALOGEN NUCLEOPHILE (*Continued*)

Sulfoxide	Nucleophile	Reaction Conditions	Product(s) and Yield(s) (%)	Refs.
C_9				
$CH_3S(O)CH_2COC_6H_5$	$SOCl_2$	CH_2Cl_2, 4 h	$CH_3SCHClCOC_6H_5$ (—)	105, 254
	CH_3SOCl		" (87)	254
4-$S(O)CH_3$-3,5-dimethylphenol (S(O)CH₃, OH)	AcCl	C_6H_6, ~1 h	2-Cl-4-SCH_3-3,5-dimethylphenol (SCH_3, Cl, OH) (18)[c]	261
$S(O)C_6H_5$-cyclopropane	$SOCl_2$	CH_2Cl_2, reflux 1 h	1-Cl-1-SC_6H_5-cyclopropane (quant)	256, 369
	C_2H_5COCl	CH_2Cl_2, reflux 1 h	" (quant)	256, 369
3-$S(O)CH_3$-cinnolin-4(1H)-one (O, $S(O)CH_3$, N, N, H)	$SOCl_2$	Reflux 6 h	4-Cl-3-SCH_2Cl-cinnoline (Cl, SCH_2Cl, N=N) (92)	194
C_{10}				
$CH_3S(O)CH(CH_3)COC_6H_5$	$SOCl_2$	CH_2Cl_2, 2 h	$CH_3SCH(CH_2Cl)COC_6H_5$ (85) + $CH_3SCH(CH_3)COC_6H_5$ (10)	239
2-$S(O)CH_3$-indane-1,3-dione (O, O, $S(O)CH_3$)	HCl[b]	1. $NaHCO_3$, H_2O 2. H_2O	2-Cl-2-SCH_3-indane-1,3-dione (O, Cl, SCH_3, O) (66)	270
1-CH_3-3-$S(O)CH_3$-cinnolin-4-one (O, $S(O)CH_3$, N, N, CH_3)	$SOCl_2$		4-Cl-3-SCH_2Cl-cinnoline (Cl, SCH_2Cl, N=N) (—)	194

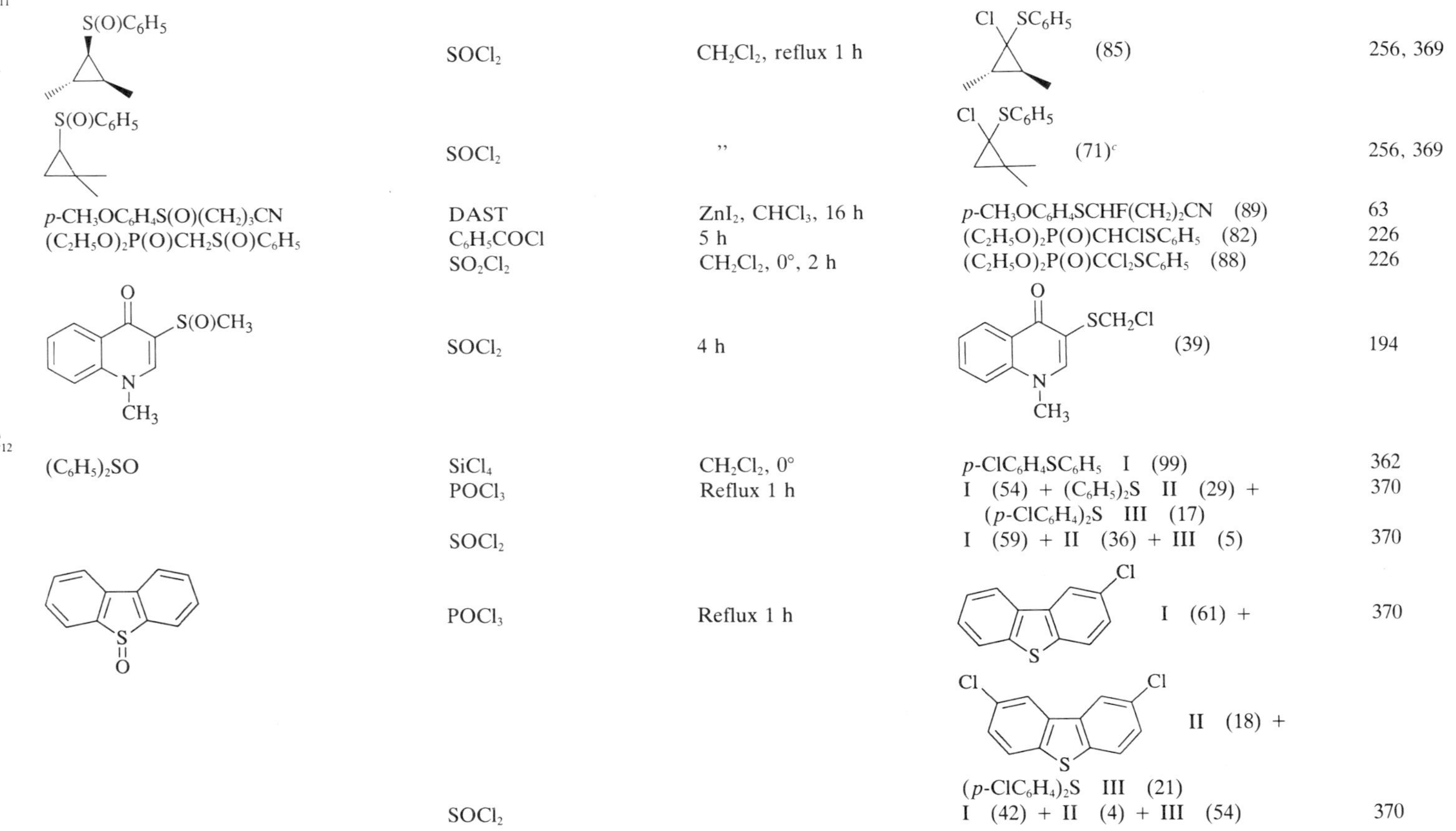

	Reactant	Reagent	Conditions	Product(s) and Yield(s) (%)	Refs.
C_{11}	$S(O)C_6H_5$ (cyclopropane)	$SOCl_2$	CH_2Cl_2, reflux 1 h	Cl, SC_6H_5 (cyclopropane) (85)	256, 369
	$S(O)C_6H_5$ (cyclopropane)	$SOCl_2$	"	Cl, SC_6H_5 (cyclopropane) (71)[c]	256, 369
	$p\text{-}CH_3OC_6H_4S(O)(CH_2)_3CN$	DAST	ZnI_2, $CHCl_3$, 16 h	$p\text{-}CH_3OC_6H_4SCHF(CH_2)_2CN$ (89)	63
	$(C_2H_5O)_2P(O)CH_2S(O)C_6H_5$	C_6H_5COCl	5 h	$(C_2H_5O)_2P(O)CHClSC_6H_5$ (82)	226
		SO_2Cl_2	CH_2Cl_2, 0°, 2 h	$(C_2H_5O)_2P(O)CCl_2SC_6H_5$ (88)	226
	$S(O)CH_3$ (quinolone, N-CH_3)	$SOCl_2$	4 h	SCH_2Cl (quinolone, N-CH_3) (39)	194
C_{12}	$(C_6H_5)_2SO$	$SiCl_4$	CH_2Cl_2, 0°	$p\text{-}ClC_6H_4SC_6H_5$ I (99)	362
		$POCl_3$	Reflux 1 h	I (54) + $(C_6H_5)_2S$ II (29) + ($p\text{-}ClC_6H_4)_2S$ III (17)	370
		$SOCl_2$		I (59) + II (36) + III (5)	370
	S=O (dibenzothiophene oxide)	$POCl_3$	Reflux 1 h	Cl (dibenzothiophene) I (61) + Cl, Cl (dibenzothiophene) II (18) + $(p\text{-}ClC_6H_4)_2S$ III (21)	370
		$SOCl_2$		I (42) + II (4) + III (54)	370

TABLE V. SULFOXIDES WITH A HALOGEN NUCLEOPHILE (*Continued*)

Sulfoxide	Nucleophile	Reaction Conditions	Product(s) and Yield(s) (%)	Refs.
Thianthrene S-oxide	$POCl_3$	Reflux 1 h	2-Chlorothianthrene (60) + 2,7-dichlorothianthrene (10) + thianthrene (30)	370
Phenoxathiin S-oxide	$POCl_3$	Reflux 1 h	Chlorophenoxathiin (58) + dichlorophenoxathiin (15) + phenoxathiin (26)	370

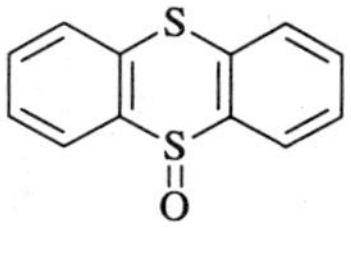

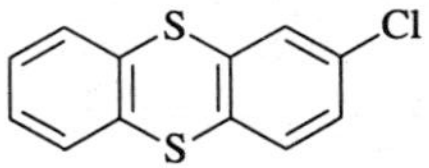

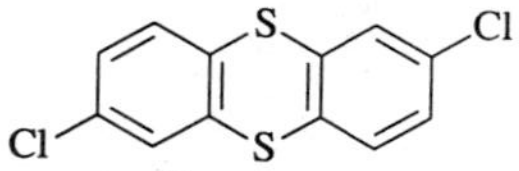

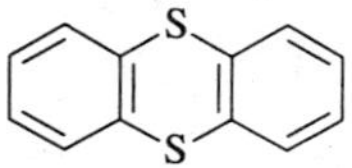

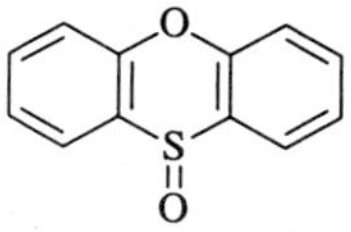

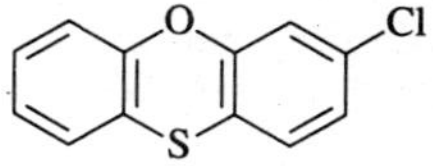

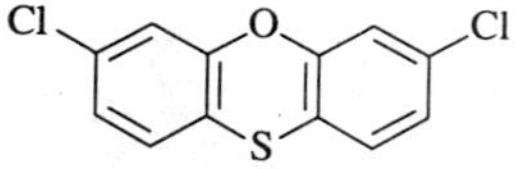

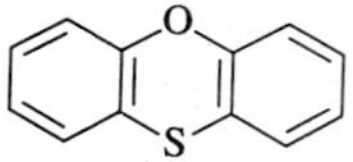

C_{13}				
$C_6H_5S(O)CH_2SC_6H_5$	$SOCl_2$	CH_2Cl_2, reflux 1 h	$C_6H_5SCH_2Cl$ (90) + $(C_6H_5S)_2$ (48)	367
$C_6H_5S(O)CH_2C_6H_5$	DAST	1. ZnI_2, $CHCl_3$, 63 h 2. mCPBA	$C_6H_5S(O)CHFC_6H_5$ (44)	63
p-$CH_3OC_6H_4S(O)(CH_2)_3CO_2C_2H_5$	DAST	ZnI_2, $CHCl_3$, 16 h	p-$CH_3OC_6H_4SCHF(CH_2)_2CO_2C_2H_5$ (79)	63
C_{15}				
$C_6H_5S(O)(CH_2)_3C_6H_5$	DAST	ZnI_2, $CHCl_3$, 16 h	$C_6H_5SCHF(CH_2)_2C_6H_5$ (100)	63
p-$CH_3OC_6H_4S(O)(CH_2)_2C_6H_5$	DAST	1. ZnI_2, $CHCl_3$, 18 h 2. mCPBA	p-$CH_3OC_6H_4S(O)CHFCH_2C_6H_5$ (86)	63
C_{16}				
p-$CH_3OC_6H_4S(O)(CH_2)_3C_6H_5$	DAST	$CHCl_3$, 18 h	p-$CH_3OC_6H_4SCHF(CH_2)_2C_6H_5$ (85)	63
C_{19}				
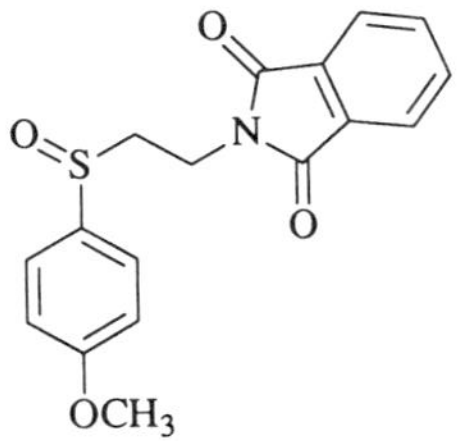	DAST	ZnI_2, $CHCl_3$, 72 h	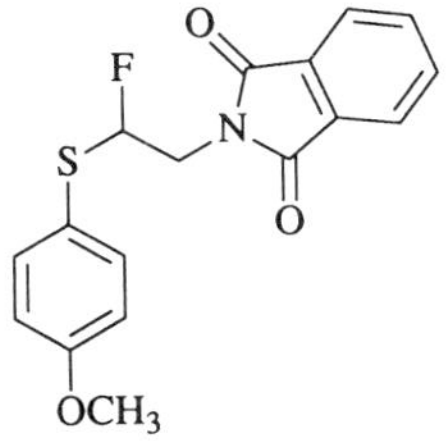 (91)	63
C_{23}				
p-$CH_3OC_6H_4S(O)C_{16}H_{33}$-$n$	DAST	$CHCl_3$, 16 h	p-$CH_3OC_6H_4SCHFC_{15}H_{31}$-$n$ (8)	63
	DAST	ZnI_2, $CHCl_3$, 16 h	" (94)	63

[a] The product reacted further with $(CH_3)_2SO$ and/or the reagent.
[b] The nucleophile was added in the second step.
[c] Other non-Pummerer products were also formed.

TABLE VI. DIRECT FORMATION OF CARBONYL COMPOUNDS AND THIOLS

Sulfoxide	Reaction Conditions	Product(s) and Yield(s) (%)	Refs.
C_2			
$(CH_3)_2SO$	1. AcOH, "prolonged heating" 2. $HgCl_2$, CH_3OH	$(CH_3S)_2Hg$ (—)	371
	1. C_6H_5COCl, CCl_4, H_2O 2. DNPH	$CH_2{=}NNHC_6H_3(NO_2)_2$-2,4 (28–30)	359
C_3			
$CH_3S(O)CH_2CN$	TsNSO, C_6H_6, 80°, 6.5 h	$NCCH(NHTs)_2$ (61) + $NCCH(SCH_3)_2$ (53) + CH_3SCH_2CN (4)	320
C_4			
$C_2H_5S(O)CH_2CO_2H$	HCl (2 N), heat 3 h	$(C_2H_5S)_2CHCO_2H$ (—)	69
$CH_3S(O)CH_2CO_2CH_3$	TsNSO, C_6H_6, 80°, 7.5 h	$(TsNH)_2CHCO_2CH_3$ (66) + $CH_3SCH(NHTs)CO_2CH_3$ (3)	320
$(HO_2CCH_2)_2SO$	HCl (anh), C_2H_5OH, reflux	HO_2CCH_2SH (—) + $[OHCCO_2H]$ (—)	35
C_5			
i-$C_3H_7S(O)CH_2CO_2H$	HCl (2 N), heat	$(i$-$C_3H_7S)_2CHCO_2H$ (—)	69
$CH_3S(O)(CH_2)_2CH(NH_2)CO_2H$	1. Ac_2O, reflux 1 h 2. HCl, H_2O, reflux 1 h 3. CH_2O, 80°, 30 min	$CH_2[S(CH_2)_2CH(NH_2)CO_2H]_2$ (—)	203
C_6			
t-$C_4H_9S(O)CH_2CO_2H$	HCl (4 N)	t-C_4H_9SH (60) + $[OHCCO_2H]$ (—)	372
C_7			
$C_6H_5S(O)CH_3$	$CHCl_2COCl$, THF, 25°, 2 h	$CH_2(SC_6H_5)_2$ (12) + $(C_6H_5S)_2$ (3)	67
	1. Ac_2O, 120° 2. HO^-	C_6H_5SH (—)[a]	373, 374[a]
$C_6H_5S(O)CD_3$	"	" (—)[a]	373, 374[a]
p-$XC_6H_4S(O)CH_3$	"	p-XC_6H_4SH (—)[a] X = Cl, NO_2	373, 374[a]
p-$O_2NC_6H_4S(O)CD_3$	"	p-$O_2NC_6H_4SH$ (—)[a]	373, 374[a]
cyclopropyl-$COCH(CH_3)S(O)CH_3$	HCl, $(CH_3)_2SO$, 1.5 h	cyclopropyl-$COCOCH_3$ (37)	239

C_8			
$C_6H_5CH_2S(O)CH_3$	Ac_2O, 100°, 20 h	$C_6H_5CH(SCH_3)_2$ (57)	182
	HCl, CCl_4, reflux 5 min	C_6H_5CHO (60)	375
p-$CH_3C_6H_4S(O)CH_3$	1. Ac_2O, 120° 2. HO^-	p-$CH_3C_6H_4SH$ (—)[a]	373, 374[a]
	TFAA, reflux 30 min	" (97)	114
p-$HOCC_6H_4S(O)CH_3$	TFAA, reflux 30 min	p-$[(CF_3CO_2)_2CH]C_6H_4SH$ (100)	114
p-$HOC_6H_4S(O)CH_3$	"	p-$(CF_3CO_2CH_2)C_6H_4SH$ (86)	114
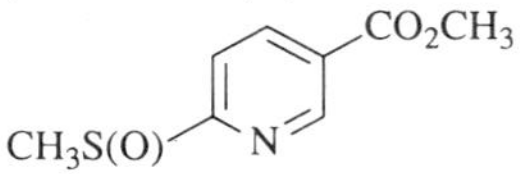 $CH_3S(O)$ N CO_2CH_3	"	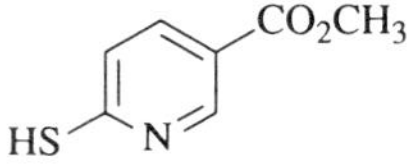 HS N CO_2CH_3 (80)	114
n-$C_5H_{11}COCH_2S(O)CH_3$	TsNSO, C_6H_6, 80°, 5 h	n-$C_5H_{11}COCH(NHTs)_2$ (58)	320
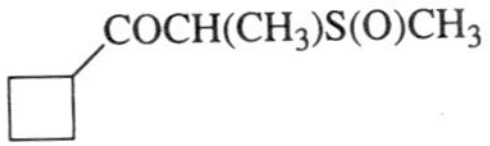 $COCH(CH_3)S(O)CH_3$	HCl, $(CH_3)_2SO$, 1.5 h	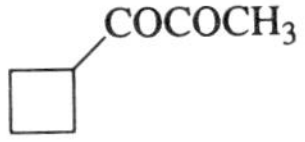 $COCOCH_3$ (39)	239
$C_6H_5SCH_2CN$	H_2O_2, H_2SO_4, reflux 1 h	$(C_6H_5S)_2$ (51) + $C_6H_5SO_2CH_2CN$ (8)	376
	Br_2, AcOH, 4 d	$(p$-$BrC_6H_4S)_2$ (—)	376
$C_6H_5SCH_2CO_2H$	H_2O_2 (30%), reflux 5 min	$C_6H_5S(O)CH_2CO_2H$ I (50) + $C_6H_5SCHOHCO_2H$ II (14) + C_6H_5SH III (7) + $(C_6H_5S)_2$ (9)	31
	H_2O_2 (30%), 40°	I (95) + II (5)	31
	1. H_2O_2 (30%), H_2SO_4, reflux 2. Zn	III (80)	31
$C_6H_5S(O)CH_2CO_2H$	"Mineral acids," heat	C_6H_5SH I (—) + $OHCCO_2H$ II (—)	15
	AcOH, reflux overnight	$(C_6H_5S)_2CHCO_2H$ III (90)	30
	H_2O, 6 d	I (7)[b] + III (—)[b] + $(C_6H_5S)_2$ IV (4)[b] + $C_6H_5SCHOHCO_2H$ V (—)[b]	
	H_2O, 3 months	I (47) + III (22) + IV (14) + V (8)	31
	H_2O, 100°, 4 h	III (91) + V (8)	31
	C_6H_5SH, AcOH, H_2O, reflux 2 h	III (40)	31
p-$ClC_6H_4S(O)CH_2CO_2H$	AcOH, reflux overnight	$(p$-$ClC_6H_4S)_2CHCO_2H$ (76)	30
	H_2SO_4 (6 N), "warm"	" (—)	30

TABLE VI. DIRECT FORMATION OF CARBONYL COMPOUNDS AND THIOLS (*Continued*)

Sulfoxide	Reaction Conditions	Product(s) and Yield(s) (%)	Refs.
$p\text{-BrC}_6\text{H}_4\text{SCH}_2\text{CO}_2\text{H}$	HNO_3, H_2O, reflux 1.5 h	$(p\text{-BrC}_6\text{H}_4\text{S})_2$ I (—) + $p\text{-BrC}_6\text{H}_4\text{SO}_3\text{SC}_6\text{H}_4\text{Br-}p$ II (—)	376
	H_2SO_4, KMnO_4, H_2O, reflux	I (38) + II (—)	376
$p\text{-O}_2\text{NC}_6\text{H}_4\text{S(O)CH}_2\text{CO}_2\text{H}$	AcOH, reflux overnight	$(p\text{-O}_2\text{NC}_6\text{H}_4\text{S})_2$ (—)	30
	H_2SO_4(6 N), "warm"	" (—)	30
$\text{HO}_2\text{CCH(CH}_3\text{)CH}_2\text{S(O)C(CH}_3\text{)}_2\text{CO}_2\text{H}$	HCl	$\text{HO}_2\text{CC(CH}_3\text{)}_2\text{SH}$ (—) + $\text{HO}_2\text{CCH(CH}_3\text{)CHO}$ (—)	377
$[\text{HO}_2\text{CCH(CH}_3\text{)CH}_2]_2\text{SO}$	HCl	$\text{HO}_2\text{CCH(CH}_3\text{)CH}_2\text{SH}$ (—) + $\text{HO}_2\text{CCH(CH}_3\text{)CHO}$ (—)	377
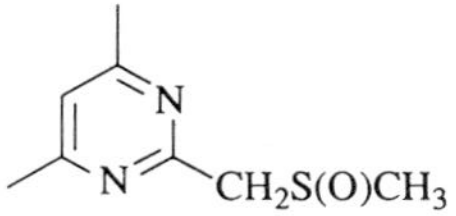 N, N, $\text{CH}_2\text{S(O)CH}_3$	I_2, CH_3OH, reflux 6 h	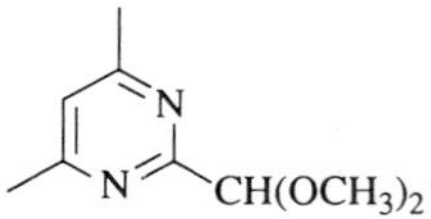 N, N, $\text{CH(OCH}_3\text{)}_2$ (81)	378
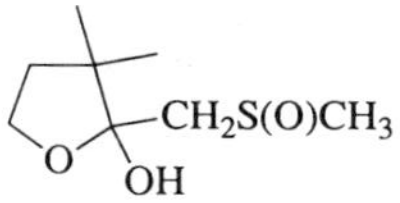 O, OH, $\text{CH}_2\text{S(O)CH}_3$	I_2, CH_3OH	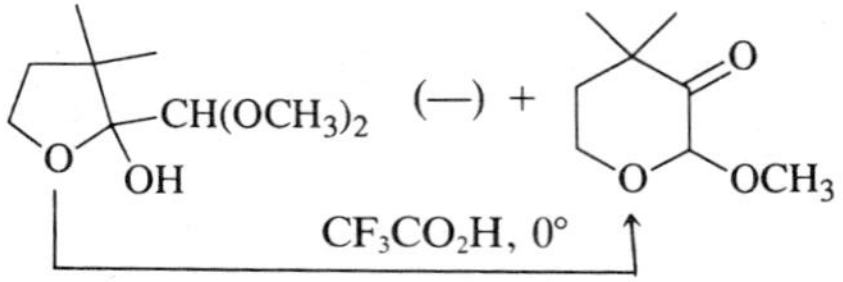 O, OH, $\text{CH(OCH}_3\text{)}_2$ (—) + O, O, OCH_3 (40); $\text{CF}_3\text{CO}_2\text{H}$, 0°	117
C_9			
$\text{C}_6\text{H}_5\text{SCH}_2\text{COCH}_3$	Ac_2O, H_2O_2, H_2SO_4, reflux	$(\text{C}_6\text{H}_5\text{S})_2$ (62)	376
	1. Br_2, AcOH, CCl_4 2. H_2O_2	" (53)	376
$p\text{-O}_2\text{NC}_6\text{H}_4\text{S(O)CH}_2\text{COCH}_3$	1. TsN_3, $\text{C}_2\text{H}_5\text{OH}$, H_2O, $(\text{C}_2\text{H}_5)_3\text{N}$, 2 d	$(p\text{-O}_2\text{NC}_6\text{H}_4\text{S})_2$ (78) + $\text{CH}_3\text{C(CO}_2\text{CH}_3\text{)=NNHC}_6\text{H}_3\text{(NO}_2\text{)}_2\text{-2,4}$ (51)	379
$\text{C}_6\text{H}_5\text{S(CH}_2\text{)}_2\text{CO}_2\text{H}$	H_2O_2, H_2SO_4, reflux	$(\text{C}_6\text{H}_5\text{S})_2$ (9) + $\text{C}_6\text{H}_5\text{SO}_2\text{(CH}_2\text{)}_2\text{CO}_2\text{H}$ (—)	376
$\text{C}_6\text{H}_5\text{CH}_2\text{S(O)CH}_2\text{CO}_2\text{H}$	H_2SO_4, steam distil, 0.5 h	$(\text{C}_6\text{H}_5\text{CH}_2\text{S})_2\text{CHCO}_2\text{H}$ I (—) + $(\text{C}_6\text{H}_5\text{CH}_2\text{S})_2$ (—)	69
	HCl, 100°, 0.5 h	I (—) + $\text{C}_6\text{H}_5\text{CH}_2\text{SCHOHCO}_2\text{H}$ (—)	69
$p\text{-CH}_3\text{C}_6\text{H}_4\text{SCH}_2\text{CO}_2\text{H}$	1. H_2O_2, H_2SO_4, reflux 2. Zn	$p\text{-CH}_3\text{C}_6\text{H}_4\text{SH}$ (75)	31
$p\text{-CH}_3\text{C}_6\text{H}_4\text{S(O)CH}_2\text{CO}_2\text{H}$	AcOH, reflux overnight	$(p\text{-CH}_3\text{C}_6\text{H}_4\text{S})_2\text{CHCO}_2\text{H}$ (82)	30

Substrate	Conditions	Product(s) (Yield, %)	Refs.
p-$CH_3OC_6H_4CH_2S(O)CH_3$	HCl, CCl_4, reflux 15 min	p-$CH_3OC_6H_4CH_2Cl$ (100)	375
o-$(CF_3CONH)C_6H_4S(O)CH_3$	TFAA, reflux 30 min	o-$(CF_3CONH)C_6H_4SH$ (90)	114
$C_6H_5COCH_2S(O)CH_3$	1. $SOCl_2$, CH_2Cl_2, 20 min 2. CH_3OH, 30 min	$C_6H_5COCH(OCH_3)_2$ (—)	105
	I_2, CH_3OH, reflux	" (88)	60
	1. $SOCl_2$, CH_2Cl_2 2. $HS(CH_2)_3SH$, 3 h	C_6H_5CO–[1,3-dithian-2-yl] (51)	105
	1. $SOCl_2$, CH_2Cl_2 2. $HS(CH_2)_3SH$	C_6H_5CO–[1,3-dithiolan-2-yl] (76)	105
o-$O_2NC_6H_4COCH_2S(O)CH_3$	HCl, H_2O, "warm"	[indigo structure] (39)	380
[cyclopentyl]–$COCH(CH_3)S(O)CH_3$	HCl, $(CH_3)_2SO$, 1.5 h	[cyclopentyl]–$COCOCH_3$ (74)	239
$C_6H_5S(O)CH(CH_3)CO_2H$	H_2SO_4, reflux 4 h	C_6H_5SH (—)	16
n-$C_6H_{13}COCH_2S(O)CH_3$	[2,5,6-triamino-4-hydroxypyrimidine structure: OH, NH_2, H_2N, NH_2], AcOH, AcONa, reflux, 1 h	[2-amino-4-hydroxypteridine structure: OH, H_2N, C_6H_{13}-n] (15) + CH_3SH (—)	381
C_{10}			
$(i$-$C_5H_{11})_2SO$	HCl, C_2H_5OH, reflux	i-$C_5H_{11}SH$ (—) + s-C_4H_9CHO (—)	35
s-$C_4H_9S(O)C_6H_5$	1. Ac_2O, pyr, 120°, 4 h 2. NaOH, H_2O 3. DNPH, 24 h	i-$C_3H_7CH{=}NNHC_6H_3(NO_2)_2$-2,4 (79)	382
	1. Ac_2O, pyr, 120°, 6 h 2. NaOH, H_2O 3. NaCN, 24 h 4. NH_3, H_2O, 60°, 4 h	i-$C_3H_7CH(CN)NH_2$ (80)	382

TABLE VI. DIRECT FORMATION OF CARBONYL COMPOUNDS AND THIOLS (*Continued*)

Sulfoxide	Reaction Conditions	Product(s) and Yield(s) (%)	Refs.
p-(*i*-$C_3H_7CO)C_6H_4S(O)CH_3$	TFAA, reflux 30 min	*p*-(*i*-$C_3H_7CO)C_6H_4SH$ (98)	114
p-$[NC(CH_2)_2CO]C_6H_4S(O)CH_3$	"	*p*-$[NC(CH_2)_2CO]C_6H_4SH$ (100)	114
$CH_3COCH_2S(O)CH_2C_6H_5$	1. TsN_3, C_2H_5OH, H_2O, $(C_2H_5)_3N$, 2 d 2. DNPH, CH_3OH, H_2SO_4	$(C_6H_5CH_2S)_2$ (73) + $CH_3C(CO_2CH_3){=}NNHC_6H_3(NO_2)_2$-2,4 (59)	379
$C_6H_{11}COCH(CH_3)S(O)CH_3$	1. HCl, $(CH_3)_2SO$, 25°, 36 h 2. SiO_2	$C_6H_{11}COCOCH_3$ (63)	239
$RCH_2COCH_2S(O)CH_3$	AcOH, AcONa, reflux 1 h, 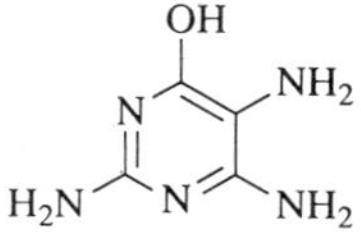 2-amino-4-hydroxy-5,6-diaminopyrimidine (structure: OH, NH_2, H_2N, NH_2)	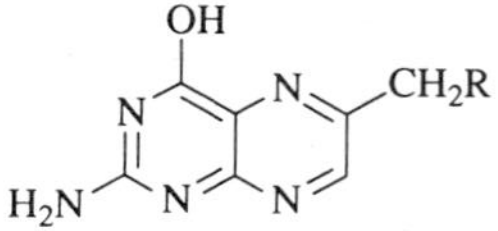 2-amino-4-hydroxy-6-(CH_2R)pteridine (structure: OH, H_2N, CH_2R)	
R = C_6H_{11}		(9) + CH_3SH (—)	381
R = C_6H_{13}-*n*		(17) + CH_3SH (—)	381
$C_6H_{11}COS(O)C_3H_7$-*n*	$C_6H_{11}COSC_3H_7$-*n*, 18 h	$(C_6H_{11}CO)_2O$ (85) + (*n*-$C_3H_7S)_2$ (84)	383
$C_6H_{11}COS(O)C_3H_7$-*i*	$C_6H_{11}COSC_3H_7$-*i*, 18 h	" (97) + (*i*-$C_3H_7S)_2$ (86)	383
$C_6H_5CH(CH_3)S(O)CH_2CO_2H$	H_2SO_4(1 N), distill	$C_6H_5CH(CH_3)SH$ (—) + $[OHCCO_2H]$ (—)	384, 385
$C_6H_5CH_2S(O)CO_2C_2H_5$	$SOCl_2$, CH_2Cl_2, −10°	$(C_6H_5CH_2S)_2$ (18) + $C_6H_5CH_2SO_nSCH_2C_6H_5$ n = 1 (10), n = 2 (11) + $C_6H_5CH_2SCO_2C_2H_5$ (32)[c] + $C_2H_5O_2CCl$ (20)	383
	$C_6H_5CH_2SCO_2C_2H_5$, CH_2Cl_2, 40°, 20 h	$(C_6H_5CH_2S)_2$ (—)[b]	383
p-$CH_3COC_6H_4SCH_2CO_2H$	H_2O_2 (1.5 M), H_2SO_4, reflux 2 h	*p*-$CH_3COC_6H_4SH$ I (26)[b] + (*p*-$CH_3COC_6H_4S)_2$ II (27)[b]	386
	H_2O_2 (2 M), H_2SO_4, reflux 90 min	I (—) + II (84)	386
o-$AcNHC_6H_4CH_2S(O)CH_3$	HCl, $(ClCH_2)_2$, 55°, 24 min	*o*-$AcNHC_6H_4CHO$ (49)	375

	Reactant	Conditions	Product(s) (Yield, %)	Refs.
	5-$S(O)CH_3$-2-C_6H_5-1,3,2-dioxaborinane	1. TFAA, CH_2Cl_2, −14° to rt, 2 h 2. DNPH	$(HOCH_2)_2CH{=}NNHC_6H_3(NO_2)_2$-2,4 (35)	219
	2-methyl-2-[$C(CH_3)_2COCH_2S(O)CH_3$]-1,3-dioxolane	I_2, CH_3OH, reflux 4 h	2,5-dimethoxy-4,4,5-trimethyldihydrofuran-3-one (CH_3O, OCH_3) (—)	61
C_{11}	$C_6H_{11}COS(O)C_4H_9$-*n*,	$C_6H_{11}COSC_4H_9$-*n*, 18 h	$(C_6H_{11}CO)_2O$ I (94) + $(n\text{-}C_4H_9S)_2$ (84)	383
	$C_6H_{11}COS(O)C_4H_9$-*s*,	$C_6H_{11}COSC_4H_9$-*s*, 18 h	I (91) + $(s\text{-}C_4H_9S)_2$ (84)	383
	n-$C_5H_{11}S(O)C_6H_5$	1. Ac_2O, AcONa, reflux 3 h 2. $NaHCO_3$, 2 d 3. DNPH, overnight	*n*-$C_4H_9CH{=}NNHC_6H_3(NO_2)_2$-2,4 (66)	382
	p-$AcC_6H_4SCH_2COCH_3$	H_2O_2, H_2SO_4, reflux 1 h	$(p\text{-}AcC_6H_4S)_2$ (53)	376
	p-$CH_3C_6H_4SC(CH_3)_2CO_2H$	"	$(C_6H_5S)_2$ (9) + *p*-$CH_3C_6H_4SO_2C(CH_3)_2CO_2H$ (—)	376
	$C_6H_5(CH_2)_2COCH_2S(O)CH_3$	TFAA, CH_2Cl_2, reflux	$C_6H_5(CH_2)_2COCH(SCH_3)_2$ (42)	341
		TFAA, CH_3CN, reflux 1 h	" (42)	37
	2-methyl-4-$COCH_3$-1-SCH_2CO_2H-benzene	H_2O_2 (1.5 M), H_2SO_4, reflux 2 h	2-methyl-4-$COCH_3$-benzenethiol (SH) (17)[b] + $[4\text{-}CH_3CO\text{-}2\text{-}CH_3C_6H_3S]_2$ I (53)[b]	386

TABLE VI. DIRECT FORMATION OF CARBONYL COMPOUNDS AND THIOLS (*Continued*)

Sulfoxide	Reaction Conditions	Product(s) and Yield(s) (%)	Refs.
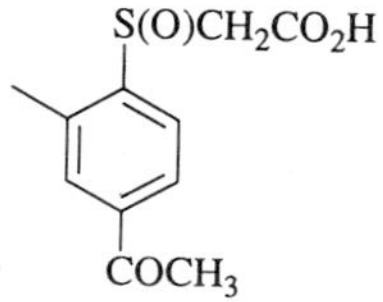 $S(O)CH_2CO_2H$ / $COCH_3$ (ring-substituted)	H_2O_2 (2 M), H_2SO_4, reflux 90 min	I (80)	386
	H_2SO_4 (6%), reflux 5–6 h	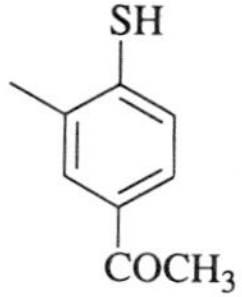 SH / $COCH_3$ (ring-substituted) (43)	386
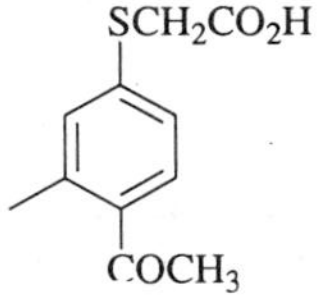 SCH_2CO_2H / $COCH_3$ (ring-substituted)	H_2O_2 (1.5 M), H_2SO_4, reflux 2 h	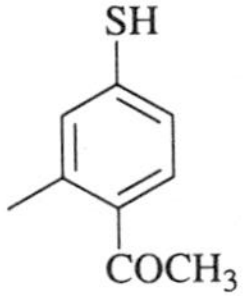 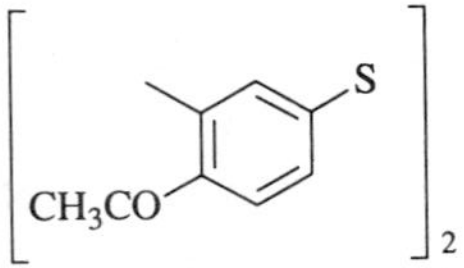 SH / $COCH_3$ (ring-substituted) (18)[b] + [CH_3CO / S (ring-substituted)]$_2$ I (40)[b]	386
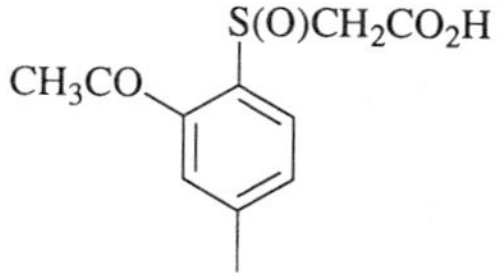 CH_3CO / $S(O)CH_2CO_2H$ (ring-substituted)	H_2O_2 (2 M), H_2SO_4, reflux 90 min	I (82)	386
	H_2SO_4 (6%), reflux 5–6 h	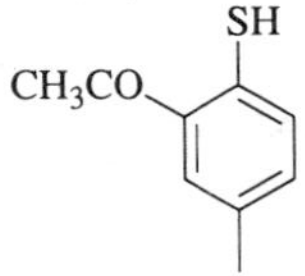 CH_3CO / SH (ring-substituted) (45)	386
$C_6H_5COCH_2S(O)(CH_2)_2CO_2H$	H_2SO_4 (dil), steam distill	$C_6H_5COCH[S(CH_2)_2CO_2H]_2$ (—)	68
	H_2SO_4 (dil), steam distill, $HgCl_2$	C_6H_5COCHO (—) + $ClHgS(CH_2)_2CO_2H$ (—)	68
	HCl (5 N), $HgCl_2$, heat	C_6H_5COCHO (—) + $HSCH{=}CHCO_2H$ (—) + $C_6H_5COCHOHS(CH_2)_2CO_2H$ (—)	68

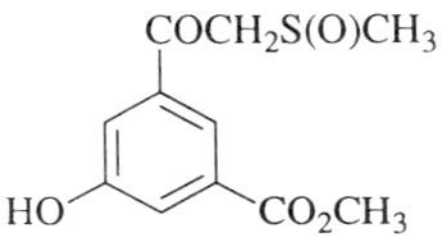 $COCH_2S(O)CH_3$, HO, CO_2CH_3	HCl, CH_3OH, reflux 2 h	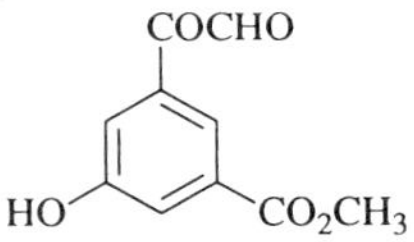 COCHO, HO, CO_2CH_3 (quant)	387
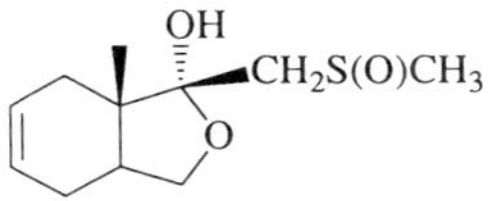 OH, $CH_2S(O)CH_3$, O	I_2, CH_3OH, reflux	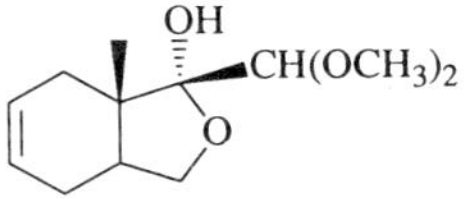 OH, $CH(OCH_3)_2$, O (—)	117
o-$AcNHC_6H_4COCH_2S(O)CH_3$	HCl (1 N), 90°, 15 min	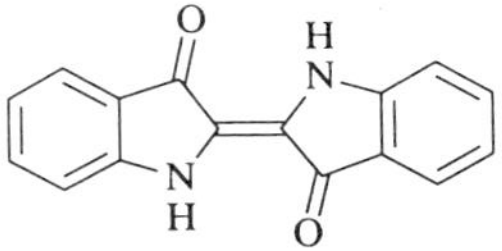 O, H, N, N, H, O (77)	380
$C_6H_5(CH_2)_2COCH_2S(O)CH_3$	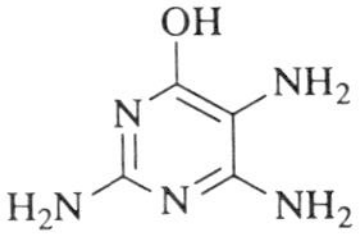 OH, NH_2, N, H_2N, N, NH_2 AcOH, AcONa, 0.5 h, reflux 1 h	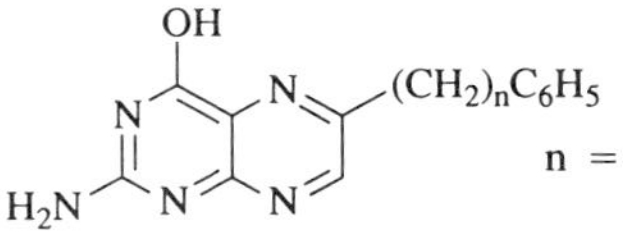 OH, N, $(CH_2)_nC_6H_5$, N, H_2N, N, N n = 2 (17) + CH_3SH (—)	381
$C_6H_5(CH_2)_3COCH_2S(O)CH_3$	"	" n = 3 (14) + CH_3SH (—)	381
n-$C_9H_{19}COCH_2S(O)CH_3$	"	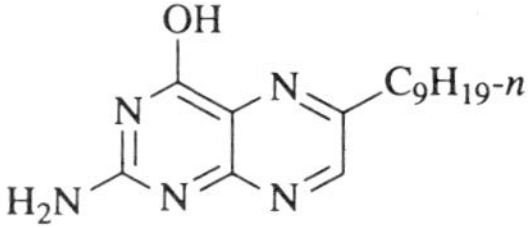 OH, N, C_9H_{19}-n, N, H_2N, N, N (9) + CH_3SH (—)	381
	I_2, CH_3OH, reflux 1.5 h	n-$C_9H_{19}COCH(OCH_3)_2$ (85)	60
	Br_2, H_2SO_4 (cat.), CH_3OH, reflux 30 min	" (85)	60
p-$C_4H_9C_6H_4S(O)CH_2CO_2H$	AcOH, reflux overnight	(p-$C_4H_9C_6H_4S)_2CHCO_2H$ (87)	30

TABLE VI. DIRECT FORMATION OF CARBONYL COMPOUNDS AND THIOLS (*Continued*)

Sulfoxide	Reaction Conditions	Product(s) and Yield(s) (%)	Refs.
$(CH_3)_2C(NO_2)CH(CH_3)CH_2S(O)C_6H_5$	1. TFAA, 2,6-lutidine, CH_3CN, 0°, 10 min 2. $NaHCO_3$, H_2O, 4 h	$(CH_3)_2C(NO_2)CH(CH_3)CHO$ (—)	388
$C_2H_5CH(NO_2)CH(CH_3)CH_2S(O)C_6H_5$	"	$C_2H_5CH(NO_2)CH(CH_3)CHO$ (—)	388
$C_2H_5C(CH_3)(NO_2)(CH_2)_2S(O)C_6H_5$	"	$C_2H_5C(CH_3)(NO_2)CH_2CHO$ (—)	388
C_{13}			
$C_6H_5S(O)CH_2C_6H_5$	1. Ac_2O, AcOH, reflux 1 h 2. DNPH, 5 h	$C_6H_5CH{=}NNHC_6H_3(NO_2)_2$-2,4 (90)	58, 382
	1. TFAA, 2,6-lutidine, CH_3CN, 0°, 10 min 2. $NaHCO_3$, H_2O, 2 h	C_6H_5CHO (85)	115
	$Cl_2CHCOCl$, THF, $(C_2H_5)_3N$, 25°, overnight	$C_6H_5CH(SC_6H_5)_2$ I (38) + $C_6H_5CH_2SC_6H_5$ (18) + $C_6H_5CHClS(O)C_6H_5$ (16)	67
	Ac_2O, *p*-xylene, 140°, 6 h	I (80)	67
$C_6H_5S(O)CH_2C_6H_4Cl$-*p*	AcOH, H_2O, H_2SO_4, reflux 24 h	*p*-ClC_6H_4CHO (30)	389
$C_6H_5SCH_2C_6H_4NO_2$-*p*	AcOH, H_2O_2, H_2SO_4, reflux	*p*-$O_2NC_6H_4CHO$ (90)	389
$[C_6H_5S(O)]_2CH_2$	AcOH, reflux 48 h	$(C_6H_5S)_2$ (80)	30
	H_2SO_4, heat 24 h	" (77)	30
1-nitro-1-[$(CH_2)_2S(O)C_6H_5$]cyclopentane	1. TFAA, 2,6-lutidine, CH_3CN, 0°, 10 min 2. $NaHCO_3$, H_2O, 4 h	1-nitro-1-(CH_2CHO)cyclopentane (—)	388
$CH_2{=}CH(CH_2)_8COCH_2S(O)CH_3$	I_2, CH_3OH, reflux	$CH_2{=}CH(CH_2)_8COCH(OCH_3)_2$ (92)	60
$C_6H_{11}COS(O)C_6H_{13}$-*n*	$C_6H_{11}COSC_6H_{13}$-*n*, 18 h	$(C_6H_{11}CO)_2O$ I (82) + $(n\text{-}C_6H_{13}S)_2$ (82)	383
C_{14}			
$C_6H_{11}COS(O)CH_2C_6H_{11}$	$C_6H_{11}COSCH_2C_6H_{11}$, 18 h	I (90) + $(C_6H_{11}CH_2S)_2$ (85)	383
n-$C_8H_{17}S(O)C_6H_5$	1. TFAA, 2,6-lutidine, CH_3CN, 0°, 10 min 2. $NaHCO_3$, H_2O, 2 h	*n*-$C_7H_{15}CHO$ (72)	115
	1. TFAA, 2,6-lutidine, CH_3CN, 0°, 10 min	" (74)	115

	2. $CuCl_2$, H_2O, 2 h		
	1. TFAA, 2,6-lutidine, CH_3CN, 0°, 10 min 2. $HgCl_2$, H_2O, 2 h	” (86)	115
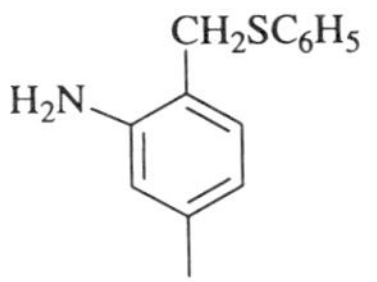 H_2N, $CH_2SC_6H_5$	1. Ac_2O, 15 min 2. H_2O_2, overnight 3. AcOH, reflux 1.5 h 4. Na_2CO_3, 1.5 h	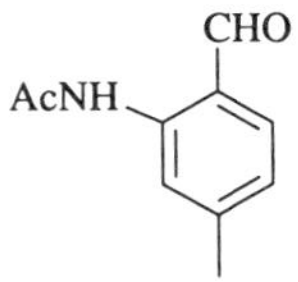 AcNH, CHO (73)	382
$(C_6H_5CH_2)_2SO$	$Cl_2CHCOCl$, THF, $(C_2H_5)_3N$, 25°, overnight	$C_6H_5CH(SCH_2C_6H_5)_2$ I (91)	67
	$(C_6H_5)_2CHCOCl$, THF, $(C_2H_5)_3N$, 25°, overnight	I (27)	67
	AcCl, THF, $(C_2H_5)_3N$, 25°, overnight	I (5) + $(C_6H_5CH_2)_2S$ II (88)	67
	Ac_2O, 100°, 20 h	I (55) + II (4) + $C_6H_5CH_2SCOCH_3$ (7)	67
	$(Cl_2CHCO)_2O$, THF, 25°, 2 h	I (76) + II (21)	67
	H_2O, reflux or heat	C_6H_5CHO I (—)	390
	HCl[d]	I (1–3) + $(C_6H_5CH_2S)_2$ II (15–48) + $(C_6H_5CH_2)_2S$ III (0.5–30) + $C_6H_5CH_2Cl$ (1–43) + $C_6H_5CH_2S(O)_2SCH_2C_6H_5$ (0.4–32) + $C_6H_5CH_2SH$ IV (—) + $C_6H_5CHOHSCH_2C_6H_5$ V + $C_6H_5CH(SCH_2C_6H_5)_2$ VI V + VI (0–8)	34
	C_6H_5COCl, CCl_4, reflux	I (—) + II (—) + III (—) + IV (—) + VI (—)	34
	Ac_2O, 140°, 2 h	I (42) + $C_6H_5CH_2SAc$ (46)	201
	1. PPSE, $(CH_2Cl)_2$, 80°, 3 h 2. NaOH, H_2O 3. DNPH, H_3PO_4, C_2H_5OH	$C_6H_5CHNNHC_6H_3(NO_2)_2$-2,4 (91)	58
$C_6H_5CH_2S(O)SCH_2C_6H_5$	TFAA, CCl_4, −10°	$(C_6H_5CH_2S)_2$ (50) + $C_6H_5CH_2SO_2COCF_3$ (50)	391
	Ac_2O, 60°, 2 h	$C_6H_5CH_2S(O)CH(SAc)C_6H_5$ I (37)	285
$C_6H_5CH_2S(O)SCD_2C_6H_5$	Ac_2O, 60°, 2 h	I + $C_6H_5CH_2S(O)CD(SAc)C_6H_5$ 6:4	285
p-$O_2NC_6H_4CH_2S(O)CH_2C_6H_5$	AcOH, reflux 24 h	*p*-$O_2NC_6H_4CH(SCH_2C_6H_5)_2$ I (73) + *p*-$O_2NC_6H_4CHO$ (97)	392
	HCl, ether, 22 h	I (45) + (*p*-$O_2NC_6H_4CH_2S)_2CHC_6H_5$ (55)	392

TABLE VI. DIRECT FORMATION OF CARBONYL COMPOUNDS AND THIOLS (*Continued*)

Sulfoxide	Reaction Conditions	Product(s) and Yield(s) (%)	Refs.
$C_6H_5S(O)CH_2C_6H_4CN$-*p*	AcOH, H_2O, H_2SO_4, reflux 24 h	*p*-NCC_6H_4CHO (85)	389
$C_6H_5S(O)CH_2C_6H_4OCH_3$-*p*	"	C_6H_5CHO (tr)	389
$C_6H_5S(O)CH_2COC_6H_5$	AcOH, reflux 11 h	$C_6H_5SCH(OAc)COC_6H_5$ (45) + $(C_6H_5S)_2CHCOC_6H_5$ I (22)	30
	1. PPSE, $(CHCl_2)_2$, 80°, 3 h 2. NaOH, H_2O	$C_6H_5SCOCOC_6H_5$ (22) + I (30)	58
	1. PPSE, 150° 2. NaOH, H_2O	" (36) + I (8)	58
s-$C_4H_9C(CH_3)(NO_2)(CH_2)_2S(O)C_6H_5$	1. TFAA, 2,6-lutidine, CH_3CN, 0°, 10 min 2. $NaHCO_3$, H_2O, 4 h	*s*-$C_4H_9C(CH_3)(NO_2)CH_2CHO$ (—)	388
1-NO_2-1-$(CH_2)_2S(O)C_6H_5$-cyclohexane (structure)	"	1-NO_2-1-CH_2CHO-cyclohexane (structure) (—)	388
$CH_2S(O)C_6H_4CH_3$-*p* substituted cyclopenta-isoxazoline (structure)	1. TFAA, 2,6-lutidine, CH_3CN, 0° 30 min, 30° 30 min 2. TsOH, $Hg(OAc)_2$, CH_3OH, 1 h	$CH(OCH_3)_2$ substituted cyclopenta-isoxazoline (structure) (75)	62
C_{15}			
$C_6H_5CH_2S(O)CO_2CH_2C_6H_5$	Ac_2O, 60°, 2 h	$C_6H_5CH_2S(O)CH(SAc)C_6H_5$ (65) + $C_6H_5CH_2OAc$ (44) + $C_6H_5CH_2S(O)SCH_2C_6H_5$ (tr)	285
p-$C_6H_5COC_6H_4SCH_2CO_2H$	H_2O_2 (1.5 M), H_2SO_4, reflux 2 h	*p*-$C_6H_5COC_6H_4SH$ (27)[b] + (*p*-$C_6H_5COC_6H_4S)_2$ I (35)[b]	386
	H_2O_2 (2 M), H_2SO_4, reflux 90 min	I (86)	386

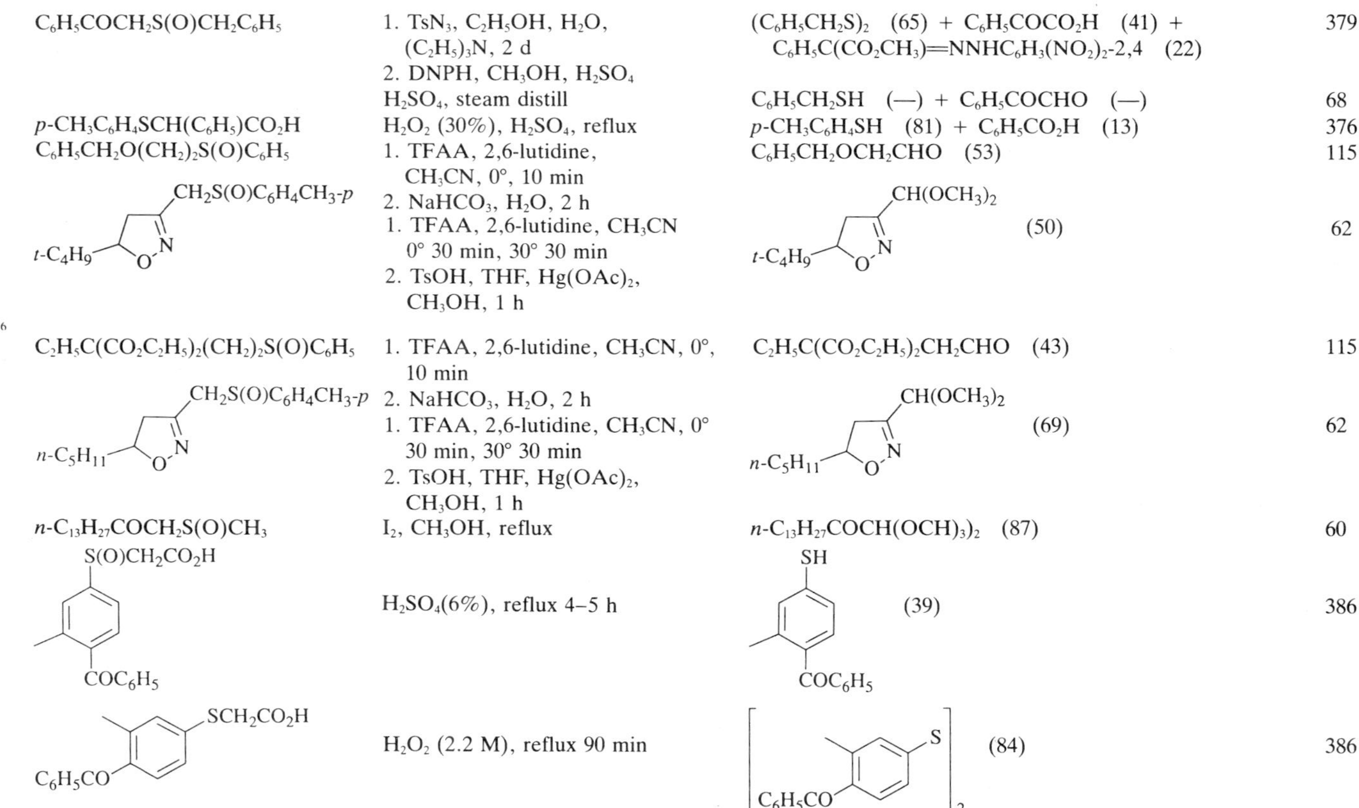

$C_6H_5COCH_2S(O)CH_2C_6H_5$	1. TsN_3, C_2H_5OH, H_2O, $(C_2H_5)_3N$, 2 d 2. DNPH, CH_3OH, H_2SO_4	$(C_6H_5CH_2S)_2$ (65) + $C_6H_5COCO_2H$ (41) + $C_6H_5C(CO_2CH_3)\!=\!NNHC_6H_3(NO_2)_2$-2,4 (22)	379
	H_2SO_4, steam distill	$C_6H_5CH_2SH$ (—) + C_6H_5COCHO (—)	68
p-$CH_3C_6H_4SCH(C_6H_5)CO_2H$	H_2O_2 (30%), H_2SO_4, reflux	p-$CH_3C_6H_4SH$ (81) + $C_6H_5CO_2H$ (13)	376
$C_6H_5CH_2O(CH_2)_2S(O)C_6H_5$	1. TFAA, 2,6-lutidine, CH_3CN, 0°, 10 min 2. $NaHCO_3$, H_2O, 2 h	$C_6H_5CH_2OCH_2CHO$ (53)	115
t-C_4H_9 … $CH_2S(O)C_6H_4CH_3$-p (O, N)	1. TFAA, 2,6-lutidine, CH_3CN 0° 30 min, 30° 30 min 2. TsOH, THF, $Hg(OAc)_2$, CH_3OH, 1 h	t-C_4H_9 … $CH(OCH_3)_2$ (O, N) (50)	62
C_{16}			
$C_2H_5C(CO_2C_2H_5)_2(CH_2)_2S(O)C_6H_5$	1. TFAA, 2,6-lutidine, CH_3CN, 0°, 10 min 2. $NaHCO_3$, H_2O, 2 h	$C_2H_5C(CO_2C_2H_5)_2CH_2CHO$ (43)	115
n-C_5H_{11} … $CH_2S(O)C_6H_4CH_3$-p (O, N)	1. TFAA, 2,6-lutidine, CH_3CN, 0° 30 min, 30° 30 min 2. TsOH, THF, $Hg(OAc)_2$, CH_3OH, 1 h	n-C_5H_{11} … $CH(OCH_3)_2$ (O, N) (69)	62
n-$C_{13}H_{27}COCH_2S(O)CH_3$	I_2, CH_3OH, reflux	n-$C_{13}H_{27}COCH(OCH)_3)_2$ (87)	60
$S(O)CH_2CO_2H$ … COC_6H_5	H_2SO_4(6%), reflux 4–5 h	SH … COC_6H_5 (39)	386
C_6H_5CO … SCH_2CO_2H	H_2O_2 (2.2 M), reflux 90 min	[C_6H_5CO … S]$_2$ (84)	386

TABLE VI. DIRECT FORMATION OF CARBONYL COMPOUNDS AND THIOLS (*Continued*)

Sulfoxide	Reaction Conditions	Product(s) and Yield(s) (%)	Refs.
C_6H_5CO–$C_6H_3(CH_3)$–SCH_2CO_2H	H_2O_2(2.5 M), reflux 90 min	C_6H_5CO–$C_6H_3(CH_3)$–SH (23) +	386
		[C_6H_5CO–$C_6H_3(CH_3)$–S]$_2$ (21)	
$[HOCH_2C(CH_3)_2CHOHCONH(CH_2)_2]_2SO$	Stand 3 months	$[HOCH_2C(CH_3)_2CHOHCONH(CH_2)_2S]_2$ (>60)	393
Phenothiazine-N–$CH_2COCH_2S(O)CH_3$	1. TsOH, THF, 65°, 15 min 2. Heat or SiO_2	Phenothiazine-N–CH=COHCHO (—) +	294
		Phenothiazine-N–$CH_2COCH(SCH_3)_2$ (—) +	
		Phenothiazine-N–$CH_2COCH_2SCH_3$ (—)	

C_{18}	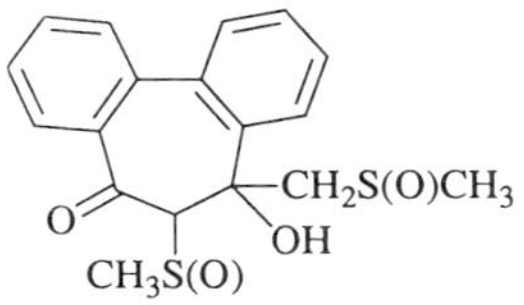	HCl, $(CH_3)_2SO$, 25°, 24 h	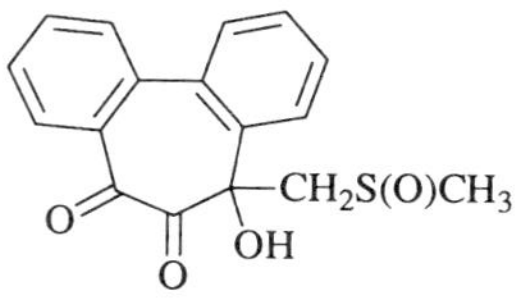 (—)	365
	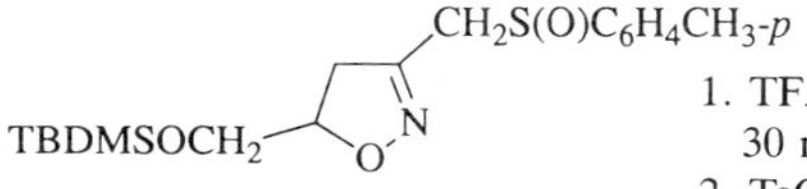	1. TFAA, 2,6-lutidine, CH_3CN, 0° 30 min, 30° 30 min 2. TsOH, THF, $Hg(OAc)_2$, CH_3OH, 1 h	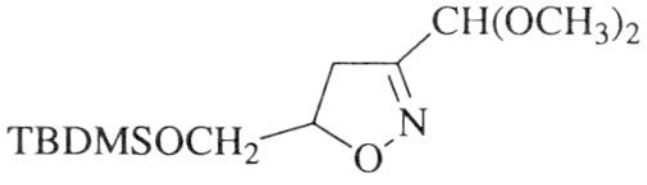 (67)	62
C_{19}	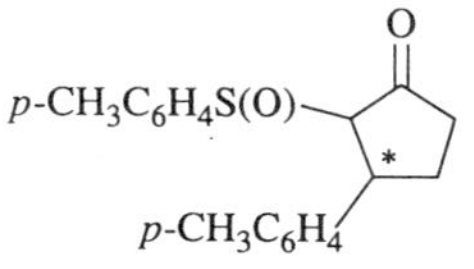	$(CH_2OH)_2$, TsOH	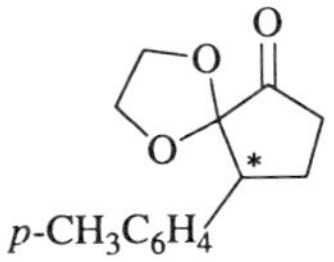 (94)	394
	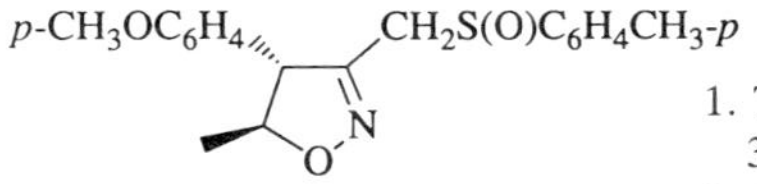	1. TFAA, 2,6-lutidine, CH_3CN, 0° 30 min, 30° 30 min 2. TsOH, THF, $Hg(OAc)_2$, CH_3OH, 1 h	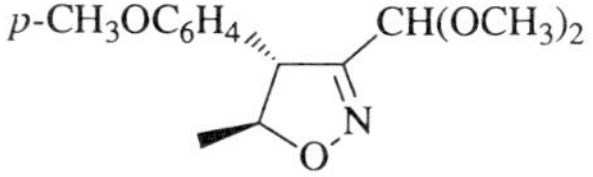 (45)	62
	$CH_2{=}CH(CH_2)_2C^*H(OCH_2C_6H_5)CH_2S(O)C_6H_5$	1. "Pummerer conditions" 2. DIBAH	$CH_2{=}CH(CH_2)_2C^*H(OCH_2C_6H_5)CHO$ (73)	395
C_{20}				

TABLE VI. Direct Formation of Carbonyl Compounds and Thiols (*Continued*)

Sulfoxide	Reaction Conditions	Product(s) and Yield(s) (%)	Refs.
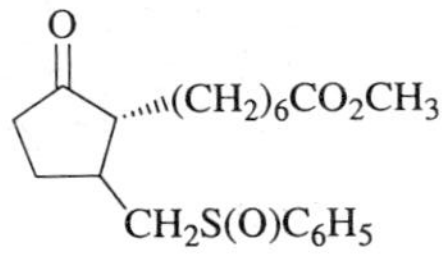 2-[$(CH_2)_6CO_2CH_3$]-3-[$CH_2S(O)C_6H_5$]cyclopentanone	1. Ac_2O, AcONa 2. CH_3OH, H_2O, H_2SO_4, $HgCl_2$	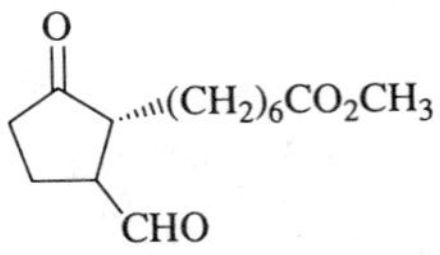 2-[$(CH_2)_6CO_2CH_3$]-3-CHO-cyclopentanone (32) 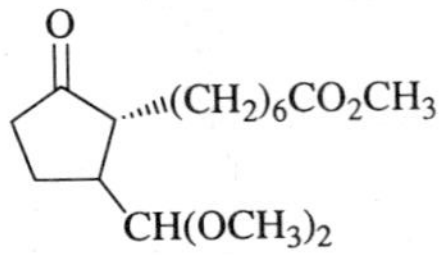 + 2-[$(CH_2)_6CO_2CH_3$]-3-[$CH(OCH_3)_2$]cyclopentanone (32)	396
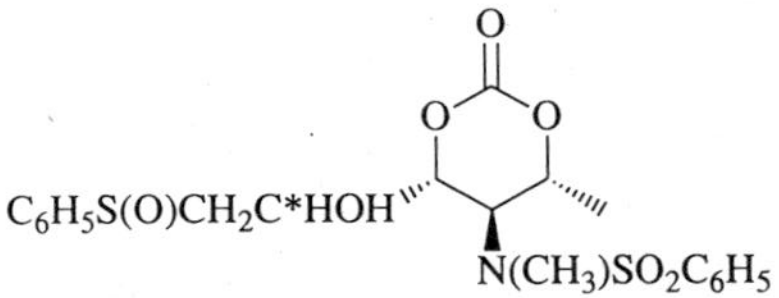 $C_6H_5S(O)CH_2C^*HOH$-(cyclic carbonate, $N(CH_3)SO_2C_6H_5$)	1. TFAA, CH_2Cl_2, 0° 30 min, rt 30 min 2. CH_3OH, $Hg(OAc)_2$, CH_3SO_3H, 10 min	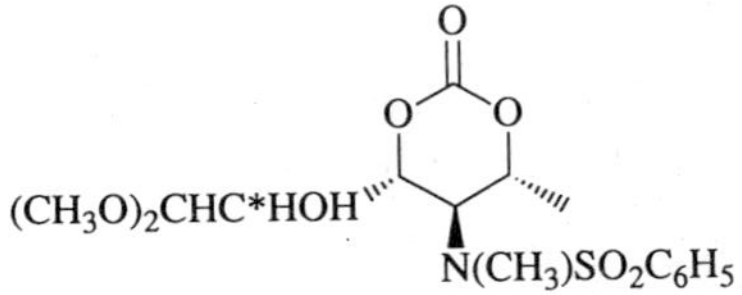 $(CH_3O)_2CHC^*HOH$-(cyclic carbonate, $N(CH_3)SO_2C_6H_5$) (78)	397
(*Z*)-*n*-$C_8H_{17}CH{=}CH(CH_2)_7COCH_2S(O)CH_3$	I_2, CH_3OH, reflux	(*Z*)-*n*-$C_8H_{17}CH{=}CH(CH_2)_7COCH(OCH_3)_2$ (100)	60
C_{22}			
n-$C_{16}H_{33}S(O)C_6H_5$	1. TFAA, 2,6-lutidine, CH_3CN, CH_2Cl_2, 0° 10 min 2. $NaHCO_3$, H_2O, 2 h	*n*-$C_{15}H_{31}CHO$ (68)	115
C_{23}			
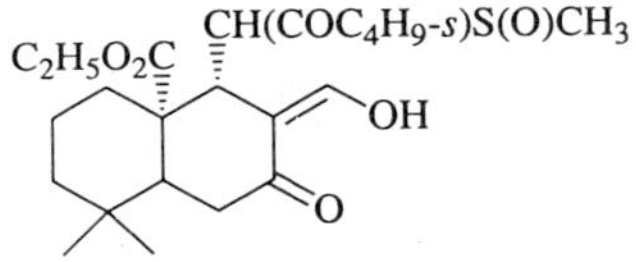 Decalone bearing $C_2H_5O_2C$, $CH(COC_4H_9\text{-}s)S(O)CH_3$, =CHOH	AcOH, H_2O, reflux 4 h	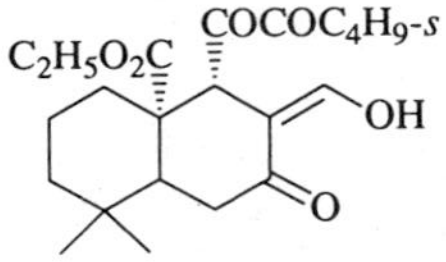 Decalone bearing $C_2H_5O_2C$, $COCOC_4H_9\text{-}s$, =CHOH (100)	116

Reactant	Conditions	Product(s) and Yield(s) (%)	Refs.
CF_3CO-N (azabicyclic) with $CH_2S(O)C_6H_5$ and $CH=CH(CH_2)_3CO_2CH_3$ side chains	1. TFAA 2. $NaHCO_3$, H_2O	CF_3CO-N (azabicyclic) with CHO and $CH=CH(CH_2)_3CO_2CH_3$ side chains (—)	398
C_{25}			
$CH_3CO_2(CH_2)_4CH(O_2CC_6H_5)CH=CHCH=CHCH_2S(O)C_6H_5$	1. Ac_2O, TFAA, 2,6-lutidine, AcONa, 0° 80 min, 25° 20 min 2. $HgCl_2$, $CaCO_3$, CH_3CN, H_2O, 0°, 2.5 h	$CH_3CO_2(CH_2)_4CH(O_2CC_6H_5)CH=CHCH=CHCHO$ (65)	146
C_{26}			
2,2-dimethyl-1,3-dioxolane with $CH_2S(O)C_6H_5$ and $CH_2OCH(C_6H_5)_2$	Ac_2O, AcONa, rt to reflux 0.5 h, reflux 3 h	2,2-dimethyl-1,3-dioxolane with CHO and $CH_2OCH(C_6H_5)_2$ (84–93)	153
C_{31}			
bis(2,2-dimethyl-1,3-dioxolane) with $CH_2OCH(C_6H_5)_2$ and $C_6H_5S(O)CH_2$	"	bis(2,2-dimethyl-1,3-dioxolane) with $CH_2OCH(C_6H_5)_2$ and CHO (71–90)	153
C_{32}			
t-$C_4H_9(CH_3)_2SiOCH(CH_2C{\equiv}C(CH_2)_7S(O)C_6H_5)CH=CHCH=CHC_5H_{11}$-$n$	1. TFAA, pyr 2. $NaHCO_3$, H_2O	t-$C_4H_9(CH_3)_2SiOCH(CH_2C{\equiv}C$-$(CH_2)_6CHO)CH=CHCH=CHC_5H_{11}$-$n$ (—)	399

[a] The reaction was studied from a kinetic point of view.
[b] The reaction was carried out to partial conversion.
[c] The product was not formed when $AgBF_4$ was present in the reaction mixture.
[d] The range in yield reflects different solvents and reaction conditions.

TABLE VII. DIRECT FORMATION OF VINYL SULFIDES

Sulfoxide	Reaction Conditions	Product(s) and Yield(s) (%)	Refs.
C_4			
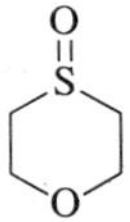 O, S, O	Ac_2O, TsOH, toluene, reflux 3 h	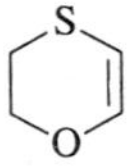 S, O (15) 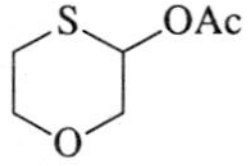 + S, OAc, O (4) 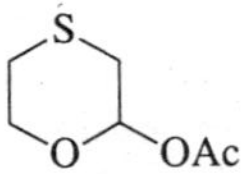 + S, O, OAc (6)	400
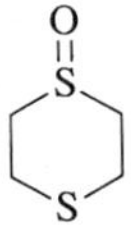 O, S, S	Ac_2O, 100°, 65 h	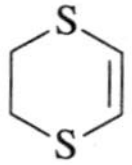 S, S (53)	223
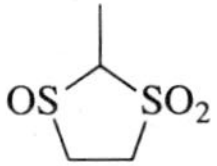 OS, SO_2	TFAA, $(C_2H_5)_3N$, − 15° to rt	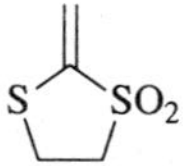 S, SO_2 (58–67)	401
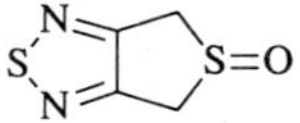 N, S, N, S=O	Ac_2O, reflux	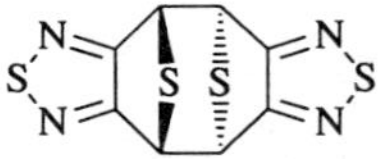 N, S, N, S, S, N, S, N (—)	402
	Ac_2O, 100°, $CH_3O_2CC{\equiv}CCO_2CH_3$	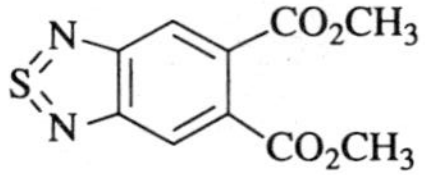 N, S, N, CO_2CH_3, CO_2CH_3 (70)	402
	Ac_2O, 100°, 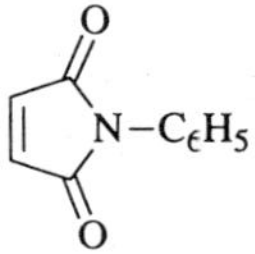 O, N−C_6H_5, O	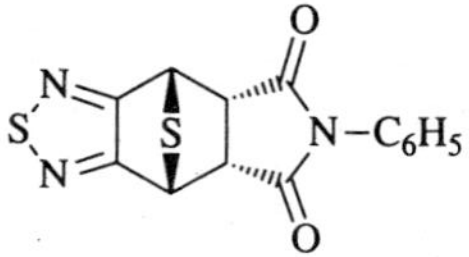 N, S, N, S, O, N−C_6H_5, O (46)	402

	Reactant	Conditions	Product(s) and Yield(s) (%)	Refs.
C_5				
	[structure]	$(C_6H_5CO)_2O$, C_6H_6, reflux 14 h	 [structure] (54)	403
	[structure]	$(CH_3)_3SiCl$, CH_2Cl_2, 12 h	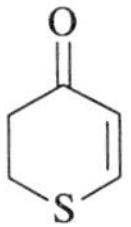 [structure] (86)[a,b]	54, 241
	[structure]	$(CH_3)_2SO$, 100°, 65 h	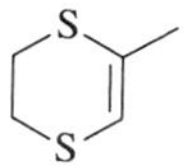 [structure] (80)	129
C_6				
	$t\text{-}C_4H_9S(O)CHClCH_3$	$(CH_3)_3SiOSO_2CF_3$, $(C_2H_5)_3N$, ether, 0°, 15 min	$t\text{-}C_4H_9SCCl{=}CH_2$ (78)	404
	[structure]	H_2O_2 (30%), AcOH, acetone, 50°, 24 h	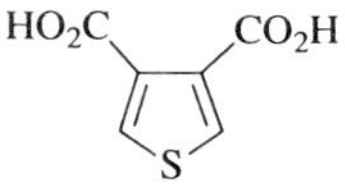 [structure] (68)	405
	[structure]	"	" (97)	405
	[structure]	H_2O, heat	" (89)	405

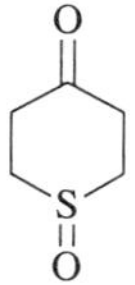

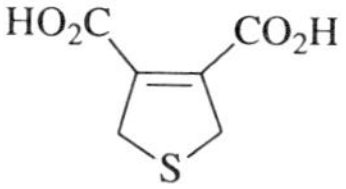

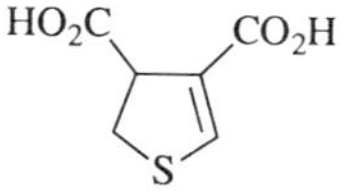

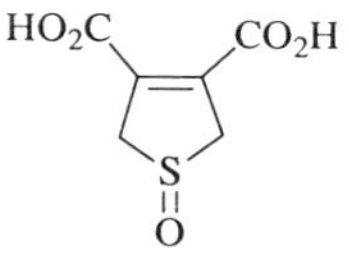

TABLE VII. DIRECT FORMATION OF VINYL SULFIDES (*Continued*)

Sulfoxide	Reaction Conditions	Product(s) and Yield(s) (%)	Refs.
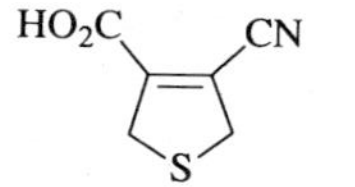 HO_2C, CN (structure)	H_2O_2, (30%), AcOH, acetone, 25°, 38 h	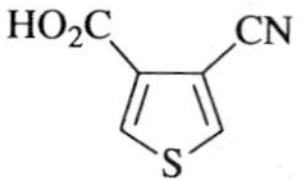 HO_2C, CN (structure) (92)	405
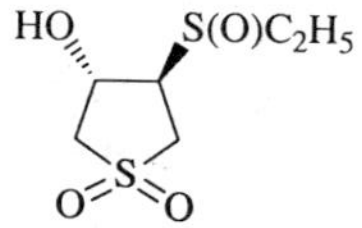 HO, $S(O)C_2H_5$ (structure)	Ac_2O, AcOH	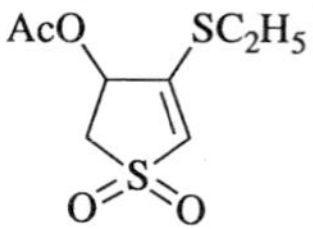 AcO, SC_2H_5 (structure) (—)	406
C_7			
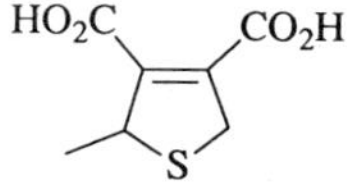 HO_2C, CO_2H (structure)	H_2O_2 (30%), Ac_2O, acetone, 25°, 38 h	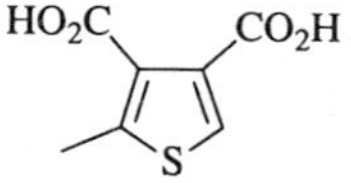 HO_2C, CO_2H (structure) (90)	405
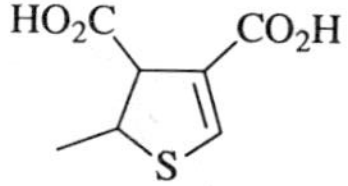 HO_2C, CO_2H (structure)	"	" (83)	405
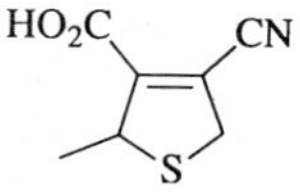 HO_2C, CN (structure)	"	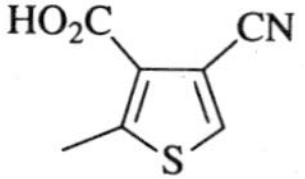 HO_2C, CN (structure) (92)	405
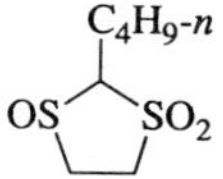 C_4H_9-*n*, OS, SO_2 (structure)	TFAA, pyr, −15° to rt	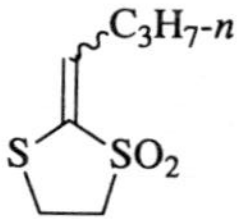 C_3H_7-*n*, S, SO_2 (structure) (58–67)	401
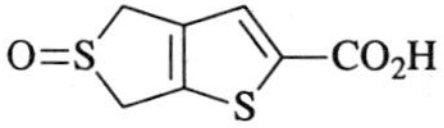 O=S, CO_2H (structure)	Ac_2O, reflux 1.5 h	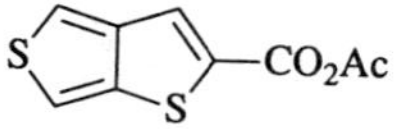 S, CO_2Ac (structure) (95)	126

Reactant	Conditions	Product(s) and Yield(s) (%)	Refs.
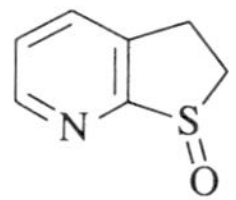	1. Ac_2O, reflux 24 h 2. Evaporate Ac_2O 3. Na_2CO_3, acetone, H_2O, 40°, 2 h	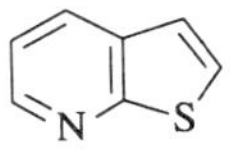 (64)	407
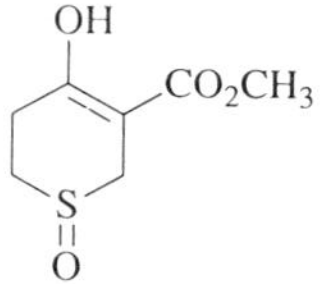		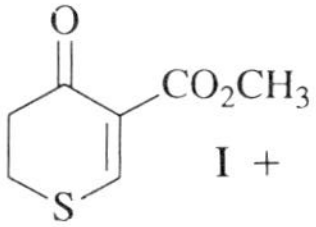 I + 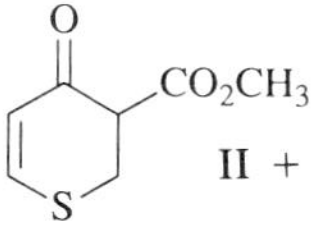 II + 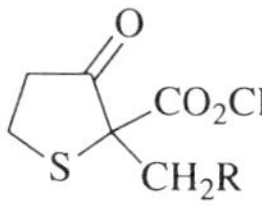 III + 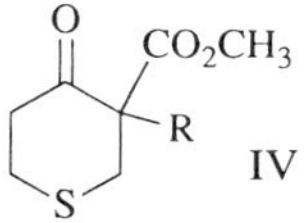 IV	54, 159
	Ac_2O, reflux 2 h	III + IV R = OAc, 5:1 (87)	
	$(CH_3)_3SiCl$ (5 eq), CCl_4, reflux 10 min	I (72) + II (16) + III + IV R = Cl, 5:1 (7)	
	$(CH_3)_3SiCl$ (2.2 eq), CH_2Cl_2, reflux 10 min	I (74) + II (13) + III + IV R = Cl, 5:1 (12)	
	$(CH_3)_3SiCl$ (2.2 eq), CH_2Cl_2, 10 min	I (54) + II (4) + III + IV R = Cl, 5:1 (40)	
	$(CH_3)_3SiCl$ (2.2 eq), $(i\text{-}C_3H_7)_2NC_2H_5$, CH_2Cl_2, 60 h	III + IV R = Cl, 5:1 (41)	
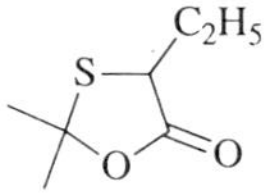	AcOH, H_2O_2 (30°), 0° to rt, overnight	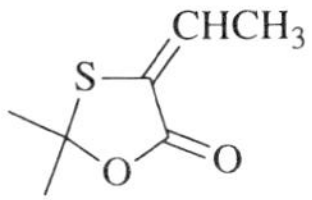 (low)[a,b]	36
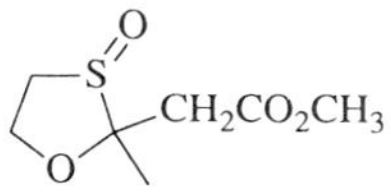	TsOH, C_6H_6-DMF (1:1), 50°, 60 h	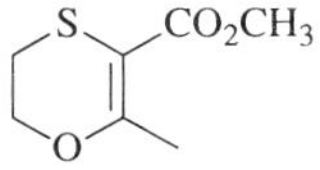 (90)	408

TABLE VII. Direct Formation of Vinyl Sulfides (*Continued*)

	Sulfoxide	Reaction Conditions	Product(s) and Yield(s) (%)	Refs.
	(spiro sulfoxide, S=O, S)	TsOH, C_6H_6, reflux with water removal, 18 h	(S, S bicyclic) (96)	130
C_8	$(n\text{-}C_4H_9)_2SO$	Ac_2O, $CHCl_3$, 25°, 4 d	$n\text{-}C_4H_9SCH{=}CHC_2H_5$ (68)	201
		Ac_2O, C_6H_6, 80°, 6 h	" (60)	201
		$(C_6H_5CO)_2O$, C_6H_6, 80°, 6 h	" (96)	201
	$C_2H_5S(O)(CH_2)_2PO(OC_2H_5)_2$	Ac_2O, reflux 2 h	$(E)\text{-}C_2H_5SCH{=}CHPO(OC_2H_5)_2$ (77)	409
		AcCl, 4 h	" (65)	409
		$SOCl_2$, C_6H_6, 3 h	" (68)	409
		TFAA, CH_2Cl_2, −78 to −20°, 15 min	" (78)	409
	$C_6H_5S(O)CHClCH_3$	$(CH_3)_3SiOSO_2CF_3$, $(C_2H_5)_3N$, ether, 25°, 1 h	$C_6H_5SCCl{=}CH_2$ (86)	404
	$C_6H_5S(O)CHBrCH_3$	$(CH_3)_3SiOSO_2CF_3$, $(C_2H_5)_3N$, ether, 0°, 30 min	$C_6H_5SCBr{=}CH_2$ (77)	404
	CH_3O_2C, CO_2CH_3, S	H_2O_2 (30%), Ac_2O, acetone, 25°, 38 h	CH_3O_2C, CO_2CH_3, S (23)[b]	405
	CH_3O_2C, CO_2CH_3, S	"	" (10)[b]	405
	S, S=O	Ac_2O, (O, N−C_6H_5, O) reflux 4 h	S, S, O, N−C_6H_5, O (67) *exo*:*endo* = 2.7:1	120, 410

S=O	Al_2O_3 (neutral), 100–125°, 20 torr	S (94)	127
	Al_2O_3 (neutral), 120–130°, 25 torr	" (15)	411
	Ac_2O, O, O, O, reflux 15 h	O, S, O, O (24–25)	127, 411
O, N–C_6H_5, O	Ac_2O, O, N–C_6H_5, O, reflux 2 h	O, S, N–C_6H_5, O (86) *exo*:*endo* = 1:1.2	127, 411
	Ac_2O, 220°	" *exo*:*endo* = 2:1 (72)	411
S, S=O	Ac_2O, 150°	S, S (50)	412
F, O, S, N, NH, O, O	Ac_2O, 115°, 10.5 h	F, O, S, N, NH, O, O (24)	400
O, S, C_4H_9-*t*, S	DMF, 100°, 15 h	S, S, C_4H_9-*t* (80)	129

TABLE VII. Direct Formation of Vinyl Sulfides (*Continued*)

	Sulfoxide	Reaction Conditions	Product(s) and Yield(s) (%)	Refs.
	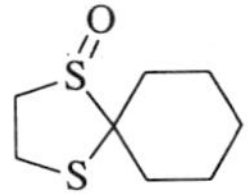	TsOH, C_6H_6, reflux with water removal, 18 h	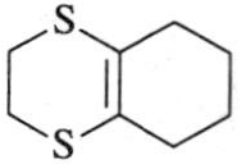 (93)	130, 134
	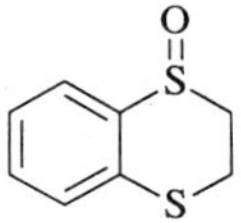	Ac_2O, reflux 10 min	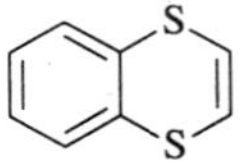 (49)	134
C_9	$C_6H_5S(O)CH(CN)CH_3$	$(CH_3)_3SiOSO_2CF_3$, $[(CH_3)_3Si]_2NH$, ether, 25° 3 h	$C_6H_5SC(CN){=}CH_2$ (84)	413
	$CH_3S(O)CH_2CHOHC_6H_5$	1. $SOCl_2$, CH_2Cl_2, 10 h 2. KOH, C_2H_5OH, reflux overnight	$CH_3SC{\equiv}CC_6H_5$ (63)	414
	$CH_3SCH[PO(OC_2H_5)_2]CH_2CH{=}CH_2$	Ac_2O, CH_3SO_2OH, CH_2Cl_2, 15 h	$CH_3SC[PO(OC_2H_5)_2]{=}CHCH{=}CH_2$ E:Z = 3:1 (67)	233
	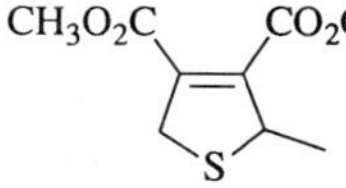	H_2O_2 (30%), Ac_2O, acetone, 25°, 38 h	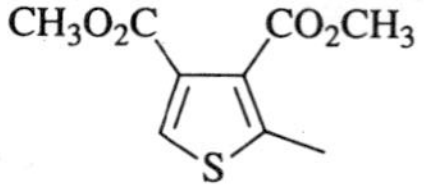 (13)[b]	405
	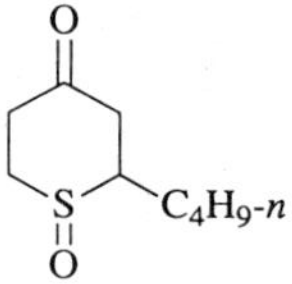	$(CH_3)_3SiCl$, CH_2Cl_2, 15 h	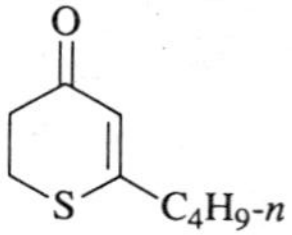 (49) + 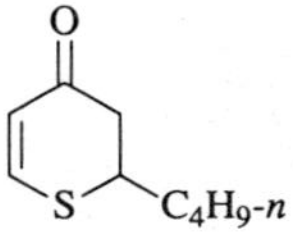 (15) + 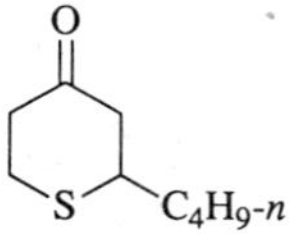 (13)	54, 241

	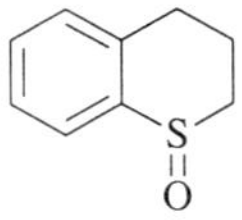	Ac_2O, 90°, 5 h	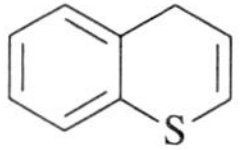 (78)	415
	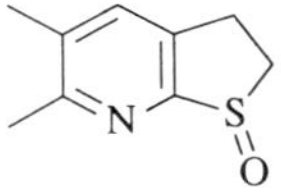	1. Ac_2O, reflux 24 h 2. Evaporate Ac_2O 3. Na_2CO_3, acetone, H_2O, 40°, 2 h	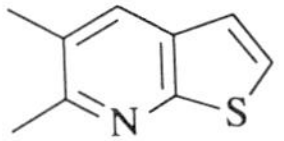 (89)	407
	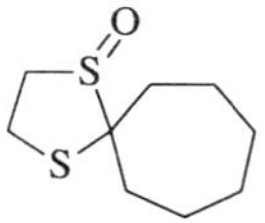	TsOH, C_6H_6, reflux with water removal, 18 h	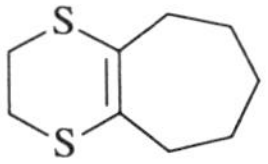 (95)	130
C_{10}	$C_6H_5S(O)CH(COCH_3)CH_3$	Ac_2O, CH_3SO_3H, CH_2Cl_2, 40°, 2 h	$C_6H_5SC(COCH_3){=}CH_2$ (—)	416
	$C_6H_5S(O)CH(CN)C_2H_5$	$Sn(OSO_2CF_3)_2$, DABCO, $(CH_3)_3SiCl$, CH_2Cl_2, 0°	$C_6H_5SC(CN){=}CHCH_3$ (66)	122
	$C_6H_5S(O)CHClC_3H_7\text{-}i$	$(CH_3)_3SiOSO_2CF_3$, $(C_2H_5)_3N$, ether, 0°, 30 min	$C_6H_5SCCl{=}C(CH_3)_2$ (92)	404
	$C_6H_5S(O)CHBrC_3H_7\text{-}i$	$(CH_3)_3SiOSO_2CF_3$, $(C_2H_5)_3N$, ether, 0°, 15 min	$C_6H_5SCBr{=}C(CH_3)_2$ (89)	404
	$C_6H_5S(O)CHClC_3H_7\text{-}n$	$(CH_3)_3SiOSO_2CF_3$, $(C_2H_5)_3N$, ether, 0°, 30 min	$C_6H_5SCCl{=}CHC_2H_5$ $Z{:}E$ = 1:1 (91)	404
	$C_6H_5S(O)CHBrC_3H_7\text{-}n$	$(CH_3)_3SiOSO_2CF_3$, $(C_2H_5)_3N$, ether, 0°, 15 min	$C_6H_5SCBr{=}CHC_2H_5$ $Z{:}E$ = 5:4 (89)	404
	$C_6H_5S(O)CHClCH_2CH{=}CH_2$	$(CH_3)_3SiOSO_2CF_3$, $(C_2H_5)_3N$, ether, 25°, 1 h	$C_6H_5SCCl{=}CHCH{=}CH_2$ $Z{:}E$ = ~2:1 (72)	404
	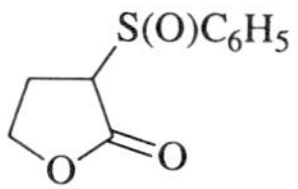	Ac_2O, 60–70° overnight, reflux 2 h	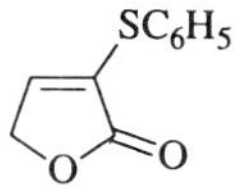 (86)	417, 418

TABLE VII. DIRECT FORMATION OF VINYL SULFIDES (*Continued*)

Sulfoxide	Reaction Conditions	Product(s) and Yield(s) (%)	Refs.
	Ac_2O, CH_3SO_2OH, CH_2Cl_2, 16 h	" (97)	121
	1. PPSE, $(CH_2Cl)_2$, 80°, 3 h 2. NaOH, H_2O	" (93)	58
	1. $SOCl_2$, CCl_4, 0°, 2 h 2. THF, LiBr, Li_2CO_3, reflux 30 min	" (93)	418
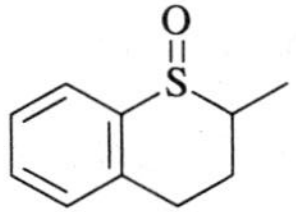	Ac_2O, reflux	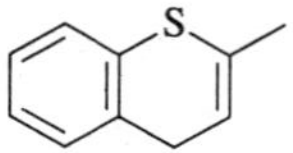 (—)	419
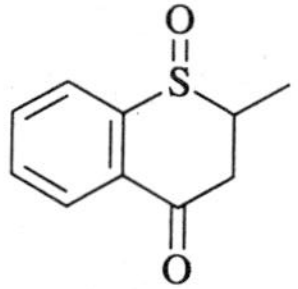	Ac_2O, reflux	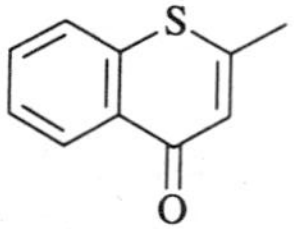 (—)	419
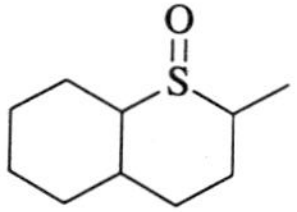	Ac_2O	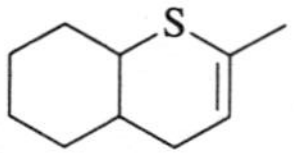 (—)	420
	Ac_2O	(—) + OAc (—)	420

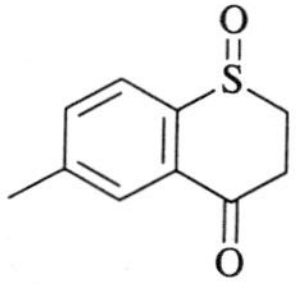

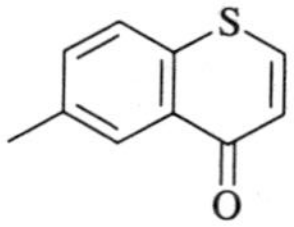

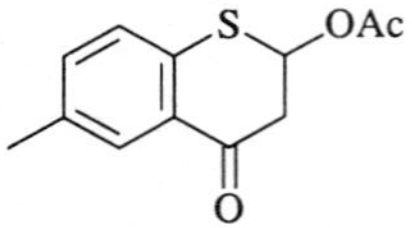

Reactant	Conditions	Product(s) (Yield, %)	Refs.
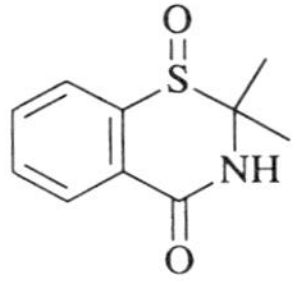 O, S, NH, O	Ac_2O, AcONa, reflux	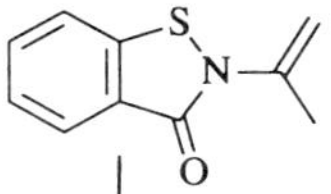 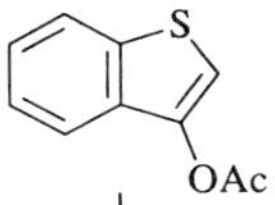 S, N, O (—) + S, OAc (—); 3:2; Ac_2O, AcONa	421
Ac, N, S, O	Ac_2O, reflux 3 h	Ac, N, S (39)	278
CH_3O_2C, S, S=O, CH_3O_2C	Ac_2O, O, N$-C_6H_5$, O; reflux 3 h	CH_3O_2C, S, S, O, N$-C_6H_5$, O, CH_3O_2C (67); *exo*:*endo* = 4.2:1	410, 422
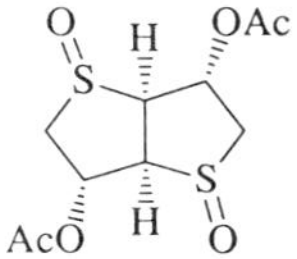 O, S, H, OAc, S, O, H, AcO	Ac_2O, 110°, 20 h	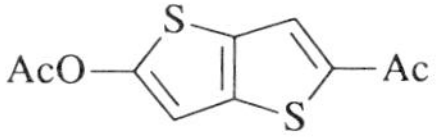 AcO—, S, S, —Ac (27) 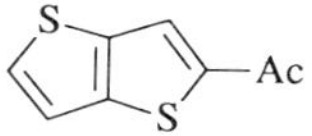 + S, S, —Ac (6)	423
	Ac_2O, $NaIO_4$ or $NaIO_3$, 110°, 12 h	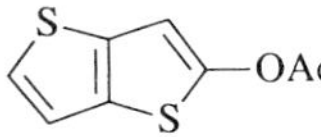 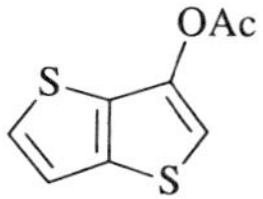 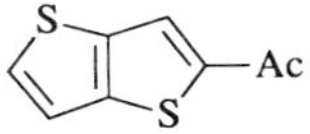 S, S, —OAc I + OAc, S, S II + S, S, —Ac III	423

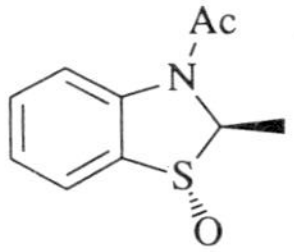

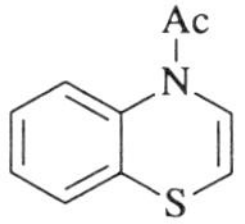

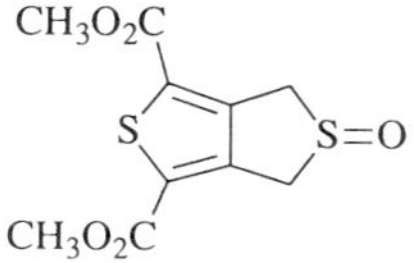

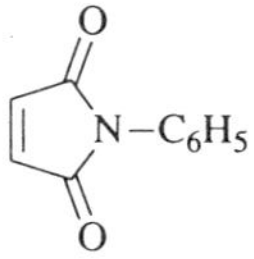

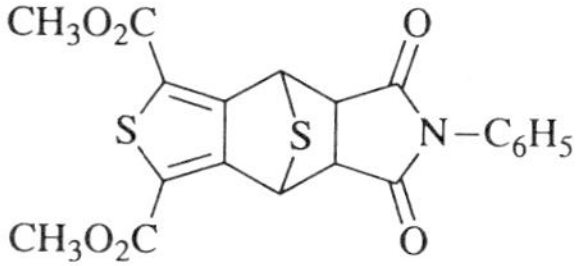

TABLE VII. Direct Formation of Vinyl Sulfides (*Continued*)

Sulfoxide	Reaction Conditions	Product(s) and Yield(s) (%)	Refs.
		I + II (51)	423
	Ac_2O, $FeCl_3$, 110°, 12 h	I + II (9) + III (28)	423
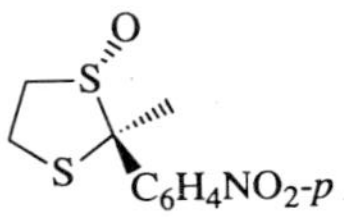 $C_6H_4NO_2$-*p*	$(CH_3)_2SO$, 100°, 15 h	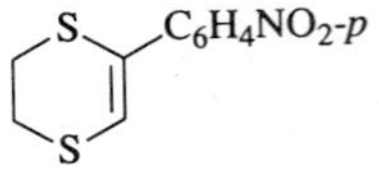 $C_6H_4NO_2$-*p* (50)	129
C_{11}			
$C_6H_5S(O)CH(CO_2C_2H_5)CH_3$	Ac_2O, CH_3SO_2OH, CH_2Cl_2, 16 h	$C_6H_5SC(CO_2C_2H_5){=}CH_2$ (92)	121
	Ac_2O, CH_3SO_2OH, CH_2Cl_2, 40°, few h	" (83)	416
	$(CH_3)_3SiOSO_2CF_3$, HMDS, ether, 25°, 6 h	" (83)	424
$C_6H_5S(O)CH(COCH_3)C_2H_5$	Ac_2O, CH_3SO_2OH, CH_2Cl_2, 40°, 8 h	$C_6H_5SC(COCH_3){=}CHCH_3$ $Z{:}E = 30{:}1$ (89)	416
$C_6H_5S(O)CH(CN)C_3H_7$-*n*	$Sn(OSO_2CF_3)_2$, DABCO, TMSCl, CH_2Cl_2, 0°	$C_6H_5SC(CN){=}CHC_2H_5$ (65)	122
	$(CH_3)_3SiOSO_2CF_3$, HMDS, ether, 25°, 3 h	" (74)	424
$C_6H_5S(O)CH(CN)C_3H_7$-*i*	$Sn(OSO_2CF_3)_2$, DABCO, TMSCl, CH_2Cl_2, 0°, overnight	$C_6H_5SC(CN){=}C(CH_3)_2$ I (78) + $C_6H_5SCl(CN)C_3H_7$-*i* II (8)	122, 425
	$Sn(OSO_2CF_3)_2$, 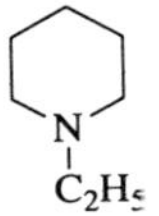 N–C_2H_5	I (40)	425
	$(CH_3)_3SiOSO_2CF_3$, CH_2Cl_2, 0°, overnight		
	$Sn(OSO_2CF_3)_2$, TMSCl, CH_2Cl_2, 0°, overnight	I (17) + II (63)	425

Reactant	Conditions	Product(s) and Yield(s) (%)	Refs.
	$(CH_3)_3SiOSO_2CF_3$, HMDS, ether, 35°, 3 h	I (87)	424
5-Methyl-3-$S(O)C_6H_5$-butyrolactone	Ac_2O, 60–70°, overnight, reflux 2 h	5-Methyl-3-SC_6H_5-butenolide (—)	417, 418
3-$CH_2S(O)C_6H_5$-butyrolactone	TFAA, Ac_2O, 2.5 h	3-($=CHSC_6H_5$)-butyrolactone (—) + 3-$CH(OAc)SC_6H_5$-butyrolactone (—)	426
2-$S(O)C_6H_5$-cyclopentanone	Ac_2O, CH_3SO_2OH, CH_2Cl_2, 16 h	2-SC_6H_5-cyclopent-2-enone (73)	121
Tetrahydro-dihydrothieno[2,3-b]quinoline S-oxide	1. Ac_2O, reflux 24 h 2. Evaporate Ac_2O 3. Na_2CO_3, acetone, H_2O, 40°, 2 h	Tetrahydrothieno[2,3-b]quinoline (74)	407
2,2-Dimethylthiochroman S-oxide	$h\nu$, (Vycor filter), C_6H_6	2-C_3H_7-*i*-benzothiophene (—)	65
	Ac_2O, reflux	2-Methyl-2-CH_2OAc-thiochroman (80)	419
2,2-Dimethylthiochroman-4-one S-oxide	Ac_2O, reflux	2-Isopropylidene-benzothiophen-3-one (—) + 3-OAc-2-isopropenylbenzothiophene (—)	419

TABLE VII. DIRECT FORMATION OF VINYL SULFIDES (*Continued*)

Sulfoxide	Reaction Conditions	Product(s) and Yield(s) (%)	Refs.
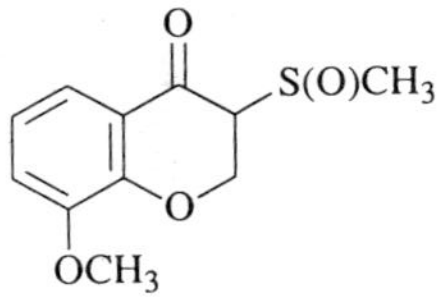 O, $S(O)CH_3$, O, OCH_3	Ac_2O, 100°, 4 h	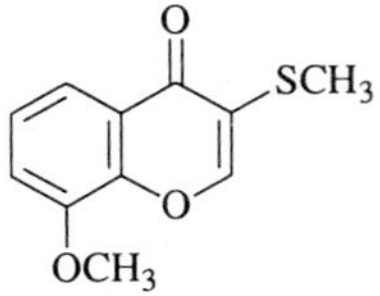 O, SCH_3, O, OCH_3 (85)	194
	$SOCl_2$	" (—)	194
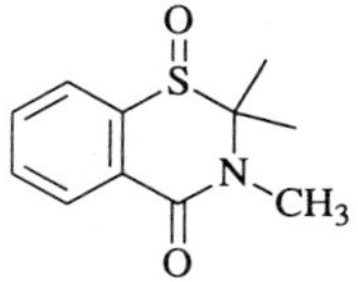 O, S, N, CH_3, O	Ac_2O, AcONa, reflux	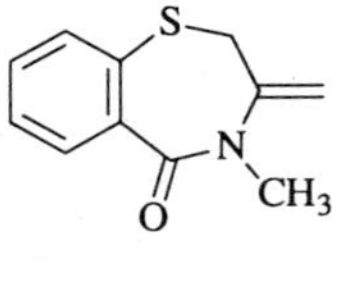 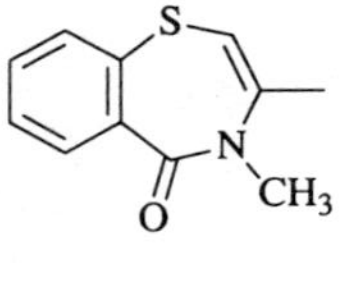 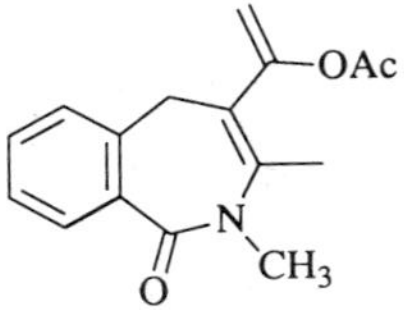 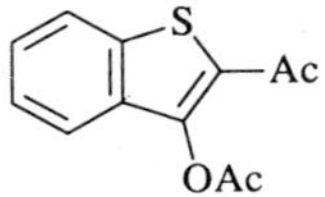 S, N, CH_3, O (—) + S, N, CH_3, O (—) +	421
		OAc, N, CH_3, O (50) + S, Ac, OAc (10)	
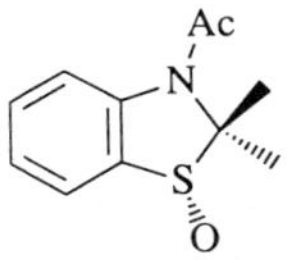 Ac, N, S, O	TFAA, CH_2Cl_2, 0–25°, 24 h	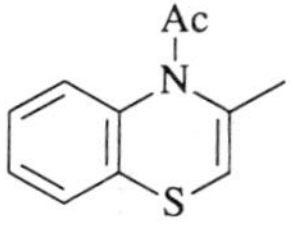 Ac, N, S (32)	278, 427
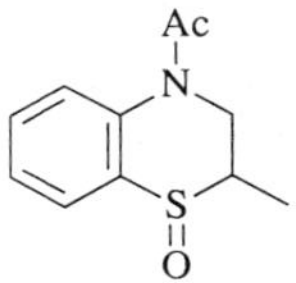 Ac, N, S, O	Ac_2O, reflux 100 min	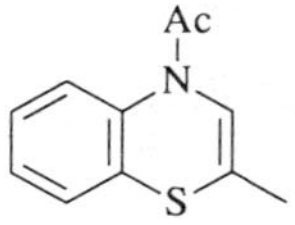 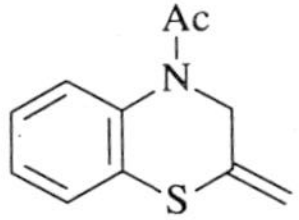 Ac, N, S + Ac, N, S 1:1 mixture (—)	278, 427

	Reactant	Conditions	Product(s) and Yield(s) (%)	Refs.
	2-$S(O)C_3H_7$-*i*-$C_6H_4CONHCH_3$	Ac_2O, 80°, 15 h	2-$SC(CH_3)=CH_2$-$C_6H_4CONRCH_3$ R = H (84) + R = Ac (6) + 2,2-dimethyl-3-methyl-benzothiazin-4-one (S, N–CH_3, C=O) (5)	267
	2-$C_6H_4OCH_3$-*p*-1,3-dithiolane 1-oxide (S=O)	$(CH_3)_2SO$, 100°, 15 h	5-$C_6H_4OCH_3$-*p*-2,3-dihydro-1,4-dithiin (81)	129
C_{12}	$C_6H_5S(O)CH(CO_2C_2H_5)C_2H_5$	Ac_2O, CH_3SO_2OH, CH_2Cl_2	$C_6H_5SC(CO_2C_2H_5)=CHCH_3$ $Z:E = 4:1$ (67)	428
		Ac_2O, CH_3SO_2OH, CH_2Cl_2, 40°, 8 h	" $Z:E = 4:1$ (89)	112
	$C_6H_5S(O)CH(CO_2CH_3)C_3H_7$-*n*	"	$C_6H_5SC(CO_2CH_3)=CHC_2H_5$ $Z:E = 3:1$ (74–82)	112, 428
	$C_6H_5S(O)CH(CO_2CH_3)C_3H_7$-*i*	"	$C_6H_5SC(CO_2CH_3)=C(CH_3)_2$ (90)	112, 428
	$C_6H_5S(O)CH(CN)C_4H_9$-*s*	$Sn(OSO_2CF_3)_2$, $(CH_3)SiCl$, DABCO, CH_2Cl_2, 0°	$C_6H_5SC(CN)=CHC_3H_7$-*i* (57)	122
	$C_6H_5S(O)CH(CN)CH_2CH(OCH_3)_2$	"	$C_6H_5SC(CN)=CHCH(OCH_3)_2$ (25)	122
	$C_6H_5S(O)CH(CO_2CH_3)CH_2CO_2CH_3$	Ac_2O, CH_3SO_2OH, 16 h	$C_6H_5SC(CO_2CH_3)=CHCO_2CH_3$ $E:Z = 3:2$ (—)[b]	429
	$C_6H_{11}S(O)C_6H_5$	TFAA, $(C_2H_5)_3N$, CH_2Cl_2, 0°, few min	1-SC_6H_5-cyclohexene (97)	430
	2-$S(O)C_6H_5$-cyclohexanone	Ac_2O, CH_3SO_2OH, CH_2Cl_2, 16 h	2-SC_6H_5-2-cyclohexen-1-one (86)	121, 416
	3-$S(O)C_6H_5$-6-methyltetrahydropyran-2-one	TFAA, CH_2Cl_2, 1 h	3-SC_6H_5-6-methyl-5,6-dihydropyran-2-one (71)	179

TABLE VII. DIRECT FORMATION OF VINYL SULFIDES (*Continued*)

Sulfoxide	Reaction Conditions	Product(s) and Yield(s) (%)	Refs.
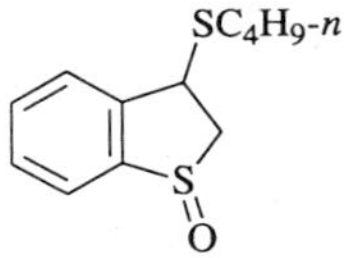 SC_4H_9-*n*; S=O		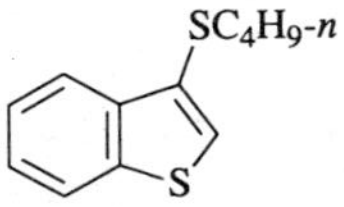 SC_4H_9-*n*; S (—)	431
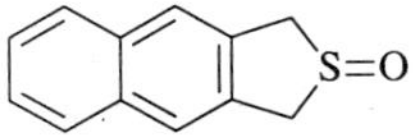 S=O	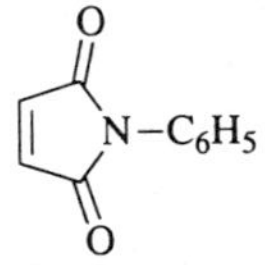 Ac_2O, N–C_6H_5; reflux 4 h, rt 12 h	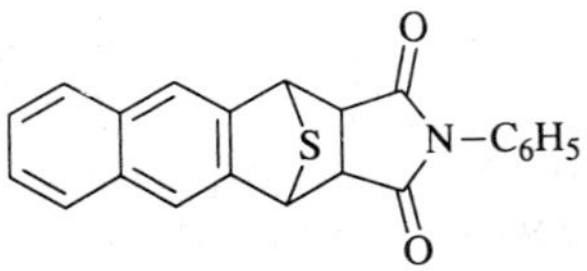 S; N–C_6H_5 (48); *exo*:*endo* = 3.3:1; 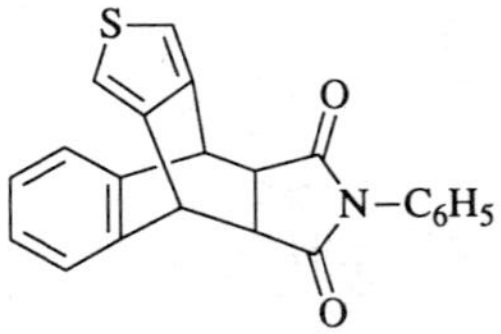 + S; N–C_6H_5 (8)	411
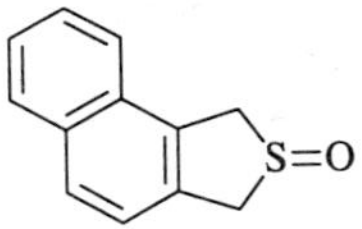 S=O	Al_2O_3 (neutral), 160–180°, 25 torr, 1 h	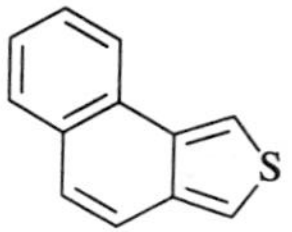 S (47)	411
	Ac_2O, 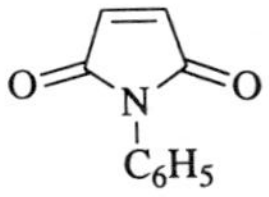 N; C_6H_5; reflux 5 h, rt 4 d	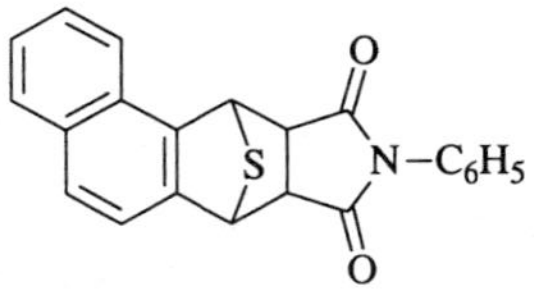 S; N–C_6H_5 (72); *exo*:*endo* = 1.3:1	411

Reactant	Conditions	Product(s) and Yield(s) (%)	Refs.
	Ac_2O, reflux, N-C_6H_5-maleimide	" (89)	127
	"	S-bridged N-C_6H_5 imide adduct (55)	125
	"	" (70)	124
	Ac_2O, pyr, 1,4-naphthoquinone, reflux	S-bridged bis-OAc adduct (51)	125
N-Ac, C_2H_5	Ac_2O, reflux 1.5 h	N-Ac, =CHCH_3 (28) + N-Ac, C_2H_5 (15) + N-Ac (25) + N-Ac (7)	278, 427
N-CH_3, C_2H_5	Ac_2O, reflux 2.5 h	N-CH_3, C_2H_5 I (10) + N-CH_3, =CHCH_3 (25) + N-CH_3, =CHCH_3, OAc (23) + N-CH_3 (10)	278, 427

TABLE VII. DIRECT FORMATION OF VINYL SULFIDES (*Continued*)

Sulfoxide	Reaction Conditions	Product(s) and Yield(s) (%)	Refs.
	TsOH, C_6H_6, reflux 1 h	I (15) 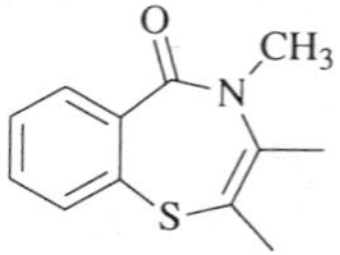 + [benzothiazepinone structure: O, N, CH_3, S] (2)	278, 427
 [oxathiolane S-oxide structure: O, S, O] $CH_2CONHC_6H_5$	TsOH, C_6H_6-DMF (1:1), 50°, 24 h	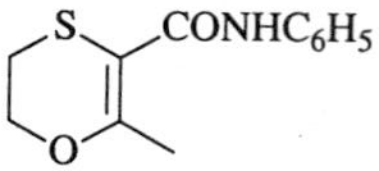 [oxathiin structure: S, O] $CONHC_6H_5$ (90)	186
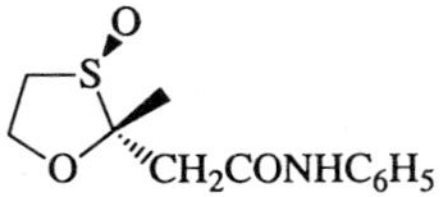 [oxathiolane S-oxide structure: O, S, O] $CH_2CONHC_6H_5$	"	" (84)	186
	DMF, 100°, 7 d	" (—)	186
C_{13}			
$C_6H_5S(O)CH(COC_3H_7\text{-}n)C_2H_5$	Ac_2O, CH_3SO_2OH, CH_2Cl_2, 16 h	$C_6H_5SC(COC_3H_7\text{-}n){=}CHCH_3$ (65)	121
	$Sn(OSO_2CF_3)_2$, 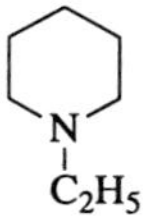 [piperidine structure: N–C_2H_5]	" (31)	122, 425
	$(CH_3)_2SiOSO_2CF_3$, CH_2Cl_2, 0°, overnight		
$C_6H_5S(O)CH(CO_2CH_3)C_4H_9\text{-}n$	Ac_2O, CH_3SO_2OH, CH_2Cl_2, 40°, 8 h	$C_6H_5SC(CO_2CH_3){=}CHC_3H_7\text{-}n$ $Z:E$ = 2:1 (93)	416, 428
$C_6H_5S(O)CH(CO_2CH_3)C_4H_9\text{-}s$	"	$C_6H_5SC(CO_2CH_3){=}CHC_3H_7\text{-}i$ $Z:E$ = 1:1 (79–98)	416, 428
$C_6H_5S(O)CH(CO_2C_2H_5)C_3H_7\text{-}n$	$(CH_3)_2SiOSO_2CF_3$, $[(CH_3)_3Si]_2NH$, ether, 25°, 3 h	$C_6H_5SC(CO_2C_2H_5){=}CHC_2H_5$ $Z:E$ = 6:94 (84)	424
$C_6H_5S(O)CH(CO_2C_2H_5)C_3H_7\text{-}i$	$(CH_3)_2SiOSO_2CF_3$, $[(CH_3)_3Si]_2NH$, ether, 35°, 3 h	$C_6H_5SC(CO_2C_2H_5){=}C(CH_3)_2$ (84)	424

Reactant	Conditions	Product(s) and Yield(s) (%)	Refs.
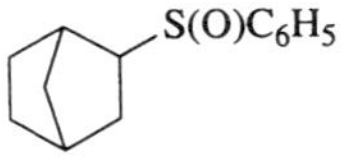 $S(O)C_6H_5$	TFAA (excess), $(C_2H_5)_3N$, CH_2Cl_2, 0°, 1 h	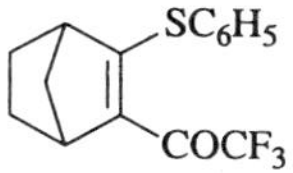 SC_6H_5, $COCF_3$ I (70) 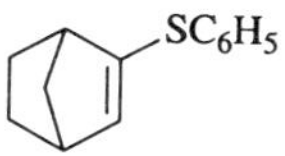 + SC_6H_5 II (—)	430
	TFAA (1 eq), $(C_2H_5)_3N$, CH_2Cl_2, 0°, 1 h	I (70) + II (15)	430
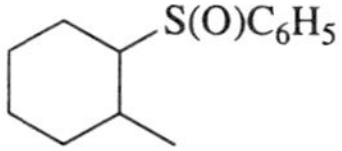 $S(O)C_6H_5$	”	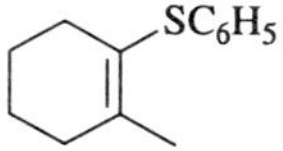 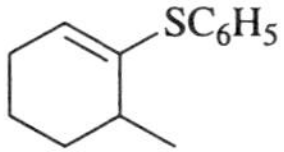 SC_6H_5 + SC_6H_5 11:8 (91)	430
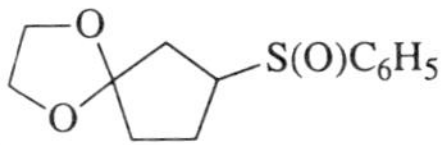 O, O, $S(O)C_6H_5$	TFAA, $(C_2H_5)_3N$, CH_2Cl_2, 5°, 2.5 h	 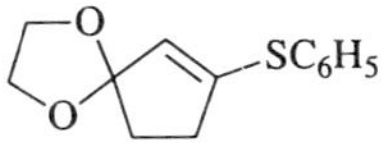 O, O, SC_6H_5 + O, O, SC_6H_5 1:4.6 (85)	430
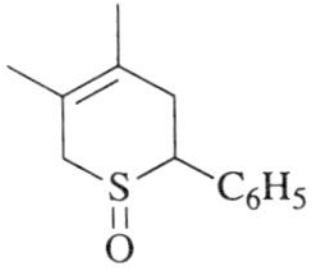 S, O, C_6H_5	Ac_2O, reflux 3 h	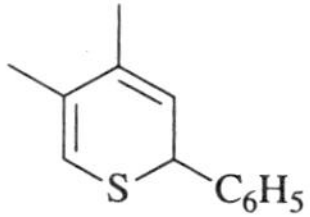 S, C_6H_5 (75)	432
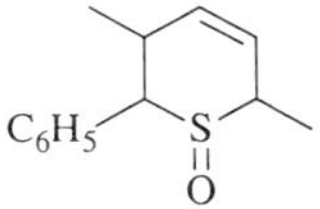 C_6H_5, S, O	Ac_2O, reflux 1 h	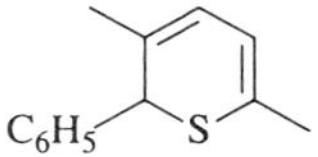 C_6H_5, S (13)	432
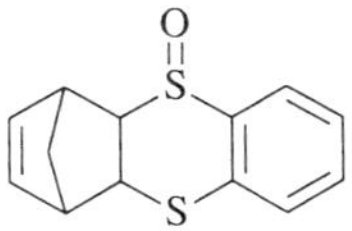 O, S, S	Ac_2O, reflux 6.5 h	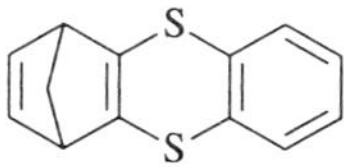 S, S (—)	134
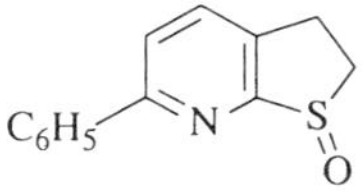 C_6H_5, N, S, O	1. Ac_2O, reflux 24 h 2. Evaporate Ac_2O 3. Na_2CO_3, acetone, H_2O, 40°, 2 h	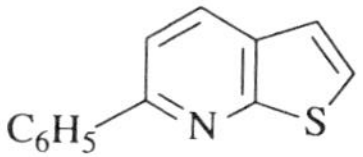 C_6H_5, N, S (62)	407
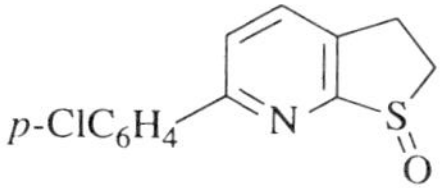 p-ClC_6H_4, N, S, O	”	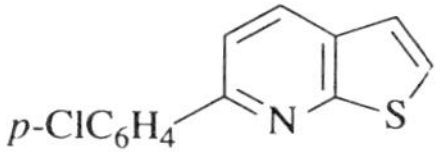 p-ClC_6H_4, N, S (68)	407

TABLE VII. DIRECT FORMATION OF VINYL SULFIDES (*Continued*)

Sulfoxide	Reaction Conditions	Product(s) and Yield(s) (%)	Refs.
C_{14}			
$C_6H_5S(O)CHClCH_2C_6H_5$	$(CH_3)_3SiOSO_2CF_3$, $(C_2H_5)_3N$, ether, 0°, 30 min	$C_6H_5SCCl{=}CHC_6H_5$ (—)	404
$C_6H_5S(O)CHBrCH_2C_6H_5$	"	(*Z*)-$C_6H_5SCBr{=}CHC_6H_5$ (86)	404
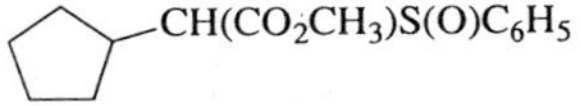 [cyclopentyl]$CH(CO_2CH_3)S(O)C_6H_5$	Ac_2O, CH_3SO_2OH, CH_2Cl_2, 40°, 8 h	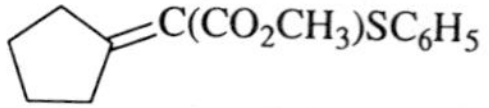 [cyclopentylidene]$=C(CO_2CH_3)SC_6H_5$ (69)	416, 428
$C_6H_5S(O)(CH_2)_2SC_6H_5$	Ac_2O	$C_6H_5SCH{=}CHSC_6H_5$ (56)	433
$C_6H_5S(O)(CH_2)_2S(O)C_6H_5$	Ac_2O	" (—)	433
n-$C_5H_{11}S(O)(CH_2)_4SC_5H_{11}$-$n$	Ac_2O	n-$C_5H_{11}SCH{=}CH(CH_2)_2SC_5H_{11}$-$n$ (36)	433
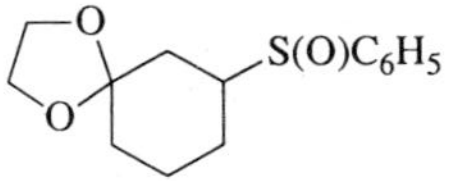 1,4-dioxaspiro[4.5]decan-7-yl $S(O)C_6H_5$	TFAA, lutidine, CH_2Cl_2, 5°, 1.15 h	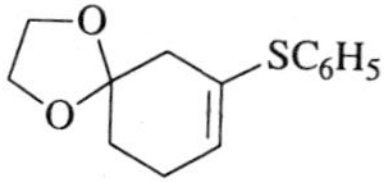 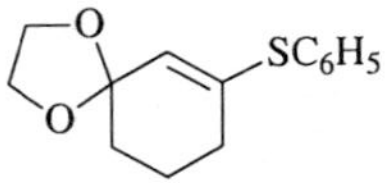 7-SC_6H_5-1,4-dioxaspiro[4.5]dec-7-ene + 7-SC_6H_5-1,4-dioxaspiro[4.5]dec-6-ene 1.3:1 (89)	430
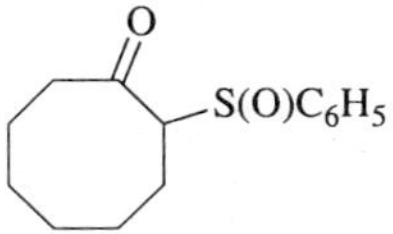 2-$S(O)C_6H_5$-cyclooctanone	Ac_2O, CH_3SO_2OH, CH_2Cl_2, 16 h	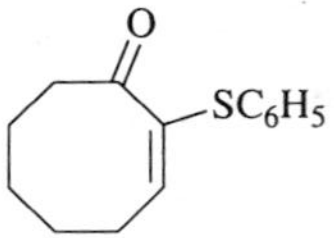 2-SC_6H_5-cyclooct-2-enone (75)	121
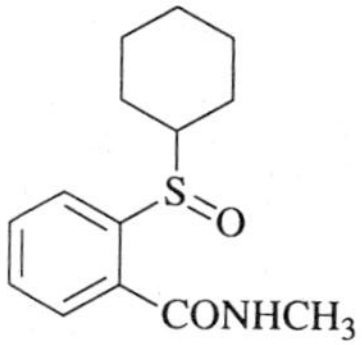 2-(cyclohexyl-S(=O))-$C_6H_4CONHCH_3$	Ac_2O, 80°, 7 h	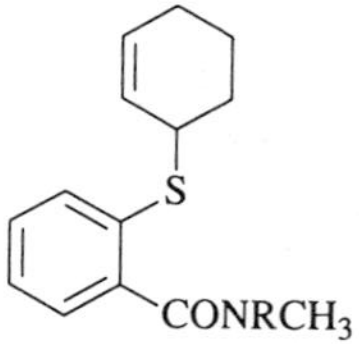 2-(cyclohex-2-enyl-S)-$C_6H_4CONRCH_3$ R = H (78) + R = Ac (4) 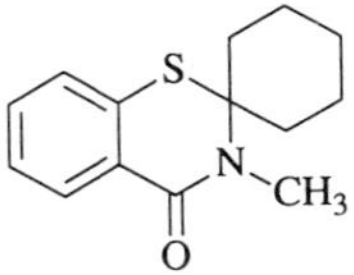 + 3-methylspiro[benzothiazine-2,1′-cyclohexan]-4-one (4)	267

SC_6H_{13}-*n*, S, O		SC_6H_{13}-*n*, S (—)	431
$S(O)C_6H_5$, O, O	TFAA, CH_2Cl_2, 1 h	SC_6H_5, O, O (quant)	179
S=O	Ac_2O, pyr or AcONa, or $(C_2H_5)_3N$	S (30)	123
	Ac_2O, O, N, O, C_6H_5 reflux	O, S, N–C_6H_5, O (45)	124
S=O	,,	O, S, N–C_6H_5, O (60)	123
O, S, N, CO_2CH_3, COC_6H_5	TFAA, $CDCl_3$, 25° (−30°)	S, N, CO_2CH_3, COC_6H_5 (quant)	242

TABLE VII. DIRECT FORMATION OF VINYL SULFIDES (*Continued*)

Sulfoxide	Reaction Conditions	Product(s) and Yield(s) (%)	Refs.
[structure: 3-Ac-2-methyl-2-$CH_2CO_2C_2H_5$-benzothiazoline S-oxide]	Ac_2O, reflux 1.5 h	[structure, $CO_2C_2H_5$] I (42) + [structure, $CO_2C_2H_5$] II (42) + [structure, $CH_2CO_2C_2H_5$] III (3)	278, 427
	TFAA, CH_2Cl_2, 0°, 2 h	I (26) + III (16)	278, 427
[structure: diastereomeric S-oxide, $CH_2CO_2C_2H_5$]	Ac_2O, reflux 1.5 h	I (26) + II (23) + III (17)	278, 427
[structure: $C_6H_5SO_2$, C_6H_5, N=N, S=O]	Stand at 0°	[structure: $C_6H_5SO_2$–thiadiazole–C_6H_5] (53)	434
[structure: $(CH_2)_7$ spiro dithiolane S-oxide]	TsOH, C_6H_6, reflux with water removal	[structure: $(CH_2)_7$ fused dithiin] (75)	130
	"	" (96)	131

	Reactant	Conditions	Product(s) and Yield(s) (%)	Refs.
C_{15}	$C_6H_5S(O)CHClCH(CH_3)C_6H_5$	$(CH_3)_3SiOSO_2CF_3$, $(C_2H_5)_3N$, ether,	$C_6H_5SCCl{=}C(CH_3)C_6H_5$ $Z{:}E = 5{:}4$ (92)	404
	(Z)-$C_6H_5S(O)C[Si(CH_3)_3]{=}CH(CH_2)_2OAc$	1. C_6H_6, reflux 12 h 2. CH_3OH, TsOH, (cat.)	$C_6H_5S(O)CH{=}CH(CH_2)_2OAc$ I (15) + $C_6H_5SC{\equiv}C(CH_2)_2OAc$ II (41) + $C_6H_5SCO(CH_2)_3OAc$ III (35)	173
	(E)-$C_6H_5S(O)C[Si(CH_3)_3]{=}CH(CH_2)_2OAc$	"	I (12) + II (40) + III (33)	173
		$(CH_3)_3SiOSO_2CF_3$, $[(CH_3)_3Si]_2NH$, ether, 25°, 3 h	$C_6H_5SC(CO_2C_4H_9\text{-}t){=}CHC_2H_5$ $Z{:}E = 13{:}87$ (76)	424
	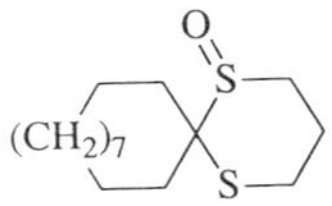 O, S, S, $(CH_2)_7$	TsOH, C_6H_6, reflux with water removal, 10 h	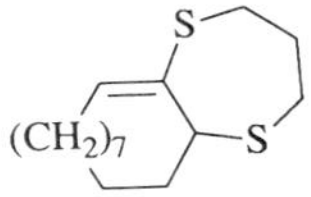 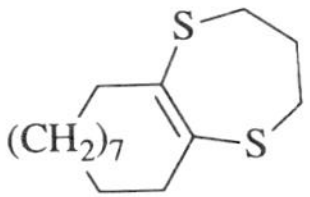 S, S, $(CH_2)_7$ I (11) + S, S, $(CH_2)_7$ (3)	131
		TsCl, pyr, 110–120°, 4 h	I (99)	131
	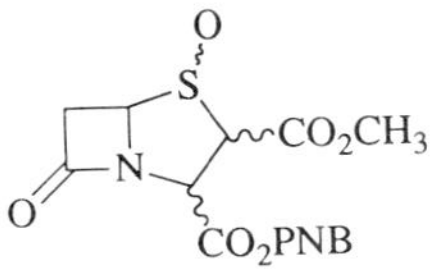 O, S, N, O, CO_2CH_3, CO_2PNB	TFAA, lutidine, 25° overnight	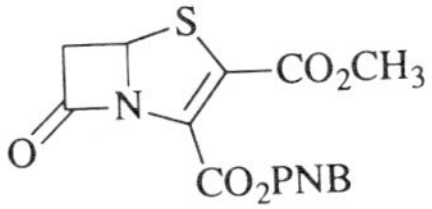 S, N, O, CO_2CH_3, CO_2PNB (37)	435
C_{16}	$C_6H_5S(O)CH(CO_2C_2H_5)C_6H_{11}$	$(CH_3)_3SiOSO_2CF_3$, $[(CH_3)_3Si]_2NH$, ether, 35°, 4 h	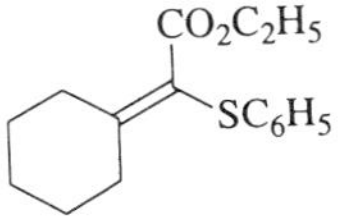 $CO_2C_2H_5$, SC_6H_5 (73)	424
	$C_6H_5S(O)CH(COC_6H_5)C_2H_5$	$Sn(OSO_2CF_3)_2$ (2.8 eq), $(CH_3)_3SiOSO_2CF_3$ (3 eq), CH_2Cl_2, 0° overnight; N-C_2H_5 (7 eq);	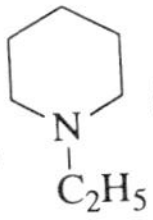 (E)-$C_6H_5SC(COC_6H_5){=}CHCH_3$ (86)[a]	425

TABLE VII. DIRECT FORMATION OF VINYL SULFIDES (*Continued*)

Sulfoxide	Reaction Conditions	Product(s) and Yield(s) (%)	Refs.
p-$CH_3C_6H_4SO_2$, p-$CH_3OC_6H_4$ (N=N, S→O)	Silica gel, ether, or stand	p-$CH_3C_6H_4SO_2$ (N-N, S) $C_6H_4OCH_3$-p (65)	434
$(CH_2)_7$ (O=S, S)	TsCl, pyr, 110–120°, 2.5 h	$(CH_2)_7$ (S, S) (89)	131
Ac, N, C_6H_5, S, O	TFAA, CH_2Cl_2, 0° 1 h, rt 2 h	Ac, N, C_6H_5, S (38)	278, 427
O, N, CH_3, C_6H_5, S, O	Ac_2O, reflux 2 h	O, N, CH_3, C_6H_5, S (32)[b]	278, 427
		+ O, N, CH_3, C_6H_5, S, Ac (40)[b]	

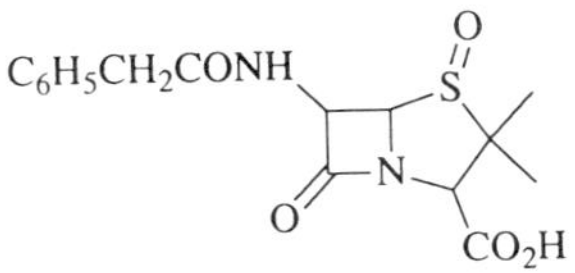

	Reactant	Conditions	Product(s) and Yield(s) (%)	Refs.
	$C_6H_5CH_2CONH$ (penicillin sulfoxide; CO_2H)	$(CH_3)_3SiCl$, 2-methylpyridine, $CHCl_3$, 83°, 20 h	$C_6H_5CH_2CONH$ (cephem; CO_2H) I (55) + $C_6H_5CH_2CONH$ (exo-methylene cepham) (—)	141
		$CH_3C(OTMS){=}NTMS$, 2-methylpyridinium bromide (N^+ H, Br^-), dioxane, CH_2Cl_2, 102°, 6 h	I (78)	141
	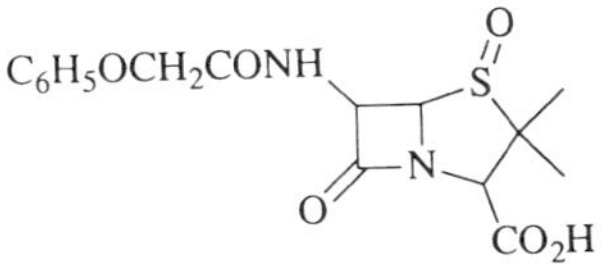 $C_6H_5OCH_2CONH$ (penicillin sulfoxide; CO_2H)	Xylene, mineral acid, reflux	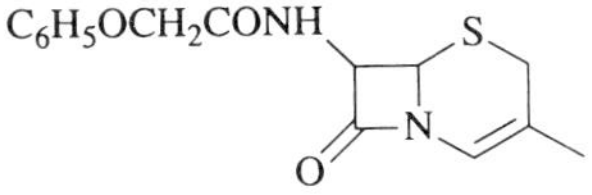 $C_6H_5OCH_2CONH$ (cephem) (—)	139
C_{17}	$C_6H_5S(O)CH(COC_6H_5)C_3H_7$-$n$	Ac_2O, CH_3SO_2OH, CH_2Cl_2, 40°, 8 h	$C_6H_5SC(COC_6H_5){=}CHC_2H_5$ $Z{:}E = 2{:}1$ (69)	416
		Ac_2O, AcOH, CH_3SO_2OH	" Z only (—)	436
		$Sn(OSO_2CF_3)_2$, N-ethylpiperidine (C_2H_5), $(CH_3)_3SiOSO_2CF_3$, CH_2Cl_2, 0° overnight	" (65)	122, 425
	$C_6H_5S(O)CH(C_6H_{13}$-$n)(CH_2)_2COCH_3$	TFAA, pyr, CH_2Cl_2, 75 min	$C_6H_5SC[(CH_2)_2COCH_3]{=}CHC_5H_{11}$-$n$ (—) + $C_6H_5SC(C_6H_{13})$-$n){=}CHCH_2COCH_3$ (—)	437

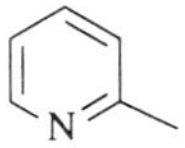

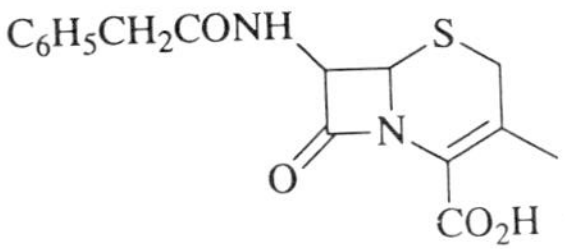

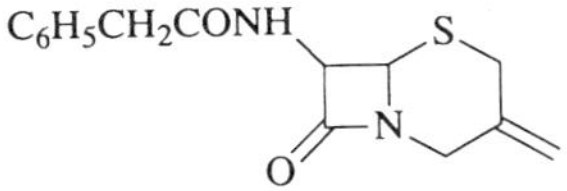

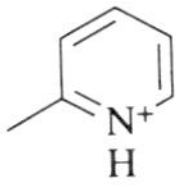

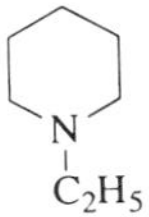

TABLE VII. DIRECT FORMATION OF VINYL SULFIDES (*Continued*)

Sulfoxide	Reaction Conditions	Product(s) and Yield(s) (%)	Refs.
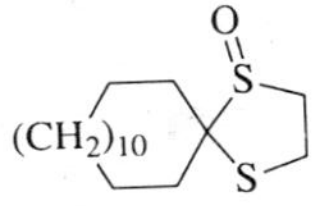 $(CH_2)_{10}$ S=O, S	TsOH, C_6H_6, reflux with water removal, 18 h	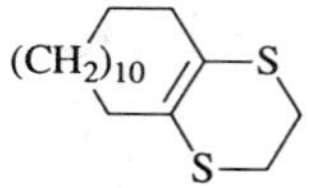 $(CH_2)_{10}$ S, S (85)	130
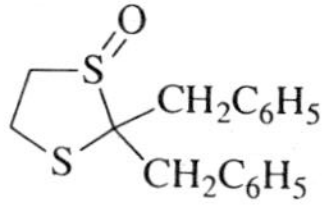 S=O, S, $CH_2C_6H_5$, $CH_2C_6H_5$	$(CH_3)_2SO$, 100°, 6 h	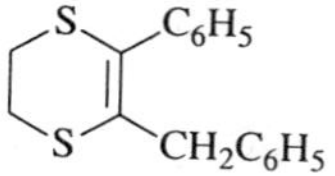 S, S, C_6H_5, $CH_2C_6H_5$ (85)	129
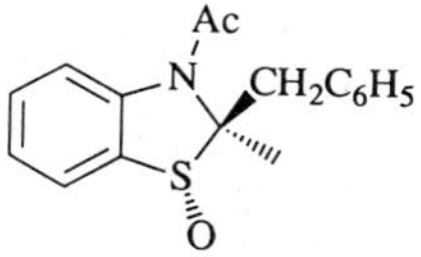 Ac, N, $CH_2C_6H_5$, S, O	Ac_2O, reflux 1.5 h	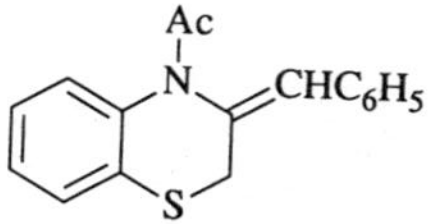 Ac, N, CHC_6H_5, S (35)	278, 427
		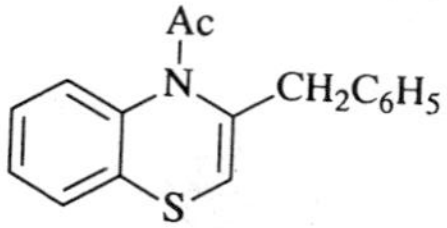 + Ac, N, $CH_2C_6H_5$, S (5)	
		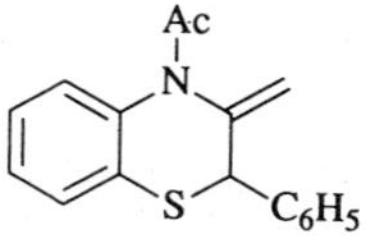 + Ac, N, S, C_6H_5 (23)	
		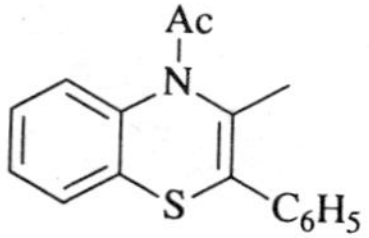 + Ac, N, S, C_6H_5 (9)	

	Reactant	Conditions	Product(s) (Yield, %)	Refs.
	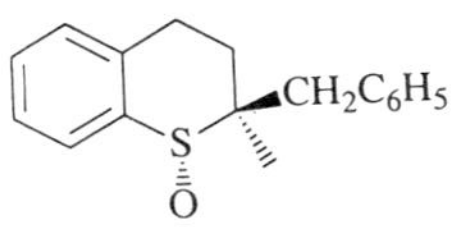 $CH_2C_6H_5$, S, O	TsOH, xylene, reflux 45 min	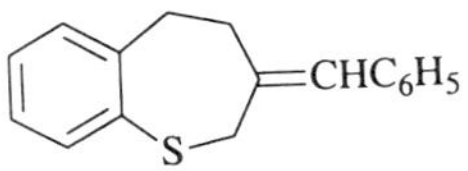 $=CHC_6H_5$, S (27)	278
			+ $-CH_2C_6H_5$, S (20)	
	$C_6H_5CH_2$, S, O	,,	S, CHC_6H_5 (43)	278
			+ S, $CH_2C_6H_5$ (52)	
		Ac_2O, reflux 45 min	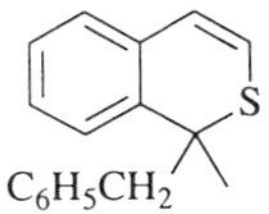 S, $C_6H_5CH_2$ (43)	278
	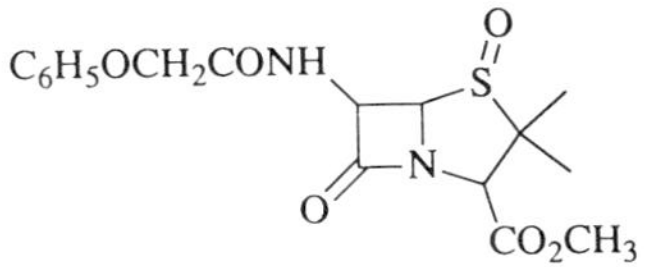 $C_6H_5OCH_2CONH$, O, S, N, O, CO_2CH_3	TsOH, xylene, reflux 1 h	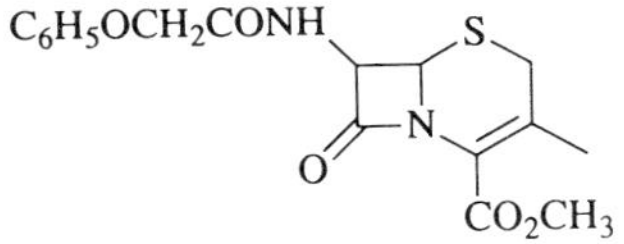 $C_6H_5OCH_2CONH$, S, N, O, CO_2CH_3 (15)	139, 140
C_{18}	(*Z*)-$C_6H_5S(O)CSi(CH_3)_3{=}CHC_7H_{15}$-*n*	1. C_6H_6, reflux 4.5 h 2. CH_3OH, TsOH (cat.)	$C_6H_5S(O)CH{=}CHC_7H_{15}$-*n* I (5) + $C_6H_5SC{\equiv}CC_7H_{15}$-*n* II (38) + $C_6H_5SCOC_8H_{17}$-*n* (48)	173
	(*E*)-$C_6H_5S(O)CSi(CH_3)_3{=}CHC_7H_{15}$-*n*	,,	I (6) + II (46) + III (36)	173

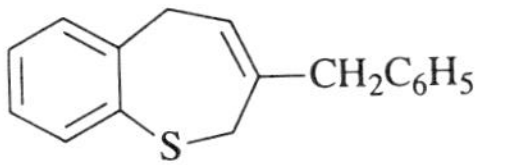

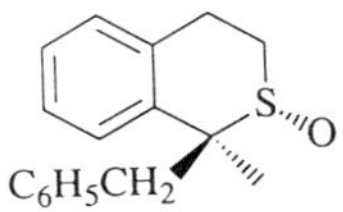

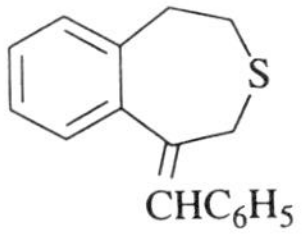

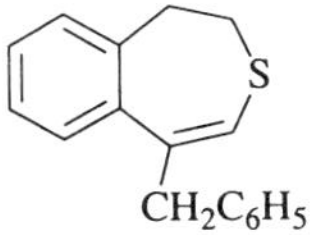

TABLE VII. Direct Formation of Vinyl Sulfides (*Continued*)

Sulfoxide	Reaction Conditions	Product(s) and Yield(s) (%)	Refs.
C_{19}			
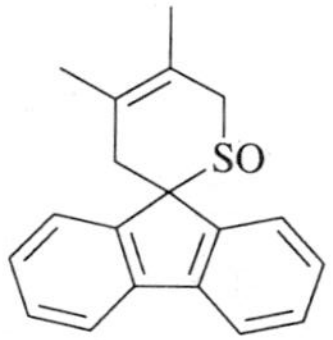 SO	Ac_2O, 48 h	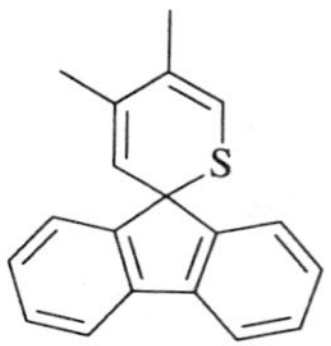 S (99)	432, 438
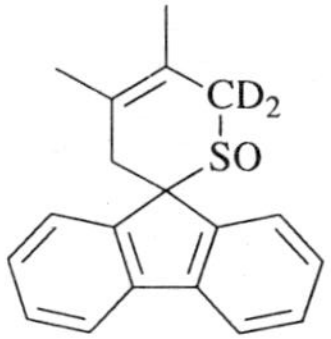 CD_2 SO	Ac_2O, 48 h	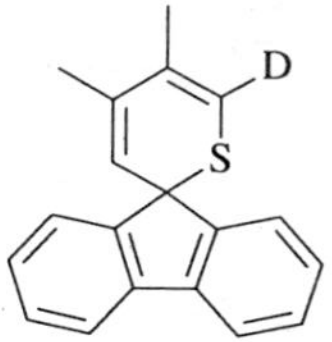 D S (49)	432
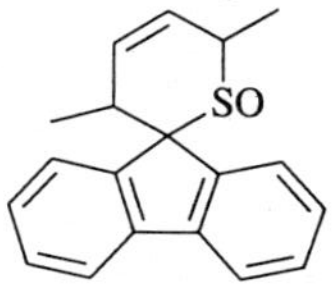 SO	Ac_2O, 120°, 30 min	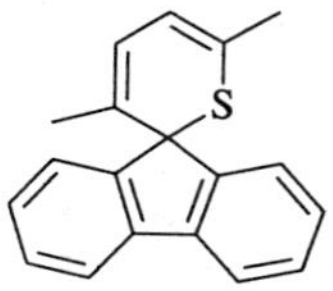 S (54)	432
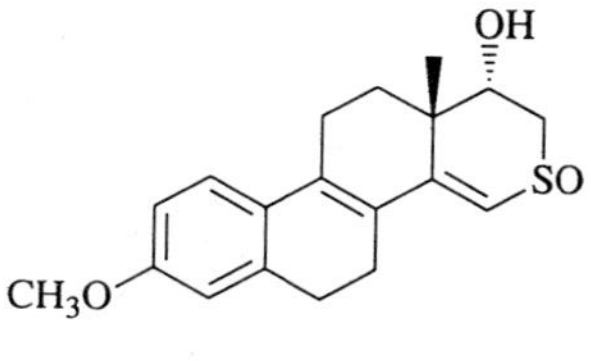 OH SO CH_3O	1. Ac_2O, TsOH, 1.5 h 2. $LiAlH_4$	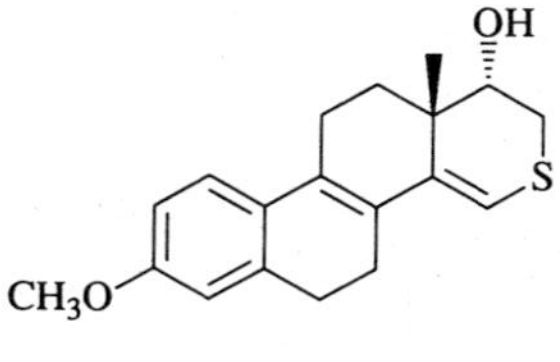 OH S CH_3O (10)	439

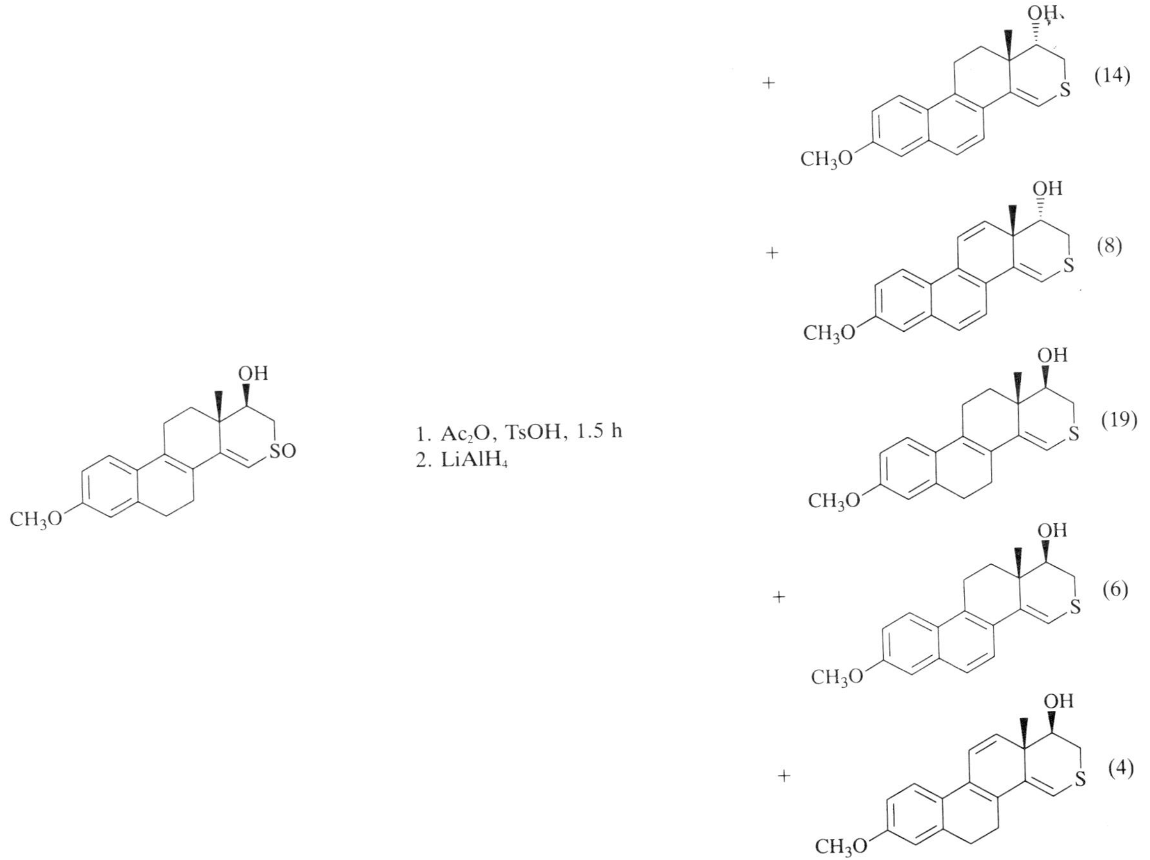

CH_3O ... OH ... SO

1. Ac_2O, TsOH, 1.5 h
2. $LiAlH_4$

CH_3O ... OH ... S (14)

\+ CH_3O ... OH ... S (8)

\+ CH_3O ... OH ... S (19)

\+ CH_3O ... OH ... S (6)

\+ CH_3O ... OH ... S (4)

439

TABLE VII. DIRECT FORMATION OF VINYL SULFIDES (*Continued*)

Sulfoxide	Reaction Conditions	Product(s) and Yield(s) (%)	Refs.
		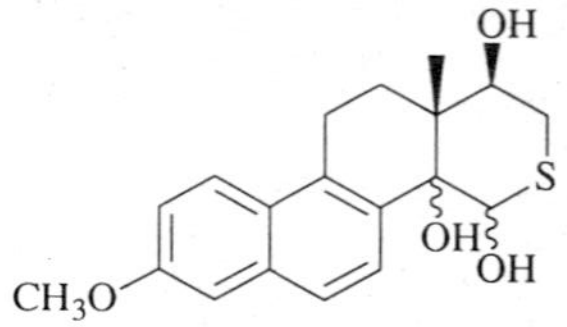 + CH_3O … OH, OH, OH, S (34)	
		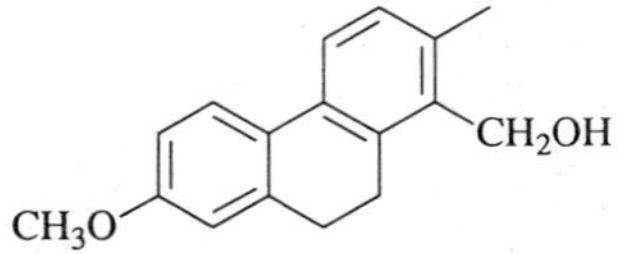 + CH_3O … CH_2OH (8)	
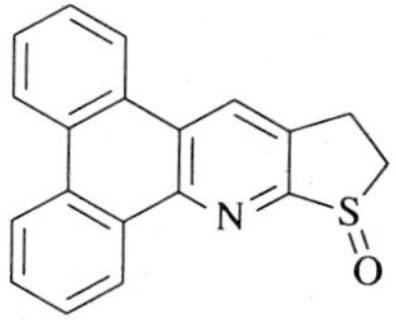 N, S=O	1. Ac_2O, reflux 24 h 2. Evaporate Ac_2O 3. Na_2CO_3, C_2H_5OH, CH_2Cl_2, H_2O, 20 h	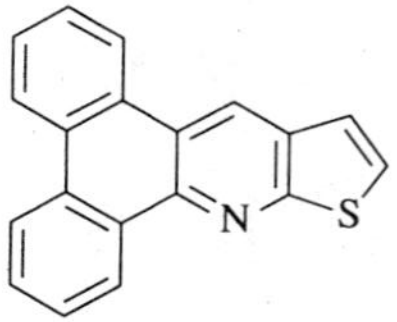 N, S (41)	407
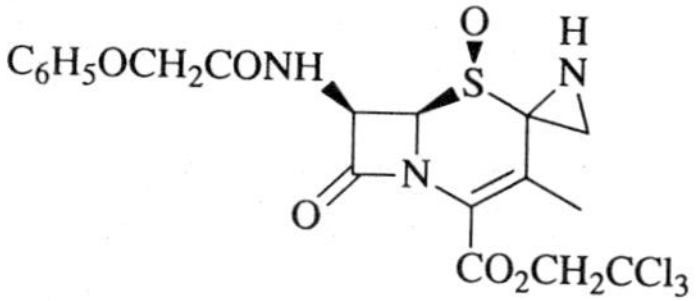 $C_6H_5OCH_2CONH$, O, S, H, N, O, N, $CO_2CH_2CCl_3$	PCl_3, DMF	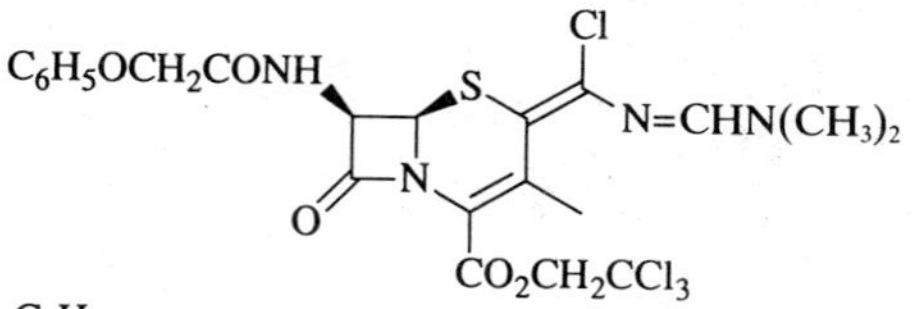 $C_6H_5OCH_2CONH$, S, Cl, N=CHN$(CH_3)_2$ (37), O, N, $CO_2CH_2CCl_3$	440
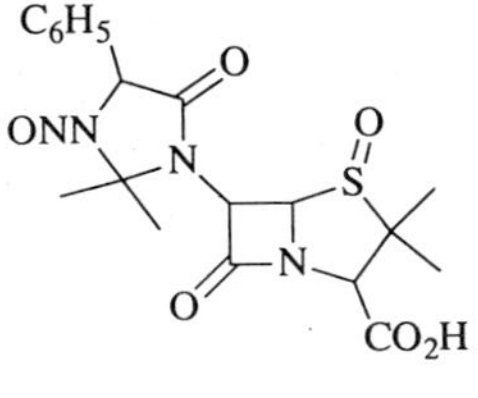 C_6H_5, O, ONN, N, O, S, O, N, CO_2H	TsOH, $[(CH_3)_2N]_2CO$, 135°, 2 h	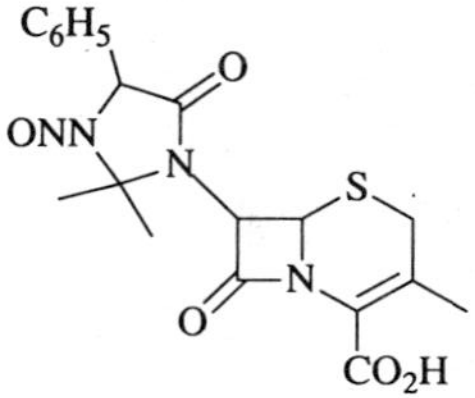 C_6H_5, O, ONN, N, S, O, N, CO_2H (32)	441

	p-ClC$_6$H$_4$S(O)NHCH(C$_6$H$_4$Cl-p)$_2$	Ac$_2$O, 45 h	p-ClC$_6$H$_4$SN═C(C$_6$H$_4$Cl-p)$_2$ I (55) + p-ClC$_6$H$_4$SNHCH(C$_6$H$_4$Cl-p)$_2$ II (25)	136
		Ac$_2$O, 70 h	I (58) + II (38)	136
		SOCl$_2$, pyr	I (64)	136
C$_{20}$	C$_6$H$_5$S(O)CH(COC$_6$H$_5$)C$_2$H$_5$	Sn(OSO$_2$CF$_3$)$_2$, N-ethylpiperidine (ring N–C$_2$H$_5$), (CH$_3$)$_3$SiOSO$_2$CF$_3$, CH$_2$Cl$_2$, 0° overnight	(E)-C$_6$H$_5$SC(COC$_6$H$_5$)═CHCH$_3$ (86)	122
	C$_6$H$_5$S(O)CH(C$_6$H$_{13}$-n)-(CH$_2$)$_2$COCH$_2$CO$_2$C$_2$H$_5$	TFAA, pyr, CH$_2$Cl$_2$	C$_6$H$_5$SC[(CH$_2$)$_2$COCH$_2$C$_2$H$_5$]═CHC$_5$H$_{11}$-n I + C$_6$H$_5$SC(C$_6$H$_{13}$-n)═CHCH$_2$COCH$_2$CO$_2$C$_2$H$_5$ II I + II (83)	437
	(structure: 1,4-bis(1,3-benzodithiol-2-yl)benzene S-oxide, S, S═O)	DBN, CH$_2$Cl$_2$, pyr, reflux 10 min	(structure: quinoid bis(1,3-benzodithiol-2-ylidene) S-oxide, S, S═O) (37)	442
	C$_6$H$_5$S(O)CH[PO(C$_6$H$_5$)$_2$]CH$_3$	Ac$_2$O, CH$_3$SO$_2$OH, CH$_2$Cl$_2$, 25°, 8 d	C$_6$H$_5$SC[PO(C$_6$H$_5$)$_2$]═CH$_2$ (65)	443, 444
	p-CH$_3$C$_6$H$_4$S(O)NHCH(C$_6$H$_4$Cl-p)$_2$	Ac$_2$O, 45 h	p-CH$_3$C$_6$H$_4$SN═C(C$_6$H$_4$Cl-p)$_2$ I (45) + p-CH$_3$C$_6$H$_4$SNHCH(C$_6$H$_4$Cl-p)$_2$ II (26)	136
		Ac$_2$O, 70 h	I (48) + II (43)	136
		SOCl$_2$, pyr	I (55)	136
	p-CH$_3$C$_6$H$_4$S(O)NHCH(C$_6$H$_5$)$_2$	Ac$_2$O, 45 h	p-CH$_3$C$_6$H$_4$SN═C(C$_6$H$_5$)$_2$ I (23) + p-CH$_3$C$_6$H$_4$SNHCH(C$_6$H$_5$)$_2$ II (30)	136
		Ac$_2$O, 70 h	I (32) + II (42)	136
		SOCl$_2$, pyr	I (19)	136

TABLE VII. DIRECT FORMATION OF VINYL SULFIDES (*Continued*)

Sulfoxide	Reaction Conditions	Product(s) and Yield(s) (%)	Refs.
9-[$NHS(O)C_6H_4CH_3$-*p*]fluorene	Ac_2O, 45 h	9-($NSC_6H_4CH_3$-*p*)fluorene I (25) + 9-($NHSC_6H_4CH_3$-*p*)fluorene II (35)	136
	Ac_2O, 70 h	I (33) + II (48)	136
	$SOCl_2$, pyr	I (23)	136
5,7-di-C_6H_5-thieno-benzothiadiazole S=O	Ac_2O, 140°	5,7-di-C_6H_5-thieno-benzothiadiazole (—)	445
C_{21}			
$C_6H_5S(O)CH[PO(C_6H_5)_2]C_2H_5$	Ac_2O, CH_3SO_2OH, CH_2Cl_2	(*Z*)-$C_6H_5SC[PO(C_6H_5)_2]{=}CHCH_3$ (76)	444
2-C_6H_5-3-$S(O)C_6H_5$-chroman-4-one	Acetone, reflux 64 h	2-C_6H_5-3-SC_6H_5-chromone I (26) + 2-C_6H_5-chromone II (5)	446

Reactant	Conditions	Product(s) and Yield(s) (%)	Refs.
	Ac_2O, H_2SO_4, 24 h	I (61) + II (7) + [structure] SC_6H_5, OAc, C_6H_5 III (8) + [structure] C_6H_5, OH (3)	446
	Ac_2O, 95 h	I (10) + II (74) + III 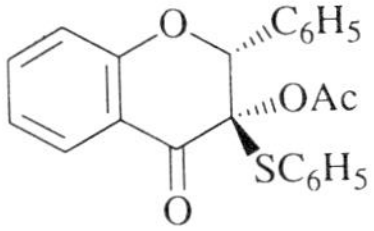 + [structure] C_6H_5, OAc, SC_6H_5 IV; III:IV = 2:1 (16)	446
[structure] C_6H_5, C_6H_5, S=O	"Pummerer conditions"	[structure] C_6H_5, C_6H_5, S (37)	447
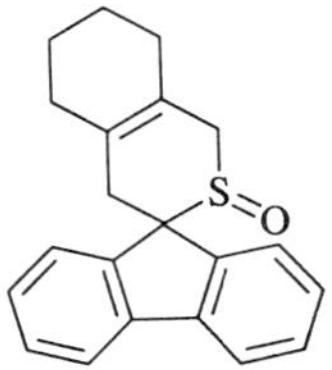 [structure] S=O	Ac_2O, 110°, 1.5 h	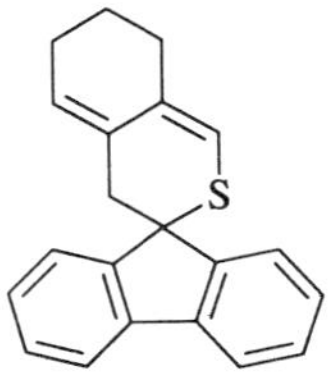 [structure] S (38)	432

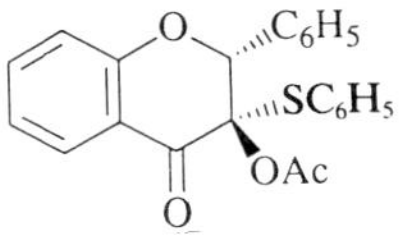

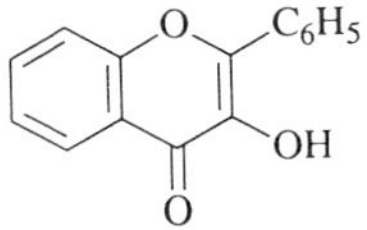

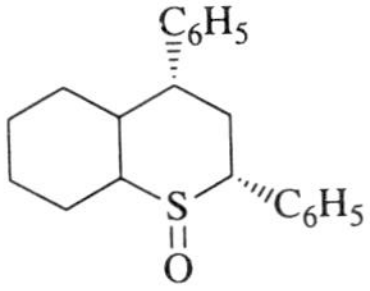

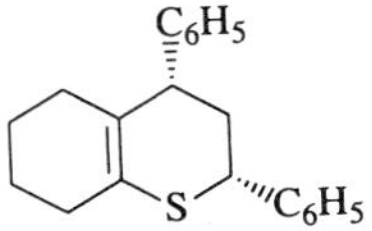

TABLE VII. Direct Formation of Vinyl Sulfides (*Continued*)

Sulfoxide	Reaction Conditions	Product(s) and Yield(s) (%)	Refs.
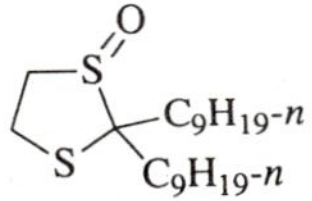 [structure: 1-oxo-1,3-dithiolane, 2,2-di-C_9H_{19}-n]	$(CH_2)_2SO$, 100°, 72 h	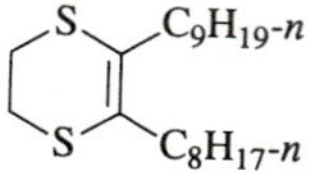 [structure: dihydro-1,4-dithiin, C_9H_{19}-n, C_8H_{17}-n] (83)	129
C_{22}			
$C_6H_5S(O)CH(COC_6H_5)(CH_2)_2C_6H_5$	$Sn(OSO_2CF_3)_2$, 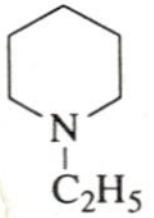 N-C_2H_5-piperidine, $(CH_3)_3SiOSO_2CF_3$, CH_2Cl_2, 0° overnight	$C_6H_5SC(COC_6H_5){=}CHCH_2C_6H_5$ + $C_6H_5SCH(COC_6H_5)CH{=}CHC_6H_5$ 1:1 (71)	122, 425
p-$CH_3C_6H_4S(O)NHCH(C_6H_4CH_3$-$p)_2$	Ac_2O, 45 h	p-$CH_3C_6H_4SN{=}C(C_6H_4CH_3$-$p)_2$ I (10) + p-$CH_3C_6H_4SNHCH(C_6H_4CH_3$-$p)_2$ II (25)	136
	Ac_2O, 70 h	I (14) + II (37)	136
	$SOCl_2$, pyr	I (16)	136
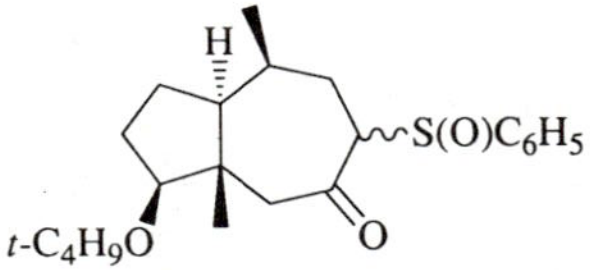 [structure: H, t-C_4H_9O, ~$S(O)C_6H_5$, O]	Ac_2O, CH_3SO_2OH, 25°, 4 h	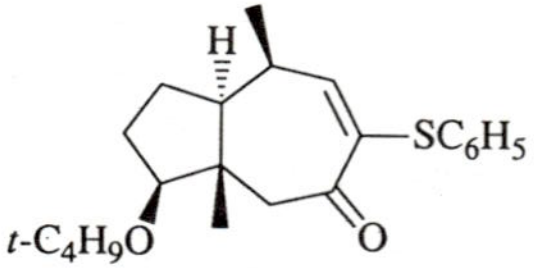 [structure: H, t-C_4H_9O, SC_6H_5, O] (80), (100)[a]	149, 391
C_{23}			
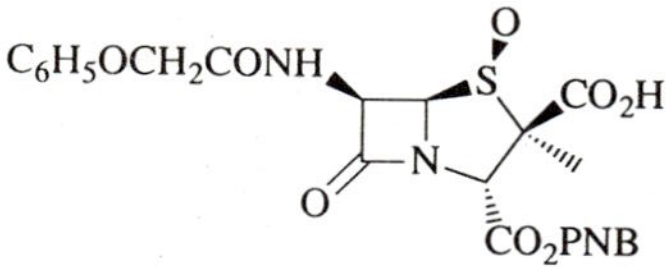 [structure: $C_6H_5OCH_2CONH$, S→O, CO_2H, CO_2PNB]		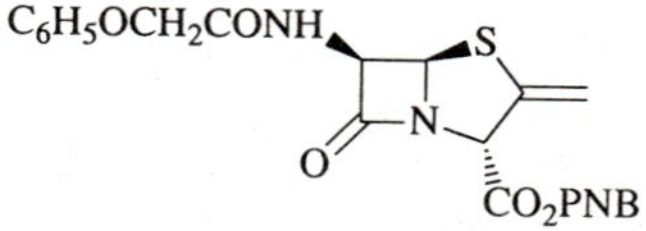 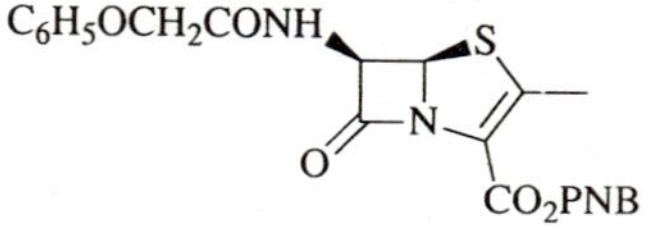 [structure: $C_6H_5OCH_2CONH$, S, CO_2PNB] I + [structure: $C_6H_5OCH_2CONH$, S, CO_2PNB] II	

Substrate	Conditions	Product(s) (Yield, %)	Refs.
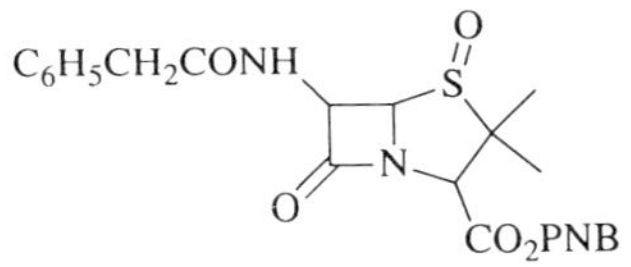 $C_6H_5CH_2CONH$ … CO_2PNB (S=O)	$(COCl)_2$, pyr, CH_2Cl_2 or THF, 16 h	I + II = 4:1 (18)	448
	$(COCl)_2$, molecular sieves, CH_2Cl_2 or THF	I (44) $C_6H_5CH_2CONH$ … CO_2PNB I + $C_6H_5CH_2CONH$ … Cl … CO_2PNB II + O … $NHCOCH_2C_6H_5$ … N … S … O_2CPNB III	448
	Imidazole or *N*-methylimidazole, CH_2Cl_2	III (—)	449
	4-Picoline	I + II 1:1 (—)	449
	Pyr, $(CH_2Cl)_2$, reflux 29 h	I (45) + III (22)	449
	Pyr, CH_3CN or dioxane	III (—)	449
	Pyr, $Cl_2C{=}CHCl$	III (quant)	449
	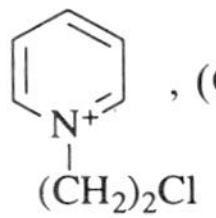 N^+–$(CH_2)_2Cl$ (pyridinium), $(CH_2Cl)_2$, reflux 21 h	II (83)	449

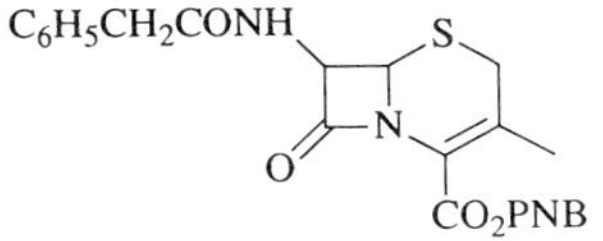

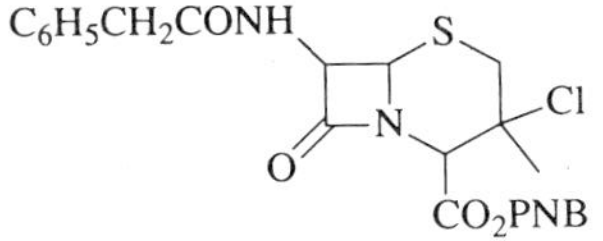

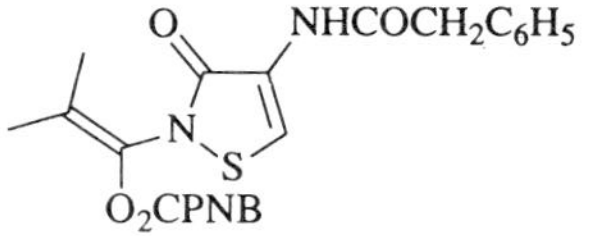

TABLE VII. Direct Formation of Vinyl Sulfides (*Continued*)

Sulfoxide	Reaction Conditions	Product(s) and Yield(s) (%)	Refs.
	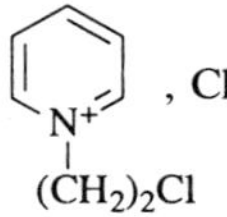 N-(2-chloroethyl)pyridinium [$(CH_2)_2Cl$], $Cl_2C{=}CHCl$	II : III = 2 : 3 (—)	449
	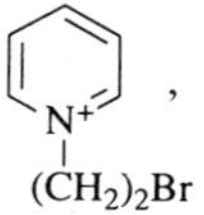 N-(2-bromoethyl)pyridinium [$(CH_2)_2Br$], $(CH_2Br)_2$, 90°, 10 h	II (26)	449
	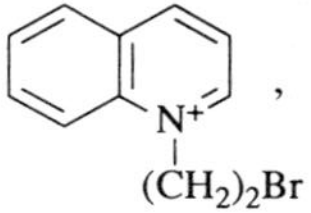 N-(2-bromoethyl)quinolinium [$(CH_2)_2Br$], $(CH_2Br)_2$, 85°, 16 h	II (48)	449
C_{25}			
1,3-Diphenyl-2,3-dihydro-1H-naphtho[2,1-b]thiopyran S-oxide [C_6H_5, C_6H_5, S=O]	Ac_2O, 140°, 6 h	1,3-Diphenyl-1H-naphtho[2,1-b]thiopyran [C_6H_5, C_6H_5, S] (80)	450
	$h\nu$, Pyrex filter, C_6H_6	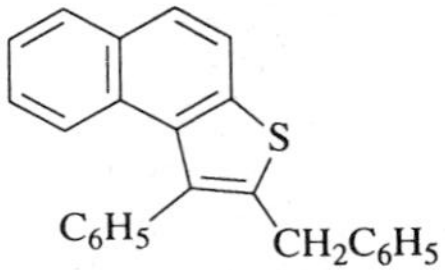 1-Phenyl-2-benzylnaphtho[2,1-b]thiophene [C_6H_5, $CH_2C_6H_5$, S] (74)	450, 451

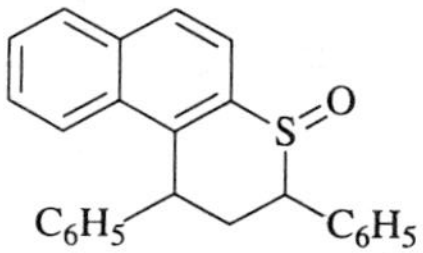

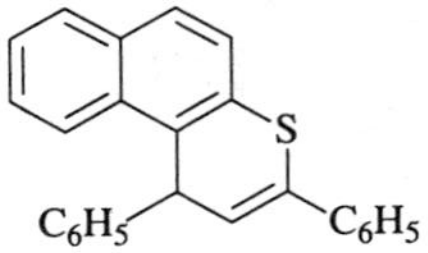

Reactant	Conditions	Product(s) and Yield(s) (%)	Refs.
	$h\nu$, Pyrex filter, AcOH	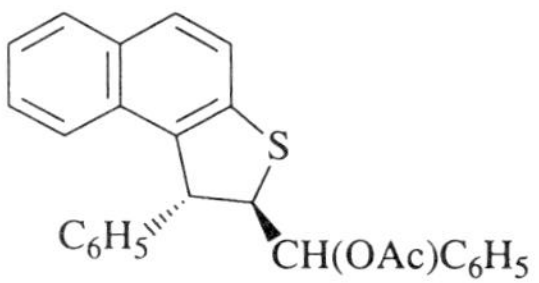 C_6H_5 / $CH(OAc)C_6H_5$ (75)	451
	$h\nu$, Pyrex filter, C_6H_6-CH_3OH (1:1)	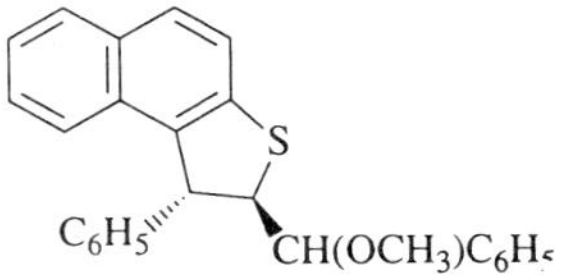 C_6H_5 / $CH(OCH_3)C_6H_5$ (85) 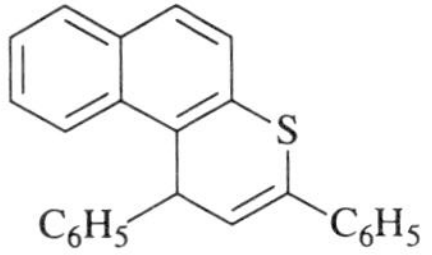 + C_6H_5 / C_6H_5 (5)	451
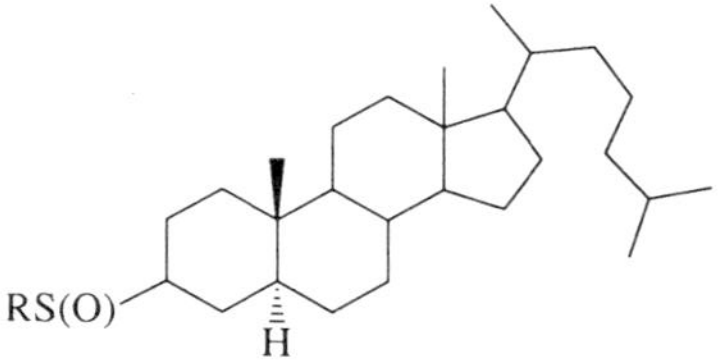 C_{28} RS(O) / H		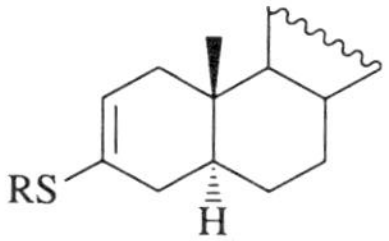 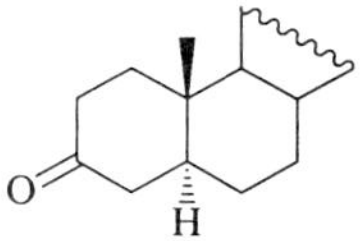 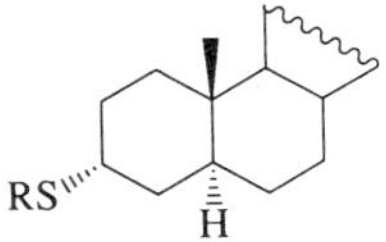 RS / H I; O / H II; RS / H III	
3α: R = CH_3	Ac_2O, 80°, 4 h	I (26) + III, R = $AcOCH_2$ (56)	19
R = C_3H_7-*i*	Ac_2O, C_6H_6, 80°, 16 h	I (40)	19
R = $CH_2C_6H_5$	Ac_2O, C_6H_6, 80°, 16 h	II (28) + III, R = H (35)	19

TABLE VII. DIRECT FORMATION OF VINYL SULFIDES (*Continued*)

Sulfoxide	Reaction Conditions	Product(s) and Yield(s) (%)	Refs.
3β: R = CH_3	Ac_2O, 80°, 4 h	I (4) 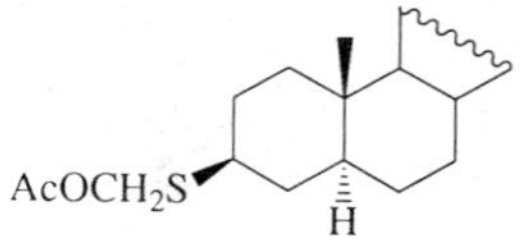 + [structure: $AcOCH_2S$ … H] (70)	19
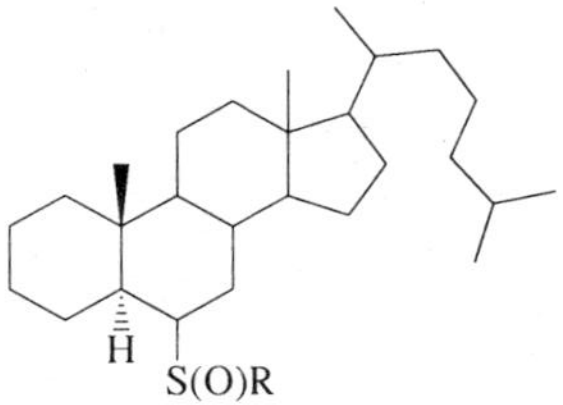 [structure: S(O)R, H]		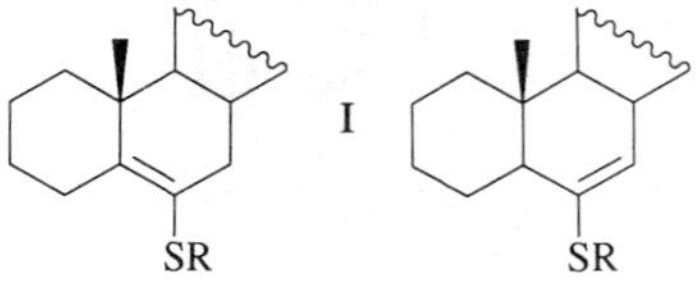 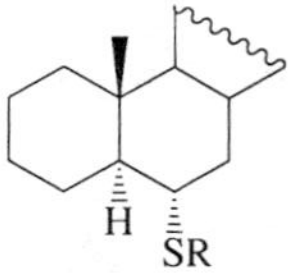 [structures I, II (SR); III (H, SR)]	
6α: R = CH_3	Ac_2O, 80°, 4–16 h	I (15) + III, R = CH_2OAc (68)	19
C_2H_5	"	III, R = $CH{=}CH_2$ (23)	19
C_3H_7-*i*	"	I + II (30) + III, R = H (59)	19
C_4H_9-*n*	"	III, R = $CH{=}CHC_2H_5$ (90)	19
$CH_2C_6H_5$	"	I (9) + III, R = H (73)	19
6β: R = CH_3	"	I (73)	19
C_2H_5	"	I (81)	19
C_3H_7-*i*	"	I:II = 2:1 (58)	19
C_4H_9-*n*	"	I (79)	19
$CH_2C_6H_5$	"	I (69)	19

Reactant	Conditions	Product(s) and Yield(s) (%)	Refs.
C_{30}			
(structure: C_6H_5, C_6H_5, S, S···O, C_6H_5, C_6H_5)	Ac_2O, reflux 4 h	(structure: C_6H_5, C_6H_5, S, S, C_6H_5, C_6H_5) (87)	422, 452
(structure: $C_6H_5S(O)(CH_2)_2-N$, O, H, CH_3O, N, CO_2CH_3)	TFAA, toluene, 0° to reflux	(structure: $C_6H_5SCH{=}CH-N$, O, H, CH_3O, N, CO_2CH_3) (26)[b]	356
C_{43}			
(structure: $(C_6H_5)_3CNH$, O, S, $CH_2CO_2CH_3$, N, O, $CO_2CH(C_6H_5)_2$)	1. TFAA, Ac_2O, 4 h 2. 2,6-Lutidine, 15 h	(structure: CO_2CH_3, $(C_6H_5)_3CNH$, S, N, O, $CO_2CH(C_6H_5)_2$) (58)	453
C_{45}			
(structure: $(C_6H_5)_3CNH$, O, S, $CH_2CO_2CH_3$, N, O, CH_2OAc, $CO_2CH(C_6H_5)_2$)	"	(structure: CO_2CH_3, $(C_6H_5)_3CNH$, S, N, O, CH_2OAc, $CO_2CH(C_6H_5)_2$) (49)	453

[a] The reaction was carried out to partial conversion.
[b] Other non-Pummerer products were also formed.

TABLE VIII. SULFILIMINES

	Sulfilimine	Reaction Conditions	Product(s) and Yield(s) (%)	Refs.
C_4				
	$(CH_3)_2SNAc$	AcOH, reflux 3 h	CH_3SCH_2OAc (—)	454
		Ac_2O, 70–75°	" (63) + Ac_2NH (58)	455
		AcCl, CH_2Cl_2, 25–30°	CH_3SCH_2Cl (48) + AcNH (52)	455
		C_6H_5COCl, CH_2Cl_2, 25–30°	" (39) + $C_6H_5CONHAc$ (40)	455
C_8				
	$C_6H_5S(NCONH_2)CH_3$	CH_3OH, reflux 10 h	$C_6H_5SCH_2OCH_3$ (14)	456
	$(CH_3)_2SNC_6H_5$	$(C_2H_5)_3N$, toluene, reflux 4 h	o-$CH_3SCH_2C_6H_4NH_2$ (>90)	457
		o-$CH_3C_6H_4OH$, 120–130°	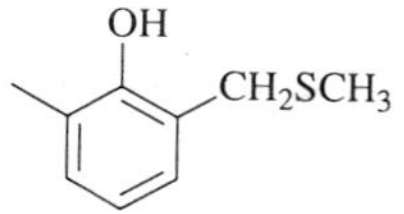 OH, CH_2SCH_3 (39)[a]	458
	$(CH_3)_2SNC_6H_4Cl$ (o, m, or p)	$(C_2H_5)_3N$, toluene, reflux 4 h	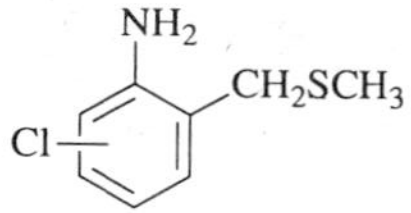 NH_2, Cl, CH_2SCH_3 (90–95)	457
		o-$CH_3C_6H_4OH$, 120–130°, 3 h	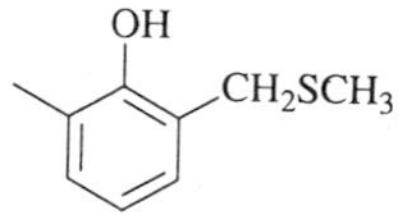 OH, CH_2SCH_3 (35)[a]	458
	$(CH_3)_2SNC_6H_3(NO_2)_2$-2,4	"	" (95)	458
C_9				
	$(CH_3)_2SNC_6H_4CH_3$-p	o-$CH_3C_6H_4OH$, 120–130°, 5 h	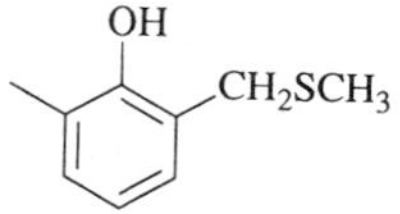 OH, CH_2SCH_3 (78)	458

Reactant	Conditions	Product(s) (Yield, %)	Refs.
	o-$CH_3OC_6H_4OH$, 120–130°, 5 h	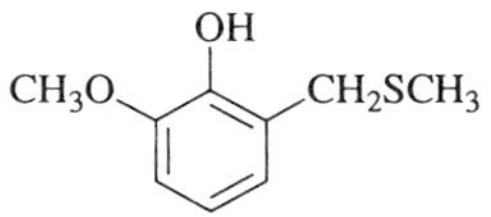 OH, CH_3O, CH_2SCH_3 (35)[a]	458
	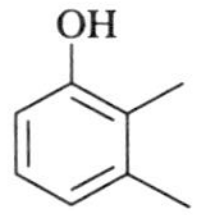 OH , 120–130°, 7 h	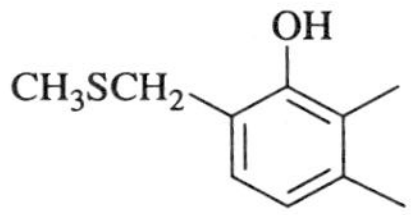 OH, CH_3SCH_2 (69)[a]	458
	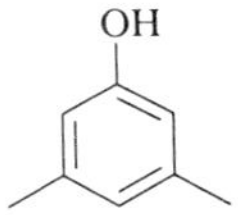 OH , 120–130°, 7 h	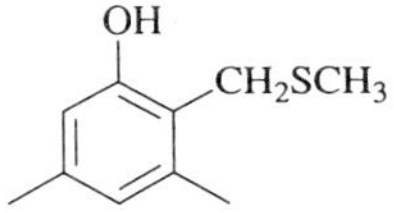 OH, CH_2SCH_3 (58) 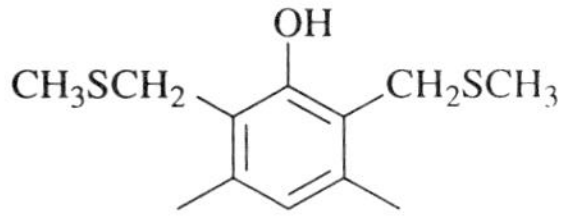 + OH, CH_3SCH_2, CH_2SCH_3 (35)	458
	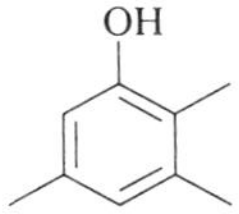 OH , 120–130°, 7 h	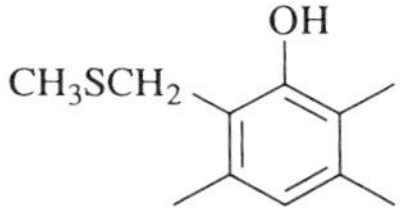 OH, CH_3SCH_2 (37)[a]	458
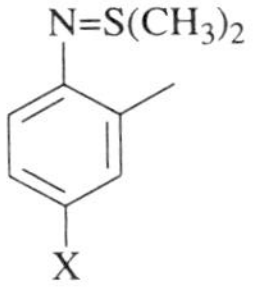 $N{=}S(CH_3)_2$, X	$(C_2H_5)_3N$, toluene, reflux 4 h	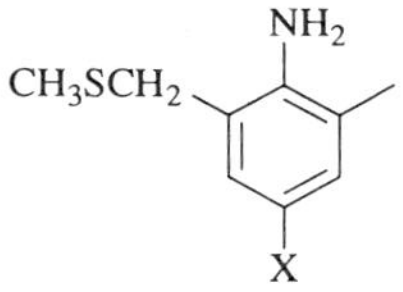 NH_2, CH_3SCH_2, X (>90) X = Cl, Br	457
$C_6H_5S(NCONH_2)C_2H_5$	CH_3OH, reflux 10 h	$C_6H_5SCH(OCH_3)CH_3$ (17)	456

TABLE VIII. SULFILIMINES (*Continued*)

Sulfilimine	Reaction Conditions	Product(s) and Yield(s) (%)	Refs.
C_{11}			
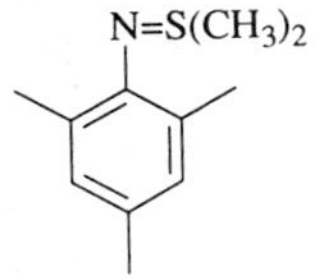 $N{=}S(CH_3)_2$	o-$CH_3C_6H_4OH$, 120–130°	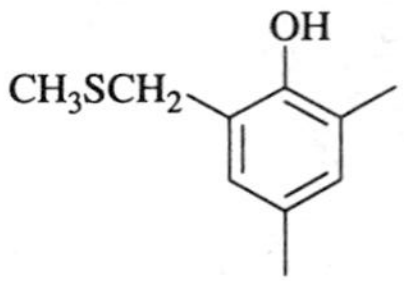 OH, CH_3SCH_2 (62)	458
C_{12}			
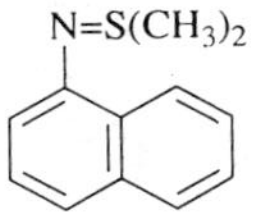 $N{=}S(CH_3)_2$	$(C_2H_5)_3N$, toluene, reflux 4 h	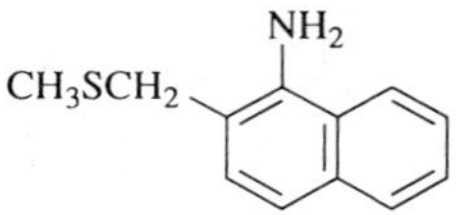 NH_2, CH_3SCH_2 (>90)	457
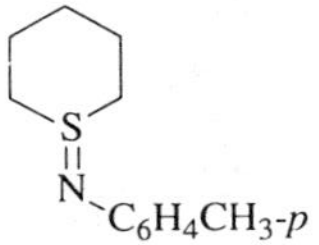 S, N, $C_6H_4CH_3$-p	t-C_4H_9OK, C_6H_6, 15 h	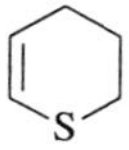 S (50)	459
C_{13}			
p-$XC_6H_4S(NSO_2C_6H_4Y)CH_3$ X = H, CH_3, OCH_3, Cl, NO_2 Y = H, p-CH_3, p-Br, m-NO_2	Ac_2O, 120°, 10 h	p-$XC_6H_4SCH_2OAc$ (—)[b,c]	460
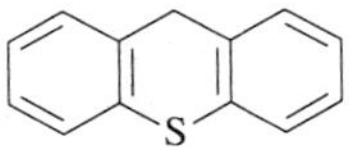 S		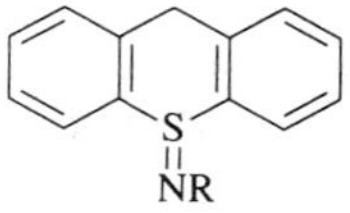 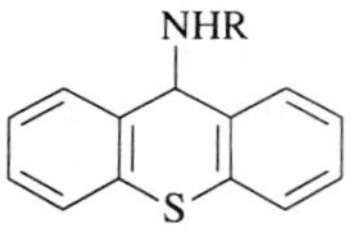 S, NR I; NHR, S II	
	TsNNaCl, CH_3OH, CH_2Cl_2, AcOH, 60 min	I, R = Ts (38) + II, R = Ts (26)	461
	$C_6H_5SO_2NNaCl$, CH_3OH, CH_2Cl_2, AcOH, 60 min	I, R = $SO_2C_6H_5$ (55) + II, R = $SO_2C_6H_5$ (40)	461

		TsNNaCl, CH_3OH, CH_2Cl_2, 2 h	I, R = Ts (5)[a,b] + II, R = Ts (11)[a,b] + 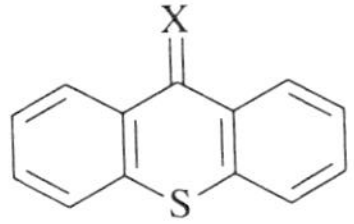 X = O (1); X = NTs (7)[a,b]	461
C_{15}	$C_6H_5CH_2S(NSO_2C_6H_5)CH_2CO_2H$	1. HCl (2 N), heat 2. I_2	$(C_6H_5CH_2S)_2CHCO_2H$ (—)	69
	$C_6H_5S(NC_6H_4CH_3\text{-}p)C_2H_5$	$t\text{-}C_4H_9OK$, C_6H_6, 15 h	$C_6H_5SCH{=}CH_2$ (27)	459
C_{16}	$C_6H_5S(NC_6H_4CH_3\text{-}p)C_3H_7\text{-}n$	,,	$C_6H_5SCH{=}CHCH_3$ (60)	459
	$C_6H_5S(NC_6H_4CH_3\text{-}p)C_3H_7\text{-}i$	,,	$C_6H_5SC(CH_3){=}CH_2$ (80)	459
	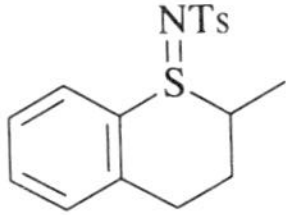 NTs	Ac_2O, AcONa, reflux 30 min	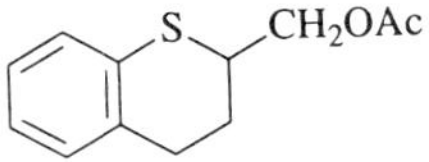 CH_2OAc (90) 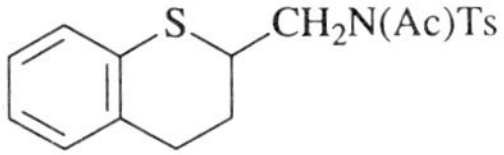 + $CH_2N(Ac)Ts$ (9)	255
	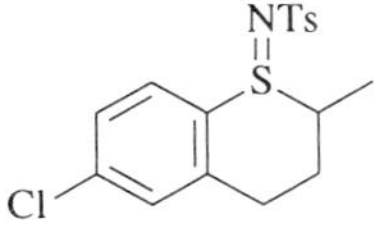 NTs, Cl	,,	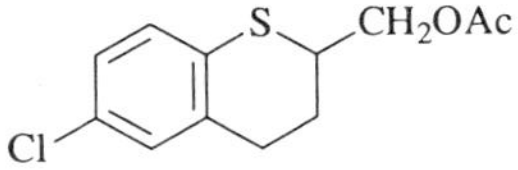 CH_2OAc, Cl (73)	255
	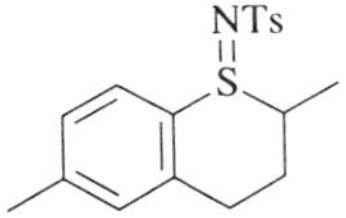 NTs	,,	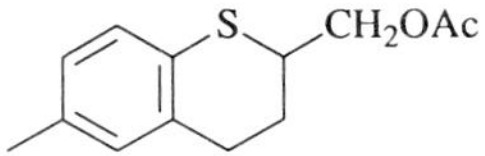 CH_2OAc (90)	255
	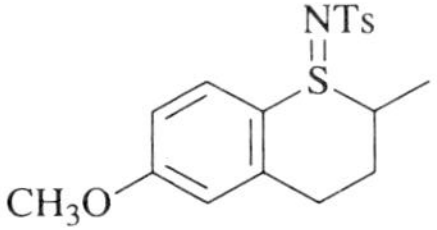 NTs, CH_3O	,,	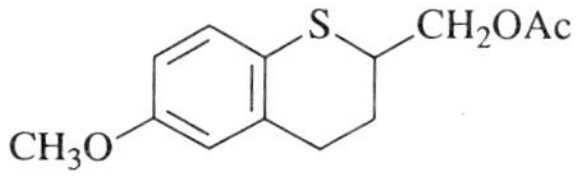 CH_2OAc, CH_3O (90)	255

TABLE VIII. SULFILIMINES (*Continued*)

Sulfilimine	Reaction Conditions	Product(s) and Yield(s) (%)	Refs.
[structure: 2,2-dimethylthiochroman S-NTs sulfilimine]	Ac_2O, reflux 30 min	[structure: 2-(acetoxymethyl)-2-methylthiochroman, CH_2OAc] I (41) + [structure: 2-methylthiochroman with $CH_2N(R)Ts$] II R = H (10) + R = Ac (28)	255
	Ac_2O, AcONa, reflux 30 min	I (72) + II, R = Ac (3)	255
	AcOH, reflux 30 min	I (87)	255
	C_6H_6, reflux 1 h	II, R = $SC_6H_4[(CH_2)_2C(CH_3){=}CH_2]$-*o* (quant)	462, 463
C_{19}			
$C_6H_5S(NC_6H_4CH_3\text{-}p)C_6H_{11}$	$t\text{-}C_4H_9OK$, C_6H_6, 15 h	[structure: 1-(phenylthio)cyclohexene, SC_6H_5] (83)	459

C_{20}	S=NTs	HCl, C_6H_6, reflux 5 h	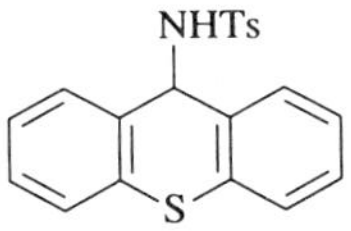 NHTs, S (76)	461
		AcOH, H_2O, reflux	" (76)	461
		DBU, C_6H_6, 4 h	" (84)	461
		TsNNaCl, CH_3OH, reflux 4.5 h	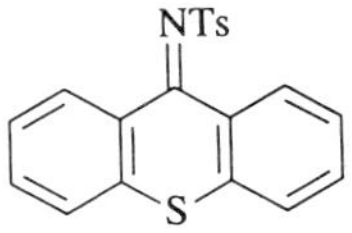 NTs, S (61)	461
C_{21}	$C_6H_5S(NC_6H_4CH_3\text{-}p)CH(CH_3)C_6H_5$	$t\text{-}C_4H_9OK$, C_6H_6, 15 h	$C_6H_5SC(C_6H_5){=}CH_2$ (76)	459

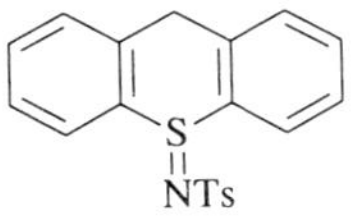

[a] The reaction was carried out to partial conversion.
[a] Other non-Pummerer products were also formed.
[c] The reaction was studied from a kinetic point of view.

TABLE IX. THE SELENO–PUMMERER REACTION

Substrate	Reaction Conditions	Product(s) and Yield(s) (%)	Refs.
C_2			
$(CH_3)_2SeO$	AcOH, 60°, 24 h	CH_3SeCH_2OAc (29)	464
C_4			
$CH_3Se(O)CH_2OAc$	Ac_2O	$(AcOCH_2)_2Se$ (—)	464
C_8			
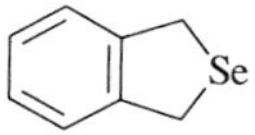 Se	AcOH, H_2O_2, 0–5°, 30 min	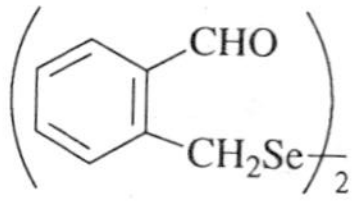 CHO, CH_2Se, 2 (58)	465
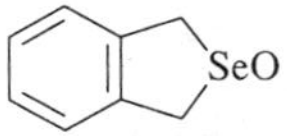 SeO	NaOH (40%), H_2O, hexane	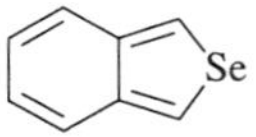 Se (—)	466
$C_6H_5SeCH_2CN$	mCPBA, $CHCl_3$, 0° to rt, 2–3 h	$C_6H_5SeCH(O_2CC_6H_6Cl\text{-}m)CN$ (45)	467
	AcO_2H, $CHCl_3$, 0° to rt, 2–3 h	$C_6H_5SeCH(OAc)CN$ (58)	467
$C_6H_5Se(O)CH{=}CH_2$	Cl_3CCOCl, Zn-Cu	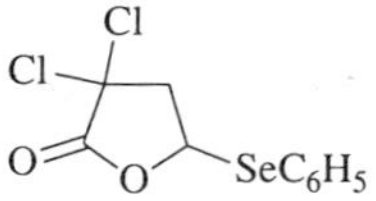 Cl, Cl, O, O, SeC_6H_5 (50)	468
C_9			
$C_6H_5SeCH_2CO_2CH_3$	mCPBA, $CHCl_3$, 0° to rt, 2–3 h	$C_6H_5SeCH(O_2CC_6H_4Cl\text{-}m)CO_2CH_3$ (37)	467
$C_6H_5Se(O)C(CH_3){=}CH_2$	Cl_3CCOCl, Zn-Cu	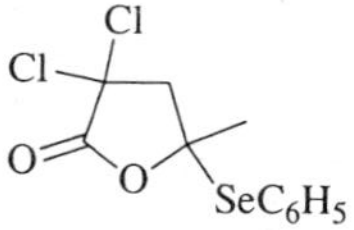 Cl, Cl, O, O, SeC_6H_5 (55)	468
	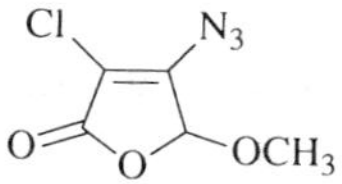 Cl, N_3, O, O, OCH_3, toluene, reflux 2 h	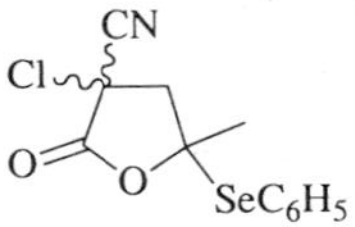 CN, Cl, O, O, SeC_6H_5 (30)[a]	468

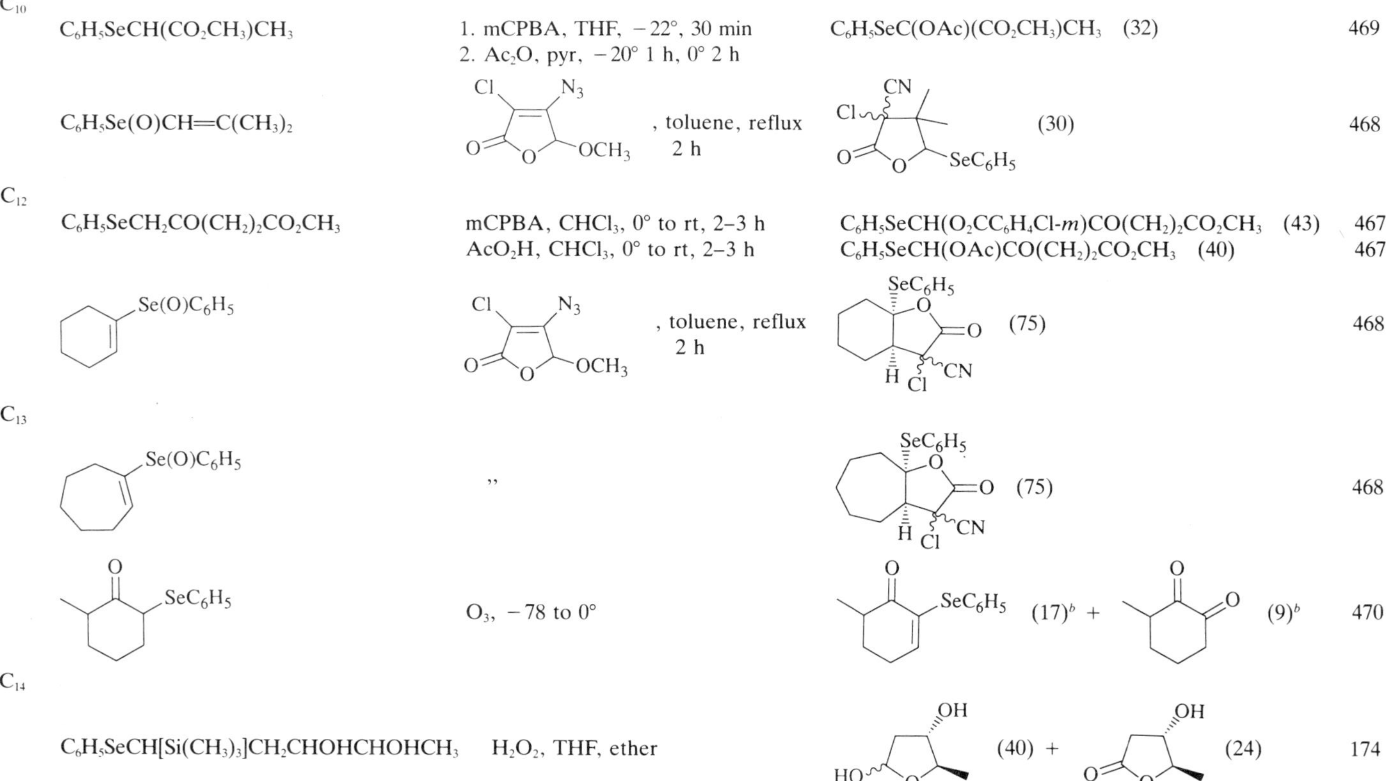

	Reactant	Conditions	Product(s) and Yield(s) (%)	Refs.
C_{10}				
	$C_6H_5SeCH(CO_2CH_3)CH_3$	1. mCPBA, THF, −22°, 30 min 2. Ac_2O, pyr, −20° 1 h, 0° 2 h	$C_6H_5SeC(OAc)(CO_2CH_3)CH_3$ (32)	469
	$C_6H_5Se(O)CH{=}C(CH_3)_2$	, toluene, reflux 2 h	(30)	468
C_{12}				
	$C_6H_5SeCH_2CO(CH_2)_2CO_2CH_3$	mCPBA, $CHCl_3$, 0° to rt, 2–3 h	$C_6H_5SeCH(O_2CC_6H_4Cl\text{-}m)CO(CH_2)_2CO_2CH_3$ (43)	467
		AcO_2H, $CHCl_3$, 0° to rt, 2–3 h	$C_6H_5SeCH(OAc)CO(CH_2)_2CO_2CH_3$ (40)	467
		, toluene, reflux 2 h	(75)	468
C_{13}				
		"	(75)	468
		O_3, −78 to 0°	(17)[b] + (9)[b]	470
C_{14}				
	$C_6H_5SeCH[Si(CH_3)_3]CH_2CHOHCHOHCH_3$	H_2O_2, THF, ether	(40) + (24)	174

TABLE IX. The Seleno–Pummerer Reaction (*Continued*)

	Substrate	Reaction Conditions	Product(s) and Yield(s) (%)	Refs.
	O, SeC_6H_5	H_2O_2, CH_2Cl_2, pyr, 25°, 25 min	O, SeC_6H_5 I $(2)^b$ +	471
			O, O II $(13)^b$	
		O_3, CH_2Cl_2, −78°	I $(80^b$ + II $(43)^b$	471
		1. O_3, CH_2Cl_2, −78° 2. $(i\text{-}C_3H_7)_2NH$	I $(14)^b$ + II $(<2)^b$	471
	O, $SeC_6H_4CF_3$-*m*	H_2O_2, CH_2Cl_2, pyr	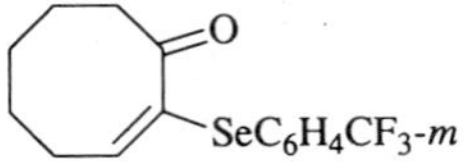 O, $SeC_6H_4CF_3$-*m* III $(0)^b$ + II $(5)^b$	471
		O_3, CH_2Cl_2, −78°	III $(14)^b$ + II $(32)^b$	471
		1. O_3, CH_2Cl_2, −78° 2. $(i\text{-}C_3H_7)_2NH$	III $(22)^b$ + II $(<2)^b$	471
C_{15}	$C_6H_5SeCH(C_6H_5)Si(CH_3)_2H$	H_2O_2, CH_2Cl_2, 3 h	C_6H_5CHO (—)	175
	$C_6H_5SeCH[Si(CH_3)_3]C_5H_{11}$-*n*	H_2O_2 (30%), THF, ether, 0 to 25°	*n*-$C_5H_{11}CHO$ (80)	472
C_{16}	$C_6H_5SeCH[Si(CH_3)_3]C_6H_{13}$-*n*	"	*n*-$C_6H_{13}CHO$ (80)	472
	$C_6H_5Se(O)CH(COC_6H_5)C_2H_5$	H_2SO_4, CH_3OH, −40 to 0°	$C_6H_5COC(OCH_3)_2C_2H_5$ (22)	471
	SeC_6H_5, O	H_2O_2, CH_2Cl_2, 0°, 10 min	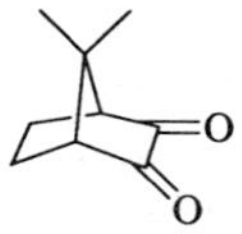 O, O (65)	473
	O, HO, Se, OH, O	$NaIO_4$, CH_3OH, 30 min	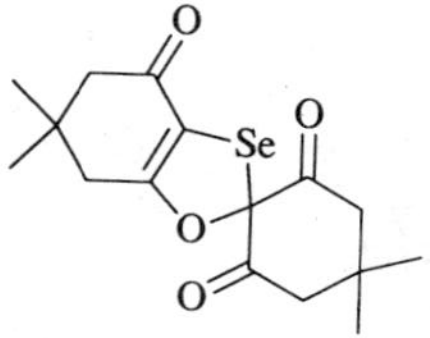 O, Se, O, O, O (quant)	474

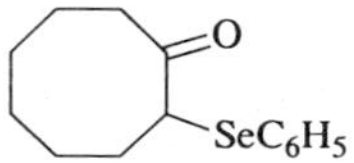

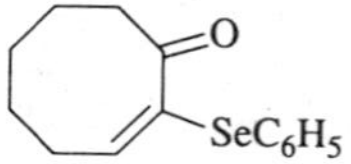

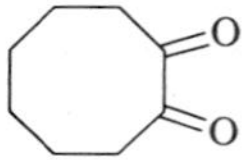

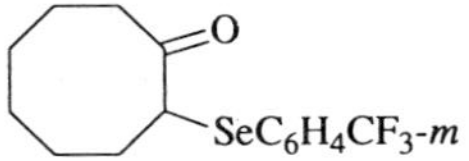

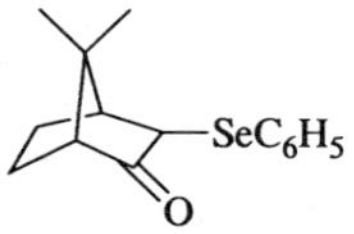

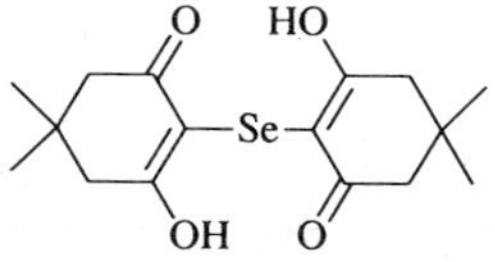

C_{17}				
	$C_6H_5SeCH[Si(CH_3)_3]CH_2C_6H_5$	1. mCPBA, −10° 2. CCl_4, heat	(*E*)-$C_6H_5CH{=}CH[Si(CH_3)_3]$ (47) + C_6H_5CHO (24)	175
	[structure: bicyclic, $Se(O)C_6H_5$]	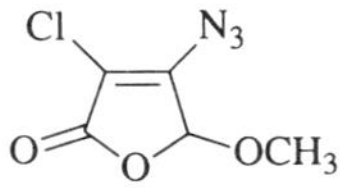 [structure: Cl, N_3, OCH_3 butenolide], toluene, reflux 2 h	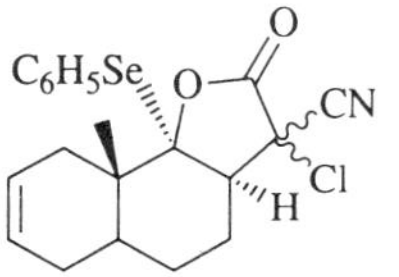 [structure: C_6H_5Se, O, CN, Cl, H] (30)	468
	$C_6H_5Se(O)C(CH_3)(C_6H_5)Si(CH_3)_3$		$C_6H_5SeC(CH_3){=}CH_2$ I + $C_6H_5SeC(CH_3)(C_6H_5)OSi(CH_3)_3$ II + $C_6H_5C[Si(CH_3)_3]{=}CH_2$ III	
		CCl_4, 80°	I (6) + II (6) + III (64)	175
		CCl_4, $(i\text{-}C_3H_7)_2NH$, 80°	I (7) + II (7) + III (55)	175
		C_2Cl_4 $(i\text{-}C_3H_7)_2NH$, 100°	I (10) + II (10) + III (55)	175
		THF, $(i\text{-}C_3H_7)_2NH$, 25°	I (24) + II (22) + III (30)	175
		CH_3OH, $(i\text{-}C_3H_7)_2NH$, 40°	I (12) + II (8) + III (51)	175
		Acetone, $(i\text{-}C_3H_7)_2NH$, 56°	I (17) + II (20) + III (49)	175
C_{18}				
	m-$CF_3C_6H_4Se(O)C(CH_3)(C_6H_5)Si(CH_3)_3$		*m*-$CF_3C_6H_4SeC(C_6H_5){=}CH_2$ I + *m*-$CF_3C_6H_4SeC(CH_3)(C_6H_5)OSi(CH_3)_3$ II + $C_6H_5C[Si(CH_3)_3]{=}CH_2$ III	
		CCl_4, 80°	I (<3) + II (<3) + III (66)	175
		THF, $(i\text{-}C_3H_7)_2NH$, 25°	I (<2) + II (11) + III (58)	175
	m-$CF_3C_6H_4SeCH[Si(CH_3)_3]CH_2C_6H_5$	1. mCPBA, −10° 2. CCl_4, heat	(*E*)-$C_6H_5CH{=}CHSi(CH_3)_3$ (47) + $C_6H_5CH_2CHO$ (24)	175
C_{19}				
	$C_6H_5SeCH[Si(CH_3)_3](CH_2)_3C_6H_5$	H_2O_2 (30%), THF, ether, 0 to 25°	$C_6H_5(CH_2)_3CHO$ (75)	472
C_{20}				
	$C_6H_5SeC[Si(CH_3)_3]_2C_6H_5$	H_2O_2, CH_2Cl_2, 3 h	$C_6H_5COSi(CH_3)_3$ (46) + C_6H_5CHO (8)	175

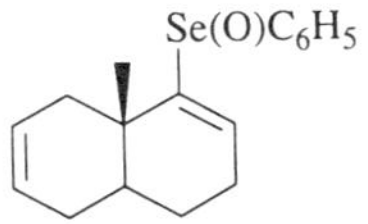

TABLE IX. THE SELENO–PUMMERER REACTION (*Continued*)

Sulfoxide	Reaction Conditions	Product(s) and Yield(s) (%)	Refs.
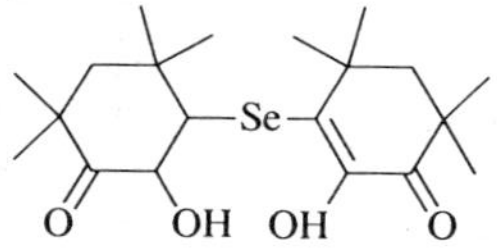 Se, OH, OH, O, O	$NaIO_4$, CH_3OH, reflux 45 min	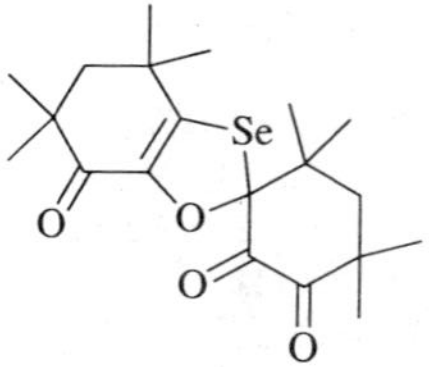 Se, O, O, O, O (35)	474
	SeO_2, dioxane, reflux 4–6 h	" (—)	475
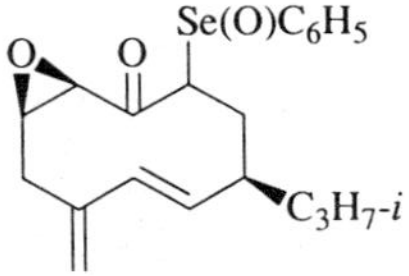 $Se(O)C_6H_5$, O, O, C_3H_7-*i*	1. Ac_2O, AcONa, THF 2. K_2CO_3, CH_3OH, H_2O	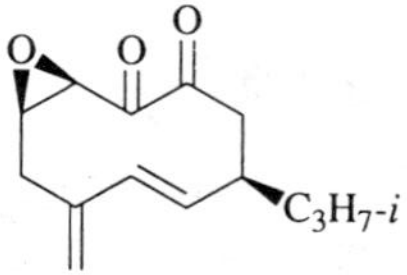 O, O, O, C_3H_7-*i* (60)	159
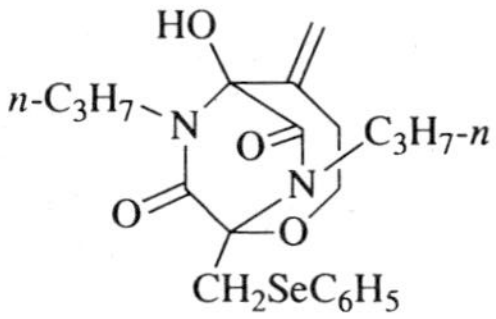 HO, *n*-C_3H_7–N, O, N, C_3H_7-*n*, O, O, $CH_2SeC_6H_5$	1. mCPBA 2. Ac_2O, AcONa, 50°	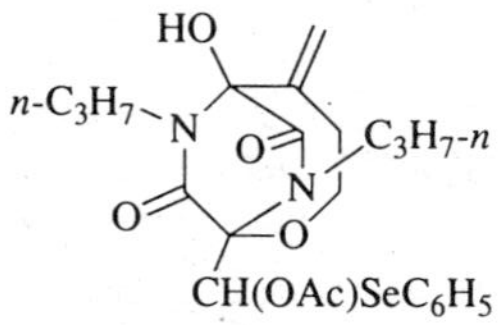 HO, *n*-C_3H_7–N, O, N, C_3H_7-*n*, O, O, $CH(OAc)SeC_6H_5$ (85)	157
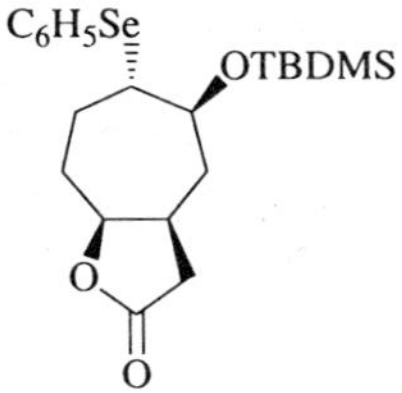 C_6H_5Se, OTBDMS, O, O	1. mCPBA, THF, −78°, 20 min 2. Ac_2O, AcONa, THF, 65°, 3 h	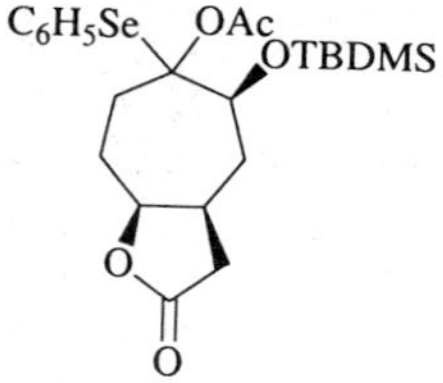 C_6H_5Se, OAc, OTBDMS, O, O (80)	158

Reactant	Conditions	Product(s) and Yield(s) (%)	Refs.
C_{24}			
$C_6H_5SeCH(TMS)C_{14}H_{29}$-*n*	H_2O_2 (30%), THF, ether, 0° 10 min, 25° 1 h	*n*-$C_{14}H_{29}CHO$ (90)	472
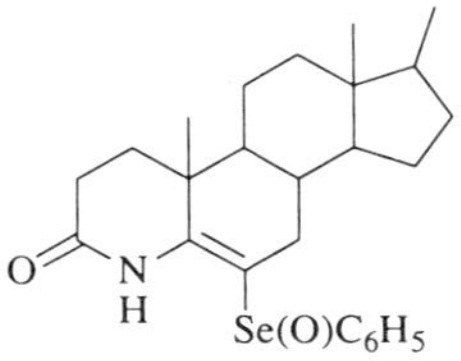 (steroidal lactam with Se(O)C_6H_5 substituent)		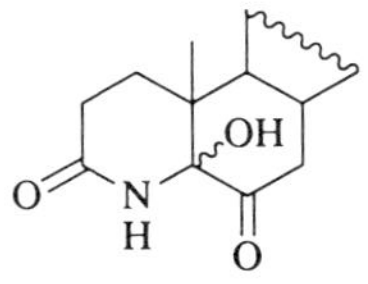 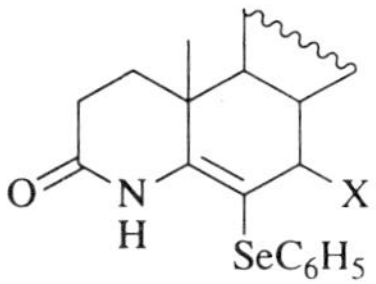 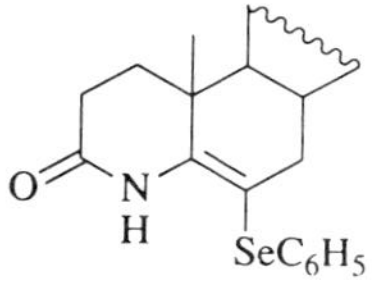 (I) + (II, with SeC$_6$H$_5$ and X) + (III, with SeC$_6$H$_5$)	
R = OAc	$(C_6H_5SeO)_2O$, CH_2Cl_2, 18 h	I (40) + II, X = α-OH (28) + II, X = β-OH + III (16) (12)	476, 477
R = OH	"	I (9) + II, X = α-OH (30) + II, X = β-OH + III (18) (9)	477
C_{27}			
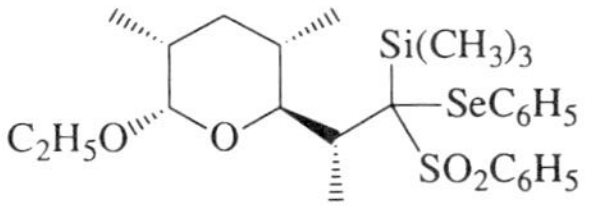 (tetrahydropyran with C_2H_5O, Si$(CH_3)_3$, SeC$_6$H$_5$, SO$_2$C$_6$H$_5$)	H_2O_2, H_3O^+, THF, 30 min	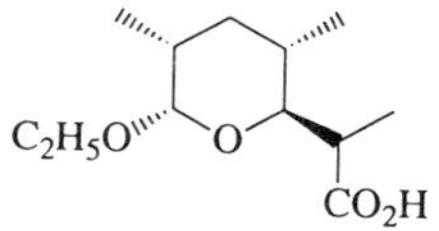 (tetrahydropyran with C_2H_5O and CH(CH$_3$)CO$_2$H) (68)	478

[a] The yield is of the dechlorinated ketone.
[a] Other non-Pummerer products were also formed.

TABLE X. THE SILA–PUMMERER REACTION

Substrate	Reaction Conditions	Product(s) and Yield(s) (%)	Refs.
C_4			
$(CH_2)_4SO$	LDA, $(CH_3)_3SiCl$, −10 to 25°	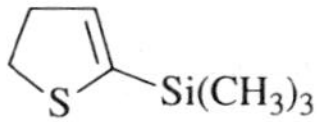 $Si(CH_3)_3$ (60)	479
C_6			
t-$C_4H_9S(O)C_2H_5$	"	t-$C_4H_9SC[Si(CH_3)_3]{=}CH_2$ (75)	479
C_7			
$C_6H_5S(O)CH_3$	$TBDMSOC(OCH_3){=}CH_2$, ZnI_2, CH_3CN, 12 h	$C_6H_5SCH_2OTBDMS$ (60)	55, 344
C_8			
$(n$-$C_4H_9)_2SO$	$(CH_3)_3SiI$, $(i$-$C_3H_7)_2NC_2H_5$, CH_2Cl_2, 25°, 24 h	n-$C_4H_9SCH{=}CHC_2H_5$ (75)	57
n-$C_4H_9S(O)CH(CH_3)C_2H_5$	$(CH_3)_3SiI$, $(i$-$C_3H_7)_2NC_2H_5$, CH_2Cl_2, 25°, 14 h	$C_2H_5CH{=}CHSCH(CH_3)C_2H_5$ (81)	57
$C_6H_5S(O)C_2H_5$	LDA, THF, $(CH_3)_3SiCl$, −10 to 25°	$C_6H_5SC[Si(CH_3)_3]{=}CH_2$ (75)	479
$C_6H_5S(O)CHClCH_3$	1. LDA, THF, −78° 2. $(CH_3)_3SiCl$, −78 to 60°	$C_6H_5SCOCH_3$ (63)	165
C_9			
$C_6H_5S(O)CHClC_2H_5$	"	$C_6H_5SCOC_2H_5$ (62)	165
$C_6H_5S(O)CH_2CH{=}CH_2$	$TBDMSOC(OCH_3){=}CH_2$, ZnI_2, CH_3CN, 15 min	$C_6H_5SCH[OTBDMS]CH{=}CH_2$ (54)	480
	$TMSOC(OCH_3){=}C(CH_3)_2$, ZnI_2, CH_3CN	$C_6H_5SCH(CH{=}CH_2)C(CH_3)_2CO_2CH_3$ I + (E)-$C_6H_5SCH{=}CHCH_2C(CH_3)_2CO_2CH_3$ II I + II (47)	480
$C_6H_5S(O)CH_2C{\equiv}CH$	$TBDMSOC(OCH_3){=}CH_2$, CH_3CN, 1 h	$C_6H_5SCH[OTBDMS]C{\equiv}CH$ (36) + $C_6H_5SCH(CH_2CO_2CH_3)C{\equiv}CH$ (24)	480
(Z)-$CH_2S(O)CH{=}CHC_6H_5$	$TBDMSOC(OCH_3){=}CH_2$, ZnI_2, CH_3CN, 11 h	(Z)-$C_6H_5CH{=}CHSCH_2OTBDMS$ (43)	480
C_{10}			
$C_6H_5SCH_2Si(CH_3)_3$	1. mCPBA, CH_2Cl_2, −23°, 2 h 2. C_6H_6, reflux 4 h	$C_6H_5SCH_2OSi(CH_3)_3$ (89)	169
$C_6H_5S(O)CH_2Si(CH_3)_3$	60°	" (72)	171
	60°, 1 h	" (79)	481

	Reactant	Conditions	Product(s) and Yield(s) (%)	Refs.
	p-$ClC_6H_4S(O)CH_2Si(CH_3)_3$	TFAA or CF_3CO_2H, CH_2Cl_2, 0°	p-$ClC_6H_4SCH_2O_2CCF_3$ (quant)	482
	$C_6H_5S(O)C_4H_9$-n	LDA, THF, $(CH_3)_3SiCl$, −10 to 25°	$C_6H_5SC[Si(CH_3)_3]{=}CHC_2H_5$ $E{:}Z$ = 2:1 (80)	479
	$C_6H_5S(O)CH_2CH{=}CHCH_3$	"	$C_6H_5SC[Si(CH_3)_3]{=}CHCH{=}CH_2$ $E{:}Z$ = 3:2 (70)	479
		$(CH_3)_3SiI$, $(i$-$C_3H_7)_2NC_2H_5$, 25°, 2 h	" E >95% (85)	57
	$C_6H_5S(O)CH_2CO_2C_2H_5$	$TBDMSOC(OCH_3){=}CH_2$, CH_3CN, 70°, 14 h	$C_6H_5SCH(CO_2C_2H_5)OTBDMS$ (79)	55, 344
		$TBDMSOC(OCH_3){=}CH_2$, ZnI_2, CH_3CN, 12 h	" (42)	344
C_{11}	$C_6H_5SCH[Si(CH_3)_3]CH_3$	1. mCPBA, CH_2Cl_2, −23°, 2 h 2. C_6H_6, reflux 4 h	$C_6H_5SCH(CH_3)OSi(CH_3)_3$ (97)	169
	$C_6H_5S(O)CH[Si(CH_3)_3]CH_3$	80°	" (—)[a]	171
		C_6H_6, reflux 2 h	" (68)	483
	$C_6H_5S(O)CH{=}CHC_3H_7$-n	$(CH_3)_3SiI$, $(i$-C_3H_7-$n)_2NC_2H_5$, CH_2Cl_2, −25°, 2 h	$C_6H_5S(CH{=}CH)_2CH_3$ (91)	57
C_{12}	$C_6H_5SCH[Si(CH_3)_3]C_2H_5$	1. mCPBA, CH_2Cl_2, −23°, 2 h 2. C_6H_6, reflux 4 h	$C_6H_5SCH(C_2H_5)OSi(CH_3)_3$ (88)	169
	$C_6H_5S(O)C_6H_{13}$-n	$TBDMSOC(OCH_3){=}CH_2$, ZnI_2, CH_3CN, 24 h	$C_6H_5SCH(C_5H_{11}$-$n)OTBDMS$ (42)	344
	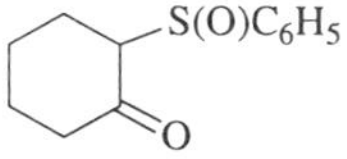 [structure: $S(O)C_6H_5$, O]	$TBDMSOC(OCH_3){=}CH_2$, ZnI_2, CH_3CN, rt 1 h, 70° 14 h	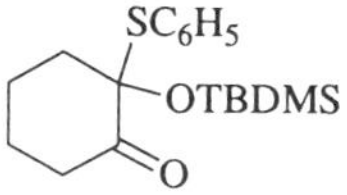 [structure: SC_6H_5, OTBDMS, O] (75)	55, 344
	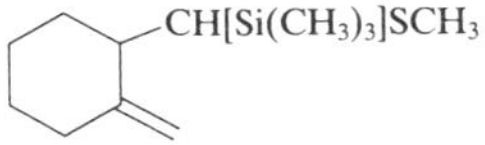 [structure: $CH[Si(CH_3)_3]SCH_3$]	1. mCPBA, CH_2Cl_2, −40° to rt 2. THF, $(CO_2H)_2$, 20°, 18 h	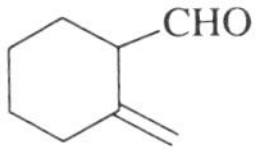 [structure: CHO] (40–60)	167
	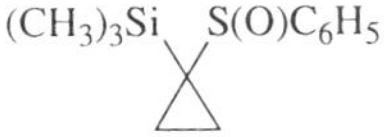 [structure: $(CH_3)_3Si$, $S(O)C_6H_5$]	C_6H_6, reflux 22 h	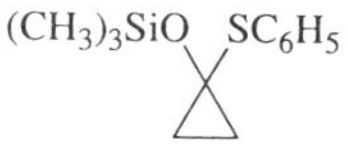 [structure: $(CH_3)_3SiO$, SC_6H_5] (75)	484
		CH_3OH, reflux 14 d	" (3) + [structure: CH_3O, SC_6H_5] (42)	484

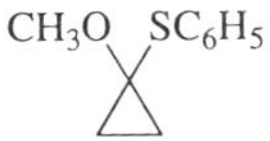

TABLE X. THE SILA–PUMMERER REACTION (*Continued*)

	Substrate	Reaction Conditions	Product(s) and Yield(s) (%)	Refs.
C_{13}	$C_6H_5SCH[Si(CH_3)_3]C_3H_7$-*n*	1. mCPBA, CH_2Cl_2, −23°, 2 h 2. C_6H_6, reflux 4 h	$C_6H_5SCH(C_3H_7\text{-}n)OSi(CH_3)_3$ (85)	169
	$C_6H_5SCH[Si(CH_3)_3]CH_2CH{=}CH_2$	"	$C_6H_5SCH(CH_2CH{=}CH_2)OSi(CH_3)_3$	169
	$C_6H_5S(O)CHClC_6H_5$	1. LDA, THF, −78° 2. $(CH_3)_3SiCl$, −78 to 0°	$C_6H_5SCOC_6H_5$ (74)	165
	$C_6H_5S(O)CHClC_6H_{11}$	"	$C_6H_5SCOC_6H_{11}$ (58)	165
	1-($CH_2{=}CH$)-1-[$CH(Si(CH_3)_3)SCH_3$]cyclohexane	1. mCPBA, CH_2Cl_2, −40° to rt 2. THF, $(CO_2H)_2$, 20°, 18 h	1-($CH_2{=}CH$)-1-CHO-cyclohexane (40–60)	167
C_{14}	$C_6H_5SCH[Si(CH_3)_3]C_4H_9$-*n*	1. mCPBA, CH_2Cl_2, −23°, 2 h 2. C_6H_6, reflux 4 h	$C_6H_5SCH(C_4H_9\text{-}n)OSi(CH_3)_3$ (83)	169
	$C_6H_5SCH[Si(CH_3)_3]C_4H_9$-*s*	"	$C_6H_5SCH(C_4H_9\text{-}s)OSi(CH_3)_3$ (83)	169
	$C_6H_5S(O)C_8H_{17}$-*n*	$(CH_3)_3SiI$, $(i\text{-}C_3H_7)_2NC_2H_5$, CH_2Cl_2, 25°, 1.5 h	$C_6H_5SCH{=}CHC_6H_{13}$-*n* (90)	57
	$C_6H_5S(O)CH[Si(CH_3)_3]CH{=}C(CH_3)_2$	rt	$C_6H_5SCH(OSi(CH_3)_3)CH{=}C(CH_3)_2$ (—)	166
	t-$C_4H_9S(O)CH[Si(CH_3)_3]C_6H_5$	rt or 68°	*t*-$C_4H_9SCH(OSi(CH_3)_3)C_6H_5$ (76)[a] + $C_6H_5CH(SC_4H_9\text{-}t)_2$ (4)	171
	p-$CH_3C_6H_4S(O)CH_2Si(C_2H_5)_3$	60°, 1 h	*p*-$CH_3C_6H_4SCH_2OSi(C_2H_5)_3$ (50)	481
	$CH_3SCH[Si(CH_3)_3]C(CH_3)(CH{=}CH_2)$-$CH_2CH{=}C(CH_3)_2$	1. mCPBA, CH_2Cl_2, −40° to rt 2. THF, $(CO_2H)_2$, 20°, 18 h	$OHCC(CH_3)(CH{=}CH_2)CH_2CH{=}C(CH_3)_2$ (40–60)	68
	2,2-dimethyl-6-methylene-1-[$CH(Si(CH_3)_3)SCH_3$]cyclohexane	"	2,2-dimethyl-6-methylenecyclohexanecarbaldehyde (CHO) (40–60)	167
	$C_6H_5S(O)CH_2CH(C_2H_5)C_4H_9$-*n*	$TBDMSOC(OCH_3){=}CH_2$ ZnI_2, CH_3CN, 24 h	$C_6H_5SCH(OTBDMS)CH(C_2H_5)C_4H_9$-*n* (42)	55, 344
	$C_6H_5S(O)(CH_2)_2C_6H_5$	$TBDMSOC(OCH_3){=}CH_2$, ZnI_2, CH_3CN, 20 h	$C_6H_5SCH(CH_2C_6H_5)OTBDMS$ (55) + $C_6H_5SCH{=}CHC_6H_5$ $Z{:}E$ = ~1:2 (16)	55, 344
		LDA, THF, $(CH_3)_3SiCl$, −10 to 25°	$C_6H_5SC[Si(CH_3)_3]{=}CHC_6H_5$ $E{:}Z$ = 3:1 (75)	479

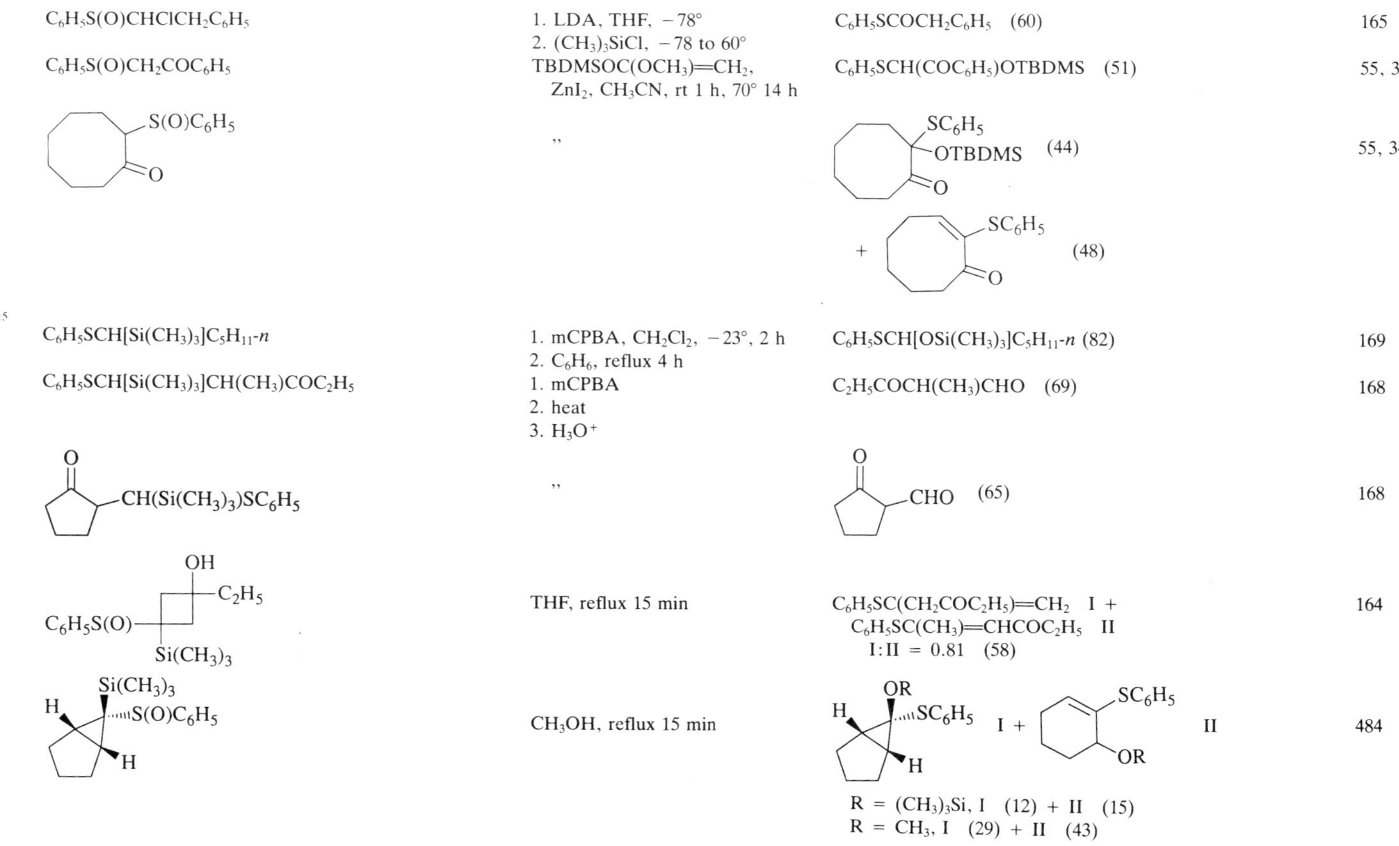

	Reactant	Conditions	Product(s) and Yield(s) (%)	Refs.
	$C_6H_5S(O)CHClCH_2C_6H_5$	1. LDA, THF, −78° 2. $(CH_3)_3SiCl$, −78 to 60°	$C_6H_5SCOCH_2C_6H_5$ (60)	165
	$C_6H_5S(O)CH_2COC_6H_5$	$TBDMSOC(OCH_3){=}CH_2$, ZnI_2, CH_3CN, rt 1 h, 70° 14 h	$C_6H_5SCH(COC_6H_5)OTBDMS$ (51)	55, 344
	$S(O)C_6H_5$, O	"	SC_6H_5, OTBDMS, O (44) + SC_6H_5, O (48)	55, 344
C_{15}	$C_6H_5SCH[Si(CH_3)_3]C_5H_{11}$-*n*	1. mCPBA, CH_2Cl_2, −23°, 2 h 2. C_6H_6, reflux 4 h	$C_6H_5SCH[OSi(CH_3)_3]C_5H_{11}$-*n* (82)	169
	$C_6H_5SCH[Si(CH_3)_3]CH(CH_3)COC_2H_5$	1. mCPBA 2. heat 3. H_3O^+	$C_2H_5COCH(CH_3)CHO$ (69)	168
	O, $CH(Si(CH_3)_3)SC_6H_5$	"	O, CHO (65)	168
	OH, C_2H_5, $C_6H_5S(O)$, $Si(CH_3)_3$	THF, reflux 15 min	$C_6H_5SC(CH_2COC_2H_5){=}CH_2$ I + $C_6H_5SC(CH_3){=}CHCOC_2H_5$ II I:II = 0.81 (58)	164
	$Si(CH_3)_3$, H, $S(O)C_6H_5$, H	CH_3OH, reflux 15 min	OR, H, SC_6H_5, H I + SC_6H_5, OR II R = $(CH_3)_3Si$, I (12) + II (15) R = CH_3, I (29) + II (43)	484

TABLE X. THE SILA–PUMMERER REACTION (*Continued*)

Substrate	Reaction Conditions	Product(s) and Yield(s) (%)	Refs.
C_{16}			
$C_6H_5SCH(TMS)C_6H_{13}$-*n*	1. mCPBA, CH_2Cl_2, $-23°$, 2 h 2. C_6H_6, reflux 4 h	$C_6H_5SCH[OSi(CH_3)_3]C_6H_{13}$ (85)	169
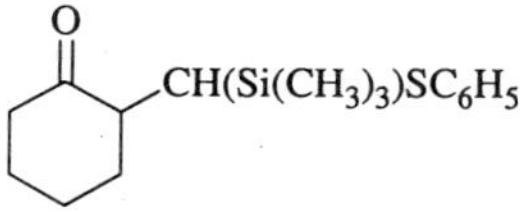 O; $CH(Si(CH_3)_3)SC_6H_5$	1. mCPBA 2. heat 3. H_3O^+	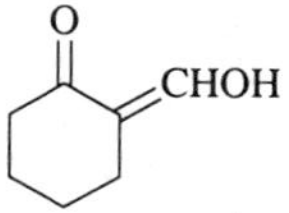 O; CHOH (72)	168
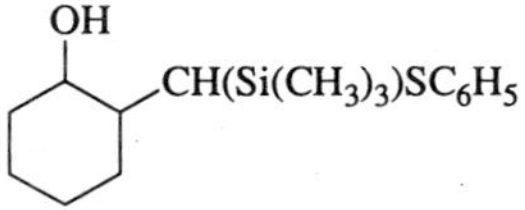 OH; $CH(Si(CH_3)_3)SC_6H_5$	"	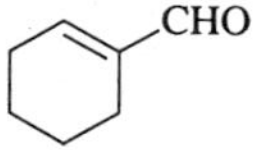 CHO (42)	168
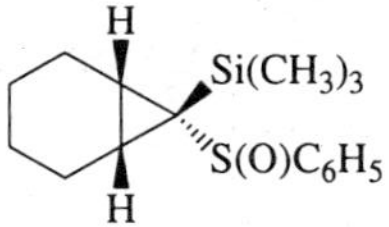 H; $Si(CH_3)_3$; $S(O)C_6H_5$; H	C_6H_6, reflux 30 min	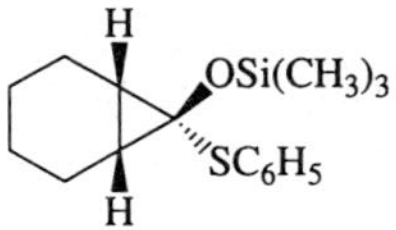 H; $OSi(CH_3)_3$; SC_6H_5; H I (86)	484
	CH_3OH, reflux 17 h	I (15) 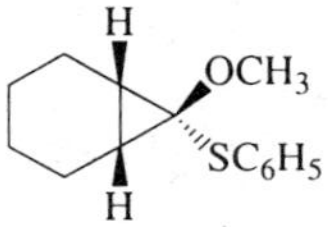 + H; OCH_3; SC_6H_5; H (68)	484
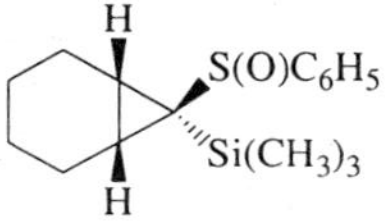 H; $S(O)C_6H_5$; $Si(CH_3)_3$; H	C_6H_6, reflux 15 min	I 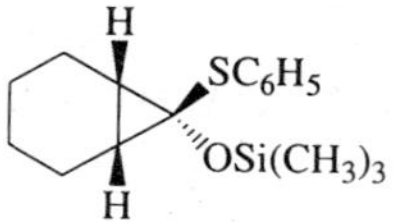 + H; SC_6H_5; $OSi(CH_3)_3$; H 84:16 (69)	484

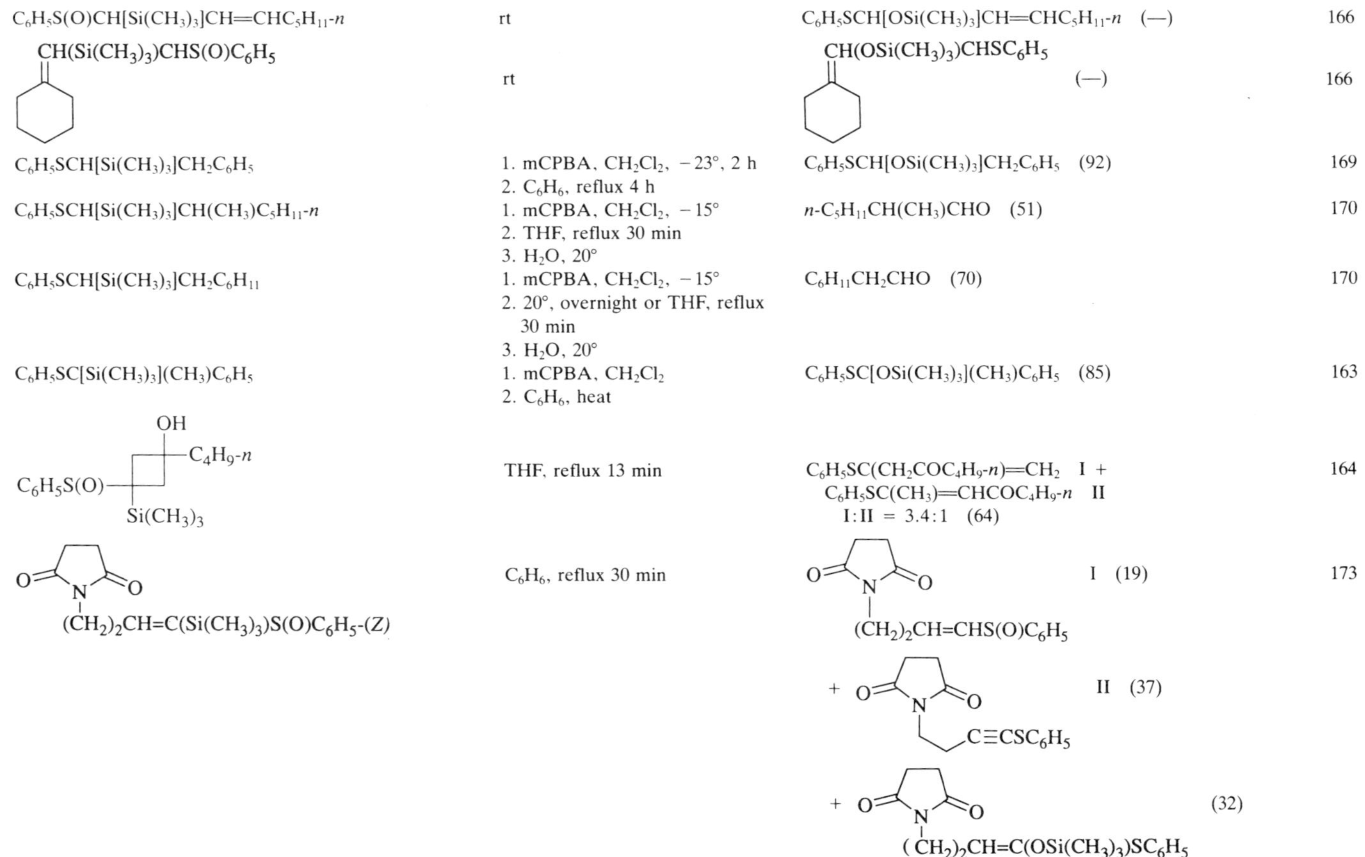

C_{17}			
$C_6H_5S(O)CH[Si(CH_3)_3]CH{=}CHC_5H_{11}$-$n$	rt	$C_6H_5SCH[OSi(CH_3)_3]CH{=}CHC_5H_{11}$-$n$ (—)	166
$CH(Si(CH_3)_3)CHS(O)C_6H_5$ (cyclohexylidene)	rt	$CH(OSi(CH_3)_3)CHSC_6H_5$ (cyclohexylidene) (—)	166
$C_6H_5SCH[Si(CH_3)_3]CH_2C_6H_5$	1. mCPBA, CH_2Cl_2, −23°, 2 h 2. C_6H_6, reflux 4 h	$C_6H_5SCH[OSi(CH_3)_3]CH_2C_6H_5$ (92)	169
$C_6H_5SCH[Si(CH_3)_3]CH(CH_3)C_5H_{11}$-$n$	1. mCPBA, CH_2Cl_2, −15° 2. THF, reflux 30 min 3. H_2O, 20°	n-$C_5H_{11}CH(CH_3)CHO$ (51)	170
$C_6H_5SCH[Si(CH_3)_3]CH_2C_6H_{11}$	1. mCPBA, CH_2Cl_2, −15° 2. 20°, overnight or THF, reflux 30 min 3. H_2O, 20°	$C_6H_{11}CH_2CHO$ (70)	170
$C_6H_5SC[Si(CH_3)_3](CH_3)C_6H_5$	1. mCPBA, CH_2Cl_2 2. C_6H_6, heat	$C_6H_5SC[OSi(CH_3)_3](CH_3)C_6H_5$ (85)	163
$C_6H_5S(O)$–cyclobutane with OH, C_4H_9-n, $Si(CH_3)_3$	THF, reflux 13 min	$C_6H_5SC(CH_2COC_4H_9\text{-}n){=}CH_2$ I + $C_6H_5SC(CH_3){=}CHCOC_4H_9$-$n$ II I:II = 3.4:1 (64)	164
succinimido-$(CH_2)_2CH{=}C(Si(CH_3)_3)S(O)C_6H_5$-(Z)	C_6H_6, reflux 30 min	succinimido-$(CH_2)_2CH{=}CHS(O)C_6H_5$ I (19) + succinimido-$CH_2CH_2C{\equiv}CSC_6H_5$ II (37) + succinimido-$(CH_2)_2CH{=}C(OSi(CH_3)_3)SC_6H_5$ (32)	173

TABLE X. THE SILA–PUMMERER REACTION (*Continued*)

Substrate	Reaction Conditions	Product(s) and Yield(s) (%)	Refs.
	1. C_6H_6, reflux 30 h 2. CH_3OH, TsOH (cat.)	I (20) + II (36) + N-[$(CH_2)_3COSC_6H_5$]succinimide (32)	173
C_{18}			
$C_6H_5SC[Si(CH_3)_3](C_2H_5)C_6H_5$	1. mCPBA, CH_2Cl_2 2. C_6H_6, heat	$C_6H_5SC[OSi(CH_3)_3](C_2H_5)C_6H_5$ (78)	163
$C_6H_5SCH[Si(CH_3)_3]C_8H_{17}$-*n*	1. mCPBA, CH_2Cl_2, −15° 2. THF, reflux 30 min 3. H_2O, 20°	*n*-$C_8H_{17}CHO$ (68)	170
$C_6H_5SCH[Si(CH_3)_3]CH_2COC_6H_5$	1. mCPBA 2. heat 3. H_3O^+	$C_6H_5COCH_2CHO$ (78)	168
2-[$CH(Si(CH_3)_3)SC_6H_5$]cyclohexanone ethylene ketal	"	2-CHO-cyclohexanone ethylene ketal (65)	168
C_{19}			
$C_6H_5SC[Si(CH_3)_3](C_6H_5)C_3H_7$-*n*	1. mCPBA, CH_2Cl_2 2. C_6H_6, heat	$C_6H_5SC[OSi(CH_3)_3](C_6H_5)C_3H_7$-*n* (77)	163
$C_6H_5SC[Si(CH_3)_3](C_6H_5)C_3H_7$-*i*	"	$C_6H_5SC[OSi(CH_3)_3](C_6H_5)C_3H_7$-*i* (—)	163
$C_6H_5SC[Si(CH_3)_3](C_6H_5)CH_2CH{=}CH_2$	"	$C_6H_5SC[OSi(CH_3)_3](C_6H_5)CH_2CH{=}CH_2$ (79)	163
$C_6H_5S(O)CH[Si(CH_3)_3]CH{=}C(CH_3)(CH_2)_2CH{=}C(CH_3)_2$		$C_6H_5SCH[OSi(CH_3)_3]CH{=}C(CH_3)(CH_2)_2CH{=}C(CH_3)_2$ (—)	485
1-$C_6H_5S(O)$-1-$Si(CH_3)_3$-3-OH-3-C_6H_5-cyclobutane	Ether, reflux 47 min	$C_6H_5SC(CH_2COC_6H_5){=}CH_2$ I + $C_6H_5SC(CH_3){=}CHCOC_6H_5$ II I:II = 0.83 (59)	164
C_{20}			
$C_6H_5SC[Si(CH_3)_3](C_6H_5)C_4H_9$-*n*	1. mCPBA, CH_2Cl_2 2. C_6H_6, heat	$C_6H_5SC[OSi(CH_3)_3](C_6H_5)C_4H_9$-*n* (81)	163
C_{21}			
$C_6H_5SCH[Si(CH_3)_3]C_{12}H_{25}$-*n*	1. mCPBA, CH_2Cl_2, −23°, 2 h 2. C_6H_6, reflux 4 h	$C_6H_5SCH[OSi(CH_3)_3]C_{12}H_{25}$-*n* (88)	169

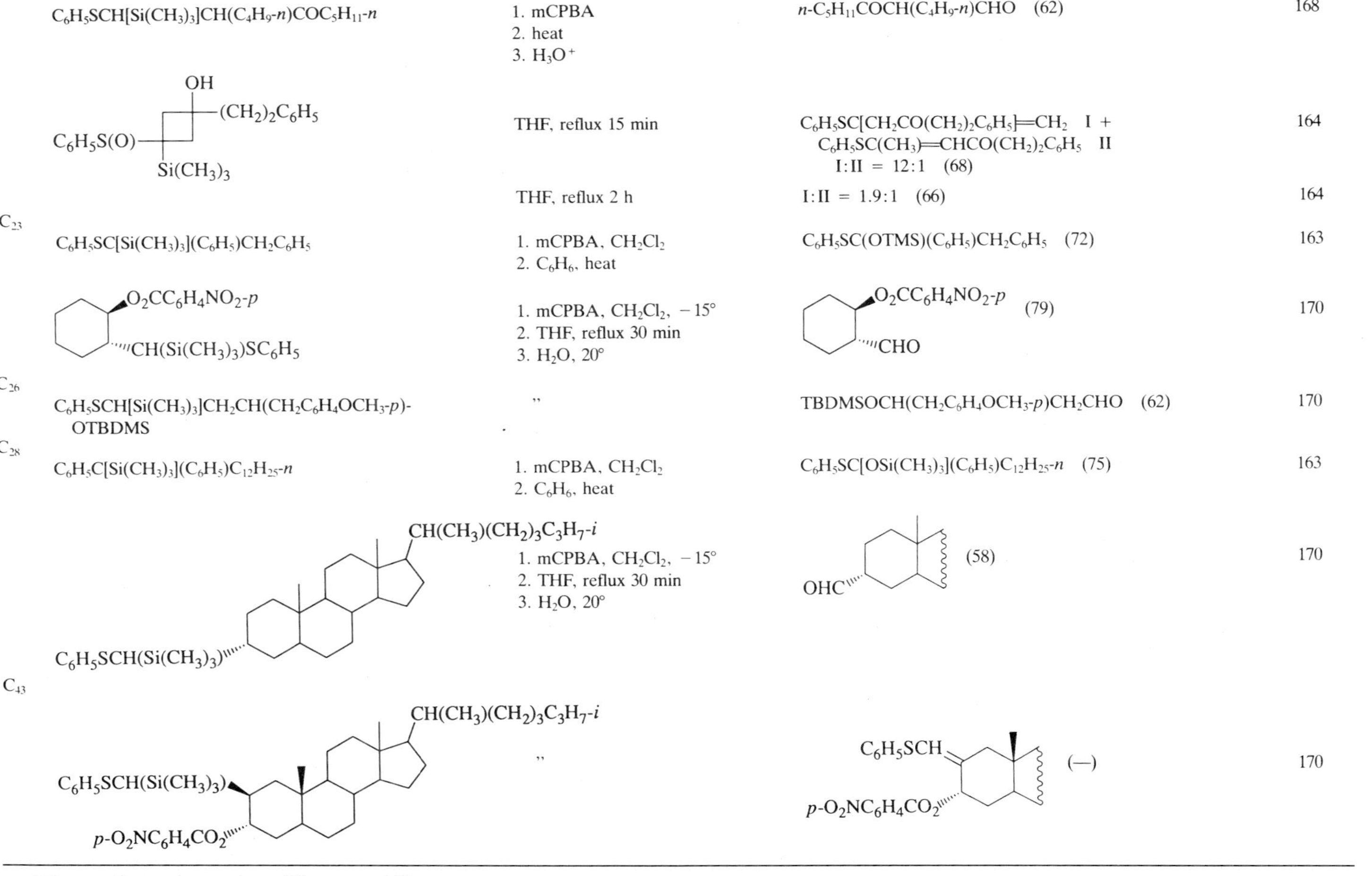

	Reactant	Conditions	Product(s) and Yield(s) (%)	Refs.
	$C_6H_5SCH[Si(CH_3)_3]CH(C_4H_9\text{-}n)COC_5H_{11}\text{-}n$	1. mCPBA 2. heat 3. H_3O^+	$n\text{-}C_5H_{11}COCH(C_4H_9\text{-}n)CHO$ (62)	168
	$C_6H_5S(O)$–cyclobutane bearing OH, $(CH_2)_2C_6H_5$, $Si(CH_3)_3$	THF, reflux 15 min	$C_6H_5SC[CH_2CO(CH_2)_2C_6H_5]{=}CH_2$ I + $C_6H_5SC(CH_3){=}CHCO(CH_2)_2C_6H_5$ II I:II = 12:1 (68)	164
		THF, reflux 2 h	I:II = 1.9:1 (66)	164
C_{23}	$C_6H_5SC[Si(CH_3)_3](C_6H_5)CH_2C_6H_5$	1. mCPBA, CH_2Cl_2 2. C_6H_6, heat	$C_6H_5SC(OTMS)(C_6H_5)CH_2C_6H_5$ (72)	163
	cyclohexane with $O_2CC_6H_4NO_2\text{-}p$ and $CH(Si(CH_3)_3)SC_6H_5$	1. mCPBA, CH_2Cl_2, $-15°$ 2. THF, reflux 30 min 3. H_2O, 20°	cyclohexane with $O_2CC_6H_4NO_2\text{-}p$ and CHO (79)	170
C_{26}	$C_6H_5SCH[Si(CH_3)_3]CH_2CH(CH_2C_6H_4OCH_3\text{-}p)$-OTBDMS	"	$TBDMSOCH(CH_2C_6H_4OCH_3\text{-}p)CH_2CHO$ (62)	170
C_{28}	$C_6H_5C[Si(CH_3)_3](C_6H_5)C_{12}H_{25}\text{-}n$	1. mCPBA, CH_2Cl_2 2. C_6H_6, heat	$C_6H_5SC[OSi(CH_3)_3](C_6H_5)C_{12}H_{25}\text{-}n$ (75)	163
	steroid with $C_6H_5SCH(Si(CH_3)_3)$ and $CH(CH_3)(CH_2)_3C_3H_7\text{-}i$	1. mCPBA, CH_2Cl_2, $-15°$ 2. THF, reflux 30 min 3. H_2O, 20°	OHC-substituted structure (58)	170
C_{43}	steroid with $C_6H_5SCH(Si(CH_3)_3)$, $p\text{-}O_2NC_6H_4CO_2$ and $CH(CH_3)(CH_2)_3C_3H_7\text{-}i$	"	structure with $C_6H_5SCH{=}$ and $p\text{-}O_2NC_6H_4CO_2$ (—)	170

[a] The two diastereoisomers have different reactivities.

TABLE XI. VINYLOGOUS AND ADDITIVE PUMMERER REACTIONS

	Substrate	Reaction Conditions	Product(s) and Yield(s) (%)	Refs.
C_5	(*Z*)-$CH_3S(O)C(SCH_3){=}C(NH_2)CH_3$	Ac_2O, pyr, CH_2Cl_2, 4 h	$CH_3SCOC(SCH_3)(NHAc)CH_3$ (87)	183
C_7	(*Z*)-$CH_3S(O)C(SCH_3){=}C(NH_2)C_3H_7$-*i*	"	$CH_3SCOC(SCH_3)(NHAc)C_3H_7$-*i* (87)	183
	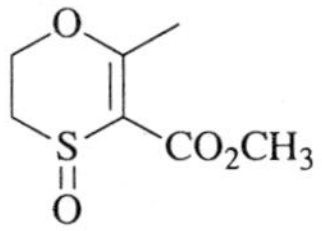 [structure: 1,4-oxathiin S-oxide, 2-CH₃, 3-CO₂CH₃]	Ac_2O, AcOH, C_6H_6, reflux 50 min	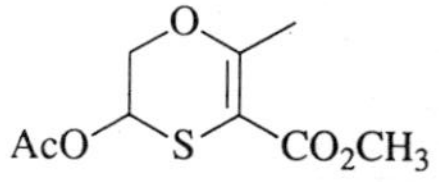 [structure: AcO-oxathiin, CH₃, CO₂CH₃] (78) 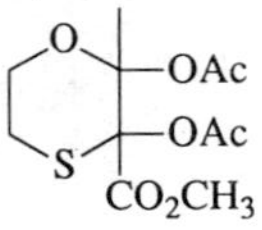 + [structure: oxathiane, OAc, OAc, CO₂CH₃] (9) 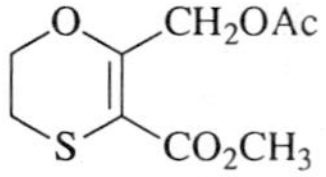 + [structure: oxathiin, CH₂OAc, CO₂CH₃] (10)	178
		TFAA, C_6H_6	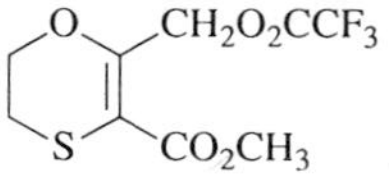 [structure: oxathiin, $CH_2O_2CCF_3$, CO_2CH_3] (—)	178
	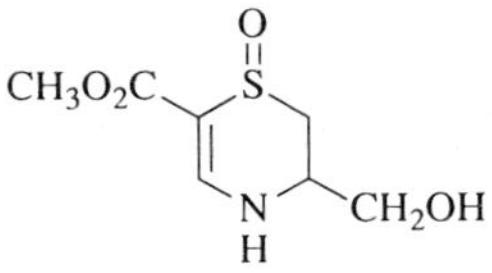 [structure: CH_3O_2C-thiazine S-oxide, N-H, CH_2OH]	AcCl, CH_3CN	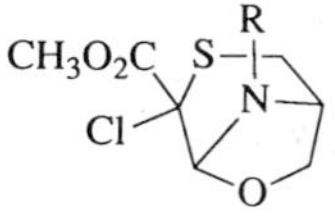 [structure: bicyclic, CH_3O_2C, Cl, S, N-R, O] R = Ac (72)	486
		AcCl, CH_3CN, overnight	" R = H (71)	485
C_8	$C_6H_5S(O)CH{=}CH_2$	$SOCl_2$ (5 eq), CH_2Cl_2, −5 to 25°, 30 min	$C_6H_5SCHClCH_2Cl$ (85–95)	177
		$CH_3C(S)SH$	$C_6H_5SCH[SC(S)CH_3]CH_2SC(S)CH_3$ (quant)	487
		$AcOC(CH_3){=}CH_2$, TsOH, CH_3CN, reflux	$C_6H_5SCH_2CHO$ (40)[a]	51

Reactant	Conditions	Product(s) (Yield, %)	Refs.
	$AcOC(CH_3)=CH_2$, TsOH, AcOH, CH_3CN, reflux	$C_6H_5SCH_2CH(OAc)_2$ (quant)	51
	Cl_3CCOCl, Zn, ether, reflux	3,3-dichloro-5-(phenylthio)dihydrofuran-2-one (SC_6H_5) (51)	195
	$Cl_2CHCOCl$, $(C_2H_5)_3N$, ether	" (40)	195
	$Cl_2CHCOCl$, $(C_2H_5)_3N$, CH_2Cl_2	" (15)	195
	$TBDMSOC(OCH_3)=CH_2$, ZnI_2, CH_3CN, 5 h	$C_6H_5SCH(OTBDMS)(CH_2)_2CO_2CH_3$ (65)	480
	$TBDMSOC(OCH_3)=CHCH_3$, ZnI_2, CH_3CN, 12 h	$C_6H_5SCH(OTBDMS)CH_2CH(CH_3)CO_2CH_3$ (66)	480
	$TMSOC(OCH_3)=CHCH_3$, ZnI_2, CH_3CN, 5 min	$C_6H_5SCH[CH(CH_3)CO_2CH_3]CH_2CH(CH_3)CO_2CH_3$ (58)	480
	$TMSOC(OCH_3)=C(CH_3)_2$, ZnI_2, CH_3CN, 80°, 5 h	$C_6H_5SCH[C(CH_3)_2CO_2CH_3]CH_2C(CH_3)_2CO_2CH_3$ (45)	480
1,4-benzodithiin 1-oxide (S=O)	Ac_2O	2,3-diacetoxy-2,3-dihydro-1,4-benzodithiin (OAc, OAc) 1:1 (—)	134
C_9			
$C_6H_5S(O)CH=CHCH_3$	$TBDMSOC(OCH_3)=CH_2$, ZnI_2, CH_3CN, 15 h	$C_6H_5SCH(OTBDMS)CH(CH_3)CH_2CO_2CH_3$ 1:1 diastereomeric mixture (40)	480
(Z)-$CH_3S(O)CH=CHC_6H_5$	PCl_5, CH_2Cl_2, 1 h	$CH_3SCH_2COC_6H_5$ I (25) + $C_6H_5CH(SCH_3)CHO$ II (23) + $CH_3SCH=CClC_6H_5$ III (42)	488
	$SOCl_2$, CCl_4, reflux 12 h	I (42) + II (53)	488
(E)-$CH_3S(O)CH=CHC_6H_5$	PCl_5, CH_2Cl_2, 1 h	I (30) + II (10) + III (27)	488

TABLE XI. VINYLOGOUS AND ADDITIVE PUMMERER REACTIONS (*Continued*)

Substrate	Reaction Conditions	Product(s) and Yield(s) (%)	Refs.
$CH_3S(O)CH{=}CHC_6H_5$	$SOCl_2$, CH_2Cl_2	$CH_3SCH{=}CClC_6H_5$ (—)	254, 414
$CH_3S(O)CD{=}CHC_6H_5$	"	$CH_3SCD{=}CClC_6H_5$ (—)	254, 414
$CH_3S(O)CD_2CHODC_6H_5$	"	" (—)	414
(*Z*)-$CH_3S(O)C(SCH_3){=}C(NH_2)(CH_2)_2CH(OCH_3)_2$	Ac_2O, pyr, CH_2Cl_2	$CH_3SCOC(SCH_3)(NHAc)(CH_2)_2CH(OCH_3)_2$ (73)	183
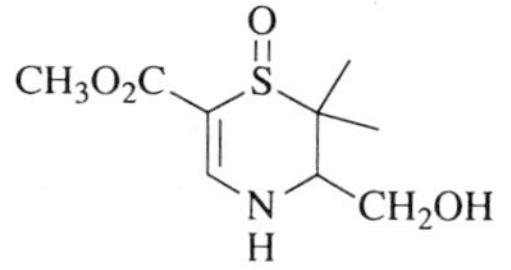 CH_3O_2C, S(=O), N–H, CH_2OH (ring structure)	AcCl, CH_3CN	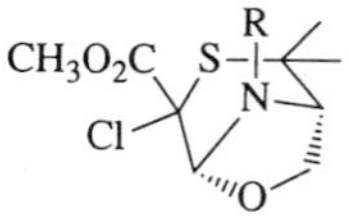 CH_3O_2C, Cl, S–R, N, O (ring structure) R = H, (—)	486
	AcCl, CH_3CN, 3 h	" R = Ac, (80)	489
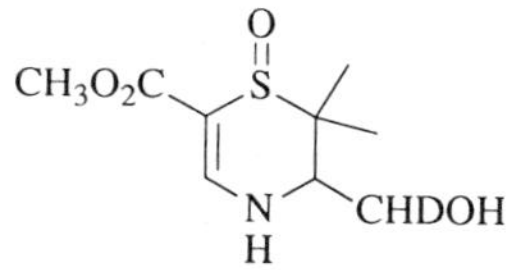 CH_3O_2C, S(=O), N–H, CHDOH (ring structure)	AcCl, CH_3CN	" R = Ac, (80)	486
C_{10}			
(*Z*)-$CH_3S(O)C(SCH_3){=}C(NH_2)C_6H_5$	Ac_2O, pyr, CH_2Cl_2, 4 h	$CH_3SCOC(SCH_3)(NHAc)C_6H_5$ (88)	183
(*E*)-$C_6H_5S(O)CH{=}CHCO_2CH_3$	$TBDMSOC(OCH_3){=}CH_2$, ZnI_2, CH_3CN, 36 h	$C_6H_5SCH(OTBDMS)CH(CO_2CH_3)CH_2CO_2CH_3$ 1:1 diastereomeric mixture (53)	480
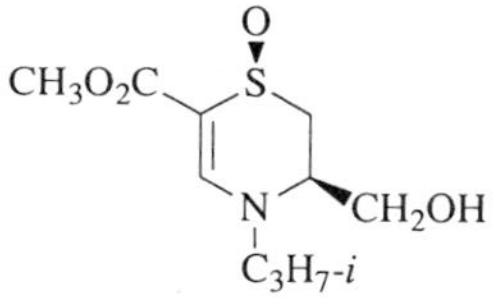 CH_3O_2C, S(O), N–C_3H_7-*i*, CH_2OH (ring structure)	AcCl, CH_3CN, −15°, 25 min	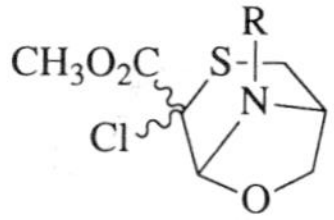 CH_3O_2C, Cl, S–R, N, O (ring structure) R = C_3H_7-*i* 4:1 diastereomeric mixture (40)	485

Reactant	Conditions	Product(s) (Yield, %)	Refs.
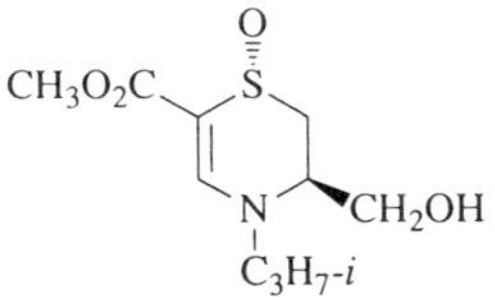 [structure: 4-isopropyl-3-(hydroxymethyl)-6-methoxycarbonyl-1,4-thiazine S-oxide; labels CH_3O_2C, O, S, N, CH_2OH, C_3H_7-*i*]	AcCl, CH_3CN, 10 min	" R = C_3H_7-*i* 4:1 diasteromeric mixture (74)	485
C_{11}			
$C_6H_5S(O)C(CH_3)=C(CH_3)_2$	$AcOC(CH_3)=CH_2$, TsOH, CH_3CN, reflux	$C_6H_5SC(CH_3)_2COCH_3$ (55)[a,b]	51
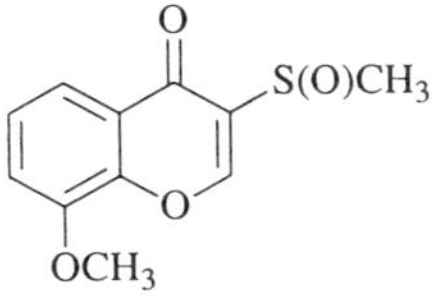 [structure: 8-methoxy-3-(methylsulfinyl)chromone; labels O, $S(O)CH_3$, O, OCH_3]	Ac_2O, reflux 10 h	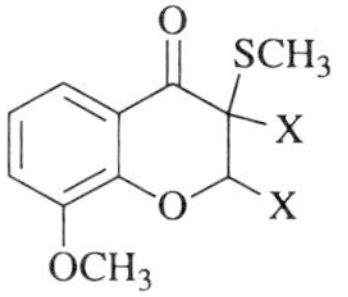 [structure: 8-methoxy-3-(methylthio)-2,3-X-chroman-4-one; labels O, SCH_3, X, O, X, OCH_3] X = OAc (63)	194
	$SOCl_2$, 3 h	" X = Cl (97)	194, 268
C_{12}			
$C_6H_5S(O)C(CO_2CH_3)=CHC_2H_5$	H_2SO_4, dioxane, reflux	[structure: 5-methyl-3-(phenylthio)-2(5H)-furanone; labels SC_6H_5, O, O] (10)	179
$C_6H_5S(O)C(CO_2CH_3)=C(CH_3)_2$	"	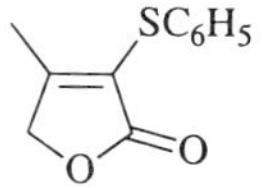 [structure: 4-methyl-3-(phenylthio)-2(5H)-furanone; labels SC_6H_5, O, O] (24)	179
	AcCl, CH_2Cl_2, 30 min	$C_6H_5SC(OAc)(CO_2CH_3)C(CH_3)_2Cl$ (67)	179
	$SOCl_2$, CH_2Cl_2, 30 min	$C_6H_5SCCl(CO_2CH_3)C(CH_3)_2Cl$ (quant)	179
	TFAA, CH_2Cl_2, 20 min	$C_6H_5SC(CH_3)_2COCO_2CH_3$ (35)	179
$C_6H_5S(O)C(CO_2CH_3)=CHC_2H_5$	AcCl, CH_2Cl_2, 30 min	$C_6H_5SC(OAc)(CO_2CH_3)CHClC_2H_5$ 4:3 diastereomeric mixture (70)	179
(*E*)-$C_6H_5S(O)C(CO_2CH_3)=C(CH_3)C_2H_5$	AcCl, CH_2Cl_2, 30 min	$C_6H_5SC(OAc)(CO_2CH_3)CCl(CH_3)C_2H_5$ (66)	179
(*Z*)-	"	" (71)	179

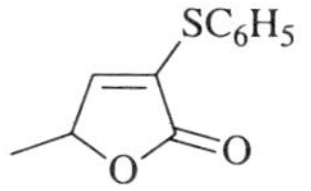

TABLE XI. VINYLOGOUS AND ADDITIVE PUMMERER REACTIONS (*Continued*)

Substrate	Reaction Conditions	Product(s) and Yield(s) (%)	Refs.
$CH_3S(O)C(SCH_3){=}CHC_6H_3(OCH_3)_2$-3,4	HCl, *t*-C_4H_9OH, 2 h	$CH_3SCOCH(SCH_3)C_6H_3(OCH_3)_2$-3,4 (67)	184
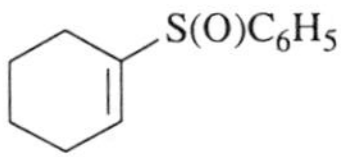 $S(O)C_6H_5$	Cl_3CCOCl, Zn, ether, reflux	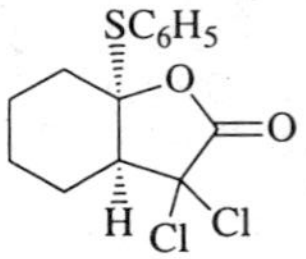 SC_6H_5 O O H Cl Cl (72)	195
	$Cl_2CHCOCl$, $(C_2H_5)_3N$, ether	" (41)	195
	$AcOC(CH_3){=}CH_2$, TsOH, CH_3CN, reflux	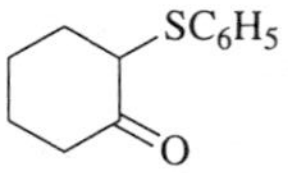 SC_6H_5 O (90)[a]	51
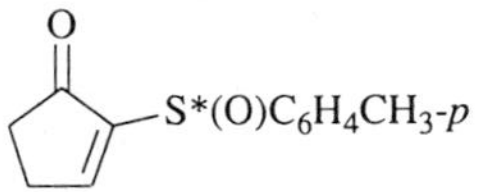 O $S^*(O)C_6H_4CH_3$-*p*	$Cl_2C{=}C{=}O$, ether, reflux 15 min	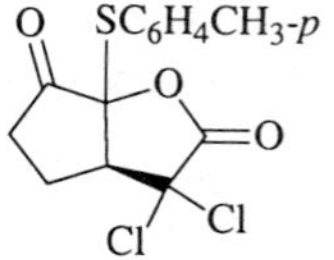 O $SC_6H_4CH_3$-*p* O O Cl Cl (—)	177
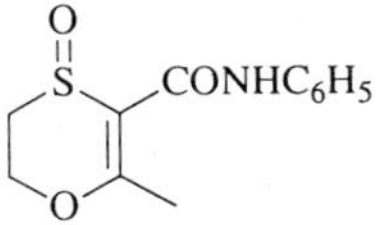 O S $CONHC_6H_5$ O	TFAA, C_2H_5OAc, 3° 15 min	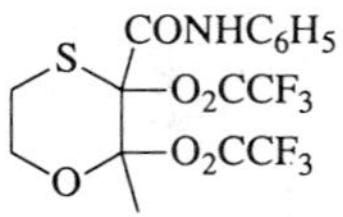 S $CONHC_6H_5$ O_2CCF_3 O_2CCF_3 O (—)	231
	TFAA, C_6H_6, 15 min	" (90)	490
	Ac_2O, AcOH, C_6H_6, reflux 80 min	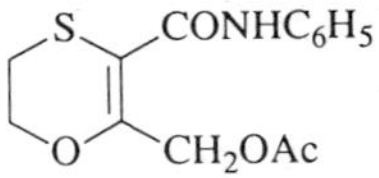 S $CONHC_6H_5$ O CH_2OAc (56)  + S O $COCONHC_6H_5$ (14)	178

Reactant	Conditions	Product(s) and Yield(s) (%)	Refs.
		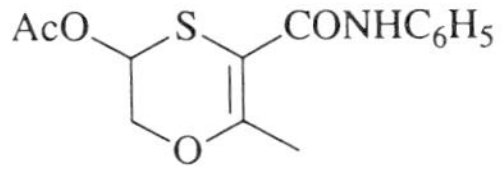 + AcO-(dihydro-1,4-oxathiin)-$CONHC_6H_5$ (16)	
		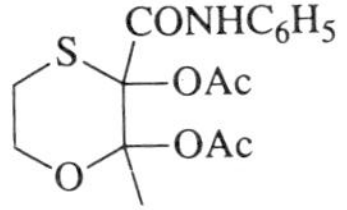 + (1,4-oxathiane)-$CONHC_6H_5$, OAc, OAc (12)	
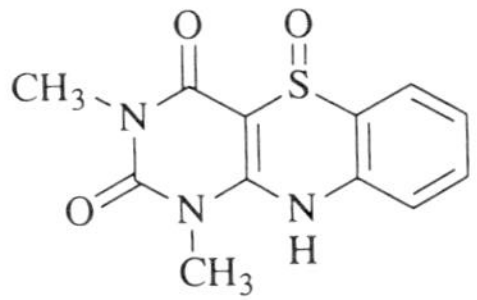 CH_3-N, N-CH_3 pyrimido[5,4-b][1,4]benzothiazine-2,4-dione S-oxide (N-H)	CH_3OH, $(C_2H_5)_3N$, reflux 20 h	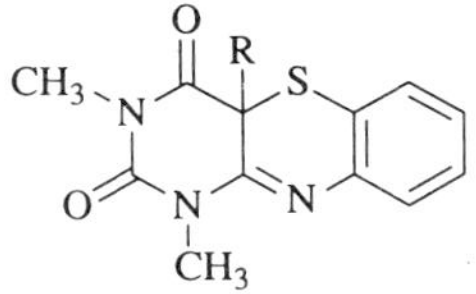 R = OCH_3 (~70)	491
	C_2H_5OH, $(C_2H_5)_3N$, reflux 20 h	" R = OC_2H_5 (~70)	491
	$SOCl_2$	" R = Cl (—)	491
	Ac_2O, 100°, 2 h	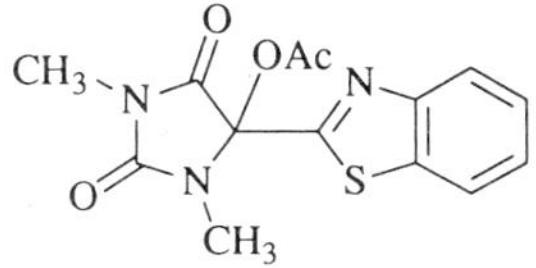 1,3-dimethyl-5-OAc-5-(2-benzothiazolyl)hydantoin (~85)	491
C_{13}			
(*E*)-$C_6H_5S(O)C(CO_2CH_3)$=$C(CH_3)C_2H_5$	H_2SO_4, dioxane, reflux	C_2H_5, SC_6H_5 butenolide I + SC_6H_5 butenolide II I:II = 2.2:1 (32)	179
(*Z*)-	"	I:II = 1:1 (33)	179
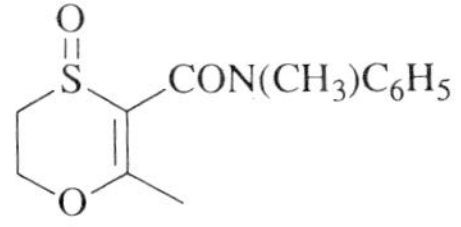 (1,4-oxathiin S-oxide)-$CON(CH_3)C_6H_5$	TFAA, C_6H_6, 15 min	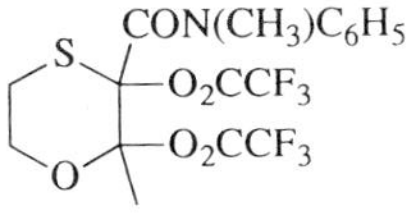 (1,4-oxathiane)-$CON(CH_3)C_6H_5$, O_2CCF_3, O_2CCF_3 (—)	490

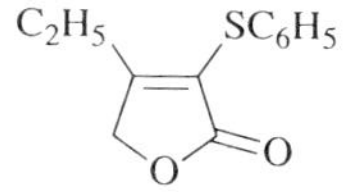

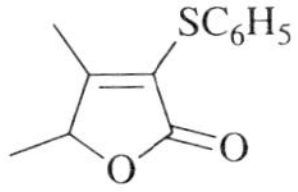

TABLE XI. VINYLOGOUS AND ADDITIVE PUMMERER REACTIONS (*Continued*)

Substrate	Reaction Conditions	Product(s) and Yield(s) (%)	Refs.
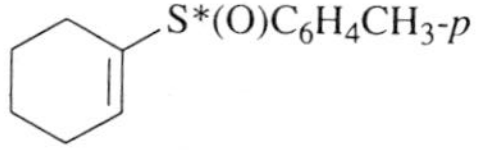 $S^*(O)C_6H_4CH_3$-*p* (cyclohexenyl) (*R*)-(+)	$Cl_2CHCOCl$, Zn-Cu, ether, reflux	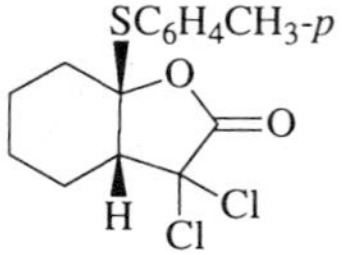 $SC_6H_4CH_3$-*p*, H, Cl, Cl, O (70)	181
(*S*)-(—)	"	" (68)	181
(*R*)-(+)	$ClCH_2COCl$, Zn-Cu, ether, reflux	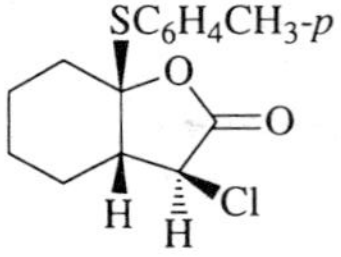 $SC_6H_4CH_3$-*p*, H, H, Cl, O (60)	181
O, $S^*(O)C_6H_4CH_3$-*p* (*S*)-(+)	"	O, $SC_6H_4CH_3$-*p*, H, Cl, Cl, O (60)	181
AcNH, O, S, N, O, $CO_2CH_2CCl_3$	PX_3, DMF	AcNH, S, N, O, X, $CO_2CH_2CCl_3$ X = Cl (70–80) X = Br (52)	492 492
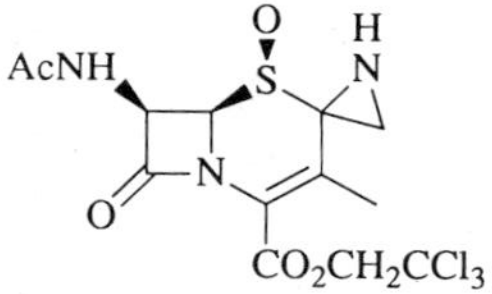 AcNH, O, S, H, N, N, O, $CO_2CH_2CCl_3$	PCl_3, DMF	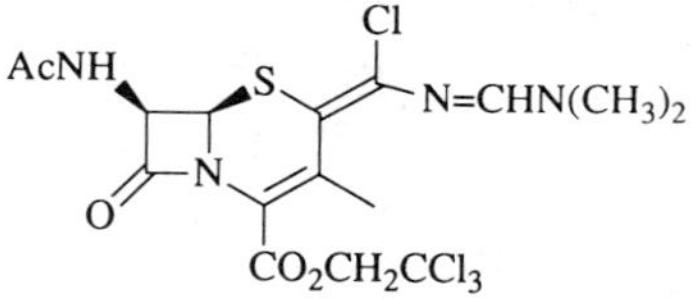 AcNH, S, N, O, Cl, $N{=}CHN(CH_3)_2$, $CO_2CH_2CCl_3$ (47)	440

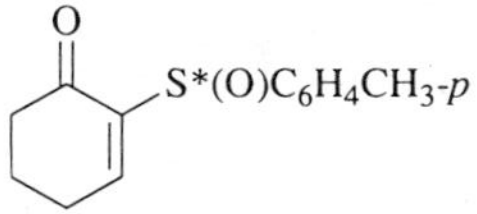

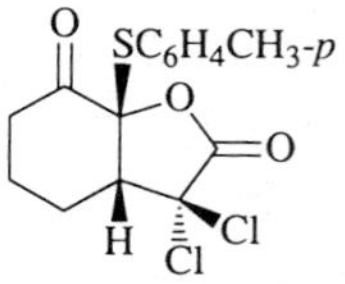

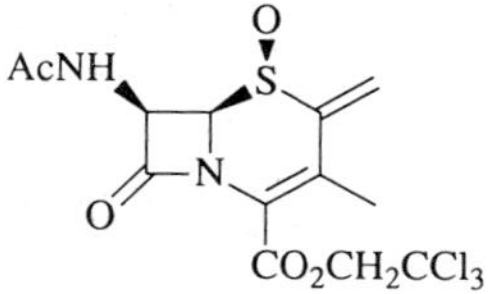

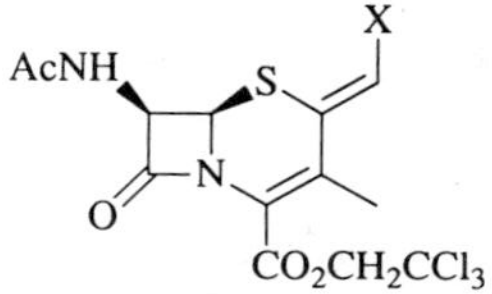

C_{14}

(*E*)-$C_6H_5S(O)CH{=}CHC_6H_5$	Cl_3CCOCl, Zn, ether, reflux 15 + 15 min	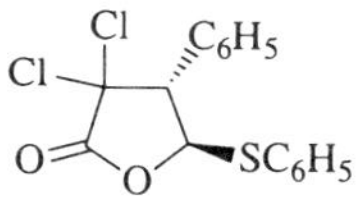 (65)	195
	$Cl_2CHCOCl$, $(C_2H_5)_3N$, ether	" (54)	195
	$Cl_2CHCOCl$, $(C_2H_5)_3N$, CH_2Cl_2	" (30)	195
	$TBDMSOC(OCH_3){=}CH_2$, ZnI_2, CH_3CN, 15 h	$C_6H_5SCH(OTBDMS)CH(C_6H_5)CH_2CO_2CH_3$ 1:1.6 mixture of diastereoisomers (41)	480
(*Z*)-$C_6H_5S(O)CH{=}CHC_6H_5$	Cl_3CCOCl, Zn, ether, reflux	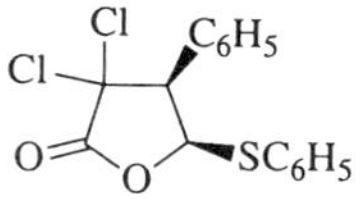 (25)	195
	$Cl_2CHCOCl$, $(C_2H_5)_3N$, ether	" (20)	195
	PCl_5, CH_2Cl_2, 1 h	$C_6H_5SCH_2COC_6H_5$ I (20) + $C_6H_5SCH(C_6H_5)CHO$ II (18) + $C_6H_5SCH{=}CClC_6H_5$ (53)	488
	$SOCl_2$, CCl_4, reflux 12 h	I (29) + II (38)	488
	$TBDMSOC(OCH_3){=}CH_2$, ZnI_2, CH_3CN, 70°, 5 h	$C_6H_5SCH(OTBDMS)CH(C_6H_5)CH_2CO_2CH_3$ 1:1.7 mixture of diastereoisomers (32)	480
(*E*)-*p*-$CH_3C_6H_4S(O)CH{=}C(CH_3)C_4H_9$-*n*	Cl_3CCOCl, Zn-Cu, ether, reflux	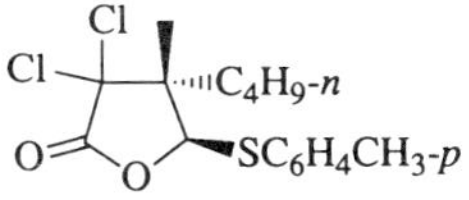 (70)	493
(*E*)-*p*-$CH_3C_6H_4S^*(O)C(C_4H_9$-*n*$){=}CHCH_3$	"	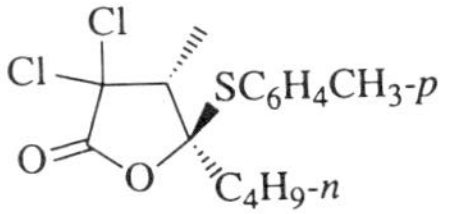 (75)	493
$C_6H_5S(O)CH{=}CHC_6H_5$	$SOCl_2$ (5 eq), CH_2Cl_2, −5 to 25°, 30 min	$C_6H_5SCHClC_6H_5$ (85–95)	177

TABLE XI. VINYLOGOUS AND ADDITIVE PUMMERER REACTIONS (*Continued*)

Substrate	Reaction Conditions	Product(s) and Yield(s) (%)	Refs.
$C_6H_5S(O)CH{=}CHC_6H_{13}$-*n*	"	$C_6H_5SCHClC_6H_{13}$-*n* (85–95)	177
$S(O)C_6H_5$, CO_2CH_3	Ac_2O, 75°, 3 h	SC_6H_5, CO_2CH_3, OAc (50) + OAc, SC_6H_5, CO_2CH_3 (15) + SC_6H_5, CO_2CH_3 (19) I	494
	Ac_2O, reflux	I (quant)	179
	TFAA, 0°, 30 min	X, $C(Y)(CO_2CH_3)SC_6H_5$; X = Y = O_2CCF_3 (quant)	179
	$SOCl_2$, 0°, 30 min	X = Y = Cl (quant)	179
	AcCl, CH_2Cl_2, 0° to rt, 15 min	X = Cl, Y = OAc (90)	179
	Ac_2O, pyr, overnight	OAc, CO_2CH_3 (83)	179, 494
	H_2SO_4, H_2O, dioxane, reflux 3 h	SC_6H_5, O, O (53) + OH, SC_6H_5, CO_2CH_3 (tr)	494
$S(O)C_6H_5$, O, O	H_2SO_4, dioxane, reflux 4 h	SC_6H_5, O, O I (18) + SC_6H_5, O, O, OH (24)	179

Reactant	Conditions	Product(s) and Yield(s) (%)	Refs.
	Ac_2O, CH_3SO_2OH, reflux 1 h	I (44)	179
	Ac_2O, reflux 3.5 h	I (13) 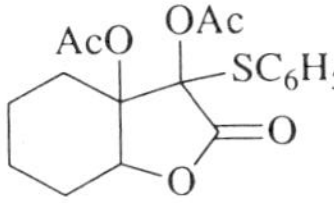 + [structure: AcO, OAc, SC_6H_5 lactone] (20)	179
	Ac_2O, pyr, overnight	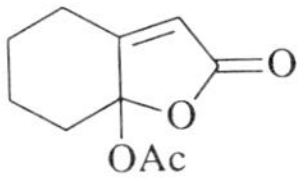 [structure: OAc butenolide] (45)	179
	AcCl, CH_2Cl_2, 0° to rt, 1.5 h	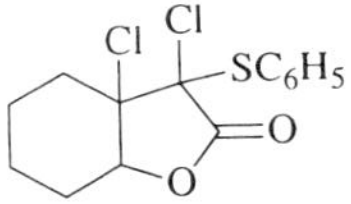 [structure: Cl, Cl, SC_6H_5 lactone] (64) 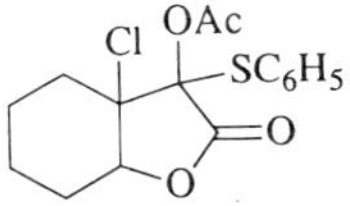 + [structure: Cl, OAc, SC_6H_5 lactone] (9) 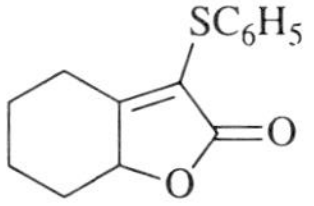 + [structure: SC_6H_5 butenolide] (12)	179
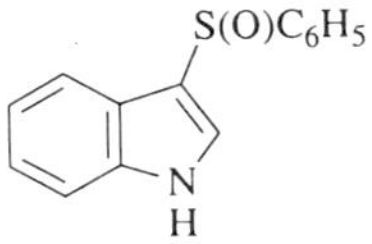 [structure: 3-$S(O)C_6H_5$-indole]	$SOCl_2$, $NaHCO_3$, 0°	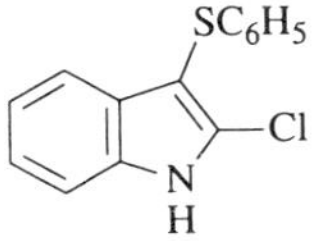 [structure: 3-SC_6H_5-2-Cl-indole] (58)	495
C_{15}			
$CH_3S(O)CH{=}C(C_6H_5)_2$	$SOCl_2$, 3 h	$CH_3SCHClC(C_6H_5)_2Cl$ (26)	194
(Z)-$C_6H_5S(O)C(CO_2CH_3){=}CHC_3H_7$-$i$	AcCl, CH_2Cl_2, 30 min	$C_6H_5SC(OAc)(CO_2CH_3)CHClC_3H_7$-$i$ (64)	179
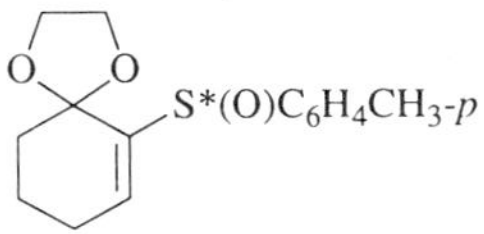 [structure: dioxolane cyclohexene $S^*(O)C_6H_4CH_3$-p]	$Cl_2CHCOCl$, Zn-Cu, ether, reflux	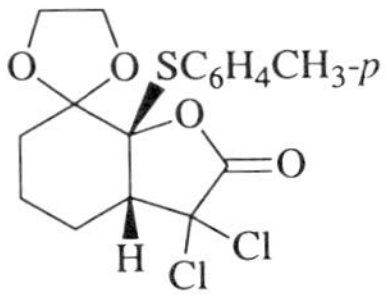 [structure: $SC_6H_4CH_3$-p, dichlorolactone, H] (25)	181

TABLE XI. VINYLOGOUS AND ADDITIVE PUMMERER REACTIONS (*Continued*)

Substrate	Reaction Conditions	Product(s) and Yield(s) (%)	Refs.
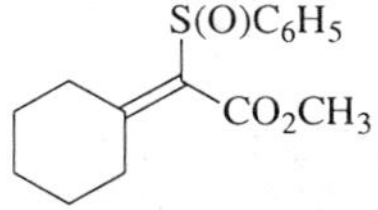 $S(O)C_6H_5$ / CO_2CH_3	H_2SO_4, dioxane, reflux	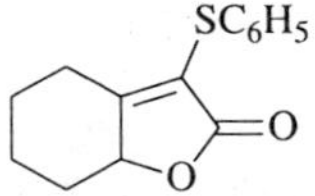 SC_6H_5 / =O (21)	179
$S(O)C_6H_5$ / CO_2CH_3	"	SC_6H_5 / =O (27)	179
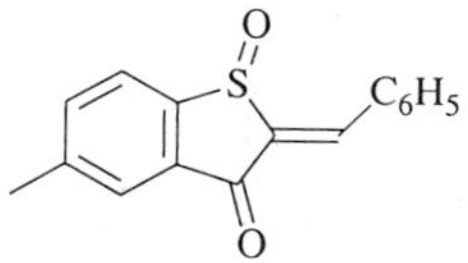 C_6H_5	HNO_3, H_2O, CH_3OH	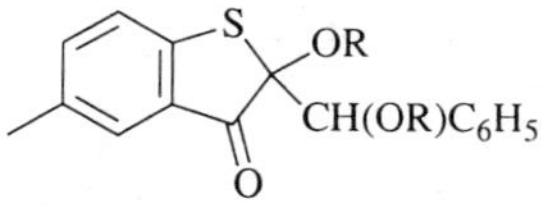 OR / $CH(OR)C_6H_5$ R = CH_3 (53)	180
	HCl, CH_3OH, 23 h	" R = CH_3 (59 + 25)[c]	180
	HCl, C_2H_5OH	" R = C_2H_5 (—)	180
C_{16}			
C_6H_5 / C_6H_5	HCl, dioxane, H_2O, 1 h	C_6H_5 / COC_6H_5 I (32)	496
		+ C_6H_5 / C_6H_5 / Cl II (3)	
		+ C_6H_5 / C_6H_5 III (45)	

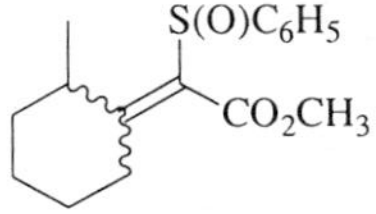

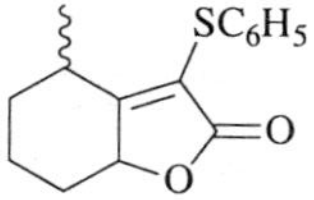

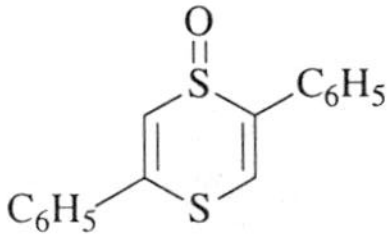

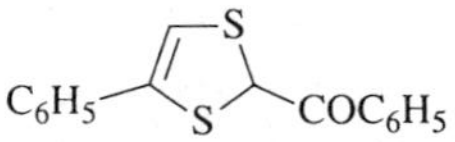

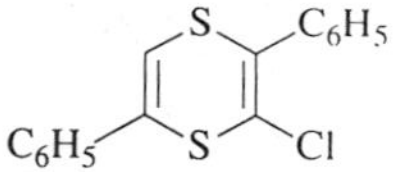

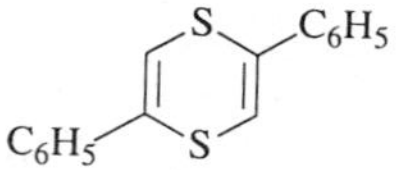

Substrate	Conditions	Product(s) and Yield(s) (%)	Refs.
p-ClC6H4-(1-oxo-1,4-dithiin)-C6H4Cl-p	HCl (anh), dioxane	I (34) + II (22) + III (25)	496
	AcCl, CH_2Cl_2	II (53) + III (21)	496
	HCl (anh), CH_3OH	I (26) + II (3) + [dithiin: C_6H_5, C_6H_5, OCH_3, Cl] (42)	496
	HCl, dioxane, H_2O	p-ClC_6H_4-(1,3-dithiole)-COC_6H_4Cl-p (68) + p-ClC_6H_4-(1,4-dithiin)-C_6H_4Cl-p (32)	496
C_{17} 3-$S(O)C_6H_5$-1-CH_2CN-indole	$SOCl_2$, $NaHCO_3$, 0°	3-SC_6H_5-2-Cl-1-CH_2CN-indole (60–97)	495
3-$S(O)C_6H_5$-1-$CH_2CO_2CH_3$-indole	"	3-SC_6H_5-2-Cl-1-$CH_2CO_2CH_3$-indole (60–97)	495
p-$CH_3C_6H_4S(O)$C(=CHCH_3)-3,4-$(OCH_3)_2C_6H_3$	Cl_3CCOCl, Zn-Cu, ether, reflux	dichlorolactone with $SC_6H_4CH_3$-p, 3,4-dimethoxyphenyl (60)	493

TABLE XI. VINYLOGOUS AND ADDITIVE PUMMERER REACTIONS (*Continued*)

Substrate	Reaction Conditions	Product(s) and Yield(s) (%)	Refs.
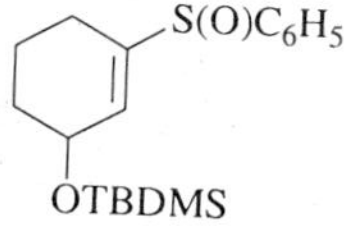 [3-OTBDMS-cyclohexenyl] $S(O)C_6H_5$, OTBDMS	Cl_3CCOCl, Zn, ether, reflux	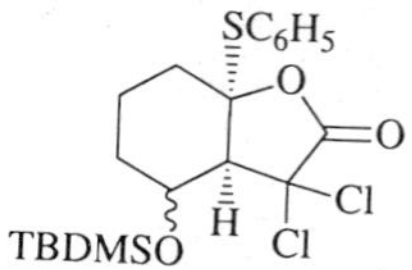 SC_6H_5, O, =O, Cl, Cl, H, TBDMSO (80)	195
	$Cl_2CHCOCl$, $(C_2H_5)_3N$, ether	" (20)	195
One diastereoisomer	Cl_3CCOCl, Zn, ether, reflux	" 9:1 mixture (80)	195
Another diastereoisomer	"	" 19:1 mixture (95)	195
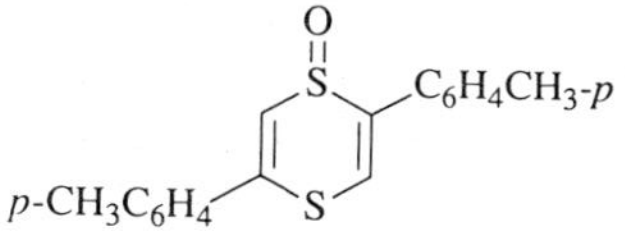 [dithiin S-oxide] O, S, S, $C_6H_4CH_3$-*p*, *p*-$CH_3C_6H_4$	HCl, dioxane, H_2O	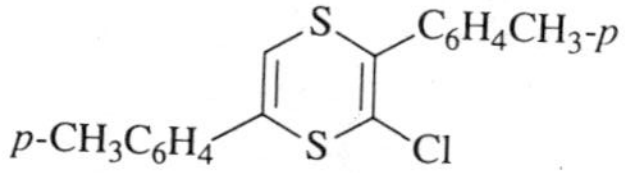 [dithiin] S, S, $C_6H_4CH_3$-*p*, Cl, *p*-$CH_3C_6H_4$ (21) 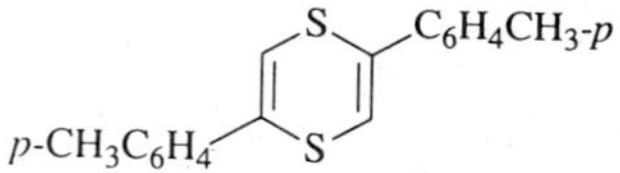 + [dithiin] S, S, $C_6H_4CH_3$-*p*, *p*-$CH_3C_6H_4$ (79)	496
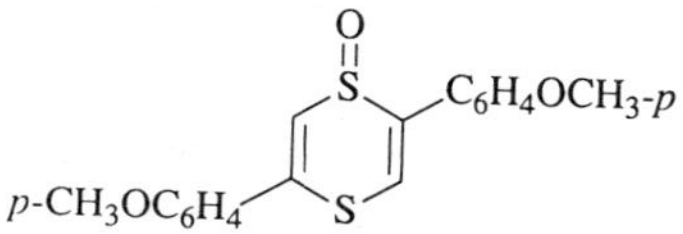 [dithiin S-oxide] O, S, S, $C_6H_4OCH_3$-*p*, *p*-$CH_3OC_6H_4$	"	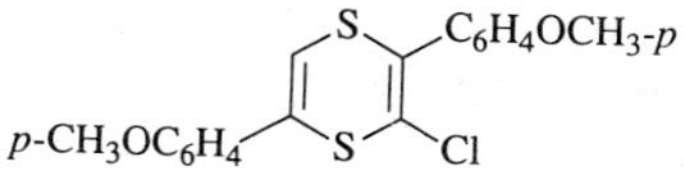 [dithiin] S, S, $C_6H_4OCH_3$-*p*, Cl, *p*-$CH_3OC_6H_4$ (61) 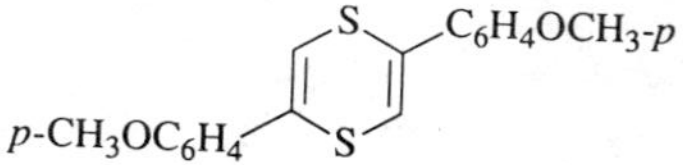 + [dithiin] S, S, $C_6H_4OCH_3$-*p*, *p*-$CH_3OC_6H_4$ (39)	496

Reactant	Conditions	Product(s) and Yield(s) (%)	Refs.
C_{19}			
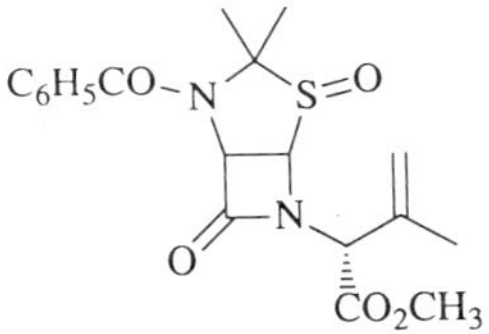 $C_6H_5CO{-}N$ … $S{=}O$ … CO_2CH_3	H_2SO_4, C_6H_6, $AcN(CH_3)_2$, 105° 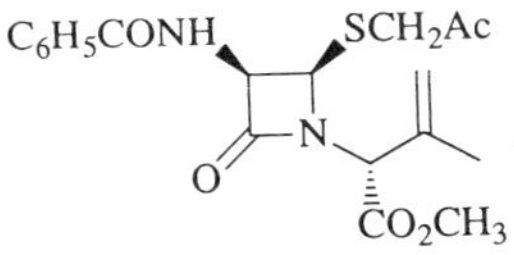	C_6H_5CONH … SCH_2Ac … CO_2CH_3 (41)	497
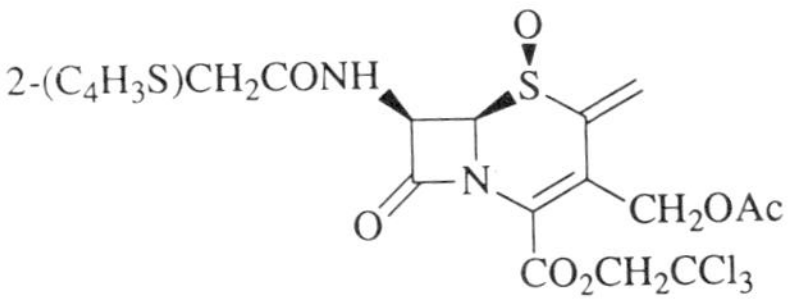 2-$(C_4H_3S)CH_2CONH$ … CH_2OAc … $CO_2CH_2CCl_3$	PCl_3, DMF 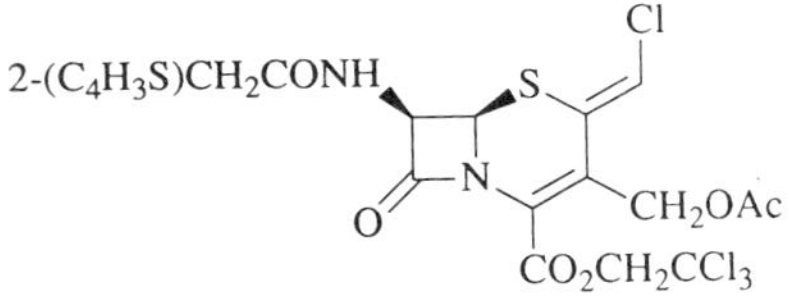	2-$(C_4H_3S)CH_2CONH$ … Cl … CH_2OAc … $CO_2CH_2CCl_3$ (50)	492
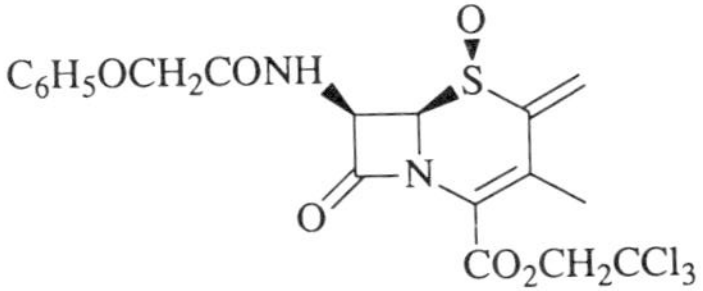 $C_6H_5OCH_2CONH$ … $CO_2CH_2CCl_3$	PX_3, DMF	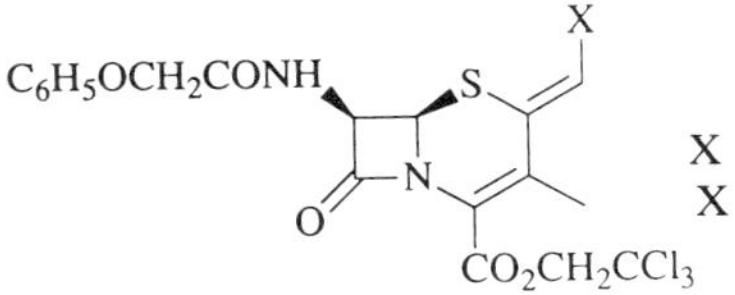 $C_6H_5OCH_2CONH$ … X … $CO_2CH_2CCl_3$ X = Cl (70–75) X = Br (82)	492
C_{21}			
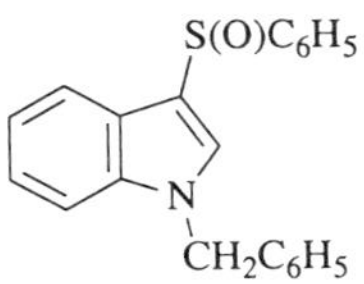 $S(O)C_6H_5$ … N … $CH_2C_6H_5$	$SOCl_2$, $NaHCO_3$, 0°	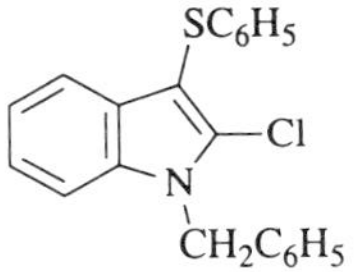 SC_6H_5 … Cl … N … $CH_2C_6H_5$ (60–97)	495
C_{22}			
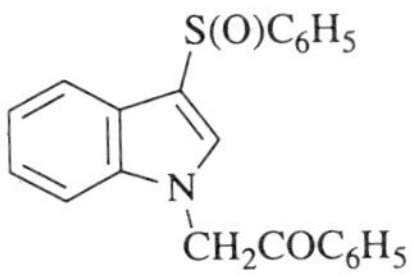 $S(O)C_6H_5$ … N … $CH_2COC_6H_5$	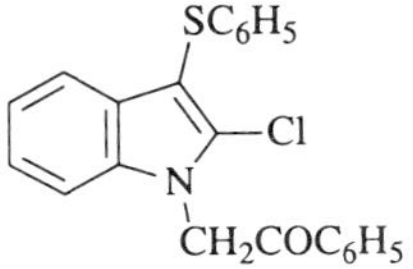 ,,	SC_6H_5 … Cl … N … $CH_2COC_6H_5$ (60–97)	495

TABLE XI. VINYLOGOUS AND ADDITIVE PUMMERER REACTIONS (*Continued*)

Substrate	Reaction Conditions	Product(s) and Yield(s) (%)	Refs.
C_{24}			
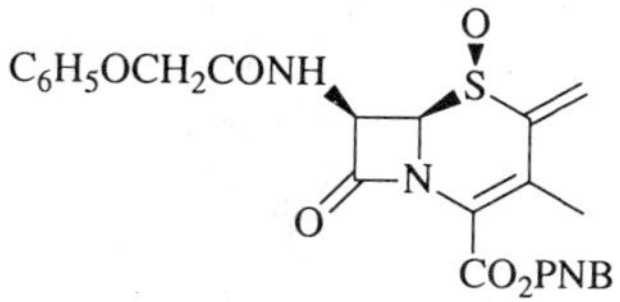 $C_6H_5OCH_2CONH$ … CO_2PNB	PX_3, DMF	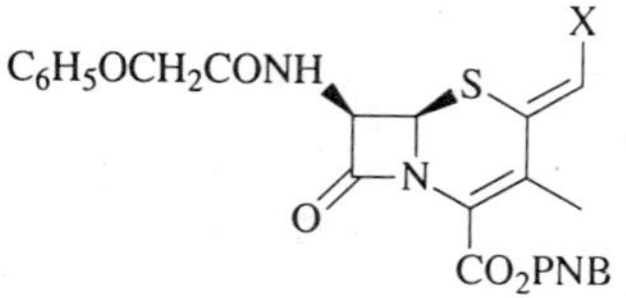 $C_6H_5OCH_2CONH$ … X … CO_2PNB X = Cl (60–65) X = Br (74)	492
C_{30}			
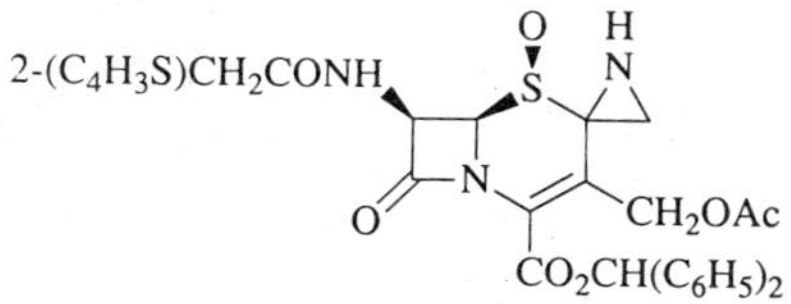 $2\text{-}(C_4H_3S)CH_2CONH$ … CH_2OAc $CO_2CH(C_6H_5)_2$	PCl_3, DMF	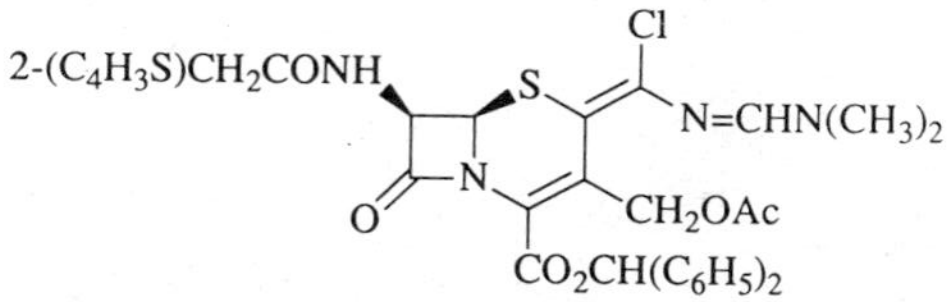 $2\text{-}(C_4H_3S)CH_2CONH$ … Cl $N{=}CHN(CH_3)_2$ CH_2OAc $CO_2CH(C_6H_5)_2$ (52)	440

[a] The reaction was carried out to partial conversion.
[b] Other non-Pummerer products were also formed.
[c] The yields are of the two diastereoisomers formed in the reaction.

REFERENCES

[1] S. Oae and T. Numata, *Isot. Org. Chem.*, **5**, 45 (1980) [*C.A.*, **94**, 46279s (1981)].
[2] S. Oae, T. Numata, and T. Yoshimura, in *The Chemistry of the Sulphonium Group*, C. J. M. Stirling and S. Patai (Eds.), Wiley, New York, 1981, p. 571.
[3] S. Oae in "*Top. Organic Sulphur Chem., Plenary Lect., Int. Symp. 8th*," M. Tishler (Ed.), Univ. Press, Ljubljana 1978, p. 289 [*C.A.*, **91**, 107279d (1979)].
[4] T. Numata and S. Oae, *Yuki Gosei Kagaku Kyokaishi*, **35**, 726 (1977) [*C.A.*, **88**, 5639s (1978)].
[5] T. Numata, *Yuki Gosei Kagaku Kyokaishi*, **36**, 845 (1978) [*C. A.*, **90**, 136847x (1979)].
[6] G. A. Russell and G. J. Mikol, in *Mechanism of Molecular Migrations*, B. S. Thyagarajan (Ed.), Vol. 1, Wiley-Interscience, New York, 1968, p. 157.
[6a] D. Grierson, *Org. React.*, **39**, 85 (1990).
[7] E. Block, *Reactions of Organosulfur Compounds*, Academic Press, New York, 1978, p. 175.
[8] C. L. Jenkins, *Diss. Abstr. Int. B*, **39**, 1289 (1978) [*C.A.*, **90**, 38359v (1979)].
[9] J. G. Tillett, *Chem. Rev.*, **76**, 747 (1976).
[10] T. Durst, *Adv. Org. Chem.*, **6**, 285 (1969).
[11] S. Warren, *Chem. Ind. (London)*, **1980**, 824.
[12] P. Welzel, *Nachr. Chem., Tech. Lab.*, **31**, 892 (1983) [*C.A.*, **100**, 33732b (1984)].
[13] S. W. Schneller, *Int. J. Sulfur Chem.*, **8**, 485 (1973).
[14] G. Kresze in *Houben-Weyl: Methoden der Organischen Chemie*, D. Klamann (Ed.), Band E 11, G. Thieme, Stuttgart-New York, 1985, p. 872.
[15] R. Pummerer, *Ber.*, **42**, 2282 (1909).
[16] R. Pummerer, *Ber.*, **43**, 1401 (1910).
[17] L. Horner, *Justus Liebigs Ann. Chem.*, **631**, 198 (1960).
[18] M. Lounasmaa and A. Koskinen, *Heterocycles*, **22**, 1591 (1984).
[19] D. Neville-Jones, E. Helmy, and R. D. Whitehouse, *J. Chem. Soc., Perkin Trans. 1*, **1972**, 1329.
[20] H. Fujihara, J.-J. Chiu, and N. Furukawa, *J. Chem. Res. (S)*, **1987**, 204.
[21] V. Ruffato and U. Miotti, *Gazz. Chim. It.*, **108**, 92 (1978); T. Numata and S. Oae, *Int. J. Sulfur Chem. A*, **1**, 6 (1971).
[22] B. Stridsberg and S. Allenmark, *Acta Chem. Scand., Ser. B*, **28**, 591 (1974).
[23] B. Stridsberg and S. Allenmark, *Acta Chem. Scand., Ser. B*, **B30**, 219 (1976).
[24] T. Numata and S. Oae, *Tetrahedron Lett.*, **1977**, 1337.
[25] S. Glue, I. T. Kay, and M. R. Kipps, *J. Chem. Soc., Chem. Commun.*, **1970**, 1158.
[26] T. Numata, O. Ito, and S. Oae, *Tetrahedron Lett.*, **1979**, 1869.
[27] O. Itoh, T. Numata, T. Yoshimura, and S. Oae, *Bull. Chem. Soc. Jpn.*, **56**, 266 (1983).
[28] T. Masuda, T. Numata, N. Furukawa, and S. Oae, *Chem. Lett.*, **1977**, 703.
[29] T. Masuda, T. Numata, N. Furukawa, and S. Oae, *J. Chem. Soc., Perkin Trans. 2*, **1978**, 1302.
[30] W. J. Kenney, J. A. Walsh, and D. A. Davenport, *J. Am. Chem. Soc.*, **83**, 4019 (1961).
[31] D. Walker and J. Leib, *Can. J. Chem.*, **40**, 1242 (1962).
[32] H.-D. Becker, G. J. Mikol, and G. A. Russell, *J. Am. Chem. Soc.*, **85**, 3410 (1963).
[33] G. A. Russell and G. J. Mikol, *J. Am. Chem. Soc.*, **88**, 5498 (1966).
[34] J. A. Smythe, *J. Chem. Soc.*, **1909**, 349.
[35] T. P. Hilditch, *Ber.*, **44**, 3583 (1911).
[36] J. M. McIntosh and R. K. Leavitt, *Can. J. Chem.*, **63**, 3313 (1985).
[37] Y. Oikawa and O. Yonemitsu, *Tetrahedron*, **30**, 2653 (1974).
[38] Y. Oikawa and O. Yonemitsu, *J. Org. Chem.*, **41**, 1118 (1976).
[39] H. Ishibashi, M. Okada, H. Komatsu, M. Ikeda, and Y. Tamura, *Synthesis*, **1985**, 643.
[40] Y. Tamura, H. -D. Choi, H. Shindo, J. Uenishi, and H. Ishibashi, *Tetrahedron Lett.*, **22**, 81 (1981).
[41] See for example: M. Watanabe, S. Nakamori, H. Hasegawa, K. Shirai, and T. Kumamoto, *Bull. Chem. Soc. Jpn.*, **54**, 817 (1981).
[42] R. Tanikaga, Y. Yabuki, N. Ono, and A. Kaji, *Tetrahedron Lett.*, **1976**, 2257.

[43] C. E. Adams, F. J. Walker, and K. B. Sharpless, *J. Org. Chem.*, **50**, 420 (1985).
[44] A. K. Sharma and D. Swern, *Tetrahedron Lett.*, **1974**, 1503.
[45] R. Tanikaga, Y. Hiraki, N. Ono, and A. Kaji, *J. Chem. Soc., Chem. Commun.*, **1980**, 41.
[46] Y. Hiraki, M. Kamiya, R. Tanikaga, N. Ono, and A. Kaji, *Bull. Chem. Soc. Jpn.*, **50**, 447 (1977).
[47] Y. Tamura, H. Maeda, H. D. Choi, and H. Ishibashi, *Synthesis*, **1982**, 56.
[48] I. K. Stamos, *Tetrahedron Lett.*, **23**, 2787 (1985).
[49] See also: T. H. Kim and D. Y. Oh, *Tetrahedron Lett.*, **27**, 1165 (1986); H. Ishibashi, H. Nakatani, Y. Umei, W. Yamamoto, and M. Ikeda, *J. Chem. Soc., Perkin Trans. 1*, **1987**, 589; H. Ishibashi, T. Sato, M. Irie, M. Ito, and M. Ikeda, *Ibid.*, **1987**, 1095.
[50] R. P. Hatch, J. Shringarpure, and S. M. Weinreb, *J. Org. Chem.*, **43**, 4172 (1978).
[51] O. De Lucchi, G. Marchioro, and G. Modena, *J. Chem. Soc., Chem. Commun.*, **1984**, 513.
[52] A. Dossena, R. Marchelli, and G. Casnati, *J. Chem. Soc., Chem. Commun.*, **1979**, 370.
[53] A. Dossena, R. Marchelli, and G. Casnati, *J. Chem. Soc., Perkin Trans. 1*, **1981**, 2737.
[54] S. Lane, S. J. Quick, and R. J. K. Taylor, *J. Chem. Soc., Perkin Trans. 1*, **1984**, 2549.
[55] K. Kita, H. Yasuda, O. Tamura, F. Itoh, and Y. Tamura, *Tetrahedron Lett.*, **25**, 4681 (1984).
[56] N. Tokitoh, Y. Igarashi, and W. Ando, *Tetrahedron Lett.*, **28**, 5903 (1987).
[57] R. D. Miller and D. R. McKean, *Tetrahedron Lett.*, **24**, 2619 (1983).
[58] M. Kakimoto and Y. Imai, *Chem. Lett.*, **1984**, 1831.
[59] M. Shimizu, T. Akiyama, and T. Mukaiyama, *Chem. Lett.*, **1984**, 1531.
[60] T. L. Moore, *J. Org. Chem.*, **32**, 2786 (1967).
[61] T. Matsumoto, H. Shirahama, A. Ichihara, H. Shin, and S. Kagawa, *Bull. Chem. Soc. Jpn.*, **45**, 1144 (1972).
[62] F. Cozzi, C. Gobbi, L. Raimondi, and A. Restelli, *Gazz. Chim. It.*, **116**, 717 (1986).
[63] J. R. McCarthy, N. P. Peet, M. E. LeTourneau, and M. Inbasekaran, *J. Am. Chem. Soc.*, **107**, 735 (1985).
[64] R. K. Marat and A. F. Janzen, *Can. J. Chem.*, **55**, 3031 (1977).
[65] R. A. Archer and B. S. Kitchell, *J. Am. Chem. Soc.*, **68**, 3462 (1966).
[66] R. G. Petrova and R. K. Freidlina, *Bull. Acad. Sci. USSR, Div. Chem. Soc. (Engl. Transl.)*, **1966**, 1797.
[67] T. D. Harris and V. Boekelheide, *J. Org. Chem.*, **41**, 2770 (1976).
[68] B. Holmberg, *Archiv Kemi, Mineral. Geol.*, **14A**, No. 9 (1940) [*C.A.*, **35**, 2129(7) (1941)].
[69] A. Tananger, *Archiv Kemi, Mineral. Geol.*, **24A**, No. 10 (1947) [*C.A.*, **42**, 8786a (1948)].
[70] H. E. Zimmerman in *Molecular Rearrangements*, P. de Mayo, Ed., Vol. 1, 1963, p. 381.
[71] C. R. Johnson and W. G. Phillips, *J. Org. Chem.*, **32**, 1926 (1967).
[72] R. Annunziata, M. Cinquini, and S. Colonna, *J. Chem. Soc., Perkin Trans. 1*, **1973**, 1231.
[73] E. Vilsmaier, R. Bayer, I. Laengenfelder, and U. Welz, *Chem. Ber.*, **111**, 1136 (1978).
[74] B. M. Dilworth and M. A. McKervey, *Tetrahedron*, **42**, 3731 (1986).
[75] L. A. Paquette, *Org. React.*, **25**, 1 (1977).
[76] T. P. Dawson and W. E. Lawson, *J. Chem. Soc.*, **49**, 3119 (1927).
[77] T. P. Dawson and W. E. Lawson, *J. Chem. Soc.*, **49**, 3125 (1927).
[78] H. Richtzenhain and B. Alfredsson, *Chem. Ber.*, **86**, 142 (1953); F. Boberg, G. Winter, and J. Moos, *Justus Liebigs Ann. Chem.*, **616**, 1 (1958).
[79] G. E. Wilson, Jr. and R. Albert, *Tetrahedron Lett.*, **1968**, 6271.
[80] P. Bakuzis, M. L. F. Bakuzis, C. C. Fortes, and R. Santos, *J. Org. Chem.*, **41**, 2769 (1976).
[81] L. A. Paquette, W. D. Klobucar, and R. A. Snow, *Synth. Commun.*, **6**, 575 (1978).
[82] N. Kunieda, Y. Fujiwara, A. Suzuki, and M. Kinoshita, *Phosphorus Sulfur*, **16**, 223 (1983).
[83] A. J. Mura, Jr., D. A. Bennett, and T. Cohen, *Tetrahedron Lett.*, **1975**, 4433.
[84] R. Raetz and O. J. Sweeting, *Tetrahedron Lett.*, **1963**, 529.
[85] T. Takata, L. Huang, and W. Ando, *Chem. Lett.*, **1985**, 1705.
[86] T. Takata, K. Hoshino, E. Takeuchi, Y. Tamura, and W. Ando, *Tetrahedron Lett.*, **25**, 4767 (1984).
[87] W. Adam, O. De Lucchi, K. Hill, E.-M. Peters, K. Peters, and H. G. von Schnering, *Chem. Ber.*, **118**, 3070 (1985).

[88] Y. Nagao, M. Ochiai, K. Kanedo, A. Maeda, K. Watanabe, and E. Fujita, *Tetrahedron Lett.*, **1977**, 1345.
[89] Y. Nagao, K. Kaneko, and E. Fujita, *Tetrahedron Lett.*, **1978**, 4115.
[90] B. M. Trost and G. S. Massiot, *J. Am. Chem. Soc.*, **99**, 4405 (1977).
[91] T. Yagihara and S. Oae, *Tetrahedron*, **28**, 2759 (1972).
[92] W. A. Pryor and H. T. Bickley, *J. Org. Chem.*, **37**, 2885 (1972).
[93] G. Sosnovsky, *J. Org. Chem.*, **26**, 281 (1961).
[94] L. Horner and E. Juergens, *Justus Liebigs Ann. Chem.*, **602**, 135 (1957).
[95] G. Sosnovsky, *Tetrahedron*, **18**, 15 (1962).
[96] G. Sosnovsky and S. -O. Lawesson, *Angew. Chem. Int. Ed. Engl.*, **3**, 269 (1964).
[97] K. Almdal and O. Hammerich, *Sulfur Lett.*, **2**, 1 (1984).
[98] J. Nokami, M. Hatate, S. Wakabayashi, and R. Okawara, *Tetrahedron Lett.*, **21**, 2557 (1980).
[99] J. Nokami, K. Nishiuchi, S. Wakabayashi, and R. Okawara, *Tetrahedron Lett.*, **21**, 4455 (1980).
[100] J. Nokami, M. Kawada, and R. Okawara, *Tetrahedron Lett.*, **1979**, 1045.
[101] R. Oda and K. Yamamoto, *J. Org. Chem.*, **26**, 4679 (1961).
[102] M. Hojo, R. Masuda, T. Saeki, K. Fujimori, and S. Tsutsumi, *Tetrahedron Lett.*, **1977**, 3883.
[103] P. Manya, A. Sekera, and A. Rumpf, *Tetrahedron*, **26**, 467 (1970).
[104] N. Kunieda, J. Nokami, and M. Kinoshita, *Chem. Lett.*, **1974**, 369.
[105] G. A. Russell and L. A. Ochrymowycz, *J. Org. Chem.*, **34**, 3618 (1969).
[106] S. Iriuchijima, K. Maniwa, and G. Tsuchihashi, *J. Am. Chem. Soc.*, **96**, 4280 (1974).
[107] S. Iriuchijima, K. Maniwa, and G. Tsuchihashi, *J. Am. Chem. Soc.*, **97**, 596 (1975).
[108] K. Ogura, N. Yahata, J. Watanabe, K. Takahashi, and H. Iida, *Bull. Chem. Soc. Jpn.*, **56**, 3543 (1983).
[109] J. Otera, *Synthesis*, **1988**, 95.
[110] E. Block and M. Aslam, *Tetrahedron*, **44**, 281 (1988).
[111] E. J. Corey and M. G. Bock, *Tetrahedron Lett.*, **1975**, 2643.
[112] P. M. Pojer and S. J. Angyal, *Tetrahedron Lett.*, **1976**, 3067.
[113] P. M. Pojer and S. J. Angyal, *Austr. J. Chem.*, **31**, 1031 (1978).
[114] R. M. Young, J. Y. Gauthier, and W. Coombs, *Tetrahedron Lett.*, **25**, 1753 (1984).
[115] H. Sugihara, R. Tanikaga, and A. Kaji, *Synthesis*, **1978**, 881.
[116] W. L. Meyer, R. A. Manning, E. Schindler, R. S. Schroeder, and D. Craig Shew, *J. Org. Chem.*, **41**, 1005 (1976).
[117] B. M. Trost and G. H. Miller, *J. Am. Chem. Soc.*, **97**, 7182 (1975).
[118] T. Sakamoto, K. Tanji, S. Niitsuma, T. Ono, and H. Yamanaka, *Chem. Pharm. Bull.*, **28**, 3362 (1980).
[119] M. von Strandtman, D. Connor, and J. Shavel, Jr. *J. Heterocycl. Chem.*, **9**, 175 (1972).
[120] M. P. Cava and N. M. Pollack, *J. Am. Chem. Soc.*, **89**, 3639 (1967).
[121] H. L. Monteiro and A. L. Gemal, *Synthesis*, **1975**, 437.
[122] T. Akiyama, H. Iwakiri, M. Shimizu, and T. Mukaiyama, *Chem. Lett.*, **1984**, 1843.
[123] R. H. Schlessinger and I. S. Ponticello, *Tetrahedron Lett.*, **1967**, 4057.
[124] R. H. Schlessinger and I. S. Ponticello, *J. Am. Chem. Soc.*, **89**, 3641 (1967).
[125] M. P. Cava, N. M. Pollack, and D. A. Repella, *J. Am. Chem. Soc.*, **89**, 3640 (1967).
[126] H. Wynberg and D. J. Zwanenburg, *Tetrahedron Lett.*, **1967**, 761.
[127] M. P. Cava and N. M. Pollack, *J. Am. Chem. Soc.*, **88**, 4112 (1966).
[128] G. J. Harner, L. E. Saris, M. V. Lakshmikantham, and M. P. Cava, *Tetrahedron Lett.*, **1976**, 2581.
[129] C. H. Chen, *Tetrahedron Lett.*, **1976**, 25.
[130] C. H. Chen and B. A. Donatelli, *J. Org. Chem.*, **41**, 3053 (1976).
[131] A. Nickon, A. D. Rodriguez, V. Shirhatti, and R. Ganguly, *J. Org. Chem.*, **50**, 4218 (1985).
[132] Z. Majerski, Z. Marinic, and R. Sarac-Arneri, *J. Org. Chem.*, **48**, 5109 (1983).
[133] C. G. Francisco, R. Freire, R. Hernandez, J. A. Salazar, and E. Suarez, *Tetrahedron Lett.*, **25**, 1621 (1984).
[134] E. Wenkert and C. A. Broka, *Finn. Chem. Lett.*, **1984**, 126.

[135] T. Kobayashi, K. Iino, and T. Hiraoka, *J. Am. Chem. Soc.*, **99**, 5505 (1977).
[136] M. Isola, E. Ciuffarin, L. Sagramora, and C. Niccolai, *Tetrahedron Lett.*, **23**, 1381 (1982).
[137] R. J. Stoodley, *Tetrahedron*, **31**, 2321 (1975).
[138] R. D. G. Cooper and D. O. Spry in *Cephalosporins and Penicillins*, E. H. Flynn (Ed.), Academic Press, New York, 1972, p. 183.
[139] R. B. Morin, B. G. Jackson, R. A. Mueller, E. R. Lavagnino, W. B. Scanlon, and S. L. Andrews, *J. Am. Chem. Soc.*, **85**, 1896 (1963).
[140] R. B. Morin, B. G. Jackson, R. A. Mueller, E. R. Lavagnino, W. B. Scanlon, and S. L. Andrews, *J. Am. Chem. Soc.*, **91**, 1401 (1969).
[141] J. J. de Koning, H. J. Kooreman, H. S. Tan, and J. Verweij, *J. Org. Chem.*, **40**, 1346 (1975).
[142] S. Wolfe, P. M. Kazmaier, and H. Auksi, *Can. J. Chem.*, **57**, 2412 (1979).
[143] T. Kaneko, *J. Am. Chem. Soc.*, **107**, 5490 (1985).
[144] T. Kaneko, Y. Akamoto, and K. Hatada, *J. Chem. Soc., Chem. Commun.*, **1987**, 1511.
[145] S. F. Dyke and E. P. Tiley, *Tetrahedron*, **31**, 561 (1975).
[146] E. J. Corey and D. J. Hoover, *Tetrahedron Lett.*, **23**, 3463 (1982).
[147] T. Matsumoto, H. Shirahama, F. Sakan, and K. Takigawa, *Bull. Chem. Soc. Jpn.*, **50**, 325 (1977).
[148] T. Matsumoto, H. Shirahama, H. Ichihara, H. Shin, S. Kagawa, F. Sakan, S. Matsumoto, and S. Nishida, *J. Am. Chem. Soc.*, **90**, 3280 (1968).
[149] Y. Oikawa and O. Yonemitsu, *J. Chem. Soc. Perkin Trans. 1*, **1976**, 1479.
[150] P. T. Lansbury, A. K. Serelis, J. W. Hengeveld, and D. G. Hangauer, Jr., *Tetrahedron*, **36**, 2701 (1980).
[151] A. W. M. Lee, V. S. Martin, S. Masamune, K. B. Sharpless, and F. J. Walker, *J. Am. Chem. Soc.*, **104**, 3515 (1982).
[152] P. Magnus, T. Gallagher, P. Brown, and P. Pappalardo, *Acc. Chem. Res.*, **17**, 35 (1984).
[153] S. Y. Ko, A. W. M. Lee, S. Masamune, L. A. Reed, III, K. B. Sharpless, F. J. Walker, *Science (Washington D.C.)*, **220**, 949 (1983).
[154] D. L. J. Clive, *Tetrahedron*, **34**, 1049 (1978).
[155] H. J. Reich, *Acc. Chem. Res.*, **12**, 22 (1979).
[156] H. J. Reich in *Oxidation in Organic Chemistry, Part C*, W. S. Thrahanovsky (Ed.), Academic Press, New York, 1978, pp. 1–130.
[157] T. Fukuyama, B. D. Robins, and R. A. Sachleben, *Tetrahedron Lett.*, **22**, 4155 (1981).
[158] J. A. Marshall and R. D. Royce, Jr., *J. Org. Chem.*, **47**, 693 (1982).
[159] S. L. Schreiber and C. Santini, *J. Am. Chem. Soc.*, **106**, 4038 (1984).
[160] A. G. Brook, *Acc. Chem. Res.*, **7**, 77 (1974).
[161] M. Kise, M. Murase, M. Kitano, and M. Murai, *Tetrahedron Lett.*, **1976**, 4355.
[162] E. W. Colvin, *Silicon in Organic Synthesis*, Butterworth, London, 1981, p. 33. A. G. Brook and A. R. Bassindale in *Rearrangements in Ground and Exited State*, P. De Mayo (Ed.), Vol. 2, Academic Press, New York, 1980, pp. 149–227.
[163] D. J. Ager, *Tetrahedron Lett.*, **21**, 4759 (1980).
[164] T. Takeda, T. Tsuchida, K. Ando, and T. Fujiwara, *Chem. Lett.*, **1983**, 549.
[165] K. M. More and J. Wemple, *J. Org. Chem.*, **43**, 2713 (1978).
[166] I. Cutting and P. J. Parsons, *Tetrahedron Lett.*, **22**, 2021 (1981).
[167] P. J. Kocienski, *J. Chem. Soc., Chem. Commun.*, **1980**, 1096.
[168] D. J. Ager, *Tetrahedron Lett.*, **24**, 419 (1983).
[169] D. J. Ager and R. C. Cookson, *Tetrahedron Lett.*, **21**, 1677 (1980).
[170] P. J. Kocienski, *Tetrahedron Lett.*, **21**, 1559 (1980).
[171] E. Vedejs and M. Mullins, *Tetrahedron Lett.*, **1975**, 2017.
[172] M. Aono, C. Hyodo, Y. Terao, and K. Achiwa, *Tetrahedron Lett.*, **27**, 4039 (1986).
[173] D. J. Hart and Y. M. Tsai, *Tetrahedron Lett.*, **24**, 4387 (1983).
[174] J. D. White, M. Knag, and B. G. Sheldon, *Tetrahedron Lett.*, **24**, 4539 (1983).
[175] H. J. Reich and S. K. Shah, *J. Org. Chem.*, **42**, 1773 (1977).
[176] K. Ogura, Y. Ito, and G. Tsuchihashi, *Synthesis,* **1980**, 736.
[177] G. H. Posner, E. Asirvatham, and S. F. Ali, *J. Chem. Soc., Chem. Commun.*, **1985**, 542.
[178] R. R. King, *J. Org. Chem.*, **45**, 5347 (1980).

179 H. Kosugi, H. Uda, and S. Yamagiwa, *J. Chem. Soc., Chem. Commun.*, **1976**, 71. S. Yamagiwa, H. Sato, N. Hoshi, H. Kosugi, and H. Uda, *J. Chem. Soc., Perkin Trans. 1*, **1979**, 570.
180 L. S. S. Reamonn and W. I. O'Sullivan, *J. Chem. Soc., Chem. Commun.*, **1976**, 642.
181 J. P. Marino and A. D. Perez, *J. Am. Chem. Soc.*, **106**, 7643 (1984).
182 W. E. Parham and L. D. Edwards, *J. Org. Chem.*, **33**, 4150 (1968).
183 K. Ogura and G. Tsuchihashi, *J. Am. Chem. Soc.*, **96**, 1960 (1974).
184 K. Ogura, Y. Ito, and G. Tsuchihashi, *Bull. Chem. Soc. Jpn.*, **52**, 2013 (1979).
185 T. W. Green, *Protective Groups in Organic Synthesis*, Wiley, New York, 1981; Y. Wolman in *The Chemistry of the Thiol Group*, S. Patai (Ed.), Wiley, London, 1974, p. 669.
186 E. Jonsson, *Acta Chem. Scand.*, **21**, 1277 (1967).
187 S. Oae and N. Furukawa in *Chemistry of Sulfilimines and Related Derivatives*, A.C.S. Monograph, 1983.
188 See for example: J. Drabowicz, T. Numata, and S. Oae, *Org. Prep. Proc. Int.*, **9**, 63 (1977). C. Tenca, A. Dossena, R. Marchelli, and G. Casnati, *Synthesis*, **1981**, 141. G. A. Olah, A. P. Fung, B. G. Gupta, and S. C. Narang, *Synthesis*, **1980**, 221. G. A. Olah, R. Malhotra, and S. C. Narang, *ibid.*, **1979**, 58.
189 K. Naumann, G. Zon, and K. Mislow, *J. Am. Chem. Soc.*, **91**, 7012 (1969).
190 A. J. Mancuso and D. Swern, *Synthesis,* **1981**, 165.
191 R. F. Butterworth and S. Hanessian, *Synthesis,* **1971**, 70.
192 W. W. Epstein and F. W. Sweat, *Chem. Rev.*, **67**, 247 (1967).
193 T. D. Connor, P. A. Young, and M. von Strandtmann, *Synthesis*, **1978**, 208.
194 D. T. Connor, P. A. Young, and M. von Strandtmann, *J. Heterocycl. Chem.*, **15**, 115 (1978).
195 J. P. Marino and M. Neisser, *J. Am. Chem. Soc.*, **103**, 7687 (1981).
196 See for example: J. D. Albright and L. Goldman, *J. Am. Chem. Soc.*, **87**, 4214 (1965).
197 J. B. Albright and L. Goldman, *J. Am. Chem. Soc.*, **89**, 2416 (1967).
198 J. G. Moffatt in *Oxidation. Techniques and Applications in Organic Synthesis*, R. L. Augustine and D. J. Trecker, Eds., Vol. 2, Dekker, New York, 1971, pp. 1–63.
199 S. L. Huang, K. Omura, and D. Swern, *Synthesis,* **1978**, 297.
200 K. Kondo and A. Negishi, *Chem. Lett.*, **1974**, 1525.
201 L. Horner and P. Kaiser, *Justus Liebigs Ann. Chem.*, **626**, 19 (1959).
202 S. Oae, T. Kitao, S. Kawamura, and Y. Kitaoka, *Tetrahedron*, **19**, 817 (1963).
203 S. Oae, T. Kitao, and S. Kawamura, *Tetrahedron*, **19**, 1783 (1963).
204 K. Omura, A. K. Sharma, and D. Swern, *J. Org. Chem.*, **41**, 957 (1976).
205 A. K. Sharma, T. Ku, A. D. Dawson, and D. Swern, *J. Org. Chem.*, **40**, 2758 (1975).
206 S. Iwanami, S. Arita, and K. Takeshita, *Yuki Gosei Kagaku Kyokay Shi,* **26**, 375 (1968) [*C.A.*, **69**, 35641m (1968)].
207 W. R. Sorenson, *J. Org. Chem.*, **24**, 978 (1959).
208 K. Onodera, S. Hirano, N. Kashimura, and T. Yajima, *Tetrahedron Lett.*, **1965**, 4327.
209 V. Lerch and J. G. Moffatt, *J. Org. Chem.*, **36**, 3391 (1971).
210 A. Dossena, R. Marchelli, and G. Casnati, *J. Chem. Soc., Perkin Trans. 1*, **1983**, 1141.
211 M. G. Burdon and J. G. Moffatt, *J. Am. Chem. Soc.*, **89**, 4725 (1967).
212 M. G. Burdon and J. G. Moffatt, *J. Am. Chem. Soc.*, **87**, 4656 (1965).
213 A. H. Fenselau, E. H. Hamamura, and J. G. Moffatt, *J. Org. Chem.*, **35**, 3546 (1970).
214 A. Dossena, G. Palla, R. Marchelli, and T. Lodi, *Int. J. Pept. Protein Res.*, **23**, 198 (1984).
215 K. Omura and D. Swern, *Tetrahedron*, **34**, 1651 (1978).
216 J. H. Jones, D. W. Thomas, R. M. Thomas, and M. E. Wood, *Synth. Commun.*, **16**, 1607 (1986).
217 J. A. Schwindeman and P. D. Magnus, *Tetrahedron Lett.*, **22**, 4925 (1981).
218 M. J. Prior and G. H. Whitham, *J. Chem. Soc., Perkin Trans. 1*, **1986**, 683.
219 J. E. McCormick and R. S. McElhinney, *J. Chem. Res. (S)*, **1981**, 12.
220 E. Block and J. O'Connor, *J. Am. Chem. Soc.*, **96**, 3929 (1974).
221 M. G. Burdon and J. G. Moffatt, *J. Am. Chem. Soc.*, **88**, 5855 (1966).
222 W. A. Szarek, D. M. Vyas, and B. Achmatowicz, *J. Heterocycl. Chem.*, **12**, 123 (1975).
223 W. Parham and M. D. Bhavsar, *J. Org. Chem.*, **28**, 2686 (1963).

[224] C. R. Johnson, J. C. Sharp, and W. G. Phillips, *Tetrahedron Lett.*, **1967**, 5299.
[225] J. E. McCormick and R. S. McElhinney, *J. Chem. Soc., Perkin Trans. 1*, **1976**, 2533.
[226] M. Mikolajczyk, A. Zatorski, S. Grzejszczak, B. Costisella, and W. Midura, *J. Org. Chem.*, **43**, 2518 (1978).
[227] Y. Tamura, H. -D. Choi, H. Maeda, and H. Ishibashi, *Tetrahedron Lett.*, **22**, 1343 (1981).
[228] P. Calinaud and J. Gelas, *Can. J. Chem.*, **61**, 2103 (1983).
[229] M. Gannon, J. E. McCormick, and R. S. McElhinney, *J. Chem. Res. (S)*, **1979**, 52.
[230] J. Buter, P. W. Raynolds, and R. M. Kellogg, *Tetrahedron Lett.*, **1974**, 2901.
[231] R. Greenhalgh, R. R. King, and W. D. Marshall, *J. Agric. Food Chem.*, **26**, 475 (1978).
[232] S. E. Dinizo and D. S. Watt, *Synthesis*, **1977**, 181.
[233] H. Ishibashi, T. Sato, M. Irie, M. Ito, and M. Ikeda, *J. Chem Soc., Perkin Trans. 1*, **1987**, 1095.
[234] P. Claus, *Monatsh. Chem.*, **99**, 1034 (1968).
[235] N. Furukawa, A. Kawada, T. Kawai, and H. Fujihara, *J. Chem. Soc., Chem. Commun.*, **1985**, 1266.
[236] K. Morihara, *Bull. Chem. Soc. Jpn.*, **37**, 1785 (1964).
[237] O. Itoh, T. Numata, T. Yoshimura, and S. Oae, *Bull. Chem. Soc. Jpn.*, **56**, 343 (1983).
[238] H. Ishibashi, H. Komatsu, K. Maruyama, and M. Ikeda, *Tetrahedron Lett.*, **26**, 5791 (1985).
[239] G. A. Russell and G. Hamprecht, *J. Org. Chem.*, **35**, 3007 (1970).
[240] Y. Tamura, H. Maeda, S. Akai, K. Ishiyama, and H. Ishibashi, *Tetrahedron Lett.*, **22**, 4301 (1981).
[241] S. Lane, S. J. Quick, and R. J. K. Taylor, *Tetrahedron Lett.*, **25**, 1039 (1984).
[242] S. Wolfe and P. M. Kazmaier, *Can. J. Chem.*, **57**, 2388 (1979).
[243] T. Fuchigami, Y. Nakagawa, and T. Nonaka, *Tetrahedron Lett.*, **27**, 3869 (1986).
[244] S. Wolfe and P. M. Kazmaier, *Can. J. Chem.*, **57**, 2397 (1979).
[245] T. Numata, O. Itoh, T. Yoshimura, and S. Oae, *Bull. Chem. Soc. Jpn.*, **56**, 257 (1983).
[246] S. Oae and T. Numata, *Tetrahedron*, **30**, 2641 (1974).
[247] T. Numata and S. Oae, *Chem. Ind. (London)*, **1972**, 726.
[248] T. Numata and S. Oae, *Tetrahedron*, **32**, 2699 (1976).
[249] I. Saito and S. Fukui, *J. Vitaminol (Kyoto)*, **12**, 244 (1966) [*C.A.*, **66**, 2502k (1967)].
[250] J. E. McCormick and R. S. McElhinney, *J. Chem. Soc., Chem. Commun.*, **1969**, 171.
[251] J. Kuszmann and P. Sohar, *Acta Chim. Acad. Sci. Hung.*, **88**, 167 (1976) [*C.A.*, **85**, 33309r (1976)].
[252] B. Lindberg and H. Lundstrom, *Acta Chem. Scand.*, **22**, 1861 (1968).
[253] S. Iriuchijima, K. Maniwa, and S. Tsuchihashi, *Agric. Biol. Chem.*, **40**, 2389 (1976).
[254] G. A. Russell and E. T. Sabourin, *J. Org. Chem.*, **34**, 2336 (1969).
[255] M. Kise, M. Murase, M. Kitano, and H. Murai, *Tetrahedron Lett.*, **1976**, 4355.
[256] T. Masuda, N. Furukawa, and S. Oae, *Bull. Chem. Soc. Jpn.*, **51**, 2659 (1978).
[257] T. Masuda, T. Numata, N. Furukawa, and S. Oae, *Chem. Lett.*, **1977**, 745.
[258] M. Bhupathy and T. Cohen, *Tetrahedron Lett.*, **28**, 4787 (1987).
[259] Y. Kita, O. Tamura, T. Miki, and Y. Tamura, *Tetrahedron Lett.*, **28**, 6479 (1987).
[260] D. T. Connor and M. von Strandtmann, *J. Org. Chem.*, **39**, 1594 (1974).
[261] R. R. King, *J. Org. Chem.*, **43**, 3784 (1987).
[262] R. Greenhalgh, W. D. Marshall, and R. R. King, *J. Agric. Food Chem.*, **24**, 266 (1976).
[263] S. Oae, O. Itoh, T. Numata, and T. Yoshimura, *Bull. Chem. Soc. Jpn.*, **56**, 270 (1983).
[264] P. Clayton, J. H. C. Nayler, M. J. Pearson, and R. Southgate, *J. Chem. Soc., Perkin Trans. 1*, **1974**, 22.
[265] T. Numata, O. Itoh, and S. Oae, *Chem. Lett.*, **1977**, 909.
[266] D. T. Connor and R. J. Sorenson, *J. Heterocycl. Chem.*, **18**, 587 (1981).
[267] Y. Uchida and S. Oae, *Gazz. Chim. Ital.*, **117**, 649 (1987).
[268] D. T. Connor and M. Von Strandtmann, *J. Heterocycl. Chem.*, **15**, 113 (1978).
[269] L. Pichat and J. Tostain, *J. Labelled Compd. Radiopharm.*, **13**, 587 (1977) [*C.A.*, **88**, 136255m (1978)].
[270] H. -D. Becker, *J. Org. Chem.*, **29**, 1358 (1964).

[271] B. M. Pinto, D. M. Vyas, and W. A. Szarek, *Can. J. Chem.*, **55**, 937 (1977).
[272] D. O. Spry, *J. Chem. Soc., Chem. Commun.*, **1973**, 259.
[273] J. E. McCormick and R. S. McElhinney, *J. Chem. Res. (S)*, **1981**, 310.
[274] E. S. Schroeder and R. M. Dodson, *J. Am. Chem. Soc.*, **84**, 1904 (1962).
[275] R. Lett and Y. Kuroki, *Tetrahedron Lett.*, **23**, 5541 (1982).
[276] T. Numata, O. Itoh, and S. Oae, *Tetrahedron Lett.*, **1979**, 161.
[277] D. O. Spry, *J. Am. Chem. Soc.*, **92**, 5006 (1970).
[278] H. Shimizu, N. Ueda, T. Katahoka, and M. Hori, *Chem. Pharm. Bull.*, **32**, 2590 (1984).
[279] M. Bhupathy and T. Cohen, *Tetrahedron Lett.*, **28**, 4797 (1987).
[280] P. J. R. Nederlof, M. J. Moolenaar, E. R. de Waard, and H. O. Huisman, *Tetrahedron*, **34**, 2205 (1978).
[281] J. -S. Dung, R. W. Armstrong, O. P. Anderson, and R. M. Williams, *J. Org. Chem.*, **48**, 3592 (1983).
[282] T. P. Hilditch and S. Smiles, *J. Chem. Soc.*, **1911**, 145.
[283] G. E. Fr. Wilson and C. J. Strong, *J. Org. Chem.*, **37**, 2376 (1972).
[284] N. Furukawa, T. Morishita, T. Akasaka, and S. Oae, *Tetrahedron Lett.*, **1978**, 1567.
[285] N. Furukawa, T. Morishita, T. Akasaka, and S. Oae, *Tetrahedron Lett.*, **1977**, 1653.
[286] F. M. Hauser, S. R. Ellenberger, J. P. Glusker, C. J. Smart, and H. L. Carrell, *J. Org. Chem.*, **51**, 50 (1986).
[287] S. Wolfe, P. M. Kazmaier, and H. Auksi, *Can. J. Chem.*, **57**, 2404 (1979).
[288] K. Sindelar and M. Protiva, *Collect. Czech. Chem. Commum.*, **35**, 3328 (1970).
[289] E. Kameyama, S. Inokuma, R. Oda, and T. Kuwamura, *Kogyo Kagaku Zasshi*, **74**, 87 (1971) [*C.A.*, **74**, 124718n (1971)].
[290] H. L. Yale, *J. Heterocycl. Chem.*, **15**, 331 (1978).
[291] R. B. Gammil, *J. Org. Chem.*, **49**, 5035 (1984).
[292] Sanwa Chemical K. K. Pat. No. 82 46974 [*C.A.*, **97**, 55677j (1982)].
[293] A. de Groot and B. J. M. Jansen, *J. Org. Chem.*, **49**, 2034 (1984).
[294] L. C. Vishwakarma and A. R. Martin, *J. Heterocycl. Chem.*, **19**, 103 (1982).
[295] T. Katsuki, A. W. M. Lee, P. Ma, V. S. Martin, S. Masamune, K. B. Sharpless, T. T. Tuddenham, and S. J. Walker, *J. Org. Chem.*, **47**, 1373 (1982).
[296] T. Ikeda and C. R. Hutchinson, *J. Org. Chem.*, **49**, 2837 (1984).
[297] D. H. R. Barton, F. Comer, D. G. T. Greig, P. G. Sammes, C. M. Cooper, G. Hewitt, and W. G. E. Underwood, *J. Chem. Soc. (C)*, **1971**, 3540.
[298] D. H. Bremner and M. M. Campbell, *J. Chem. Soc., Chem. Commun.*, **1976**, 538.
[299] D. H. Bremner and M. M. Campbell, *J. Chem. Soc., Perkin Trans. 1*, **1977**, 2298.
[300] H. D. Huisman, *Pure Appl. Chem.*, **49**, 1307 (1977).
[301] T. Matsumoto, H. Shirahama, A. Ichihara, H. Shin, S. Kagawa, N. Ito, T. Hisamitsu, T. Kamada, F. Sakan, K. Saito, S. Nishida, and S. Matsumoto, *Tetrahedron Lett.*, **1968**, 1925.
[302] T. Terasawa and T. Okada, *Heterocycles*, **11**, 171 (1978).
[303] J. -P. Corbet and C. Benezra, *Tetrahedron Lett.*, **21**, 2061 (1980).
[304] G. A. Koppel and L. J. McShane, *J. Am. Chem. Soc.*, **100**, 288 (1978).
[305] T. Mandai, M. Kawada, and J. Otera, *J. Org. Chem.*, **48**, 5183 (1983).
[306] R. E. Brown, D. M. Lustgarten, and R. J. Stanback, *J. Org. Chem.*, **34**, 3694 (1969).
[307] F. Theil, C. Lindig, and K. Repke, *Z. Chem.*, **20**, 414 (1980).
[308] J. M. Akkerman, H. de Koning, and H O Huisman, *Heterocycles*, **15**, 797 (1981).
[309] J. M. Akkerman, H. de Koning, and H. O. Huisman, *J. Chem. Soc., Perkin Trans. 1*, **1979**, 2124.
[310] N. M. Lucey and R. S. McElhinney, *J. Chem. Res. (S)*, **1985**, 240.
[311] T. Gallagher, P. Magnus, and J. C. Huffman, *J. Am. Chem. Soc.*, **104**, 1140 (1982).
[312] K. Perlman, H. K. Schnoes, Y. Tanaka, H. F. DeLuca, Y. Kabayashi, and T. Taguchi, *Biochemistry*, **23**, 5041 (1984).
[313] E. Block and J. O'Connor, *J. Am. Chem. Soc.*, **95**, 5048 (1973).
[314] H. Ishibashi, S. Harada, M. Okada, M. Ikeda, K. Ishiyama, H. Yamashita, and Y. Tamura, *Synthesis*, **1986**, 847.

[315] Y. D. Vankar and C. T. Rao, *Tatrahedron*, **41**, 3405 (1985).
[316] A. F. Janzen, G. N. Lypka, and R. E. Wasylishen, *J. Heterocycl. Chem.*, **16**, 415 (1979).
[317] S. Rakhit, M. Georges, and J. F. Bagli, *Can. J. Chem.*, **57**, 1153 (1979).
[318] W. Czuba, T. Kowalska, H. Poradowska, and P. Kowalski, *Pol. J. Chem.*, **58**, 1221 (1984).
[319] G. M. Brooke and J. A. K. Jamie Ferguson, *J. Chem. Soc., Perkin Trans. 1*, **1987**, 2091.
[320] T. Minami, Y. Tsumori, K. Yoshida, and T. Agawa, *J. Org. Chem.*, **39**, 3412 (1974).
[321] I. Nagakura, H. Oka, and Y. Nitta, *Heterocycles*, **3**, 453 (1975).
[322] Y. Uchida, Y. Kobayashi, and S. Kozuka, *Bull. Chem. Soc. Jpn.*, **54**, 1781 (1981).
[323] R. McCrindle, A. J. Alees, and D. K. Stephenson, *J. Chem. Soc., Perkin Trans. 1*, **1981**, 3070.
[324] Y. Hayashi and R. Oda, *J. Org. Chem.*, **32**, 457 (1967).
[325] P. Claus, *Monatsh. Chem.*, **102**, 913 (1971).
[326] K. Sato, S. Inoue, K. Ozawa, T. Kobayashi, T. Ota, and M. Tazaki, *J. Chem. Soc., Perkin Trans. 1*, **1987**, 1753.
[327] K. Sato, S. Inoue, K. Ozawa, and M. Tazaki, *J. Chem. Soc., Perkin Trans. 1*, **1984**, 2715.
[328] J. P. Marino, K. E. Pfitzner, and R. A. Olofson, *Tetrahedron*, **27**, 4181 (1971).
[329] K. E. Pfitzner, J. P. Marino, and R. A. Olofson, *J. Am. Chem. Soc.*, **87**, 4658 (1965).
[330] P. Claus, *Monatsh. Chem.*, **99**, 1034 (1968).
[331] G. R. Pettit and T. H. Brown, *Can. J. Chem.*, **45**, 1306 (1967).
[332] I. K. Stamos, *Tetrahedron Lett.*, **26**, 477 (1985).
[333] I. K. Stamos, *Tetrahedron Lett.*, **27**, 6261 (1986).
[334] T. H. Kim, D. Y. Kim, and D. Y. Oh, *Synth. Commun.*, **17**, 755 (1987).
[335] H. Ishibashi, H. Komatsu, and M. Ikeda, *J. Chem Res. (S)*, **1987**, 296.
[336] R. Hunter and C. D. Simon, *Tetrahedron Lett.*, **27**, 1385 (1986).
[337] Y. Oikawa, O. Setayama, and O. Yonemitsu, *Heterocycles*, **2**, 21 (1974).
[338] Y. Tamura, H. Maeda, S. Akai, and H. Ishibashi, *Tetrahedron Lett.*, **23**, 2209 (1982).
[339] L. N. Mander and P. H. C. Mundill, *Synthesis*, **1981**, 620.
[340] H. Ishibashi, H. Ozeki, and M. Ikeda, *J. Chem. Soc., Chem. Commun.*, **1986**, 654.
[341] Y. Oikawa and O. Yonemitsu, *Tetrahedron Lett.*, **1972**, 3396.
[342] J. G. Cannon, T. Lee, H. D. Goldman, and B. Costall, *J. Med. Chem.*, **20**, 1111 (1977).
[343] Y. Oikawa and O. Yonemitsu, *J. Chem. Soc., Chem. Commun.*, **1971**, 555.
[344] Y. Kita, O. Tamura, H. Yasuda, F. Itoh, and Y. Tamura, *Chem. Pharm. Bull.*, **33**, 4235 (1985).
[345] I. A. Blair, L. N. Mander, and P. H. C. Mundill, *Aust. J. Chem.*, **34**, 1235 (1981).
[346] Y. Oikawa, M. Tanaka, H. Hirasawa, and O. Yonemitsu, *Heterocycles*, **15**, 207 (1981).
[347] M. Hori, T. Katahoka, M. Shimizu, and A. Tamoto, *Tetrahedron Lett.*, **22**, 3629 (1981).
[348] T. Kataoka, A. Tomoto, H. Shimizu, and M. Hori, *J. Chem. Soc., Perkin Trans. 1*, **1983**, 2913.
[349] Y. Oikawa and O. Yonemitsu, *Heterocycles*, **5**, 233 (1975).
[350] P. Magnus and P. M. Cairns, *J. Am. Chem. Soc.*, **108**, 217 (1986).
[351] S. G. Pyne, *Tetrahedron Lett.*, **28**, 4737 (1987).
[352] M. Ladlow, P. M. Cairus, and P. Magnus, *J. Chem. Soc., Chem. Commun.*, **1986**, 1756.
[353] T. Gallagher, P. Magnus, and J. C. Huffman, *J. Am. Chem. Soc.*, **105**, 4750 (1983).
[354] T. Gallagher and P. D. Magnus, *J. Am. Chem. Soc.*, **105**, 2086 (1983).
[355] P. Magnus, T. Gallagher, P. Brown, and J. C. Huffman, *J. Am. Chem. Soc.*, **106**, 2105 (1984).
[356] K. Cardwell, B. Hewitt, and P. Magnus, *Tetrahedron Lett.*, **28**, 3303 (1987).
[357] P. Magnus, P. M. Cairns, and C. Sook Kim, *Tetrahedron Lett.*, **26**, 1963 (1985).
[358] P. D. Magnus and P. Pappalardo, *J. Am. Chem. Soc.*, **105**, 6525 (1983).
[359] R. Michelot and M. B. Tchoubar, *Bull. Soc. Chim. Fr.*, **1966**, 3039.
[360] E. H. Amonoo-Neizer, S. K. Ray, R. A. Shaw, and B. C. Smith, *J. Chem. Soc.*, **1965**, 6250.
[361] R. E. Boyle, *J. Org. Chem.*, **31**, 3880 (1966).
[362] M. F. Lappert and J. K. Smith, *J. Chem. Soc.*, **1961**, 3224.
[363] R. Raetz and O. J. Sweeting, *J. Org. Chem.*, **28**, 1612 (1963).

[364] J. S. Grossert, W. R. Hardstaff, and R. F. Langler, *J. Chem. Soc., Chem. Commun.*, **1973**, 50.
[365] G. A. Russell, E. T. Sabourin, and G. Hamprecht, *J. Org. Chem.*, **34**, 2339 (1969).
[366] H. -D. Becker and G. A. Russell, *J. Org. Chem.*, **28**, 1896 (1963).
[367] B. Wladislaw, L. Marzorati, and M. A. Carvalho Andrade, *An. Acad. Bras. Cienc.*, **52**, 11 (1980) [*C.A.*, **94**, 29885u (1981)].
[368] J. J. A. vanAsten and R. Louw, *Tetrahedron Lett.*, **1975**, 671.
[369] T. Masuda, N. Furukawa, and S. Oae, *Chem. Lett.*, **1977**, 1103.
[370] C. W. Bird, *J. Chem. Soc. C*, **1968**, 1230.
[371] S. Oae, M. Yokoyama, and M. Kise, *Bull. Chem. Soc. Jpn.*, **41**, 1221 (1968).
[372] N. Hellstroem and T. Lauritzson, *Chem. Ber.*, **69**, 2003 (1963).
[373] M. Kise and S. Oae, *Bull. Chem. Soc. Jpn.*, **43**, 1426 (1970).
[374] S. Oae and M. Kise, *Tetrahedron Lett.*, **1968**, 2261.
[375] J. P. Chupp, T. M. Balthazor, M. J. Miller, and M. J. Pozzo, *J. Org. Chem.*, **49**, 4711 (1984).
[376] D. Walker, *J. Org. Chem.*, **31**, 835 (1966).
[377] E. Larson and K. Joensson, *Ber.*, **67**, 1263 (1934).
[378] T. Sakamoto, K. Tanji, S. Niitsuma, T. Ono, and H. Yamanaka, *Chem. Pharm. Bull.*, **28**, 3362 (1980).
[379] D. Hodson and G. Holt, *J. Chem. Soc. C*, **1968**, 1602.
[380] J. Gosteli, *Helv. Chim. Acta*, **60**, 1980 (1977).
[381] A. Rosowsky and K. K. N. Chen, *J. Org. Chem.*, **38**, 2073 (1973).
[382] S. Iriuchijiama, T. Sakakibara, and G. Tsuchihashi, *Agric. Biol. Chem.*, **40**, 1369 (1976).
[383] H. J. Chaves das Neves and M. F. Machete, *Tetrahedron Lett.*, **1977**, 187.
[384] B. Holmberg, *Ark. Kemi, Mineral. Geol.*, **13A** No 8 (1939) [*C.A.*, **33**, 6278 (1939)].
[385] B. Holmberg, *Ark. Kemi, Mineral. Geol.*, **12A**, No 14 (1937) [*C.A.*, **31**, 4292 (1937)].
[386] D. Walker and J. Leib, *J. Org. Chem.*, **28**, 3077 (1963).
[387] C. F. Schwender, B. R. Sunday, J. Shavel, Jr., and R. E. Giles, *J. Med. Chem.*, **17**, 1112 (1974).
[388] N. Ono, H. Miyake, R. Tanikaga, and A. Kaji, *J. Org. Chem.*, **47**, 5017 (1982).
[389] G. A. Russell and J. M. Pecoraro, *J. Org. Chem.*, **44**, 3990 (1979).
[390] H. Rheinboldt and E. Giesbrecht, *J. Am. Chem. Soc.*, **69**, 644 (1947).
[391] N. Furukawa, T. Morishita, T. Akasaka, and S. Oae, *Tetrahedron Lett.*, **1979**, 3973.
[392] D. A. Davenport, D. B. Moss, J. E. Rhodes, and J. A. Walsh, *J. Org. Chem.*, **34**, 3353 (1969).
[393] J. Barnett, *J. Chem. Soc.*, **1944**, 5.
[394] G. H. Posner and T. P. Kogan, *J. Chem. Soc., Chem. Commun.*, **1983**, 1481.
[395] T. Itoh, A. Yoshinaka, T. Sato, and T. Fujisawa, *Chem. Lett.*, **1985**, 1679.
[396] K. Kondo, E. Hiro, and D. Tunemoto, *Tetrahedron Lett.*, **1976**, 4489.
[397] P. J. Maurer, C. G. Knudsen, A. D. Palkowitz, and H. Rapoport, *J. Org. Chem.*, **50**, 325 (1985).
[398] S. Kosuge, M. Hayashi, and N. Hamanaka, *Tetrahedron Lett.*, **23**, 4027 (1982).
[399] C. Sato, S. Ikeda, H. Shirahama, and T. Matsumoto, *Tetrahedron Lett.*, **23**, 2099 (1982). C. Sato, S. Ikeda, H. Shirahama, and T. Matsumoto, *Tennen Yuki Kagobutsu Toronkai Koen Yoshishu*, 24th, 410 (1981) [*C.A.*, **96**, 199361d (1982)].
[400] L. J. J. Hronowski and W. A. Szarek, *J. Med. Chem.*, **25**, 522 (1982).
[401] M. Behforouz and R. Benrashid, *Tetrahedron Lett.*, **1979**, 4493.
[402] E. Shaumann, *Bull. Soc. Chim. Belg.*, **95**, 995 (1986).
[403] W. E. Parham, L. Christensen, S. H. Groen, and R. M. Dodson, *J. Org. Chem.*, **29**, 2211 (9164).
[404] R. D. Miller and R. Haessig, *Synth. Commun.*, **14**, 1285 (1984).
[405] T. Taya, S. Kosaka, Y. Otsuji, and E. Imoto, *Bull. Chem. Soc. Jpn.*, **41**, 2086 (1968).
[406] N. N. Novitskaya, B. V. Flekhter, L. V. Spirikhin, and G. A. Tolstikov, *Khim. Geterotsikl. Soedin.*, **1979**, 563 [*C.A.*, **91**, 39234h (1979)].

[407] E. C. Taylor and J. E. Macor, *J. Org. Chem.*, **52**, 4280 (1987).
[408] W. S. Lee, H. G. Hahan, and K. D. Nam, *J. Org. Chem.*, **51**, 2789 (1986).
[409] M. Mikolajczyk, B. Costisella, and Ş. Grzejszczak, *Tetrahedron*, **39**, 1189 (1983).
[410] M. P. Cava, N. M. Pollack, and G. A. Dieterle, *J. Am. Chem. Soc.*, **95**, 2558 (1973).
[411] M. P. Cava, N. M. Pollack, O. A. Mamer, and M. J. Mitchell, *J. Org. Chem.*, **36**, 3932 (1971).
[412] R. S. Schlessinger and G. S. Ponticello, *J. Am. Chem. Soc.*, **89**, 7138 (1967).
[413] S. Torii, H. Tanaka, and N. Sayo, *J. Org. Chem.*, **44**, 2938 (1979).
[414] G. A. Russell, E. Sabourin, and G. J. Mikol, *J. Org. Chem.*, **31**, 2854 (1966).
[415] W. E. Parham and R. Konkos, *J. Am. Chem. Soc.*, **83**, 4034 (1961).
[416] J. Durman, J. I. Grayson, P. G. Hunt, and S. Warren, *J. Chem. Soc., Perkin Trans. 1*, **1986**, 1939.
[417] K. Iwai, K. Hiroshi, and H. Uda, *Chem. Lett.*, **1974**, 1237.
[418] K. Iwai, H. Kosugi, H. Uda, and M. Kawai, *Bull. Chem. Soc. Jpn.*, **50**, 242 (1977).
[419] R. B. Morin, D. O. Spry, and R. A. Mueller, *Tetrahedron Lett.*, **1969**, 849.
[420] E. N. Karaulova, G. D. Gal'pern, V. D. Nikitina, I. V. Cherepanova, and L. R. Barykina, *Neftekhimiya*, **12**, 104 (1972) [*C.A.*, **76**, 163506p (1972)].
[421] R. B. Morin and D. Spry, *J. Chem. Soc., Chem. Commun.*, **1970**, 335.
[422] M. P. Cava, M. Behforouz, G. E. M. Husbands, and M. Srinivasan, *J. Am. Chem. Soc.*, **95**, 2561 (1973).
[423] J. Kuszmann, P. Sohar, and G. Horvath, *Tetrahedron*, **27**, 5055 (1971).
[424] R. D. Miller and R. Haessig, *Tetrahedron Lett.*, **26**, 2395 (1985).
[425] T. Akiyama, H. Iwakiri, M. Shimizu, and T. Mukaiyama, *Chem. Lett.*, **1984**, 1843.
[426] J. -P. Corbet and C. Benezra, *Can. J. Chem.*, **57**, 213 (1979).
[427] M. Hori, T. Kataoka, M. Shimizu, and N. Ueda, *Tetrahedron Lett.*, **22**, 1701 (1981).
[428] J. Durnam, P. G. Hunt, and S. Warren, *Tetrahedron Lett.*, **24**, 2113 (1983).
[429] J. A. Kaydos and D. L. Smith, *J. Org. Chem.*, **48**, 1096 (1983).
[430] P. Bakuzis and M. L. F. Bakuzis, *J. Org. Chem.*, **50**, 2569 (1985).
[431] I. M. Nasyrov, U. K. Karimov, I. U. Numanov, and S. S. Dzhalolov, *Khim. Tadzh.*, **1973**, 104 [*C.A.*, **84**, 89929 (1976)].
[432] K. Praefcke and C. Weichsel, *Justus Liebigs Ann. Chem.*, **1979**, 784.
[433] G. M. Prokhorov, V. I. Dronov, R. V. Zainullina, E. E. Zaev, N. S. Lyubopytova, V. I. Khvostenko, and E. G. Galkin, *Zh. Org. Khim.*, **10**, 1852 (1974) [*C.A.*, **81**, 151715k (1974)].
[434] B. Zwanenburg and A. Wagenaar, *Tetrahedron Lett.*, **1973**, 5009.
[435] C. U. Kim, P. F. Misco, and D. N. McGregor, *J. Org. Chem.*, **47**, 171 (1982).
[436] P. Blatcher and S. Warren, *J. Chem. Soc., Perkin Trans. 1*, **1985**, 1055.
[437] P. J. Brown, D. N. Jones, M. A. Khan, and N. A. Meanwell, *Tetrahedron Lett.*, **24**, 405 (1983).
[438] K. Praefcke and C. Weichsel, *Tetrahedron Lett.*, **1976**, 2229.
[439] T. Terasawa and T. Okada, *Heterocycles,* **11**, 181 (1978).
[440] D. O. Spry, *Tetrahedron Lett.*, **1977**, 3611.
[441] W. J. Gottstein, P. F. Misco, and L. C. Cheney, *J. Org. Chem.*, **37**, 2765 (1972).
[442] M. V. Lakshmikantham and M. P. Cava, *J. Org. Chem.*, **43**, 82 (1978).
[443] S. Warren and A. T. Zaslona, *Tetrahedron Lett.*, **23**, 4167 (1982).
[444] J. I. Grayson, S. Warren, and A. T. Zaslona, *J. Chem. Soc., Perkin Trans. 1*, **1987**, 967.
[445] J. D. Bower and R. H. Schlessinger, *J. Am. Chem. Soc.*, **91**, 6891 (1969).
[446] T. Patonay, E. Patonay-Pely, and G. Litkei, *Tetrahedron*, **43**, 1827 (1987).
[447] T. V. Stolbova, S. K. Klimenko, A. A. Shcherbakov, G. G. Aleksandrov, Y. T. Struchkov, and V. G. Kharchenko, *Khim. Gerterotsikl. Soedin.*, **1980**, 1056 [*C.A.*, **94**, 47081h (1981)].
[448] J. E. Baldwin, A. K. Forrest, S. Ko, and L. N. Sheppard, *J. Chem. Soc., Chem. Commun.*, **1987**, 81.
[449] A. Nudelman and R. J. McCaully, *J. Org. Chem.*, **42**, 2887 (1977).
[450] A. G. Schultz and R. H. Schlessinger, *Tetrahedron Lett.*, **1973**, 491.
[451] A. G. Schultz and R. H. Schlessinger, *Tetrahedron Lett.*, **1973**, 4787.
[452] M. P. Cava and G. E. M. Husbands, *J. Am. Chem. Soc.*, **91**, 3952 (1969).

[453] C. U. Kim, P. F. Misco, U. J. Haynes, and D. N. McGregor, *J. Med. Chem.*, **27**, 1225 (1984).
[454] H. Kise, G. F. Whitfield, and D. Swern, *J. Org. Chem.*, **37**, 1125 (1972).
[455] H. Kise, G. F. Whitfield, and D. Swern, *Tetrahedron Lett.*, **1971**, 4839.
[456] S. Oae, T. Masuda, K. Tsujihara, and N. Furukawa, *Bull. Chem. Soc. Jpn.*, **45**, 3586 (1972).
[457] P. Claus and W. Vykudilik, *Tetrahedron Lett.*, **1968**, 3607.
[458] T. Yamamoto and M. Okavara, *Bull. Chem. Soc. Jpn.*, **51**, 2443 (1978).
[459] N. Furukawa, S. Oae, and T. Masuda, *Chem. Ind. (London)*, **1975**, 396.
[460] N. Furukawa, T. Yoshimura, and S. Oae, *Phosphorus Sulfur*, **3**, 277 (1977).
[461] Y. Tamura, Y. Nishikawa, K. Sumoto, M. Ikeda, M. Murase, and M. Kise, *J. Org. Chem.*, **42**, 3226 (1977).
[462] M. Kise, M. Murase, M. Kitano, T. Tomita, and H. Murai, *Tetrahedron Lett.*, **1976**, 691.
[463] M. Kise, M. Murase, T. Tomita, M. Kitano, and H. Murai, *Hukusokan Kagaku Toronkai Koen Yoshishu*, 8th, 109 (1975) [*C.A.*, **85**, 5453v (1976)].
[464] N. Miyoshi, S. Murai, and N. Sonoda, *Tetrahedron Lett.*, **1977**, 851.
[465] B. E. Norcross, J. M. Lansinger, and R. L. Martin, *J. Org. Chem.*, **42**, 369 (1977).
[466] L. E. Saris and M. P. Cava, *J. Am. Chem. Soc.*, **98**, 867 (1976).
[467] G. Galambos and V. Simonidesz, *Tetrahedron Lett.*, **23**, 4371 (1982).
[468] J. P. Marino and M. W. Kim, *Tetrahedron Lett.*, **28**, 4925 (1987).
[469] N. Ikota and B. Ganem, *J. Org. Chem.*, **43**, 1607 (1978).
[470] H. J. Reich, J. M. Renga, and I. L. Reich, *J. Org. Chem.*, **39**, 2133 (1974).
[471] H. J. Reich, J. M. Renga, and I. L. Reich, *J. Am. Chem. Soc.*, **97**, 5434 (1975).
[472] K. Sachdev and H. S. Sachdev, *Tetrahedron Lett.*, **1976**, 4223.
[473] F. C. Brown, D. G. Morris, and A. M. Murray, *Tetrahedron*, **34**, 1845 (1978).
[474] T. Laitalainen, T. Simonen, R. Kivekaes, and M. Klinga, *J. Chem. Soc., Perkin Trans. 1*, **1983**, 333.
[475] T. Laitalainen, T. Simonen, and R. Kivekaes, *Tetrahedron Lett.*, **1978**, 3079.
[476] T. G. Back and N. Ibrahim, *Tetrahedron Lett.*, **1979**, 4931.
[477] T. G. Back, N. Ibrahim, and D. J. McPhee, *J. Org. Chem.*, **47**, 3283 (1982).
[478] M. Isobe, Y. Ichikawa, and T. Goto, *Tetrahedron Lett.*, **22**, 4287 (1981).
[479] R. D. Miller and R. Haessig, *Tetrahedron Lett.*, **25**, 5351 (1984).
[480] Y. Kita, O. Tamura, S. Itoh, M. Yasuda, T. Miki, and Y. Tamura, *Chem. Pharm. Bull.*, **35**, 562 (1987).
[481] A. G. Brook and D. G. Anderson, *Can. J. Chem.*, **46**, 2115 (1968).
[482] H. Ishibashi, H. Nakatani, K. Maruyama, K. Minami, and M. Ikeda, *J. Chem. Soc., Chem. Commun.*, **1987**, 1443.
[483] F. A. Carey and O. Hernandez, *J. Org. Chem.*, **38**, 2670 (1973).
[484] M. Bhupathy and T. Cohen, *Tetrahedron Lett.*, **28**, 4793 (1987).
[485] J. Kitchin and R. J. Stoodley, *J. Chem. Soc., Perkin Trans. 1*, **1973**, 2464.
[486] J. Kitchin and R. J. Stoodley, *J. Chem. Soc., Chem. Commun.*, **1972**, 959.
[487] S. Oae, T. Yagihara, and T. Okabe, *Tetrahedron*, **28**, 3203 (1972).
[488] N. Miyamoto, D. Fukuoka, K. Utimoto, and H. Nozaki, *Bull. Chem. Soc. Jpn.*, **47**, 1817 (1974).
[489] J. Kitchin and R. J. Stoodley, *J. Chem. Soc., Perkin Trans. 1*, **1973**, 22.
[490] R. R. King, R. Greenhalgh, and W. D. Marshall, *J. Org. Chem.*, **43**, 1262 (1978).
[491] Y. Maki and T. Hiramitsu, *Chem. Pharm. Bull.*, **24**, 3135 (1976).
[492] D. D. Spry, *Tetrahedron Lett.*, **21**, 1289 (1980).
[493] J. P. Marino and R. F. de la Pradilla, *Tetrahedron Lett.*, **26**, 5381 (1985).
[494] H. Kosugi, H. Uda, and S. Yamagiwa, *J. Chem. Soc., Chem. Commun.*, **1975**, 192.
[495] J. Garcia, R. Greenhouse, J. M. Muchowski, and J. A. Ruiz, *Tetrahedron Lett.*, **26**, 1827 (1985).
[496] K. Kobayashi and K. Mutai, *Chem. Lett.*, **1981**, 1105.
[497] J. E. Baldwin, M. A. Christie, S. B. Haber, and L. I. Kruse, *J. Am. Chem. Soc.*, **98**, 3045 (1976).

CHAPTER 4

THE CATALYZED NUCLEOPHILIC ADDITION OF ALDEHYDES TO ELECTROPHILIC DOUBLE BONDS*

HERMANN STETTER AND HEINRICH KUHLMANN

Technische Hochschule Aachen, West Germany

CONTENTS

* Translated from the German by E. Ciganek, Medical Products Department, E. I. duPont de Nemours & Co., Wilmington, Delaware.

Organic Reactions, *Vol. 40*, Edited by Leo A. Paquette et al.
ISBN 0-471-53841-8 Published by John Wiley & Sons, Inc.

INTRODUCTION

The cyanide-ion-catalyzed dimerization of aromatic and heterocyclic aldehydes to α-hydroxy ketones, also known as the benzoin condensation, is one of the oldest reactions in organic chemistry.[1] A key intermediate in this reversible reaction is a nitrile-stabilized carbanion. In 1973 it was shown that these carbanions add irreversibly to α,β-unsaturated ketones, esters, and nitriles to give 1,4-diketones, 4-ketocarboxylic esters, and 4-ketocarbonitriles, respectively.[2] The reaction succeeds only in aprotic solvents, preferably dimethylformamide. Benzoins and aldehydes can be used interchangeably as starting materials since they are in rapid equilibrium. Cyanide ion catalysis fails with aliphatic aldehydes because they resinify under the strongly basic conditions. Vitamin B_1 (thiamine) has long been known to convert aliphatic

$$\mathrm{RCHO} + \mathrm{R^1CH{=}CHCOR^2} \xrightarrow{\mathrm{CN^-}} \mathrm{RCOCHR^1CH_2COR^2}$$

$$\mathrm{RCHO} + \mathrm{R^1CH{=}CHCO_2R^2} \xrightarrow{\mathrm{CN^-}} \mathrm{RCOCHR^1CH_2CO_2R^2}$$

$$\mathrm{RCHO} + \mathrm{R^1CH{=}CHCN} \xrightarrow{\mathrm{CN^-}} \mathrm{RCOCHR^1CH_2CN}$$

aldehydes into acyloins in buffered aqueous solution.[3] The catalytic effect of vitamin B_1 is due to the presence of a thiazolium cation. Thiazolium salts in general as well as other azolium salts[4] also catalyze the formation of acyloins. The optimum conditions for thiazolium salt catalysis have been worked out.[5] It was shown subsequently that thiazolium salts, in combination with bases, also catalyze the addition of aliphatic, aromatic, and heterocyclic aldehydes to α,β-unsaturated carbonyl compounds.[6] The 1,4-diketones, 4-ketocarboxylic esters, and 4-ketocarbonitriles are usually formed in good to excellent yields. Cyanide ion and thiazolium salt catalysis thus complement each other.

This chapter covers the literature of the catalyzed nucleophilic addition of aldehydes to electrophilic double bonds up to 1988. The reaction has been reviewed previously.[2c]

MECHANISM

The benzoin condensation is a reversible, thermodynamically controlled reaction. The addition of the intermediate anion **1** to a Michael acceptor is irreversible and kinetically controlled. The mechanism of the cyanide-ion-catalyzed reactions is summarized in Eq. 1.

$$\mathrm{RCHO} + \mathrm{CN^-} \rightleftharpoons \mathrm{RCH(O^-)(CN)} \rightleftharpoons \underset{\mathbf{1}}{\mathrm{RC^-(OH)(CN)}}$$

$$\mathbf{1} + \mathrm{RCHO} \rightleftharpoons \mathrm{RC(OH)(CN)\text{-}CH(O^-)R} \rightleftharpoons \mathrm{RC(O^-)(CN)\text{-}CH(OH)R} \rightleftharpoons \mathrm{RCOCH(OH)R} + \mathrm{CN^-}$$

$$\mathbf{1} + \mathrm{R^1CH{=}CHCOR^2} \longrightarrow \mathrm{RC(OH)(CN)CHR^1CH{=}C(O^-)R^2} \rightleftharpoons$$

$$\mathrm{RC(O^-)(CN)CHR^1CH{=}C(OH)R^2} \longrightarrow \mathrm{RCOCHR^1CH_2COR^2} + \mathrm{CN^-} \quad \text{(Eq. 1)}$$

With thiazolium salts, the catalytic species is the ylide **2**, formed by proton abstraction under the influence of a base (Eq. 2); otherwise, the mechanism resembles that of the cyanide-ion-catalyzed reaction.

R^2, R^3, N^+, Cl^-, R^1, S + $(C_2H_5)_3N$ ⇌ R^2, R^3, N^+, R^1, S, – + $(C_2H_5)_3NH^+ Cl^-$

2

2 + RCHO ⇌ R^2, R^3, N^+, O^-, CHR, R^1, S ⇌ R^2, R^3, N^+, OH, C^-, R, R^1, S

3

3 + RCHO ⇌ R^2, R^3, N^+, HO O^-, C–CHR, R, R^1, S ⇌ R^2, R^3, N^+, O^- OH, C–CHR, R, R^1, S

⟶ RCOCHR(OH) + **2**

3 + CH_2=CHCOR^4 ⟶ R^2, R^3, N^+, HO, O^-, CCH$_2$CH=CR^4, R, R^1, S ⇌

R^2, R^3, N^+, O^-, OH, CCH$_2$CH=CR^4, R, R^1, S ⇌ R^2, R^3, N^+, O^-, CCH$_2$CH$_2$COR^4, R, R^1, S

⟶ RCO(CH$_2$)$_2$COR^4 + **2** (Eq. 2)

SCOPE AND LIMITATIONS

Catalysts

For cyanide-catalyzed additions, sodium or potassium cyanide and aprotic solvents such as dimethylformamide (DMF) or dimethyl sulfoxide (DMSO) are used. The reaction temperatures are in the range 30–35°. Any thiazolium salt, including thiamine (vitamin B_1), can be employed in the thiazolium-salt-catalyzed additions. However, quaternary salts of 5-(2-hydroxyethyl)-4-methyl-1,3-thiazole, an inexpensive intermediate in the industrial vitamin B_1 synthesis, are used in practice almost without exception. Among the reagents available for quaternization are methyl iodide, ethyl bromide, and benzyl chloride. 3-Benzyl-5-(2-hydroxyethyl)-4-methyl-1,3-thiazolium chloride (cat-

(Structure: CH_3 and $HO(CH_2)_2$ substituted thiazolium ring, N–R, X^-)

	R	X
Catalyst a	$CH_2C_6H_5$	Cl
b	CH_3	I
c	C_2H_5	Br
d	$(CH_2)_2OC_2H_5$	Br

alyst a) is best for the addition of aliphatic aldehydes,[2c,7] whereas 5-(2-hydroxyethyl-3,4-dimethyl-1,3-thiazolium iodide (catalyst b)[8] and 3-ethyl-5-(2-hydroxyethyl)-4-methyl-1,3-thiazolium bromide (catalyst c)[2c,7,9] are the catalysts of choice for the addition of aromatic aldehydes. Any one of these three catalysts is suitable for additions involving heterocyclic aldehydes. Catalyst d [3-(2-ethoxyethyl)-5-(2-hydroxyethyl)-4-methyl-1,3-thiazolium bromide] is particularly well suited for additions to α,β-unsaturated esters.[10] Another catalyst that has been proposed, 3-hydroxyethyl-1,3-thiazolium bromide,[11] does not appear to offer any special advantages. Attempts to attach the catalyst to a polymer have not always produced satisfactory results. The most important among these is the catalyst that is obtained by quaternization of 5-(2-hydroxyethyl)-4-methyl-1,3-thiazole with Merrifield's resin, which is a chloromethylated copolymer of styrene and divinylbenzene.[12a–c] This catalyst is well suited for the synthesis of acyloins,[13] but in additions of aldehydes it usually gives lower yields than soluble thiazolium salts.[12] In addition, these catalysts can be regenerated only partially so they rapidly lose their effectiveness.

Addition of metal salts or metal carbonyls either prevents the reaction or leads to different reactions.[14] Treatment of the products with chromium trioxide on silica gel may be used to remove any unpleasant odor caused by traces of the catalyst or its degradation products.[15]

Preparation of 1,4-Diketones

The Aldehyde Component. The thiazolium-salt-catalyzed addition of aliphatic aldehydes to α,β-unsaturated ketones usually proceeds in high yields. With straight-chain aldehydes, yields are in the range of 60 to 80%, but they decrease with α-branched aliphatic aldehydes such as isobutyraldehyde. Unsaturated aldehydes with both conjugated and isolated double bonds may be employed. The former lead to the little-studied class of δ,ε-unsaturated γ-diketones (Eq. 3).[16]

$$(CH_3)_2C{=}CHCHO + CH_2{=}CHCOCH_3 \xrightarrow[C_2H_5OH,\ 80^\circ]{\text{Cat. a}} (CH_3)_2C{=}CHCO(CH_2)_2COCH_3 \quad (52\%) \qquad \text{(Eq. 3)}$$

Acrolein cannot be added directly to α,β-unsaturated ketones. However, the same net result is achieved by employing the mixture of *exo*- and *endo*-

2-formyl-5-norbornenes which is readily obtained by Diels–Alder addition of acrolein to cyclopentadiene. The intermediate adduct is then subjected to retro diene cleavage (Eq. 4).[17]

$$\text{(2-formyl-5-norbornene, CHO)} + CH_2{=}CHCOCH_3 \xrightarrow{\text{Cat. a, }80^\circ} \text{(norbornenyl)}CO(CH_2)_2COCH_3\ (70\%)$$

$$\xrightarrow[\text{gas phase}]{500^\circ,} \text{(cyclopentadiene)} + CH_2{=}CHCO(CH_2)_2COCH_3\ (95\%) \qquad \text{(Eq. 4)}$$

Aldehydes with isolated triple bonds add readily to α,β-unsaturated ketones,[18] as do alkoxy-substituted aldehydes.[19,20] Other aliphatic aldehydes that have been used successfully include 3,4-dihydro-2*H*-pyran-2-carboxaldehyde, tetrahydropyran-2-carboxaldehyde, and tetrahydropyran-3-carboxaldehyde.[20] Attempts to employ aldoses as aldehyde components have been unsuccessful. On the other hand, D-glyceraldehyde can be used provided the hydroxy groups are masked by ketalization.[21]

Glyoxal, the simplest dialdehyde, has so far resisted thiazolium-salt-catalyzed addition; however, the monoacetal of glyoxal, 2,2-diethoxyacetaldehyde can be used successfully.[22] Higher aliphatic dialdehydes, from 1,4-butanedial to 1,10-decanedial, readily give the corresponding tetraketones.[8] Examples of ketoaldehydes include levulinic aldehyde[8] and 5-acetoxy-4-oxoheptanal.[23] Aliphatic ester aldehydes, such as methyl 4-formylbutanoate, methyl 5-formylpentanoate, and methyl 8-formyloctanoate add readily.[24] Esters of glyoxylic acid cannot be used since they suffer hydrolysis under the reaction conditions. Amides of glyoxylic acid, such as the dimethylamide or the pyrrolidide, present no problems.[25]

The successful employment of phthalimidoaldehydes is of particular importance since it leads to nitrogen-containing 1,4-diketones (Eq. 5).[26]

$$\text{PhthN}(CH_2)_nCHO + CH_2{=}CHCOR \longrightarrow \text{PhthN}(CH_2)_nCO(CH_2)_2COR \qquad \text{(Eq. 5)}$$

n = 1,2,3,5

The addition of aromatic aldehydes to α,β-unsaturated ketones is best carried out with cyanide ion catalysis in dimethylformamide, although thiazolium salt catalysis can, in principle, be used as well. Substituents on the aromatic ring do not prevent the addition. Exceptions are nitro-substituted aromatic aldehydes, which do not react, and *ortho*-substituted aromatic aldehydes, for which thiazolium salt catalysis must be used and which add only to vinyl ketones. Thus 2-chlorobenzaldehyde reacts with 3-buten-2-one under thiazolium ion catalysis to give the adduct in 56% yield; cyanide ions do not catalyze this reaction.[2c,27a,b] Alkoxy- and aryloxy-substituted benzaldehydes are another class for which cyanide catalysis often fails and for which thiazolium salt catalysis usually gives better results. Additions of 2,6-disubstituted benzaldehydes have not been reported, probably because of steric hindrance.

Either cyanide ion or thiazolium salt catalysis works well with heterocyclic aldehydes such as those of furan, thiophene, and pyridine. 2-Pyrrolecarboxaldehyde does not add, but dimethyl 2-formyl-4-methylpyrroledicarboxylate and 1-benzoylpyrrole-2-carboxaldehyde do.[28] With furan aldehydes, thiazolium salt catalysis is preferred since the products are obtained in high purity, whereas cyanide ion catalysis leads to partial resinification which makes the isolation and purification more difficult. Cyanide ion catalysis is preferred with thiophene- and pyridinealdehydes. An exception is pyridine-2-carboxaldehyde which gives better yields of adducts with thiazolium ion catalysis.[29] The reason probably lies in the stability imparted to 2-pyridoin by strong hydrogen bonding, which makes regeneration of the carbanion more difficult.

Vitamin B_1 (thiamine) catalyzes the conversion of pyruvic acid into acetoin in buffered aqueous solution.[30a–d] The decarboxylation is catalyzed by the thiazolium salt present in the vitamin. Both aliphatic and aromatic α-keto acids may therefore be substituted for the corresponding aldehydes in thiazolium salt catalyzed addition to α,β-unsaturated ketones (Eq. 6).[31]

$$n\text{-}C_3H_7COCO_2H \;+\; CH_2{=}CHCOCH_3 \xrightarrow[(C_2H_5)_3N,\ 80^\circ]{\text{Cat. a, } C_2H_5OH,} \underset{(79\%)}{n\text{-}C_3H_7CO(CH_2)_2COCH_3} \;+\; CO_2 \qquad \text{(Eq. 6)}$$

Acceptors. Most α,β-unsaturated ketones can serve as acceptors in the synthesis of 1,4-diketones. Vinyl ketones give especially good yields. A particular advantage in the preparation of unsymmetrical 1,4-diketones lies in the possibility of introducing the different group either via the aldehyde or the α,β-unsaturated ketone. This is illustrated in Eq. 7 for the synthesis of 1-(bicyclo[2.2.1]hept-5-enyl)-1,4-hexanedione by paths a[17] and b.[32]

Aromatic and heterocyclic α,β-unsaturated ketones are particularly well-suited acceptors since they are readily prepared by aldol condensation. They

$$\text{(5-norbornen-2-yl)CHO} + CH_2{=}CHCOC_2H_5 \xrightarrow{a} \text{(5-norbornen-2-yl)}CO(CH_2)_2COC_2H_5$$

$$C_2H_5CHO + \text{(5-norbornen-2-yl)}COCH{=}CH_2 \xrightarrow{b} \text{(5-norbornen-2-yl)}CO(CH_2)_2COC_2H_5 \quad \text{(Eq. 7)}$$

may be used to prepare 1,4-diketones with all possible permutations of substituent groups (Eqs. 8,[33] 9,[33] and 10[29]).

$$C_6H_5CHO + C_6H_5CH{=}CHCOCH_3 \xrightarrow[\text{DMF}]{\text{CN}^-,} C_6H_5COCH(C_6H_5)CH_2COCH_3 \quad \text{(Eq. 8)}$$

(80%)

$$p\text{-}ClC_6H_4CHO + C_6H_5CH{=}CHCOCH_3 \xrightarrow[\text{DMF}]{\text{CN}^-,} p\text{-}ClC_6H_4COCH(C_6H_5)CH_2COCH_3 \quad \text{(Eq. 9)}$$

(98%)

$$\text{3-pyridyl-CHO} + C_6H_5CH{=}CHCO\text{-(2-furyl)} \xrightarrow[\text{DMF}]{\text{CN}^-,} \text{3-pyridyl-}COCH(C_6H_5)CH_2CO\text{-(2-furyl)} \quad \text{(Eq. 10)}$$

(80%)

Cyclopentenone and cyclohexenone are poor acceptors in the catalyzed addition of aldehydes; they either do not react at all or give very poor yields.[34] This is surprising since they are normally excellent Michael acceptors. α,β-Unsaturated cyclic ketones with exocyclic double bonds, on the other hand, react very well.

The reaction is not suited for the conversion of α,β-unsaturated aldehydes into the corresponding 1,4-cyclohexanediones. The only known example of such a dimerization is mentioned in the section on γ-keto acids.

The sodium salts of 3-aroylacrylic acids, which are readily obtained by Friedel–Crafts addition of maleic anhydride to aromatic hydrocarbons,[35] can also serve as acceptors. The intermediate β-keto acid salts spontaneously decarboxylate to give the 1,4-diketones directly (Eq. 11).[36] The fact that

$$\text{2-thienyl-CHO} + NaO_2CCH{=}CHCO\text{-}C_6H_4\text{-}CH_3\text{-}p \longrightarrow \text{2-thienyl-}CO(CH_2)_2CO\text{-}C_6H_4\text{-}CH_3\text{-}p \quad (82\%) \qquad \text{(Eq. 11)}$$

3-aroylacrylic acids are in general more readily accessible than the corresponding vinyl ketones makes this variation particularly valuable.

In Michael additions, Mannich bases may often be used in place of α,β-unsaturated ketones. The same holds true for the catalyzed addition of aldehydes. Thus aliphatic aldehydes react with Mannich bases under thiazolium salt catalysis; the best yields are obtained by carrying out the addition in dimethylformamide at 80–90° (Eq. 12).[37] The optimum conditions for the

$$n\text{-}C_3H_7CHO + (CH_3)_2N(CH_2)_2COC_6H_5 \xrightarrow[-\ HN(CH_3)_2]{\text{Cat. b}} n\text{-}C_3H_7CO(CH_2)_2COC_6H_5 \quad (69\%) \qquad \text{(Eq. 12)}$$

corresponding reaction of aromatic and heterocyclic aldehydes are cyanide ion catalysis, dimethylformamide as the solvent, and temperatures in the range 35–100° (Eq. 13).[37]

$$C_6H_5CHO + (CH_3)_2N(CH_2)_2COC_3H_7\text{-}i \xrightarrow[-\ HN(CH_3)_2]{CN^-} C_6H_5CO(CH_2)_2COC_3H_7\text{-}i \quad (60\%) \qquad \text{(Eq. 13)}$$

The alkaloid gramine, which is readily prepared by the Mannich reaction from indole, formaldehyde, and dimethylamine, reacts with benzaldehyde under cyanide ion catalysis to give 2-(3-indolyl)acetophenone in 52% yield (Eq. 14).[37]

$$C_6H_5CHO + \text{3-}[(CH_3)_2NCH_2]\text{indole (N–H)} \xrightarrow[-\ HN(CH_3)_2]{CN^-} \text{3-}(C_6H_5COCH_2)\text{indole (N–H)} \quad (52\%) \qquad \text{(Eq. 14)}$$

An example involving a heterocyclic aldehyde is given in Eq. 15.[38]

$$\text{2-thienyl-CHO} + (CH_3)_2N(CH_2)_2CO\text{-2-thienyl} \xrightarrow[-\ HN(CH_3)_2]{CN^-} \text{2-thienyl-}CO(CH_2)_2CO\text{-2-thienyl}\ (75\%) \quad \text{(Eq. 15)}$$

Of particular interest in this context are thiazolium-salt-catalyzed additions involving Mannich bases of unsaturated ketones, which lead to the same unsaturated 1,4-diketones that are obtained from unsaturated aldehydes. In these reactions it is sometimes necessary to use the quaternary salts rather than the free Mannich bases.[37]

The thiazolium-salt-catalyzed addition of aldehydes to vinyl sulfones gives equal amounts of 1,4-diketones and 1,4-disulfones (Eq. 16).[39]

$$RCHO + CH_2{=}CHSO_2R^1 \longrightarrow RCO(CH_2)_2SO_2R^1$$

$$RCO(CH_2)_2SO_2R^1 \xrightarrow{\text{Base}} RCOCH{=}CH_2 \ \ \text{and} \ \ BH^+ + R^1SO_2^-$$

$$RCOCH{=}CH_2 \xrightarrow{RCHO} RCO(CH_2)_2COR \qquad BH^+ + R^1SO_2^- \xrightarrow{CH_2=CHSO_2R^1} R^1SO_2(CH_2)_2SO_2R^1 \quad \text{(Eq. 16)}$$

The reaction can be used to prepare symmetrical 1,4-diketones. The separation of the 1,4-diketones from the 1,4-disulfones is often difficult. It is best to use the commercially available divinyl sulfone, for which separation of the reaction products does not pose a problem. An added advantage is that only 0.5 molar equivalent of divinyl sulfone is required (Eq. 17).[39]

$$2\ n\text{-}C_{10}H_{21}CHO + CH_2{=}CHSO_2CH{=}CH_2 \xrightarrow[C_2H_5OH]{\text{Cat. b, }NaO_2CCH_3} n\text{-}C_{10}H_{21}CO(CH_2)_2COC_{10}H_{21}\text{-}n\ (69\%) + \text{polysulfones} \quad \text{(Eq. 17)}$$

Preparation of 4-Ketonitriles and 4,7-Diketonitriles

Aromatic and heterocyclic aldehydes readily add to α,β-unsaturated nitriles under cyanide ion catalysis to give 4-ketonitriles in high yields. Acrylonitrile, crotononitrile, methacrylonitrile, and cinnamonitrile are among the acceptors that have been used successfully. Aliphatic aldehydes require thiazolium salt catalysis; yields are in the range of 30–60%. Thiazolium salt catalysis is not recommended for aromatic or heterocyclic aldehydes since satisfactory results are obtained only with acrylonitrile and the yields are distinctly lower than the 50–90% typically obtained with cyanide ion catalysis (Eq. 18).

$$\text{2-furyl-CHO} + CH_2{=}CHCN \xrightarrow[\text{Cat. a}]{CN^-} \text{2-furyl-}CO(CH_2)_2CN \quad (63\text{-}68\%)^{40}\ (48\%)^{7} \qquad \text{(Eq. 18)}$$

4,7-Diketonitriles are obtained from the adduct of 5-norbornene-2-carboxaldehyde to acrylonitrile. Thermal cleavage gives a vinyl ketonitrile which is subjected to another addition of an aliphatic or aromatic aldehyde (Eq. 19).[41]

$$\text{5-norbornene-2-CHO} + CH_2{=}CHCN \xrightarrow{\text{Cat. a}} \text{5-norbornene-2-}CO(CH_2)_2CN\ (50\%)$$

$$\xrightarrow{500^\circ} \text{cyclopentadiene} + CH_2{=}CHCO(CH_2)_2CN \xrightarrow[\text{Cat. a}]{n\text{-}C_3H_7CHO} n\text{-}C_3H_7CO(CH_2)_2CO(CH_2)_2CN\ (72\%) \qquad \text{(Eq. 19)}$$

Preparation of 4-Ketocarboxylic Esters

The conditions for the cyanide-ion-catalyzed addition of aromatic and heterocyclic aldehydes to α,β-unsaturated esters are similar to those employed for α,β-unsaturated ketones. Acceptors include acrylates, methacrylates, crotonates, and cinnamates. Yields are usually lower than in the corresponding reactions that lead to 1,4-diketones. The reason is a side reaction that consumes the catalyst (Eq. 20);[42] it can be made the main reaction path by leaving out the aldehyde and using more than catalytic amounts of cyanide.[43a–c]

$$C_6H_5CH{=}CHCO_2C_2H_5 \underset{}{\overset{CN^-}{\rightleftharpoons}} C_6H_5\overset{CN}{\overset{|}{C}}H\overset{-}{C}HCO_2C_2H_5 \rightleftharpoons$$

$$C_6H_5\overset{CN}{\overset{|}{\underline{C}}}CH_2CO_2C_2H_5 \xrightarrow{C_6H_5CH=CHCO_2C_2H_5}$$

$$\begin{matrix} CN \\ | \\ C_6H_5CCH_2CO_2C_2H_5 \\ | \\ C_6H_5CH\overset{-}{C}HCO_2C_2H_5 \end{matrix} \xrightarrow{-\,C_2H_5O^-} \text{3-oxo-4-cyano-4,5-diphenylcyclopentanecarboxylate } (C_6H_5, NC, C_6H_5, =O, CO_2C_2H_5) \quad (72\%) \qquad \text{(Eq. 20)}$$

Yields in the cyanide-ion-catalyzed addition to α,β-unsaturated esters can be improved by using isopropyl or *tert*-butyl esters in place of methyl or ethyl esters. Thus, the yield is 19% higher in the addition of benzaldehyde to *tert*-butyl crotonate than in the reaction involving the corresponding ethyl ester (Eq. 21).[44]

$$C_6H_5CHO + CH_3CH{=}CHCO_2R \longrightarrow C_6H_5CH(CH_3)CH_2CO_2R \qquad \text{(Eq. 21)}$$

$$R = C_2H_5 \quad (33\%)$$
$$t\text{-}C_4H_9 \quad (52\%)$$

Yields in the addition of aliphatic, aromatic, and heterocyclic aldehydes to α,β-unsaturated esters catalyzed by thiazolium salts are generally lower. Catalyst d is best for this purpose; yields in the additions of aliphatic aldehydes to acrylates then are slightly above 50%.[10] α,β-Unsaturated esters with a single alkyl or aryl substituent in the β position do not normally react under thiazolium salt catalysis. Exceptions are the additions of glyoxylic acid pyrrolidide to crotonic, methacrylic, and cinnamic esters where yields in excess of 50% are obtained.[25] Other special cases are additions of aldehydes to fumarates, which proceed in good yield (Eq. 22).[10]

$$n\text{-}C_6H_{13}CHO + \begin{matrix} C_2H_5O_2C \diagdown \\ CH \\ \| \\ CH \\ \diagdown CO_2C_2H_5 \end{matrix} \xrightarrow{\text{Cat. a}} \begin{matrix} n\text{-}C_6H_{13}COCHCO_2C_2H_5 \\ | \\ CH_2CO_2C_2H_5 \end{matrix} \quad (73\%) \qquad \text{(Eq. 22)}$$

The reaction fails with maleic esters. Alkylidene- or arylidenemalonates, on the other hand, react readily (Eq. 23).[36] The adducts are easily converted into 4-ketocarboxylic acids by hydrolysis followed by decarboxylation. This method thus permits access to 4-ketocarboxylic acids that cannot be made by direct addition.

2-Furyl-CHO + $C_6H_5CH{=}C(CO_2C_2H_5)_2$ —Cat. c→ 2-Furyl-$COCH(C_6H_5)CH(CO_2C_2H_5)_2$ (Eq. 23)

(56%)

α-Methylene-γ-butyrolactone and α-methylenevalerolactone are excellent acceptors (Eq. 24).[10]

CH_3CHO + α-methylene-γ-butyrolactone —Cat. b→ α-(CH_3COCH_2)-γ-butyrolactone (Eq. 24)

(56%)

The only example of an intramolecular addition is a key step in the total synthesis of hirsutic acid (Eq. 25);[45] the reaction requires unusually large amounts of catalyst.

CN, RO_2C, OHC —Cat. b→ CN, RO_2CCH_2, O (Eq. 25)

Amides of 2,5-diketocarboxylic acids are obtained by thiazolium-salt-catalyzed addition of glyoxylamides to α,β-unsaturated ketones. Thus addition of the pyrrolidide of glyoxylic acid to methyl vinyl ketone gives the pyrrolidide of 2,5-diketohexanoic acid (Eq. 26); the latter is readily converted into the

(pyrrolidine)NCOCHO + $CH_2{=}CHCOCH_3$ —Cat. a→ (pyrrolidine)$NCOCO(CH_2)_2COCH_3$ (Eq. 26)

(83%)

ethyl ester.[25] Another example is the addition of dihydrocitronellal to 2-oxo-3-methyl-3-butenoates, which gives derivatives of 2,5-dioxododecanoic acid.[46]

Diketocarboxylic esters can be obtained by two routes. One involves the addition of formyl esters to vinyl ketones (Eq. 27),[24] and the other uses aldehydes and vinyl ketoesters as substrates (Eq. 28).[24] In this way, a large number of 5,8- and 9,12-diketocarboxylic esters are obtained from methyl 4-formylbutanoate and methyl 8-formyloctanoate.[24]

$$CH_3O_2C(CH_2)_2CHO + CH_2{=}CHCOCH_3 \xrightarrow{\text{Cat. a}} CH_3O_2C(CH_2)_2CO(CH_2)_2COCH_3 \text{ (70\%)} \quad \text{(Eq. 27)}$$

$$n\text{-}C_3H_7CHO + CH_2{=}CHCO(CH_2)_3CO_2CH_3 \xrightarrow{\text{Cat. a}} n\text{-}C_3H_7CO(CH_2)_2CO(CH_2)_3CO_2CH_3 \text{ (38\%)} \quad \text{(Eq. 28)}$$

Branched diketocarboxylic acids are obtained from the thiazolium-salt-catalyzed addition of aldehydes to α-alkylidene- or α-arylidene-β-ketoesters (Eq. 29).[36]

$$\text{2-furyl-CHO} + C_6H_5CH{=}C(COCH_3)(CO_2C_2H_5) \xrightarrow{\text{Cat. b}} \text{2-furyl-}COCH(C_6H_5)CH(COCH_3)(CO_2C_2H_5) \text{ (81\%)} \quad \text{(Eq. 29)}$$

Preparation of Tri- and Polyketo Compounds

Derivatives of 1,2,5-triketo compounds are obtained by thiazolium-salt-catalyzed addition of diethoxyacetaldehyde to α,β-unsaturated ketones (Eq. 30).[22]

$$(C_2H_5O)_2CHCHO + CH_2{=}CHCOCH_3 \xrightarrow[\text{dioxane}]{\text{Cat. a}} (C_2H_5O)_2CHCO(CH_2)_2COCH_3 \text{ (78\%)} \quad \text{(Eq. 30)}$$

1,3,6-Triketones are formed by thiazolium-salt-catalyzed addition of aldehydes to the readily available[47] monoacetals of vinyl-1,3-diketones followed by hydrolysis (Eq. 31).[47] Cyclic 1,3,6-triketones are accessible by the same method (Eq. 32).[47]

$$C_2H_5CHO + CH_2{=}CHCOCH_2\text{-(2,4,5-trimethyl-1,3-dioxolan-2-yl)} \xrightarrow[\text{dioxane}]{\text{Cat. a}}$$

$$C_2H_5CO(CH_2)_2COCH_2\text{-(2,4,5-trimethyl-1,3-dioxolan-2-yl)}\ (78\%) \xrightarrow{H^+} C_2H_5CO(CH_2)_2COCH_2COCH_3\ (78\%) \quad \text{(Eq. 31)}$$

$$C_6H_5CHO + CH_2{=}CHCO\text{-(4,5-dimethyl-1,3-dioxolane-2-spirocyclohexyl)} \xrightarrow{\text{Cat. b}}$$

$$C_6H_5CO(CH_2)_2CO\text{-(4,5-dimethyl-1,3-dioxolane-2-spirocyclohexyl)}\ (67\%) \xrightarrow{H^+} C_6H_5CO(CH_2)_2CO\text{-(2-oxocyclohexyl)}\ (71\%) \quad \text{(Eq. 32)}$$

There are several routes to 1,4,7-triketones. Formaldehyde adds to 2 molecules of methyl vinyl ketone under thiazolium salt catalysis to give 2,5,8-nonanetrione, albeit in low yields (Eq. 33).[48a,b] Levulinic aldehyde, which is an intermediate in this reaction, adds to α,β-unsaturated ketones in much better yields (Eq. 34).[8] A better route to symmetrical 1,4,7-triketones involves addition of aldehydes to divinyl ketone. Optimum yields are obtained with aliphatic aldehydes when no solvent is used (Eq. 35); with aromatic and heterocyclic aldehydes, dimethylformamide is the solvent of choice.[8]

$$CH_2O + 2\,CH_2{=}CHCOCH_3 \xrightarrow{\text{Cat. a}} CH_3CO(CH_2)_2CO(CH_2)_2COCH_3\ (27\%) \quad \text{(Eq. 33)}$$

$$CH_3CO(CH_2)_2CHO + CH_2{=}CHCOCH_3 \xrightarrow[\text{dioxane}]{\text{Cat. a}} CH_3CO(CH_2)_2CO(CH_2)_2COCH_3\ (60\%) \quad \text{(Eq. 34)}$$

$$2\ n\text{-}C_5H_{11}CHO + CH_2{=}CHCOCH{=}CH_2 \xrightarrow[\text{DMF}]{\text{Cat. a}} n\text{-}C_5H_{11}CO(CH_2)_2CO(CH_2)_2COC_5H_{11}\text{-}n \quad (65\%) \qquad \text{(Eq. 35)}$$

Dibenzylideneacetone (Eq. 36) and difurylideneacetone can be used in place of divinyl ketone.[8] Interestingly, aliphatic aldehydes add only once to these ketones, even when an excess of aldehyde is used. Addition of 2-furancarboxaldehyde to the reaction mixture leads to an unsymmetrical triketone (Eq. 37).[8]

$$2\ \text{(2-furyl)CHO} + C_6H_5CH{=}CHCOCH{=}CHC_6H_5 \xrightarrow[\text{DMF}]{\text{Cat. a}} \text{(2-furyl)}COCH(C_6H_5)CH_2COCH_2CH(C_6H_5)CO\text{(2-furyl)} \quad (54\%) \qquad \text{(Eq. 36)}$$

$$CH_3CHO + C_6H_5CH{=}CHCOCH{=}CHC_6H_5 \xrightarrow[\text{DMF}]{\text{Cat. a}} CH_3COCH(C_6H_5)CH_2COCH{=}CHC_6H_5 \xrightarrow{\text{(2-furyl)CHO}} CH_3COCH(C_6H_5)CH_2COCH_2CH(C_6H_5)CO\text{(2-furyl)} \quad (45\%) \qquad \text{(Eq. 37)}$$

Monoadditions to divinyl ketone produce vinyl-1,4-diketones in unsatisfactory yields. A much better route to unsymmetrical 1,4,7-triketones involves thermolysis of the adduct of 5-norbornene-2-carboxaldehyde to methyl vinyl ketone followed by a second aldehyde addition to the vinyl 1,4-diketone so obtained (Eq. 38).[17]

Branched triketones are formed in the thiazolium-salt-catalyzed addition of aldehydes to α-alkylidene-β-diketones (Eq. 39).[36] Another route starts with [bicyclo[2.2.1]hept-5-ene-2-carboxaldehyde] + $CH_2{=}CHCOCH_3$ $\xrightarrow{\text{Cat. a}}$

[bicycloheptenyl]$CO(CH_2)_2COCH_3$ (70%) $\xrightarrow{500^\circ}$ $CH_2{=}CHCO(CH_2)_2COCH_3$ (95%)

$\xrightarrow[\text{Cat. a}]{\text{2-furyl-CHO}}$ 2-furyl-$CO(CH_2)_2CO(CH_2)_2COCH_3$ (70%) (Eq. 38)

2-furyl-CHO + $C_6H_5CH{=}C(COCH_3)_2$ $\xrightarrow[C_2H_5OH]{\text{Cat. a}}$

2-furyl-$COCH(C_6H_5)CH(COCH_3)_2$ (63%) (Eq. 39)

the condensation of a 1,4-diketone monoketal with formaldehyde, followed by catalyzed addition of an aldehyde and hydrolysis of the ketal (Eq. 40).[49]

[4,5-dimethyl-2-methyl-1,3-dioxolane-2-yl]$(CH_2)_2COC_6H_5$ $\xrightarrow[-H_2O]{CH_2O}$ [4,5-dimethyl-2-methyl-1,3-dioxolane-2-yl]$CH_2C({=}CH_2)COC_6H_5$ (67%) $\xrightarrow[\text{Cat. a}]{\text{2-furyl-CHO}}$

[4,5-dimethyl-2-methyl-1,3-dioxolane-2-yl]$CH_2CH(COC_6H_5)CH_2CO$-2-furyl (57%) $\xrightarrow{H^+}$ $C_6H_5COCH(CH_2CO\text{-2-furyl})CH_2COCH_3$ (80%) (Eq. 40)

1,4,8-Triketones are obtained by the route outlined in Eq. 41;[50] this method can be adapted to the synthesis of 1,4,9-triketones.[50]

$$\text{(5-norbornene-2-COCl)} + \text{ClMg(CH}_2)_3\text{-(2,4,5-trimethyl-1,3-dioxolan-2-yl)} \longrightarrow$$

$$\text{(norbornenyl)CO(CH}_2)_3\text{-(dioxolanyl)}\ (66\%) \xrightarrow{H^+} \text{(norbornenyl)CO(CH}_2)_3\text{COCH}_3\ (92\%)$$

$$\xrightarrow{500^\circ} CH_2{=}CHCO(CH_2)_3COCH_3\ (90\%)$$

$$\xrightarrow[\text{Cat. a}]{CH_3CHO} CH_3CO(CH_2)_2CO(CH_2)_3COCH_3\ (78\%) \quad \text{(Eq. 41)}$$

A general method for the synthesis of tetraketones consists of thiazolium-salt-catalyzed addition of aliphatic, aromatic, or heterocyclic dialdehydes to vinyl ketones (Eq. 42).[51] Attempts to use glyoxaldehyde in this reaction have been unsuccessful. The expected 1,4,5,8-tetraketones are formed as side products in the addition of diethoxyacetaldehyde to vinyl ketones (Eq. 30).[22]

$$\text{HOC-(2,5-furandiyl)-CHO} + 2\ CH_2{=}CHCOCH_3 \xrightarrow{\text{Cat. a}}$$

$$CH_3CO(CH_2)_2CO\text{-(2,5-furandiyl)-}CO(CH_2)_2COCH_3\ (58\%) \quad \text{(Eq. 42)}$$

Symmetrical 1,4,7,10-tetraketones are accessible by the method outlined in Eq. 43; unsymmetrical tetraketones of this type are obtained in the same way.[32]

The synthesis of polyketones can be approached in several ways. Reaction of 2 equivalents of levulinic aldehyde with one equivalent of divinyl ketone gives 2,5,8,11,14-pentadecanepentaone (Eq. 44).[8] This method has been extended to the synthesis of a heptaketone (Eq. 45).[8] Unsymmetrical polyketones are synthesized by addition of 5-norbornene-2-carboxaldehyde to vinyl

CHO + $COCH{=}CH_2$ $\xrightarrow{\text{Cat. a}}$

$CO(CH_2)_2CO$ (61%) $\xrightarrow{500°}$ $CH_2{=}CHCO(CH_2)_2COCH{=}CH_2$ (100%)

$\xrightarrow[\text{Cat. a}]{2\ n\text{-}C_3H_7CHO}$ $n\text{-}C_3H_7CO(CH_2)_2CO(CH_2)_2CO(CH_2)_2COC_3H_7\text{-}n$ (69%) (Eq. 43)

$2\ CH_3CO(CH_2)_2CHO + CH_2{=}CHCOCH{=}CH_2 \xrightarrow[\text{DMF}]{\text{Cat. a}}$

$CH_3(COCH_2CH_2)_4COCH_3$ (40%) (Eq. 44)

CH_3–(furan)–$(CH_2)_2CHO$ + $CH_2{=}CHCOCH{=}CH_2 \xrightarrow[\text{DMF}]{\text{Cat. a}}$

CH_3–(furan)–$(CH_2)_2(COCH_2CH_2)_3$–(furan)–CH_3 $\xrightarrow{CH_3CO_2H,\ H_2SO_4}$

$CH_3(COCH_2CH_2)_6COCH_3$ (70%) (Eq. 45)

ketones, thermal cleavage of the adduct, another addition of 5-norbornene-2-carboxaldehyde, repetition of this sequence until the desired vinylpolyketone is obtained, and finally a last addition of an aldehyde. This sequence is illustrated with the preparation of 1-phenyl-1,4,7,10-undecanetetraone in Eq. 46.[52] The same principle allows preparation of triketocarboxylic esters and triketonitriles.

Branched triketo esters are obtained by thiazolium-salt-catalyzed addition of aldehydes to 3,3-diacylacrylates (Eq. 47); the latter are readily accessible from ethyl diethoxyacetate and β-diketones.[53] Branched triketo esters and diketo diesters are made in an analogous manner.

$$\text{(bicyclo[2.2.1]hept-5-ene-2-carbaldehyde)} + CH_2{=}CHCOCH_3 \xrightarrow{\text{Cat. a}} \text{(bicycloheptenyl)}CO(CH_2)_2COCH_3 \; (70\%)$$

$$\xrightarrow{500°} CH_2{=}CHCO(CH_2)_2COCH_3 \; (95\%) \xrightarrow[\text{Cat. a}]{\text{(bicycloheptenyl)CHO}}$$

$$\text{(bicycloheptenyl)}CO(CH_2)_2CO(CH_2)_2COCH_3 \xrightarrow{500°}$$

$$CH_2{=}CHCO(CH_2)_2CO(CH_2)_2COCH_3 \; (95\%) \xrightarrow[\text{Cat. b}]{C_6H_5CHO}$$

$$C_6H_5CO(CH_2)_2CO(CH_2)_2CO(CH_2)_2COCH_3 \; (89\%) \quad \text{(Eq. 46)}$$

$$\text{(2-furyl)}CHO + C_2H_5O_2CCH{=}C(COCH_3)_2 \xrightarrow{\text{Cat. a}}$$

$$\text{(2-furyl)}COCH(CO_2C_2H_5)CH(COCH_3)_2 \; (57\%) \quad \text{(Eq. 47)}$$

SYNTHETIC UTILITY

The 1,4-diketones accessible by the catalyzed addition of aldehydes to α,β-unsaturated ketones readily undergo intramolecular aldol condensation to give substituted cyclopentenones. This is particularly important because many natural products, such as jasmine-type fragrances and the prostaglandins, either contain cyclopentane rings or can be derived from this structural element. A synthesis of the natural product *cis*-jasmone is shown in Eq. 48.[9] Since many, even complex, 1,4-diketones are now easily accessible, a large number of substituted cyclopentenones can be prepared. Some are excellent perfumes. Thus, 3-(3,4-dihydro-2*H*-pyran-2-yl)-2-*n*-pentyl-2-cyclopenten-1-one, whose synthesis is shown in Eq. 49, has a very intense, jasmone-like fragrance.[20]

$$C_2H_5CH{=}CH(CH_2)_2CHO + CH_2{=}CHCOCH_3 \xrightarrow{\text{Cat. a}}$$

$$C_2H_5CH{=}CH(CH_2)_2CO(CH_2)_2COCH_3 \ (76\%) \xrightarrow{OH^-} \text{cyclopentenone product} \ (81\%) \quad \text{(Eq. 48)}$$

$$\text{dihydropyran-CHO} + CH_2{=}CHCOCH_2C_5H_{11}\text{-}n \xrightarrow{\text{Cat. a}}$$

$$\text{dihydropyranyl-}CO(CH_2)_2COCH_2C_5H_{11}\text{-}n \ (59\%) \longrightarrow \text{cyclopentenone product } (n\text{-}C_5H_{11}) \ (50\%) \quad \text{(Eq. 49)}$$

There are two ways that 1,4,8-triketones can undergo intramolecular aldol condensation. However, formation of the cyclohexenone clearly predominates over cyclization leading to five-membered rings (Eq. 50).[50]

$$n\text{-}C_3H_7CO(CH_2)_2CO(CH_2)_3COCH_3 \xrightarrow{OH^-} \text{2-}(n\text{-}C_3H_7COCH_2)\text{-3-}CH_3\text{-cyclohexenone} \ (74\%) \quad \text{(Eq. 50)}$$

Base-catalyzed cyclization of γ,ϵ-unsaturated 1,4-diketones is accompanied by double-bond isomerization (Eq. 51).[54]

$$CH_3CO(CH_2)_2COCH{=}C(CH_3)_2 \xrightarrow{OH^-} [CH_3CO(CH_2)_2C(O^-){=}CH{-}C(CH_3){=}CH_2] \longrightarrow \text{2-}(CH_2{=}C(CH_3))\text{-3-}CH_3\text{-cyclopentenone} \ (58\%) \quad \text{(Eq. 51)}$$

Ethers of 2-hydroxy-2-cyclopenten-1-one are obtained by aldol condensation of the readily accessible 1-alkoxy-2,5-alkanediones (Eq. 52).[20]

$$n\text{-}C_4H_9OCH_2CO(CH_2)_2COCH_3 \xrightarrow{OH^-} \text{2-}(n\text{-butoxy})\text{-3-methyl-2-cyclopenten-1-one}\ (70\%) \quad \text{(Eq. 52)}$$

2-Hydroxycyclopentenones, some of which are important flavoring agents, may be prepared as shown in Eq. 53.[55] Other 2-hydroxy-2-cyclopenten-1-ones with a variety of substituents in the 3 position can be derived from 1-acetoxy-3-buten-2-one, which is easily prepared from 2-butyne-1,4-diol. Examples of the application of the catalyzed addition of aldehydes to the synthesis of prostaglandins are found in references 11, 56, and 57.

$$CH_3CHO + CH_2{=}CHCOCH_2OAc \xrightarrow{\text{Cat. a}} CH_3CO(CH_2)_2COCH_2OAc$$

$$\xrightarrow{OH^-} \text{2-hydroxy-3-methyl-2-cyclopenten-1-one}\ (66\%) \quad \text{(Eq. 53)}$$

Another important use of 1,4-diketones is in the synthesis of heterocycles such as furans, pyrroles, thiophenes, and pyridazines. Examples are given in Eqs. 54,[58] 55,[51] and 56.[59]

Catalytic reduction of 4-ketonitriles leads to pyrrolines and pyrrolidines. Thus, depending on the reaction conditions, 4-oxo-4-(3-pyridyl)butyronitrile is converted into either nornicotine[60] or myosmine[61] (Eq. 57).

$$\text{3-thienyl-CHO} + \text{2-thienyl-}CH{=}CHCOC_6H_5 \xrightarrow{\text{Cat. b}}$$

$$\text{(2-thienyl)}CH(CO\text{-3-thienyl})CH_2COC_6H_5 \xrightarrow{CH_3NH_3^+\,Cl^-} \text{1-methyl-2-(3-thienyl)-3-(2-thienyl)-5-phenylpyrrole}\ (98\%) \quad \text{(Eq. 54)}$$

$$2\ \text{(2-furyl)CHO} \xrightarrow[\text{Cat. a}]{CH_2{=}CHSO_2CH{=}CH_2} \text{(2-furyl)}CO(CH_2)_2CO\text{(2-furyl)}\ (51\%) \xrightarrow[(CH_3CO)_2O]{H_2SO_4,} \text{2,2':5',2''-terfuran}\ (65\%) \quad \text{(Eq. 55)}$$

$$\text{(2-thienyl)CHO} + (CH_3)_2N(CH_2)_2CO\text{(2-thienyl)} \xrightarrow{CN^-} \text{(2-thienyl)}CO(CH_2)_2CO\text{(2-thienyl)}\ (70\%) \xrightarrow[NaHCO_3]{P_2S_{10},} \text{2,2':5',2''-terthiophene}\ (85\%) \quad \text{(Eq. 56)}$$

$$\text{(3-pyridyl)}CO(CH_2)_2CN \xrightarrow[C_2H_5OH,\ 25°]{H_2,\ \text{Raney Ni},\ NH_3,} \begin{cases} \text{24 h} \rightarrow \text{3-(2-pyrrolidinyl)pyridine}\ (60\%) \\ \text{20 h} \rightarrow \text{3-(1-pyrrolin-2-yl)pyridine}\ (69\%) \end{cases} \quad \text{(Eq. 57)}$$

Pyrrolidines are also accessible by reductive amination of 1,4-diketones. An example is the prepartion of *cis*- and *trans*-2-ethyl-4-*n*-pentylpyrrolidines, which are toxins of the Pharaoh ant (Eq. 58).[62]

$$C_2H_5CHO + CH_2{=}CHCOC_5H_{11}\text{-}n \longrightarrow C_2H_5CO(CH_2)_2COC_5H_{11}\text{-}n\ (50\%) \xrightarrow[\text{2. }NaBH_4]{\text{1. }Na(CN)BH_3, NH_4O_2CCH_3,\ KOH,\ CH_3OH}$$

$$\textit{cis}\text{-}C_2H_5\text{-}(C_4H_7NH)\text{-}C_5H_{11}\text{-}n + \textit{trans}\text{-}C_2H_5\text{-}(C_4H_7NH)\text{-}C_5H_{11}\text{-}n\ (70\%) \quad \text{(Eq. 58)}$$

Reductive amination of 1,4,7-triketones analogously leads to pyrrolizidines (Eq. 59).[23] The product is an ant toxin; other such toxins are obtained in a similar way.

$$n\text{-}C_7H_{15}CO(CH_2)_2CHO + CH_2{=}CHCOCH_3 \xrightarrow{\text{Cat. a}}$$

$$n\text{-}C_7H_{15}CO(CH_2)_2CO(CH_2)_2COCH_3 \xrightarrow[NH_4O_2CCH_3]{Na(CN)BH_3} \quad \text{(Eq. 59)}$$

H, CH_3, C_7H_{15}-n, N

The adducts of 2-pyridinecarboxaldehyde to α,β-unsaturated ketones are readily converted into quinolizidines by hydrogenation (Eq. 60).[29] Another

$$+ \; CH_2{=}CHCOCH_3 \xrightarrow{\text{Cat. c}}$$

N, CHO, N, $CO(CH_2)_2COCH_3$

$$\xrightarrow[CH_3CO_2H]{H_2,\ PtO_2,} \quad \text{(Eq. 60)}$$

OH, H, N, CH_3

(83%)

interesting approach to heterocycles is the reaction of 1,4,7-triketones with hydrazine in aqueous acetic acid to give dihydropyrrolo[1,2-*b*]pyridazines (Eq. 61).[63] Since the starting triketones are readily available by catalyzed aldehyde addition, this sequence constitutes the most expeditious synthesis of this ring system. Dehydrogenation with chloranil affords the fully aromatic pyrrolo-[1,2-*b*]pyridazines.

$$CH_3CO(CH_2)_2CO(CH_2)_2COCH_3 \xrightarrow[CH_3CO_2H]{N_2H_4,\ H_2O,} \quad \text{(Eq. 61)}$$

CH_3, N, N, CH_3

(83%)

The addition of phthalimidoaldehydes to α,β-unsaturated ketones leads to phthalimido-1,4-diketones, which are precursors to a number of condensed bicyclic and tricyclic pyrroles. Thus, reaction with 2-aminobenzophenone followed by hydrazinolysis gives a derivative of 4*H*-pyrrole[1,2-*a*]-[1,4]-benzodiazepine (Eq. 62).[64] The same ring system is obtained with anthranilates.[64] Urethane leads to the pyrrolo[1,2-*c*]imidazole ring system (Eq. 63),[64] and glycine esters to pyrrolo[1,2-*a*]pyrazines (Eq. 64).[64] The pyrrolo[1,2-*c*]-pyrimidine and pyrrolo[2,1-*d*]-[1,4]-diazepine ring systems are accessible by similar methods.[64]

$NCH_2CO(CH_2)_2COCH_3$ + NH_2, COC_6H_5

p-$CH_3C_6H_4SO_3H$, diglyme, reflux, 8 h → NCH_2, N, CH_3, COC_6H_5

N_2H_4, H_2O → CH_3, N, N, C_6H_5 (78%) (Eq. 62)

$NCH_2CO(CH_2)_2COCH_3$

1. $H_2NCO_2C_2H_5$ 2. N_2H_4 → HN, N, O (Eq. 63)

1. $H_2NCH_2CO_2R$ 2. N_2H_4 → HN, N, O (Eq. 64)

COMPARISON WITH OTHER METHODS

Many methods exist for the synthesis of 1,4-dicarbonyl compounds.[65] They are, almost without exception, noncatalytic. The much smaller number of methods that use aldehydes as starting materials also are, with the exception of radical-induced reactions, noncatalytic. Most involve a reversal of polarity (*umpolung*) of the carbonyl group. The best known of these employ anions of cyanohydrin derivatives and of α-dialkylaminonitriles.[66]

The most commonly used derivatives of cyanohydrins are the ethoxyethyl ethers obtained by acid-catalyzed addition to ethyl vinyl ether,[67] and the trimethylsilyl ethers formed by reaction of aldehydes with trimethylsilyl cyanide.[68] Generation of the anions requires organolithium reagents or lithium amides. Depending on the conditions, the reaction of these anions with α,β-unsaturated ketones leads to either 1,2 or 1,4 addition, and the carbonyl groups are regenerated in the last step by acid-catalyzed hydrolysis. Although this method is more circuitous, it does have advantages over the catalyzed addition of aldehydes in certain cases. For instance, mesityl oxide is not amenable to the latter, and cyclohexenones usually give unsatisfactory yields.

The more convenient method for *umpolung* of carbonyl groups employs α-dialkylaminonitriles, which are easily prepared from aldehydes by the Strecker reaction. The most commonly used secondary amine is morpholine. The anions, which are generated with sodium methoxide or with potassium hydroxide in ethanol, readily undergo Michael addition to acrylates and acrylonitriles. This method is superior to the catalyzed aldehyde addition in reactions involving substituted acrylates, as illustrated in Eqs. 65[44] and 66.[66]

$$C_6H_5CHO + CH_3CH{=}CHCO_2C_2H_5 \xrightarrow[DMF]{CN^-} \underset{(33\%)}{C_6H_5COCH(CH_3)CH_2CO_2C_2H_5} \quad \text{(Eq. 65)}$$

$$\underset{\text{(morpholino)}C_6H_5CHCN}{} + CH_3CH{=}CHCO_2CH_3 \xrightarrow{CH_3ONa,\ THF}$$

$$\underset{(98\%)}{\text{(morpholino)}C_6H_5C(CN)CH(CH_3)CH_2CO_2CH_3} \xrightarrow{CH_3CO_2H} \underset{(64\%)}{C_6H_5COCH(CH_3)CH_2CO_2CH_3} \quad \text{(Eq. 66)}$$

Michael additions of α-dialkylaminonitrile anions to α,β-unsaturated ketones, on the other hand, are less advantageous. Thus, reaction of α-dimethylaminophenylacetonitrile with benzalacetophenone in tetrahydrofuran with sodium methoxide as the base gives the diketone in only 38% yield. The yield can be increased by employing potassium amide in liquid ammonia.[69] A less commonly used method involves 1,4 addition of lithiodithioacetals to α,β-unsaturated ketones; cleavage of the thioacetals furnishes 1,4-diketones.[70]

The radical-catalyzed addition of aldehydes to olefins frequently proceeds with excellent yields, but additions to α,β-unsaturated esters and ketones are often unsatisfactory.[71] Acrylates and vinyl ketones cannot be employed since they are polymerized or telomerized under the reaction conditions. Radical catalyzed additions of aldehydes proceed well with maleates but not with fumarates.[72] The situation is reversed in the thiazolium-salt-catalyzed addition where only fumarates give satisfactory results. There are only a few examples of radical additions of aldehydes to α,β-unsaturated ketones; remarkable among these are additions to mesityl oxide, which proceed in good yield. By comparison, cyanide-ion or thiazolium-salt-catalyzed additions fail with α,β-unsaturated carbonyl compounds that have two alkyl substituents in the β position.

In summary, the cyanide-ion and thiazolium-salt-catalyzed addition of aldehydes stands out among all the other methods mentioned here by its simplicity and much wider range of application. It is inferior in only a few special cases. Its advantages are well illustrated by the two-step synthesis of the perfume dihydrojasmone (Eq. 67).[15] The process is easy to carry out and proceeds from starting materials to product with loss of only one water molecule. By comparison, other methods require more steps and employ auxiliary groups of varying complexity. The radical addition of aldehydes fails with vinyl ketones.

$$n\text{-}C_5H_{11}CH_2CHO + CH_2{=}CHCOCH_3 \xrightarrow[C_2H_5OH,\ 80°]{\text{Cat. a, } N(C_2H_5)_3,}$$

$$n\text{-}C_5H_{11}CH_2CO(CH_2)_2COCH_3 \ (75\%) \xrightarrow[-H_2O]{OH^-} \text{3-methyl-2-}(n\text{-}C_5H_{11})\text{cyclopent-2-enone} \ (88\%) \quad \text{(Eq. 67)}$$

EXPERIMENTAL CONDITIONS

Aprotic solvents are required for the cyanide-ion-catalyzed addition of aromatic and heterocyclic aldehydes to α,β-unsaturated ketones, esters, and nitriles. Dimethylformamide (DMF) is the solvent of choice; others include dimethyl sulfoxide (DMSO) and hexamethylphosphoramide (HMPA; *CAUTION*: this solvent is a suspected carcinogen and must be handled with utmost care). Dry, analytically pure sodium cyanide, and occasionally potassium cyanide, are used as catalysts. The quantity of catalyst is usually 0.1 equivalent although larger amounts may be advantageous in certain reactions. The reaction mixture is stirred under nitrogen at about 35° for generally 1–4 hours. The method of isolation varies with the reaction product.

The thiazolium-salt-catalyzed addition can be carried out in either protic or aprotic solvents. Ethanol, dioxane, and dimethylformamide are most commonly used. Often it is best to use no solvent at all, although it may prove difficult to control the exothermic reaction on a large scale. The best bases are triethylamine and sodium acetate. The amount of catalyst is generally 0.1 equivalent but it can be lowered to half that quantity in some cases. Reaction times are in the range of 6 to 16 hours at 60–80°. A stirrer, reflux condenser, and an inert nitrogen atmosphere are required.

EXPERIMENTAL PROCEDURES

A. Cyanide-Ion-Catalyzed Additions

4-(3-Pyridyl)-4-oxobutyronitrile (Cyanide-Ion-Catalyzed Addition of a Heterocyclic Aldehyde to an α,β-Unsaturated Nitrile). This preparation is described in *Organic Syntheses*.[73]

Ethyl 3-(4-Chlorobenzoyl)propionate (Cyanide-Ion-Catalyzed Addition of an Aromatic Aldehyde to an α,β-Unsaturated Ester).[44] A mixture of 28.1 g (200 mmol) of 4-chlorobenzaldehyde, 1.96 g (40 mmol) of sodium cyanide, and 160 mL of dimethylformamide (DMF) was stirred at room temperature for 1 hour, and 15 g (150 mmol) of ethyl acrylate in 80 mL of DMF was then added dropwise during 30 minutes. The mixture was stirred at room temperature for 2 hours, treated with 600 mL of water, and extracted several times with chloroform. The combined extracts were washed with dilute sulfuric acid, dilute aqueous sodium bicarbonate, and water. Removal of the solvent and distillation of the residue gave 24.5 g (68%) of ethyl 3-(4-chlorobenzoyl)propionate, bp 132–134° (0.1 mm), mp 58–59° (isopropyl alcohol).

2,4-Diphenyl-1-(2-thienyl)-1,4-butanedione (Cyanide-Ion-Catalyzed Addition of a Heterocyclic Aldehyde to an α,β-Unsaturated Ketone).[38] A solution of 14.5 g (125 mmol) of freshly distilled 2-thiophenecarboxaldehyde in 40 mL of DMF was added at room temperature during 15 minutes to a stirred mixture of 0.5 g (10 mmol) of sodium cyanide and 40 mL of DMF. After stirring another 15 minutes, a solution of 20.8 g (100 mmol) of benzylideneacetone in 100 mL of DMF was added dropwise at room temperature. The mixture was stirred another 2 hours, treated with 500 mL of water, and extracted several times with chloroform. The extracts were washed repeatedly with water until neutral. Removal of the solvent and crystallization of the residue from ethanol gave 29.6 g (90%) of 2,4-diphenyl-1-(2-thienyl)-1,4-butanedione, mp 142°.

1,4-Diphenyl-1,4-butanedione (Cyanide-Ion-Catalyzed Addition of an Aromatic Aldehyde to a Mannich Base).[37] Sodium cyanide (0.98 g, 20 mmole) and DMF (20 mL) were placed in a 250-mL three-necked flask equipped with a stirrer, reflux condenser with drying tube (KOH), pressure-equalizing addition funnel, and nitrogen inlet. A solution of 21.2 g (200 mmol) of benzaldehyde in 50 mL of DMF was added dropwise during 1.5 hours at a bath temperature of 35°. After stirring another 30 minutes, the temperature was raised to 100° and a solution of 35.4 g (200 mmol) of β-(dimethylamino)-propiophenone in 50 mL of DMF was added during 2 hours. The mixture was stirred another hour at 100°, poured into 500 mL of water and acidified with dilute hydrochloric acid to remove any unreacted Mannich base as the water-soluble hydrochloride. The mixture was extracted with four 100-mL portions of chloroform, the combined extracts were washed with dilute aqueous sodium bicarbonate and dried (Na_2SO_4). Removal of the solvent and recrystallization of the residue from petroleum ether gave 27.4 g (64%) of 1,4-diphenyl-1,4-butanedione, mp 144–146°.

B. Thiazolium-Salt-Catalyzed Additions

3-Benzyl-5-(2-hydroxyethyl)-4-methyl-1,3-thiazolium Chloride. The synthesis of this commercially available catalyst is described in *Organic Syntheses* (Note 1).[5b]

5-(2-Hydroxyethyl)-3,4-dimethyl-1,3-thiazolium Iodide.[8] A mixture of 143.2 g (1.00 mol) of 5-(2-hydroxyethyl)-4-methyl-1,3-thiazole, 142.0 g (1.00 mole) of iodomethane and 500 mL of dry acetonitrile was heated under reflux with exclusion of moisture (KOH drying tube) for 24 hours. The solvent was removed under vacuum and the residue was dissolved in 200 mL of isopropyl alcohol. Ether was added until the solution just started to become cloudy, and crystallization was induced by seeding or scratching. When the crystallization was complete, the product was collected by filtration, washed with ether, and dried under aspirator vacuum to give 225.3 g (79%) of 5-(2-hydroxyethyl)-3,4-dimethyl-1,3-thiazolium iodide, mp 86°.

The synthesis of **2,5-Undecanedione** by thiazolium-salt-catalyzed addition and its cyclization to **Dihydrojasmone** are described in *Organic Syntheses*.[15]

7,11-Dimethyl-10-dodecene-2,5-dione (Thiazolium-Salt-Catalyzed Addition of an Aliphatic Aldehyde to a Vinyl Ketone).[16] A mixture of 154.2 g (1 mol) of citronellal, 87.6 g (1.25 mole) of methyl vinyl ketone, 27.0 g (0.1 mol) of 3-benzyl-5-(2-hydroxyethyl)-4-methyl-1,3-thiazolium chloride (catalyst a), 32.8 g (0.4 mol) of anhydrous sodium acetate, and 500 mL of ethanol was stirred under nitrogen at a bath temperature of 80° for 15–17 hours. The cooled mixture was concentrated under aspirator vacuum, ether was added, and the organic phase was washed with very dilute sulfuric acid, dilute sodium bicarbonate, and water. Each aqueous phase was extracted with ether, and the combined organic layers were dried with magnesium sulfate and concentrated. The residue was distilled to give 179.4 g (80%) of 7,11-dimethyl-10-dodecen-2,5-dione, bp 97° (0.26 mm).

1-(2-Furyl)-2,4-diphenyl-1,4-butanedione (Thiazolium-Salt-Catalyzed Addition of a Heterocyclic Aldehyde to an α,β-Unsaturated Ketone).[7] A mixture of 104.2 g (0.5 mol) of benzylideneacetophenone, 48 g (0.5 mol) of 2-furancarboxaldehyde, 6.3 g of 3-ethyl-5-(2-hydroxyethyl)-4-methyl-1,3-thiazolium bromide (catalyst c),* 15.2 g (0.15 mol) of triethylamine, and 250 mL of ethanol was heated under reflux under nitrogen for 12 hours. The mixture was allowed to cool to room temperature with stirring, the precipitate was collected by filtration, washed with ethanol, and dried to give 138.2 g (91%) of 1-(2-furyl)-2,4-diphenyl-1,4-butanedione, mp 114–115°.

4,7,10-Tridecanetrione (Thiazolium-Salt-Catalyzed Double Addition of an Aliphatic Aldehyde to Divinylketone in the Absence of a Solvent).[8] A mixture of 21.6 g (300 mmol) of butanal, 10.3 g (125 mmol) of divinyl ketone, 6.7 g (25 mmol) of 3-benzyl-5-(2-hydroxyethyl)-4-methyl-1,3-thiazolium chloride (catalyst a), and 30.4 g (300 mmol) of triethylamine was heated to 65° under nitrogen for 6 hours. The mixture was cooled, dissolved in chloroform, and washed once with aqueous sodium bicarbonate solution and brine. The aqueous phases were each extracted with chloroform, and the combined chloroform phases were dried ($MgSO_4$) and concentrated. Distillation of the residue gave 18.4 g (65%) of 4,7,10-tridecanetrione, bp 125° (0.3 mm), mp 78°.

1-(Bicyclo[2.2.1]hept-5-en-2-yl)-1,4-octanedione and 1-Decene-3,6-dione (Thiazolium-Salt-Catalyzed Addition of a Masked α,β-Unsaturated Aldehyde to an α,β-Unsaturated Ketone Followed by Thermal Cleavage to an Unsaturated Diketone).[17] A mixture of 122.2 g (1 mol) of bicyclo[2.2.1]hept-5-ene-2-carboxaldehyde, 112.2 g (1 mol) of 1-hepten-3-one, 13.5 g (0.05 mol) of 3-benzyl-5-(2-hydroxyethyl)-4-methyl-1,3-thiazolium chloride (catalyst a), and 50.6 g (0.5 mol) of triethylamine was stirred under nitrogen at 65° for 15 hours. Isolation, as described in the previous preparation, gave 192.2 g (82%) of 1-(bicyclo[2.2.1]hept-5-enyl)-1,4-octanedione, bp 107° (0.001 mm).

The thermal cleavage was carried out in a vertical quartz tube (300 × 13 mm) filled with 4 × 4 mm pieces of quartz tubing. The upper end of the tube carried a dropping funnel with a ground glass stopcock; the lower end was connected in series to a receiving flask, a reflux condenser, and a trap cooled with dry ice/acetone. The system was evacuated to 10 mm, the tube was heated to 500°, and 82 g (0.35 mol) of 1-(bicyclo[2.2.1]hept-5-enyl)-1,4-octanedione was fed into the tube at a rate of about 12 drops per minute. The product collected in the receiving flask, whereas the cyclopentadiene was condensed in the dry-ice trap. The crude product was distilled in the presence of 1% of hydroquinone to give 54.2 g (92%) of 1-decene-3,6-dione, bp 75° (0.6 mm).

Methyl 4-(Bicyclo[2.2.1]hept-5-en-2-yl)-4-oxobutanoate and Methyl 4-Oxo-5-hexenoate (Thiazolium-Salt-Catalyzed Addition of a Masked α,β-Unsaturated Aldehyde to an α,β-Unsaturated Ester Followed by Thermal Cleavage to an Unsaturated Keto Ester).[41] A mixture of 366.6 g (3 mol) of

* Catalyst c may be replaced by catalysts a or b without decrease in yields.

bicyclo[2.2.1]hept-5-ene-2-carboxaldehyde, 258.3 g (3 mol) of methyl acrylate, 40.5 g (0.15 mol) of 3-benzyl-5-(2-hydroxyethyl)-4-methyl-1,3-thiazolium iodide (catalyst a), and 101.2 g (1 mol) of triethylamine was heated under nitrogen with stirring to 65° for 15 hours. Isolation, as described above, gave 318.7 g (51%) of methyl 4-(bicyclo[2.2.1]hept-5-enyl)-4-oxo-5-hexenoate, bp 110° (0.7 mm). Pyrolysis of 50 g of this product as described in the preceding preparation gave 32.4 (95%) of methyl 4-oxo-5-hexenoate, bp 47° (0.3 mm).

1,4-Bis(2-furyl)-1,4-butanedione (Thiazolium-Salt-Catalyzed Addition of a Heterocyclic Aldehyde to Divinyl Sulfone.[39] The addition was carried out in a three-necked flask fitted with a stirrer, a pressure-equalizing addition funnel, a reflux condenser protected by a calcium-chloride tube, and a gas inlet tube. A mixture of 19.2 g (0.2 mol) of 2-furancarboxaldehyde, 4.9 g (0.06 mol) of sodium acetate, 5.7 g (0.02 mol) of 5-(2-hydroxyethyl)-3,4-dimethyl-1,3-thiazolium chloride (catalyst b) and 200 ml of ethanol was placed in the flask and heated under a slow stream of nitrogen to 80° bath temperature. A solution of 11.8 g (0.1 mol) of divinyl sulfone in 50 mL of ethanol was then added during 3 hours and the mixture was heated under reflux for another 12 hours. The mixture was filtered while hot, and the solids were extracted several times with chloroform. The combined filtrate and chloroform extracts were concentrated under aspirator vacuum and the residue was dissolved in 200 mL of chloroform. The solution was washed with aqueous sodium bicarbonate solution and water, and the aqueous phases were reextracted with chloroform. The combined organic phases were dried with sodium sulfate and concentrated. Crystallization of the residue from ethanol gave 16.4 g (75%) of 1,4-bis(2-furyl)-1,4-butanedione, mp 131°.

TABULAR SURVEY

The literature has been searched through December 1988. The tables are arranged according to product type. Tables I–IV list reactions that produce 1,4-diketones. Table I contains catalyzed additions of aldehydes to simple α,β-unsaturated ketones including their Mannich-base equivalents. Table II lists catalyzed additions of aldehydes to divinyl sulfone, a reaction that also produces 1,4-diketones. Catalyzed additions of α-keto acids to α,β-unsaturated ketones, which lead to 1,4-diketones, are collected in Table III. Table IV contains all additions that give tri- and polyketones; these include double additions of α,β-unsaturated ketones to formaldehyde, double additions of aldehydes to divinyl ketones and other (bis)-α,β-unsaturated ketones, and additions of aldehydes to α,β-unsaturated ketones containing additional ketonic carbonyl groups in other parts of the acceptor molecule. Table V lists additions that produce 4-ketocarboxylic acids, esters, and lactones; reactions leading to 4-ketonitriles are found in Table VI. Table VII contains the lone example of an addition of an aldehyde to an acceptor not covered in Tables I–VI.

In additions to Michael acceptors containing electron-withdrawing groups on both carbons of the double bond, the dominant group determines in which table the reaction is listed. Reactions involving Michael acceptors with two electron-withdrawing groups on the same carbon atom of the double bond are listed in both appropriate tables.

Within each table, entries are arranged in the order of increasing carbon count of the aldehyde; within each aldehyde listing, the entries are arranged in the order of increasing carbon number of the Michael acceptor; Mannich bases are listed under the carbon count of their Michael-acceptor equivalents.

Where no solvent is mentioned in the Conditions column, reactions were carried out neat. Reaction times have been omitted. These are usually 12–15 hours for thiazolium-salt-catalyzed reactions. Cyanide-catalyzed reactions involve three phases: stirring of the aldehyde with the catalyst, addition of the Michael acceptor, followed by continued stirring to complete the reaction. A dash (—) indicates that no yield was reported. The following abbreviations are used in the tables:

Ac	acetyl
Cat. a	3-benzyl-5-(2-hydroxyethyl)-4-methyl-1,3-thiazolium chloride
Cat. b	5-(2-hydroxyethyl)-3,4-dimethyl-1,3-thiazolium iodide
Cat. c	3-ethyl-5-(2-hydroxyethyl)-4-methyl-1,3-thiazolium bromide
Cat. d	3-(ethoxyethyl)-5-(2-hydroxyethyl)-4-methyl-1,3-thiazolium bromide
Cat. e	5-(2-hydroxyethyl)-4-methyl-3-(polystyrylmethyl)-1,3-thiazolium chloride
Cat. f	2-hydroxyethyl-1,3-thiazolium bromide
C_4H_3O	2-furyl
C_4H_3O-3	3-furyl
C_4H_3S	2-thienyl
C_4H_3S-3	3-thienyl
C_4H_8N	*N*-pyrrolidinyl
C_5H_4N	2-pyridinyl
C_5H_4N-3	3-pyridinyl
C_5H_4N-4	4-pyridinyl
DMF	dimethylformamide
EtOH	ethanol
PhThN	*N*-phthalimidyl
rt	room temperature
TEA	triethylamine
THP	2-tetrahydropyranyl

TABLE I. Catalyzed Additions of Aldehydes to α,β-Unsaturated Ketones

Aldehyde	Michael Acceptor	Conditions	Product(s) and Yield(s) (%)	Refs.
C_2				
CH_3CHO	$CH_2{=}CHCOCH_3$	Cat. a, TEA, 80°	$CH_3CO(CH_2)_2COCH_3$ (61)	48
	$CH_2{=}CHCOCH_3$	Cat. e, EtOH, reflux	" (61)	12c
	$CH_2{=}CHCO(CH_2)_2OCH_3$	Cat. a, dioxane, TEA, 90°	$CH_3CO(CH_2)_2CO(CH_2)_2OCH_3$ (80)	19
	$CH_2{=}CHCOCH_2O_2CCH_3$	Cat. a, dioxane, TEA, 80°	$CH_3CO(CH_2)_2COCH_2O_2CCH_3$ (70)	20
	"	Cat. e, i-C_3H_7OH, TEA, 82°	" (25)	74
	$I^-(CH_3)_3N^+(CH_2)_2COCH{=}C(CH_3)_2$	Cat. c, DMF, TEA, 80°	$CH_3CO(CH_2)_2COCH{=}C(CH_3)_2$ (23)	37
	$CH_2{=}CHCO(CH_2)_2OC_3H_7$-$n$	Cat. a, dioxane, TEA, 90°	$CH_3CO(CH_2)_2CO(CH_2)_2OC_3H_7$-$n$ (74)	19
	$CH_2{=}CH[CO(CH_2)_2]_2CN$	Cat. a, TEA, 65°	$CH_3[CO(CH_2)_2]_3CN$ (71)	52
	$CH_2{=}CHCO(CH_2)_3$— (1,3-dioxolan-2-yl; O, O)	Cat. a, TEA, 65°	$CH_3CO(CH_2)_2CO(CH_2)_3$— (1,3-dioxolan-2-yl; O, O) (82)	50
	$CH_2{=}CHCOCH_2$ (2-CH_3-4,5-di-CH_3-1,3-dioxolan-2-yl; CH_3, CH_3, O, O, CH_3)	Cat. a, dioxane, TEA, 80°	$CH_2{=}CHCOCH_2$ (2-CH_3-4,5-di-CH_3-1,3-dioxolan-2-yl; CH_3, CH_3, O, O, CH_3) (79)	47
	(E)-$C_6H_5CH{=}CHCOCH_3$	Cat. a, EtOH, TEA, 80°	$CH_3COCH(C_6H_5)CH_2COCH_3$ (20)	7
	"	Cat. e, EtOH, TEA, reflux	" (18)	12c
	$CH_2{=}CHCOC_6H_4OCH_3$-p	Cat. a, dioxane, TEA, 90°	$CH_3CO(CH_2)_2COC_6H_4OCH_3$-$p$ (80)	19
	$CH_2{=}CHCO(CH_2)_3$ (2-CH_3-1,3-dioxolan-2-yl; O, O, CH_3)	Cat. a, TEA, 65°	$CH_3CO(CH_2)_2CO(CH_2)_3$ (2-CH_3-1,3-dioxolan-2-yl; O, O, CH_3) (83)	50
	R, R (4,5-positions), O, O, R^1: $CH_2CH{=}C(CO_2CH_3)COCH_3$	Cat. a, EtOH, TEA, 80°	R, R (4,5-positions), O, O, R^1: $CH_2CH(COCH_3)CH(CO_2CH_3)COCH_3$	
	R = H, R′ = CH_3		(44)	49
	R = CH_3, R′ = H		(45)	49

TABLE I. CATALYZED ADDITIONS OF ALDEHYDES TO α,β-UNSATURATED KETONES (*Continued*)

Aldehyde	Michael Acceptor	Conditions	Product(s) and Yield(s) (%)	Refs.
	$CH_2{=}CHCOC_6H_3(OCH_3)_2$-2,4	Cat. a, dioxane, TEA, 90°	$CH_3CO(CH_2)_2COC_6H_3(OCH_3)_2$-2,4 (73)	19
	$CH_2{=}CHCO(CH_2)_2CO(CH_2)_2CO_2CH_3$	Cat. a, TEA, 65°	$CH_3[CO(CH_2)_2]_2CO(CH_2)_2CO_2CH_3$ (76)	52
	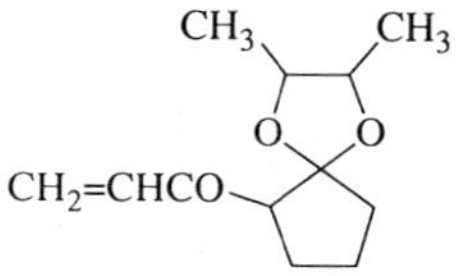 $CH_2{=}CHCO$–, CH_3, CH_3, O, O	Cat. a, dioxane, TEA, 80°	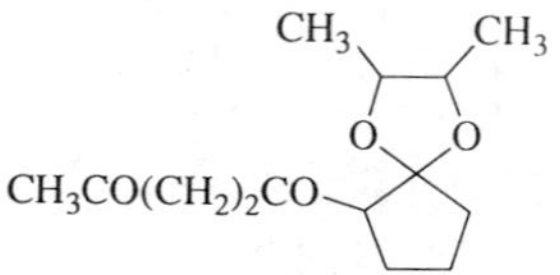 $CH_3CO(CH_2)_2CO$–, CH_3, CH_3, O, O (70)	47
	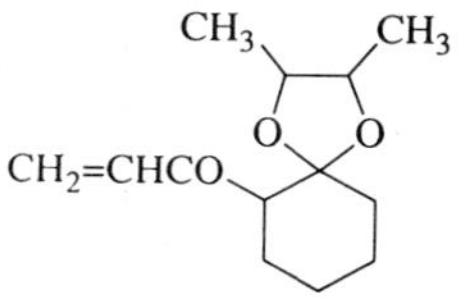 $CH_2{=}CHCO$–, CH_3, CH_3, O, O	Cat. a, dioxane, TEA, 80°	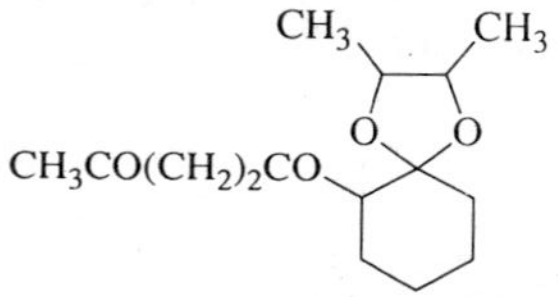 $CH_3CO(CH_2)_2CO$–, CH_3, CH_3, O, O (71)	47
	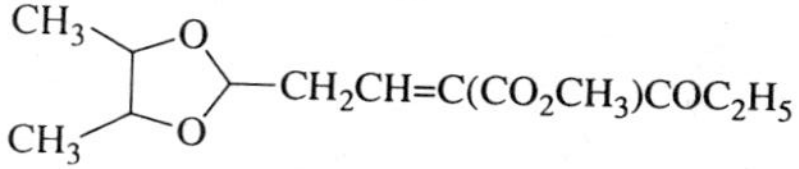 CH_3, CH_3, O, O, –$CH_2CH{=}C(CO_2CH_3)COC_2H_5$	Cat. a, EtOH, TEA, 80°	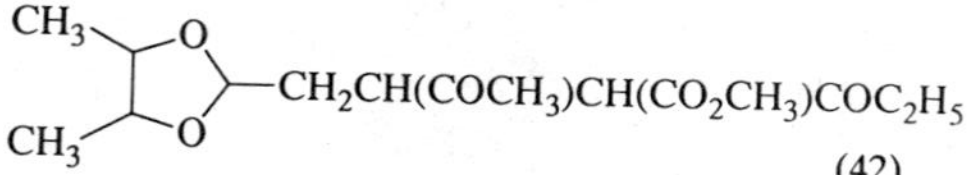 CH_3, CH_3, O, O, –$CH_2CH(COCH_3)CH(CO_2CH_3)COC_2H_5$ (42)	49
	 $CH_2{=}CRCO$–, O; R = –CH_2–, CH_3, O, O, CH_3, CH_3 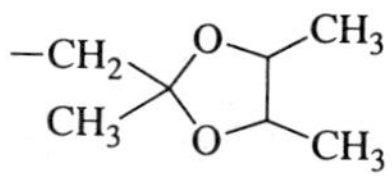	Cat. a, EtOH, TEA, 80°	$CH_3COCH_2CHRCOC_4H_3O$ (52)	49
	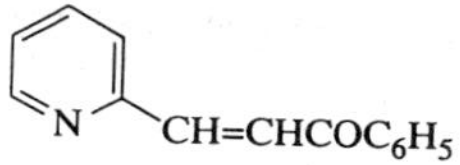 N, –$CH{=}CHCOC_6H_5$	Cat. a, dioxane, TEA, 80°	$CH_3COCH(C_5H_4N\text{-}2)CH_2COC_6H_5$ (70)	29
	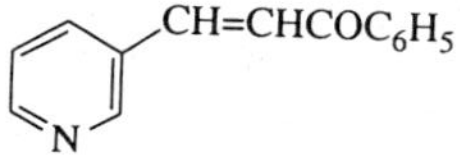 N, –$CH{=}CHCOC_6H_5$	Cat. a, DMF, TEA, 80°	$CH_3COCH(C_5H_4N\text{-}3)CH_2COC_6H_5$ (69)	29

C_3	$C_6H_5CH{=}CHCOCH{=}CHC_6H_5$	Cat. a, DMF, TEA, 80°	$CH_3COCH(C_6H_5)CH_2COCH{=}CHC_6H_5$ (85)	8
C_2H_5CHO	$CH_2{=}CHCOCH_3$	Cat. a, EtOH, TEA, 80°	$C_2H_5CO(CH_2)_2COCH_3$ (60)	48
	$CH_2{=}CHCOC_2H_5$	Cat. a, TEA, 88°	$C_2H_5CO(CH_2)_2COC_2H_5$ (48)	75
	$CH_2{=}C(CH_3)COCH_3$	Cat. e, EtOH, TEA, 80°	$C_2H_5COCH_2CH(CH_3)COCH_3$ (28)	12a
	$CH_2{=}CHCO(CH_2)_2CN$	Cat. a, TEA, 65°	$C_2H_5CO(CH_2)_2CO(CH_2)_2CN$ (72)	41
	$CH_2{=}CHCOCH_2O_2CCH_3$	Cat. a, dioxane, TEA, 80°	$C_2H_5CO(CH_2)_2COCH_2O_2CCH_3$ (72)	20
	$CH_2{=}CHCO(CH_2)_2OCH_3$	Cat. a, dioxane, TEA, 90°	$C_2H_5CO(CH_2)_2CO(CH_2)_2OCH_3$ (76)	19
	O	Cat. a, dioxane, TEA, 80°	$C_2H_5COCH_2-$ O (67)	76
	(*E*)-$C_2H_5O_2CCH{=}CHCOCH_3$	Cat. a, dioxane, TEA, 80°	$C_2H_5COCH(CO_2C_2H_5)CH_2COCH_3$ (71)	36
	$CH_2{=}CHCO(CH_2)_2CO_2CH_3$	Cat. a, dioxane, TEA, 100°	$C_2H_5[CO(CH_2)_2]_2CO_2CH_3$ (50)	24
	$CH_2{=}CHCOC_5H_{11}$-*n*	Cat. a, TEA, 90°	$C_2H_5CO(CH_2)_2COC_5H_{11}$-*n* (50)	62
	$CH_3CH{=}C(CO_2C_2H_5)COCH_3$	Cat. a, dioxane, TEA, 80°	$C_2H_5COCH(CH_3)CH(CO_2C_2H_5)COCH_3$ (49)	36
	$C_2H_5O_2CCH{=}C(COCH_3)_2$	Cat. a, dioxane, TEA, 80°	$C_2H_5COCH(CO_2C_2H_5)CH(COCH_3)_2$ (43)[a]	53
	O	Cat. a, EtOH, TEA, 80°	$C_2H_5COCH_2-$ O (74)	76
	$CH_2{=}CHCO$	Cat. a, dioxane, TEA, 65°	$C_2H_5CO(CH_2)_2CO$ (72)	32
	CH_3 CH_3 O O $CH_2{=}CHCOCH_2$ CH_3	Cat. a, dioxane, TEA, 80°	CH_3 CH_3 O O $C_2H_5CO(CH_2)_2COCH_2$ CH_3 (78)	47
	$C_2H_5O_2CCH{=}C(CO_2C_2H_5)COCH_3$	Cat. a, dioxane, TEA, 80°	$C_2H_5COCH(CO_2C_2H_5)CH(CO_2C_2H_5)COCH_3$ (43)[a]	53
	$CH_2{=}CH[CO(CH_2)_2]_2CO_2CH_3$	Cat. a, TEA, 65°	$C_2H_5[CO(CH_2)_2]_3CO_2CH_3$ (73)	52

TABLE I. CATALYZED ADDITIONS OF ALDEHYDES TO α,β-UNSATURATED KETONES (*Continued*)

Aldehyde	Michael Acceptor	Conditions	Product(s) and Yield(s) (%)	Refs.
	$CH_2{=}CHCOC_6H_4OCH_3$-*p*	Cat. a, dioxane, TEA, 100°	$C_2H_5CO(CH_2)_2COC_6H_4OCH_3$-*p* (64)	19
	$CH_2{=}CHCOC_6H_3(OCH_3)_2$-3,4	Cat. a, dioxane, TEA, 100°	$C_2H_5CO(CH_2)_2COC_6H_3(OCH_3)_2$-3,4 (60)	19
	$CH_2{=}CHCOC_6H_3(OCH_3)_2$-2,4	Cat. a, dioxane, TEA, 100°	$C_2H_5CO(CH_2)_2COC_6H_3(OCH_3)_2$-2,4 (60)	19
	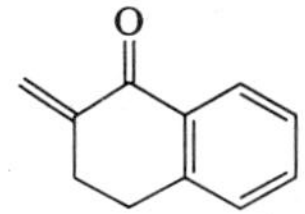 O (structure)	Cat. a, EtOH, TEA, 80°	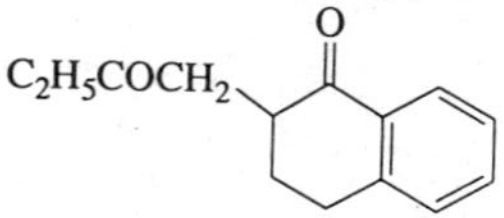 $C_2H_5COCH_2$ O (structure) (63)	76
	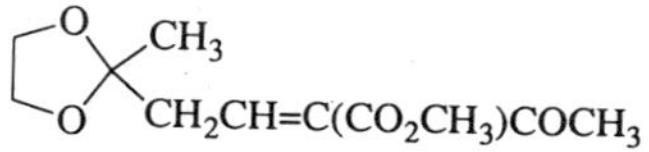 O, O, CH_3, $CH_2CH{=}C(CO_2CH_3)COCH_3$ (structure)	Cat. a, EtOH, TEA, 80°	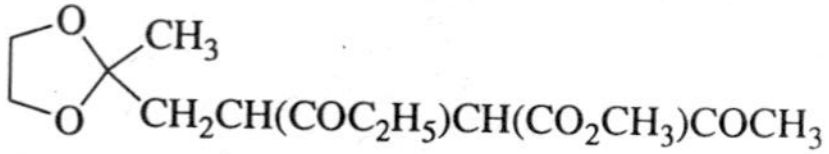 O, O, CH_3, $CH_2CH(COC_2H_5)CH(CO_2CH_3)COCH_3$ (structure) (44)	49
	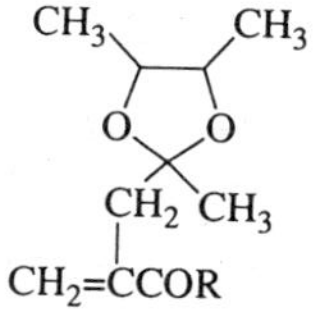 CH_3, CH_3, O, O, CH_2, CH_3, $CH_2{=}CCOR$ (structure)	Cat. a, EtOH, TEA, 80°	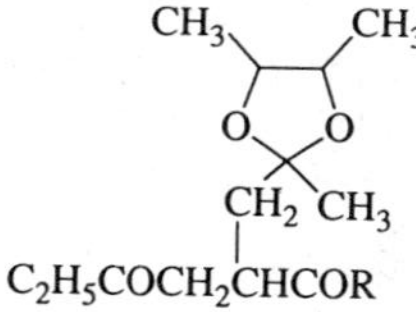 CH_3, CH_3, O, O, CH_2, CH_3, $C_2H_5COCH_2CHCOR$ (structure)	
	R			
	$COCH_3$		(30)	49
	COC_6H_4Cl-*p*		(48)	49
	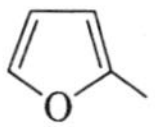 O (structure)		(53)	49
	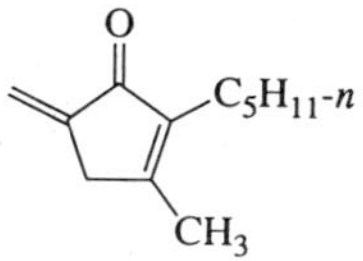 O, C_5H_{11}-*n*, CH_3 (structure)	Cat. a, EtOH, TEA 80°	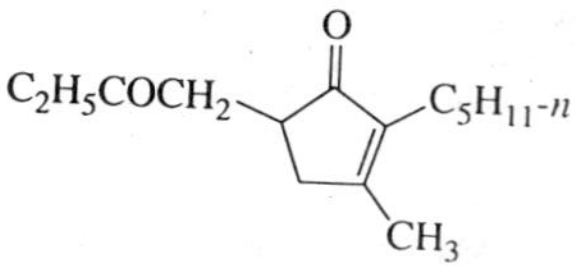 $C_2H_5COCH_2$, O, C_5H_{11}-*n*, CH_3 (structure) (70)	76

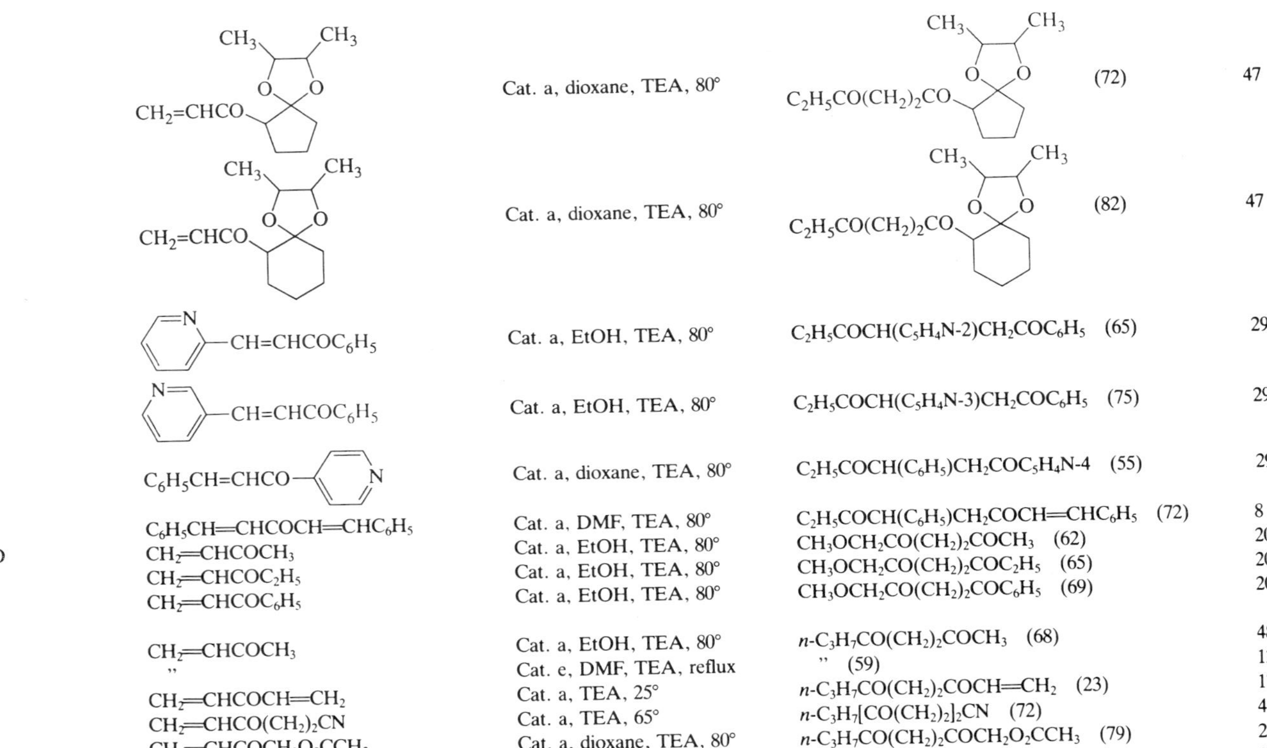

	Aldehyde	Acceptor	Conditions	Product(s) (Yield, %)	Refs.
		$CH_2{=}CHCO$-(2-substituted 2,3-dimethyl-1,4-dioxaspiro[4.4]nonane; structure)	Cat. a, dioxane, TEA, 80°	$C_2H_5CO(CH_2)_2CO$-(2,3-dimethyl-1,4-dioxaspiro[4.4]nonane; structure) (72)	47
		$CH_2{=}CHCO$-(2-substituted 2,3-dimethyl-1,4-dioxaspiro[4.5]decane; structure)	Cat. a, dioxane, TEA, 80°	$C_2H_5CO(CH_2)_2CO$-(2,3-dimethyl-1,4-dioxaspiro[4.5]decane; structure) (82)	47
		2-pyridyl-$CH{=}CHCOC_6H_5$	Cat. a, EtOH, TEA, 80°	$C_2H_5COCH(C_5H_4N\text{-}2)CH_2COC_6H_5$ (65)	29
		3-pyridyl-$CH{=}CHCOC_6H_5$	Cat. a, EtOH, TEA, 80°	$C_2H_5COCH(C_5H_4N\text{-}3)CH_2COC_6H_5$ (75)	29
		$C_6H_5CH{=}CHCO$-4-pyridyl	Cat. a, dioxane, TEA, 80°	$C_2H_5COCH(C_6H_5)CH_2COC_5H_4N\text{-}4$ (55)	29
		$C_6H_5CH{=}CHCOCH{=}CHC_6H_5$	Cat. a, DMF, TEA, 80°	$C_2H_5COCH(C_6H_5)CH_2COCH{=}CHC_6H_5$ (72)	8
	CH_3OCH_2CHO	$CH_2{=}CHCOCH_3$	Cat. a, EtOH, TEA, 80°	$CH_3OCH_2CO(CH_2)_2COCH_3$ (62)	20
		$CH_2{=}CHCOC_2H_5$	Cat. a, EtOH, TEA, 80°	$CH_3OCH_2CO(CH_2)_2COC_2H_5$ (65)	20
		$CH_2{=}CHCOC_6H_5$	Cat. a, EtOH, TEA, 80°	$CH_3OCH_2CO(CH_2)_2COC_6H_5$ (69)	20
C_4	$n\text{-}C_3H_7CHO$	$CH_2{=}CHCOCH_3$	Cat. a, EtOH, TEA, 80°	$n\text{-}C_3H_7CO(CH_2)_2COCH_3$ (68)	48
		"	Cat. e, DMF, TEA, reflux	" (59)	12c
		$CH_2{=}CHCOCH{=}CH_2$	Cat. a, TEA, 25°	$n\text{-}C_3H_7CO(CH_2)_2COCH{=}CH_2$ (23)	17
		$CH_2{=}CHCO(CH_2)_2CN$	Cat. a, TEA, 65°	$n\text{-}C_3H_7[CO(CH_2)_2]_2CN$ (72)	41
		$CH_2{=}CHCOCH_2O_2CCH_3$	Cat. a, dioxane, TEA, 80°	$n\text{-}C_3H_7CO(CH_2)_2COCH_2O_2CCH_3$ (79)	20
		$CH_2{=}CHCO(CH_2)_2OCH_3$	Cat. a, dioxane, TEA, 90°	$n\text{-}C_3H_7CO(CH_2)_2CO(CH_2)_2OCH_3$ (76)	19
		$CH_3CH{=}C(CO_2C_2H_5)COCH_3$	Cat. a, dioxane, TEA, 80°	$n\text{-}C_3H_7COCH(CH_3)CH(CO_2C_2H_5)COCH_3$ (43)	36
		$I^-(CH_3)_3N^+(CH_2)_2COC(CH_3){=}CHCH_3$	Cat. a, DMF, TEA, 80°	$n\text{-}C_3H_7CO(CH_2)_2COC(CH_3){=}CHCH_3$ (41)	37

TABLE I. CATALYZED ADDITIONS OF ALDEHYDES TO α,β-UNSATURATED KETONES (*Continued*)

Aldehyde	Michael Acceptor	Conditions	Product(s) and Yield(s) (%)	Refs.
	$(CH_3)_2N(CH_2)_2COCH{=}C(CH_3)_2$	Cat. a, DMF, TEA, 80°	n-$C_3H_7CO(CH_2)_2COCH{=}C(CH_3)_2$ (53)	37
	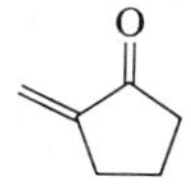	Cat. a, dioxane, TEA, 80°	n-$C_3H_7COCH_2$– 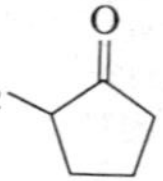 (63)	76
	(E)-$C_2H_5O_2CCH{=}CHCOCH_3$	Cat. a, dioxane, TEA, 80°	n-$C_3H_7COCH(CO_2C_2H_5)CH_2COCH_3$ (74)	36
	$CH_2{=}CHCO(CH_2)_2CO_2CH_3$	Cat. a, dioxane, TEA, 100°	n-$C_3H_7[CO(CH_2)_2]_2CO_2CH_3$ (52)	24
	$CH_2{=}CHCO(CH_2)_3CO_2CH_3$	Cat. a, dioxane, TEA, 100°	n-$C_3H_7CO(CH_2)_2CO(CH_2)_3CO_2CH_3$ (38)	24
	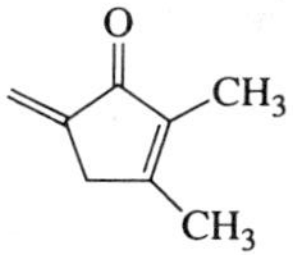 CH_3, CH_3	Cat. a, dioxane, TEA, 80°	n-$C_3H_7COCH_2$– 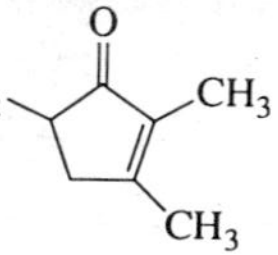 CH_3, CH_3 (59)	76
	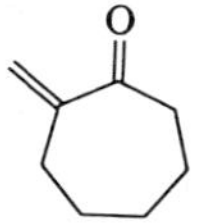	Cat. a, EtOH, TEA, 80°	n-$C_3H_7COCH_2$– 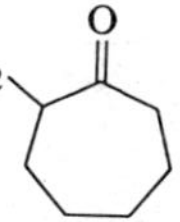 (32)	76
	$(CH_3)_3\overset{+}{N}(CH_2)_2CO$– 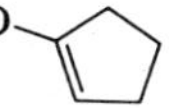 I^-	Cat. a, DMF, TEA, 80°	n-$C_3H_7CO(CH_2)_2CO$– 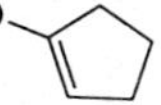 (48)	37
	(E)-$C_6H_5CH{=}CHCOCH_3$	Cat. a, EtOH, TEA, 70°	n-$C_3H_7COCH(C_6H_5)CH_2COCH_3$ (40)	6a
	$CH_2{=}CH[CO(CH_2)_2]_2CN$	Cat. a, TEA, 65°	n-$C_3H_7[CO(CH_2)_2]_3CN$ (66)	52
	$(CH_3)_3\overset{+}{N}(CH_2)_2CO$– 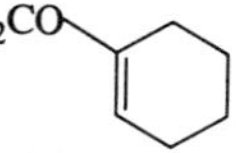 I^-	Cat. a, DMF, TEA, 80°	n-$C_3H_7CO(CH_2)_2CO$– 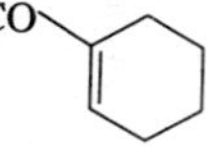 (61)	37
	$CH_2{=}CHCO(CH_2)_3$– 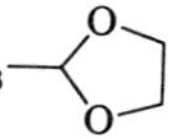	Cat. a, TEA, 65°	n-$C_3H_7CO(CH_2)_2CO(CH_2)_3$– 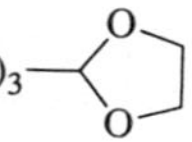 (73)	50

$C_2H_5O_2CCH{=}C(COCH_3)_2$	Cat. a, dioxane, TEA, 80°	n-$C_3H_7COCH(CO_2C_2H_5)CH(COCH_3)_2$ (42)[a]	53
"	Cat. a, dioxane, TEA, 80°	" (21)	53
$HO_2CCH{=}CHCOC_6H_5$	Cat. a, EtOH, TEA, 80°	$C_6H_5CO(CH_2)_2COC_3H_7$-n (80)	36
$(CH_3)_2N(CH_2)_2COC_6H_5$	Cat. a, DMF, TEA, 80°	" (69)	37
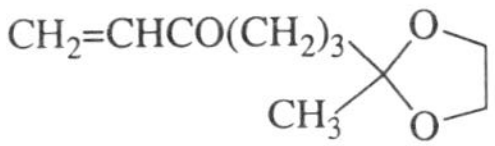 $CH_2{=}CHCO(CH_2)_3$ CH3 O O	Cat. a, TEA, 65°	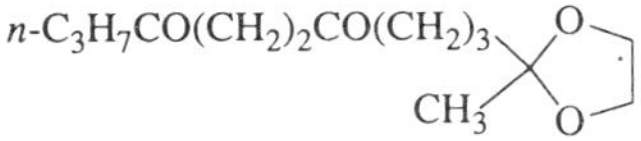 n-$C_3H_7CO(CH_2)_2CO(CH_2)_3$ CH3 O O (77)	50
$(CH_3)_2NCH_2CH(CH_3)COC_6H_5$	Cat. a, DMF, TEA, 80°	n-$C_3H_7COCH_2CH(CH_3)COC_6H_5$ (35)	37
 $CH_2{=}CHCO$	Cat. a, dioxane, TEA, 65°	 n-$C_3H_7CO(CH_2)_2CO$ (63)	32
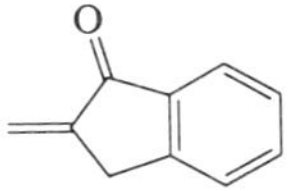 O	Cat. a, EtOH, TEA, 80°	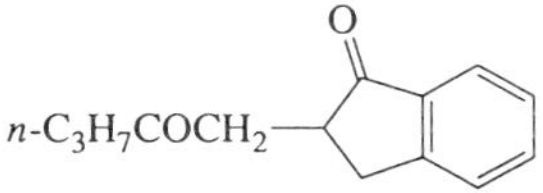 n-$C_3H_7COCH_2$ O (76)	76
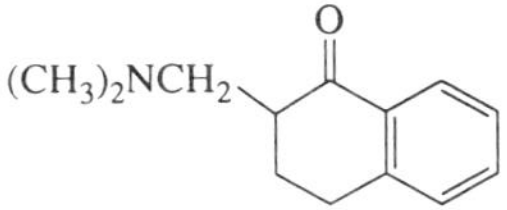 $(CH_3)_2NCH_2$ O	Cat. a, DMF, TEA, 80°	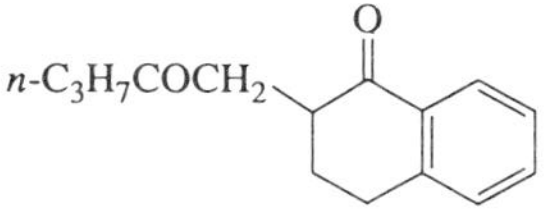 n-$C_3H_7COCH_2$ O (72)	37
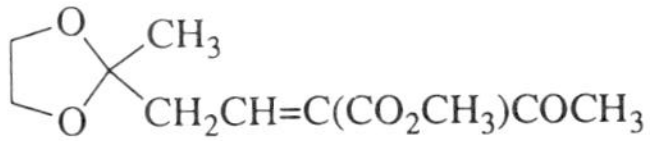 O O CH3 $CH_2CH{=}C(CO_2CH_3)COCH_3$	Cat. a, EtOH, TEA, 80°	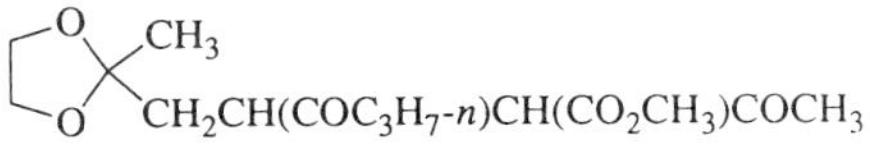 O O CH3 $CH_2CH(COC_3H_7$-$n)CH(CO_2CH_3)COCH_3$ (46)	49
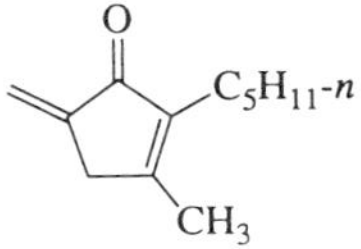 O C_5H_{11}-n CH_3	Cat. a, EtOH, TEA, 80°	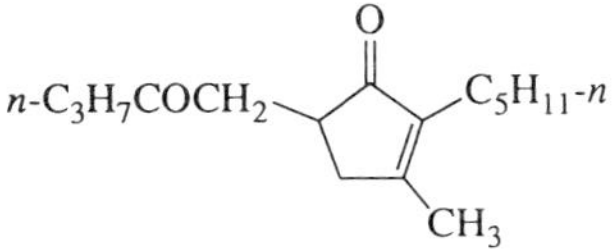 n-$C_3H_7COCH_2$ O C_5H_{11}-n CH_3 (80)	76
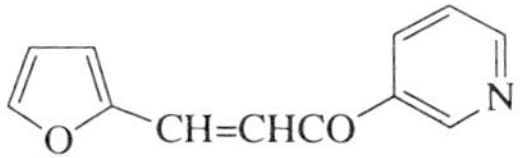 O CH=CHCO N	Cat. b, EtOH, TEA, 80°	n-$C_3H_7COCH(C_4H_3O)CH_2COC_5H_4N$-3 (88)	29

TABLE I. CATALYZED ADDITIONS OF ALDEHYDES TO α,β-UNSATURATED KETONES (*Continued*)

Aldehyde	Michael Acceptor	Conditions	Product(s) and Yield(s) (%)	Refs.
	$C_2H_5O_2CCH{=}C(CO_2C_2H_5)COCH_3$	Cat. a, dioxane, TEA, 80°	n-$C_3H_7COCH(CO_2C_2H_5)CH(CO_2C_2H_5)COCH_3$ (44)[a]	53
	$C_6H_5CH{=}C(CO_2C_2H_5)COCH_3$	Cat. a, EtOH, TEA, 80°	n-$C_3H_7COCH(C_6H_5)CH(CO_2C_2H_5)COCH_3$ (52)	36
	$C_6H_5CH{=}CHCO$-(2-thienyl)	Cat. b, EtOH, TEA, 80°	n-$C_3H_7COCH(C_6H_5)CH_2COC_4H_3S$ (77)	29
	$C_6H_5CH{=}CHCO$-(2-furyl)	Cat. b, DMF, TEA, 80°	n-$C_3H_7COCH(C_6H_5)CH_2COC_4H_3O$ (80)	29
	(2-furyl)-$CH{=}CHCOC_6H_5$	Cat. b, TEA, 80°	n-$C_3H_7COCH(C_4H_3O)CH_2COC_6H_5$ (92)	29
	(2-furyl)-$CH{=}CHCO$-(2-furyl)	Cat. b, TEA, 80°	n-$C_3H_7COCH(C_4H_3O)CH_2COC_4H_3O$ (84)	29
	$CH_2{=}CRCOC_6H_5$ R = $-CH_2$-(2-methyl-1,3-dioxolan-2-yl)	Cat. a, EtOH, TEA, 80°	n-$C_3H_7COCH_2CHRCOC_6H_5$ (50)	49
	$C_6H_5CH{=}CHCO$-(3-pyridyl)	Cat. b, EtOH, TEA, 80°	n-$C_3H_7COCH(C_6H_5)CH_2COC_5H_4N$-3 (89)	29
	(3-pyridyl)-$CH{=}CHCO$-(3-pyridyl)	Cat. a, EtOH, TEA, 80°	n-$C_3H_7COCH(C_5H_4N$-3$)CH_2COC_5H_4N$-3 (84)	29
	(3-pyridyl)-$CH{=}CHCOC_6H_5$	Cat. a, EtOH, TEA, 80°	n-$C_3H_7COCH(C_5H_4N$-3$)CH_2COC_6H_5$ (95)	29
	$C_6H_5CH{=}CHCOC_6H_5$	Cat. a, EtOH, TEA, 80°	n-$C_3H_7COCH(C_6H_5)CH_2COC_6H_5$ (70)	7
	$(CH_3)_2NCH_2CH(C_6H_5)COC_6H_5$	Cat. a, DMF, TEA, 80°	n-$C_3H_7COCH_2CH(C_6H_5)COC_6H_5$ (47)	37

	Aldehyde	Acceptor	Conditions	Product(s) (Yield, %)	Refs.
	i-C_3H_7CHO	$CH_2{=}CHCOCH_3$	Cat. a, EtOH, TEA, 80°	i-$C_3H_7CO(CH_2)_2COCH_3$ (41)	48
		$CH_2{=}CHCOCH_2OCH_3$	Cat. a, dioxane, TEA, 90°	i-$C_3H_7CO(CH_2)_2COCH_2OCH_3$ (73)	19
		$CH_2{=}CHCOCH_2O_2CCH_3$	Cat. a, dioxane, TEA, 80°	i-$C_3H_7CO(CH_2)_2COCH_2O_2CCH_3$ (72)	20
		$CH_2{=}CHCO(CH_2)_2CO_2CH_3$	Cat. a, dioxane, TEA, 100°	i-$C_3H_7CO(CH_2)_2CO(CH_2)_2CO_2CH_3$ (32)	24
		(E)-$C_2H_5O_2CCH{=}CHCOCH_3$	Cat. a, dioxane, TEA, 80°	i-$C_3H_7COCH(CO_2C_2H_5)CH_2COCH_3$ (46)	36
		$CH_2{=}CHCO(CH_2)_2OC_3H_7$-$n$	Cat. a, dioxane, TEA, 90°	i-$C_3H_7CO(CH_2)_2CO(CH_2)_2OC_3H_7$-$n$ (66)	19
		$CH_2{=}CHCO$	Cat. a, dioxane, TEA, 65°	i-$C_3H_7CO(CH_2)_2CO$ (46)	32
	(E,Z)-$CH_3CH{=}CHCHO$	$CH_2{=}CHCOCH_3$	Cat. a, EtOH, TEA, 80°	(E,Z)-$CH_3CH{=}CHCO(CH_2)_2COCH_3$ (28)	16
		$CH_2{=}CHCOC_2H_5$	Cat. a, EtOH, NaOAc, 80°	(E,Z)-$CH_3CH{=}CHCO(CH_2)_2COC_2H_5$ (21)	16
		$CH_2{=}CHCOC_6H_5$	Cat. a, EtOH, NaOAc, 80°	(E,Z)-$CH_3CH{=}CHCO(CH_2)_2COC_6H_5$ (25)	16
C_5	n-C_4H_9CHO	$CH_2{=}CHCOCH_3$	Cat. a, EtOH, TEA, 80°	n-$C_4H_9CO(CH_2)_2COCH_3$ (68)	48
		(E)-$C_2H_5O_2CCH{=}CHCOCH_3$	Cat. a, dioxane, TEA, 80°	n-$C_4H_9COCH(CO_2C_2H_5)CH_2COCH_3$ (57)	36
		$CH_2{=}CHCO(CH_2)_2CO_2CH_3$	Cat. a, dioxane, TEA, 100°	n-$C_4H_9CO(CH_2)_2CO(CH_2)_2CO_2CH_3$ (61)	24
		$CH_3CH{=}C(CO_2C_2H_5)COCH_3$	Cat. a, dioxane, TEA, 80°	n-$C_4H_9COCH(CH_3)CH(CO_2C_2H_5)COCH_3$ (32)	36
		$CH_2{=}CHCOC_5H_{11}$-n	Cat. a, TEA, 90°	n-$C_4H_9CO(CH_2)_2COC_5H_{11}$-$n$ (72)	77
			Cat. a, EtOH, TEA, 80°	n-$C_4H_9COCH_2$ (52)	76
		$HO_2CCH{=}CHCOC_6H_5$	Cat. a, EtOH, TEA, 80°	n-$C_4H_9CO(CH_2)_2COC_6H_5$ (83)	36
		$C_2H_5O_2CCH{=}C(COCH_3)_2$	Cat. a, dioxane, TEA, 80°	n-$C_4H_9COCH(CO_2C_2H_5)CH(COCH_3)_2$ (30)[a]	53
		$CH_2{=}CHCOC_7H_{15}$-n	Cat. a, TEA, 90°	n-$C_4H_9CO(CH_2)_2COC_7H_{15}$-$n$ (61)	77
		(E)-$C_6H_5CH{=}CHCOCH_3$	Cat. c, EtOH, TEA, 80°	n-$C_4H_9COCH(C_6H_5)CH_2COCH_3$ (35)	7
		$C_2H_5O_2CCH{=}C(CO_2C_2H_5)COCH_3$	Cat. a, dioxane, TEA, 80°	n-$C_4H_9CO[CH(CO_2C_2H_5)]_2COCH_3$ (36)[a]	53
		$C_6H_5CH{=}CHCOCH{=}CHC_6H_5$	Cat. a, DMF, TEA, 80°	n-$C_4H_9COCH(C_6H_5)CH_2COCH{=}CHC_6H_5$ (65)	8
	$CH_3O(CH_2)_3CHO$	$CH_2{=}CHCOCH_3$	Cat. a, dioxane, TEA, 90°	$CH_3O(CH_2)_3CO(CH_2)_2COCH_3$ (68)	21
		$CH_2{=}CHCOC_2H_5$	Cat. a, dioxane, TEA, 90°	$CH_3O(CH_2)_3CO(CH_2)_2COC_2H_5$ (71)	21
		$CH_2{=}CHCOC_6H_5$	Cat. a, dioxane, TEA, 90°	$CH_3O(CH_2)_3CO(CH_2)_2COC_6H_5$ (68)	21

TABLE I. CATALYZED ADDITIONS OF ALDEHYDES TO α,β-UNSATURATED KETONES (*Continued*)

Aldehyde	Michael Acceptor	Conditions	Product(s) and Yield(s) (%)	Refs.
$i\text{-}C_3H_7OCH_2CHO$	$CH_2{=}CHCOCH_3$	Cat. a, EtOH, TEA, 80°	$i\text{-}C_3H_7OCH_2CO(CH_2)_2COCH_3$ (63)	20
	$CH_2{=}CHCOC_2H_5$	Cat. a, EtOH, TEA, 80°	$i\text{-}C_3H_7OCH_2CO(CH_2)_2COC_2H_5$ (67)	20
	$CH_2{=}CHCOC_6H_5$	Cat. a, EtOH, TEA, 80°	$i\text{-}C_3H_7OCH_2CO(CH_2)_2COC_6H_5$ (61)	20
$(CH_3)_2C{=}CHCHO$	$CH_2{=}CHCOCH_3$	Cat. a, EtOH, NaOAc, 80°	$(CH_3)_2C{=}CHCO(CH_2)_2COCH_3$ (52)	16
	$CH_2{=}CHCOC_2H_5$	Cat. a, EtOH, NaOAc, 80°	$(CH_3)_2C{=}CHCO(CH_2)_2COC_2H_5$ (55)	16
	$CH_2{=}CHCOCH{=}C(CH_3)_2$	Cat. a, EtOH, NaOAc, 80°	$(CH_3)_2C{=}CHCO(CH_2)_2COCH{=}C(CH_3)_2$ (58)	16
	$I^-(CH_3)_3N^+(CH_2)_2COCH{=}C(CH_3)_2$	Cat. c, DMF, TEA, 80°	$(CH_3)_2C{=}CHCO(CH_2)_2COCH{=}C(CH_3)_2$ (16)	37
	$CH_2{=}CHCOC_6H_5$	Cat. a, EtOH, NaOAc, 80°	$(CH_3)_2C{=}CHCO(CH_2)_2COC_6H_5$ (62)	16
$CH_3O_2C(CH_2)_2CHO$	$CH_2{=}CHCOR$	Cat. a, dioxane, TEA, 90°	$CH_3O_2C(CH_2)_2CO(CH_2)_2COR$	24
	R			
	CH_3		(70)	
	$C_3H_7\text{-}n$		(64)	
	$C_4H_9\text{-}n$		(65)	
	$C_5H_{11}\text{-}n$		(65)	
	$C_6H_{13}\text{-}n$		(69)	
	$C_7H_{15}\text{-}n$		(71)	
	$C_8H_{17}\text{-}n$		(69)	
	$C_9H_{19}\text{-}n$		(67)	
2-Furyl-CHO	$CH_2{=}CHCOCH_3$	Cat. a, EtOH, NaOAc, 80°	$C_4H_3OCO(CH_2)_2COCH_3$ (80)	48
	"	Cat. a, EtOH, NaOAc, reflux	" (66)	28
	"	Cat. e, EtOH, TEA, 80°	" (50)	12a
	"	Cat. b, EtOH, TEA, 80°	" (98)	51
	$CH_3CH{=}CHCOCH_3$	Cat. c, EtOH, TEA, 80°	$C_4H_3OCOCH(CH_3)CH_2COCH_3$ (34)	7
	$CH_2{=}CHCO(CH_2)_2CN$	Cat. b, TEA, 65°	$C_4H_3OCO(CH_2)_2CO(CH_2)_2CN$ (45)	41
	$CH_2{=}CHCOCH_2O_2CCH_3$	Cat. a, dioxane, TEA, 80°	$C_4H_3OCO(CH_2)_2COCH_2O_2CCH_3$ (55)	20
	$CH_2{=}CHCO(CH_2)_2OCH_3$	Cat. b, dioxane, TEA, 90°	$C_4H_3OCO(CH_2)_2CO(CH_2)_2OCH_3$ (76)	19
	$CH_2{=}CHCO(CH_2)_2CO_2CH_3$	Cat. b, dioxane, TEA, 100°	$C_4H_3OCO(CH_2)_2CO(CH_2)_2CO_2CH_3$ (20)	24
	$HO_2CCH{=}CHCOC_6H_5$	Cat. b, EtOH, TEA, 80°	$C_4H_3OCO(CH_2)_2COC_6H_5$ (82)	36

(*E*)-$C_6H_5CH{=}CHCOCH_3$	NaCN, DMF, 35°	$C_4H_3OCOCH(C_6H_5)CH_2COCH_3$ (44)	78
"	Cat. c, EtOH, TEA, 80°	" (80)	7
$C_2H_5O_2CCH{=}C(COCH_3)_2$	Cat. a, dioxane, TEA, 80°	$C_4H_3OCOCH(CO_2C_2H_5)CH(COCH_3)_2$ (57)	53
$CH_2{=}CHCOC_6H_4OCH_3$-*p*	Cat. b, dioxane, TEA, 90°	$C_4H_3OCO(CH_2)_2COC_6H_4OCH_3$-*p* (70)	19
$CH_2{=}CH[CO(CH_2)_2]_2CO_2CH_3$	Cat. b, TEA, 65°	$C_4H_3O[CO(CH_2)_2]_3CO_2CH_3$ (78)	52
$C_2H_5O_2CCH{=}C(CO_2C_2H_5)COCH_3$	Cat. a, dioxane, TEA, 80°	$C_4H_3OCOCH(CO_2C_2H_5)CH(CO_2C_2H_5)COCH_3$ (33)	53
$CH_2{=}CHCOCH_2$-(2-methyl-4,5-dimethyl-1,3-dioxolan-2-yl)	Cat. b, dioxane, TEA, 80°	$C_4H_3OCO(CH_2)_2COCH_2$-(2-methyl-4,5-dimethyl-1,3-dioxolan-2-yl) (64)	47
2-methylene-1-indanone	Cat. b, EtOH, TEA, 80°	2-($C_4H_3OCOCH_2$)-1-indanone (69)	76
2-methylene-1-tetralone	Cat. b, EtOH, TEA, 80°	2-($C_4H_3OCOCH_2$)-1-tetralone (61)	76
$CH_2{=}CHCOC_6H_3(OCH_3)_2$-3,4	Cat. b, dioxane, TEA, 90°	$C_4H_3OCO(CH_2)_2COC_6H_3(OCH_3)_2$-3,4 (62)	19
$CH_2{=}C(COR^1)CH_2$-(2-R_2-4,5-dimethyl-1,3-dioxolan-2-yl)		$C_4H_3OCOCH_2CH(COR^1)CH_2$-(2-$R_2$-4,5-dimethyl-1,3-dioxolan-2-yl)	
R^1 / R^2			
CH_3 / CH_3	Cat. b, EtOH, TEA, 80°	(55)	49
C_6H_4Cl-*p* / CH_3	Cat. b, EtOH, TEA, 80°	(78)	49
C_6H_5 / CH_3	Cat. c, EtOH, TEA, 80°	(57)	49
C_6H_5 / C_2H_5	Cat. b, EtOH, TEA, 80°	(74)	49
2-furyl-$CH{=}CHCO$-2-furyl	NaCN, DMF, 35°	$C_4H_3OCOCH(C_4H_3O)CH_2COC_4H_3O$ (45)	29

TABLE I. CATALYZED ADDITIONS OF ALDEHYDES TO α,β-UNSATURATED KETONES (*Continued*)

Aldehyde	Michael Acceptor	Conditions	Product(s) and Yield(s) (%)	Refs.
	CH_3 CH_3 O O CH_2=CHCO 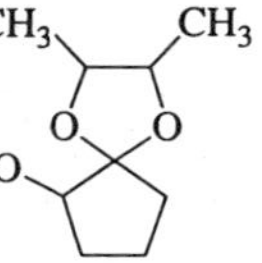	Cat. b, dioxane, TEA, 80°	CH_3 CH_3 O O $C_4H_3OCO(CH_2)_2CO$ 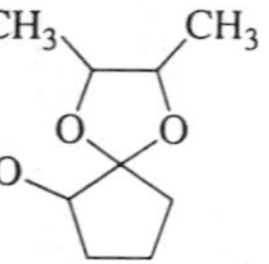 (82)	47
	$C_2H_5O_2CCH$=$CHCOC_6H_5$	Cat. b, EtOH, TEA, 80°	$C_4H_3OCOCH(CO_2C_2H_5)CH_2COC_6H_5$ (70)	36
	CH_3 CH_3 O O CH_2=CHCO 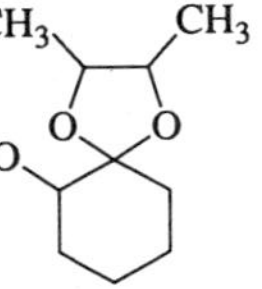	Cat. b, dioxane, TEA, 80°	CH_3 CH_3 O O $C_4H_3OCO(CH_2)_2CO$ 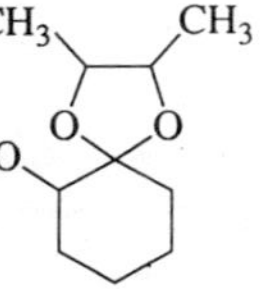 (75)	47
	C_6H_5CH=$C(CO_2C_2H_5)COCH_3$	Cat. b, EtOH, TEA, 80°	$C_4H_3OCOCH(C_6H_5)CH(CO_2C_2H_5)COCH_3$ (81)	36
	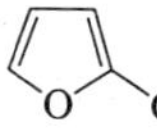 O CH=$CHCOC_6H_5$	Cat. c, EtOH, TEA, 80°	$C_4H_3OCOCH(C_4H_3O)CH_2COC_6H_5$ (63)	29
	"	NaCN, DMF, 35°	" (61)	29
	$C_2H_5O_2CCH$=$CCOCH_3$ | COC_6H_5	Cat. a, dioxane, TEA, 80°	$C_4H_3OCOCH(CO_2C_2H_5)CHCOCH_3$ | COC_6H_5 (59)	53
	C_6H_5CH=$CHCOC_6H_5$	NaCN, DMF, 35°	$C_4H_3OCOCH(C_6H_5)CH_2COC_6H_5$ (93)	33
	"	Cat. c, EtOH, TEA, 80°	" (91)	7
	$C_2H_5O_2CCH$=$CH(COC_6H_5)_2$	Cat. a, dioxane, TEA, 80°	$C_4H_3OCOCH(CO_2C_2H_5)CH(COC_6H_5)_2$ (74)	53
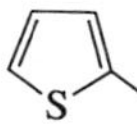 S CHO	CH_2=$CHCOCH_3$	NaCN, DMF, 20°	$C_4H_3SCO(CH_2)_2COCH_3$ (80)	38
	CH_2=$CHCO(CH_2)_2OCH_3$	Cat. b, dioxane, TEA, 90°	$C_4H_3SCO(CH_2)_2CO(CH_2)_2OCH_3$ (71)	19
	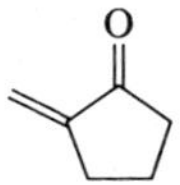 O	Cat. b, dioxane, TEA, 80°	O $C_4H_3SCOCH_2$ 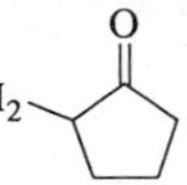 (60)	76

$CH_2{=}CH[CO(CH_2)_2]_2CN$	Cat. b, EtOH, TEA, 80°	$C_4H_3S[CO(CH_2)_2]_3CN$ (67)	52
$I^-(CH_3)_3N^+(CH_2)_2COCH{=}C(CH_3)_2$	Cat. c, DMF, TEA, 80°	$C_4H_3SCO(CH_2)_2COCH{=}C(CH_3)_2$ (45)	37
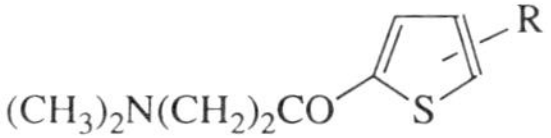	NaCN, DMF, rt	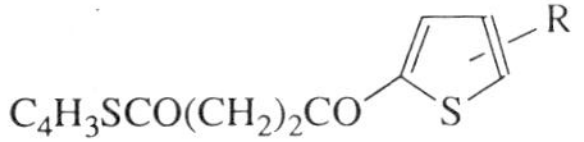	
R			
H		(75)	38
3-CH_3		(60)	59
4-CH_3		(62)	59
5-CH_3		(58)	59
(*E*)-$C_6H_5CH{=}CHCOCH_3$	NaCN, DMF, 20°	$C_4H_3SCOCH(C_6H_5)CH_2COCH_3$ (77)	38
$HO_2CCH{=}CHC_6H_4CH_3$-*p*	Cat. b, EtOH, TEA, 80°	$C_4H_3SCO(CH_2)_2COC_6H_4CH_3$-*p* (82)	36
$(CH_3)_2N(CH_2)_2COC_6H_4X$-*p*	NaCN, DMF, rt	$C_4H_3SCO(CH_2)_2COC_6H_4X$-*p*	59
X			
CH_3		(60)	
o-Cl		(0)	
p-Cl		(57)	
CH_3O		(33)	
C_6H_5O		(86)	
C_6H_5		(90)	
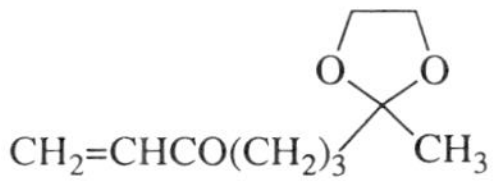	Cat. b, TEA, 65°	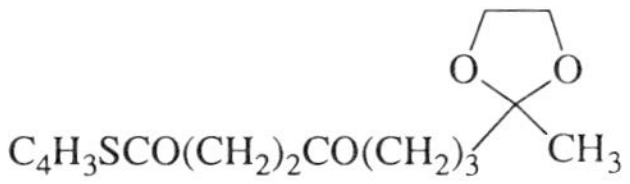 (71)	50
$CH_2{=}CH[CO(CH_2)_2]_2CO_2CH_3$	Cat. b, TEA, 65°	$C_4H_3S[CO(CH_2)_2]_3CO_2CH_3$ (76)	52
	Cat. b, DMF, TEA, 80°	$C_4H_3SCOCH(C_4H_3O)CH(CO_2C_2H_5)COCH_3$ (57)	36
	Cat. b, dioxane, TEA, 80°	$C_4H_3SCOCH(C_4H_3S)CH(CO_2C_2H_5)COCH_3$ (71)	36
$C_6H_5CH{=}C(CO_2CH_3)COCH_3$	Cat. b, EtOH, TEA, 80°	$C_4H_3SCOCH(C_6H_5)CH(CO_2C_2H_5)COCH_3$ (51)	36

TABLE I. CATALYZED ADDITIONS OF ALDEHYDES TO α,β-UNSATURATED KETONES (*Continued*)

	Aldehyde	Michael Acceptor	Conditions	Product(s) and Yield(s) (%)	Refs.
		$C_2H_5O_2CCH{=}CHCOC_6H_5$	Cat. b, EtOH, TEA, 80°	$C_4H_3SCOCH(CO_2C_2H_5)CH_2COC_6H_5$ (46)	36
		$C_6H_5CH{=}CHCO$-(2-thienyl)	NaCN, DMF, 20°	$C_4H_3SCOCH(C_6H_5)CH_2COC_4H_3S$ (80)	38
		(3-thienyl)-$CH{=}CHCOC_6H_5$	Cat. b, EtOH, TEA, reflux	$C_4H_3SCOCH(C_4H_3S\text{-}3)CH_2COC_6H_5$ (91)	58
		(2-thienyl)-$CH{=}CHCOC_6H_5$	Cat. b, EtOH, TEA, reflux	$C_4H_3SCOCH(C_4H_3S)CH_2COC_6H_5$ (94)	58
		$C_6H_5CH{=}CHCOC_6H_5$	NaCN, DMF, 20°	$C_4H_3SCOCH(C_6H_5)CH_2COC_6H_5$ (90)	38
	thiophene-3-CHO	(2-thienyl)-$CH{=}CHCOC_6H_5$	Cat. b, EtOH, TEA, reflux	$3\text{-}C_4H_3SCOCH(C_4H_3S)CH_2COC_6H_5$ (81)	58
		(3-thienyl)-$CH{=}CHCOC_6H_5$	Cat. b, EtOH, TEA, reflux	$3\text{-}C_4H_3SCOCH(C_4H_3S\text{-}3)CH_2COC_6H_5$ (72)	58
	5-Cl-thiophene-2-CHO	$(CH_3)_2N(CH_2)_2COC_6H_4CH_3$-*p*	NaCN, DMF, rt	$5\text{-}ClC_4H_2SCO(CH_2)_2COC_6H_4CH_3$-*p* (64)	59
		$(CH_3)_2N(CH_2)_2COC_6H_4Cl$-*p*	NaCN, DMF, rt	$5\text{-}ClC_4H_2SCO(CH_2)_2COC_6H_4Cl$-*p* (42)	59
C_6	*n*-$C_5H_{11}CHO$	$CH_2{=}CHCOCH_3$	Cat. a, EtOH, TEA, 80°	*n*-$C_5H_{11}CO(CH_2)_2COCH_3$ (68)	48
		(*E*)-$C_2H_5O_2CCH{=}CHCOCH_3$	Cat. a, dioxane, TEA, 80°	*n*-$C_5H_{11}COCH(CO_2C_2H_5)CH_2COCH_3$ (46)	36
		$CH_2{=}CHCO(CH_2)_2CO_2CH_3$	Cat. a, dioxane, TEA, 100°	*n*-$C_5H_{11}[CO(CH_2)_2]_2CO_2CH_3$ (67)	24
		$CH_2{=}CHCO(CH_2)_3CO_2CH_3$	Cat. a, dioxane, TEA, 100°	*n*-$C_5H_{11}CO(CH_2)_2CO(CH_2)_3CO_2CH_3$ (36)	24

Aldehyde	Acceptor	Conditions	Product(s) and Yield(s) (%)	Refs.
	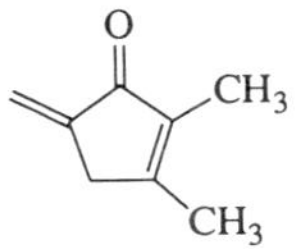 O, CH_3, CH_3	Cat. a, EtOH, TEA, 80°	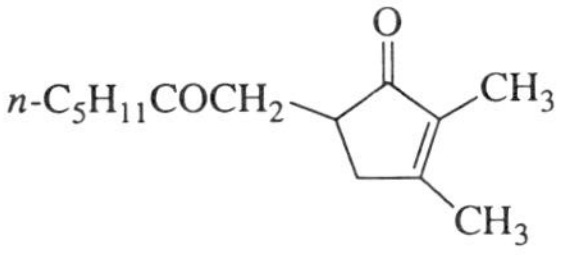 n-$C_5H_{11}COCH_2$–, O, CH_3, CH_3 (68)	76
	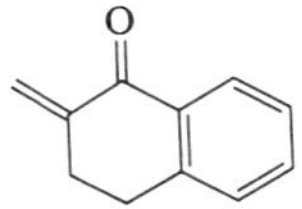 O	Cat. a, EtOH, TEA, 80°	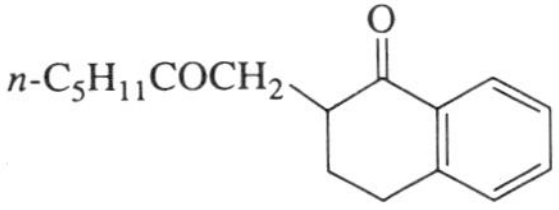 n-$C_5H_{11}COCH_2$–, O (81)	76
n-$C_3H_7CH(CH_3)CHO$	$CH_2{=}CHCOCH_3$	Cat. a, EtOH, TEA, 80°	n-$C_3H_7CH(CH_3)CO(CH_2)_2COCH_3$ (38)	48
	$CH_2{=}CHCO(CH_2)_2CO_2CH_3$	Cat. a, dioxane, TEA, 100°	n-$C_3H_7CH(CH_3)[CO(CH_2)_2]_2CO_2CH_3$ (21)	24
$(C_2H_5)_2CHCHO$	$CH_2{=}CHCOCH{=}CH_2$	Cat. a, TEA, 65°	$(C_2H_5)_2CHCO(CH_2)_2COCH{=}CH_2$ (25)	17
(E)-n-$C_3H_7CH{=}CHCHO$	$CH_2{=}CHCOCH_3$	Cat. a, EtOH, NaOAc, 80°	(E)-n-$C_3H_7CH{=}CHCO(CH_2)_2COCH_3$ (32)	16
(E,E)-$CH_3(CH{=}CH)_2CHO$	$CH_2{=}CHCOCH_3$	Cat. a, EtOH, NaOAc, 80°	(E,E)-$CH_3(CH{=}CH)_2CO(CH_2)_2COCH_3$ (35)	16
	$CH_2{=}CHCOC_6H_5$	Cat. a, EtOH, NaOAc, 80°	(E,E)-$CH_3(CH{=}CH)_2CO(CH_2)_2COC_6H_5$ (36)	16
$(C_2H_5O)_2CHCHO$	$CH_2{=}CHCOCH_3$	Cat. a, dioxane, TEA, 80°	$(C_2H_5O)_2CHCO(CH_2)_2COCH_3$ (71)	22
	$CH_2{=}CHCOC_2H_5$	Cat. a, dioxane, TEA, 80°	$(C_2H_5O)_2CHCO(CH_2)_2COC_2H_5$ (65)	22
	$CH_2{=}CHCOC_5H_{11}$-n	Cat. a, dioxane, TEA, 80°	$(C_2H_5O)_2CHCO(CH_2)_2COC_5H_{11}$-$n$ (58)	22
	$CH_2{=}CHCOC_6H_{13}$-n	Cat. a, dioxane, TEA, 80°	$(C_2H_5O)_2CHCO(CH_2)_2COC_6H_{13}$-$n$ (67)	22
n-$C_4H_9OCH_2CHO$	$CH_2{=}CHCOCH_3$	Cat. a, EtOH, TEA, 80°	n-$C_4H_9OCH_2CO(CH_2)_2COCH_3$ (68)	20
	$CH_2{=}CHCOC_2H_5$	Cat. a, EtOH, TEA, 80°	n-$C_4H_9OCH_2CO(CH_2)_2COC_2H_5$ (64)	20
	$CH_2{=}CHCOC_6H_5$	Cat. a, EtOH, TEA, 80°	n-$C_4H_9OCH_2CO(CH_2)_2COC_6H_5$ (65)	20
$CH_3O(CH_2)_4CHO$	$CH_2{=}CHCOCH_3$	Cat. a, dioxane, TEA, 90°	$CH_3O(CH_2)_4CO(CH_2)_2COCH_3$ (67)	21
	$CH_2{=}CHCOC_2H_5$	Cat. a, dioxane, TEA, 90°	$CH_3O(CH_2)_4CO(CH_2)_2COC_2H_5$ (65)	21
	$CH_2{=}CHCOC_6H_5$	Cat. a, dioxane, TEA, 90°	$CH_3O(CH_2)_4CO(CH_2)_2COC_6H_5$ (63)	21
$C_2H_5O(CH_2)_3CHO$	$CH_2{=}CHCOCH_3$	Cat. a, dioxane, TEA, 90°	$C_2H_5O(CH_2)_3CO(CH_2)_2COCH_3$ (63)	21
	$CH_2{=}CHCOC_2H_5$	Cat. a, dioxane, TEA, 90°	$C_2H_5O(CH_2)_3CO(CH_2)_2COC_2H_5$ (63)	21
	$CH_2{=}CHCOC_6H_5$	Cat. a, dioxane, TEA, 90°	$C_2H_5O(CH_2)_3CO(CH_2)_2COC_6H_5$ (64)	21
$CH_3CO_2(CH_2)_3CHO$	$CH_2{=}CHCOCH_3$	Cat. a, dioxane, TEA, 90°	$CH_3CO_2(CH_2)_3CO(CH_2)_2COCH_3$ (52)	21
$CH_3O_2C(CH_2)_3CHO$	$CH_2{=}CHCOCH_3$	Cat. a, dioxane, TEA, 90°	$CH_3O_2C(CH_2)_3CO(CH_2)_2COCH_3$ (62)	24
	$CH_2{=}CHCOC_6H_{13}$-n	Cat. a, dioxane, TEA, 90°	$CH_3O_2C(CH_2)_3CO(CH_2)_2COC_6H_{13}$-$n$ (69)	24
	$CH_2{=}CHCOC_9H_{19}$-n	Cat. a, dioxane, TEA, 90°	$CH_3O_2C(CH_2)_3CO(CH_2)_2COC_9H_{19}$-$n$ (67)	24
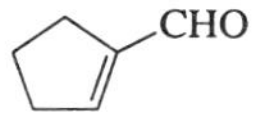 CHO	$CH_2{=}CHCOCH_3$	Cat. a, EtOH, TEA, 80°	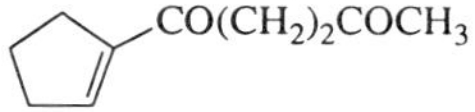 $CO(CH_2)_2COCH_3$ (27)	16

TABLE I. CATALYZED ADDITIONS OF ALDEHYDES TO α,β-UNSATURATED KETONES (*Continued*)

Aldehyde	Michael Acceptor	Conditions	Product(s) and Yield(s) (%)	Refs.
	CH_2=$CHCOC_6H_5$	Cat. a, EtOH, NaOAc, 80°	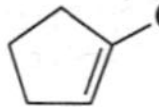 $CO(CH_2)_2COC_6H_5$ (28)	16
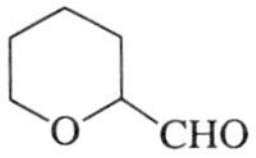 CHO	CH_2=$CHCOCH_3$	Cat. a, EtOH, TEA, 80°	$THPCO(CH_2)_2COCH_3$ (83)	20
	CH_2=$CHCOC_2H_5$	Cat. a, EtOH, TEA, 80°	$THPCO(CH_2)_2COC_2H_5$ (71)	20
	CH_2=$CHCOC_5H_{11}$-*n*	Cat. a, EtOH, TEA, 80°	$THPCO(CH_2)_2COC_5H_{11}$-*n* (59)	20
	CH_2=$CHCOC_6H_{13}$-*n*	Cat. a, EtOH, TEA, 80°	$THPCO(CH_2)_2COC_6H_{13}$-*n* (57)	20
	CH_2=$CHCOC_6H_5$	Cat. a, EtOH, TEA, 80°	$THPCO(CH_2)_2COC_6H_5$ (64)	20
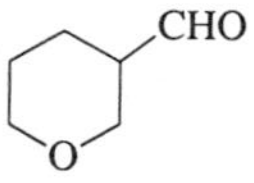 CHO	CH_2=$CHCOCH_3$	Cat. a, EtOH, TEA, 80°	3-$THPCO(CH_2)_2COCH_3$ (65)	20
	CH_2=$CHCOC_2H_5$	Cat. a, EtOH, TEA, 80°	3-$THPCO(CH_2)_2COC_2H_5$ (61)	20
	CH_2=$CHCOC_5H_{11}$-*n*	Cat. a, EtOH, TEA, 80°	3-$THPCO(CH_2)_2COC_5H_{11}$-*n* (68)	20
	CH_2=$CHCOC_6H_{13}$-*n*	Cat. a, EtOH, TEA, 80°	3-$THPCO(CH_2)_2COC_6H_{13}$-*n* (67)	20
	CH_2=$CHCOC_6H_5$	Cat. a, EtOH, TEA, 80°	3-$THPCO(CH_2)_2COC_6H_5$ (46)	20
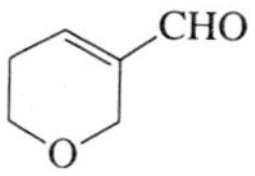 CHO	CH_2=$CHCOCH_3$	Cat. a, EtOH, TEA, 80°	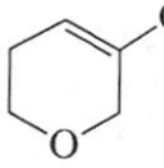 $CO(CH_2)_2COCH_3$ (73)	20
	CH_2=$CHCOC_2H_5$	Cat. a, EtOH, TEA, 80°	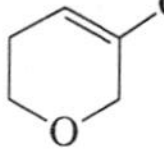 $CO(CH_2)_2COC_2H_5$ (58)	20
	CH_2=$CHCOC_5H_{11}$-*n*	Cat. a, EtOH, TEA, 80°	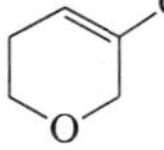 $CO(CH_2)_2COC_5H_{11}$-*n* (65)	20
	CH_2=$CHCOC_6H_{13}$-*n*	Cat. a, EtOH, TEA, 80°	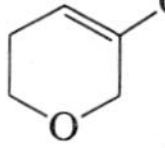 $CO(CH_2)_2COC_6H_{13}$-*n* (63)	20

	$CH_2{=}CHCOC_6H_5$	Cat. a, EtOH, TEA, 80°	$CO(CH_2)_2COC_6H_5$ (O) (57)	20
CHO (O)	$CH_2{=}CHCOCH_3$	Cat. a, dioxane, TEA, 80°	$CO(CH_2)_2COCH_3$ (O) (64)	20
	$CH_2{=}CHCOC_2H_5$	Cat. a, dioxane, TEA, 80°	$CO(CH_2)_2COC_2H_5$ (O) (68)	20
	$CH_2{=}CHCOC_5H_{11}$-*n*	Cat. a, dioxane, TEA, 80°	$CO(CH_2)_2COC_5H_{11}$-*n* (O) (53)	20
	$CH_2{=}CHCOC_6H_{13}$-*n*	Cat. a, dioxane, TEA, 80°	$CO(CH_2)_2COC_6H_{13}$-*n* (O) (43)	20
	$CH_2{=}CHCOC_6H_5$	Cat. a, dioxane, TEA, 80°	$CO(CH_2)_2COC_6H_5$ (O) (83)	20
CH_3 (O) CHO	$CH_2{=}CHCOCH_3$	Cat. a, EtOH, TEA, 80°	CH_3 (O) $CO(CH_2)_2COCH_3$ (99)	51
	$CH_2{=}CHCOCH_2O_2CCH_3$	Cat. a, dioxane, TEA, 80°	CH_3 (O) $CO(CH_2)_2COCH_2O_2CCH_3$ (57)	20
CH_3 (S) CHO	$(CH_3)_2N(CH_2)_2COC_6H_5$	NaCN, DMF, rt	CH_3 (S) $CO(CH_2)_2COC_6H_5$ (0)	59
$HOCH_2$ (O) CHO	$CH_2{=}CHCOCH_3$	Cat. b, EtOH, TEA, 80°	$HOCH_2$ (O) $CO(CH_2)_2COCH_3$ (53)	51

TABLE I. CATALYZED ADDITIONS OF ALDEHYDES TO α,β-UNSATURATED KETONES (*Continued*)

Aldehyde	Michael Acceptor	Conditions	Product(s) and Yield(s) (%)	Refs.
2,2-dimethyl-1,3-dioxolane-4-CHO (CH_3, CH_3, O, O, CHO)	$CH_2{=}CHCOCH_3$	Cat. a, EtOH, TEA, 80°	2,2-dimethyl-1,3-dioxolan-4-yl-$CO(CH_2)_2COCH_3$ (37)	21
	$CH_2{=}CHCOCH_2O_2CCH_3$	Cat. a, EtOH, TEA, 80°	2,2-dimethyl-1,3-dioxolan-4-yl-$CO(CH_2)_2COCH_2O_2CCH_3$ (43)	21
	$CH_2{=}CHCOC_2H_5$	Cat. a, EtOH, TEA, 80°	2,2-dimethyl-1,3-dioxolan-4-yl-$CO(CH_2)_2COC_2H_5$ (36)	21
	$CH_2{=}CHCOC_6H_{13}$-*n*	Cat. a, EtOH, TEA, 80°	2,2-dimethyl-1,3-dioxolan-4-yl-$CO(CH_2)_2COC_6H_{13}$-*n* (28)	21
	$CH_2{=}CHCOC_7H_{15}$-*n*	Cat. a, EtOH, TEA, 80°	2,2-dimethyl-1,3-dioxolan-4-yl-$CO(CH_2)_2COC_7H_{15}$-*n* (29)	21
	$CH_2{=}CHCOC_8H_{17}$-*n*	Cat. a, EtOH, TEA, 80°	2,2-dimethyl-1,3-dioxolan-4-yl-$CO(CH_2)_2COC_8H_{17}$-*n* (28)	21
	$CH_2{=}CHCOC_6H_5$	Cat. a, EtOH, TEA, 80°	2,2-dimethyl-1,3-dioxolan-4-yl-$CO(CH_2)_2COC_6H_5$ (56)	21
2-pyridine-CHO (N, CHO)	$CH_2{=}CHCOCH_3$	NaCN, DMF, 35°	2-$C_5H_4NCO(CH_2)_2COCH_3$ (12)	29
	"	Cat. c, dioxane, TEA, 80°	" (75)	29
	$CH_2{=}CHCOC_2H_5$	Cat. c, dioxane, TEA, 80°	2-$C_5H_4NCO(CH_2)_2COC_2H_5$ (65)	29
	$CH_2{=}CHCOC_6H_5$	Cat. c, dioxane, TEA, 80°	2-$C_5H_4NCO(CH_2)_2COC_6H_5$ (76)	29
	(*E*)-$C_6H_5CH{=}CHCOCH_3$.	Cat. c, dioxane, TEA, 80°	2-$C_5H_4NCOCH(C_6H_5)CH_2COCH_3$ (45)	29
	$C_6H_5CH{=}CHCOC_6H_5$	NaCN, DMF, 35°	2-$C_5H_4NCOCH(C_6H_5)CH_2COC_6H_5$ (91)	33

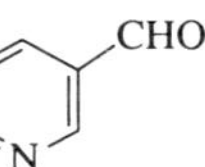	$CH_2{=}CHCOCH_3$	NaCN, DMF, 35°	$3\text{-}C_5H_4NCO(CH_2)_2CHCOCH_3$ (88)	33
	$CH_2{=}CHCO(CH_2)_2OCH_3$	Cat. b, dioxane, TEA, 90°	$3\text{-}C_5H_4NCO(CH_2)_2CO(CH_2)_2OCH_3$ (70)	19
	$CH_2{=}CHCOC_6H_5$	NaCN, DMF, 35°	$3\text{-}C_5H_4NCO(CH_2)_2COC_6H_5$ (80)	33
	$(CH_3)_2N(CH_2)_2COC_6H_5$	NaCN, DMF, 35°	$3\text{-}C_5H_4NCO(CH_2)_2COC_6H_5$ (35)	37
	$CH_2{=}CHCOC_6H_4OCH_3\text{-}p$	Cat. b, dioxane, TEA, 90°	$3\text{-}C_5H_4NCO(CH_2)_2COC_6H_4OCH_3\text{-}p$ (82)	19
	$CH_2{=}CHCO(CH_2)_2C{=}O$ with $2,4\text{-}(CH_3O)_2C_6H_3$ on C=O	Cat. b, dioxane, TEA, 90°	$3\text{-}C_5H_4NCO(CH_2)_2CO(CH_2)_2C{=}O$ with $2,4\text{-}(CH_3O)_2C_6H_3$ on C=O (70)	19
	$C_6H_5CH{=}CHCO(C_4H_3O\text{-}2)$	NaCN, DMF, 35°	$3\text{-}C_5H_4NCOCH(C_6H_5)CH_2COC_4H_3O$ (80)	29
	"	Cat. b, EtOH, TEA, 80°	" (70)	29
	$(C_4H_3O\text{-}2)CH{=}CHCOC_6H_5$	NaCN, DMF, 35°	$3\text{-}C_5H_4NCOCH(C_4H_3O)CH_2COC_6H_5$ (67)	29
	"	Cat. b, EtOH, TEA, 80°	" (65)	29
	$(C_4H_3O\text{-}2)CH{=}CHCO(C_4H_3O\text{-}2)$	NaCN, DMF, 35°	$3\text{-}C_5H_4NCOCH(C_4H_3O)CH_2COC_4H_3O$ (88)	29
	"	Cat. b, EtOH, TEA, 80°	" (67)	29
	$(C_4H_3O\text{-}2)CH{=}CHCO(C_5H_4N\text{-}3)$	Cat. b, EtOH, TEA, 80°	$3\text{-}C_6H_4NCOCH(C_4H_3O)CH_2COC_5H_4N\text{-}3$ (66)	29
	"	NaCN, DMF, 35°	" (56)	29
	$(C_5H_4N\text{-}3)CH{=}CHCOC_6H_5$	Cat. c, EtOH, TEA, 80°	$3\text{-}C_5H_4NCOCH(C_5H_4N\text{-}3)CH_2COC_6H_5$ (80)	29
	"	NaCN, DMF, 35°	" (83)	29
	$C_6H_5CH{=}CHCO(C_5H_4N\text{-}3)$	Cat. c, EtOH, TEA, 80°	$3\text{-}C_5H_4NCOCH(C_6H_5)CH_2COC_5H_4N\text{-}3$ (76)	29
	"	NaCN, DMF, 35°	" (72)	29

TABLE I. CATALYZED ADDITIONS OF ALDEHYDES TO α,β-UNSATURATED KETONES (*Continued*)

Aldehyde	Michael Acceptor	Conditions	Product(s) and Yield(s) (%)	Refs.
	2-quinolyl-CH=CHCOC$_6$H$_5$	NaCN, DMF, 35°	3–C$_5$H$_4$NCOCH(2-quinolyl)CH$_2$COC$_6$H$_5$ (63)	29
	3-pyridyl-CH=CHCO-3-pyridyl	Cat. c, EtOH, TEA, 80°	3-C$_5$H$_4$NCOCH(C$_5$H$_4$N-3)CH$_2$COC$_5$H$_4$N-3 (66)	29
	"	NaCN, DMF, 35°	" (74)	29
4-pyridyl-CHO	CH$_2$═CHCOCH$_3$	NaCN, DMF, 35°	4-C$_5$H$_4$NCO(CH$_2$)$_2$COCH$_3$ (70)	33
1-pyrrolidinyl-COCHO	CH$_2$═CHCOCH$_3$	Cat. a, dioxane, TEA, 80°	C$_4$H$_8$NCOCO(CH$_2$)$_2$COCH$_3$ (83)	25
	CH$_2$═CHCOC$_2$H$_5$	Cat. a, dioxane, TEA, 80°	C$_4$H$_8$NCOCO(CH$_2$)$_2$COC$_2$H$_5$ (80)	25
	CH$_2$═C(CH$_3$)COCH$_3$	Cat. a, dioxane, TEA, 80°	C$_4$H$_8$NCOCOCH$_2$CH(CH$_3$)COCH$_3$ (77)	25
	CH$_2$═CHCOC$_6$H$_5$	Cat. a, dioxane, TEA, 80°	C$_4$H$_8$NCOCO(CH$_2$)$_2$COC$_6$H$_5$ (40)	25
	C$_6$H$_5$CH═CHCOC$_6$H$_5$	Cat. a, dioxane, TEA, 80°	C$_4$H$_8$NCOCOCH(C$_6$H$_5$)CH$_2$COC$_6$H$_5$ (60)	25
C$_7$				
n-C$_6$H$_{13}$CHO	CH$_2$═CHCOCH$_3$	Cat. a, EtOH, TEA, 80°	*n*-C$_6$H$_{13}$CO(CH$_2$)$_2$COCH$_3$ (78)	9
	"	Thiamine · HCl, EtOH, TEA, 80°	" (58)	74
	"	Cat. e, EtOH, TEA, reflux	" (65)	12c
	"	Cat. e, *i*-C$_3$H$_7$OH, TEA, 82°	" (68)	74
	"	Cat. e, TEA, reflux	" (69)	12c
	"	Cat. a, EtOH, TEA, 80°	" (71–75)	15

Aldehyde	Acceptor	Conditions	Product (% yield)	Refs.
	$CH_2{=}CHCOCH{=}CH_2$	Cat. a, TEA, 25°	$n\text{-}C_6H_{13}CO(CH_2)_2COCH{=}CH_2$ (20)	17
	$CH_2{=}CHCO(CH_2)_2CO_2CH_3$	Cat. a, dioxane, TEA, 100°	$n\text{-}C_6H_{13}[CO(CH_2)_2]_2CO_2CH_3$ (55)	24
	$CH_2{=}CHCO(CH_2)_3CO_2CH_3$	Cat. a, dioxane, TEA, 100°	$n\text{-}C_6H_{13}CO(CH_2)_2CO(CH_2)_3CO_2CH_3$ (33)	24
	$CH_2{=}CHCOC_5H_{11}\text{-}n$	Cat. a, TEA, 90°	$n\text{-}C_6H_{13}CO(CH_2)_2COC_5H_{11}\text{-}n$ (78)	77
	$CH_2{=}CHCO(CH_2)_3$–(1,3-dioxolan-2-yl, H)	Cat. a, TEA, 65°	$n\text{-}C_6H_{13}CO(CH_2)_2CO(CH_2)_3$–(1,3-dioxolan-2-yl, H) (73)	50
	$CH_2{=}CHCOCH_2$–(2-CH_3-4,5-di-CH_3-1,3-dioxolan-2-yl)	Cat. a, dioxane, TEA, 80°	$n\text{-}C_6H_{13}CO(CH_2)_2COCH_2$–(2-$CH_3$-4,5-di-$CH_3$-1,3-dioxolan-2-yl) (67)	47
	$CH_2{=}CHCOC_6H_4OCH_3\text{-}p$	Cat. a, dioxane, TEA, 100°	$n\text{-}C_6H_{13}CO(CH_2)_2COC_6H_4OCH_3\text{-}p$ (60)	19
	$CH_2{=}CHCOCH_2$–(2-C_2H_5-4,5-di-CH_3-1,3-dioxolan-2-yl)	Cat. a, dioxane, TEA, 80°	$n\text{-}C_6H_{13}CO(CH_2)_2COCH_2$–(2-$C_2H_5$-4,5-di-$CH_3$-1,3-dioxolan-2-yl) (69)	47
	$CH_2{=}CHCOC_6H_3(OCH_3)_2\text{-}2,4$	Cat. a, dioxane, TEA, 100°	$n\text{-}C_6H_{13}CO(CH_2)_2COC_6H_3(OCH_3)_2\text{-}2,4$ (60)	19
	$CH_2{=}CHCOC_6H_3(OCH_3)_2\text{-}3,4$	Cat. a, dioxane, TEA, 100°	$n\text{-}C_6H_{13}CO(CH_2)_2COC_6H_3(OCH_3)_2\text{-}3,4$ (57)	19
$CH_2{=}CH(CH_2)_4CHO$	$CH_2{=}CHCOC_5H_{11}\text{-}n$	Cat. a, TEA, 90°	$CH_2{=}CH(CH_2)_4CO(CH_2)_2COC_5H_{11}\text{-}n$ (50)	62
	$CH_2{=}CHCOC_9H_{19}\text{-}n$	Cat. a, TEA, 90°	$CH_2{=}CH(CH_2)_4CO(CH_2)_2COC_9H_{19}\text{-}n$ (50)	62
(*Z*)-$C_2H_5CH{=}CH(CH_2)_2CHO$	$CH_2{=}CHCOCH_3$	Cat. a, EtOH, TEA, 80°	(*Z*)-$C_2H_5CH{=}CH(CH_2)_2CO(CH_2)_2COCH_3$ (76)	9
$C_2H_5C{\equiv}C(CH_2)_2CHO$	"	Thiamine · HCl, DMF, TEA, 80°	$C_2H_5C{\equiv}C(CH_2)_2CO(CH_2)_2COCH_3$ (65)	18
$C_2H_5O(CH_2)_4CHO$	"	Cat. a, dioxane, TEA, 90°	$C_2H_5O(CH_2)_4CO(CH_2)_2COCH_3$ (73)	21
$n\text{-}C_3H_7O(CH_2)_3CHO$	"	Cat. a, dioxane, TEA, 90°	$n\text{-}C_3H_7O(CH_2)_3CO(CH_2)_2COCH_3$ (61)	21
3-cyclohexenyl-CHO	$CH_2{=}CHCOCH_3$	Cat. a, EtOH, NaOAc, 80°	3-cyclohexenyl-$CO(CH_2)_2COCH_3$ (70)	16
		Cat. a, EtOH, NaOAc, 80°		16

TABLE I. CATALYZED ADDITIONS OF ALDEHYDES TO α,β-UNSATURATED KETONES (*Continued*)

Aldehyde	Michael Acceptor	Conditions	Product(s) and Yield(s) (%)	Refs.
	$CH_2{=}CHCOCH{=}C(CH_3)_2$	Cat. a, EtOH, NaOAc, 80°	cyclohexenyl–$CO(CH_2)_2COCH{=}C(CH_3)_2$ (61)	16
	$CH_2{=}CHCOC_6H_5$	Cat. a, EtOH, NaOAc, 80°	cyclohexenyl–$CO(CH_2)_2COC_6H_5$ (71)	16
C_6H_5CHO	$CH_2{=}CHCOCH_3$	Cat. c, TEA, 80°	$C_6H_5CO(CH_2)_2COCH_3$ (65)	48
	"	NaCN, DMF, 35°	" (82)	33
	"	Cat. e, DMF, TEA, reflux	" (64)	12c
	"	Cat. e, TEA, reflux	" (53)	12c
	"	Cat. a, EtOH, NaOAc, reflux	" (64)	28
	$(CH_3)_2N(CH_2)_2COCH_3$	NaCN, DMF, 40°	" (45)	37
	$CH_2{=}CHCOCH{=}CH_2$	Cat. b, TEA, 40°	$C_6H_5CO(CH_2)_2COCH{=}CH_2$ (20)	17
	$CH_2{=}CHCO(CH_2)_2CN$	Cat. b, TEA, 65°	$C_6H_5[CO(CH_2)_2]_2CN$ (62)	41
	2-methylenecyclopentanone	Cat. b, EtOH, TEA, 80°	$C_6H_5COCH_2$–(2-oxocyclopentyl) (44)	76
	$CH_2{=}CHCO(CH_2)_2OCH_3$	Cat. b, dioxane, TEA, 90°	$C_6H_5CO(CH_2)_2CO(CH_2)_2OCH_3$ (60)	19
	$CH_2{=}CHCOCH_2O_2CCH_3$	Cat. a, dioxane, TEA, 80°	$C_6H_5CO(CH_2)_2COCH_2O_2CCH_3$ (63)	20
	$CH_2{=}CHCO(CH_2)_2CO_2CH_3$	Cat. b, dioxane, TEA, 100°	$C_6H_5[CO(CH_2)_2]_2CO_2CH_3$ (33)	24
	$I^-(CH_3)_3N^+(CH_2)_2COCH{=}C(CH_3)_2$	Cat. c, DMF, TEA, 80°	$C_6H_5CO(CH_2)_2COCH{=}C(CH_3)_2$ (55)	37
	$(CH_3)_2N(CH_2)_2COC_4H_9$-*t*	NaCN, DMF, 70°	$C_6H_5CO(CH_2)_2COC_4H_9$-*t* (60)	37
	$(CH_3)_2N(CH_2)_2CO$–(5-methyl-2-thienyl)	NaCN, DMF, rt	$C_6H_5CO(CH_2)_2CO$–(5-methyl-2-thienyl) (76)	79
	$CH_2{=}CH[CO(CH_2)_2]_2CN$	Cat. b, EtOH, TEA, 80°	$C_6H_5[CO(CH_2)_2]_3CN$ (63)	52

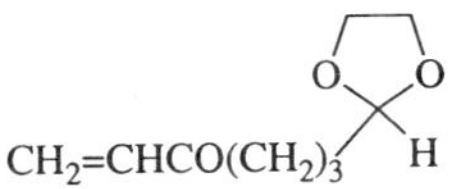 $CH_2{=}CHCO(CH_2)_3$–(1,3-dioxolan-2-yl, H)	Cat. b, TEA, 65°	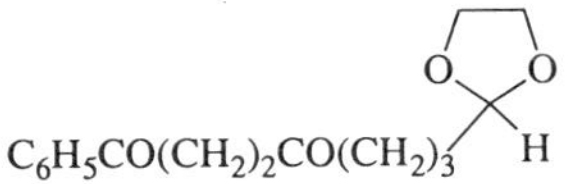 $C_6H_5CO(CH_2)_2CO(CH_2)_3$–(1,3-dioxolan-2-yl, H) (64)	50
$HO_2CCH{=}CHCOC_6H_5$	Cat. a, EtOH, TEA, 80°	$C_6H_5CO(CH_2)_2COC_6H_5$ (81)	36
$(CH_3)_2N(CH_2)_2COC_6H_5$	NaCN, DMF, 35°	" (64)	37
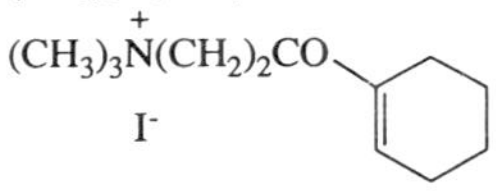 $(CH_3)_3\overset{+}{N}(CH_2)_2CO$–(cyclohexenyl) I^-	Cat. a, DMF, TEA, 80°	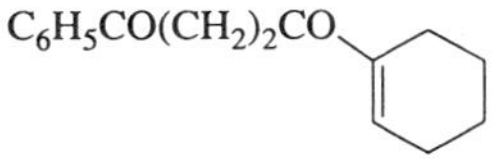 $C_6H_5CO(CH_2)_2CO$–(cyclohexenyl) (25)	37
(*E*)-$C_6H_5CH{=}CHCOCH_3$	NaCN, DMF, 35°	$C_6H_5COCH(C_6H_5)CH_2COCH_3$ (80)	33
"	Cat. e, DMF, TEA, reflux	" (82)	12c
"	Cat. e, EtOH, TEA, reflux	" (76)	12c
$CH_2{=}CH[CO(CH_2)_2]_2CO_2CH_3$	Cat. b, TEA, 65°	$C_6H_5[CO(CH_2)_2]_3CO_2CH_3$ (67)	52
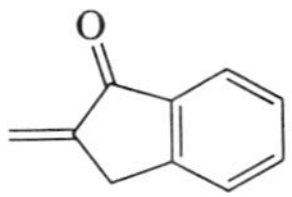 2-methylene-1-indanone (O)	Cat. b, EtOH, TEA, 80°	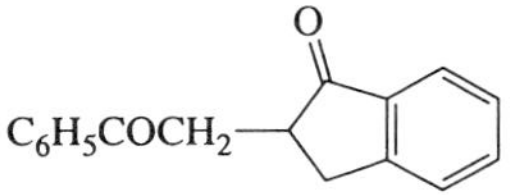 $C_6H_5COCH_2$–(1-indanon-2-yl, O) (51)	76
$(CH_3)_2NCH_2CH(CH_3)COC_6H_5$	NaCN, DMF, 100°	$C_6H_5COCH_2CH(CH_3)COC_6H_5$ (49)	37
$CH_2{=}CHCOC_6H_4OCH_3$-*p*	Cat. b, dioxane, TEA, 90°	$C_6H_5CO(CH_2)_2COC_6H_4OCH_3$-*p* (63)	19
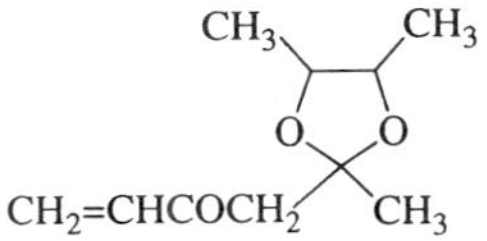 $CH_2{=}CHCOCH_2$–(2,4,5-trimethyl-1,3-dioxolan-2-yl; CH_3, CH_3, CH_3, O, O)	Cat. b, dioxane, TEA, 80°	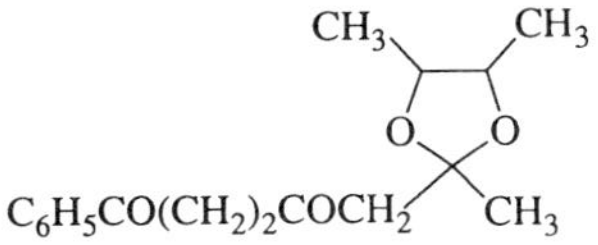 $C_6H_5CO(CH_2)_2COCH_2$–(2,4,5-trimethyl-1,3-dioxolan-2-yl; CH_3, CH_3, CH_3, O, O) (65)	47
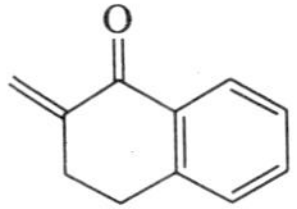 2-methylene-1-tetralone (O)	Cat. b, EtOH, TEA, 80°	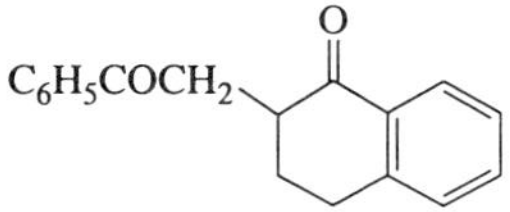 $C_6H_5COCH_2$–(1-tetralon-2-yl, O) (66)	76
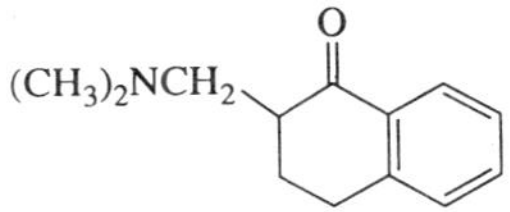 $(CH_3)_2NCH_2$–(1-tetralon-2-yl, O)	NaCN, DMF, 40°	" (63)	37

TABLE I. CATALYZED ADDITIONS OF ALDEHYDES TO α,β-UNSATURATED KETONES (*Continued*)

Aldehyde	Michael Acceptor	Conditions	Product(s) and Yield(s) (%)	Refs.
	$CH_2{=}CHCOC_6H_3(OCH_3)_2$-3,4	Cat. b, dioxane, TEA, 90°	$C_6H_5CO(CH_2)_2COC_6H_3(OCH_3)_2$-3,4 (60)	19
	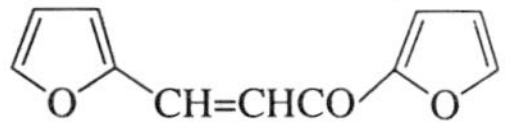 2-furyl-CH=CHCO-2-furyl	Cat. b, EtOH, TEA, 80°	$C_6H_5COCH(C_4H_3O)CH_2COC_4H_3O$ (82)	29
	"	NaCN, DMF, 35°	" (75)	29
	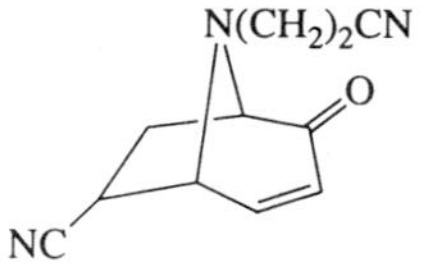 NC-substituted N(CH_2)$_2$CN bicyclic enone	Cat. c, TEA, 70°	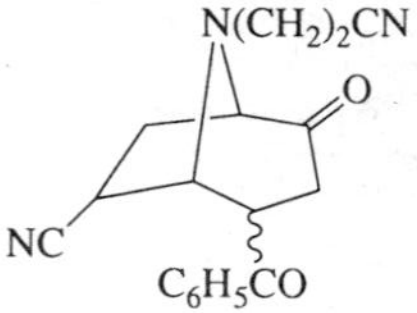 NC-substituted N(CH_2)$_2$CN bicyclic ketone bearing C_6H_5CO (2)	34
	$C_2H_5O_2CCH{=}CHCOC_6H_5$	Cat. b, EtOH, TEA, 80°	$C_6H_5COCH(CO_2C_2H_5)CH_2COC_6H_5$ (53)	36
	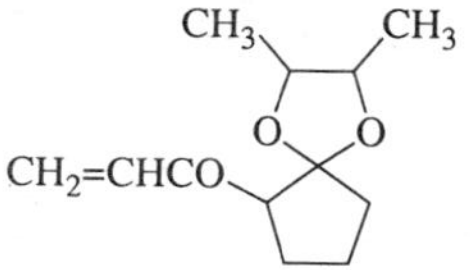 $CH_2{=}CHCO$-(dimethyl-dioxaspiro cyclopentyl)	Cat. b, dioxane, TEA, 80°	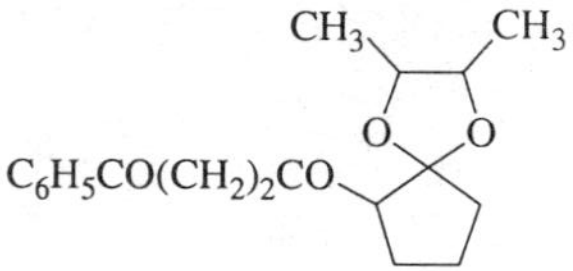 $C_6H_5CO(CH_2)_2CO$-(dimethyl-dioxaspiro cyclopentyl) (71)	47
	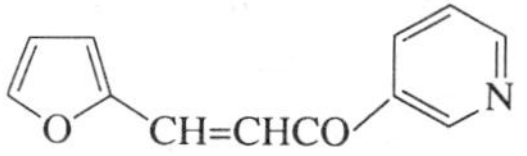 2-furyl-CH=CHCO-3-pyridyl	NaCN, DMF, 35°	$C_6H_5COCH(C_4H_3O)CH_2COC_5H_4N$-3 (53)	29
	"	Cat. b, EtOH, TEA, 80°	" (65)	29
	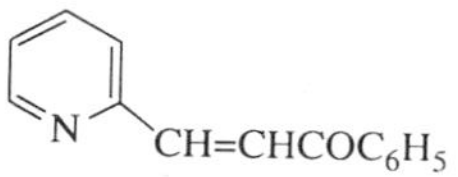 2-pyridyl-$CH{=}CHCOC_6H_5$	Cat. c, EtOH, TEA, 80°	$C_6H_5COCH(C_5H_4N)CH_2COC_6H_5$ (86)	29
	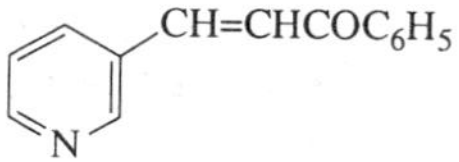 3-pyridyl-$CH{=}CHCOC_6H_5$	Cat. c, EtOH, TEA, 80°	$C_6H_5COCH(C_5H_4N\text{-}3)CH_2COC_6H_5$ (72)	29
	"	NaCN, DMF, 35°	" (87)	29

Reactant	Conditions	Product (Yield, %)	Refs.
$C_6H_5CH{=}CHCO$-(3-pyridyl)	NaCN, DMF, 35°	$C_6H_5COCH(C_6H_5)CH_2COC_5H_4N$-3 (65)	29
"	Cat. b, EtOH, TEA, 80°	" (79)	29
$CH_2{=}CHCO$-(2,3-dimethyl-1,4-dioxaspiro[4.5]dec-6-yl)	Cat. b, dioxane, TEA, 80°	$C_6H_5CO(CH_2)_2CO$-(2,3-dimethyl-1,4-dioxaspiro[4.5]dec-6-yl) (67)	47
$C_6H_5CH{=}CHCO$-(2-furyl)	Cat. c, EtOH, TEA, 80°	$C_6H_5COCH(C_6H_5)CH_2COC_4H_3O$ (71)	29
"	NaCN, DMF, 35°	" (77)	29
(2-furyl)-$CH{=}CHCOC_6H_5$	NaCN, DMF, 35°	$C_6H_5COCH(C_4H_3O)CH_2COC_6H_5$ (73)	29
"	Cat. b, TEA, 80°	" (85)	29
$C_6H_5CH{=}CHCO$-(2-thienyl)	NaCN, DMF, 35°	$C_6H_5COCH(C_6H_5)CH_2COC_4H_3S$ (75)	29
(3-pyridyl)-$CH{=}CHCO$-(3-pyridyl)	NaCN, DMF, 35°	$C_6H_5COCH(C_5H_4N\text{-}3)CH_2COC_5H_4N$-3 (48)	29
"	Cat. b, EtOH, TEA, 80°	" (47)	29
$CH_2{=}CRCO$-(2-furyl), R = $-CH_2$-(2,4,5-trimethyl-1,3-dioxolan-2-yl)	Cat. c, EtOH, TEA, 80°	$C_6H_5COCH_2CHRCOC_4H_3O$ (63)	49

TABLE I. Catalyzed Additions of Aldehydes to α,β-Unsaturated Ketones (*Continued*)

Aldehyde	Michael Acceptor	Conditions	Product(s) and Yield(s) (%)	Refs.
	$C_2H_5O_2CCHC(COC_6H_5)COCH_3$	Cat. b, dioxane, TEA, 80°	$C_6H_5COCH(CO_2C_2H_5)CH(COC_6H_5)COCH_3$ (47)	53
	$C_6H_5CH{=}CHCOC_6H_5$	NaCN, DMF, 35°	$C_6H_5COCH(C_6H_5)CH_2COC_6H_5$ (93)	33
	"	Cat. c, DMF, TEA, 80°	" (83)	7
	$(CH_3)_2NCH_2CH(C_6H_5)COC_6H_5$	NaCN, DMF, 40°	" (58)	37
	$C_2H_5O_2CCH{=}C(CO_2C_2H_5)COC_6H_5$	Cat. b, dioxane, TEA, 80°	$C_6H_5COCH(CO_2C_2H_5)CH(CO_2C_2H_5)COC_6H_5$ (66)	53
	(2-quinolyl)$CH{=}CHCOC_6H_5$	NaCN, DMF, 35°	$C_6H_5COCH($2-quinolyl$)CH_2COC_6H_5$ (48)	29
	$C_2H_5O_2CCH{=}C(COC_6H_5)_2$	Cat. b, dioxane, TEA, 80°	$C_6H_5COCH(CO_2C_2H_5)CH(COC_6H_5)_2$ (73)	53
o-ClC_6H_4CHO	$(CH_3)_2N(CH_2)_2CO$-(2-thienyl)	NaCN, DMF, rt	o-$ClC_6H_4CO(CH_2)_2COC_4H_3S$ (55)	79
p-ClC_6H_4CHO	$CH_2{=}CHCOCH_3$	NaCN, DMF, 35°	p-$ClC_6H_4CO(CH_2)_2COCH_3$ (98)	33
	$CH_2{=}CHCO(CH_2)_2CN$	Cat. b, TEA, 65°	p-$ClC_6H_4[CO(CH_2)_2]_2CN$ (50)	41
	$CH_2{=}CHCO(CH_2)_2CO_2CH_3$	Cat. a, TEA, 65°	p-$ClC_6H_4[CO(CH_2)_2]_2CO_2C_2H_5$ (60)	41
	$(CH_3)_2N(CH_2)_2CO$-(5-methyl-2-thienyl)	NaCN, DMF, rt	p-$ClC_6H_4CO(CH_2)_2CO$-(5-methyl-2-thienyl) (52)	79
	$HO_2CCH{=}CHCOC_6H_5$	Cat. b, EtOH, TEA, 80°	p-$ClC_6H_4CO(CH_2)_2COC_6H_5$ (86)	36
	$HO_2CCH{=}CHCOC_6H_4Cl$-p	Cat. b, EtOH, TEA, 80°	p-$ClC_6H_4CO(CH_2)_2COC_6H_4Cl$-$p$ (82)	36
	$HO_2CCH{=}CHCOC_6H_4OCH_3$-$p$	Cat. a, EtOH, TEA, 80°	p-$ClC_6H_4CO(CH_2)_2COC_6H_4OCH_3$-$p$ (65)	36
	$CH_2{=}CHCOC_6H_4OCH_3$-p	Cat. b, dioxane, TEA, 90°	" (77)	19
	(E)-$C_6H_5CH{=}CHCOCH_3$	NaCN, DMF, 35°	p-$ClC_6H_4COCH(C_6H_5)CH_2COCH_3$ (98)	33
	$CH_3O_2CCH{=}CHCOC_6H_5$	Cat. b, EtOH, TEA, 80°	p-$ClC_6H_4COCH(CO_2CH_3)CH_2COC_6H_5$ (70)	36
	$C_6H_5CH{=}CHCOC_6H_5$	NaCN, DMF, 35°	p-$ClC_6H_4COCH(C_6H_5)CH_2COC_6H_5$ (98)	33

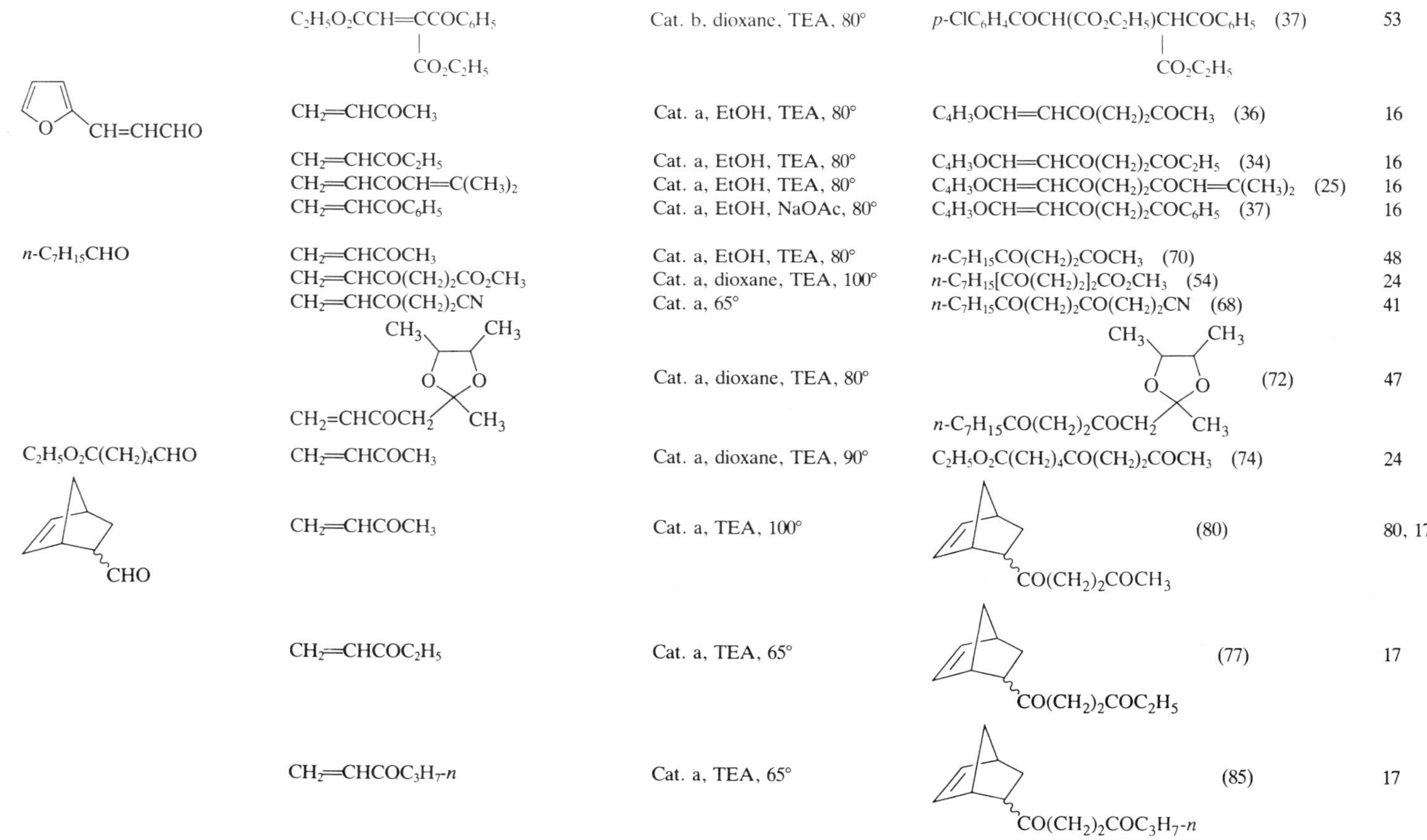

	Aldehyde	Acceptor	Conditions	Product (Yield, %)	Refs.
		$C_2H_5O_2CCH{=}C(CO_2C_2H_5)COC_6H_5$	Cat. b, dioxane, TEA, 80°	p-$ClC_6H_4COCH(CO_2C_2H_5)CH(CO_2C_2H_5)COC_6H_5$ (37)	53
	2-Furyl-CH=CHCHO	$CH_2{=}CHCOCH_3$	Cat. a, EtOH, TEA, 80°	$C_4H_3OCH{=}CHCO(CH_2)_2COCH_3$ (36)	16
		$CH_2{=}CHCOC_2H_5$	Cat. a, EtOH, TEA, 80°	$C_4H_3OCH{=}CHCO(CH_2)_2COC_2H_5$ (34)	16
		$CH_2{=}CHCOCH{=}C(CH_3)_2$	Cat. a, EtOH, TEA, 80°	$C_4H_3OCH{=}CHCO(CH_2)_2COCH{=}C(CH_3)_2$ (25)	16
		$CH_2{=}CHCOC_6H_5$	Cat. a, EtOH, NaOAc, 80°	$C_4H_3OCH{=}CHCO(CH_2)_2COC_6H_5$ (37)	16
C_8	n-$C_7H_{15}CHO$	$CH_2{=}CHCOCH_3$	Cat. a, EtOH, TEA, 80°	n-$C_7H_{15}CO(CH_2)_2COCH_3$ (70)	48
		$CH_2{=}CHCO(CH_2)_2CO_2CH_3$	Cat. a, dioxane, TEA, 100°	n-$C_7H_{15}[CO(CH_2)_2]_2CO_2CH_3$ (54)	24
		$CH_2{=}CHCO(CH_2)_2CN$	Cat. a, 65°	n-$C_7H_{15}CO(CH_2)_2CO(CH_2)_2CN$ (68)	41
		$CH_2{=}CHCOCH_2$-(2,4,5-trimethyl-1,3-dioxolan-2-yl)	Cat. a, dioxane, TEA, 80°	n-$C_7H_{15}CO(CH_2)_2COCH_2$-(2,4,5-trimethyl-1,3-dioxolan-2-yl) (72)	47
	$C_2H_5O_2C(CH_2)_4CHO$	$CH_2{=}CHCOCH_3$	Cat. a, dioxane, TEA, 90°	$C_2H_5O_2C(CH_2)_4CO(CH_2)_2COCH_3$ (74)	24
	5-Norbornene-2-CHO	$CH_2{=}CHCOCH_3$	Cat. a, TEA, 100°	5-Norbornen-2-yl-$CO(CH_2)_2COCH_3$ (80)	80, 17
		$CH_2{=}CHCOC_2H_5$	Cat. a, TEA, 65°	5-Norbornen-2-yl-$CO(CH_2)_2COC_2H_5$ (77)	17
		$CH_2{=}CHCOC_3H_7$-n	Cat. a, TEA, 65°	5-Norbornen-2-yl-$CO(CH_2)_2COC_3H_7$-n (85)	17

TABLE I. CATALYZED ADDITIONS OF ALDEHYDES TO α,β-UNSATURATED KETONES (*Continued*)

Aldehyde	Michael Acceptor	Conditions	Product(s) and Yield(s) (%)	Refs.
	$CH_2{=}CHCO(CH_2)_2CN$	Cat. a, TEA, 65°	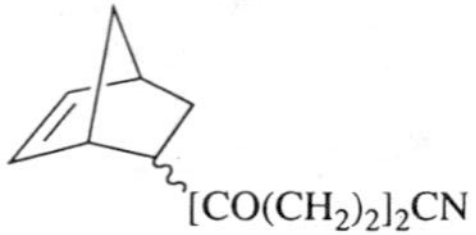 $[CO(CH_2)_2]_2CN$ (60)	52
	$CH_2{=}CHCO(CH_2)_2CO_2CH_3$	Cat. a, TEA, 65°	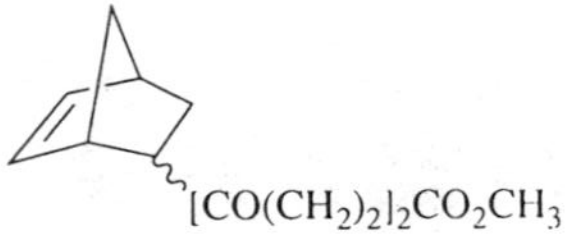 $[CO(CH_2)_2]_2CO_2CH_3$ (56)	52
	$CH_2{=}CHCOC_4H_9$-*n*	Cat. a, TEA, 65°	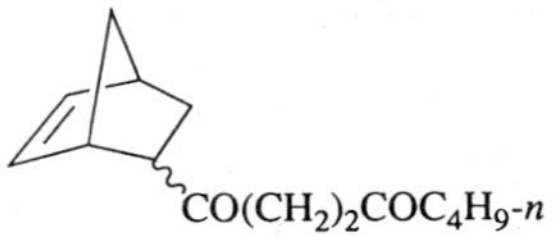 $CO(CH_2)_2COC_4H_9$-*n* (82)	17
	$CH_2{=}CHCOC_6H_{13}$-*n*	Cat. a, TEA, 65°	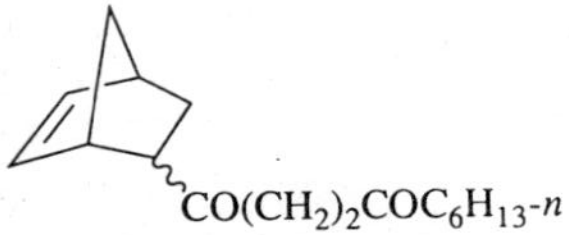 $CO(CH_2)_2COC_6H_{13}$-*n* (83)	17
	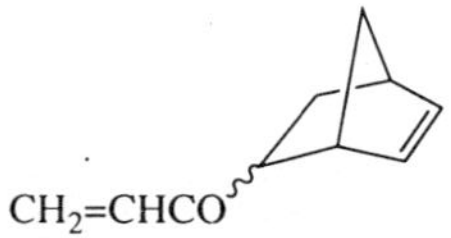 $CH_2{=}CHCO$	Cat. a, TEA, 65°	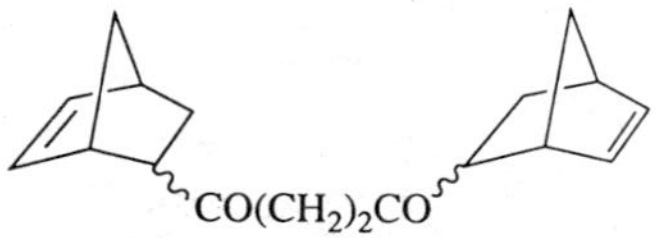 $CO(CH_2)_2CO$ (61)	32
p-$CH_3C_6H_4CHO$	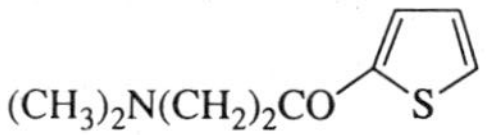 $(CH_3)_2N(CH_2)_2CO$ S	NaCN, DMF, rt	*p*-$CH_3C_6H_4CO(CH_2)_2COC_4H_3S$ (0)	79
	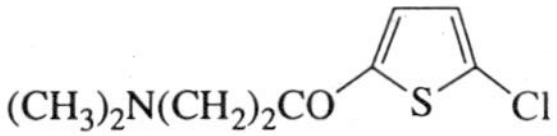 $(CH_3)_2N(CH_2)_2CO$ S Cl	NaCN, DMF, rt	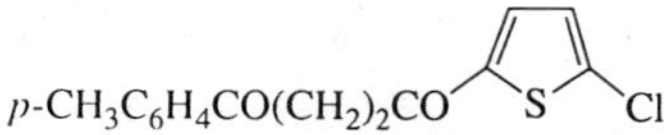 *p*-$CH_3C_6H_4CO(CH_2)_2CO$ S Cl (0)	79

	p-$CH_3OC_6H_4CHO$	$CH_2{=}CHCOCH_3$	Cat. b, TEA, 75°	p-$CH_3OC_6H_4CO(CH_2)_2COCH_3$ (42)	81
		"	Cat. c, EtOH, TEA, 80°	" (42)	48
		$HO_2CCH{=}CHCOC_6H_5$	Cat. b, EtOH, TEA, 80°	p-$CH_3OC_6H_4CO(CH_2)_2COC_6H_5$ (73)	36
		$CH_2{=}CHCOC_6H_4OCH_3$-p	Cat. b, dioxane, TEA, 90°	p-$CH_3OC_6H_4CO(CH_2)_2COC_6H_4OCH_3$-p (47)	19
	m-$CF_3C_6H_4CHO$	$(CH_3)_2N(CH_2)_2CO$-(5-methyl-2-thienyl)	NaCN, DMF, rt	m-$CF_3C_6H_4CO(CH_2)_2COC_4H_2SCH_3$-5 (40)	79
	3,4-$(OCH_2O)C_6H_3CHO$	$CH_2{=}CHCOCH_3$	Cat. c, TEA, 80°	3,4-$(OCH_2O)C_6H_3CO(CH_2)_2COCH_3$ (40)	48
C_9	n-$C_8H_{17}CHO$	$CH_2{=}CHCOCH_3$	Cat. a, EtOH, TEA, 80°	n-$C_8H_{17}CO(CH_2)_2COCH_3$ (70)	48
		$CH_2{=}CHCO(CH_2)_2CO_2CH_3$	Cat. a, dioxane, TEA, 100°	n-$C_8H_{17}[CO(CH_2)_2]_2CO_2CH_3$ (37)	24
	$C_6H_5CH_2OCH_2CHO$	$CH_2{=}CHCOC_2H_5$	Cat. a, EtOH, TEA, 80°	$C_6H_5CH_2OCH_2CO(CH_2)_2COC_2H_5$ (68)	20
		$CH_2{=}CHCOC_6H_5$	Cat. a, EtOH, TEA, 80°	$C_6H_5CH_2OCH_2CO(CH_2)_2COC_6H_5$ (65)	20
	$C_6H_5(CH_2)_2CHO$	$CH_2{=}CHCOCH_3$	Cat. a, EtOH, TEA, 80°	$C_6H_5(CH_2)_2CO(CH_2)_2COCH_3$ (61)	48
		$C_6H_5CH{=}CHCOC_6H_5$	Cat. a, EtOH, TEA, 80°	$C_6H_5(CH_2)_2COCH(C_6H_5)CH_2COC_6H_5$ (46)	7
	(E)-$C_6H_5CH{=}CHCHO$	$CH_2{=}CHCOCH_3$	Cat. a, EtOH, TEA, 80°	(E)-$C_6H_5CH{=}CHCO(CH_2)_2COCH_3$ (44)	16
		$CH_2{=}CHCOC_4H_9$-t	Cat. a, EtOH, TEA, 80°	(E)-$C_6H_5CH{=}CHCO(CH_2)_2COC_4H_9$-t (41)	16
		$CH_2{=}CHCOC_6H_5$	Cat. a, EtOH, TEA, 80°	(E)-$C_6H_5CH{=}CHCO(CH_2)_2COC_6H_5$ (39)	16
		$C_6H_5CH{=}C(CO_2C_2H_5)COCH_3$	Cat. a, EtOH, TEA, 80°	(E)-$C_6H_5CH{=}CHCOCH(C_6H_5)CH(CO_2C_2H_5)COCH_3$ (20)	36
	m- + p-$CH_2{=}CHC_6H_4CHO$	$CH_2{=}CHCOCH_3$	NaCN, DMF, 35°	m- + p-$(CH_2{=}CH)C_6H_4CO(CH_2)_2COCH_3$ (—)	80
C_{10}	n-$C_9H_{19}CHO$	$CH_2{=}CHCOCH_3$	Cat. a, EtOH, TEA, 80°	n-$C_9H_{19}CO(CH_2)_2COCH_3$ (67)	48
		"	Cat. a, TEA, 80°	" (77)	48
		$CH_2{=}CHCO(CH_2)_2CN$	Cat. a, TEA, 65°	n-$C_9H_{19}[CO(CH_2)_2]_2CN$ (70)	41
		$CH_2{=}CHCO(CH_2)_2CO_2CH_3$	Cat. a, TEA, 65°	n-$C_9H_{19}[CO(CH_2)_2]_2CO_2CH_3$ (68)	41
	i-$C_3H_7(CH_2)_3CH(CH_3)CH_2CHO$	$CH_2{=}CHCOCH_3$	Cat. a, TEA, 75°	i-$C_3H_7(CH_2)_3CH(CH_3)CH_2CO(CH_2)_2COCH_3$ (49)	81

TABLE I. CATALYZED ADDITIONS OF ALDEHYDES TO α,β-UNSATURATED KETONES (*Continued*)

Aldehyde	Michael Acceptor	Conditions	Product(s) and Yield(s) (%)	Refs.
$(CH_2)_2C{=}CH(CH_2)_2CHCH_3$ | $OHCCH_2$	$CH_2{=}CHCOCH_3$	Cat. a, EtOH, NaOAc, 80°	$(CH_3)_2C{=}CH(CH_2)_2CH(CH_3)CH_2$ | $CH_3CO(CH_2)_2C{=}O$ (80)	16
	"	Cat. a, TEA, 75°	" (47)	81
	$CH_2{=}CHCOCH{=}C(CH_3)_2$	Cat. a, EtOH, NaOAc, 80°	$(CH_3)_2C{=}CH(CH_2)_2CH(CH_3)CH_2$ | $(CH_3)_2C{=}CHCO(CH_2)_2C{=}O$ (76)	16
	$CH_2{=}CHCOC_6H_5$	Cat. a, EtOH, NaOAc, 80°	$(CH_3)_2C{=}CH(CH_2)_2CH(CH_3)CH_2$ | $C_6H_5CO(CH_2)_2C{=}O$ (73)	16
(E,Z)-$(CH_3)_2C{=}CH(CH_2)_2$ | $OHCCH{=}CCH_3$	$CH_2{=}CHCOCH_3$	Cat. a, EtOH, TEA, 80°	(E,Z)-$(CH_3)_2C{=}CH(CH_2)_2C(CH_3){=}CH$ | $CH_3CO(CH_2)_2C{=}O$ (65)	16
	$CH_2{=}CHCOC_2H_5$	Cat. a, EtOH, NaOAc, 80°	(E,Z)-$(CH_3)_2C{=}CH(CH_2)_2C(CH_3){=}CH$ | $C_2H_5CO(CH_2)_2C{=}O$ (53)	16
	$CH_2{=}CHCOCH{=}C(CH_3)_2$	Cat. a, EtOH, NaOAc, 80°	(E,Z)-$(CH_3)_2C{=}CH(CH_2)_2C(CH_3){=}CH$ | $(CH_3)_2C{=}CHCO(CH_2)_2C{=}O$ (66)	16
	$CH_2{=}CHCOC_6H_5$	Cat. a, EtOH, NaOAc, 80°	(E,Z)-$(CH_3)_2C{=}CH(CH_2)_2C(CH_3){=}CH$ | $C_6H_5CO(CH_2)_2C{=}O$ (72)	16
CH_3, CH_3, CH_3, CH_2CHO (cyclopentene structure)	$CH_2{=}CHCOCH_3$	Cat. a, EtOH, TEA, 80°	CH_3, CH_3, CH_3, $CH_2CO(CH_2)_2COCH_3$ (cyclopentene structure) (—)	82
n-C_5H_{11}, O, O, CH_2CHO (dioxolane structure)	O (cyclopentenone structure)	Cat. a, TEA, 70°	n-C_5H_{11}, O, O, CH_2CO, O (dioxolane–cyclopentanone structure) (30)	56

$i\text{-}C_4H_9(CH_2)_2CHCH_3$ | $OHCCH_2$	$CH_2{=}C(CH_3)COCO_2C_2H_5$	Cat. f, dioxane, TEA, 95°	$i\text{-}C_4H_9(CH_2)_2CH(CH_3)CH_2COCH_2$ (—) | $C_2H_5O_2CCOCHCH_3$	46
$CH_3O_2C(CH_2)_7CHO$	$CH_2{=}CHCOR$	Cat. a, dioxane, TEA, 90°	$CH_3O_2C(CH_2)_7CO(CH_2)COR$	24
	R			
	CH_3		(77)	
	C_2H_5		(74)	
	$n\text{-}C_4H_9$		(83)	
	$n\text{-}C_5H_{11}$		(85)	
	$n\text{-}C_6H_{13}$		(87)	
	$n\text{-}C_7H_{15}$		(84)	
	$n\text{-}C_8H_{17}$		(87)	
	$n\text{-}C_9H_{19}$		(86)	
$CH_3O_2C(CH_2)_7CHO$	$CH_2{=}CHCOCH_3$	Cat. f, dioxane, TEA, 100°	$CH_3O_2C(CH_2)_7CO(CH_2)_2COCH_3$ (42)	11
O NCH_2CHO O	$CH_2{=}CHCOCH_3$	Cat. a, EtOH, TEA, 80°	$PhThNCH_2CO(CH_2)_2COCH_3$ (74)	26
	$CH_2{=}CHCOC_2H_5$	Cat. a, EtOH, TEA, 80°	$PhThNCH_2CO(CH_2)_2COC_2H_5$ (67)	26
	$CH_2{=}CHCOC_6H_{13}\text{-}n$	Cat. a, EtOH, TEA, 80°	$PhThNCH_2CO(CH_2)_2COC_6H_{13}\text{-}n$ (71)	26
	$CH_2{=}CHCOC_6H_5$	Cat. a, EtOH, TEA, 80°	$PhThNCH_2CO(CH_2)_2COC_6H_5$ (75)	26
	O $CH{=}C(CO_2C_2H_5)COCH_3$	Cat. a, EtOH, TEA, 80°	$PhThNCH_2COCH(C_4H_3O)CHCOCH_3$ (47) | $CO_2C_2H_5$	26
	S $CH{=}C(CO_2C_2H_5)COCH_3$	Cat. a, EtOH, TEA, 80°	$PhThNCH_2COCH(C_4H_3S)CHCOCH_3$ (53) | $CO_2C_2H_5$	26
	S $CH{=}CHCO$ S	Cat. a, EtOH, TEA, 80°	$PhThNCH_2COCH(C_4H_3S)CH_2COC_4H_3S$ (62)	26
	$C_6H_5CH{=}C(CO_2C_2H_5)COCH_3$	Cat. a, EtOH, TEA, 80°	$PhThNCH_2COCH(C_6H_5)CHCOCH_3$ (51) | $CO_2C_2H_5$	26

TABLE I. CATALYZED ADDITIONS OF ALDEHYDES TO α,β-UNSATURATED KETONES (*Continued*)

Aldehyde	Michael Acceptor	Conditions	Product(s) and Yield(s) (%)	Refs.
	$C_6H_5CH{=}CHCOC_6H_5$	Cat. a, EtOH, TEA, 80°	$PhThNCH_2COCH(C_6H_5)CH_2COC_6H_5$ (72)	26
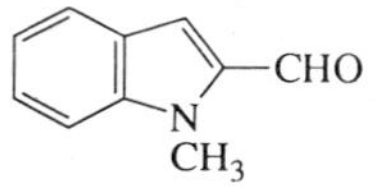 —CHO, N, CH_3	$CH_2{=}CHCOCH_3$	Cat. b, EtOH, TEA, 80°	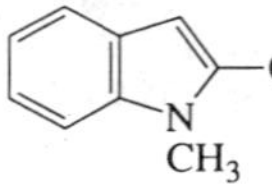 —$CO(CH_2)_2COCH_3$, N, CH_3 (57)	26
	$CH_2{=}CHCOC_2H_5$	Cat. b, EtOH, TEA, 80°	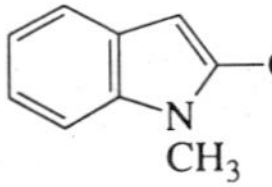 —$CO(CH_2)_2COC_2H_5$, N, CH_3 (55)	26
	$CH_2{=}CHCOC_6H_{13}$-*n*	Cat. b, EtOH, TEA, 80°	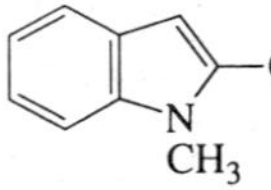 —$CO(CH_2)_2COC_6H_{13}$-*n*, N, CH_3 (66)	26
	$CH_2{=}CHCOC_6H_5$	Cat. b, EtOH, TEA, 80°	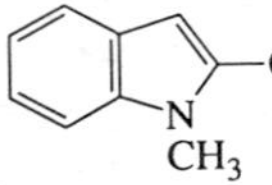 —$CO(CH_2)_2COC_6H_5$, N, CH_3 (57)	26
	$CH_2{=}CHCOC_7H_{15}$-*n*	Cat. b, EtOH, TEA, 80°	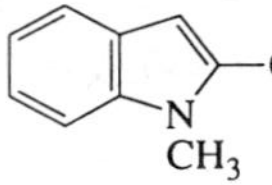 —$CO(CH_2)_2COC_7H_{15}$-*n*, N, CH_3 (59)	26
	$CH_2{=}CHCOC_8H_{17}$-*n*	Cat. b, EtOH, TEA, 80°	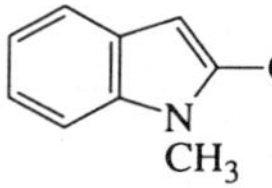 —$CO(CH_2)_2COC_8H_{17}$-*n*, N, CH_3 (62)	26
	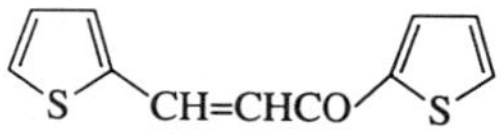 S, CH=CHCO, S	Cat. b, EtOH, TEA, 80°	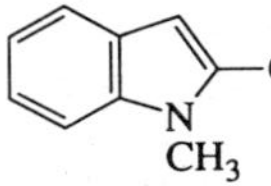 —$COCH(C_4H_3S)CH_2COC_4H_3S$, N, CH_3 (65)	26
	$C_6H_5CH{=}CHCOC_6H_5$	Cat. b, EtOH, TEA, 80°	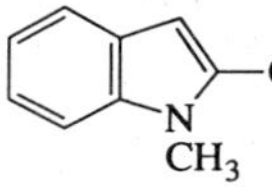 —$COCH(C_6H_5)CH_2COC_6H_5$, N, CH_3 (54)	26

	Reactant	Acceptor	Conditions	Product (Yield, %)	Refs.
C_{11}	$CH_2{=}CH(CH_2)_8CHO$	$CH_2{=}CHCOCH_3$	Cat. a, EtOH, NaOAc, 80°	$CH_2{=}CH(CH_2)_8CO(CH_2)_2COCH_3$ (81)	16
		$CH_2{=}CHCOCH{=}C(CH_3)_2$	Cat. a, EtOH, NaOAc, 80°	$CH_2{=}CH(CH_2)_8CO(CH_2)_2COCH{=}C(CH_3)_2$ (82)	16
		$CH_2{=}CHCOC_6H_5$	Cat. a, EtOH, NaOAc, 80°	$CH_2{=}CH(CH_2)_8CO(CH_2)_2COC_6H_5$ (75)	16
	$CH_3OC(CH_3)_2(CH_2)_3CHCH_3$ with $OHCCH_2$ on CH	$CH_2{=}CHCOCH_3$	Cat. a, TEA, 75°	$CH_3OC(CH_3)_2(CH_2)_3CH(CH_3)CH_2$ with $CH_3CO(CH_2)_2C{=}O$ on CH_2 (45)	81
		$CH_2{=}C(CH_3)COCO_2C_3H_7$-*i*	Cat. f, dioxane, TEA, 95°	$CH_3OC(CH_3)_2(CH_2)_3CH(CH_3)CH_2$ with *i*-$C_3H_7O_2CCOCH(CH_3)CH_2C{=}O$ on CH_2 (—)	46
	$C_6H_5CO_2(CH_2)_4CHO$	$CH_2{=}CHCOCH_3$	Cat. a, dioxane, TEA, 90°	$C_6H_5CO_2(CH_2)_4CO(CH_2)_2COCH_3$ (54)	21
	$C_6H_5(CH{=}CH)_2CHO$	$CH_2{=}CHCOCH_3$	Cat. a, EtOH, NaOAc, 80°	$C_6H_5(CH{=}CH)_2CO(CH_2)_2COCH_3$ (32)	16
	[drawn structure: O, OCH_2, O, CHO]	$CH_2{=}CHCOCH_3$	Cat. b, EtOH, TEA, 80°	[drawn structure: O, OCH_2, O, $CO(CH_2)_2COCH_3$] (95)	51
	[drawn structure: O, O, $N(CH_2)_2CHO$]	$CH_2{=}CHCOCH_3$	Cat. a, EtOH, TEA, 80°	$PhThN(CH_2)_2CO(CH_2)_2COCH_3$ (68)	26
		$CH_2{=}CHCOC_2H_5$	Cat. a, EtOH, TEA, 80°	$PhThN(CH_2)_2CO(CH_2)_2COC_2H_5$ (71)	26
		$CH_2{=}CHCOC_6H_5$	Cat. a, EtOH, TEA, 80°	$PhThN(CH_2)_2CO(CH_2)_2COC_6H_5$ (67)	26
		$C_6H_5CH{=}CHCOC_6H_5$	Cat. a, EtOH, TEA, 80°	$PhThN(CH_2)_2COCH(C_6H_5)CH_2COC_6H_5$ (74)	26
	[drawn structure: N, $C_6H_5SO_2$, CHO]	$CH_2{=}CHCOCH_3$	Cat. a, EtOH, NaOAc, reflux	[drawn structure: N, $C_6H_5SO_2$, $CO(CH_2)_2COCH_3$] (64)	28
C_{12}	*n*-$C_{11}H_{23}CHO$	$CH_2{=}CHCOCH_3$	Cat. a, EtOH, TEA, 80°	*n*-$C_{11}H_{23}CO(CH_2)_2COCH_3$ (75)	48
		$CH_2{=}CHCO(CH_2)_2CO_2CH_3$	Cat. a, dioxane, TEA, 100°	*n*-$C_{11}H_{23}[CO(CH_2)_2]_2CO_2CH_3$ (26)	24
	$C_6H_5CO_2(CH_2)_4CHO$	$CH_2{=}CHCOCH_3$	Cat. a, dioxane, TEA, 65°	$C_6H_5CO_2(CH_2)_4CO(CH_2)_2COCH_3$ (60)	21

TABLE I. CATALYZED ADDITIONS OF ALDEHYDES TO α,β-UNSATURATED KETONES (*Continued*)

Aldehyde	Michael Acceptor	Conditions	Product(s) and Yield(s) (%)	Refs.
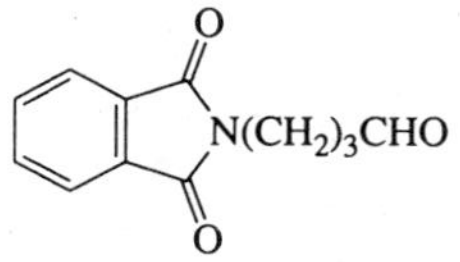 O, O, $N(CH_2)_3CHO$	$CH_2{=}CHCOCH_3$	Cat. a, EtOH, TEA, 80°	$PhThN(CH_2)_3CO(CH_2)_2COCH_3$ (73)	26
	$CH_2{=}CHCOC_6H_5$	Cat. a, EtOH, TEA, 80°	$PhThN(CH_2)_3CO(CH_2)_2COC_6H_5$ (73)	26
	$C_6H_5CH{=}CHCOC_6H_5$	Cat. a, EtOH, TEA, 80°	$PhThN(CH_2)_3COCH(C_6H_5)CH_2COC_6H_5$ (66)	26
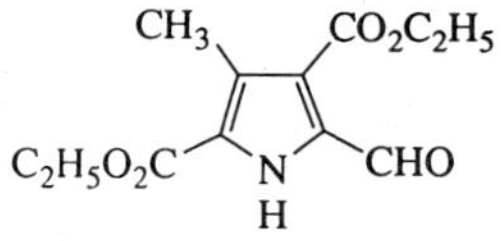 CH_3, $CO_2C_2H_5$, $C_2H_5O_2C$, N, H, CHO	$CH_2{=}CHCOCH_3$	Cat. a, EtOH, NaOAc, reflux	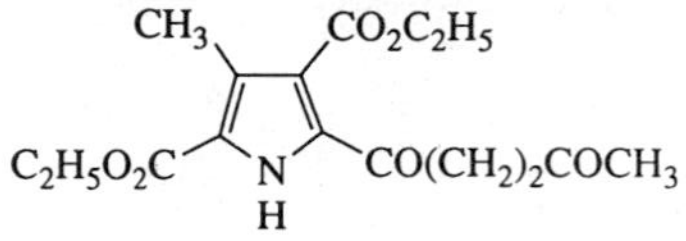 CH_3, $CO_2C_2H_5$, $C_2H_5O_2C$, N, H, $CO(CH_2)_2COCH_3$ (80)	28
C_{13}				
p-$C_6H_5OC_6H_4CHO$	$CH_2{=}CHCOCH_3$	Cat. b, TEA, 75°	p-$C_6H_5OC_6H_4CO(CH_2)_2COCH_3$ (37)	81

		$(CH_3)_2N(CH_2)_2CO$–(2-thienyl)	NaCN, DMF, rt	p-$C_6H_5OC_6H_4CO(CH_2)_2COC_4H_3S$ (2)	79
C_{14}	N-phthalimido-$(CH_2)_5CHO$	$CH_2{=}CHCOCH_3$	Cat. a, EtOH, TEA, 80°	$PhThN(CH_2)_5CO(CH_2)_2COCH_3$ (74)	26
		$CH_2{=}CHCOC_6H_5$	Cat. a, EtOH, TEA, 80°	$PhThN(CH_2)_5CO(CH_2)_2COC_6H_5$ (75)	26
		$C_6H_5CH{=}CHCOC_6H_5$	Cat. a, EtOH, TEA, 80°	$PhThN(CH_2)_5COCH(C_6H_5)CH_2COC_6H_5$ (74)	26
C_{15}	$(C_6H_5)_2C{=}CHCHO$	$CH_2{=}CHCOCH_3$	Cat. a, EtOH, TEA, 80°	$(C_6H_5)_2C{=}CHCO(CH_2)_2COCH_3$ (81)	16
		$CH_2{=}CHCOCH{=}C(CH_3)_2$	Cat. a, EtOH, NaOAc, 80°	$(C_6H_5)_2C{=}CHCO(CH_2)_2COCH{=}C(CH_3)_2$ (64)	16
		$CH_2{=}CHCOC_6H_5$	Cat. a, EtOH, NaOAc, 80°	$(C_6H_5)_2C{=}CHCO(CH_2)_2COC_6H_5$ (86)	16

[a] This product could not be isolated because it cyclized to the furan.

TABLE II. CATALYZED ADDITIONS OF ALDEHYDES TO DIVINYL SULFONE TO GIVE 1,4-DIKETONES

	Aldehyde	Michael Acceptor	Conditions	Product(s) and Yield(s) (%)	Refs.
	RCHO R	$CH_2{=}CHSO_2CH{=}CH_2$	Cat. a, EtOH, NaOAc, 80°	$RCO(CH_2)_2COR$	39
C_4					
	n-C_3H_7			(42)	
C_5					
	n-C_4H_9			(55)	
	$(CH_3)_2C{=}CH$			(25)	
	$CH_3O_2C(CH_2)_3$			(38)	
	2-furyl			(75)	
	2-thienyl			(48)	
C_6					
	n-C_5H_{11}			(53)	
C_7					
	n-C_6H_{13}			(63)	
	C_6H_5			(46)	
	m-ClC_6H_4			(40)	
	p-ClC_6H_4			(36)	
C_8					
	n-C_7H_{15}			(61)	
	m-$CH_3OC_6H_4$			(39)	
	p-$CH_3OC_6H_4$			(35)	

C_9		
	$n\text{-}C_8H_{17}$	(49)
C_{10}		
	$n\text{-}C_9H_{19}$	(42)
	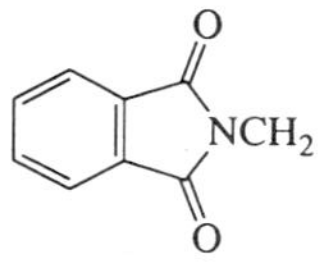 N-phthalimido-NCH_2 (structure)	(68)
C_{11}		
	$n\text{-}C_{10}H_{21}$	(69)
	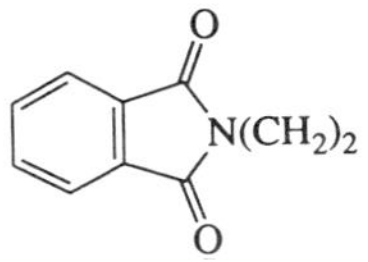 N-phthalimido-$N(CH_2)_2$ (structure)	(69)
C_{12}		
	$n\text{-}C_{11}H_{23}$	(57)
C_{15}		
	$(C_6H_5)_2C{=}CH$	(52)
	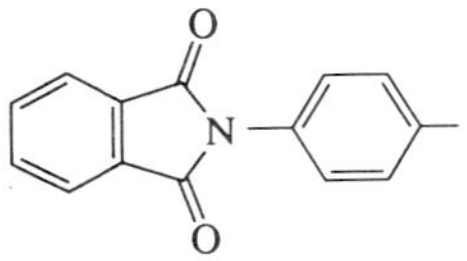 N-(p-tolyl)phthalimide (structure)	(49)

TABLE III. CATALYZED ADDITIONS OF α-KETOACIDS TO α,β-UNSATURATED KETONES

	Aldehyde	Michael Acceptor	Conditions	Product(s) and Yield(s) (%)	Refs.
	R^1COCO_2H	$CH_2{=}CHCOR^2$		$R^1CO(CH_2)_2COR^2$	
	R^1	R^2			
C_3					
	CH_3	CH_3	Cat. a, EtOH, TEA, 80°	(69)	31
C_4					
	C_2H_5	CH_3	Cat. a, EtOH, TEA, 80°	(72)	31
C_5					
	n-C_3H_7	CH_3	Cat. a, EtOH, TEA, 80°	(75)	31
	$HO_2C(CH_2)_2$	CH_3	Cat. a, dioxane, TEA, 80°	(42)	31
C_6					
	n-C_4H_9	CH_3	Cat. a, EtOH, TEA, 80°	(79)	31
	$HO_2C(CH_2)_3$	CH_3	Cat. a, dioxane, TEA, 80°	(47)	31
	[structure: 2,3-dihydrofuran-4-yl ring, O]	CH_3	Cat. a, EtOH, TEA, 80°	(79)	31
C_7					
	$HO_2C(CH_2)_4$	CH_3	Cat. a, dioxane, TEA, 80°	(41)	31
	n-C_5H_{11}	CH_3	Cat. a, EtOH, TEA, 80°	(76)	31
C_8					
	n-C_6H_{13}	CH_3	Cat. a, EtOH, TEA, 80°	(85)	31
	$HO_2C(CH_2)_5$	CH_3	Cat. a, dioxane, TEA, 80°	(74)	31
		C_2H_5	Cat. a, dioxane, TEA, 80°	(67)	31
		n-C_3H_7	Cat. a, dioxane, TEA, 80°	(76)	31
		n-C_4H_9	Cat. a, dioxane, TEA, 80°	(86)	31

C_9					
	$C_6H_5CH_2$	CH_3	Cat. a, EtOH, TEA, 80°	(63)	31
		C_2H_5	Cat. a, EtOH, TEA, 80°	(74)	31
		n-C_3H_7	Cat. a, EtOH, TEA, 80°	(78)	31
		n-C_4H_9	Cat. a, EtOH, TEA, 80°	(76)	31
		C_6H_5	Cat. a, EtOH, TEA, 80°	(79)	31
	o-$ClC_6H_4CH_2$	CH_3	Cat. a, EtOH, TEA, 80°	(73)	31
	m-$ClC_6H_4CH_2$	CH_3	Cat. a, EtOH, TEA, 80°	(66)	31
	n-C_7H_{15}	CH_3	Cat. a, EtOH, TEA, 80°	(74)	31
	$HO_2C(CH_2)_6$	CH_3	Cat. a, dioxane, TEA, 80°	(64)	31
		C_2H_5	Cat. a, dioxane, TEA, 80°	(70)	31
		n-C_3H_7	Cat. a, dioxane, TEA, 80°	(78)	31
		n-C_4H_9	Cat. a, dioxane, TEA, 80°	(78)	31
	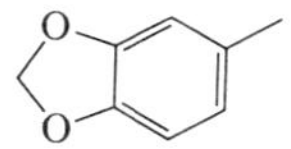 3,4-methylenedioxybenzyl (structure: O, O)	CH_3	Cat. a, EtOH, TEA, 80°	(81)	31
C_{10}					
	m-$CH_3OC_6H_4CH_2$	CH_3	Cat. a, EtOH, TEA, 80°	(81)	31
	p-$CH_3OC_6H_4CH_2$	CH_3	Cat. a, EtOH, TEA, 80°	(84)	31

TABLE IV. CATALYZED ADDITIONS OF ALDEHYDES THAT YIELD TRI- AND POLYKETONES

	Aldehyde	Michael Acceptor	Conditions	Product(s) and Yield(s) (%)	Refs.
C_1					
	HCHO	$CH_2{=}CHCOCH_3$	Cat. a, EtOH, TEA, 80°	$CH_3CO(CH_2)_2CO(CH_2)_2COCH_3$ (27)	48
C_2					
	CH_3CHO	$CH_2{=}CHCOCH{=}CH_2$	Cat. a, TEA, 65°	" (70)	8
		$CH_2{=}CHCO(CH_2)_3COCH_3$	Cat. a, TEA, 65°	$CH_3CO(CH_2)_2CO(CH_2)_3COCH_3$ (78)	50
		$CH_2{=}CHCO(CH_2)_2COC_3H_7$-$n$	Cat. a, TEA, 65°	$CH_3[CO(CH_2)_2]_2COC_3H_7$-$n$ (91)	17
		$CH_2{=}CHCO(CH_2)_4COCH_3$	Cat. a, TEA, 65°	$CH_3CO(CH_2)_2CO(CH_2)_4COCH_3$ (75)	50
		C_4H_3O–$CH{=}C(COCH_3)_2$	Cat. a, DMF, TEA, 80°	$CH_3COCH(C_4H_3O)CH(COCH_3)_2$ (43)	36
		$CH_2{=}CH[CO(CH_2)_2]_2COCH_3$	Cat. a, EtOH, TEA, 80°	$CH_3[CO(CH_2)_2]_3COCH_3$ (82)	52
		$CH_2{=}CHCO(CH_2)_2COCH(C_2H_5)_2$	Cat. a, TEA, 65°	$CH_3[CO(CH_2)_2]_2COCH(C_2H_5)_2$ (92)	17
		$CH_3{=}CHCO(CH_2)_2COC_6H_{13}$-$n$	Cat. a, TEA, 65°	$CH_3[CO(CH_2)_2]_2COC_6H_{13}$-$n$ (82)	17
		$CH_2{=}CH[CO(CH_2)_2]_3COCH_3$	Cat. a, EtOH, TEA, 80°	$CH_3[CO(CH_2)_2]_4COCH_3$ (73)	52
		$CH_2{=}CH[CO(CH_2)_2]_4COCH_3$	Cat. a, EtOH, TEA, 80°	$CH_3[CO(CH_2)_2]_5COCH_3$ (75)	52
C_3					
	C_2H_5CHO	$CH_2{=}CHCOCH{=}CH_2$	Cat. a, TEA, 65°	$C_2H_5[CO(CH_2)_2]_2COC_2H_5$ (60)	8
		$CH_2{=}CHCO(CH_2)_3COCH_3$	Cat. a, TEA, 65°	$C_2H_5CO(CH_2)_2CO(CH_2)_3COCH_3$ (71)	50
		$CH_2{=}CHCO(CH_2)_2COCH{=}CH_2$	Cat. a, dioxane, TEA, 65°	$C_2H_5[CO(CH_2)_2]_3COC_2H_5$ (53)	32
		$CH_2{=}CHCO(CH_2)_4COCH_3$	Cat. a, TEA, 65°	$C_2H_5CO(CH_2)_2CO(CH_2)_4COCH_3$ (77)	50
		$CH_2{=}CH[CO(CH_2)_2]_2COCH_3$	Cat. a, EtOH, TEA, 80°	$C_2H_5[CO(CH_2)_2]_3COCH_3$ (85)	52
		$CH_2{=}CHCO(CH_2)_2COC_6H_{13}$-$n$	Cat. a, TEA, 65°	$C_2H_5[CO(CH_2)_2]_2COC_6H_{13}$-$n$ (77)	17
		$CH_2{=}CH[CO(CH_2)_2]_4COCH_3$	Cat. a, EtOH, TEA, 80°	$C_2H_5[CO(CH_2)_2]_5COCH_3$ (71)	52
		$CH_2{=}CH[CO(CH_2)_2]_2COC_6H_5$	Cat. a, EtOH, TEA, 80°	$C_2H_5[CO(CH_2)_2]_3COC_6H_5$ (69)	52
C_4					
	n-C_3H_7CHO	$CH_2{=}CHCOCH{=}CH_2$	Cat. a, TEA, 65°	n-$C_3H_7[CO(CH_2)_2]_2COC_3H_7$-$n$ (65)	8
		$CH_2{=}CHCO(CH_2)_2COC_2H_5$	Cat. a, TEA, 65°	n-$C_3H_7[CO(CH_2)_2]_2COC_2H_5$ (74)	17
		$CH_2{=}CHCO(CH_2)_3COCH_3$	Cat. a, TEA, 65°	n-$C_3H_7CO(CH_2)_2CO(CH_2)_3COCH_3$ (76)	50
		$CH_2{=}CHCO(CH_2)_2COCH{=}CH_2$	Cat. a, dioxane, TEA, 65°	n-$C_3H_7[CO(CH_2)_2]_3COC_3H_7$-$n$ (69)	32
		$CH_2{=}CHCO(CH_2)_4COCH_3$	Cat. a, TEA, 65°	n-$C_3H_7CO(CH_2)_2CO(CH_2)_4COCH_3$ (81)	50

		$CH_2{=}CH[CO(CH_2)_2]_2COCH_3$	Cat. a, EtOH, TEA, 80°	n-$C_3H_7[CO(CH_2)_2]_3COCH_3$ (82)	52
		$CH_2{=}CH[CO(CH_2)_2]_3COCH_3$	Cat. a, EtOH, TEA, 80°	n-$C_3H_7[CO(CH_2)_2]_4COCH_3$ (69)	52
		$C_6H_5CH{=}C(COCH_3)_2$	Cat. a, EtOH, TEA, 80°	n-$C_3H_7COCH(C_6H_5)CH(COCH_3)_2$ (64)	36
		C_4H_3O-$CH{=}C(COCH_3)_2$ (2-furyl)	Cat. a, DMF, TEA, 80°	n-$C_3H_7COCH(C_4H_3O)CH(COCH_3)_2$ (60)	36
	i-C_3H_7CHO	$CH_2{=}CHCOCH{=}CH_2$	Cat. a, TEA, 65°	i-$C_3H_7[CO(CH_2)_2]_2COC_3H_7$-$i$ (62)	8
		$CH_2{=}CH[CO(CH_2)_2]_2COCH_3$	Cat. a, EtOH, TEA, 80°	i-$C_3H_7[CO(CH_2)_2]_3COCH_3$ (35)	52
	$OHC(CH_2)_2CHO$	$CH_2{=}CHCOCH_3$	Cat. a, DMF, TEA, 75°	$CH_3[CO(CH_2)_2]_3COCH_3$ (43)	8
C_5					
	n-C_4H_9CHO	$CH_2{=}CHCOCH{=}CH_2$	Cat. a, TEA, 65°	n-$C_4H_9[CO(CH_2)_2]_2COC_4H_9$-$n$ (62)	8
		$CH_2{=}CHCO(CH_2)_2COCH_3$	Cat. a, TEA, 65°	n-$C_4H_9[CO(CH_2)_2]_2COCH_3$ (72)	17
		$CH_2{=}CHCO(CH_2)_3COCH_3$	Cat. a, TEA, 65°	n-$C_4H_9CO(CH_2)_2CO(CH_2)_3COCH_3$ (76)	50
	i-C_4H_9CHO	$CH_2{=}CHCO(CH_2)_2COCH{=}CH_2$	Cat. a, dioxane, TEA, 65°	i-$C_4H_9[CO(CH_2)_2]_3COC_4H_9$-$i$ (56)	32
		$CH_2{=}CH[CO(CH_2)_2]_2COCH_3$	Cat. a, EtOH, TEA, 80°	i-$C_4H_9[CO(CH_2)_2]_3COCH_3$ (56)	52
	$CH_3CO(CH_2)_2CHO$	$CH_2{=}CHCOCH_3$	Cat. a, dioxane, TEA, 90°	$CH_3[CO(CH_2)_2]_2COCH_3$ (77)	8
		$CH_2{=}CHCOCH_2O_2CCH_3$	Cat. a, dioxane, TEA, 80°	$CH_3[CO(CH_2)_2]_2COCH_2O_2CCH_3$ (60)	20
		$CH_2{=}CHCOR$	Cat. a, dioxane, TEA, 90°	$CH_3[CO(CH_2)_2]_2COR$	8
		R			
		n-C_5H_{11}		(70)	
		n-C_6H_{13}		(77)	
		n-C_7H_{15}		(75)	
		n-C_8H_{17}		(73)	
		n-C_9H_{19}		(74)	
		n-$C_{10}H_{21}$		(40)	
	$OHC(CH_2)_3CHO$	$CH_2{=}CHCOCH_3$	Cat. a, DMF, TEA, 75°	$CH_3CO(CH_2)_2CO(CH_2)_3CO(CH_2)_2$ (33) | $COCH_3$	8
	C_4H_3O-CHO (2-furaldehyde)	$CH_2{=}CHCOCH{=}CH_2$	Cat. c, DMF, TEA, 65°	$C_4H_3O[CO(CH_2)_2]_2COC_4H_3O$ (50)	8
		$CH_2{=}CHCO(CH_2)_2COCH_3$	Cat. b, TEA, 65°	$C_4H_3O[CO(CH_2)_2]_2COCH_3$ (70)	17
		$CH_2{=}CHCO(CH_2)_3COCH_3$	Cat. b, TEA, 65°	$C_4H_3OCO(CH_2)_2CO(CH_2)_3COCH_3$ (70)	50
		$CH_2{=}CHCO(CH_2)_2COCH{=}CH_2$	Cat. a, dioxane, TEA, 65°	$C_4H_3O[CO(CH_2)_2]_3COC_4H_3O$ (62)	32

TABLE IV. CATALYZED ADDITIONS OF ALDEHYDES THAT YIELD TRI- AND POLYKETONES (*Continued*)

Aldehyde	Michael Acceptor	Conditions	Product(s) and Yield(s) (%)	Refs.
	$CH_2{=}CH[CO(CH_2)_2]_2COCH_3$	Cat. b, EtOH, TEA, 80°	$C_4H_3O[CO(CH_2)_2]_3COCH_3$ (81)	52
	$CH_2{=}CH[CO(CH_2)_2]_2COC_3H_7$-*n*	Cat. a, dioxane, TEA, 65°	$C_4H_3O[CO(CH_2)_2]_3COC_3H_7$-*n* (23)	32
	$CH_2{=}CH[CO(CH_2)_2]_3COCH_3$	Cat. b, EtOH, TEA, 80°	$C_4H_3O[CO(CH_2)_2]_4COCH_3$ (78)	52
	$CH_2{=}CH[CO(CH_2)_2]_4COCH_3$	Cat. b, EtOH, TEA, 80°	$C_4H_3O[CO(CH_2)_2]_5COCH_3$ (75)	52
	$CH_2{=}CH[CO(CH_2)_2]_2COC_6H_5$	Cat. b, EtOH, TEA, 80°	$C_4H_3O[CO(CH_2)_2]_3COC_6H_5$ (74)	52
	$C_6H_5CH{=}CHCOCH_2CH(C_6H_5)COCH_3$	Cat. c, DMF, TEA, 80°	$C_4H_3OCOCH(C_6H_5)CH_2COCH_2CH(C_6H_5)COCH_3$ (85)	8
	$C_6H_5CH{=}CHCOCH{=}CHC_6H_5$	Cat. c, EtOH, TEA, 80°	$[C_4H_3OCOCH(C_6H_5)CH_2]_2CO$ (54)	8
	2-furyl-$CH{=}C(COCH_3)_2$	Cat. b, DMF, 80°	$C_4H_3OCOCH(C_4H_3O)CH(COCH_3)_2$ (46)	36
	(2-furyl-$CH{=}CH)_2CO$	Cat. a, EtOH, TEA, 80°	$[C_4H_3OCOCH(C_4H_3O)CH_2]_2CO$ (44)	8
2-thiophenecarboxaldehyde (thiophene-2-CHO)	$CH_2{=}CHCOCH{=}CH_2$	Cat. c, DMF, TEA, 65°	$[C_4H_3SCO(CH_2)_2]_2CO$ (51)	8
	$CH_2{=}CHCO(CH_2)_2COCH_3$	Cat. b, TEA, 65°	$C_4H_3S[CO(CH_2)_2]_2COCH_3$ (43)	17
	$CH_2{=}CHCO(CH_2)_3COCH_3$	Cat. b, TEA, 65°	$C_4H_3SCO(CH_2)_2CO(CH_2)_3COCH_3$ (68)	50
	$CH_2{=}CH[CO(CH_2)_2]_2COCH_3$	Cat. b, EtOH, TEA, 80°	$C_4H_3S[CO(CH_2)_2]_3COCH_3$ (86)	52
	$CH_2{=}CH[CO(CH_2)_2]_3COCH_3$	Cat. b, EtOH, TEA, 80°	$C_4H_3S[CO(CH_2)_2]_4COCH_3$ (80)	52
	$CH_2{=}CH[CO(CH_2)_2]_4COCH_3$	Cat. b, EtOH, TEA, 80°	$C_4H_3S[CO(CH_2)_2]_5COCH_3$ (77)	52
	$CH_2{=}CH[CO(CH_2)_2]_2COC_6H_5$	Cat. b, EtOH, TEA, 80°	$C_4H_3S[CO(CH_2)_2]_3COC_6H_5$ (70)	52
	$C_6H_5CH{=}C(COCH_3)_2$	Cat. b, EtOH, TEA, 80°	$C_4H_3SCOCH(C_6H_5)CH(COCH_3)_2$ (41)	36
C_6				
n-$C_5H_{11}CHO$	$CH_2{=}CHCOCH{=}CH_2$	Cat. a, TEA, 65°	*n*-$C_5H_{11}[CO(CH_2)_2]_2COC_5H_{11}$-*n* (65)	8
	$CH_2{=}CHCO(CH_2)_2COCH_3$	Cat. a, TEA, 65°	*n*-$C_5H_{11}[CO(CH_2)_2]_2COCH_3$ (75)	17
	$CH_2{=}CHCO(CH_2)_3COCH_3$	Cat. a, TEA, 65°	*n*-$C_5H_{11}CO(CH_2)_2CO(CH_2)_3COCH_3$ (76)	50
	$CH_2{=}CHCO(CH_2)_2COCH{=}CH_2$	Cat. a, dioxane, TEA, 65°	*n*-$C_5H_{11}[CO(CH_2)_2]_3COC_5H_{11}$-*n* (54)	32
	$CH_2{=}CH[CO(CH_2)_2]_2COC_2H_5$	Cat. a, dioxane, TEA, 65°	*n*-$C_5H_{11}[CO(CH_2)_2]_3COC_2H_5$ (49)	32

		$CH_2{=}CH[CO(CH_2)_2]_2C({=}O)C_3H_7\text{-}n$	Cat. a, dioxane, TEA, 65°	$n\text{-}C_5H_{11}[CO(CH_2)_2]_3COC_3H_7\text{-}n$ (42)	32
	$(C_2H_5O)_2CHCHO$	$CH_2{=}CHCO(CH_2)_2COCH_3$	Cat. a, dioxane, TEA, 80°	$(C_2H_5O)_2CH[CO(CH_2)_2]_2COCH_3$ (59)	22
	$OHC(CH_2)_4CHO$	$CH_2{=}CHCOCH_3$	Cat. a, DMF, TEA, 75°	$CH_3CO(CH_2)_2CO(CH_2)_4C({=}O)(CH_2)_2COCH_3$ (7)	8
	OHC-(2,5-furandiyl)-CHO	$CH_2{=}CHCOCH_3$	Cat. b, EtOH, TEA, 80°	$CH_3CO(CH_2)_2CO$-(2,5-furandiyl)-$CO(CH_2)_2COCH_3$ (56)	51
	$3\text{-}C_5H_4NCHO$	$CH_2{=}CH[CO(CH_2)_2]_2COCH_3$	Cat. b, EtOH, TEA, 80°	$3\text{-}C_5H_4N[CO(CH_2)_2]_3COCH_3$ (78)	52
C_7					
	$n\text{-}C_6H_{13}CHO$	$CH_2{=}CHCOCH{=}CH_2$	Cat. a, TEA, 65°	$n\text{-}C_6H_{13}[CO(CH_2)_2]_2COC_6H_{13}\text{-}n$ (64)	8
		$CH_2{=}CHCO(CH_2)_3COCH_3$	Cat. a, TEA, 65°	$n\text{-}C_6H_{13}CO(CH_2)_2CO(CH_2)_3COCH_3$ (73)	50
		$CH_2{=}CHCO(CH_2)_2COCH{=}CH_2$	Cat. a, dioxane, TEA, 65°	$n\text{-}C_6H_{13}[CO(CH_2)_2]_3COC_6H_{13}\text{-}n$ (53)	32
		$CH_2{=}CH[CO(CH_2)_2]_2COCH_3$	Cat. a, EtOH, TEA, 80°	$n\text{-}C_6H_{13}[CO(CH_2)_2]_3COCH_3$ (76)	52
		$CH_2{=}CH[CO(CH_2)_2]_3COCH_3$	Cat. a, EtOH, TEA, 80°	$n\text{-}C_6H_{13}[CO(CH_2)_2]_4COCH_3$ (69)	52
		$CH_2{=}CHCO(CH_2)_2COC_6H_5$	Cat. a, TEA, 65°	$n\text{-}C_6H_{13}[CO(CH_2)_2]_2COC_6H_5$ (55)	17
	C_6H_5CHO	$CH_2{=}CHCOCH{=}CH_2$	Cat. c, DMF, TEA, 65°	$C_6H_5[CO(CH_2)_2]_2COC_6H_5$ (55)	8
		$CH_2{=}CHCO(CH_2)_2COCH_3$	Cat. b, TEA, 65°	$C_6H_5[CO(CH_2)_2]_2COCH_3$ (70)	17
		$CH_2{=}CHCO(CH_2)_3COCH_3$	Cat. b, TEA, 65°	$C_6H_5CO(CH_2)_2CO(CH_2)_3COCH_3$ (67)	50
		$CH_2{=}CHCO(CH_2)_4COCH_3$	Cat. b, TEA, 65°	$C_6H_5CO(CH_2)_2CO(CH_2)_4COCH_3$ (80)	50
		$CH_2{=}CH[CO(CH_2)_2]_2COCH_3$	Cat. b, EtOH, TEA, 80°	$C_6H_5[CO(CH_2)_2]_3COCH_3$ (89)	52
		$CH_2{=}CH[CO(CH_2)_2]_3COCH_3$	Cat. b, EtOH, TEA, 80°	$C_6H_5[CO(CH_2)_2]_4COCH_3$ (70)	52
		$CH_2{=}CH[CO(CH_2)_2]_4COCH_3$	Cat. b, EtOH, TEA, 80°	$C_6H_5[CO(CH_2)_2]_5COCH_3$ (68)	52
		$CH_2{=}CH[CO(CH_2)_2]_2COC_6H_5$	Cat. b, EtOH, TEA, 80°	$C_6H_5[CO(CH_2)_2]_3COC_6H_5$ (73)	52
	$p\text{-}ClC_6H_4CHO$	$CH_2{=}CHCOCH{=}CH_2$	Cat. c, DMF, TEA, 65°	$p\text{-}ClC_6H_4[CO(CH_2)_2]_2COC_6H_4Cl\text{-}p$ (60)	8
		$CH_2{=}CHCO(CH_2)_2COCH_3$	Cat. b, TEA, 65°	$p\text{-}ClC_6H_4[CO(CH_2)_2]_2COCH_3$ (70)	17
		$CH_2{=}CH[CO(CH_2)_2]_2COCH_3$	Cat. b, EtOH, TEA, 80°	$p\text{-}ClC_6H_4[CO(CH_2)_2]_3COCH_3$ (61)	52
C_8					
	$n\text{-}C_7H_{15}CHO$	$CH_2{=}CHCOCH{=}CH_2$	Cat. a, TEA, 65°	$n\text{-}C_7H_{15}[CO(CH_2)_2]_2COC_7H_{15}\text{-}n$ (68)	8

TABLE IV. CATALYZED ADDITIONS OF ALDEHYDES THAT YIELD TRI- AND POLYKETONES (*Continued*)

Aldehyde	Michael Acceptor	Conditions	Product(s) and Yield(s) (%)	Refs.
	$CH_2{=}CHCO(CH_2)_2COCH_3$	Cat. a, TEA, 65°	n-$C_7H_{15}[CO(CH_2)_2]_2COCH_3$ (51)	17
	$CH_2{=}CHCO(CH_2)_2COC_4H_9$-$n$	Cat. a, TEA, 65°	n-$C_7H_{15}[CO(CH_2)_2]_2COC_4H_9$-$n$ (72)	17
$OHC(CH_2)_6CHO$	$CH_2{=}CHCOCH_3$	Cat. a, DMF, TEA, 75°	$CH_3CO(CH_2)_2CO(CH_2)_6CO(CH_2)_2COCH_3$ (46)	8
 CHO	$CH_2{=}CHCO(CH_2)_2COCH_3$	Cat. a, TEA, 65°	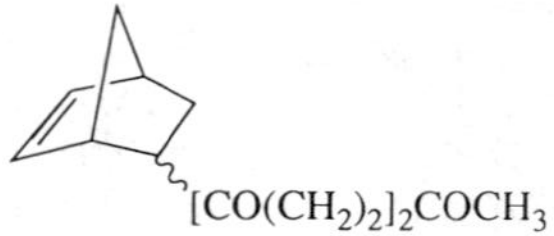 $[CO(CH_2)_2]_2COCH_3$ (65)	52
	$CH_2{=}CHCO(CH_2)_2COC_2H_5$	Cat. a, dioxane, TEA, 65°	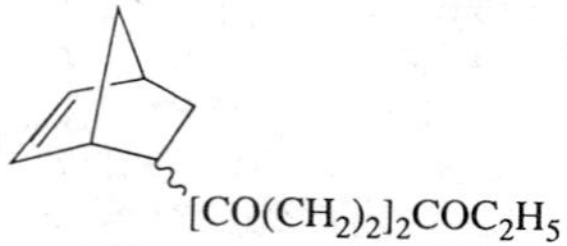 $[CO(CH_2)_2]_2COC_2H_5$ (33)	32
	$CH_2{=}CHCO(CH_2)_2COC_3H_7$-$n$	Cat. a, dioxane, TEA, 65°	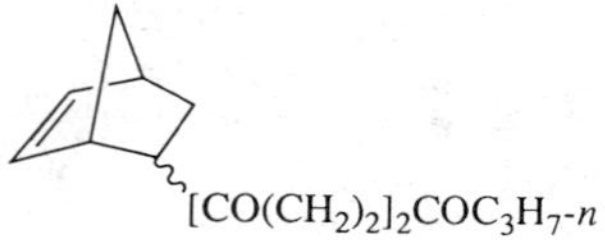 $[CO(CH_2)_2]_2COC_3H_7$-n (56)	32
	$CH_2{=}CH[CO(CH_2)_2]_2COCH_3$	Cat. a, EtOH, TEA, 80°	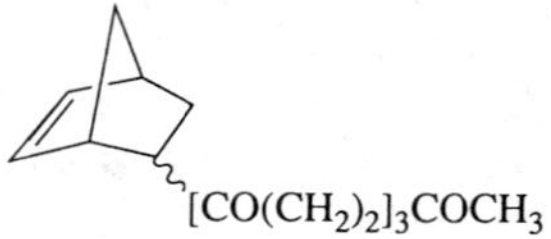 $[CO(CH_2)_2]_3COCH_3$ (63)	52
	$CH_2{=}CH[CO(CH_2)_2]_3COCH_3$	Cat. a, EtOH, TEA, 80°	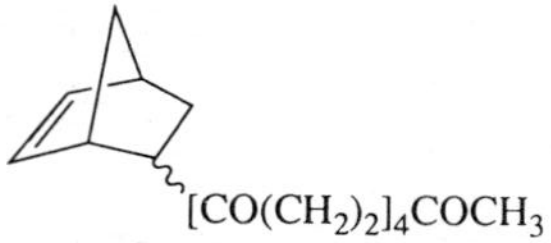 $[CO(CH_2)_2]_4COCH_3$ (52)	52
	$CH_2{=}CHCO(CH_2)_2COC_6H_5$	Cat. a, TEA, 65°	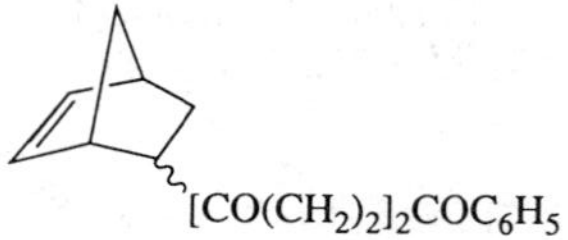 $[CO(CH_2)_2]_2COC_6H_5$ (46)	52

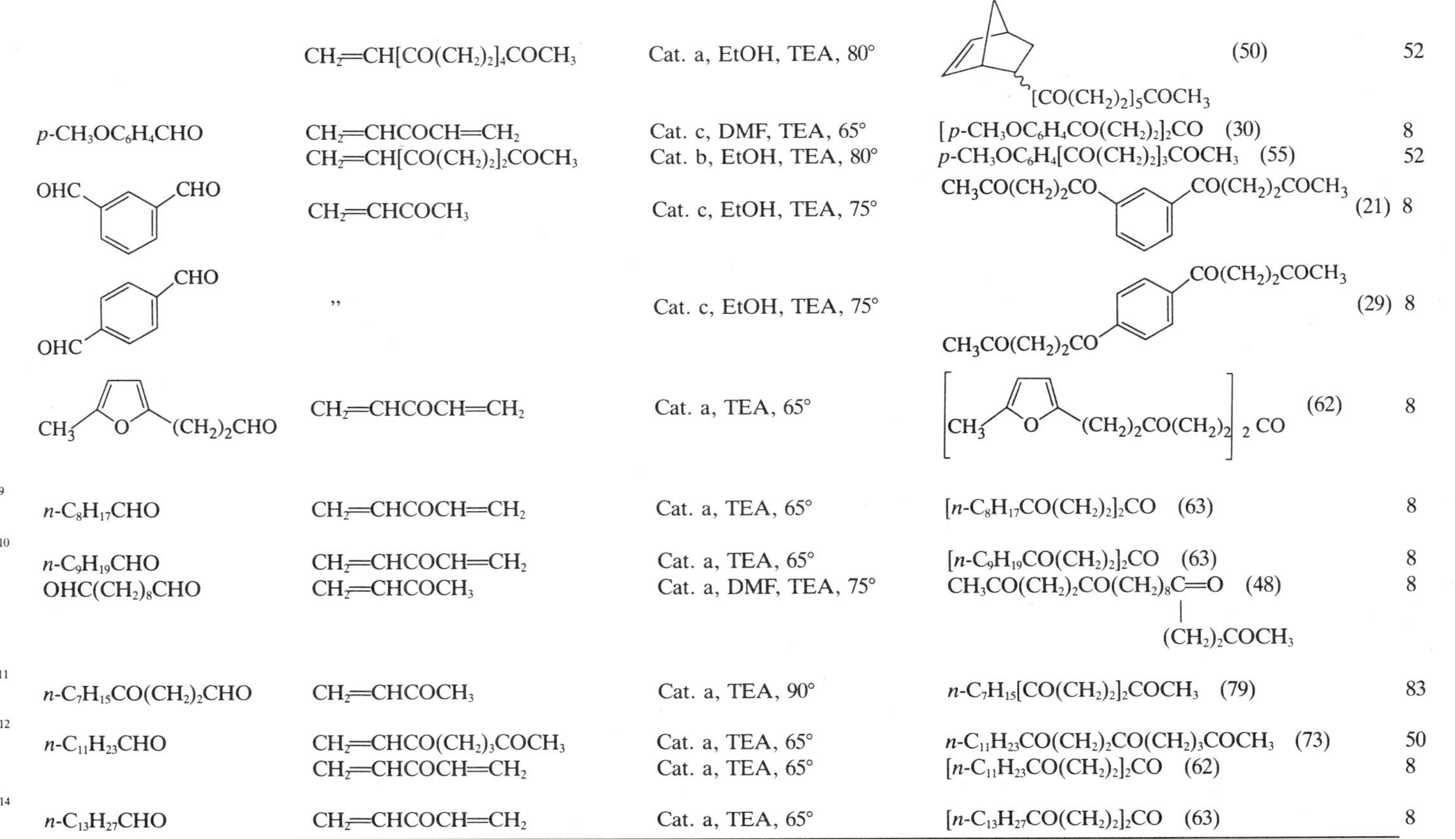

		$CH_2=CH[CO(CH_2)_2]_4COCH_3$	Cat. a, EtOH, TEA, 80°	(bicyclo[2.2.1]heptenyl)$[CO(CH_2)_2]_5COCH_3$ (50)	52
	p-$CH_3OC_6H_4CHO$	$CH_2=CHCOCH=CH_2$	Cat. c, DMF, TEA, 65°	$[p$-$CH_3OC_6H_4CO(CH_2)_2]_2CO$ (30)	8
		$CH_2=CH[CO(CH_2)_2]_2COCH_3$	Cat. b, EtOH, TEA, 80°	p-$CH_3OC_6H_4[CO(CH_2)_2]_3COCH_3$ (55)	52
	m-OHC-C_6H_4-CHO	$CH_2=CHCOCH_3$	Cat. c, EtOH, TEA, 75°	m-$C_6H_4[CO(CH_2)_2COCH_3]_2$ ($CH_3CO(CH_2)_2CO$-C_6H_4-$CO(CH_2)_2COCH_3$) (21)	8
	p-OHC-C_6H_4-CHO	"	Cat. c, EtOH, TEA, 75°	p-$C_6H_4[CO(CH_2)_2COCH_3]_2$ ($CH_3CO(CH_2)_2CO$-C_6H_4-$CO(CH_2)_2COCH_3$) (29)	8
	5-CH_3-2-furyl-$(CH_2)_2CHO$	$CH_2=CHCOCH=CH_2$	Cat. a, TEA, 65°	$[$5-CH_3-2-furyl-$(CH_2)_2CO(CH_2)_2]_2CO$ (62)	8
C_9	n-$C_8H_{17}CHO$	$CH_2=CHCOCH=CH_2$	Cat. a, TEA, 65°	$[n$-$C_8H_{17}CO(CH_2)_2]_2CO$ (63)	8
C_{10}	n-$C_9H_{19}CHO$	$CH_2=CHCOCH=CH_2$	Cat. a, TEA, 65°	$[n$-$C_9H_{19}CO(CH_2)_2]_2CO$ (63)	8
	$OHC(CH_2)_8CHO$	$CH_2=CHCOCH_3$	Cat. a, DMF, TEA, 75°	$CH_3CO(CH_2)_2CO(CH_2)_8C(=O)(CH_2)_2COCH_3$ (48)	8
C_{11}	n-$C_7H_{15}CO(CH_2)_2CHO$	$CH_2=CHCOCH_3$	Cat. a, TEA, 90°	n-$C_7H_{15}[CO(CH_2)_2]_2COCH_3$ (79)	83
C_{12}	n-$C_{11}H_{23}CHO$	$CH_2=CHCO(CH_2)_3COCH_3$	Cat. a, TEA, 65°	n-$C_{11}H_{23}CO(CH_2)_2CO(CH_2)_3COCH_3$ (73)	50
		$CH_2=CHCOCH=CH_2$	Cat. a, TEA, 65°	$[n$-$C_{11}H_{23}CO(CH_2)_2]_2CO$ (62)	8
C_{14}	n-$C_{13}H_{27}CHO$	$CH_2=CHCOCH=CH_2$	Cat. a, TEA, 65°	$[n$-$C_{13}H_{27}CO(CH_2)_2]_2CO$ (63)	8

TABLE V. Catalyzed Additions of Aldehydes to α,β-Unsaturated Acids, Esters, and Lactones

Aldehyde	Michael Acceptor	Conditions	Product(s) and Yield(s) (%)	Refs.
C_2				
CH_3CHO	(α-methylene five-membered lactone)	Cat. a, dioxane, TEA	CH_3COCH_2-substituted five-membered lactone (56)	10
	(α-methylene six-membered lactone)	Cat. a, dioxane, TEA	CH_3COCH_2-substituted six-membered lactone (58)	10
	4,5-di-CH_3-1,3-dioxolan-2-yl–$CH_2CH{=}C(CO_2CH_3)COC_2H_5$	Cat. a, EtOH, TEA, 80°	4,5-di-CH_3-1,3-dioxolan-2-yl–$CH_2CH(COCH_3)CH(CO_2CH_3)COC_2H_5$ (42)	49
	4,5-di-R-2-R^1-1,3-dioxolan-2-yl–$CH_2CH{=}C(CO_2CH_3)COCH_3$	Cat. a, EtOH, TEA, 80°	4,5-di-R-2-R^1-1,3-dioxolan-2-yl–$CH_2CH(COCH_3)CH(CO_2CH_3)COCH_3$	49
	R = H, R^1 = CH_3		(44)	
	R = CH_3, R^1 = H		(45)	
C_3				
C_2H_5CHO	(α-methylene five-membered lactone)	Cat. a, dioxane, TEA, 80°	$C_2H_5COCH_2$-substituted five-membered lactone (54)	10
	(α-methylene six-membered lactone)	Cat. a, dioxane, TEA, 80°	$C_2H_5COCH_2$-substituted six-membered lactone (56)	10
	$CH_3CH{=}C(CO_2C_2H_5)COCH_3$	Cat. a, dioxane, TEA, 80°	$C_2H_5COCH(CH_3)CH(CO_2C_2H_5)COCH_3$ (49)	36
	(E)-$C_2H_5O_2CCH{=}CHCO_2C_2H_5$	Cat. a, dioxane, TEA, 90°	$C_2H_5COCH(CO_2C_2H_5)CH_2CO_2C_2H_5$ (62)	10

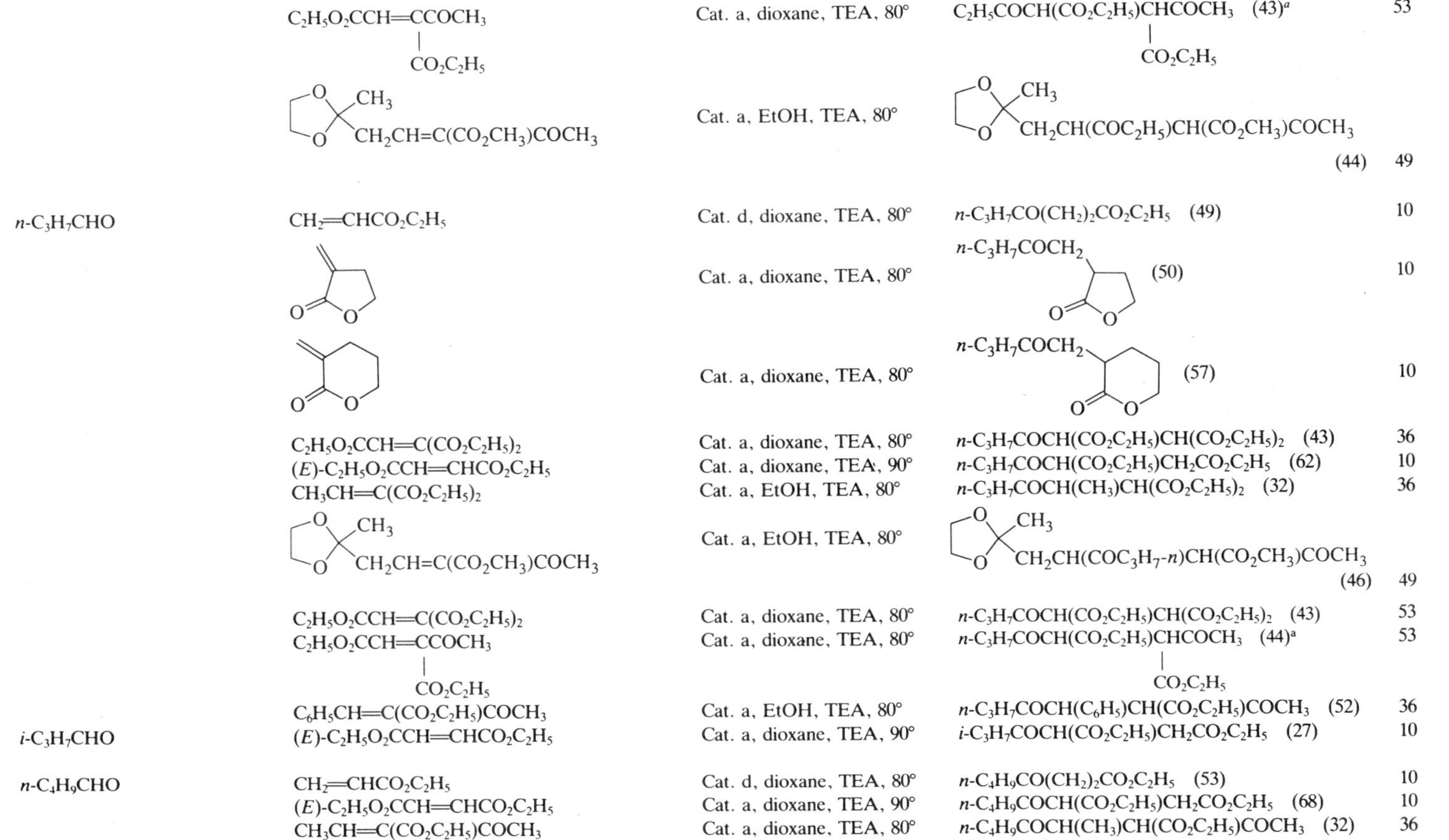

Aldehyde	Acceptor	Conditions	Product(s) and Yield(s) (%)	Refs.
	$C_2H_5O_2CCH{=}C(CO_2C_2H_5)COCH_3$	Cat. a, dioxane, TEA, 80°	$C_2H_5COCH(CO_2C_2H_5)CH(CO_2C_2H_5)COCH_3$ (43)[a]	53
	2-Methyl-1,3-dioxolan-2-yl–$CH_2CH{=}C(CO_2CH_3)COCH_3$	Cat. a, EtOH, TEA, 80°	2-Methyl-1,3-dioxolan-2-yl–$CH_2CH(COC_2H_5)CH(CO_2CH_3)COCH_3$ (44)	49
C_4 n-C_3H_7CHO	$CH_2{=}CHCO_2C_2H_5$	Cat. d, dioxane, TEA, 80°	n-$C_3H_7CO(CH_2)_2CO_2C_2H_5$ (49)	10
	(structure)	Cat. a, dioxane, TEA, 80°	n-$C_3H_7COCH_2$– (structure) (50)	10
	(structure)	Cat. a, dioxane, TEA, 80°	n-$C_3H_7COCH_2$– (structure) (57)	10
	$C_2H_5O_2CCH{=}C(CO_2C_2H_5)_2$	Cat. a, dioxane, TEA, 80°	n-$C_3H_7COCH(CO_2C_2H_5)CH(CO_2C_2H_5)_2$ (43)	36
	(E)-$C_2H_5O_2CCH{=}CHCO_2C_2H_5$	Cat. a, dioxane, TEA, 90°	n-$C_3H_7COCH(CO_2C_2H_5)CH_2CO_2C_2H_5$ (62)	10
	$CH_3CH{=}C(CO_2C_2H_5)_2$	Cat. a, EtOH, TEA, 80°	n-$C_3H_7COCH(CH_3)CH(CO_2C_2H_5)_2$ (32)	36
	2-Methyl-1,3-dioxolan-2-yl–$CH_2CH{=}C(CO_2CH_3)COCH_3$	Cat. a, EtOH, TEA, 80°	2-Methyl-1,3-dioxolan-2-yl–$CH_2CH(COC_3H_7$-$n)CH(CO_2CH_3)COCH_3$ (46)	49
	$C_2H_5O_2CCH{=}C(CO_2C_2H_5)_2$	Cat. a, dioxane, TEA, 80°	n-$C_3H_7COCH(CO_2C_2H_5)CH(CO_2C_2H_5)_2$ (43)	53
	$C_2H_5O_2CCH{=}C(CO_2C_2H_5)COCH_3$	Cat. a, dioxane, TEA, 80°	n-$C_3H_7COCH(CO_2C_2H_5)CH(CO_2C_2H_5)COCH_3$ (44)[a]	53
	$C_6H_5CH{=}C(CO_2C_2H_5)COCH_3$	Cat. a, EtOH, TEA, 80°	n-$C_3H_7COCH(C_6H_5)CH(CO_2C_2H_5)COCH_3$ (52)	36
i-C_3H_7CHO	(E)-$C_2H_5O_2CCH{=}CHCO_2C_2H_5$	Cat. a, dioxane, TEA, 90°	i-$C_3H_7COCH(CO_2C_2H_5)CH_2CO_2C_2H_5$ (27)	10
C_5 n-C_4H_9CHO	$CH_2{=}CHCO_2C_2H_5$	Cat. d, dioxane, TEA, 80°	n-$C_4H_9CO(CH_2)_2CO_2C_2H_5$ (53)	10
	(E)-$C_2H_5O_2CCH{=}CHCO_2C_2H_5$	Cat. a, dioxane, TEA, 90°	n-$C_4H_9COCH(CO_2C_2H_5)CH_2CO_2C_2H_5$ (68)	10
	$CH_3CH{=}C(CO_2C_2H_5)COCH_3$	Cat. a, dioxane, TEA, 80°	n-$C_4H_9COCH(CH_3)CH(CO_2C_2H_5)COCH_3$ (32)	36

TABLE V. CATALYZED ADDITIONS OF ALDEHYDES TO α,β-UNSATURATED ACIDS, ESTERS, AND LACTONES (*Continued*)

Aldehyde	Michael Acceptor	Conditions	Product(s) and Yield(s) (%)	Refs.
	$C_2H_5O_2CCH{=}CCOCH_3$ $\vert$ $CO_2C_2H_5$	Cat. a, dioxane, TEA, 80°	$n\text{-}C_4H_9COCH(CO_2C_2H_5)CHCOCH_3$ (36)[a] $\vert$ $CO_2C_2H_5$	53
$i\text{-}C_4H_9CHO$	$(E)\text{-}C_2H_5O_2CCH{=}CHCO_2C_2H_5$	Cat. a, dioxane, TEA, 90°	$i\text{-}C_4H_9COCH(CO_2C_2H_5)CH_2CO_2C_2H_5$ (47)	10
$CH_2{=}CH(CH_2)_2CHO$	$CH_2{=}CHCO_2CH_3$	Cat. a, TEA, 85°	$CH_2{=}CH(CH_2)_2CO(CH_2)_2CO_2CH_3$ (22)	11
	"	Cat. a, dioxane, TEA, 100°	" (30)	57
2-Furyl-CHO (C_4H_3O–CHO)	$CH_2{=}CHCO_2C_2H_5$	Cat. e, EtOH, TEA, 80°	$C_4H_3OCO(CH_2)_2CO_2C_2H_5$ (24)	12a
	"	Cat. a, EtOH, TEA, 80°	" (31)	7
	$CH_3CH{=}C(CO_2C_2H_5)_2$	Cat. b, EtOH, TEA, 80°	$C_4H_3OCOCH(CH_3)CH(CO_2C_2H_5)_2$ (43)	36
	$C_2H_5O_2CCH{=}CCOCH_3$ $\vert$ $CO_2C_2H_5$	Cat. a, dioxane, TEA, 80°	$C_4H_3OCOCH(CO_2C_2H_5)CHCOCH_3$ (33) $\vert$ $CO_2C_2H_5$	53
	$C_6H_5CH{=}C(CO_2C_2H_5)COCH_3$	Cat. b, EtOH, TEA, 80°	$C_4H_3OCOCH(C_6H_5)CH(CO_2C_2H_5)COCH_3$ (81)	36
	$C_6H_5CH{=}C(CO_2C_2H_5)_2$	Cat. b, EtOH, TEA, 80°		
2-Thienyl-CHO (C_4H_3S–CHO)	$CH_2{=}CHCO_2C_2H_5$	NaCN, DMF, 22°	$C_4H_3SCO(CH_2)_2CO_2C_2H_5$ (45)	44
	$(E,Z)\text{-}CH_3CH{=}CHCO_2C_2H_5$	NaCN, DMF, 22°	$C_4H_3SCOCH(CH_3)CH_2CO_2C_2H_5$ (47)	44
	2-Furyl-$CH{=}C(CO_2C_2H_5)COCH_3$	Cat. b, DMF, TEA, 80°	$C_4H_3SCOCH(C_4H_3O)CHCOCH_3$ (57) $\vert$ $CO_2C_2H_5$	36
	2-Thienyl-$CH{=}C(CO_2C_2H_5)COCH_3$	Cat. b, dioxane, TEA, 80°	$C_4H_3SCOCH(C_4H_3S)CHCOCH_3$ (71) $\vert$ $CO_2C_2H_5$	36
	$C_6H_5CH{=}C(CO_2CH_3)COCH_3$	Cat. b, EtOH, TEA, 80°	$C_4H_3SCOCH(C_6H_5)CH(CO_2CH_3)COCH_3$ (51)	36
C_6				
$n\text{-}C_5H_{11}CHO$	$CH_2{=}CHCO_2C_2H_5$	Cat. a, dioxane, TEA, 80°	$n\text{-}C_5H_{11}CO(CH_2)_2CO_2C_2H_5$ (52)	10
	$(E)\text{-}C_2H_5O_2CCH{=}CHCO_2C_2H_5$	Cat. a, dioxane, TEA, 90°	$n\text{-}C_5H_{11}COCH(CO_2C_2H_5)CH_2CO_2C_2H_5$ (69)	10
$(C_2H_5O)_2CHCHO$	$CH_2{=}CHCO_2C_2H_5$	Cat. a, dioxane, TEA, 80°	$(C_2H_5O)_2CHCO(CH_2)_2CO_2C_2H_5$ (24)	22

$C_4H_8NCOCHO$ (pyrrolidine N-COCHO, structure)	$CH_2{=}CHCO_2C_2H_5$	Cat. a, dioxane, TEA, 80°	$C_4H_8NCOCO(CH_2)_2CO_2C_2H_5$ (83)	25
	$CH_2{=}C(CH_3)CO_2C_2H_5$	Cat. a, dioxane, TEA, 80°	$C_4H_8NCOCOCH_2CH(CH_3)CO_2C_2H_5$ (66)	25
	(E,Z)-$CH_3CH{=}CHCO_2C_2H_5$	Cat. a, dioxane, TEA, 80°	$C_4H_8NCOCOCH(CH_3)CH_2CO_2C_2H_5$ (56)	25
	(E)-$C_2H_5O_2CCH{=}CHCO_2C_2H_5$	Cat. a, dioxane, TEA, 80°	$C_4H_8NCOCOCH(CO_2C_2H_5)CH_2CO_2C_2H_5$ (59)	25
	(E)-$C_6H_5CH{=}CHCO_2C_2H_5$	Cat. a, dioxane, TEA, 80°	$C_4H_8NCOCOCH(C_6H_5)CH_2CO_2C_2H_5$ (50)	25
3-pyridine-CHO (structure)	$CH_2{=}CHCO_2C_2H_5$	NaCN, DMF, 22°	3-$C_5H_4NCO(CH_2)_2CO_2C_2H_5$ (37)	44
C_7				
n-$C_6H_{13}CHO$	$CH_2{=}CHCO_2C_2H_5$	Cat. d, dioxane, TEA, 80°	n-$C_6H_{13}CO(CH_2)_2C_2H_5$ (54)	10
	(E)-$C_2H_5O_2CCH{=}CHCO_2C_2H_5$	Cat. a, dioxane, TEA, 90°	n-$C_6H_{13}COCH(CO_2C_2H_5)CH_2CO_2C_2H_5$ (73)	10
C_6H_5CHO	$CH_2{=}CHCO_2C_2H_5$	NaCN, DMF, 20°	$C_6H_5CO(CH_2)_2CO_2C_2H_5$ (55)	44
	(E,Z)-$CH_3CH{=}CHCO_2C_2H_5$	NaCN, DMF, 36°	$C_6H_5COCH(CH_3)CH_2CO_2C_2H_5$ (33)	44
	(E,Z)-$CH_3CH{=}CHCO_2C_3H_7$-i	NaCN, DMF, 36°	$C_6H_5COCH(CH_3)CH_2CO_2C_3H_7$-$i$ (40)	44
	(E,Z)-$CH_3CH{=}CHCO_2C_4H_9$-t	NaCN, DMF, 36°	$C_6H_5COCH(CH_3)CH_2CO_2C_4H_9$-$t$ (52)	44
	(Z)-$C_2H_5O_2CCH{=}CHCO_2C_2H_5$	NaCN, DMF, 35°	$C_6H_5COCH(CO_2C_2H_5)CH_2CO_2C_2H_5$ (32)	2a
	$C_2H_5O_2CCH{=}CCOC_6H_5$ | $CO_2C_2H_5$	Cat. b, dioxane, TEA, 80°	$C_6H_5COCH(CO_2C_2H_5)CHCOC_6H_5$ (66) | $CO_2C_2H_5$	53
p-ClC_6H_4CHO	$CH_2{=}CHCO_2C_2H_5$	NaCN, DMF, 22°	p-$ClC_6H_4CO(CH_2)_2CO_2C_2H_5$ (68)	44
	(E,Z)-$CH_3CH{=}CHCO_2CH_3$	NaCN, DMF, 22°	p-$ClC_6H_4COCH(CH_3)CH_2CO_2CH_3$ (35)	44
	E,Z-$CH_3CH{=}CHCO_2C_2H_5$	NaCN, DMF, 35°	p-$ClC_6H_4COCH(CH_3)CH_2CO_2C_2H_5$ (56)	2a
	$CH_2{=}C(CH_3)CO_2C_2H_5$	NaCN, DMF, 22°	p-$ClC_6H_4COCH_2CH(CH_3)CO_2C_2H_5$ (34)	44
	(E,Z)-$CH_3CH{=}CHCO_2C_3H_7$-i	NaCN, DMF, 22°	p-$ClC_6H_4COCH(CH_3)CH_2CO_2C_3H_7$-$i$ (49)	44
	(E,Z)-$CH_3CH{=}CHCO_2C_4H_9$-t	NaCN, DMF, 22°	p-$ClC_6H_4COCH(CH_3)CH_2CO_2C_4H_9$-$t$ (64)	44
	(Z)-$C_2H_5O_2CCH{=}CHCO_2C_2H_5$	NaCN, DMF, 35°	p-$ClC_6H_4COCH(CO_2C_2H_5)CH_2CO_2C_2H_5$ (35)	2a
	(E)-$C_6H_5CH{=}CHCO_2C_2H_5$	NaCN, DMF, 40°	p-$ClC_6H_4COCH(C_6H_5)CH_2CO_2C_2H_5$ (54)	44
	(E)-$C_6H_5CH{=}CHCO_2C_4H_9$-t	NaCN, DMF, 36°	p-$ClC_6H_4COCH(C_6H_5)CH_2CO_2C_4H_9$-$t$ (60)	44
	$C_2H_5O_2CCH{=}CCOC_6H_5$ | $CO_2C_2H_5$	Cat. b, dioxane, TEA, 80°	p-$ClC_6H_4COCH(CO_2C_2H_5)CHCOC_6H_5$ (37) | $CO_2C_2H_5$	53

TABLE V. CATALYZED ADDITIONS OF ALDEHYDES TO α,β-UNSATURATED ACIDS, ESTERS, AND LACTONES (*Continued*)

Aldehyde	Michael Acceptor	Conditions	Product(s) and Yield(s) (%)	Refs.
C_8				
n-$C_7H_{15}CHO$	$CH_2{=}CHCO_2C_2H_5$	Cat. d, dioxane, TEA, 80°	n-$C_7H_{15}CO(CH_2)_2CO_2C_2H_5$ (59)	10
	(E)-$C_2H_5O_2CCH{=}CHCO_2C_2H_5$	Cat. a, dioxane, TEA, 90°	n-$C_7H_{15}COCH(CO_2C_2H_5)CH_2CO_2C_2H_5$ (63)	10
(E)-$CH_2{=}CHCH_2CH{=}CH$ (with $OHC(CH_2)_2$ attached)	$CH_2{=}CHCO_2C_2H_5$	Cat. d, dioxane, TEA, 87°	(E)-$CH_2{=}CHCH_2CH{=}CH(CH_2)_2C{=}O$ (with $C_2H_5O_2C(CH_2)_2$ attached) (—)	84
 CHO	$CH_2{=}CHCO_2CH_3$	Cat. a, TEA, 100°	 $CO(CH_2)_2CO_2CH_3$ (55)	80, 4
$CH_3O_2C(CH_2)_5CHO$	$CH_2{=}CHCO_2CH_3$	Cat. a, dioxane, TEA, 100°	$CH_3O_2C(CH_2)_5CO(CH_2)_2COCH_3$ (32)	57
C_9				
n-$C_8H_{17}CHO$	$CH_2{=}CHCO_2C_2H_5$	Cat. d, dioxane, TEA, 80°	n-$C_8H_{17}CO(CH_2)_2CO_2C_2H_5$ (56)	10
	(E)-$C_2H_5O_2CCH{=}CHCO_2C_2H_5$	Cat. a, dioxane, TEA, 90°	n-$C_8H_{17}COCH(CO_2C_2H_5)CH_2CO_2C_2H_5$ (49)	10
(E)-$C_6H_5CH{=}CHCHO$	$C_6H_5CH{=}C(CO_2C_2H_5)COCH_3$	Cat. a, EtOH, TEA, 80°	(E)-$C_6H_5CH{=}CHCOCH(C_6H_5)CHCOCH_3$ (with $CO_2C_2H_5$ attached) (20)	36
C_{10}				
n-$C_9H_{19}CHO$	$CH_2{=}CHCO_2C_2H_5$	Cat. d, dioxane, TEA, 80°	n-$C_9H_{19}CO(CH_2)_2CO_2C_2H_5$ (51)	10
	(E)-$C_2H_5O_2CCH{=}CHCO_2C_2H_5$	Cat. a, dioxane, TEA, 90°	n-$C_9H_{19}COCH(CO_2C_2H_5)CH_2CO_2C_2H_5$ (45)	10
$(CH_3)_2C{=}CH(CH_2)_2CHCH_3$ (with $OHCCH_2$ attached)	$CH_2{=}CHCO_2C_2H_5$	Cat. d, TEA, reflux	$(CH_3)_2C{=}CH(CH_2)_2CH(CH_3)CH_2$ (with $C_2H_5O_2C(CH_2)_2C{=}O$ attached) (55)	85
$CH_3O_2C(CH_2)_7CHO$	$CH_2{=}CHCO_2CH_3$	Cat. f, dioxane, TEA, 100°	$CH_3O_2C(CH_2)_7CO(CH_2)_2CO_2CH_3$ (42)	11
(Z)-$CH_3O_2C(CH_2)_3CH{=}CH$ (with $OHC(CH_2)_2$ attached)	$CH_2{=}CHCO_2CH_3$	Cat. a, TEA, 85°	(Z)-$CH_3O_2C(CH_2)_3CH{=}CH(CH_2)_2C{=}O$ (with $CH_3O_2C(CH_2)_2$ attached) (21)	11

Reactant	Michael acceptor	Conditions	Product(s) and Yield(s) (%)	Refs.
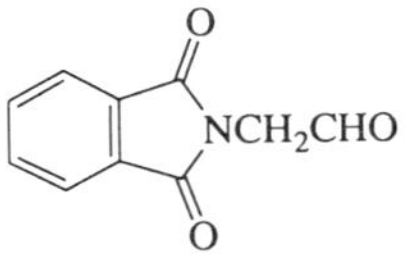 N-(formylmethyl)phthalimide (PhThNCH_2CHO)	$C_6H_5CH{=}C(CO_2C_2H_5)COCH_3$	Cat. a, EtOH, TEA, 80°	$PhThNCH_2COCH(C_6H_5)CH(CO_2C_2H_5)COCH_3$ (51)	26
	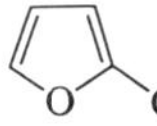 2-furyl-$CH{=}C(CO_2C_2H_5)COCH_3$	Cat. a, EtOH, TEA, 80°	$PhThNCH_2COCH(C_4H_3O)CH(CO_2C_2H_5)COCH_3$ (47)	26
	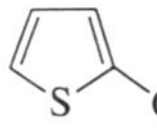 2-thienyl-$CH{=}C(CO_2C_2H_5)COCH_3$	Cat. a, EtOH, TEA, 80°	$PhThNCH_2COCH(C_4H_3S)CH(CO_2C_2H_5)COCH_3$ (53)	26
C_{11}				
n-$C_{10}H_{21}CHO$	(E)-$C_2H_5O_2CCH{=}CHCO_2C_2H_5$	Cat. a, dioxane, TEA, 90°	n-$C_{10}H_{21}COCH(CO_2C_2H_5)CH_2CO_2C_2H_5$ (49)	10
(Z)-$C_2H_5O_2C(CH_2)_3CH{=}CHCH_2CH_2CHO$ [printed as (Z)-$C_2H_5O_2C(CH_2)_3CH{=}CH$ with $OHC(CH_2)_2$ hanging below the CH]	$CH_2{=}CHCO_2C_2H_5$	Cat. a, TEA, 85°	(Z)-$C_2H_5O_2C(CH_2)_3CH{=}CH(CH_2)_2C(=O)(CH_2)_2CO_2C_2H_5$ [printed as (Z)-$C_2H_5O_2C(CH_2)_3CH{=}CH(CH_2)_2C{=}O$ with $C_2H_5O_2C(CH_2)_2$ hanging below the C=O] (—)	11
C_{13}				
Bicyclic aldehyde bearing CN, OHC and exocyclic ${=}CHCO_2CH_3$ groups (structure)		Cat. a, i-C_3H_7OH, TEA, 80°	Tricyclic ketone bearing CN and $CH_3O_2CCH_2$ groups (structure) (67)	45

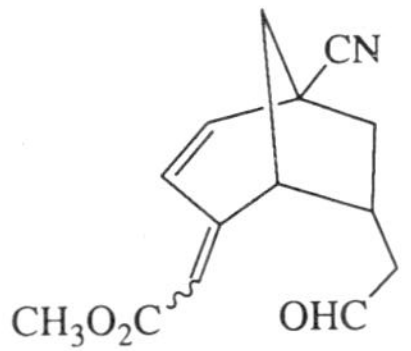

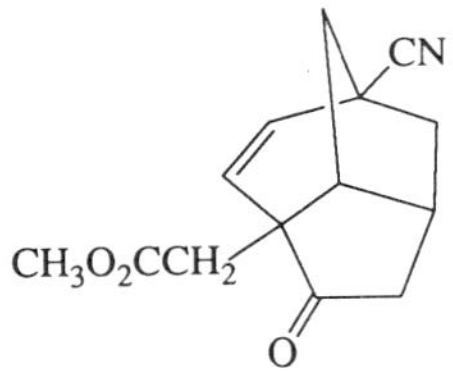

[a] This product could not be isolated because it cyclized to the furan.

TABLE VI. CATALYZED ADDITIONS OF ALDEHYDES TO α,β-UNSATURATED NITRILES

	Aldehyde	Michael Acceptor	Conditions	Product(s) and Yield(s) (%)	Refs.
C_2	CH_3CHO	$CH_2{=}CHCN$	Cat. a, EtOH, TEA, 80°	$CH_3CO(CH_2)_2CN$ (30)	7
C_3	C_2H_5CHO	"	Cat. e, EtOH, TEA, 80°	$C_2H_5CO(CH_2)_2CN$ (10)	12a
C_4	n-C_3H_7CHO	"	Cat. d, dioxane, TEA, 80°	n-$C_3H_7CO(CH_2)_2CN$ (52)	10
C_5	n-C_4H_9CHO	"	Cat. d, dioxane, TEA, 80°	n-$C_4H_9CO(CH_2)_2CN$ (55)	10
	i-C_4H_9CHO	"	Cat. d, dioxane, TEA, 80°	i-$C_4H_9CO(CH_2)_2CN$ (51)	10
	(2-furyl)CHO	"	Cat. e, EtOH, TEA, 80°	$C_4H_3OCO(CH_2)_2CN$ (13)	12a
		"	Cat. a, EtOH, TEA, 80°	" (48)	7
		"	NaCN, DMF, 30°	" (63–67)	40
		$CH_2{=}C(CH_3)CN$	NaCN, DMF, 30°	$C_4H_3OCOCH_2CH(CH_3)CN$ (60–63)	40
		$CH_3CH{=}CHCN$	NaCN, DMF, 30°	$C_4H_3OCH(CH_3)CH_2CN$ (70–75)	40
		(E)-$C_6H_5CH{=}CHCN$	NaCN, DMF, 30°	$C_4H_3OCOCH(C_6H_5)CH_2CN$ (65–70)	40
	(2-thienyl)CHO	$CH_2{=}CHCN$	NaCN, DMF, 20°	$C_4H_3SCO(CH_2)_2CN$ (85)	38
		$CH_3CH{=}CHCN$	NaCN, DMF, 20°	$C_4H_3SCOCH(CH_3)CH_2CN$ (76)	38
		(E)-$C_6H_5CH{=}CHCN$	NaCN, DMF, 20°	$C_4H_3SCOCH(C_6H_5)CH_2CN$ (71)	38

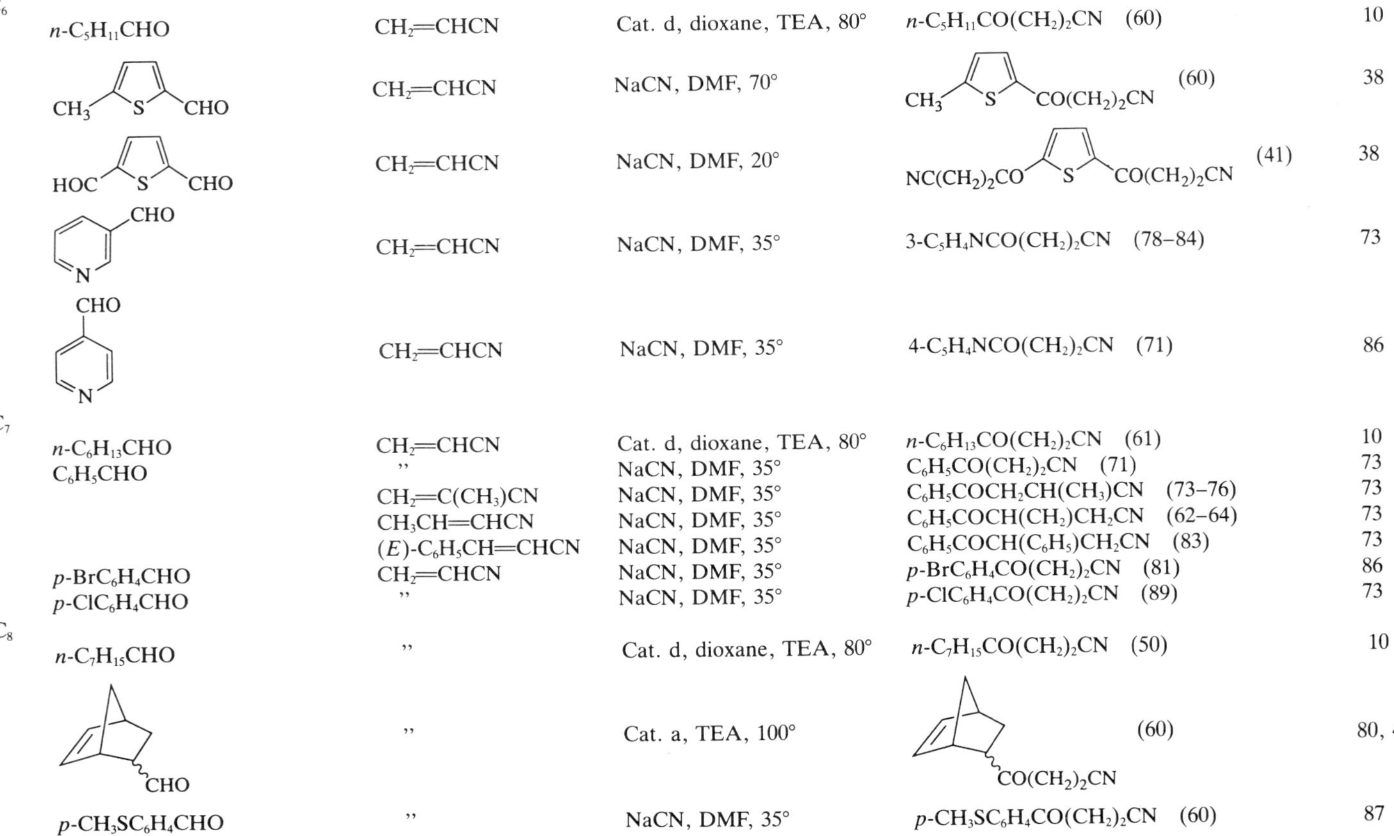

C_6	n-$C_5H_{11}CHO$	$CH_2{=}CHCN$	Cat. d, dioxane, TEA, 80°	n-$C_5H_{11}CO(CH_2)_2CN$ (60)	10
	(structure) CH_3, S, CHO	$CH_2{=}CHCN$	NaCN, DMF, 70°	(structure) CH_3, S, $CO(CH_2)_2CN$ (60)	38
	(structure) HOC, S, CHO	$CH_2{=}CHCN$	NaCN, DMF, 20°	(structure) $NC(CH_2)_2CO$, S, $CO(CH_2)_2CN$ (41)	38
	(structure) CHO, N	$CH_2{=}CHCN$	NaCN, DMF, 35°	3-$C_5H_4NCO(CH_2)_2CN$ (78–84)	73
	(structure) CHO, N	$CH_2{=}CHCN$	NaCN, DMF, 35°	4-$C_5H_4NCO(CH_2)_2CN$ (71)	86
C_7	n-$C_6H_{13}CHO$	$CH_2{=}CHCN$	Cat. d, dioxane, TEA, 80°	n-$C_6H_{13}CO(CH_2)_2CN$ (61)	10
	C_6H_5CHO	"	NaCN, DMF, 35°	$C_6H_5CO(CH_2)_2CN$ (71)	73
		$CH_2{=}C(CH_3)CN$	NaCN, DMF, 35°	$C_6H_5COCH_2CH(CH_3)CN$ (73–76)	73
		$CH_3CH{=}CHCN$	NaCN, DMF, 35°	$C_6H_5COCH(CH_3)CH_2CN$ (62–64)	73
		(E)-$C_6H_5CH{=}CHCN$	NaCN, DMF, 35°	$C_6H_5COCH(C_6H_5)CH_2CN$ (83)	73
	p-BrC_6H_4CHO	$CH_2{=}CHCN$	NaCN, DMF, 35°	p-$BrC_6H_4CO(CH_2)_2CN$ (81)	86
	p-ClC_6H_4CHO	"	NaCN, DMF, 35°	p-$ClC_6H_4CO(CH_2)_2CN$ (89)	73
C_8	n-$C_7H_{15}CHO$	"	Cat. d, dioxane, TEA, 80°	n-$C_7H_{15}CO(CH_2)_2CN$ (50)	10
	(structure) CHO	"	Cat. a, TEA, 100°	(structure) $CO(CH_2)_2CN$ (60)	80, 41
	p-$CH_3SC_6H_4CHO$	"	NaCN, DMF, 35°	p-$CH_3SC_6H_4CO(CH_2)_2CN$ (60)	87

TABLE VI. CATALYZED ADDITIONS OF ALDEHYDES TO α,β-UNSATURATED NITRILES (*Continued*)

Aldehyde	Michael Acceptor	Conditions	Product(s) and Yield(s) (%)	Refs.
C_9				
n-$C_8H_{17}CHO$	"	Cat. d, dioxane, TEA, 80°	*n*-$C_8H_{17}CO(CH_2)_2CN$ (64)	10
m- + *p*-$(CH_2{=}CH)C_6H_4CHO$	"	NaCN, DMF, 35°	*m*- + *p*-$(CH_2{=}CH)C_6H_4CO(CH_2)_2CN$ (—)	80
C_{10}				
n-$C_9H_{19}CHO$	"	Cat. d, dioxane, TEA, 80°	*n*-$C_9H_{19}CO(CH_2)_2CN$ (57)	10
C_{11}				
2-Naphthaldehyde (structure: naphthalene ring with CHO)	"	NaCN, DMF, 35°	2-$C_{10}H_7CO(CH_2)_2CN$ (81)	86
C_{13}				
m-$C_6H_5OC_6H_4CHO$	"	NaCN, DMF, 35°	*m*-$C_6H_5OC_6H_4CO(CH_2)_2CN$ (81)	88
	(*E*)-$C_6H_5CH{=}CHCN$	NaCN, DMF, 35°	*m*-$C_6H_5OC_6H_4COCH(C_6H_5)CH_2CN$ (92)	88

TABLE VII. CATALYZED ADDITIONS OF ALDEHYDES TO MISCELLANEOUS MICHAEL ACCEPTORS

Aldehyde	Michael Acceptor	Conditions	Product(s) and Yield(s) (%)	Refs.
C_7				
C_6H_5CHO	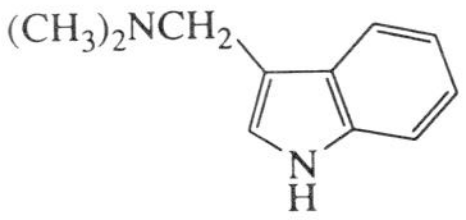 $(CH_3)_2NCH_2$-(indol-3-yl, N-H)	NacN, DMF, 70°	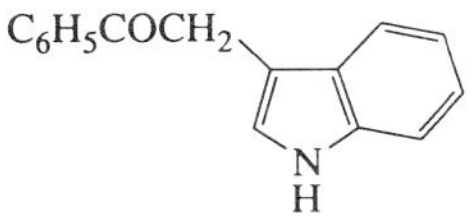 $C_6H_5COCH_2$-(indol-3-yl, N-H) (52)	37

REFERENCES

[1] W. S. Ide and J. S. Buck, *Org. React.* **4**, 269 (1948).
[2a] H. Stetter and M. Schreckenberg, *Angew. Chem.*, **85**, 89 (1973); *Angew. Chem., Int. Ed. Engl.*, **12**, 81 (1973).
[2b] H. Stetter and M. Schreckenberg (to Bayer A.-G.), Ger. Offen. 2,262,343 (1974) [*C.A.*, **81**, 105015j (1974)].
[2c] H. Stetter, *Angew. Chem.*, **88**, 695 (1976); *Angew. Chem., Int. Ed. Engl.*, **15**, 639 (1976).
[3a] R. Breslow, *J. Am. Chem. Soc.*, **80**, 3719 (1958).
[3b] N. Tagaki and H. Hara, *J. Chem. Soc., Chem. Commun.*, **1973**, 891.
[4] B. Lackmann, H. Steinmaus, and H. W. Wanzlick, *Tetrahedron*, **27**, 4085 (1971).
[5a] H. Stetter, H. Kuhlmann, and R. Y. Rämsch, *Synthesis*, **1976**, 733.
[5b] H. Stetter and H. Kuhlmann, *Org. Synth.*, **62**, 170 (1984).
[6a] H. Stetter and H. Kuhlmann, *Angew. Chem.*, **86**, 589 (1974); *Angew. Chem., Int. Ed. Engl.*, **13**, 539 (1974).
[6b] H. Stetter and H. Kuhlmann (to Bayer A.-G.) Ger. Offen. 2,437,219 (1976) [*C.A.*, **84**, 164172t (1976)].
[7] H. Stetter and H. Kuhlmann, *Chem. Ber.*, **109**, 2890 (1976).
[8] H. Stetter, W. Basse, H. Kuhlmann, A. Landscheidt, and W. Schlenker, *Chem. Ber.*, **110**, 1007 (1977).
[9] H. Stetter and H. Kuhlmann, *Synthesis*, **1975**, 379.
[10] H. Stetter, W. Basse, and J. Nienhaus, *Chem. Ber.*, **113**, 690 (1980).
[11] L. Novák, G. Baan, J. Marosfalvi, and C. Szántay, *Chem. Ber.*, **113**, 2939 (1980).
[12a] J. Castells, E. Duñach, F. Geijo, F. Pujol, and P. M. Segura, *Israel J. Chem.*, **17**, 278 (1978).
[12b] C. S. Sell and L. A. Dorman, *J. Chem. Soc., Chem. Commun.*, **1982**, 629.
[12c] B.-H. Chang and Y. L. Chang, *J. Chinese Chem. Soc.*, **30**, 55 (1983).
[13] H.-J. Krause (to Henkel K.G. aA), Ger. Offen. 2,732,714 (1979) [*C.A.* **90**, 168067j (1979)].
[14] A. F. Noels and A. J. Hubert, *J. Mol. Catal.*, **22**, 235 (1983).
[15] H. Stetter, H. Kuhlmann, and W. Haese, *Org. Synth.*, **65**, 26 (1987).
[16] H. Stetter, G. Hilboll, and H. Kuhlmann, *Chem. Ber.*, **112**, 84 (1979).
[17] H. Stetter and A. Landscheidt, *Chem. Ber.*, **112**, 1410 (1979).
[18] V. R. Mamdapur, C. S. Subramaniam, and M. S. Chadha, *J. Indian Chem. Soc. B*, **18**, 450 (1979).
[19] H. Stetter and J. Nienhaus, *Chem. Ber.*, **111**, 2825 (1978).
[20] H. Stetter, K.-H. Mohrmann, and W. Schlenker, *Chem. Ber.*, **114**, 581 (1981).
[21] H. Stetter and H. T. Leinen, *Chem. Ber.*, **116**, 254 (1983).
[22] H. Stetter and K. H. Mohrmann, *Synthesis*, **1981**, 129.
[23] T. H. Jones, R. J. Highet, A. W. Don, and M. S. Blum, *J. Org. Chem.*, **51**, 2712 (1986).
[24] H. Stetter, W. Basse, and K. Wiemann, *Chem. Ber.*, **111**, 431 (1978).
[25] H. Stetter and H. Skobel, *Chem. Ber.*, **120**, 643 (1987).
[26] H. Stetter and P. Lappe, *Chem. Ber.*, **113**, 1890 (1980).
[27a] H. Kuhlmann, Ph.D. Dissertation, Technische Hochschule Aachen, 1976.
[27b] H.-D. Jöge, Ph.D. Dissertation, Technische Hochschule Aachen, 1977.
[28] W. Hinz, R. A. Jones, S. U. Patel, and M.-H. Karatza, *Tetrahedron*, **42**, 3753 (1986).
[29] H. Stetter and J. Krasselt, *J. Heterocycl. Chem.*, **14**, 573 (1977).
[30a] S. Mizuhara, R. Tamura, and H. Arata, *Proc. Jpn. Acad.*, **27**, 302 (1951).
[30b] S. Mizuhara and P. Handler, *J. Am. Chem. Soc.*, **76**, 571 (1954).
[30c] R. Breslow and E. McNellis, *J. Am. Chem. Soc.*, **81**, 3080 (1959).
[30d] E. Jordan and Y. H. Mariam, *J. Am. Chem. Soc.*, **100**, 2534 (1978).
[31] H. Stetter and G. Lorenz, *Chem. Ber.*, **118**, 1115 (1985).
[32] H. Stetter and B. Jansen, *Chem. Ber.*, **118**, 4877 (1985).
[33] H. Stetter and M. Schreckenberg, *Chem. Ber.*, **107**, 2453 (1974).
[34] A. Katritzky, M. Abdallah, S. Bayyuk, A. M. A. Bolouri, N. Dennis, and G. J. Sabongi, *Pol. J. Chem.* **53**, 57 (1979).

[35] D. Papa and E. Schwenk, *J. Am. Chem. Soc.*, **70**, 3356 (1948).
[36] H. Stetter and F. Jonas, *Chem. Ber.*, **114**, 564 (1981).
[37] H. Stetter, P. H. Schmitz, and M. Schreckenberg, *Chem. Ber.*, **110**, 1971 (1977).
[38] H. Stetter and B. Rajh, *Chem. Ber.*, **109**, 534 (1976).
[39] H. Stetter and H. J. Bender, *Chem. Ber.*, **114**, 1226 (1981).
[40] H. Stetter and H. Kuhlmann, *Tetrahedron*, **33**, 353 (1977).
[41] H. Stetter and H. Landscheidt, *Chem. Ber.*, **112**, 2419 (1979).
[42] H. Stetter and H. Kuhlmann, *Justus Liebigs Ann. Chem.*, **1979**, 303.
[43a] H. Stetter and H. Kuhlmann, *Justus Liebigs Ann. Chem.*, **1979**, 944.
[43b] H. Stetter and H. Kuhlmann, *Justus Liebigs Ann. Chem.*, **1979**, 1122.
[43c] H. Stetter and K. Marten, *Justus Liebigs Ann. Chem.*, **1982**, 250.
[44] H. Stetter, M. Schreckenberg, and K. Wiemann, *Chem. Ber.*, **109**, 541 (1976).
[45] B. Trost, C. D. Shuey, F. DiNinno, and S. C. McElvain, *J. Am. Chem. Soc.*, **101**, 1284 (1979).
[46] G. Baan, P. Vinczer, L. Novák, and C. Szántay, *Tetrahedron Lett.*, **26**, 4261 (1985).
[47] H. Stetter and J. Nienhaus, *Chem. Ber.*, **113**, 979 (1980).
[48a] H. Stetter and H. Kuhlmann, *Tetrahedron Lett.*, **1974**, 4505.
[48b] H. Stetter and H. Kuhlmann, *Chem. Ber.*, **109**, 3426 (1976).
[49] H. Stetter and L. Simons, *Chem. Ber.*, **118**, 3172 (1985).
[50] H. Stetter and A. Mertens, *Chem. Ber.*, **114**, 2479 (1981).
[51] T. El-Haji, J. C. Martin, and G. Descotes, *J. Heterocycl. Chem.*, **20**, 233 (1983).
[52] H. Stetter and A. Mertens, *Justus Liebigs Ann. Chem.*, **1981**, 1550.
[53] H. Stetter and F. Jonas, *Synthesis*, **1981**, 626.
[54] H. Stetter and G. Hilboll, *Synthesis*, **1979**, 187.
[55] H. Stetter and W. Schlenker, *Tetrahedron Lett.*, **21**, 3479 (1980).
[56] L. Novak, B. Majoros, and C. Szantay, *Heterocycles*, **12**, 396 (1979).
[57] L. Novak, G. Baan, J. Marosfalvi, and C. Szantay, *Tetrahedron Lett.*, **1978**, 487.
[58] D. M. Perrine, J. Kagan, D. B. Huang, K. Zeng, and B. K. Theo, *J. Org. Chem.*, **52**, 2213 (1987).
[59] H. Wynberg and J. Metselaar, *Synth. Commun.*, **14**, 1 (1984).
[60] E. Leete, M. R. Chedekel, and G. B. Bodem, *J. Org. Chem.*, **37**, 4465 (1972).
[61] E. Langhals, H. Langhals, and Ch. Ruchardt, *Justus Liebigs Ann. Chem.*, **1983**, 330.
[62] T. H. Jones, J. B. Franko, M. S. Blum, and H. M. Fales, *Tetrahedron Lett.*, **21**, 789 (1980).
[63] H. Stetter and A. Landscheidt, *J. Heterocycl. Chem.*, **16**, 839 (1979).
[64] H. Stetter and P. Lappe, *Justus Liebigs Ann. Chem.*, **1980**, 703.
[65] G. Rio and A. Lecas-Nawrocka, *Bull. Soc. Chim. Fr.*, **1976**, 317.
[66] J. D. Albright, *Tetrahedron*, **39**, 3207 (1983).
[67] H. J. Sims, H. B. Parseghian, and P. L. Benneville, *J. Org. Chem.*, **23**, 724 (1958).
[68] D. A. Evans, L. K. Truesdale, and G. L. Carroll, *J. Chem. Soc., Chem. Commun.*, **1973**, 55.
[69] H. M. Taylor and C. R. Hauser, *J. Am. Chem. Soc.*, **82**, 1790 (1960).
[70] D. Seebach, *Angew. Chem.*, **91**, 259 (1979); *Angew. Chem., Int. Ed. Engl.*, **18**, 239 (1979).
[71] C. Walling and E. S. Huyser, *Org. React.*, **13**, 91 (1963).
[72] T. M. Patrick and F. B. Erickson, *Org. Synth.*, Coll. Vol. **4**, 430 (1963).
[73] H. Stetter, H. Kuhlmann, and G. Lorenz, *Org. Synth.*, **59**, 53 (1980); Coll. Vol. VI, 866 (1988).
[74] T.-L. Ho and S.-H Liu, *Synth. Commun.*, **13**, 1125 (1983).
[75] E. Bellasio, A. Campi, N. Di Mola, and E. Baldoli, *J. Med. Chem.*, **27**, 1077 (1984).
[76] H. Stetter and W. Haese, *Chem. Ber.*, **117**, 682 (1984).
[77] T. H. Jones, M. S. Blum, and H. M. Fales, *Tetrahedron Lett.*, **1979**, 1031.
[78] H. Stetter and M. Schreckenberg, *Tetrahedron Lett.*, **1973**, 1461.
[79] R. B. Phillips, S. A. Herbert, and A. J. Robichaud, *Synth. Commun.*, **16**, 411 (1986).
[80] I. S. Ponticello and P. C. Pastor, *J. Polym. Sci.*, **18**, 2293 (1980).
[81] L. Novák, J. Rohály, G. Gálik, J. Fekete, L. Varjas, and C. Szántay, *Justus Liebigs Ann. Chem.*, **1986**, 509.

[82] E.-J. Brunke and E. Klein (to Dragoco Gerberding & Co., GmBH.), Ger. 2,935,683 (1981) [*C.A.*, **95**, 42484c (1981)].

[83] T. H. Jones, M. S. Blum, H. M. Fales, and C. R. Thompson, *J. Org. Chem.*, **45**, 4778 (1980).

[84] D. J. Voaden, *Synth. Commun.*, **14**, 53 (1984).

[85] H. Dyke, R. Sauter, P. Steel, and E. J. Thomas, *J. Chem. Soc., Chem. Commun.*, **1986**, 1447.

[86] H. Stetter and M. Schreckenberg, *Chem. Ber.*, **107**, 210 (1974).

[87] R. N. Young, J. I. Gauthier, and W. Coombs, *Tetrahedron Lett.*, **25**, 1753, (1984).

[88] W. Welter, T. Hüttelmaier, K. Matterstock, and H. Mildenberger, *Z. Naturforsch.*, **37B**, 923 (1982).

AUTHOR INDEX, VOLUMES 1–40

Volume number only is designated in this index.

CHAPTER AND TOPIC INDEX, VOLUMES 1–40

Many chapters contain brief discussions of reactions and comparisons of alternative synthetic methods related to the reaction that is the subject of the chapter. These related reactions and alternative methods are not usually listed in this index. In this index, the volume number is in **boldface**, the chapter number is in ordinary type.